Dynamics and Interactions
of Galaxies

Roland Wielen (Ed.)

Dynamics and Interactions of Galaxies

Proceedings of the International Conference,
Heidelberg, 29 May – 2 June 1989

With 230 Figures

Springer-Verlag
Berlin Heidelberg New York London
Paris Tokyo Hong Kong Barcelona

Professor Dr. Roland Wielen
Astronomisches Rechen-Institut, Mönchhofstrasse 12–14,
D-6900 Heidelberg 1, Fed. Rep. of Germany

Cover picture: The interacting galaxies M51 (NGC 5194) and NGC 5195.
The photographic plate was obtained at the 2.2 m telescope of the Calar Alto Observatory in Spain of the Max-Planck-Institut für Astronomie in Heidelberg by K. Birkle and J. A. Quesada.

ISBN-13:978-3-642-75275-9 e-ISBN-13:978-3-642-75273-5
DOI: 10.1007/978-3-642-75273-5

Library of Congress Cataloging-in-Publication Data. Dynamics and interactions of galaxies: proceedings of the international conference, Heidelberg, May 29–June 2, 1989 / R. Wielen (ed.). p. cm. Papers presented at the Heidelberg Conference on Dynamics and Interactions of Galaxies, held from May 29–June 2, 1989. Includes bibliographical references. ISBN-13:978-3-642-75275-9 (U.S.) 1. Galaxies-Congresses. I. Wielen, R. (Roland), 1938 – II. Heidelberg Conference on Dynamics and Interactions of Galaxies (1989) QB857.D94 1990 523.1'12—dc20 90-9674

2156/3150-543210 – Printed on acid-free paper

Preface

Environmental effects play an important role, not only for mankind on earth, but also in the evolution of galaxies. In recent years it became more and more evident that the classical picture, in which galaxies are viewed as quasi-isolated systems, is severely incomplete. Most galaxies have probably suffered, or suffer still, from strong gravitational interactions with neighbouring galaxies. In many cases, galaxies seem to have merged. The on-going infall of gas onto galaxies from their surroundings may also be a common process, at least for spiral galaxies. The gravitational interactions of galaxies are able to trigger violent bursts of star formation in gas-rich galaxies. Even the nuclear activity of active galaxies, from Seyfert galaxies up to quasars, is probably linked to interactions between galaxies, since the interactions seem to cause the "feeding" of the central machine (probably a massive black hole) in the galactic centre.

All these problems, in addition to other problems in the dynamics of galaxies, were discussed at the Heidelberg Conference on Dynamics and Interactions of Galaxies, held from 29 May to 2 June 1989. The conference was organized by the *Sonderforschungsbereich* (SFB) 328 *Evolution of Galaxies* of the University of Heidelberg. In this special research group, five institutes in Heidelberg collaborate: the Astronomisches Rechen-Institut, the Institut für Theoretische Astrophysik der Universität, the Landessternwarte, the Max-Planck-Institut für Astronomie, and the Max-Planck-Institut für Kernphysik. The topics of the conference were mainly concerned with SFB 328 projects A (Dynamical Evolution and Structure of Galaxies) and C (Interactions of Galaxies). The SFB 328 provided the principal financial support for the conference. Additional financial support came from: Bezirkssparkasse Heidelberg, Carl Zeiss Oberkochen, Dornier GmbH Friedrichshafen, Heidelberger Druckmaschinen AG Heidelberg, and Springer-Verlag Berlin, Heidelberg. All this financial help is gratefully acknowledged.

The scientific organizing committee of the conference consisted of:

A. Toomre (USA, Chairman)
F. Bertola (Italy)
J. Einasto (USSR)
K. Freeman (Australia)
J. Fried (Fed. Rep. of Germany)
J. Kormendy (Canada)
C. Möllenhoff (Fed. Rep. of Germany)
F. Schweizer (USA)
T. van Albada (The Netherlands)
R. Wielen (Fed. Rep. of Germany)

All the members of the committee are thanked for their contributions to the planning of this conference. A special acknowledgment must go to Alar Toomre who worked extremely hard and efficiently for the success of this meeting.

It is also a pleasure for me to thank all the other persons involved from the organizing institutes and from other institutions who made the conference possible by their efforts.

Heidelberg
February 1990

Roland Wielen

Contents

Part VI **Numerical Experiments on Interacting Galaxies and Mergers**

Part VII **Elliptical Galaxies**

Part VIII **Infall**

Part IX Disk Galaxies

Part X Other Problems of Galactic Dynamics

Part XII **Nuclear Activity of Galaxies and Interactions**

Part XIII Problems of Merger Scenarios

Part I

Interactions in the Early Universe

Galaxy Mergers: Lessons from Cosmological N-Body Simulations

G. Efstathiou

Department of Astrophysics, University of Oxford, Keble Road,
Oxford, OX1 3RH, UK

We show that galaxy merging is a natural feature of models in which the structure that we see today grew by the gravitational amplification of small amplitude density fluctuations in the early universe. We argue that a proper description of the dynamics and morphologies of real galaxies must take into account the infall of material and merging that is to be expected over a Hubble time.

1. Introduction: Galaxy Formation and Initial Conditions

Astronomers have often adopted highly simplified initial conditions in their attempts to explain how galaxies formed. I can think of at least two examples where such over-simplifications have led to results that disagree with observations. (A) Gott & Thuan (1976) modelled the formation of elliptical galaxies by the dissipationless collapse of spherical uniform density clouds in solid body rotation. Numerical simulations of collapse and virialization from these initial conditions showed that the ellipticity of the final object was related to the angular momentum as predicted for the classical fluid Maclaurin spheroids. Observations have since shown that luminous ellipticals are flattened by velocity anisotropies (Binney 1978) and rotate much more slowly than predicted by Gott & Thuan's models. (B) Larson (1974, 1975) modelled the formation of elliptical galaxies by the dissipative collapse of spherical uniform density clouds in solid body rotation. In his models, discs tended to form in the inner regions, producing objects in which the ellipticity increased rapidly towards the centre, in conflict with what is seen in most ellipticals. Larson produced more plausible ellipticity profiles by invoking large amounts of angular momentum transport by 'turbulent viscosity', though the end products of these simulations were still flattened by rotation rather than velocity anisotropy.

Had these authors adopted less symmetric initial conditions, their conclusions would have been different. For example, Binney (1976) and Aarseth and Binney (1978) showed that the final equilibrium objects in a dissipationless collapse retain a memory of their initial conditions. Thus a flattened slowly rotating galaxy might form from a flattened slowly rotating protocloud. It is easy to envisage even more radical departures from spherical symmetry, in which galaxies form by the merging of a heirarchy of subunits rather than from the coherent collapse of an initial protocloud with a well defined centre. Such merging would clearly have major effects on the morphological and dynamical properties of the remnant. This was emphasized by Toomre (1977), who argued that elliptical galaxies might have formed from mergers between disc galaxies. Clearly, if we are to explain the properties of present day galaxies we must understand what constitute reasonable initial condition.

There is good evidence to doubt whether galaxies formed from highly ordered initial conditions. When we look at the distribution of galaxies we see something resembling

Dynamics and Interactions of Galaxies Editor: R. Wielen

a fractal dust (see *e.g.* Peebles 1980) indicating that cosmic structure formed from fractal-like initial conditions. As a specific example, we consider the cold dark matter (CDM) theory (see *e.g.* Blumenthal *et. al.* 1984) in which the universe is assumed to be dominated by massive weakly interacting particles and Gaussian, adiabatic, fluctuations are assumed to have been generated in the early universe with the Harrison-Zeldovich scale invariant form. After recombination, the power-spectrum of the matter fluctuations $|\delta_k|^2$ may be approximated by

$$|\delta_k|^2 = Ak/[1 + (ak + (bk)^{3/2} + (ck)^2)^\nu]^{2/\nu}, \tag{1}$$

where A is a normalization constant, $a = 6.4(\Omega h^2)^{-1}$ Mpc, $b = 3.0(\Omega h^2)^{-1}$ Mpc, $c = 1.7(\Omega h^2)^{-1}$ Mpc and $\nu = 1.13$ (Bond & Efstathiou 1984). Throughout this article h denotes the Hubble constant $H_\circ$ in units of 100 km s^{-1} Mpc^{-1} and Ω denotes the mean mass density of the universe in units of the critical density $\rho_c = 3H_\circ^2/(8\pi G)$.

On scales corresponding to large galaxies, $k^{-1} \sim 1$Mpc, the power-spectrum (1) is roughly a power-law $|\delta_k|^2 \propto k^{n_{eff}}$ with a spectral index $n_{eff} \approx -2$. As the linear density perturbations in this model are a Gaussian random field, we would not expect any special symmetry properties at the dense spots where galaxies would form and thus a developing protogalaxy should be extremely inhomogeneous. This is borne out by the cosmological N-body simulations described in the next section.

2. Merging in N-body Simulations

Figure 1 shows a projection of the particle distribution in an N-body simulation of an $\Omega = 1$ CDM universe (see Frenk *et. al.* 1988, for further details). In this simulation, the length of the computational box is 14 Mpc ($h = 0.5$) and at the time shown in Fig. 1 the *rms* mass density fluctuation in spheres of radius $x_\circ = 8h^{-1}$Mpc is $\sigma_\rho = 0.40$.

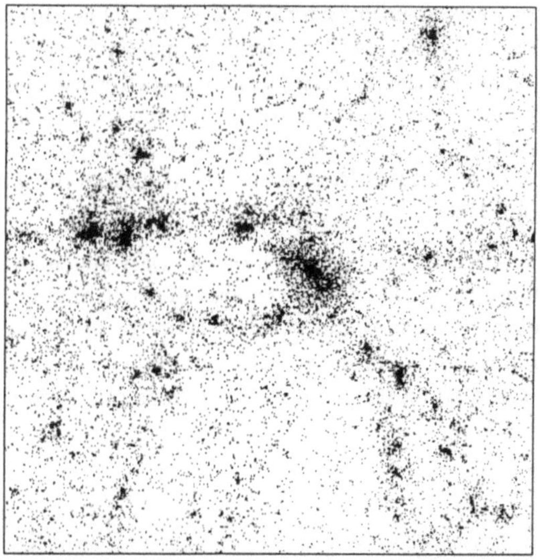

Figure 1. Projection of particle distribution in CDM simulation with box length $L_b = 14$ Mpc.

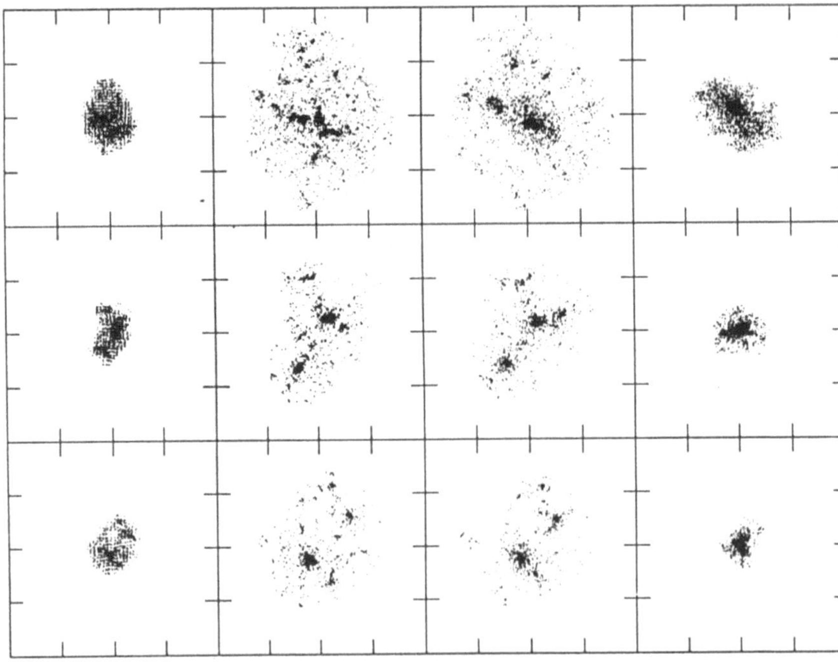

Figure 2 Evolution of the particles belonging to the three largest groups selected from Figure 1. The evolution proceeds from left to right showing the groups at redshifts of 6, 0.98, 0.41 and 0. The particles are plotted in physical coordinates and the tick marks show a separation of $1h^{-1}$Mpc.

Clumps in the simulation are located by linking together particles that lie within $b = 0.2$ times the mean interparticle spacing. This algorithm finds groups which correspond roughly to regions interior to isodensity contours at an overdensity of ~ 300. The three largest clumps from Figure 1 contain 3489, 1869 and 1482 particles respectively. The formation history of these clumps is shown in Figure 2. The clumps form by a series of merging events indicating that merging plays an important role in galaxy formation. The behaviour shown here was first pointed out for the CDM model by Frenk *et. al.* (1985) but is characteristic of the evolution of structures from scale-free Gaussian initial conditions (see Figure 2 of Efstathiou *et. al.* 1988).

Figure 2 shows the evolution of the dark matter; at this stage we can only speculate on how a dissipative baryonic component might behave. (This is an area of much ongoing research by A. Evrard, N. Katz, L. Hernquist, our group and others). It seems plausible that a considerable amount of dissipation and star formation could have taken place in the subunits. The subsequent merging of these subunits could then produce bulges and ellipticals, much as envisaged by Toomre (1977).

3. Merging and Angular Momentum

In hierarchical clustering theories like the CDM model, clumps acquire angular mometum through tidal interactions with their neighbours. This process has been investigated in detail by Barnes & Efstathiou (1987) and others. Here we summarize two important results, the consequences of which have still to be carefully explored. The following table lists the specific angular momenta, j_T, of the groups shown in Figure 2 as a function of the cosmological scale factor a. (We use units in which the box length $L_b = 1$, $G = 1$ and the mass of each particle $m_i = 1$). At early times, j_T grows $\propto a^{3/2}$, but then saturates and even decreases slightly when the groups achieve high overdensities (*cf* Figure 6 of Barnes & Efstathiou). The columns labelled j_1 list the specific angular momenta of the dominant subgroup, which we found by running the group finding algorithm with $b = 0.05$ at the final epoch ($a = 6.95$) shown in Figure 1. At the final time, the subgroups have about 64 times the mean overdensity of the groups identified with $b = 0.2$. The dominant subgroups in each of the three massive groups shown in Figure 2 contain 604, 597 and 243 particles respectively.

a	j_T	j_1	j_T	j_1	j_T	j_1
1.00	0.17	0.14	0.19	0.43	0.31	0.09
2.00	0.47	0.41	0.52	1.04	0.91	0.24
3.49	1.07	0.80	1.08	2.59	2.04	0.18
4.94	1.41	0.34	1.49	2.66	2.61	0.31
5.94	1.26	0.44	1.92	3.28	1.99	0.12
6.95	1.15	0.13	1.99	1.32	3.18	0.27

As for j_T, j_1 grows in proportion to $a^{3/2}$ at early times. However, when the subgroups become highly non-linear, j_1 can decrease substantially from its maximum value. This effect arises from the large asymmetries present within each group. As substructures merge, much of the angular momentum invested in the orbital motions of subclumps is lost to material in the outer parts of the group. This process has been pointed out by Frenk *et. al.* (1985) and Barnes & Efstathiou and plays a crucial role in establishing the radial variation of specific angular momentum in virialized groups. This effect may explain why the specific angular momenta of ellipticals and the bulges of spirals are smaller than the specific angular momenta of discs.

The second interesting result concerns the variation in the orientation of the angular momentum vector as we proceed outwards from the centre of a group. Figure 3 shows examples of groups and subgroups from N-body simulations which illustrate that large misalignments of the angular mometum vectors between high and low density regions are common. If we analyse the angular momenta of material in spherical shells around the centre of mass of groups found in CDM simulations, we find (Frenk *et. al.* 1988) that in about 50% of the groups the angular momenta within 100kpc are aligned to within about 45° of those of material within 100 – 200kpc but in some cases the angular momenta are antialigned. In the tidal torque picture we must expect that the direction of the angular momentum vector of infalling material will vary in a complex way with distance from the centre of a protogalaxy. The infall of material and the merging of blobs with misaligned angular momenta may have important consequences for the understanding of warps, ripples, rings and dynamically 'distinct' nuclei in galaxies (see the contributions by Binney and Bender in these proceedings).

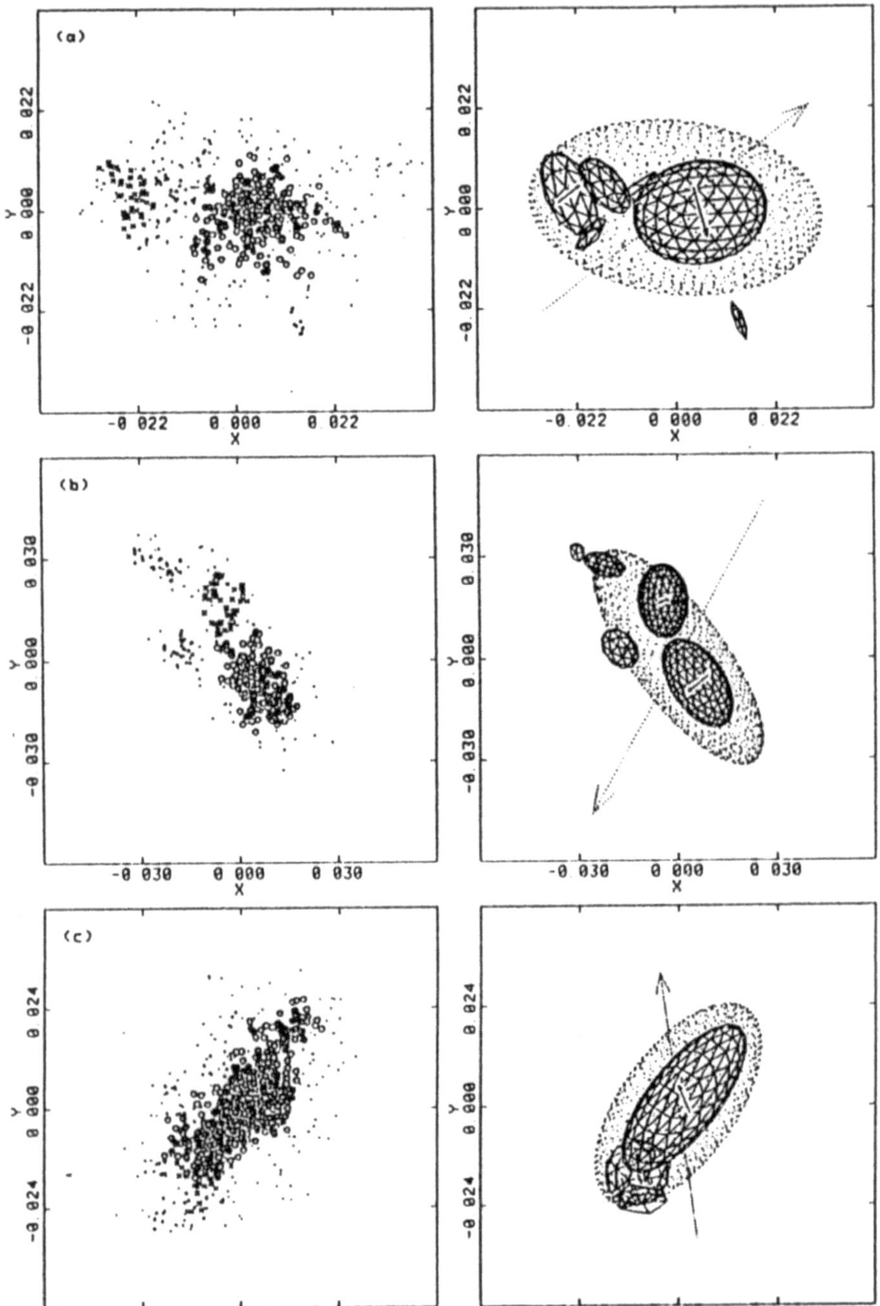

Figure 3. Internal structure of several massive groups (from N-body simulations by Barnes and Efstathiou, 1987). The first and second ranked subgroups are shown as open circles and crosses in the left hand panels while remaining particles are shown as dots. The panels to the right show smoothed representations of the groups in which they are represented by ellipsoids. In each case we have indicated the orientation of the angular momentum of the group and of the two dominant subgroups.

4. Merging and Biases

A number of authors have presented arguments against the idea that elliptical galaxies formed by the mergers of disk galaxies. For example, Ostriker (1980) argues that ellipticals are more tightly bound than present day spirals and obey a colour luminosity relation, while van den Bergh (these proceedings, see also van den Bergh & Harris, 1982) presents evidence that ellipticals have more globular clusters per unit mass than spirals.

However, it is quite plausible that the progenitors of present day ellipticals differed from present day spirals. In hierarchical clustering models, protogalaxies in a region destined to form a rich cluster of galaxies will have collapsed at higher redshift than objects of similar mass in the field. We must therefore expect 'biases' in the properties and distribution of galaxies which are modulated by large-scale density perturbations.

The formation of bound objects by hierarchical clustering and the origin of some of these biases can be described by the following simple analytic model (Bond, Cole, Efstathiou & Kaiser 1990). We imagine filtering the density field by applying a sharp low pass filter in k space and we denote the variance of the mass fluctuations of the filtered density field by σ^2. As the filtering scale is decreased, the density at any point in space obeys a random walk and defines a trajectory. The probability $Q(\delta,\sigma^2)$ that a trajectory lies between δ and $\delta + d\delta$ when the variance of the density field is σ^2 obeys a diffusion equation

$$\frac{\partial Q}{\partial \sigma^2} = \frac{1}{2}\frac{\partial^2 Q}{\partial \delta^2}. \tag{2}$$

We now wish to calculate the probability distribution of trajectories that reach (δ,σ) without exceeding a critical value δ_c/a for smaller σ. This is equivalent to solving equation (2) with an absorbing barier at $\delta = \delta_c/a$,

$$Q(\delta,\sigma^2) = \frac{1}{(2\pi)^{1/2}\sigma}\left[\exp\left(-\frac{\delta^2}{2\sigma^2}\right) - \exp\left(-\frac{(\delta - 2\delta_c/a)^2}{2\sigma^2}\right)\right]. \tag{3}$$

(Chandrasekhar, 1943). The fraction of mass $dF(M)$ that collapses into non-linear objects of mass M (corresponding to filtering scale $\sigma^2(M)$) and $M + dM$, when the universe has scale factor a, is easily calculated from (3)

$$dF(M) = \left(\frac{2}{\pi}\right)^{1/2}\frac{\delta_c}{a\sigma^2}\left(-\frac{\partial\sigma}{\partial M}\right)\exp\left(-\frac{\delta_c^2}{2a^2\sigma^2}\right)dM, \tag{4}$$

and corresponds exactly to the Press-Schechter (1974) form. Notice that $\int_0^\infty dF(M) = 1$, so unlike the Press-Schechter derivation we do not have to introduce an ad-hoc factor of two to ensure that $dF(M)$ is correctly normalized (*i.e.* our formalism accounts for the 'cloud within cloud' problem inherent in Press & Schechter's derivation). For a power-law initial fluctuation spectrum, $|\delta_k|^2 \propto k^n$, $\sigma^2(M) \propto M^{-(n+3)/3}$, and we can define a characteristic mass $M^*(a) = M[2a^2\sigma^2/\delta_c^2]^{3/(n+3)}$. The function $dF(M)$ is sometimes called the multiplicity function and the functional form of equation (4) has been found to agree extremely well with the results of N-body simulations (Efstathiou *et. al.* 1988).

Our formalism also allows us to investigate various statistical aspects of the merger history of clumps. For example, we can calculate the fraction of the mass at epoch a_1 in non-linear objects of mass M_1 to $M_1 + dM_1$ subject to the constraint that the mass ends up in groups more massive than M_c at epoch a_2. The solution is a straightforward generalization of the problem of one absorbing barrier to two absorbing barriers,

$$dF_b(M_1,a_1) = \int_{max(M_1,M_c)}^{\infty} dF(M_2,a_2)\, G(M_1,a_1,M_2,a_2) \Big/ \int_{M_c}^{\infty} dF(M_2,a_2), \quad (5)$$

where

$$G(M_1,a_1,M_2,a_2) = \left(\frac{2}{\pi}\right)^{1/2} \frac{\delta_c\sigma_1(a_2-a_1)}{a_1a_2(\sigma_1^2-\sigma_2^2)^{3/2}} \left(-\frac{\partial\sigma_1}{\partial M_1}\right) \exp\left(-\frac{\delta_c^2(a_2-a_1)^2}{2a_1^2a_2^2(\sigma_1^2-\sigma_2^2)}\right).$$

In Figure 4, we compare equation (5) with the results from N-body simulations of scale-free initial conditions ($n=-1$, $\Omega=-1$), where we have chosen $\delta_c = 1.9$ to match the predictions to non-linear groups with overdensities of a few hundred. The analytic model is in excellent agreement with the simulations and shows that the function dF_b differs markedly from the multiplicity function of an average region of space. As expected, massive groups at late times are built up by mergers of massive groups at earlier times. As Figure 4 shows, these biases can be quite large and can persist over large changes in M^* (*cf* also Figure 13 of Efstathiou *et. al.* 1988).

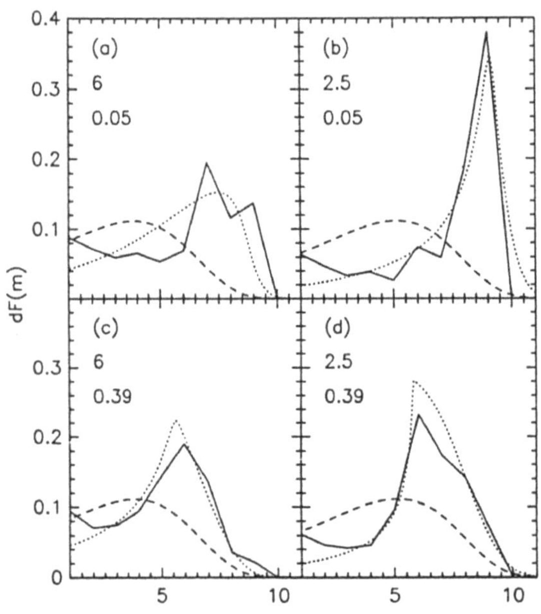

Figure 4. Multiplicity functions of particles that end up in massive groups in $n=-1$ scale-invariant simulations compared with the solutions of the two absorbing barrier problem (equation 5). In each case we located groups at the final output time (with $b=0.2$) and selected particles in the most massive groups containing 0.05 (figures a, b) and 0.39 (figures c,d) of the total mass. The multiplicity functions for these subsets of particles are shown at earlier times when the characteristic mass M^* was smaller by a factor of 2.5 (figures b, d) and by a factor of 6 (figure a, c). The predictions of the two absorbing barier problem are shown as the dotted lines. The multiplicity functions for an average region of space (equation 4) are shown by the dashed lines. (The multiplicity m of a group of N particles is defined by $2^m \leq N < 2^{(m+1)}$ particles.)

The discussion in this section shows that the progenitors of present day ellipticals could have differed significantly from present day field galaxies. Galaxies in protocluster regions will have collapsed at higher redshifts than galaxies of the same mass in the field, and consequently will have higher binding energies and circular speeds. It is, of course, a matter of speculation at present whether these biases could have affected star formation and globular cluster formation in protogalaxies.

5. Conclusions

Many speakers at this meeting have concentrated on properties of galaxies - ripples, warps, tails, activity, *etc* - that can only be understood in terms of interactions and mergers between galaxies. I have argued that merging is inevitable in hierarchical clustering models of the formation of structure and that we must expect merging and extended accretion to have played crucial roles in structuring galaxies.

Stellar dynamicists have for too long assumed that galaxies are 'island universes'. A wealth of observations show that this is not the case and it is now clear that we must understand the cosmological context in which galaxies form if we are to fully understand their properties.

The simulations of dissipationless matter described here provide a crude guide of the processes that might be relevant to galaxy formation. We can hope that our picture of galaxy formation will sharpen over the next few years as more effort is devoted to the much more complex problem of how the luminous (dissipative) components of galaxies formed. As we have emphasised in this article, it is important that studies of the latter problem are based on *realistic* initial conditions.

I thank all of my collaborators who have contributed to various aspects of this work.

REFERENCES

Aarseth, S.J. & Binney, J., 1978. *Mon. Not. R. astr. Soc.*, **185**, 227.
Barnes, J. & Efstathiou, G., 1987. *Astrophys. J.*, **319**, 575.
Binney, J., 1976. *Mon. Not. R. astr. Soc.*, **177**, 19.
Binney, J., 1978. *Mon. Not. R. astr. Soc.*, **183**, 501.
Blumenthal, G.R., Faber, S.M., Primack, J.R. & Rees, M.J., 1984. *Nature*, **311**, 517.
Bond, J.R. & Efstathiou, G., 1984. *Astrophys. J.*, **285**, L45.
Bond, J.R., Cole, S. Efstathiou, G. & Kaiser, N., 1990. *In preparation.*
Chandrasekhar, S., 1943. *Rev. Mod. Phys.*, **15**, 2.
Efstathiou, G., Frenk, C.S., White, S.D.M. & Davis, M., 1988. *Mon. Not. R. astr. Soc.*, **235**, 715.
Frenk, C.S., White, S.D.M., Efstathiou, G., & Davis, M. 1985. *Nature*, **317**, 595.
Frenk, C.S., White, S.D.M., Davis, M. & Efstathiou, G., 1987. *Astrophys. J.*, **327**, 507.
Gott, J.R. & Thuan, T.X., 1976. *Astrophys. J.*, **204**, 649.
Larson, R.B., 1974. *Mon. Not. R. astr. Soc.*, **166**, 585.
Larson, R.B., 1975. *Mon. Not. R. astr. Soc.*, **173**, 671.
Ostriker, J.P., 1980. *Comm. Astrophys.*, **8**, 177.
Peebles, P.J.E., 1980. *The Large-Scale Structure of the Universe*, Princeton University Press.
Press, W.H. & Schechter, P., 1974. *Astrophys. J.*, **187**, 425.
Toomre, A., 1977. In *The Evolution of Galaxies and Stellar Populations*, eds. B.M. Tinsley and R.B. Larson, Yale University Press. p401.
van den Bergh, S. & Harris, W.E., 1982. *Astron. J.*, **87**, 494.

The Formation of Halos via Mergers. The Organized and Organizing Dynamics of Mergers

P.J. Quinn[1], *W.H. Zurek*[2], *J.K. Salmon*[3], *and M. Warren*[3]

[1]Mount Stromlo and Siding Spring Observatories, The Australian National University, Woden Post Office, ACT2606, Canberra, Australia
[2]Theoretical Astrophysics, Los Alamos National Laboratory
[3]Caltech Concurrent Computing Project

1. Introduction

N-body models of the growth of large and small scale structures in the Universe have proved to be valuable tools in the quest for the values of fundamental cosmological parameters. From the shapes of halo rotation curves (Quinn, Salmon and Zurek, 1986, QSZ) and the form of the two-point spatial correlation function (Davis et. al., 1985), it would appear that small and intermediate scale structures imply initial density perturbations with power spectra (P(k)) that decline as a function of wavenumber (k) faster than k^{-1} ($P=Ak^n$, $n<-1$). The mass perturbations associated with these spectra imply a spherical collapse time $t_{coll} \propto M^{(n+3)/4}$. For the case $n > -3$, small mass perturbations collapse before large ones and the evolution proceeds as a cascade of mergers from small to large scales. Protogalactic complexes destined to form galaxies similar to the Milky Way must have collapsed for the first time at redshifts less than 4, ie. when the Universe was one collapse time old for these systems (Fall, 1989). This onset of major merging and agglomeration of subsystems seems to have continued for a large number of galaxies up until redshifts of less than 0.5. New studies of Butcher-Oemler clusters by Lavery and Henry (1988) suggest that the source of the blue galaxies in these clusters is bursts of star formation triggered by galactic encounters. At even smaller redshifts the strong deceleration of colliding spirals and the transient nature of tidal tails pointed out by the Toomres (1972), the number of ellipticals will shells (Schweizer, 1982), the abundance of ellipticals with counterrotating cores (Balcells and Quinn, 1989) and the old thick disks of spirals (Hernquist and Quinn, 1989) all point to a long merger related history for a large fraction of currently isolated, bright galaxies. In this article we will examine the dynamics and effects of the merger process on the dark and luminous components of galaxies at high and low redshift. As we shall see, the same dynamical processes acting in cosmological and nearby mergers have imparted a detailed memory of formation to galaxies that will be useful in probing the initial conditions and evolution of protogalactic systems.

2. Halo Formation

The formation of individual, dissipationless, dark matter halos has been modelled by several groups (Frenk et. al., 1985, QSZ). In all of the cases studied, halos are built up by the collapse of a clumpy protogalactic region into a single, virialized system via mergers of substructure. This collapse appears quite chaotic in configuration space and the notion of violent relaxation would suggest that halos largely forget about their initial conditions. However, QSZ have shown that there is a close connection between the shape of the mass distribution in the final, virialized halo and the shape of the initial density perturbation spectrum. In fact, the final halos have the mass distributions one would predict from the

Dynamics and Interactions of Galaxies Editor: R. Wielen

theory of secondary infall (Gunn and Gott, 1972, Hoffman and Shaham, 1985). In this paradigm, we imagine self-gravitating systems forming by the collapse of a set of spherically symmetric shells, each initially expanding with the Hubble flow then eventually turning around, starting with the innermost shell. This picture looks nothing like the apparently chaotic merger of lumps in highly asymmetric protogalaxies. Zurek, Quinn and Salmon (1988, ZQS) have pointed out that the success of the paradigm is not due to the behaviour of the system in configuration space but to its evolution in "binding energy-angular momentum" (Lindblad) space. Due to dynamical friction, lumps of protogalactic material sink to the centre of the forming galaxy. Hence the centres of protogalactic lumps are predestined to becomes part of the centre of the final galaxy. Similarly the diffuse components of the protogalaxy take up angular momentum and energy and become the diffuse component in the collapsed system. This preservation of the ranking in binding energy of the protogalactic soup results in a memory that contributes to the connection between the linear density perturbation field and the mass profile of the final halo. The loss of angular momentum by lumps in protogalaxies and the current low specific angular momentum content of ellipticals compared to spirals (Fall, 1982), implies that proto-ellipticals may have been more clumpy than proto-spirals (ZQS). Several galaxies found at high redshift (z~4) by Chambers and Miley (1989) do indeed look very clumpy and posses powerful radio jets like current epoch ellipticals.

This memory resulting from organized collapse goes further than the mass profile. Quinn and Zurek (1988, QZ) have pointed out that initially disorganized protogalaxies can have their internal angular momenta aligned and their mean motions curves made to look like circular orbit velocity curves by the collapse process. In this sense the sequence of mergers has an organizing influence as well as itself being an organized process. This organizing influence of mergers has been clearly demonstrated by some recent N-body studies of nearby mergers.

3. Dynamical Processes in Nearby Mergers.

Salmon, Quinn and Warren (1989, SQW) have begun a detailed study of the fully selfconsistent formation of shells around ellipticals using a Barnes-Hut treecode adapted for a parallel supercomputer. Some of their most interesting findings are not those pertaining to shells but rather to the effect a shell-forming merger has on the primary galaxy. An initially spherical, non-rotating primary can be brought into rotation about an axis perpendicular to the initial orbit plane by the collision. Dynamical friction imparts angular momentum to the primary particles in the direction of the secondary's orbital motion and, at the same time, tidal forces remove mass from the secondary and deposit it in the orbital plane. The result is a final primary that is elliptical (E1-E2) with its structural and kinematic axes aligned by the action of friction and tides (Figure 1). The fact that friction tries to spin up primary particles in the sense of the orbital motion is precisely why counterrotating cores can be formed when the orbit is retrograde with respect to the initial mean motion of the primary (Balcells and Quinn, 1989, BQ). The final mean motion curves of the primary produced by SQW are similar in shape to the circular orbit curves for the primary (Figure 1), which reminds us of the effects seen in halo formation by QZ. This implies that the deposited angular momentum knows about the underlying mass distribution. Similarly SQW and BQ find that the deposited mass profiles are the same shape as the initial mass profiles (see Figure 1) over more than five decades in density in high resolution models. This preservation of the mass profile in the deposited material is again seen in the cosmological models where small halos, and the larger halos they

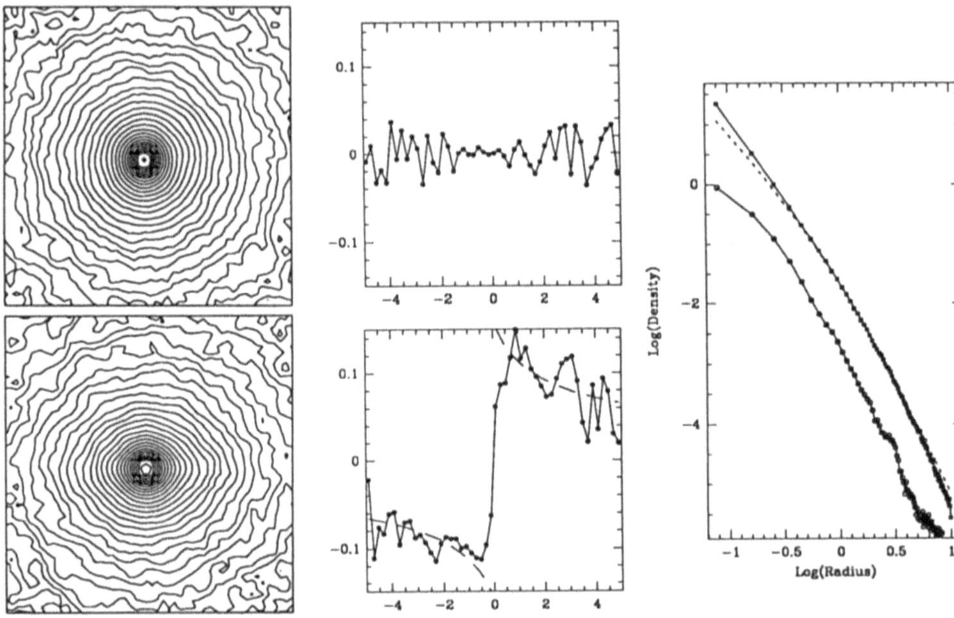

Figure 1 : **Left** : Projected log density contours of an initially nonrotating, spherical Jaffe model before (top) and after merging with a 10% mass, denser Jaffe companion (box is +/- 2 half-light radii). **Centre:** Projected radial velocity curves for a slit in the orbit (horizontal) plane before (top) and after the merger. Dotted lines show scaled down circular orbit curves. **Right:** Final density profiles of primary (upper) and deposited companion material.

build up, have similar mass distributions. It would also naturally account for the fact that the diffuse light from shells seems to track the underlying galaxy light (Pence, 1986).

Franx, Illingworth and Heckman (1989) have pointed out that more than 60% of ellipticals seem to have aligned kinematic and structural axes. These is not intrinsic need for such a correlation if ellipticals are not supported by their mean motion. Similarly, Illingworth has pointed out on several occasions that ellipticals have symmetric mean motion profiles that resemble circular orbit curves for the underlying luminous galaxy. Again, there is no need for the mean motion to be directly related to the underlying potential if it does nothing to support the figure of the galaxy. The organizing effects of the types of mergers that produce shells and counterrotating cores naturally give rise to these kinematic and structural correlations which represents new evidence of a merger history for the majority of large ellipticals.

4. A Theory of Merging : Nonviolent Relaxation.

If the dynamical processes associated with mergers, both at high and low redshift, give rise to orderly (nonviolent) mass, angular momentum and energy depositing phenomena then it may, in principle, be possible to construct a quantitative theory of dissipationless merging. With such a theory we could examine the structure of dark matter halos, ellipticals and spheroidal systems in general and critically access the thesis that they owe their current properties to dissipationless processes. Based on experience with cosmological and nearby mergers, we have conducted a series of experiments to isolate dynamical processes. From these experiments it is clear that mergers of systems described by single power law mass distributions retain their mass profiles throughout a sequence of mergers. The underlying physical reason for this is that tidal and frictional forces both depend linearly on density. The mass removed from the secondary (dm) at a radius "r" in the secondary and added to the primary (dM) at a radius "R" is:

$$dM(R,R+\Delta R) = dm(r,r+dr) = \frac{dm}{dr}\frac{dr}{dR}dR$$

If the mass profile of secondary and primary are similar ($M(R)=M_0R^{\gamma}$) and tides give rise to a Roche criteria equating the mean densities of the primary and secondary at the stripping point :

$$\frac{M_o R^{\gamma}}{R^3} = \frac{m_o r^{\gamma}}{r^3}\,[\text{Roche}]$$

$$\therefore \frac{dr}{dR} = \left(\frac{M_o}{m_o}\right)^{\frac{1}{\gamma-3}} \rightarrow \Delta M(R) = m_o\left(\frac{M_o}{m_o}\right)^{\frac{\gamma}{\gamma-3}} R^{\gamma}$$

where $\Delta M(R)$ is the deposited mass profile which has the same radial scaling as the underlying mass. A similar argument applied to the deposited velocity shows it tracks the shape of the circular orbit curves.

A complete theory based on these concepts and extended to include multicomponent systems is being developed (Quinn and Zurek, 1989a). One application of the simple, one component concepts is to ask where the core radii of dark matter halos come from? If we collide isothermal systems with no cores (ie. pure power laws) then we would expect, on the basis of the arguments given above, final halos without cores. This has been verified by experiments. Since the halos implied by decompositions of spiral rotation curves (Carignan and Freeman, 1985) have core radii several times that of the luminous spiral disks, then these cores were presumably inherited and carry useful information on the structure of the initial density perturbation spectrum (Quinn and Zurek, 1989b). It is hoped that a full understanding of the dynamics of dissipationless mergers may help isolate kinematic and structural features in galaxies and halos that have resulted from dissipative effects. In this way we may be able constrain the transition from dissipative to dissipationless dynamics and the onset of star formation in the formation of galaxies.

5. References

Balcells, M. and Quinn, P.J., 1989, *Ap.J.*, in press.
Carignan, C. and Freeman, K.C., 1985, *Ap.J.*, **294**, 494.
Chambers, K. and Miley, G.K, 1989, preprint.
Davis, M., Efstathiou, G., Frenk, C.S. and White, S.D.M., 1985, *Ap.J.*, **292**, 371.
Fall, S.M., 1982, in IAU Symposium 100, Ed.E.Athanassoula, D.Reidel, 391.
Fall, S.M., 1989, preprint.
Franx, M., Illingworth, G. and Heckman, T., 1989, *Ap.J.*, in press.
Frenk, C.S., White, S.D.M., Efstathiou, G. and Davis, M., 1985, *Nature*, **317**, 595.
Gunn, J.E and Gott, J.R., 1972, *Ap.J.*, **176**, 1.
Hernquist, L. and Quinn, P.J., 1989, in preparation.
Hoffman, Y. and Shaham, J., 1985, *Ap.J.*, **297**, 16.
Lavery, R.J. and Henry, J.P., 1988, *Ap.J.*, **330**, 596.
Pence, W.D., 1986, *Ap.J.*, **310**, 597.
Quinn, P.J. and Zurek, W.H., 1988, *Ap.J.*, **331**, 1.
Quinn, P.J. and Zurek, W.H., 1989a, in preparation.
Quinn, P.J. and Zurek, W.H., 1989b, in preparation.
Quinn, P.J., Salmon, J.K. and Zurek, W.H., 1986, *Nature*, **322**, 392.
Salmon, J.K., Quinn, P.J., and Warren, M., 1989, *this volume*.
Schweizer, F., 1982, in IAU Symposium 100, Ed. E.Athanassoula, D.Reidel, 319.
Toomre,A. and Toomre, J., 1972, *Ap.J.*, **178**, 623.
Zurek, W.H., Quinn, P.J. and Salmon, J.K., 1988, *Ap.J.*, **330**, 519.

Are High Redshift Galaxies Interacting?

G.K. Miley[1,2] *and K.C. Chambers*[2,3]

[1]Sterrewacht, Postbus 9513, NL-2300RA Leiden, The Netherlands
[2]Space Telescope Science Institute, 3700 San Martin Drive, Baltimore, MD21218, USA
[3]Dept. of Physics and Astronomy, Johns Hopkins University, Baltimore, MD21218, USA

1. INTRODUCTION

I am uncomfortable. I feel in the position of an agnostic who has been asked to praise God close to the beginning of a religious revival meeting. Unfortunately, as far as mergers at high redshifts is concerned, I cannot bring myself to do this. Instead, I propose to present you with evidence that peculiar phenomena are occurring in the most distant galaxies we can study. According to your degree of faith in the "Book of Mergers", you can decide whether and to what extent the processes we observe reflect interactions between galaxies and protogalaxies. Sticking to the catholic sense of the title assigned to me, I shall however show that some high-redshift galaxies are definitely interacting with themselves.

Throughout this talk we shall restrict ourselves to radio galaxies, because they are the only galaxies so far observable above a redshift of 2. We must bear in mind that the properties of these radio monsters are unlikely to be representative of the population of galaxies as a whole.

We shall begin by describing some searches for high-redshift radio galaxies. Then we shall show you some examples of galaxies found in our search. We shall review several properties which indicate that interaction between the radio sources and their host galaxies becomes more vigorous at large redshifts. Finally, we shall summarize observational results which suggest that galaxy merging is associated with nuclear activity in nearby radio galaxies, which we argue is weak evidence that mergers may also trigger activity in the case of our high-redshift radio galaxies.

2) RADIO SOURCES AS BEACONS

It is now about forty years since the identification of Cygnus A almost overnight changed radio astronomy from being the hobby of a few talented electrical engineers into one of the most powerful tools for

Dynamics and Interactions of Galaxies Editor: R. Wielen

studying the distant Universe.Radio sources are still among the most important cosmological probes because they can have enormous luminosities, usually have spatially extended morphologies and are intense sources of optical line emission. One of the most important applications of radio sources to cosmology is indirect, namely to use radio sources to pinpoint the most distant galaxies and quasars, for subsequent studies at non-radio wavelengths. This will be the dominant approach which we will take here.

Until recently most work in this area was directed at the 3C Catalogue, i.e. the strongest few hundred sources in the northern sky. Over the years painstaking attempts have been made to examine the optical objects associated with every 3C source, most notably by Hy Spinrad, Pat McCarthy Wil van Breugel, George Djorgovski and their colleagues at the University of California at Berkeley. The identification of this sample is now virtually complete, with more than 30 of its members associated with galaxies having redshifts in excess of 1.

During the last decade CCD optical detectors have revolutionized this subject, enabling much fainter galaxies to be imaged and their redshifts to be measured spectroscopically. Moreover, from the theoretical standpoint, the search for and study of galaxies having redshifts in excess of 2 or 3 became increasingly important as theoretical arguments based on the canonical "cold dark matter" cosmologies indicated that the majority of galaxies may have been formed during these epochs.

Since it is impossible to carry out deep optical observations on the tens of thousands of known radio sources which are fainter than the 3C objects, it is desirable to find some characteristic property of a radio source which will maximize the chance of it being a distant galaxy. One such technique which has been developed by Ken Chambers, Wil van Breugel and myself, has been to use the spectrum of the continuum radio emission as a selection criterion. The most distant galaxies tend to have steeper radio spectra than nearby ones (e.g. Tielens et al. 1979, Blumenthal and Miley 1979). 1988 was a bumper year for harvesting distant galaxies. Before that the galaxy distance record was held by 3C326.1 at a redshift of 1.8, measured by the Berkeley group. During last year several new records were set, first to redshift 2.3 by our team, then to 3.4 by Simon Lilly at the Institute of Astronomy of the University of Hawaii and subsequently to 3.8 by us. Now more than a dozen galaxies having redshifts in excess of 2 are known.

The ultra-steep spectrum selection criterion is the most efficient method so far developed for pinpointing galaxies at high redshift. Our spectroscopic survey of 4C sources shows that about half of our ultrasteep spectral candidates have emission lines and are definitely at redshifts > 0.6 and 20% (8 radio sources) are identified with optically extended objects, presumably galaxies, with z > 2 (see Chambers et al. 1988a, 1989).

3) EXAMPLES

Let us now examine two cases of galaxies with redshifts > 2.
4C40.36 (Chambers et al 1988a) is at a redshift of 2.3 and has spatially resolved optical line and continuum emission extended along the radio source by about 30 kpc.

4C41.17 is our star attraction to date. It is the object in our sample with the highest redshift and the one which we have studied in most detail. 4C41.17 was identified with a faint extended optical object, whose spectrum showed two emission lines corresponding to Lyman alpha and CIV redshifted by 3.8.

The extent of the optical continuum emission in 4C41.17 (~30 kpc) together with the large equivalent width of Lyman alpha (~1000 A) indicate that 4C41.17 is a galaxy with properties similar to those of other distant radio galaxies and the most distant galaxy known. The line-emitting gas is elongated along the radio source and a Lyman alpha halo extends by more than 100 kpc, considerably beyond the radio source. The gas is clumpy and has a turbulent velocity field , with a gradient of ~2000 km/s and a half-width of ~1200 km/s. The inner gas kinematics may be influenced by entrainment in the radio source. Mass estimates range from 10^9 to 10^{11} M_0 for the ionized gas and an upper limit of 10^{12} M_0 can be set on the total mass of the galaxy from the gas kinematics.

The structure of the optical/infrared continuum of 4C41.17 also appears to be related to radio morphology. Its spectral energy distribution (SED), like those of many other high redshift radio galaxies, resembles the SED of nearby spiral galaxies. We shall consider the implications of these observations later in the talk.

4) EPOCH DEPENDANCE

During the past two or three years, CCD imaging has revealed that distant radio galaxies have several bizarre properties. Previously, it was supposed incorrectly that the presence of radio emission had little effect on the optical properties of the host galaxies. The recent observations show that at high redshift, increased interaction takes

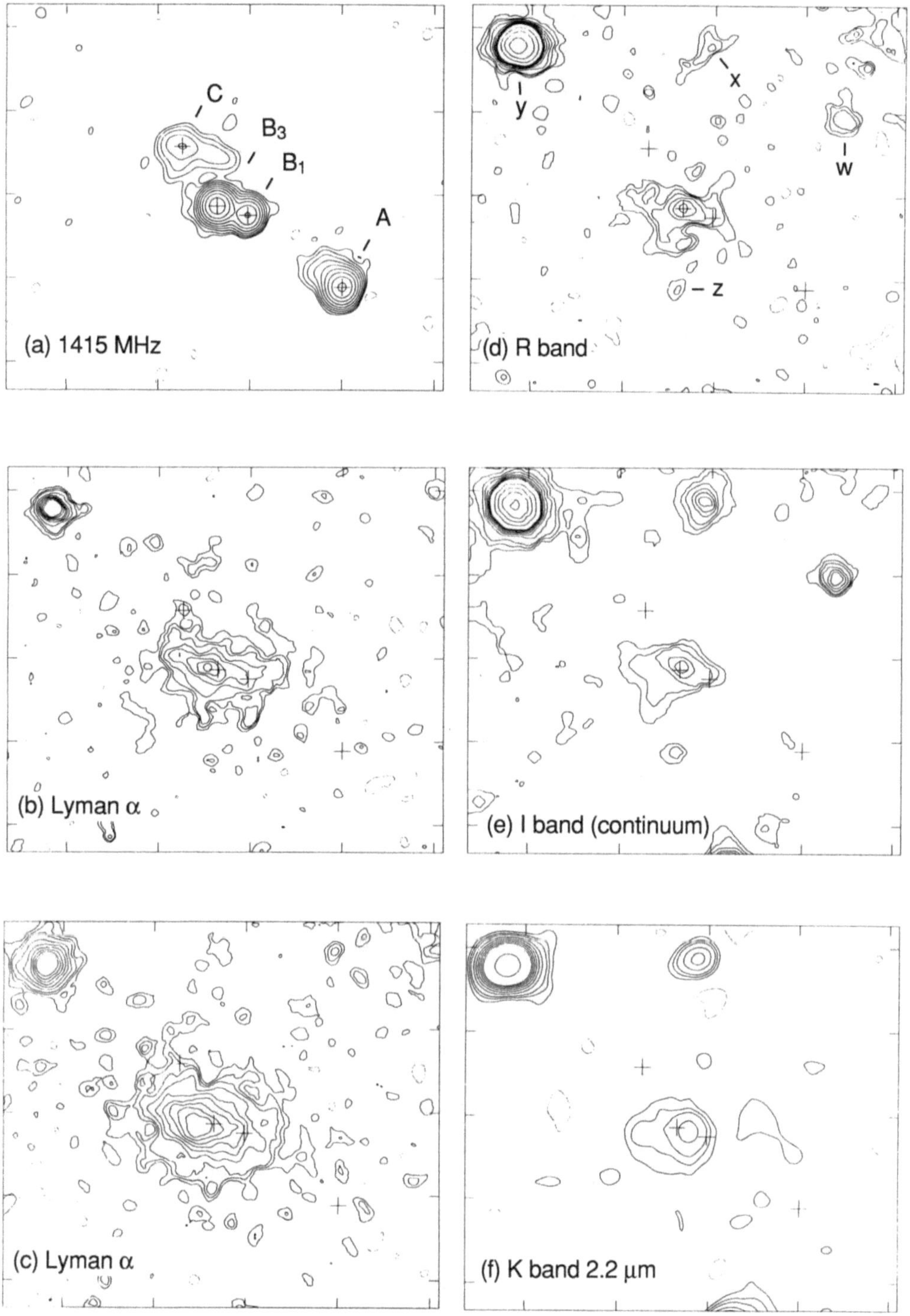

Fig. 1. Radio infrared and optical morphologies of 4C 41.17 at z = 3.8, taken from Chambers, Miley and van Breugel (1989). Tick marks are at intervals of 5" = 45 kpc for H_0 = 50 km/s/Mpc and q_0 = 0.5.

place between the relativistic plasma responsible for the radio emission and the intergalactic medium of the parent galaxies. We can list several properties that indicate such a progression with redshift. However, a word of caution! Since most of the work has been carried out on flux-limited samples of radio sources, redshift and radio luminosity are strongly correlated and in most cases it is difficult at present to disentangle redshift dependance from luminosity dependance. Let us consider each of the (apparent) redshift-dependant effects in turn.

4.1 Optical Line Orientations. During the last decade it was unexpectedly discovered several nearby radio galaxies have spatially extended optical line emission which is morphologically associated with the radio continuum emission (e.g. Miley et al, 1980, Heckman et al, 1982, Miley, 1984, van Breugel, 1986, Fosbury 1986). This was found to be present in only a minority of bright radio sources (e.g. Baum and Heckman 1989). It indicated that in some galaxies vigorous interaction could take place between the radio sources and an ambient medium. It is a two-way interaction. The line-emitting gas can bend, decollimate depolarize and enhance the radio emission, while the radio jet seems to be able to affect the ionization and kinematics (entrainment) of the gas (van Breugel et al 1985).

By analyzing at the difference in angle between the radio continuum and optical line extensions and comparing his data for high-redshift galaxies with those of Heckman and Baum at lower redshift, McCarthy (1989) has recently demonstrated very beautifully that the incidence of optical/line radio continuum associations increases dramatically at large redshifts. Practically all radio galaxies at $z > 0.6$ have luminous extended regions of line emission which are morphologically associated with the radio sources.

4.2 Radio Bending The second redshift-dependant property suggesting an increase in interaction at high redshifts is the radio morphologies. It is now more than twenty years since the first glimmerings appeared, which suggested that the structures of steep-spectrum (lobe-dominated) radio sources associated with quasars change systematically as a function of redshift.Two effects have been observed:

(i) The component sizes and separations are smaller at large redshifts.(e.g. Miley 1968; Barthel and Miley 1988).

(ii) The morphologies tend to be more bent at high redshifts. (Barthel and Miley 1988),

The implication which is a likely, but not exclusive explanation of both of these effects is (Miley 1986, Barthel and Miley 1988) that steep-spectrum radio sources associated with quasars interact more vigorously with their surroundings at large redshifts.

Evidence that radio bending in high-redshift quasars is indeed due to interaction of the radio source with gas associated with the quasars comes from work by Heckman, van Breugel and myself(GM), as yet unpublished. In a Lyman alpha survey of 20 steep spectrum radio quasars with $z > 1.9$, we found 6 cases where there was clearly huge Lyman alpha halos These appear to be associated morphologically with the radio bends. The increased bending of the quasar radio sources as a function of redshift is therefore consistent with a more vigorous interaction of the radio sources with the surrounding gas at earlier epochs.

The radio bending could be explained by collision of the radio jet with ambient clouds of each of mass $\sim 10^8$ M_0. This should be compared with typical total masses of ionized gas calculated from the Lyman alpha luminosities of typically $\sim 10^{10}$ M_0 for filling factors $\sim 10^{-4}$. The corresponding "smeared-out" densities of ~ 0.1 cm^{-3}, together with the equipartition magnetic fields calculated using canonical synchrotron theory (e.g. Miley 1980) would then be sufficient to depolarize the radio emission.

<u>4.3 Optical/IR Continuum Alignment.</u> One of the most unexpected properties of high-redshift radio galaxies was the alignment of their optical and infrared morphologies along the associated radio sources. This effect was discovered independently for two complementary samples of radio galaxies by Chambers et al. (1987) and McCarthy et al (1987).

In exhibiting the alignment effect, powerful high-redshift radio galaxies differ from radio galaxies at low redshifts (e.g. Baum and Heckman 1989). In his thesis McCarthy (1989) demonstrates that this alignment effect begins to be important above a redshift of about 0.6 and is present for almost all the galaxies with redshifts greater than 1. The alignment is also present in the near infrared. Figure 2 shows a 2.2 micron picture of the prototype object, 3C368, taken from Chambers et al 1988.

5. THE OPTICAL/INFRARED CONTINUUM EMISSION

The spectral energy distribution (SED) of high-redshift radio galaxies typically have a sharp rise redwards of an emitted wavelengrth of 3000 A (a "red bump") and are flat or rise slightly shortwards of this wavelength (a "blue shoulder"). One of the surprising properties of this emission emphasized by Lilly (1989) is the continuity in so-called "infrared Hubble plots" of the 2.2 micron observed flux density versus

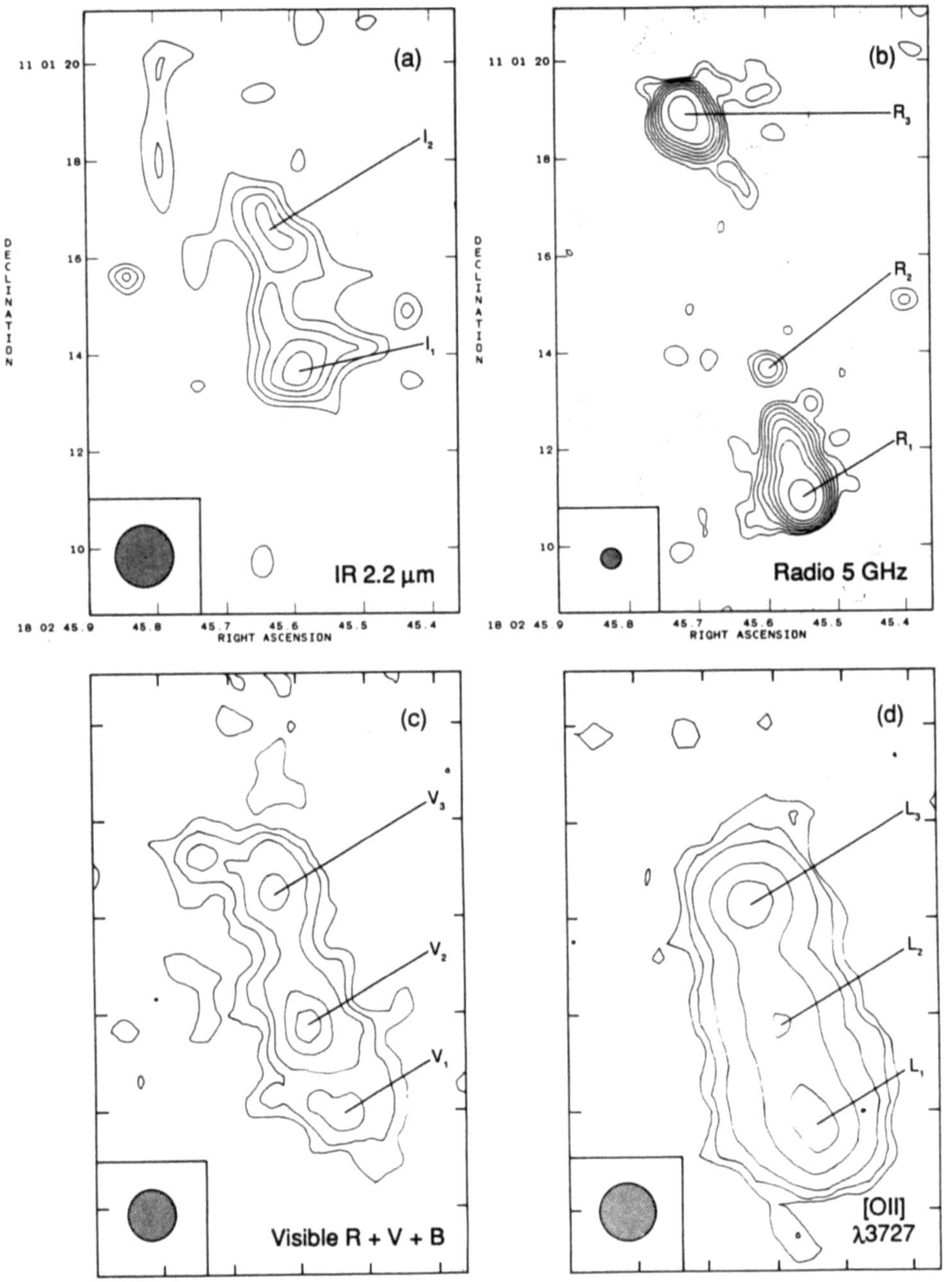

Figure 2. Images of 3C 368 in the near-IR (top left), radio (top right) and optical (bottom) on the same scale and positioned to < 1" (from Chambers, Miley and Joyce 1988).

redshift. The scatter in this diagram is low (about 0.4 mag). Moreover, there is a remarkable continuity and fairly constant scatter in this diagram from the lowest to the highest redshifts, despite the redshift-dependant changes in the properties of the radio galaxies described above.

Any explanation of the origin of the optical-infrared continuum radiation has to account for:

(i) the shape of the spectral energy distribution,
(ii) the redshift-dependant alignment with the radio in both the optical and infrared and
(iii) the continuity and low scatter in the infrared Hubble diagram.

Various mechanisms have been concocted, none of which are entirely satisfactory and some of which can be ruled out (e.g. Chambers et al. 1988). These include (with some objections indicated in parentheses):

(i) gravitational lensing (Le Fevre et al. 1987) (velocities change across source, continuity in IR Hubble diagram),
(ii) synchrotron emission (IR/optical emission much too large and not coincident with radio)
(iii) inverse Compton scattering of the microwave background by the radio-emitting relativistic electrons (Chambers et al 1988b) (IR/optical emission not coincident with radio emission)
(iv) anisotropic dust scattering (alignment seen in IR as well as blue).
(v) electron scattering of QSO nucleus (Fabian 1989, Rawlings and Eales 1989) (extended continuum does not have SED of quasar).
(vi) old stellar population (no viable process to keep such huge mass (10^{11} M_0 of stars aligned with radio, only marginally enough time to produce stars since Big Bang)
(vii) young stars formed by an interaction of the radio jet with the interstellar medium (Chambers et al 1987, McCarthy et al 1987) (contrived IMF necessary to produce observed SED, Continuity in IR Hubble diagram).

Considerable attention is now being devoted to investigating the plausibility of jet-induced star formation (e.g. Daly, 1989, Rees, 1989, De Young, 1989, Begelman and Cioffi, 1989) and although at present this mechanism appears to be an odds on favourite, there may be some dark horses still in the stable!

6. ARE WE SEEING GALAXY MERGERS?

In the spirit of this conference, I shall now proceed to consider the question whether galaxy mergers are playing some role in our high-redshift radio galaxies. First let us examine whether galaxy encounters are associated with nearby radio sources and active galaxies, which we can study in greater detail than the high-redshift galaxies. If we can implicate merging as having something to do with nuclear activity per se, then it would be a logical step to suppose that these processes also occur in the case of our radio galaxies at high redshift.

In their original paper reporting the identification, Baade and Minkowski reported that the galaxy "is a very curious object which seems to defy classification. The clue to a proper interpretation lieas in the fact that it has two nuclei that are tidally distorted and that hence we are dealing with the superimposed images of two galaxies." They concluded that Cygnus A consists of a pair of colliding galaies and that "the source of energy for the radio emission may be the relative kinetic energy of the colliding nebulae". Baade and Minkowski therefore believed that galaxy collisions were the direct source of the radio emission in Cygnus A, being unaware of the large body of data that we now have indicating that the origin of the radio emission lies in activity in galactic nuclei.

However, the wheel has come full circle in recent years and there is now considerable evidence which may implicate galaxy encounters in triggering the nuclear activity and therefore as an indirect producer of the radio emission. The observational data for this is the following:

(i) Although the party line used to be that radio galaxies are associated with elliptical galaxies, we now know that many of them (e.g. Cygnus A) have messy optical morphologies, with tails, fans, bridges, dust features etc. and the degree of messiness increases with the radio power (Heckman et al. 1986). The question is whether the messiness is produced by galaxy encounters, or some other processes more related to nuclear activity.

(ii) The IRAS far-infrared luminosities of radio galaxies increase significantly with radio power (Golombek et al. 1987). We know that large IR luminosities are often associated with multiple and presumably interacting galaxies.

(iii) Powerful radio galaxies appear to be in regions of significantly higher galaxy density than low-powered radio galaxies (Heckman et al. 1985, Yates et al 1989).

Not only the presence of radio emission, but also radio-quiet Seyfert activity is linked to merging. Keel et al. (1985) have argued that galaxy pairs are more likely to have Seyfert nuclei, while Dahari (1984) gave some evidence that Seyfert galaxies are more likely to have close neighbours.

From these results together with the clumpiness and turbulence of the Lyman alpha halos observed in 4C41.17, it is tempting to conclude that radio galaxies at large redshifts are undergoing mergers. However, it is clear that this chain of logic is riddled with circumstantial evidence and questionable assumptions. At this stage, we have a suspect, but do not have a case that would stand up in court.

7. FUTURE DREAMS

7.1 Increasing the Numbers. The establishment of the existence of a population of galaxies at epochs corresponding to $z > 3$ opens up exciting opportunities for studying the properties of the early Universe. For the first time we can measure the spatial distribution of brightness and velocity for gas in objects out to at least a redshift of 4. This will provide unique information about possible interactions in young and forming galaxies. In addition statistics of the space density of these objects and its redshift dependance is capable of providing considerable leverage on the geometry and evolution of the Universe.

Our original 4C sample covered less than a third of the sky and included only the brightest radio sources. One of the most promising areas suggested by the work until now is to extend our search technique to larger and fainter samples of radio sources. The spatial density distribution of these objects and its dependance on redshift is unknown, but a topic of great interest. Lyman alpha emission and radio emission from 4C41.17 could easily be seen if it were at a redshift of 6, above which Lyman alpha would be shifted outside optically observable wavelengths. We and others have begun extending our ultrasteep-spectrum search criterion to cover the whole sky and to fainter samples of radio sources.

7.2 Questions for Study. Enlarging the sample of and studying the properties of distant radio galaxies should allow several critical aspects of evolution and activity in galaxies to be investigated. Some pertain to cosmological problems and others to the "nature of the beasts". Some questions relevant to galaxy mergers are:

1. What are the (optical and infrared) spectral energy distributions and how do they vary across the galaxies? How do the morphologies differ from those of low-redshift radio galaxies and normal giant ellipticals? What do they tell us about the populations (and ages) of the stellar components?

2. How are the morphologies of gas, stars and relativistic plasma related? Are they clumpy? If so on what scale and are the clumps related to galaxy formation and possible merging, either by galaxies or by smaller condensations (e.g. protoglobular clusters)?

3. What are the dynamical properties of the gas in the z > 2 galaxies and what constraints can be placed on the masses of the various components? How do they differ from the dynamical properties of low-redshift radio galaxies and normal giant ellipticals?

4. What is the reason for the optical/radio alignment effect? Is the presence of the alignment connected more with the high redshifts of the 4C sample or with their large radio luminosities?

5. How do the properties of intermediate-redshift radio galaxies differ from the properties of nonactive galaxies in the same clusters? Are additional galaxies detectable close to our z > 2 galaxies? Neighbors might be expected if our objects are involved in mergers or if they are located in forming clusters or in regions of the universe which are preferentially conducive to galaxy formation.

During the next decade high-redshift radio galaxies will be the subject of intensive investigation with an arsenal of instrumentation spanning the complete electromagnetic spectrum, and will occupy the pencils and supercomputers of many leading theoreticians. At the end of this period, we should all be in a much better position to judge whether they are isolated asocial outcasts "wrapped up in their own selfishness" or whether they are indeed engaged in primeval cosmic courtship.

We am grateful to many friends for entertaining discussions, particularly Wil van Breugel, Mike Fall, Colin Norman and Peter Quinn. The work reported here was partially supported by a NATO travel grant and the Visitor program of the Space Telescope Science Institute.

REFERENCES

Baade, W. and Minkowski, R., 1954, Ap. J. 119, 206.
Barthel, P. D., and Miley, G. K., 1988, Nature 333, 2.
Baum, S. and Heckman, T., 1989, Ap. J., in press.
Begelman, M. C. and Cioffi, D. F., 1989, Submitted to Ap. J. Lett.
Blumenthal, G., Miley, G. 1979, Astronomy and Astrophysics 80, 13.
Chambers, K. C., Miley, G. K., van Breugel, W. 1987 Nature 329, 604.

Chambers, K. C., Miley, G. K., van Breugel, W. 1988a Astrophys.J. (Lett.) 327, L47.
Chambers, K. C., Miley, G. K., Joyce, R. R. 1988b, Astrophys.J. (Lett.) 329, L75
Chambers, K. C., Miley, G. K., van Breugel, W. 1989 to be submitted to Astrophys. J.
Dahari, O., 1984, Astron. J., 89, 966.
Daly, R. A. 1989, Preprint, Princeton University
De Young, D. 1989, Ap. J. Lett, in press.
Eales, S., Alexander, P. and Duncan, W. D., 1989, Mon. Not. Roy. Astro. Soc. , in press.
Fabian, A. 1989, Mon. Not. Roy. Astro. Soc., in press.
Golombek, D., Miley, G. K. and Neugebauer, G., 1988, Astron. J., 95, 26.
Keel, W. C., Kennicutt, R. C., Jr., Hummel, E., van der Hulst, J. M., 1985, A.J., 90, 708.
Heckman, T. M., Miley, G. K., Balick, B., van Breugel, W.J.M., and Butcher, H.R., 1982, Ap. J. 262, 529.
Heckman, T. M., Carty, T. J., and Bothun, G. D., 1985, Ap. J., 288, 122.
Heckman, T.M., Smith, E. P. Baum, S. A., van Breugel, W. J. M., Miley, G. K., Illingworth, G., D., Bothun, G. D., and Balick, B., 1986, Ap. J. 311, 526.
Lilly, S. J., Longair, M. S. 1984, Monthly Notices Royal Astron. Soc. 211, 833
McCarthy, P., et al., 1987a Astrophys.J. (Lett.), 319, L39
McCarthy, P. J., Van Breugel, W., Spinrad, H., Djorgovski, S. 1987, Astrophys. J. (Lett.) 321, L29
McCarthy, P. J. 1989, Ph.D Thesis, University of California.
Miley, G. K., Heckman, T. M., Butcher, H. R., and van Breugel, W. J. M., 1980, Ap. J. Lett. 247, L5.
Miley, G. K., 1988, Proc. IAU Symp. 124 on Observational Cosmology, eds. A. Hewitt, G. Burbidge and L. Fang, p267.
Miley, G. K. 1984, Proc. ESA Workshop on Quasat.
Rees, M., 1989, preprint, Univ. of Cambridge
Tielens, Miley, Willis, 1979, Astronomy and Astrophysics Suppl. 35, 153.
van Breugel, W.J.M., Miley, G. K., Heckman, T., Butcher, H. and Bridle, A., 1985, Ap. J. 290, 496.
van Breugel, W. J. M. 1986, Can J. Phys., 64, p 392.
Yates, M. G., Miller, L., and Peacock, J. A., 1989, Mon. Not. Roy. Astron. Soc., in press.

Morphology and Kinematics of Ly α in Distant Radio Galaxies

P.J. McCarthy

The Observatories of the Carnegie Institution of Washington

1. Introduction

The recent identification of large samples of faint radio sources has resulted in the discovery of a number of high luminosity galaxies at redshifts larger than 1. These objects are highly unusual in their morphological and kinematic properties. Given that there is a fairly well established connection between interactions, and possibly mergers, and nuclear activity in nearby radio sources, it seems plausible that tne high redshift radio galaxies are systems in which substantial interaction is occurring. I will discuss the morhpological and kinematic properties of these systems with the aim of determining if there is evidence for such interactions. My conclusion is that, while galaxy interactions may be occurring and may provide an abundant source of debris, interactions between nuclear outflow (both photons and particles) dominates the observed properties.

2. Emission Line Morphology

A recently completed emission line imaging survey of high redshift radio galaxies (McCarthy, Spinrad, and van Breugel *in prep*) has resulted in the discovery of a number of extemely large ($\sim$ 100 kpc) nebulae associated with galaxies at $z > 1.0$. These extented emission line regions have a wide range of morphologies and often distributed in long curving filaments or lumpy cloud like structures. The common underlying characteristic of these emission line regions is their close alignment with the axes of their double lobed radio sources (McCarthy *et al.* 1987). The degree of alignment between the emission line gas and the radio sources increases with redshift (Figure 1) until, for $z > 1$, nearly all of the nebulae are aligned within 15° of the radio axis.

Quantitative spectrophotometry of these objects strongly suggests that the gas is photoionized by a non-stellar source (McCarthy *et al.* 1988). The existance of photoionized gas along the radio axis, but *beyond* the radio hot spots is taken as evidence for ionization by a central source (van Breugel and McCarthy 1988). The lack of sufficient ionizing photons from the observed nucleus to account for the ioinzation of the gas, leads us to suppose that the central ionizing source is obscured, and preferentially emits ionizing photons into a small range of solid

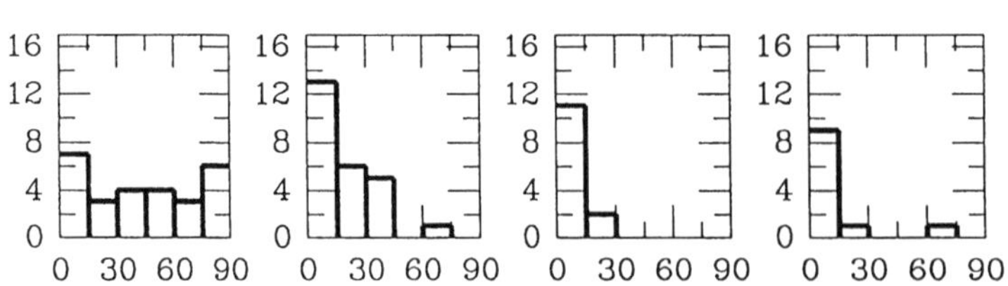

Figure 1. The position angle difference between the radio source axis and the emission line axis for 3C radio galaxies in four redshift bins. The redshift bins are, from left to right, $0.0 - 0.1, 0.1 - 0.6, 0.6 - 1.0, 1.0 - 2.0$. Some of the data in the lowest redshift bin are taken from Baum and Heckman 1989.

Dynamics and Interactions of Galaxies Editor: R. Wielen

angles. Such "anisotropic ionization sources" models have been proposed for a wide class of active galaxies (e.g. Miller and Antonucci 1987; Ward, Wilson, and Haniff 1988; Pogge 1989; Fosbury 1989). In this picture, then, the nucleus emits a beam of ionizing radiation that simply lights up whatever material it encounters.

If this picture of the ioinzation of the exended gas is correct, then one clearly needs a large supply of thermal gas at large distances from the central source. The total amount of ionized gas associated with typical 3CR nebulae at $z \sim 1$ probably lies in the range $10^7 - 10^9$ $M_\odot$. Since the ioinzing beam only illuminates a fraction of the surrounding medium, the total amount of thermal gas surrounding the galaxies will be roughly an order of magnitude larger; a significant fraction of the gas mass of a present day massive galaxy. There are a number of possible sources for such a large mass of cool gas (e.g. cooling flows; Crawford *et al.* 1988; galactic winds). One of the most natural sources, however, is the tidal debris of a strong interaction with a neighboring galaxy. While the ionization by the nucleus maybe the dominate process in determining the morphology of the gas, it is telling us that there is a great deal of material distributed around these galaxies, and tidal interactions seem a likely source of uch debree.

3. Kinematics

From long-slit spectra of the emission lines gas we can deduce it's velocity field. The velocity fields often have large amplitudes, in extreme cases with peak to peak differences of 1500 km s^{-1}. This is in marked contrast to low redshift radio galaxies which usually show low amplitudes (Baum 1989), although some show large amplitudes at the sites of strong interactions with the radio source. The velocity field of the Ly α emission associated with 3C 294 (z=1.79) is shown in Figure 2. The amplitude of the velocity field is large for a galaxy-galaxy interaction, but not unprecedanted (e.g.). Like many of the high redshift radio galaxies (e.g. 3C 324; Spinrad and Djorgovski 1984) the velocity field resembles solid body rotation. The total mass implied by the velocity field is quite large, $\sim 10^{12}$ $M_\odot$, comparable to the total mass of a giant elliptical at a similar radius.

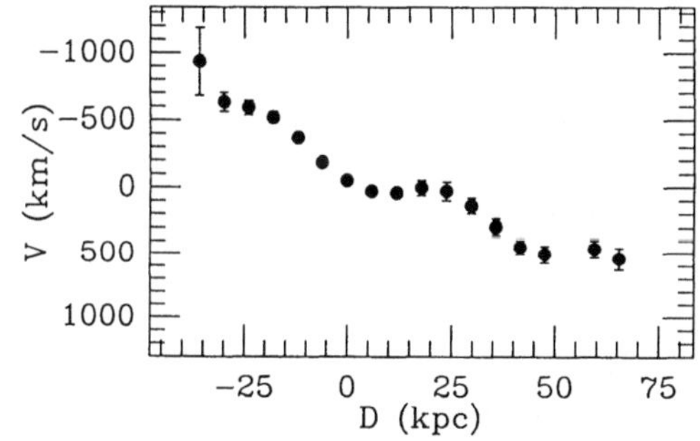

Figure 2. The velocity field of Ly α emission in the radio galaxy 3C 294 at z=1.79, from McCarthy *et al.* 1989. The zero point on the distance scale is the location of the radio core.

The high degree of alignment of the emission line gas and the radio source leads us to ask whether or not interactions with the radio plasma could be responsible for the large velocity fields. There is strong evidence for entrainment of gas to large velocities in nearby radio galaxies (e.g. van Breugel *et al.* 1986). While the emission line regions in our high redshift sample are aligned with the radio axis, in most cases they do not appear to be coincident with bright features in the radio. We have looked for correlations between the radio source parameters and the velocity field amplitude. We find good correlations with the radio luminosity and the linear size of the source, in the sense that more luminious and smaller radio sources have larger amplitude velocity fields. Both of these properties, however, are correlated with redshift, making an unambigious distinction between correlations with the radio and cosmic epoch difficult, if not impossible. It seems natural that smaller sources, which are more strongly confined, would live in deeper potential wells and consequently would show larger velocity fields. Until we are able to make similar observations of lower luminosity objects at similar redshifts, however, we will have difficulty in drawing any conclusions from the kinematic data.

4. Emission Line and Radio Source Asymmetries

One aspect of the morphology of the emission line regions is their high degree of asymmetry. This is particularly striking for $z > 0.3$. For redshifts $0.3 - 1.0$, 13/20 objects have emission lines detected only on one side of the nucleus. The origin of this high degree of asymmetry in the emission lines in unclear, but may be related to the source of the gas.

The radio sources also become markedly more asymmetric in their lobe properties at large redshift. The ratios of the distances to the far radio lobe from the core, to that of the close lobe differ significantly from unity for radio galaxies with $z > 0.7$. Further details of these results are given by McCarthy and van Breugel (1989). A similar increase in the bending of the radio structure of quasars at high redshift was found by Barthel and Miley (1987).

When comparing the asymmetry of the radio souces with that of the emission lines we find a strong correlation in the sense that the sources with extended emission lines are more likely to have large lobe asymmetries. Most remarkably, we find that in essentially all cases the side of the closer radio lobe is also the side of the high surface brightness emission lines. This is strong evidence for a causal link between the radio source asymmetry and the presence of dense clouds of thermal gas. As with the gas photoinized by the nuclues, one can think of the radio sources as unbiased probes of the surrounding medium on scales of 10 kpc to 1 Mpc. These probes tell us that the medium is very clumpy at large redshifts. Further evidence for a clumpy environment surrounding radio galaxies at large redshifts comes from the distribution of depolarizing material (Pedelty *et al.* 1989). Again a natural source of distributed material is the tidal debries from galaxy interactions.

5. Mis-aligned Radio Sources

Finaly, we consider the small fraction of radio galaxies at large redshifts that have large mis-alignments between the radio axis and the emission line gas. One example of this is the radio galaxy 3C 294 (z=1.79) whose 170 kpc cloud of Ly α emission is better aligned with its inner radio source structure than its radio lobes. An even large misalignemt is found for 3C 437 (z=1.45). A simple interpretation of this, in the context of the central ionization model, is that the collimation axis has changed in the time between the formation of the radio hot spots ($10^7 - 10^8$ yr) and the ionization of the emission line gas ($10^5 - 10^6$ yr). Precession or torqueing of the black hole spin axis could result from a merger leading to the formation of a binary black hole. Binary black holes are amoung the longest lived fossil remnants of a merger event involving two compact nuclei (Begelman, Blandford and Rees 1980; Roos 1988).

6. Conclusions

While there is little direct evidence for galaxy interactions and mergers at associated with high redshift radio galaxies, there is some indirect evidence. We should not be alarmed or surprised at the lack of direct evidence, considering that $1''$ corresponds to ~ 8 kpc, and that the luminosities of these objects are more than an order of magnitude larger than the most spectacular starbursts today. Any direct evidence for interactions may well be lost in the glare of the strong interactions between the environment and the particles and photons eminating from the nucleus. The nuclear emissions can, however, act as probes of the medium on modest scales. These probes tell us that the medium was populated with large amounts of poorly organized material. We suggest that this material may well be the remains of galaxies involved in strong interactions and possibly merger events.

Much of the work reported here was carried out in collaboration with Wil van Breugel and Hy Spinrad at the University of California.

References

Antonucci, R. R. J., and Miller, J. S. 1985, *Ap. J.*, **297**, 621.
Barthel, P. D., and Miley, G. K. 1988, *Nature*, **333**, 319.
Baum, S. A., and Heckman, T. M. 1989, *Ap. J.*, **336**, 702.
Baum, S. A. 1989, *in Extranuclear Activity in Galaxies*, Kluwer, eds. R. A. E. Fosbury and E. Meurs, *in press.*
Begelman, M. C., Blandford, R. D., and Rees, M. J. 1980, *Nature*, **287**, 307.
Crawford, C., Fabian, A. C., Johnstone, R. and Thomas, P. 1988, *M. N. R. A. S.*, **235**, 183.
Fosbury, R. A. E. 1989, *in Extranuclear Activity in Galaxies*, Kluwer, eds. R. A. E. Fosbury and E. Meurs, *in press.*
McCarthy, P. J., van Breugel, W. J. M., Spinrad, H., and Djorgovski, S. G. 1987*b*, *Ap. J. Letters*, **321**, L29.
McCarthy, P. J. *et al.* 1988, in *Cooling Flows in Clusters and Galaxies*, ed. A. C. Fabian, Kluwer Academic Press, p. 325.
McCarthy, P. J., and van Breugel, W. J. M. 1989*b*, *Ap. J.*, *submitted.*
McCarthy, P. J. *et al.* 1989, *Ap. J.*, *submitted.*
Pedelty, J. A., Rudnick, L., McCarthy, P. J., and Spinrad, H. 1989*a*, *A. J.*, **97**, 647.
Pogge, R. W. 1988, *Ap. J.*, **328**, 519.
Robinson, A., Binetta, L., Fosbury, R. A. E., and Tadhunter, C. N. 1987, *M. N. R. A. S.*, **227**, 97.
Roos, N. 1989, *Ap. J.*, **334**, 95.
van Breugel, W. J. M., and McCarthy, P. J. 1988, *in* 5th *Annual George Mason Workshop on Active Galactic Nuclei*, eds. H. R. Miller and P. Wiita, Spinrad-Verlag.
van Breugel, W. J. M., Heckman, T. M., Miley, G. K., and Filippenko, A. V. 1986. Astrophys. J. **311**, 58.
Wilson, A. S., Ward, M. J., and Haniff, C. A. 1988, *Ap. J.*, **334**, 121.

Interactions in "Butcher-Oemler" Clusters

R.J. Lavery

Mount Stromlo and Siding Spring Observatories, The Australian National University, Woden Post Office, ACT2606, Canberra, Australia

In their pioneering photometric study on the color evolution of galaxies in rich clusters, Butcher and Oemler (1978) found that two Coma-like clusters of galaxies at redshifts near 0.4 contained a much larger population of "blue" galaxies than expected. This suggested rapid color evolution of galaxies in 1/3 of a Hubble time. In order to put this on a more quantitative level, Butcher and Oemler (1984a) undertook a photometric study of 33 clusters of galaxies in the redshift range of 0.0 to 0.5. This study confirmed their original result, showing that the cores of a number of concentrated clusters at redshifts greater than 0.2 had a blue galaxy fraction of order 20%, whereas such clusters nearby had a blue galaxy fraction of only 3%. Still, this result was not readily accepted until spectroscopic observations were made of the clusters containing this large blue galaxy population.

There now exists substantial evidence, from spectroscopic observations, to support the existence of this large blue galaxy fraction in distant clusters. The two original "BO" clusters, the cluster associated with the 3C 295 radio galaxy ($z = 0.46$) and Cl 0024+1654 ($z = 0.39$), were observed by Dressler and Gunn (1983) and Dressler, Gunn and Schneider (1985). Though the case for 3C 295 was unclear, Cl 0024+1654 exhibited a large blue galaxy population. Spectroscopic observations of a "loose-type" cluster, Cl 1447+2619 ($z = 0.37$), by Butcher and Oemler (1984b), confirmed a large blue population existed in this cluster. At the lower redshift of 0.2, Lavery and Henry (1986) confirmed the large blue galaxy population for three clusters; A223, A963 and A2111. Henry and Lavery (1987) observed the last cluster at a redshift of 0.4 studied by Butcher and Oemler, A370, and confirmed the large blue galaxy population in this cluster. Additional spectroscopic observations by Couch and Sharples (1987) of three clusters at redshift 0.3, by Newberry, Kirshner, and Boroson (1988) of several "BO" clusters, and by Dressler and Gunn (1989) of a sample of clusters at redshifts near 0.45, all support the presence of a large blue galaxy population existing in distant clusters.

Photometric studies of Cl 0016+16 ($z = 0.54$) and II Zw 1305.4+2941 ($z = 0.24$) by Koo (1981) and Koo et al. (1988), which suggest no substantial blue population exists in these clusters, have been used to argue that the "BO" effect is not a universal phenomenon. But, spectroscopic observations of Cl 0016+16 by Dressler and Gunn (1986) have shown the existence of a red "post-starburst" population in this cluster. They suggest that this cluster may have had a very different galaxy color distribution if observed 1 Gyr. earlier.

Dynamics and Interactions of Galaxies Editor: R. Wielen

This large blue galaxy population in distant clusters can be explained by the presence of three different types of galaxies: 1) galaxies containing an active galactic nucleus, 2) galaxies with continuous star formation, "normal" spirals, and 3) galaxies with enhanced star formation, "burst" galaxies. Although each of these types of blue galaxies have been found in the above mentioned spectroscopic work, there is now strong evidence that the predominant type of galaxy is the "burst" galaxy (Lavery and Henry 1986; Couch and Sharples 1988), with some variation from cluster to cluster.

Two important questions on the "Butcher-Oemler" effect remain to be answered.

1) What is the mechanism inducing these bursts of star formation?

2) What is the resulting morphology of these burst galaxies?

At the present time, the two strongest candidate mechanisms are ram pressure-induced star formation, the shocking of the galactic ISM into star formation by the hot intracluster gas, and galaxy-galaxy interactions, in which gravitational disturbances induce the star formation. Each of these mechanisms was originally suggested by Dressler and Gunn (1983) and discussed by Bothun and Dressler (1986) in order to explain the "post-starburst" galaxies in the 3C 295 cluster. Both preferred the ram pressure-induced star formation mechanism because of the high velocity dispersions (1000 km/s) of the clusters. Interactions that are to produce strong gravitational effects on the galaxies require a much lower relative velocity for the interacting galaxies (Toomre and Toomre 1972).

In order to determine which mechanism is inducing the star formation, images of the blue galaxies are essential. It was thought that such imaging observations would require resolution capable only with the Hubble Space Telescope. But, the excellent observing conditions of the Mauna Kea Observatory, which regularly produce sub-arcsecond seeing, have made it possible to start to address this question from the ground.

Initial imaging observations of several of these blue galaxies were made by Thompson (1988). But, having observed only four galaxies made drawing any substantial conclusions impossible.

The first statistical evidence, based on imaging observations of the blue galaxies at the redshift of 0.2, to help distinguish between the two possible mechanisms was presented by Lavery and Henry (1988). Through the use of a nearest-neighbor analysis for ten emission line galaxies and a carefully chosen sample of 20 red galaxies, they found the blue emission line galaxies tended to have much closer "companion" galaxies than the red galaxy sample. These closer neighbor galaxies support the hypothesis of galaxy-galaxy interactions being the primary mechanism for the enhanced star formation, as did the appearance of several of the galaxies themselves. Yet, the image quality of these observations were not the best obtainable from the Mauna Kea Observatory, so additional observations of these galaxies were undertaken.

In a collaborative program with J. P. Henry (IfA/Univ. of Hawaii), we now have deep B and R CCD images of all the confirmed blue cluster members in A963 (9 galaxies), A2111 (9 galaxies), and A2125 (1 galaxy). These images were obtained on the University of Hawaii 2.2-meter Telescope of the Mauna Kea Observatory under seeing conditions of 0.8 arcseconds.

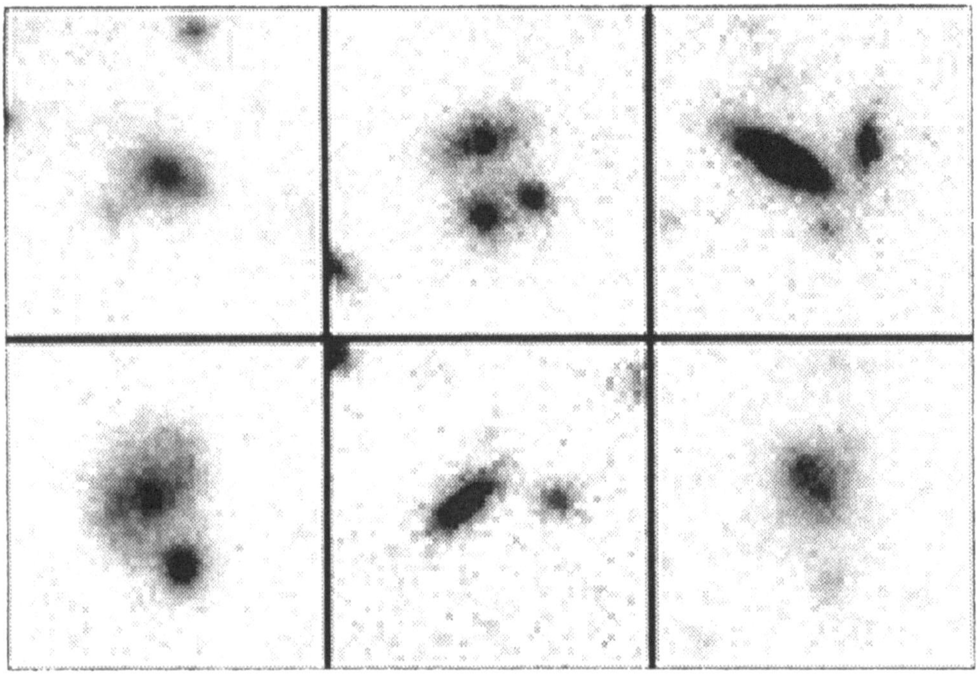

Figure 1. Six of the more interesting blue galaxies. From left to right, first row: Abell 963 BOW #60; Abell 2111 BOW #218; Abell 2125 BOW #31. Second row: Abell 2111 BOW #48; Abell 2111 BOW #76; Cl 1447+2619 BOW #7. Each box is 15 arcseconds on a side with north up and east to the left.

Exposures were 30 minutes for the B band and 15 minutes for the R band. In addition, we have a single deep R band image of two emission line galaxies in Cl 1447+2619 at a redshift of 0.37. Grey-scale images of six of the more interesting blue galaxies are shown in Figure 1. The galaxy identification is that used by Butcher, Oemler and Wells (1983; BOW).

A major point to emphasis about these galaxies is that all but one (#218) were originally identified as single objects, though there are actually multiple components to each. The first two galaxies (#60 and #218) have spectra that show extremely strong [O II] emission, equivalent widths of order 80 Angstroms, suggestive of a recent strong burst of star formation. BOW #60 is almost certainly experiencing an interaction with the faint galaxy to the north with a counter-arm being produced to the south-east (see the contour plot in Lavery and Henry, 1988). No such conclusion can be reached for BOW #218. The next galaxy, BOW #31, also an emission line galaxy, shows obvious spiral structure with a "companion" galaxy to the north at the end of the north-west spiral arm. The next two galaxies (#48 and #218) are both "triple" systems of rather complex nature. The spectra of BOW #48 shows no emission lines, whereas BOW #76 shows both [O II] and [O III]. Finally, BOW #7 is the strongest emission line galaxy found in the Cl 1447+2619 cluster of galaxies. It, too, has a faint galaxy to the north-west and is extended towards this faint galaxy, suggesting an interaction between these objects.

A more detailed analysis of all the images will be presented in a paper under preparation, but it is still possible to make several comments based on the images presented here. First, there seems to be little doubt that, at least, some of the blue galaxies are the result of galaxy-galaxy interactions. The best cases are BOW #60 and BOW #48, with BOW #31, BOW #76 and BOW #7 being highly likely examples. Second, the final morphology of some of the burst galaxies is a spiral galaxy. BOW #31 shows obvious spiral arms and BOW #60 will likely evolve into a spiral. Such structure is only obvious as these galaxies are observed face-on. The major components of both BOW #218 and BOW #48 are likely disk systems, but lack any detail due to their edge-on orientation.

To further test the hypothesis of galaxy-galaxy interactions as the inducing mechanism, two types of observations would be extremely useful. First, it is necessary to find more galaxies with spectra similar to BOW #60 and BOW #218. These galaxies, with strong [O II] emission, are in the very early stage of the burst of star formation and should provide the best clues to the burst mechanism. The fact that both galaxies have a close neighbor, and with BOW #60 undoubtedly being in an interacting system, certainly supports the galaxy-galaxy interaction burst mechanism. Second, high-resolution imaging observations of the clusters at redshifts of 0.4 are needed. Such imaging should be obtainable from Mauna Kea, especially with a simple image-stabilizing camera, such as is now available at the CFHT. There is no reason to wait for observations with the Hubble Space Telescope when the potential for such observations already exists with ground-based telescopes.

REFERENCES

Bothun, G. D., and Dressler, A., 1986. Ap. J., 301, 57.
Butcher, H. and Oemler, A., 1978. Ap. J., 219, 18.
Butcher, H. and Oemler, A., 1984a. Ap. J., 285, 426.
Butcher, H. and Oemler, A., 1984b. Nature, 310, 31.
Butcher, H., Oemler, A., and Wells, D. C., 1983. Ap. J. Supp., 52, 183.
Couch, W. J., and Sharples, R. M., 1987. M.N.R.A.S., 229, 423.
Dressler, A., and Gunn, J. E., 1983. Ap. J., 270, 7.
Dressler, A., and Gunn, J. E., 1986. Presented at STScI Symposium No. 1, Stellar Populations.
Dressler, A., and Gunn, J. E., 1989. In preparation.
Dressler, A., Gunn, J. E., and Schneider, D., 1985. Ap. J., 297, 70.
Henry, J. P., and Lavery, R. J., 1987. Ap. J., 323, 473.
Koo, D. C., 1981. Ap. J. (Letters), 251, L75.
Koo, D. C., Kron, R. G., Nanni, D., Trevese, D., and Vignato, A., 1988. Ap. J., 333, 586.
Lavery, R. J., and Henry, J. P., 1986. Ap. J. (Letters), 304, L5.
Lavery, R. J., and Henry, J. P., 1988. Ap. J., 330, 596.
Newberry, M. V., Kirshner, R. P., and Boroson, T. A., 1988. Ap. J., 335, 629.
Thompson, L. A., 1988. Ap. J., 324, 122.
Toomre, A., and Toomre, J., 1972. Ap. J., 178, 623.

Part II

Our Galaxy and Interactions

Our Fossil Galaxy

K.C. Freeman

Mount Stromlo and Siding Spring Observatories, The Australian National University, Woden Post Office, ACT 2606, Canberra, Australia

1. Introduction

The main components of our Galaxy are (**1**) the disk (thin and thick), which is rapidly rotating; (**2**) the bulge; (**3**) the metal weak halo, which rotates very slowly, and (**4**) the dark corona, which provides most of the mass. Near the sun, there is no rotationally intermediate population.

The halo and the bulge are supported mainly by their velocity dispersion. In a famous and influential paper, Eggen, Lynden-Bell and Sandage (1962) proposed that these pressure supported components formed before the disk, through the rapid collapse of part of the protogalaxy. This proposal was based on evidence at that time that the orbital and chemical properties of halo stars are tightly related, in the sense that the most metal weak halo stars have the largest orbital eccentricities and the lowest systemic rotation. In this picture, the stellar abundances and the kinematics are both measures of the time elapsed since the beginning of the collapse: as the collapse proceeds, the mean chemical abundance of the interstellar medium increases, and so does the mean systemic rotation of the stars forming at each stage of the collapse.

An alternative view of the formation of the metal weak halo and bulge developed during the 1970's, and crystallized at the Yale conference in 1977. Toomre (1977) commented:

> "It seems almost inconceivable that there wasn't a great deal of merging of sizeable bits and pieces (including quite a few lesser galaxies) early in the career of every major galaxy, no matter what it now looks like. The process would obviously have yielded halos from the stars already born, whereas any leftover gas would have settled quickly into new disks embedded within such piles of stars. More or less simultaneously though, any compact and presumably more metal rich partners or galaxy cores should have sunk quite deep into this brew before losing their separate identities, thereby contributing to what we call central bulges."

At about the same time, Searle and Zinn (1978: SZ) showed that there was no radial chemical gradient among the outer galactic globular clusters (*i.e.* the chemical and dynamical properties of the globular clusters are apparently decoupled), and they proposed that the galactic halo formed by the accretion of small metal weak "fragments", such as gas rich dwarf galaxies. SZ then showed that the shape of the chemical abundance distribution expected in this accretion picture agrees well with the distribution observed for the galactic globular clusters. Mass

Dynamics and Interactions of Galaxies Editor: R. Wielen

loss from these dwarf galaxies, which halts their star formation and chemical evolution, is important for the predicted abundance distribution. Other major contributors to this general idea include Tinsley and Larson (1979), Silk and Norman (1979), Rodgers *et al.* (1981) and Roos (1981.

If the galactic spheroid (*i.e.* the halo and bulge) formed in this way, then the dynamical properties of the spheroid would depend on the dynamics of accretion and merger events, and not on the dynamics of monolithic collapse. In particular, the kinematical and chemical properties of the halo objects would depend on the orbital and chemical properties of the infalling fragments, so we should no longer expect a tight correlation of kinematics and chemical abundance among the halo stars.

Accretion events would affect the galactic disk, because disks are easily heated by such events (*e.g.* Hernquist and Quinn, 1989). Our galaxy has a substantial thin disk, so most of these spheroid-forming accretion events must have occurred early in the Galaxy's life, while the galactic disk was mainly gaseous and could therefore still resettle. We should also note that many disk galaxies have small or negligible bulges, so whatever the bulge-forming process is, a significant fraction of disk galaxies are able to avoid it.

Most of this talk will be an elaboration of the ideas of Toomre and SZ about the formation of the halo and bulge. Many of the properties of these components can be readily interpreted in terms of an accretion and merger picture for their formation. The first topic provides some fairly direct evidence for a recent accretion event.

2. The High Velocity A Stars

Rodgers (1971) discovered that many of the halo A stars at distances of a few kpc in the direction of the SGP are young main sequence stars with abundances that are close to solar and a velocity dispersion of about 60 km s^{-1}. Further work by Rodgers *et al.* (1981) and Lance (1988) confirmed these conclusions. From Lance's study, there are 21 known main sequence A stars towards the SGP, at distances between 1 and 11 kpc from the galactic plane, with [Ca/H] in the range 0 to -0.5 and with a velocity dispersion of 62 km s^{-1}. These stars are clearly main sequence stars, not horizontal branch stars, and their ages are less than 6.10^8 y. (For comparison, the disk A stars have ages up to about 20.10^8 y.) The restricted age range of these halo A stars means that they are unlikely to be blue stragglers or normal disk stars randomly ejected from the galactic plane. Rodgers *et al.* and Lance argued that the presence of these stars is evidence for a recent accretion event (about 6.10^8 years ago, from Lance's work). They suggest that a source of relatively low abundance HI merged at high velocity with the galactic disk, producing these young stars which do not conform to the usual age-abundance-kinematics relation.

3. Some Properties of the Metal Weak Halo

Figure 1, taken from the review by Gilmore *et al.* (1989), shows the azimuthal velocity V_{rot}, relative to a nonrotating frame, against the abundance [Fe/H] for a large sample of nearby high proper motion stars observed by Laird *et al.* (1988). The thin, rapidly rotating old disk and the somewhat hotter but still rapidly rotating thick disk are seen in the top left of the figure, for [Fe/H] $\gtrsim -1.2$. The rest of the figure is occupied by stars of the slowly rotating halo, which show no trend of mean rotational velocity with [Fe/H]. A curve of mean V_{rot} against abundance for the entire sample of stars would show a gradual decline, from rapid rotation for the metal rich stars with [Fe/H] $\gtrsim -0.5$ to near zero rotation for [Fe/H] $\lesssim -1.5$. However, constructing a mean V_{rot}–abundance relation in this way would ignore the obvious presence of the separate disk and halo components and their overlap in abundance.

Here is a brief summary of the properties of the metal weak halo ([Fe/H] $\lesssim -1$): see Freeman (1987) for further references. In the solar neighborhood, the local density of the halo is about 0.1 to 0.2 percent of the disk density. Its radial density distribution falls off as $r^{-3.5}$ and its total mass is about $2.10^9\ M_\odot$. The rotational velocity of the metal weak halo is near zero (17 ± 24 km s^{-1} for a *non-kinematically* selected sample of halo stars compiled by

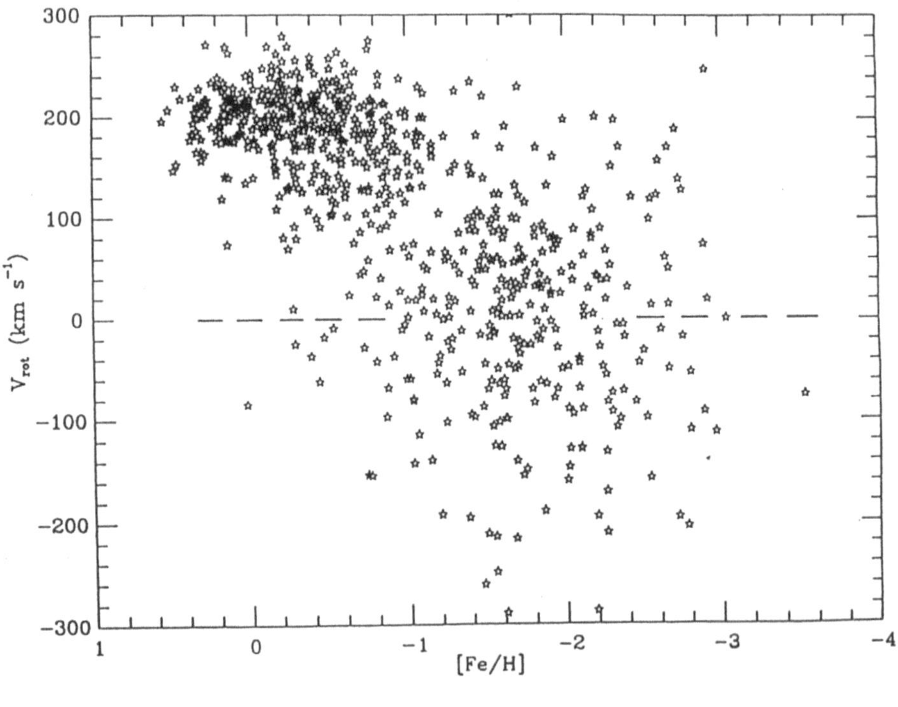

Figure 1:–The relation between rotational velocity, relative to a nonrotating frame, and [Fe/H] for the sample of high proper motion stars observed by Laird *et al.* (1988).

Morrison *et al.* 1989). This slowly rotating halo population covers a wide range in abundance, from [Fe/H] $\gtrsim -1$ to [Fe/H] $\lesssim -3$, and there is no clear evidence for a radial abundance gradient or a change of mean rotational velocity with abundance.

The presence of this slowly rotating halo population in an otherwise rapidly rotating galaxy requires some explanation. The large discontinuity in rotation between the disks and the halo, and the apparent decoupling of kinematic and chemical properties within the halo itself, suggest that the halo may not have formed as part of the simple monolithic collapse of a rotating protocloud. We now discuss the possibility that the metal weak halo formed by the accretion of small satellites or fragments during (and after) the dissipative collapse of the disk.

4. Satellite Accretion

In the SZ picture, the galactic globular clusters formed in small satellite galaxies, which were then accreted by the Galaxy. These satellites may have been like the Magellanic Clouds, which are known to be favorable sites for globular cluster formation: globular clusters have been forming in the Magellanic Clouds throughout their lives, and continue to form now. Or they may have been young versions of the present day nucleated dwarf ellipticals; the nuclei of these galaxies have about the right integrated magnitude to become globular clusters after accretion and some tidal stripping of the nuclei themselves (*e.g.* Impey *et al.* 1988); see §5. The objects which we now identify as galactic globular clusters were dense enough to survive the accretion event. The remainder of the accreted satellite would have been tidally disrupted and become part of the metal weak halo. We might then expect to find some incompletely mixed debris of disrupted satellites in the halo. This debris would now appear as moving stellar groups in the halo, and there is some evidence for the existence of such halo groups (*e.g.* Eggen, 1979; Sommer-Larsen and Christensen, 1987; Doinidis and Beers, 1989). In this picture of halo formation, the metal weak halo comes from the debris of pre-formed, short-lived metal-weak satellites. The kinematics of the halo then depends on the dynamics of sinking and merging satellites. Although the formation of satellites or fragments around a collapsing galaxy is not well understood, this picture offers some possibility of understanding the marked kinematic discontinuity between the slowly rotating metal-weak halo and the rapidly rotating disk population.

Quinn and Goodman (1986) made a preliminary dynamical study of the accretion of satellites by a disk galaxy. The main results of their study are: (**1**) Satellites with masses of a few percent of the parent disk mass, and with prograde orbits that come within about 8 disk scale lengths of the parent, are captured by dynamical friction in a few orbital periods. The capture time is longer for satellites in retrograde orbits. (**2**) If the orbits do not come within about 8 scale lengths, then the satellite survives. (**3**) If the inclination of the satellite orbit to the disk plane is less than about 60°, then the orbit is dragged down into the disk plane and then decays radially until the satellite is tidally disrupted. (**4**) The orbital energy of the satellite goes into heating the parent disk, in the plane and in the vertical direction.

If satellite accretion produced the galactic metal-weak halo, then it follows that most of the accretion events must have occurred while the disk was mainly gaseous, so that the disk could settle down again. The stellar component of this early disk, which would still be relatively metal weak, would remain puffed up, and this may be what we now recognize as the thick disk. If this identification is correct, then the thick disk should be very old, *i.e.* older than most disk stars and as old as the epoch of halo-building accretion. The ages of the oldest known disk clusters are 10 to 12 Gyr (*e.g.* Kaluzny, 1989). Unfortunately, the age of the thick disk near the sun is still in dispute. Carney *et al.* (1989) and Gilmore *et al.* (1989), using the turnoff colors as a function of abundance for the Laird *et al.* (1988) sample of nearby stars, argued that the nearby thick disk stars are as old as the globular clusters (*i.e.* at least 12 to 14 Gyr). On the other hand, Norris and Green (1988) argued that the thick disk is several Gyr younger than the globular clusters, from the apparent absence of red horizontal branch stars in a sample of thick disk stars towards the SGP.

5. The Globular Clusters

The galactic globular cluster population has two well defined components (Zinn, 1985). The clusters with [Fe/H] $\lesssim -0.8$ form a spherical, slowly rotating halo. The clusters with [Fe/H] $\gtrsim -0.8$ lie in a rapidly rotating, disk-like subsystem in the inner regions of the Galaxy. The kinematics and scale height of the disk cluster distribution are similar to those of the stellar thick disk: *i.e.* the rotational velocity $\approx$ 190 km s^{-1} and the scale height $\approx$ 1 kpc (Armandroff, 1989).

Although there is no evidence for a radial abundance gradient among the halo clusters, the second parameter effect (*i.e.* departures from a smooth dependence of horizontal branch morphology on the chemical abundance) increases with galactocentric distance. The most likely interpretation of this observation is that the outer clusters formed later and over a longer period (Zinn, 1980). There is direct evidence now for an age difference of about 3 Gyr between the second parameter pair of halo clusters NGC 288 and NGC 362, in the sense expected from their horizontal branch morphologies (Bolte, 1989). It is not yet clear whether the disk and halo globular clusters have similar ages.

We know that globular clusters can form now in the disks of late type galaxies such as the Magellanic Clouds and M33, but apparently not now in the disk of our Galaxy. Despite this clue, the formation of globular clusters remains poorly understood. We have already mentioned the SZ picture for the formation of the halo clusters in "fragments"; see Larson (1988) for an interesting discussion of this process. A related view comes from Rodgers and Paltoglou (1984) who made rotation solutions for galactic globular clusters in several abundance intervals. For most intervals, the mean rotation V_{rot} was in the range 40 to 100 km s^{-1}. However, the 30 clusters with $-1.3 >$ [Fe/H] > -1.7 gave a retrograde $V_{rot} = -70$ km s^{-1}. They suggested that these retrograde clusters may have come from a common (and fairly large) accreted satellite galaxy, like the LMC. A specific mechanism for cluster formation was proposed by Fall and Rees (1985). They suggested that clusters form by fragmentation of the collapsing protogalaxy. Thermal instability of the metal weak gas leads to a two-phase medium: in the cooler phase ($T \sim 10^4 K$), the critical mass for gravitational instability is about $10^6 M_\odot$, which is typical of the mass of the larger globular clusters.

In the SZ picture, can we identify the globular-cluster-forming "fragments" with any particular kind of irregular or dwarf galaxy that is now observed ? Systems like the LMC and SMC are probably not suitable candidates: although they are now forming globular clusters, the masses of their young clusters are smaller than the masses of the typical galactic globular clusters (Lupton *et al.* 1989).

The nucleated dwarf ellipticals (dEn) may be more appropriate objects (Zinneker *et al.* 1988). From their properties, summarized below, it seems possible that the nuclei of such dwarfs could become the galactic halo clusters. The nuclei of the dEn's observed today have absolute magnitudes in the range $-8 \gtrsim M_B \gtrsim -12$, which corresponds to a few percent of their total light. The incidence of nucleation among the dE galaxies appears to decrease with decreasing integrated luminosity; only about 20 percent of the Virgo dE galaxies with integrated magnitudes fainter than $M_B \approx -12$ show nuclei (Sandage *et al.* 1985; van den Bergh, 1986). The integrated spectra of some of the nuclei are similar to the spectra of galactic globular clusters with abundances between those of M15 and 47 Tuc; others appear to be of intermediate age (Bothun and Mould, 1988; Winsall and Freeman, 1989). Their velocity dispersions are in the range 20 to 45 km s^{-1}, similar to those of the largest globular clusters in the Galaxy and M31 (Peterson, 1989; Peterson and Caldwell, 1988). The nuclei and the envelope of these systems have similar colors (Caldwell and Bothun, 1987). From their distribution of apparent shape, van den Bergh (1986) showed that the dEn's are not disklike systems. Some dEn galaxies are found near large field galaxies like the SO system NGC 3115. There are large numbers of dEn galaxies in the Virgo and Fornax clusters (Binggeli *et al.* 1987; Ferguson, 1989). The spatial distribution of dEn systems in the Virgo cluster is much more concentrated than the distribution of non-nucleated dwarf ellipticals; the center of the concentration of the dEn systems is closer to the giant elliptical M86 than to M87 (Binggeli *et al.* 1987). Is this observation related to the high specific frequency of globular clusters in M87 (*e.g.* Harris, 1986) ? The Virgo cluster dEn's have a large velocity spread, from -700 to +1700 km s^{-1} (Bothun and Mould, 1988), which may explain why so many of them survive against accretion in the cluster environment.

Dwarf galaxies are apparently still nucleating. An example is NGC 1705, which is an isolated blue compact galaxy with a bright young stellar nucleus (Melnick *et al.* 1985). The mass of this nucleus is about $2.10^6 M_\odot$ and its age is about 1.10^7 y. Meurer (1989) has shown that this galaxy is losing its gas on a timescale of about 3.10^7 y, through a spectacular bipolar gas outflow powered by the stellar nucleus. If this galaxy remains bound, it will look like a dEn within 10^9 y. NGC 1569, 3077 and 5253 are other nearby irregular dwarf galaxies which share some of the properties of NGC 1705.

Nucleated dwarf galaxies would appear to be promising building blocks for the galactic metal weak halo. The halo contains $\sim$ 100 globular clusters, with a total mass $\sim 10^7 M_\odot$. The mass of the entire metal weak halo is about $2.10^9 M_\odot$. In the dEn galaxies observed now, the nuclei provide a few percent of the total light and therefore (because of the similarity in color of the nuclei and envelopes) probably provide a few percent of the total luminous mass. If such a dEn system were accreted by the Galaxy, its diffuse envelope and the outer part of its nucleus would be stripped off tidally to become part of the galactic field halo. If the inner parts of the nucleus are dense enough to survive the accretion event, the nucleus would then continue its evolution as a halo globular cluster. Not much is known yet about the density distribution in the nuclei of nucleated dwarfs, so it is difficult to estimate how much of the nucleus would remain after a typical accretion event. We can make a rough estimate of the density of the nucleus of the blue compact galaxy NGC 1705. At a distance of about 5 Mpc, its nucleus is unresolved on good seeing images; its mass is about $2.10^6 M_\odot$, so its mean density is at least that of typical galactic globular clusters.

Four comments on the possible formation of the metal weak halo by accretion:

(**1**). The disk globular clusters are metal rich ([Fe/H] > -0.8). The absence of metal-weak *disk* clusters suggests that the metal weak clusters formed in fragile systems, like the nucleated dwarfs (which were unable to survive orbit decay), rather than in denser, more robust satellites.

(**2**). The second parameter effect, with the interpretation that the outer clusters formed later and over a longer period than the inner clusters, fits naturally into the accretion picture. The outer satellites or "fragments" survive longer against accretion; nucleation is an ongoing process (*e.g.* NGC 1705), and the outer accreted satellites can therefore include systems that nucleated later.

(**3**). As Gilmore *et al.* (1989) have emphasized, there is a significant change in the slope of the [O/Fe]-[Fe/H] relation for nearby stars at about the same value of [Fe/H] at which the change in kinematics from disk to halo occurs (this occurs at [Fe/H] $\approx$ -1: see Figure 1). For the metal weak stars with [Fe/H] < -1, the [O/Fe] value is approximately constant at about 0.5. For the more metal rich stars, [O/Fe] declines smoothly to zero at [Fe/H]$= 0$. This is usually interpreted in terms of nucleosynthesis theory: oxygen is produced by massive stars ($M \gtrsim 20 M_\odot$), in type II supernovae, while iron is produced also in type I supernovae, from stars with $M \gtrsim 3M_\odot$. The constant [O/Fe] ratio for stars with [Fe/H]< -1 then suggests that these metal weak stars formed on timescales shorter than the timescale on which low mass supernovae explode (probably a few $\times 10^8$ years). This fits in well with the nucleated dwarf accretion picture. The [O/Fe] data would require that the epoch of star formation in these dwarfs is shorter than a few $\times 10^8$ y. We recall that Meurer's (1989) observations of the bipolar gas outflow in the blue nucleated dwarf NGC 1705 show that its gas will be removed on a timescale of about 3.10^7 y.

(**4**). There is a difference in the distribution of chemical abundance for the metal weak field stars and the halo clusters, in the sense that there are no globular clusters with [Fe/H]< -2.4 while field stars with much lower [Fe/H] are known (see Carney 1988 for a recent review). This difference, which is only marginally significant statistically, is consistent with the view that the globular clusters formed as the nuclei of nucleated dwarfs. As mentioned above, the incidence of nucleation in dwarfs is lower in the fainter dwarfs. In the accretion picture, the metal weakest field stars (which are very rare) would form in the lowest mass dwarfs (which would be expected to have the lowest metallicity). These low mass dwarfs may be too small to nucleate, and would therefore produce no globular clusters of very low abundance.

6. The Two-Component Metal Weak Halo

Hartwick (1986) showed that the galactic distribution of RR Lyrae stars with [Fe/H] < -1 appears to have two components: (i) a flattened inner component, for $R \lesssim 8$ kpc and a spherical outer component. Sommer-Larsen (1986) compared the kinematics of non-kinematically selected nearby metal weak stars in two abundance intervals: $-1.2 >$ [Fe/H] > -1.5 and [Fe/H] < -1.5. He found that the radial and azimuthal velocity dispersion components for both groups were similar, at about 140 km s^{-1} and 110 km s^{-1} respectively. However the vertical (z) components of the velocity dispersion were markedly different (59 ± 8 km s^{-1} for the metal richer stars and 102 ± 7 km s^{-1} for the metal weaker star), so it seems clear that the less metal weak stars form a more flattened distribution. Also, the less metal weak stars show somewhat more systemic rotation (67 ± 19 km s^{-1} against 29 ± 11 km s^{-1}).

In the accretion picture, we could interpret this observation in terms of the present properties of dwarf galaxies. The brighter dwarf galaxies are in the mean more metal rich and denser (see Dekel and Silk, 1986 for a summary). As satellites of a denser galaxy, the brighter dwarfs would survive orbit decay and tidal disruption in to smaller galactocentric distances, and their orbits at destruction would have lower inclinations and their debris would have a smaller vertical velocity dispersion; *i.e.* the debris from more massive dwarf satellites would form an inner, flatter, metal richer distribution.

A new study of spectroscopically selected G and K giants, by Morrison *et al.* (1989; MFF), is interesting in this context. They compared the kinematics of giants in the intermediate abundance interval $-1 >$[Fe/H] > -1.6, for $|z| < 1$ kpc and $|z| > 1$ kpc (z is the distance from the galactic plane). For $|z| > 1$ kpc, the giants in this abundance range are rotating slowly, like the metal weak halo. However, for $|z| < 1$ kpc, the mean rotational velocity is rapid (170 ± 15 km s^{-1}), like the thick disk. This indicates that there is a metal weak tail of disk stars (which they call the metal weak thick disk) extending down to about -1.6 in [Fe/H], and confirms the earlier evidence by Norris *et al.* (1985). The most metal weak stars ([Fe/H] < -1.6) have halo kinematics everywhere. The important point here, in the context of the two-component halo, is that the halo and the disk overlap in abundance over the interval $-1 >$[Fe/H] > -1.6 (this overlap is less obvious for the kinematically selected sample in Figure 1, which is biased against stars with disk kinematics).

MFF estimated the velocity dispersion components for the pure metal weak halo, uncontaminated by the metal weak thick disk stars. They derived $(133 \pm 8, 98 \pm 13, 94 \pm 6)$ km s^{-1} in the usual (r, ϕ, θ) directions. This halo velocity ellipsoid has almost equal ϕ- and θ-components, which is more consistent with the nearly spherical shape of the metal weak halo than some of the earlier estimates (see Freeman, 1987, and Wyse and Gilmore, 1989 for more details).

As an added complication, MFF also examined the kinematics of the RR Lyrae stars in the abundance interval $-1 >$[Fe/H] > -1.6 and with $|z| < 1$ kpc. They showed that the disk component of these metal weak RR Lyrae stars near the plane is much weaker than for the G and K giants in the same abundance interval. MFF suggested that the metal weak thick disk G and K giants may be somewhat younger, and therefore have few or no corresponding RR Lyrae stars. However, the bulk of the conventional thick disk (*i.e.* stars in the abundance range $-0.5 >$[Fe/H] > -1.0) appears to be as old as the disk globular clusters (although we recall that there is still some disagreement on this point: see §4).

How would we interpret these results in the accretion picture ? If the thick disk is younger than the disk globular clusters, then we could interpret the entire thick disk (conventional and metal weak) as the stars of the early galactic disk, heated by the accretion events that formed the galactic spheroid. However, if the *conventional thick disk* is indeed as old as the disk clusters, then the *metal weak thick disk* stars could be the (somewhat younger) debris of circularized satellites. We note that the circularizing will be more effective once the disk has settled, so that circularized objects would preferentially be somewhat younger than the oldest disk objects.

It seems clear, from these three sets of observations, that the metal weak stars ([Fe/H] $<$ -1) do not have a simple approximately spherical distribution. However, it is not yet clear whether the data are all pointing towards a consistent picture.

Some interesting evidence for the importance of accretion by the disk comes from a study by Norris and Ryan (1989) of the angular momentum distribution of a sample of 510 nearby kinematically selected metal weak dwarfs ([Fe/H]< -1.3). They found no stars with angular momentum per unit mass $J > J_{LSR}$, where J_{LSR} is the angular momentum per unit mass of the local standard of rest. However there are several stars in retrograde orbits with $J < -J_{LSR}$. Simulations of the selection procedure showed that this asymmetry is not a selection effect. They interpreted the asymmetry in terms of accretion: satellites in *prograde* orbits of low ($< 60^o$) inclination lose orbital angular momentum and energy to the disk before destruction, while satellites in *retrograde* orbits are less affected. The high velocity tail of the resulting debris would then be asymmetrical, in the sense observed.

7. The Bulge

For a comprehensive review of the properties of the galactic bulge, see Frogel (1988). Here we will just discuss a few general aspects to do with Toomre's picture of the formation of bulges.

It seems very likely that at least some bulges formed by mergers. For example, the galaxy IC 4767, studied by Whitmore and Bell (1988), has a distinct X-shaped structure in the surface brightness distribution of its rapidly rotating bulge. The structure and rotation of the bulge were successfully modelled as the product of the merger of a disk with a small satellite on a low inclination orbit. IC 4767 is an extreme example of a sequence of bulge morphologies which includes box- and peanut-shaped bulges. Binney and Petrou (1985) have also argued that peanut-shaped bulges result from merger events. There is good evidence from the IRAS 12-μm source counts (*e.g.* Habing 1986) and infrared balloon maps of the inner parts of the galactic bulge (*e.g.* Hayakawa *et al.* 1979) that the bulge of our Galaxy has a peanut shape. Stellar photometry of bulge fields, by Rodgers and Harding (1989), indicates that the galactic bulge may also exhibit an X-shaped structure defined by a sub-population of intermediate age stars. However, Terndrup's (1988) photometry in several galactic bulge fields suggests that most of the stars of the inner bulge have ages in the range 11 to 14 Gyr.

In external galaxies with small bulges, like the edge-on system NGC 4565, the bulge rotates with a peak rotational velocity that is typically about half the circular velocity of the disk (*e.g.* Kormendy and Illingworth, 1982). Also, in peanut- and box-shaped bulges, the rotation is cylindrical (*i.e.* independent of height above the galactic plane). A recent study of the kinematics of bulge stars in the Galaxy by Lance *et al.* (1989) shows that the rotation rate of the bulge at about 4 kpc from the galactic center is about 100 km s^{-1}. As in other peanut-shaped bulges, the rotation of the galactic bulge appears to be cylindrical. In summary, the structural and kinematical evidence now available indicates that our Galaxy has a fairly typical small peanut-shaped bulge.

We should mention the existence of a few otherwise normal SO galaxies in which a largescale burst of star formation has occurred within the last $\sim 10^8$ years. Examples are NGC 5102 (Pritchet, 1979) and the interacting galaxy Arp 144. The bulges of these systems are unusually blue; one interpretation of these galaxies is that they have suffered a recent accretion event which has contributed to their bulges.

If bulges of galaxies like the Milky Way are indeed produced by mergers, then we can summarize some necessary properties of the incoming objects. Some of these properties have already been mentioned in the quote from Toomre in §1. They must be:

- Dense enough to survive the tidal field in to the center of the parent disk (this condition is probably met by most larger satellites, because the mean densities of disk galaxies brighter than $M_B \approx -16$ are all fairly similar).
- Bright enough to have [Fe/H] high enough: in the galactic bulge, the abundance is typically like that of 47 Tuc and lower, except in the innermost regions (Rodgers *et al.* 1986; Lance *et al.* 1989; Rich, 1988).
- Old enough for chemical evolution to have proceeded to this level.
- Not too old, so that the galactic disk can still recover from the merger events.

Finally, we have already mentioned the possibility that the thick disk is produced by these early spheroid-forming merger events, through the heating of the stellar component of the early thin disk. van der Kruit and Searle (1981, 1982) have pointed out that thick disks and bulges are apparently associated, in the sense that galaxies with very weak bulges do not have thick disks. (Shaw and Gilmore (1989) have an alternative view). The lack of thick disks in galaxies with negligible bulges may argue against some of the obvious mechanisms for forming the thick disk, such as the heating of the thick disk by the usual secular heating processes that are going on now in the thin disk. The association of bulges and thick disks is an important feature for the accretion picture for bulge formation, because it provides some evidence that the epoch of accretion continued to a time when the galactic disk had already settled.

References

Armandroff, T. 1989. Astron.J., **97**, 375.

Binggeli, B., Tamman, G.A., Sandage, A. 1987. Astron.J., **94**, 251.

Binney, J., Petrou, M. 1985. Mon.Not.R.Astron.Soc., **214**, 449.

Bolte, M. 1989. Astron.J., **97**, 1688.

Bothun, G.D., Mould, J.R. 1988. Astrophys.J., **324**, 123.

Caldwell, N., Bothun, G.D. 1987. Astron.J., **94**, 1126.

Carney, B.W. 1988. *The Harlow Shapley Symposium on Globular Cluster Systems in Galaxies*, ed J.E. Grindlay, A.G.Davis Philip (Dordrecht: Kluwer), p 133.

Carney, B.W., Latham, D.W., Laird, J.B. 1989. Astron.J., **97**, 423.

Dekel, A., Silk, J. 1986. Astrophys.J., **303**, 39.

Doinidis, S.P., Beers, T.C. 1989. Astrophys.J., **340**, L57.

Eggen, O.J., Lynden-Bell, D., Sandage. A. 1962. Astrophys.J., **136**, 748.

Eggen, O.J. 1979. Astrophys.J., **229**, 158.

Fall, S.M., Rees, M. 1985. Astrophys.J., **298**, 18.

Ferguson, H.C. 1989. Astron.J., in press.

Freeman, K.C. 1987. Ann.Rev.Astron.Astrophys., **25**, 603.

Frogel, J. 1988. Ann.Rev.Astron.Astrophys., **26**, 51.

Gilmore, G., Wyse, R., Kuijken, K. 1989. Ann.Rev.Astron.Astrophys., **27**, in press.

Habing, H. 1986. *The Galaxy*, ed G. Gilmore, R. Carswell (Dordrecht: Reidel), p173.

Harris, W.E. 1986. Astron.J., **91**, 822.

Hartwick, F.D.A. 1986. *The Galaxy*, ed G. Gilmore, R. Carswell (Dordrecht: Reidel), p281.

Hayakawa, S., Matsumoto, M., Murakami, H., Uyama, K., Yamagami, T., Thomas, J. 1979. Nature, **279**, 510.

Hernquist, L., Quinn, P.J. 1989. Unpublished..

Impey, C., Bothun, G., Malin, D. 1988. Astrophys.J., **330**, 634.

Kaluzny, J. 1989. Preprint.

Kormendy, J., Illingworth, G.D. 1982. Astrophys.J., **256**, 456.

Laird, J.B., Carney, B.W., Latham, D.W. 1988. Astron.J., **95**, 1843.

Lance, C.M. 1988. Astrophys.J., **334**, 927.

Lance, C.M., Rowley, G., Freeman, K.C. 1989. In preparation.

Larson, R.B. 1988. *The Harlow Shapley Symposium on Globular Cluster Systems in Galaxies*, ed J.E. Grindlay, A.G.Davis Philip (Dordrecht: Kluwer), p 311.

Lupton, R., Fall, S.M., Freeman, K.C., Elson, R.A.W. 1989. Astrophys.J., in press.

Melnick, J., Moles, M., Terlevich, R. 1985. Astron.Astrophys., **149**, L24.

Meurer, G. 1989. PhD thesis, Australian National University.

Morrison, H.L., Flynn, C., Freeman, K.C. 1989. Preprint. (MFF)

Norris, J.E., Bessell, M., Pickles, A. 1985. Astrophys.J.Suppl., **58**, 463.

Norris, J.E., Green, E.M. 1989. Astrophys.J., **337**, 272.

Norris, J.E., Ryan, S. 1989. Astrophys.J., **336**, L17.

Peterson, R. 1989. *Dynamics of Dense Stellar Systems*, ed D. Merritt, (Cambridge: Cambridge University Press), p161.

Peterson, R., Caldwell, N. 1988. Bull.Amer.Astron.Soc., **20**, 1097.

Pritchet, C. 1979. Astrophys.J., **231**, 354.

Quinn, P.J., Goodman, J. 1986. Astrophys.J., **309**, 472.

Rich, M. 1988. Astron.J., **95**, 828.

Rodgers, A.W. 1971. Astrophys.J., **165**, 581.

Rodgers, A.W., Harding, P.X., Ryan, S. 1986. Astron.J., **92**, 600.

Rodgers, A.W., Harding, P.X. 1989. Astron.J., **97**, 1036.

Rodgers, A.W., Harding, P.X., Sadler, E.M. 1981. Astrophys.J., **244**, 912.

Rodgers, A.W., Paltoglou, G. 1984. Astrophys.J., **283**, L5.

Roos, N. 1981. Astron.Astrophys., **95**, 349.

Sandage, A., Binggeli, B., Tamman, G.A. 1985. Astron.J., **90**, 1759.

Searle, L., Zinn, R. 1978. Astrophys.J., **225**, 357. (SZ)

Shaw, M., Gilmore, G. 1989. Preprint.

Silk, J., Norman, C. 1979. Astrophys.J., **234**, 86.

Sommer-Larsen, J. 1986. Licentiate thesis, Copenhagen University.

Sommer-Larsen, J., Christensen, P. 1987. Mon.Not.R.Astron.Soc., **225**, 499.

Terndrup, D.M. 1989. Astron.J., **96**, 884.

Tinsley, B.M., Larson, R.B. 1979. Mon.Not.R.Astron.Soc., **186**, 503.

Toomre, A. 1977. *The Evolution of Galaxies and Stellar Populations*, ed B.M. Tinsley and R.B. Larson (New Haven:Yale University Observatory), p420.

van den Bergh, S. 1986. Astron.J., **91**, 271.

van der Kruit, P., Searle, L. 1981. Astron.Astrophys., **95**, 105; 116.

van der Kruit, P., Searle, L. 1982. Astron.Astrophys., **110**, 61.

Whitmore, B.C., Bell, M, 1988. Astrophys.J., **324**, 741.

Winsall, M., Freeman, K.C. 1989. In preparation.

Wyse, R., Gilmore, G. 1989. Comments on Astrophysics, in press.

Zinn, R. 1980. Astrophys.J., **241**, 602.

Zinn, R. 1985. Astrophys.J., **293**, 424.

Zinneker, H., Keable, C.J., Dunlop, J.S., Cannon, R.D., Griffiths, W.K. 1988. *The Harlow Shapley Symposium on Globular Cluster Systems in Galaxies*, ed J.E. Grindlay, A.G.Davis Philip (Dordrecht: Kluwer), p603.

Are Our Globular Clusters Coeval?

R.B. Larson

Yale Astronomy Department, Box 6666, New Haven, CT06511, USA

1. INTRODUCTION

As the oldest datable objects in our Galaxy, the globular clusters are the prime fossils from which we can hope to reconstruct a record of how our Galaxy formed. A basic goal is to determine the time interval over which the Galactic halo and disk were formed, since a short time interval (< 1 Gyr) would be consistent with the picture of rapid, ordered collapse proposed by Eggen, Lynden-Bell, and Sandage (1962), whereas a much longer time interval (>> 1 Gyr) would suggest instead a relatively prolonged and chaotic process of galaxy formation, possibly involving the merging of subsystems (e.g. Searle and Zinn 1978; Tinsley and Larson 1979). In recent years, the availability of high-quality CCD data has stimulated renewed interest in the determination of globular cluster ages, and has allowed at least relative ages to be derived with a new level of precision. Despite controversy over many of the details, agreement is beginning to emerge that some very significant age differences do exist among globular clusters.

2. SOME RECENT RESULTS

Table 1 lists most of the age determinations published during the period 1987–89 for eight well-studied clusters for which ages are available from at least two sources. In order to produce an unbiased summary, no attempt has been made to correct for the different assumptions made by different authors concerning fitting parameters such as reddening, metallicity, and oxygen-to-iron ratio. Following the cluster name and metallicity [Fe/H] listed in Table 1, the first column of ages labeled "VB" gives ages found by fitting the data to the VandenBerg and Bell (1985) isochrones; in a few cases, the tabulated numbers were derived from diagrams or other information in the papers referenced, but any error so introduced is always less than the estimated uncertainty of ± 2 Gyr typically quoted for the cluster ages. The second column of ages labelled "GDK" gives the results obtained by fitting (in most cases) the same data to the isochrones of Green, Demarque, and King (1987). Finally, the third column of ages labelled "ΔV" gives ages derived using only the magnitude difference ΔV between the main-sequence turnoff and the horizontal branch, a procedure that eliminates uncertainties in cluster reddening and distance. The latter results are taken mostly from a study by Sarajedini and King (1989) of 31 clusters with measured ΔV values, and are based on the calibration of horizontal-branch luminosity given by Lee, Demarque, and Zinn (1988, 1989).

Given the controversy that has existed between the proponents of different models, procedures, and assumptions, it is remarkable that the ages listed in the three columns of Table 1 are actually in reasonable agreement, never differing by more than the estimated uncertainty of ± 2 Gyr. The total age

Dynamics and Interactions of Galaxies Editor: R. Wielen

TABLE 1. AGES OF SOME GLOBULAR CLUSTERS

Cluster	[Fe/H]	Age ± 2 Gyr					
		VB	Ref.	GDK	Ref.	ΔV	Ref.
M92	– 2.2	16.5	(4)	17	(6)	16.5	(10)
M68	– 2.1	14	(1)	16	(6)	14	(10)
M55	– 1.8	14	(5)	15	(10)	15.5	(10)
NGC288	– 1.4	17	(9)	16.5	(6)	17	(7,10)
M5	– 1.4	16	(8)	14	(6)	15	(7,10)
NGC362	– 1.3	13.5	(9)	12.5	(6)	12	(7,10)
Pal 12	– 1.1	10–12	(3,8)	10	(10)	11	(10)
47 Tuc	– 0.7	13.5	(2)	12.5	(6)	12.5	(10,11)

REFERENCES: (1) McClure et al. (1987). (2) Hesser et al. (1987). (3) Gratton and Ortolani (1988). (4) Stetson and Harris (1988). (5) Schade, VandenBerg, and Hartwick (1988). (6) King, Demarque, and Green (1988). (7) Lee, Demarque, and Zinn (1988). (8) Stetson et al. (1989). (9) Bolte (1989). (10) Sarajedini and King (1989). (11) Lee, Demarque, and Zinn (1989).

range of at least 6 Gyr indicated by Table 1 therefore seems very significant. It is noteworthy, in particular, that three groups of investigators have all reached the conclusion that the small outer-halo cluster Pal 12 is only ~ 11 Gyr old and is younger than most other globular clusters by an amount that considerably exceeds the uncertainties. Also, all studies agree that the most metal-poor cluster in Table 1, M92, is several Gyr older than the most metal-rich cluster, 47 Tuc, although there is apparently no very close correspondence between age and metallicity.

As a test of the reality of age differences between globular clusters, attention has focused on the pair of clusters NGC288 and NGC362 which are similar in reddening, metallicity, and even location in the Galaxy, and so are well suited for differential comparisons. NGC288 appears to have a fainter main-sequence turnoff than NGC362 (Pound, Janes, and Heasley 1987; Bolte 1989), and the only plausible explanation is that NGC288 is several Gyr older than NGC362; Demarque et al. (1989) estimate that the age difference is about 5 Gyr, while Bolte (1989) concludes that an age difference of at least 3 Gyr is present. Considerable weight is added to this conclusion by the fact that a similar age difference can explain why, despite their similar metallicities, NGC288 has a very blue horizontal branch, while NGC362 has a red horizontal branch: recent horizontal-branch (HB) models can reproduce this difference if NGC288 is about 6 Gyr older than NGC362 (Lee, Demarque, and Zinn 1988, 1989). These authors also note that M5, another cluster of similar metallicity, is intermediate in HB structure between NGC288 and NGC362, consistent with its intermediate age (see Table 1). Thus all of the data and models are consistent with there being an age spread of order 5 Gyr in this group of comparison clusters, supporting the hypothesis that the HB "second parameter" is indeed age, as advocated by Searle and Zinn (1978) and Zinn (1986).

Sarajedini and King (1989) have used a larger sample of 31 clusters for which ΔV measurements are available to study the relation between age and metallicity. For a subset of 17 clusters for which ages could be determined by fitting to the GDK isochrones, a clear correlation appears to exist between age and metallicity, and these authors conclude that it is significant at more than the 99% level. For the full sample of 31 clusters, the ages estimated from the ΔV measurements show essentially the same correlation with metallicity when the HB luminosity calibration of Lee, Demarque, and Zinn (1988, 1989) is used. The resulting ages typically range from ~ 15.5 Gyr for [Fe/H] = – 2 to ~ 12 Gyr for [Fe/H] = – 1.

3. IMPLICATIONS

If the above results are basically correct, we can now begin to piece together a chronology for the formation of the halo and disk of our Galaxy. The formation of halo clusters appears to have continued for a period of ~ 6 Gyr, or one-third of the Hubble time, and was not completed until ~ 11 Gyr ago. The Galactic disk in the solar vicinity apparently began to form at about the same time that halo formation ended, and with about the same metallicity. Various indicators of the age of the local Galactic disk, including the ages of old open clusters (Janes 1988), nuclear cosmochronology (Malaney and Fowler 1989), and the white-dwarf luminosity function (Iben and Laughlin 1989) all yield disk age estimates between ~ 9 and 12 Gyr, comparable to the ages of the youngest halo objects. Chemically, there also appears to be continuity between halo and disk, the transition occurring at a metallicity of [Fe/H] = – 0.8, according to Zinn (1985). Thus it appears that the formation of the local part of the Galactic disk took up where halo formation left off approximately 10 – 12 Gyr ago.

What was happening during the ~ 6 Gyr preceding the formation of a disk in the solar vicinity? The continuing formation of globular clusters in the halo, and the failure of an extended disk to form over a period of more than 10 free-fall times, suggests that conditions remained too chaotic during this period for gas to settle permanently into a disk. Understanding this phase of Galactic evolution clearly requires understanding the origin of the globular clusters themselves. Our present knowledge of cluster systems in galaxies does not indicate any sharp distinction between globular and open clusters, so there is no reason to suppose that the globular clusters formed in a fundamentally different way from open clusters whose formation we can see taking place at the present. Searle and Zinn (1978) and Larson (1988) have argued that the globular clusters did not form in isolation but rather, like open clusters, as parts of much larger star-forming complexes or subsystems. These systems may have resembled compact dwarf galaxies like the metal-poor compact galaxy I Zw 18 = Mkn 116, whose central starburst region may be making a globular-cluster-like object (Larson 1988). Some remnants of this early population of subsystems may survive around our Galaxy as the nearby dwarf spheroidal galaxies, the largest of which, Fornax, contains several apparently normal globular clusters.

A strong piece of evidence favoring this view is the internal homogeneity of the iron abundance [Fe/H] in most globular clusters, combined with the large dispersion in iron abundance between clusters. To this evidence we can now add the cluster age dispersion discussed above, which implies that some of the cluster-forming systems maintained separate identities and continued to form clusters over a period of several Gyr, and the probable age-metallicity correlation, which suggests that these systems also evolved chemically and became more metal-rich with time. In this case, the cluster-forming subsystems would have had all the essential properties of independent small galaxies.

The fact that there is not only chronological but also chemical continuity between the halo and disk of our galaxy suggests that the disk formed from gas that was left over from, and had already been enriched by, the halo subsystems. It is likely that these small galaxy-like systems would still have been quite gas-rich at the time of their disruption, perhaps by collisions with each other, several Gyr after the big bang. The stars and clusters already formed in them would then have been dispersed to form the halo of our Galaxy, while the gas that remained would eventually have settled into a disk (Toomre 1977). Thus these hypothetical galaxy-like systems may actually have been the building blocks not just of the Galactic halo, but of the entire Galaxy.

A question not directly answered by the globular cluster evidence concerns the number and typical size of the subsystems from which our Galaxy formed – were there many very small ones, or only a few relatively large ones? The fact that the Galactic halo contains almost exclusively metal-poor objects with [Fe/H] < -1 could mean that the subunits were mostly not very large. A picture in which our Galaxy formed from many small gas-rich subsystems might perhaps be regarded as a more inhomogeneous version of the clumpy and turbulent protogalaxy models of Larson (1976), and as a middle ground between these models and the still more inhomogeneous merger picture of Toomre (1977) which envisions mergers of at least some "sizeable" objects. Making such statements more precise will be a worthy goal for much future research.

REFERENCES

Bolte, M., 1989. *Astron. J.*, **97**, 1688.

Demarque, P., Lee, Y.-W., Zinn, R., and Green, E. M., 1989. In *The Abundance Spread Within Globular Clusters: Spectroscopy of Individual Stars*, eds. G. Cayrel de Strobel, M. Spite, and J. Lloyd Evans, p. 97. Observatoire de Paris, Paris.

Eggen, O. J., Lynden-Bell, D., and Sandage, A. R., 1962. *Astrophys. J.*, **136**, 748.

Gratton, R. G., and Ortolani, S., 1988. *Astron. Astrophys. Suppl.*, **73**, 137.

Green, E. M., Demarque, P., and King, C. R., 1987. *The Revised Yale Isochrones and Luminosity Functions*. Yale University Observatory, New Haven.

Hesser, J. E., Harris, W. E., VandenBerg, D. A., Allwright, J. W. B., Shott, P., and Stetson, P. B., 1987. *Publ. Astron. Soc. Pacific*, **99**, 739.

Iben, I., and Laughlin, G., 1989. *Astrophys. J.*, **341**, 312.

Janes, K. A., 1988. In *Calibration of Stellar Ages*, ed. A. G. D. Philip, p. 59. L. Davis Press, Schenectady.

King, C. R., Demarque, P., and Green, E. M., 1988. In *Calibration of Stellar Ages*, ed. A. G. D. Philip, p. 211. L. Davis Press, Schenectady.

Larson, R. B., 1976. *Mon. Not. Roy. Astron. Soc.*, **176**, 31.

Larson, R. B., 1988. In *Globular Cluster Systems in Galaxies, IAU Symposium No. 126*, eds. J. E. Grindlay and A. G. D. Philip, p. 311. Kluwer, Dordrecht.

Lee, Y.-W., Demarque, P., and Zinn, R., 1988. In *Calibration of Stellar Ages*, ed. A. G. D. Philip, p. 149. L. Davis Press, Schenectady.

Lee, Y.-W., Demarque, P., and Zinn, R., 1989. *Astrophys. J.*, submitted.

Malaney, R. A., and Fowler, W. A., 1989. *Mon. Not. Roy. Astron. Soc.*, **237**, 67.

McClure, R. D., VandenBerg, D. A., Bell, R. A., Hesser, J. E., and Stetson, P. B., 1987. *Astron. J.*, **93**, 1144.

Pound, M. W., Janes, K. A., and Heasley, J. N., 1987. *Astron. J.*, **94**, 1185.

Sarajedini, A., and King, C. R., 1989. *Astron. J.*, **98**, in press.

Schade, D., VandenBerg, D. A., and Hartwick, F. D. A., 1988. *Astron. J.*, **96**, 1632.

Searle, L., and Zinn, R., 1978. *Astrophys. J.*, **225**, 357.

Stetson, P. B., and Harris, W. E., 1988. *Astron. J.*, **96**, 909.

Stetson, P. B., VandenBerg, D. A., Bolte, M., Hesser, J. E., and Smith, G. H., 1989. *Astron. J.*, **97**, 1360.

Tinsley, B. M., and Larson, R. B., 1979. *Mon. Not. Roy. Astron. Soc.*, **186**, 503.

Toomre, A., 1977. In *The Evolution of Galaxies and Stellar Populations*, eds. B. M. Tinsley and R. B. Larson, p. 401. Yale University Observatory, New Haven.

VandenBerg, D. A., and Bell, R. A., 1985. *Astrophys. J. Suppl.*, **58**, 561.

Zinn, R., 1985. *Astrophys. J.*, **293**, 424.

Zinn, R., 1986. In *Stellar Populations*, eds. C. A. Norman, A. Renzini, and M. Tosi, p. 73. Cambridge University Press, Cambridge.

Stellar Populations and Mergers

B.W. Carney

Department of Physics and Astronomy, University of North Carolina at Chapel Hill, CB 3255, Phillips Hall, Chapel Hill, NC27599, USA

1. Introduction

Our Galaxy has four stellar populations: the disk, the thick disk, the bulge, and the halo. Did they form out of a single proto-Galactic nebula, with each population being but a different stage in the process? Or have at least some of them nothing to do with the origins and evolution of the others, as expected from mergers? In the context of this meeting, we will look particularly at the latter possibility, and will discuss briefly the data supporting this model.

2. The Thick Disk

We differentiate one population from another by dynamics or spatial distribution, age, and by chemistry. In the simplest models, the latter two are directly related, and much effort has been devoted to deriving an age-metallicity relations, even though there are very old open clusters with solar metallicities, and younger open clusters that are metal-poor. The mean metallicity of an ensemble of gas and stars of a given age depends, however, on three characteristics: the initial level of enrichment of the gas prior to the formation of the first stars, the average "yield" of the nucleo-synthesis processes [*i.e.*, the ratio of the mass of the ejecta from supernovae to the mass locked up in remnants or long-lived stars; see (1,2)], and the degree of the completion of the transformation of the gas into stars. The halo population is, on average, metal-poor not because it began as such, but rather because its yield was extraordinarily low or because it, more likely, lost most of its gas before star formation had proceeded very far toward completion. Such a different *metallicity distribution function* (mdf) is what distinguishes one population from another. It is on this basis that Zinn (3) has argued that the globular cluster system is really composed of two populations: the "classical" metal-poor, spheroidally distributed one, and a second confined to a disk and with a higher mean metallicity. This disk globular cluster population may be related to the "thick disk" identified initially by Gilmore and Reid (4). We (5) have argued that this population is not a transition between the halo and the disk populations. If such a transition had happened, we would expect to find an mdf for the third population shifting from disk metallicities to halo metallicities as we probe greater distances from the plane. Instead, we found a third mdf, with $<[m/H]> \sim -0.5$, which is *independent* of distance from the plane, as well as independent of angular momentum (or rotational velocity). All that changes is the relative

Dynamics and Interactions of Galaxies Editor: R. Wielen

densities of disk, thick disk, and halo stars. Thus the thick disk is more readily explained as a merger event, wherein an object large enough to yield a substantial number of stars yet small enough to avoid destroying the disk merged with the Galaxy. The stars belonging to this population are old, judging by the globular clusters involved and the apparent turn-off color of the field stars (5).

3. The Halo Population

If the halo population evolved homogeneously into the disk population, we might expect to find a metallicity gradient in the halo, and correlations between the age indicator, metallicity, and kinematics. Such relations have been claimed by several groups (6,7). However, we (8,9) have not been able to confirm such relations based on our large survey of high proper motion stars (10,11). Zinn (3) had already argued the halo globular clusters have a negligible radial metallicity gradient, and we have confirmed from our analyses of local proper motion stars that field stars show neither a radial nor a vertical metallicity gradient.

The metallicity *vs.* kinematics arguments take many forms, but may be conveniently summarized in Figure 1, where we show the results of the relation between mean rotational velocity of stars [$v_{rot} = V + \Theta_0$, where V is the velocity toward $\ell = 90°$, $b = 0°$ and Θ_0 is the local value of the circular orbital velocity, about 220 km s^{-1} (12)]. Sandage and Fouts (11) argued for a smooth relation, implying homogeneous evolution. Norris (13) argued for no evolution for the lowest metallicities. Our own data (9; CLL in the figure) support Norris, and in Figure 1 we have divided the lowest metallicity bin of Sandage and Fouts in two, showing they, in fact, also agree with Norris. There is an even more compelling argument against evolution of the halo into the disk present in Figure 1, however. There is essentially no net angular momentum of the metal-poor halo population, in strong contrast to the disk. How could such a transformation occur? While there are some possible explanations, the simplest is to infer that the halo did not evolve into the disk.

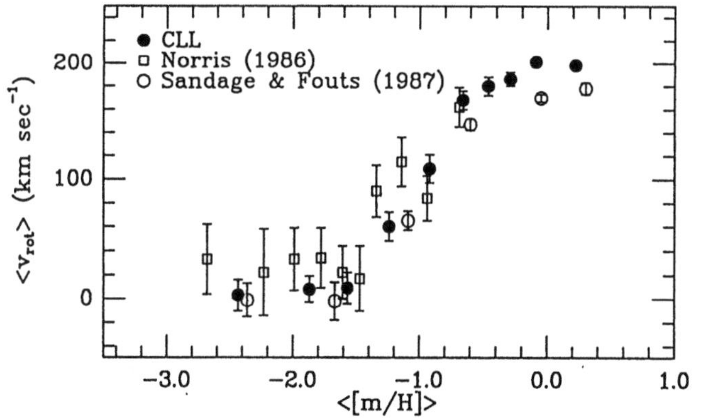

Figure 1. Mean rotational velocities binned by metallicity from three different studies

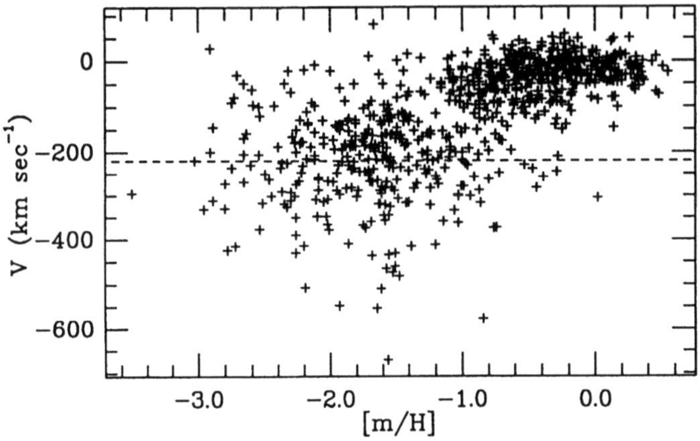

Figure 2. The individual data points from CLL

The evolution apparent in Figure 1 is also misleading. Figure 2 shows the data points used to construct our version of Figure 1. The metal-poor ([m/H] < -1.5) stars are spread out over a wide range of V velocities, but have an average value $\langle V \rangle \sim -\Theta_0$. At higher metallicities, this distribution persists, whereas a homogeneous evolution model would predict fewer low angular momentum stars as the metallicity increased. (The stars with [m/H] > -0.8 but V < -180 km s^{-1} are halo stars, not the velocity tail of the disk, judged by their velocities perpendicular to the plane.) The halo and disk populations appear to be discrete and the *apparent* evolution in Figure 1 is caused by contamination of one population by another.

The simplest explanation of these behaviors is the one offered by Searle and Zinn (2): the halo population was formed initially in gas-rich "fragments" which were randomly accreted by the Galaxy. This would explain the halo's near-zero net angular momentum, its lack of a metallicity gradient, and the lack of any obvious relations between metallicity and kinematics. Recent claims of a several Gyr age spread amongst the halo globular clusters (14,15) also support this model.

4. The Galactic Bulge

Unless the halo and disk star formation processes had heavy element yields differing by a factor of 20 to 30, the low metallicity of the halo is probably due to the loss of gas from the halo. While it is usually assumed this gas fell into the disk, this seems unlikely given the low angular momentum of the halo population. Instead, it is more likely that much of it fell into the central regions of the Galaxy, where via dissipative processes it collected and began to form stars. The Galactic bulge has about the right mass (9). Gas loss from the center of the Galaxy is a much more difficult process than for the halo, so we would expect this population to be more metal rich than the halo, since star formation could go much further toward completion. This, too, is observed (16). If this process applies to other galaxies, we also expect a crude correlation between a galaxy's integrated globular cluster light and that of its bulge, a trend which has also been found (17). Thus the halo population may have resulted from the mergers of numerous gas-rich fragments with the Galaxy. The stars and clusters from these fragments now comprise the "classical" halo, while the gas was responsible for the formation of the Galaxy's bulge.

References

1. Hartwick, F. D. A.: Astrophys. J., 209, 418 (1976).
2. Searle, L., and Zinn, R.: Astrophys. J., 225, 357 (1978).
3. Zinn, R.: Astrophys. J., 293, 424 (1985).
4. Gilmore, G., and Reid, N.: Mon. Not. Royal Astron. Soc., 202, 1025 (1983).
5. Carney, B. W., Latham, D. W., and Laird, J. B.: Astron. J., 97, 423 (1989).
6. Eggen, O. J., Lynden-Bell, D., and Sandage, A.: Astrophys. J., 136, 748 (1962).
7. Sandage, A., and Fouts, G.: Astron. J., 93, 74 (1987).
8. Carney, B. W., Latham, D. W., and Laird, J. B.: Astron. J., submitted.
9. Carney, B. W., Latham, D. W., and Laird, J. B.: Astron. J., submitted.
10. Carney, B. W., and Latham, D. W.: Astron. J., 93, 116 (1987).
11. Laird, J. B., Rupen, M. P., Carney, B. W., and Latham, D. W.: Astron. J., 95, 1843 (1988).
12. Gunn, J. E., Knapp, G. R., and Tremaine, S. D.: Astron. J., 84, 1181 (1979).
13. Norris, J.: Astrophys. J. Suppl., 67, 667 (1986).
14. Stetson, P. B., Vanden Berg, D. A., Bolte, M., Hesser, J. E., and Smith, G. H.: Astron. J., 97, 1360 (1989).
15. Bolte, M. G.: Astron. J., 97, 1688 (1989).
16. Frogel, J. A.: Ann. Rev. Astron. Astrophys., 26, 51 (1988).
17. Harris, H. G., Bothun, G. D., and Hesser, J. E.: In Globular Cluster Systems in Galaxies, ed. by J. E. Grindlay and A. G. D. Philip (Kluwer, Dordrecht), p. 613.

A New Model of the Magellanic Stream

W.L.H. Shuter

Department of Physics, University of British Columbia,
6224 Agricultural Road, Vancouver, B.C. Canada V6T2A6

Lynden-Bell (1988) has estimated the mass of the Galaxy within a Galactocentric radius of 50 kpc to be ~ 1.8×10^{12} $M_{\odot}$ from the kinematics and dynamics of the Magellanic Stream. In contrast Zaritsky et al. (1989), using the statistical procedure developed by Little and Tremaine (1987), have found this mass to be 3.6×10^{11} $M_{\odot}$ from the orbits of satellites of the Galaxy.

This contradiction is resolved if the eccentricity of the Magellanic System orbit is assumed to be ~ 2.1 , rather than ~ 0.7 required in the tidal models of Murai and Fujimoto (1980) and Lin and Lynden-Bell (1982) - i.e. the Magellanic Clouds orbit is *hyperbolic*, a view long advocated by Mathewson et al. (1974).

In the model presented here, the mass of the Galaxy interior to 50 kpc was set equal to 3.6×10^{11} $M_{\odot}$ to conform with the satellite results, the gravitational effect of M31 was included, and each galaxy was given a stiff, extensive, truncated, isothermal halo such that the total system mass maintained the timing argument of Kahn and Woltjer (1959) with a timescale of ~ 13 Byr.

In Figure 1 three time-reversed orbits for the centre of mass of the Magellanic Cloud System (CMMC) are shown, in which its presently uncertain Galactocentric tangential velocity has been set to 250, 350, and 450 km s^{-1}.

The orbit with transverse velocity 350 km s^{-1} is intriguing because its primordial conditions at T = -12.6 Byr are consistent with an origin in a radially expanding metric.

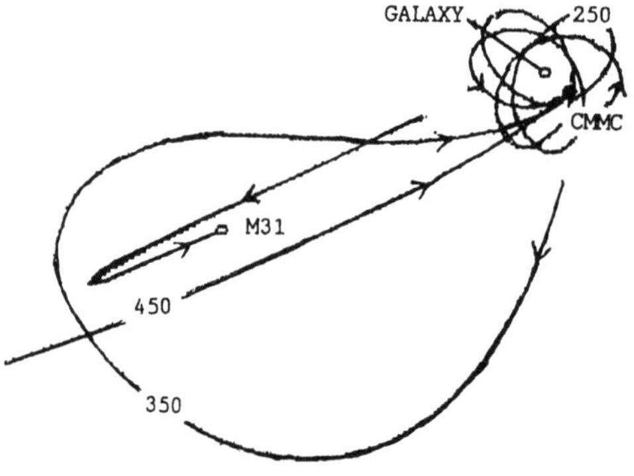

Figure 1. - The orbits of M31 and CMMC relative to that of the Galaxy, obtained by integrating back in time for 12.6 Byr. Three possible orbits for CMMC are shown, corresponding to present values for the transverse velocity of 250, 350, and 450 km s^{-1}.

Dynamics and Interactions of Galaxies Editor: R. Wielen

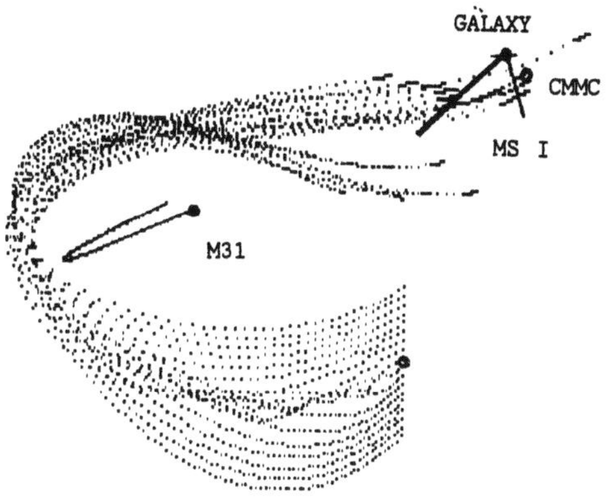

Figure 2. - The orbits of several test particles which were originally, 10 Byr ago, in retrograde circular motion about CMMC. The two diagonal lines from the Galaxy indicate the directions of MS VI and MS I, which define the extent of the Stream.

In this preliminary attempt to model the Stream a closely similar orbit (v_T = 355 km s^{-1}) has been used as a template. Test particles were placed in *retrograde* circular orbits about CMMC, in a plane defined by the Galaxy, CMMC, and M31, and their orbits were integrated forward in time from 10 Byr ago to the present. Some orbits, for which in most cases the test particles arrived with a velocity-angle distribution consistent with that observed in the Stream, are shown in Figure 2. The test particles that did not match the Stream in position or velocity tended to be scattered about a line at position angle ~ 100° in the diagram, tangent to the upper part of the test particles' orbits.

Summary

In contrast to previous models, in which the Stream is produced by recent tidal effects of the Galaxy on the Magellanic System (Toomre and Toomre 1972), in this model the constituents of the Stream are either companions or components of the Magellanic System, which had a close (~< 60 kpc) encounter with M31 about 6 Byr ago.

I thank Peter Thomas and Scott Tremaine for their interest, CITA for a Reinhardt Fellowship, and NSERC for a research grant.

References

Kahn, F. D., and Woltjer, L. 1959, *Ap. J.*, **130**, 705.
Little, B., and Tremaine, S. 1987, *Ap. J.*, **320**, 493.
Lin, D. N. C., and Lynden-Bell, D. 1982, *M.N.R.A.S.*, **198**, 707.
Lynden-Bell, D. 1988, in *The Outer Galaxy*, eds. L. Blitz and F. J. Lockman (Springer-Verlag: Berlin), p. 18.
Mathewson, D. S., Cleary, M. N., and Murray, J.D. 1974, *Ap. J.*, **190**, 291.
Murai, T., and Fujimoto, M. 1980, *Publ. Astron. Soc. Japan*, **32**, 581.
Toomre, A., and Toomre, J. 1972, *Ap. J.*, **178**, 623.
Zaritsky, D., Olszewski, E., Schommer, R., and Aaronson, M. 1989, preprint.

Part III

Recent Interactions Between Galaxies

Interactions in Our Time

F. Schweizer

Department of Terrestrial Magnetism, Carnegie Institution of Washington, Washington, DC 20015, USA

Over the past two decades, our perception of galactic collisions and mergers has changed dramatically: once thought to be interesting but rare events, they are now perceived as some of the dominant processes governing galaxy formation and evolution. I shall first briefly review the early history of the subject and our recent progress in understanding the six main types of galactic collisions. Then I shall focus on three issues: (1) the fascinating parallels that exist between the 1 Gyr old merger remnant NGC 7252 and elliptical galaxies; (2) what new light the study of young ($\lesssim$100 Myr) and old ($\gtrsim$2 Gyr) starbursts is shedding on the formation of E and S0 galaxies; and (3) the still fragmentary, but growing, evidence that even the bulges of disk galaxies may have formed through mergers. As we shall see, it now seems likely that the present position of a galaxy in Hubble's morphological sequence was determined mainly by the number and severity of mergers in its past history.

1. Interactions 1918–1978

Interacting galaxies first came to attention when Pease (1920) conducted a survey of nebulae with the Mount Wilson 60-inch reflector and published some fine photographs of binary, multiple, and distorted systems. Yet detailed observations of such systems began in earnest only with Zwicky's (1956, 1959) systematic surveys and spectroscopic work. The first half century of observations culminated in the atlases of interacting and peculiar galaxies by Vorontsov-Velyaminov (1959) and, especially, Arp (1966).

On the theoretical side, Holmberg (1941) was the first to simulate gravitational interactions between galaxies by building an ingenious analog computer. Two decades later, Pfleiderer & Siedentopf (1961) demonstrated the suitability of the electronic digital computer for the same task, and Alladin (1965) drew attention to the stickiness of galaxy collisions. Thus by 1970 the stage was set for a rash of papers describing models of galaxy interactions, including papers by Yabushita (1971), Wright (1972), Toomre & Toomre (1972), Eneev, Kozlov, & Sunyaev (1973), and others.

Of these early theoretical papers, the Toomres' is today the most often cited by far. I believe there are two main reasons for its success. First, "TT" presented not only generic models of tidal interactions, but also detailed models for four specific pairs of interacting galaxies, including M 51 and NGC 4038/39. The good agreement between the model *bridges* and *tails* and their counterparts seen on photographs of these galaxies convinced astronomers that tides were indeed responsible for much of the observed damage. Second, the Toomres did not only reemphasize that dynamical friction and orbital decay are important consequences of galactic collisions, but they also put forth the new hypothesis that mergers of disk galaxies are all but inevitable and may form remnant ellipticals.

Two papers toward the end of the period 1918–1978 deserve special mention. In 1975, Marie-Helene Ulrich (1975) discovered the first *kinematic subsystem* in NGC 2685 (= Helix galaxy), where filaments orbit the main galactic disk over its poles; I remember

Dynamics and Interactions of Galaxies Editor: R. Wielen

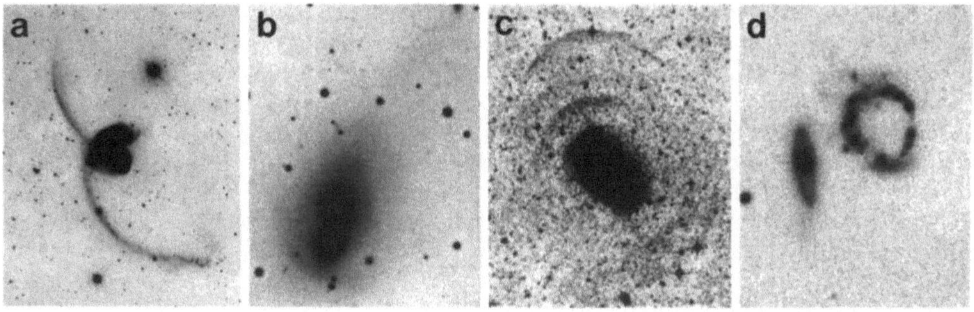

Figure 1. Four galactic collisions of different types: (*a*) NGC 4038/39 (type DD), (*b*) NGC 750/51 (EE), (*c*) NGC 3923 (Ed), and (*d*) Arp 147 (De). (Panels *b* and *d* reproduced from Arp 1966; panel *c* from Malin & Carter 1980.)

intense debate among extragalactic astronomers whether gas could indeed follow polar orbits or whether this peculiar object was a prolate spindle. And in 1978, Larson & Tinsley (1978) demonstrated that tidal interactions can trigger *bursts of star formation*, in which up to 5% of the total mass of a galaxy is converted from gas to stars.

2. A Cornucopia of Collisions

In trying to assess recent progress in our understanding of present-day interactions, I still find it helpful to order the multitude of possible collisions by the types of the involved galaxies (Schweizer 1983). I visualize interactions in a simple 3×3 matrix, in which there are only three kinds of galaxies: ellipticals (E), disk galaxies (D), and gas-rich irregulars (G).

Since more than 80% of all galaxies are disks, *DD collisions* predominate. The biggest excitement of the past decade has been the gradual discovery — fostered by IRAS — that the crunching of gas during such collisions is accompanied by spectacular fireworks (see Joseph's review in this volume). As groundbased infrared techniques continue to improve, we are increasingly able to pierce the dust that often shrouds these fireworks, and we are beginning to study spectroscopically the makeup and kinematics of the central starbursts. One popular new game is the separation of young and old stellar populations by comparing visual and near-infrared images. For example, Gatley *et al.* (1988) show beautiful 1.65-μm images of interacting galaxies that reveal where the old populations and, hence, the main luminous mass lie.

Neglected in all this excitement has been the study of velocity fields of colliding and merging galaxies, which is in great need of attention. In trying to model such systems, theoreticians often have to fall back on data that are 10–20 years old. One nice exception is NGC 7753/52, where Marcelin *et al.* (1987) have mapped the velocity field two-dimensionally with a Fabry-Perot interferometer. This close kin of M 51 now deserves serious attention from model builders, especially since there is evidence that a mass transfer from the main disk into the companion took place (Arp 1969).

Although the relatively rare *EE collisions* have long been neglected by observers for lack of striking morphological signatures, recent work by Borne *et al.* (1988, and in this volume) shows that colliding ellipticals can yield interesting information on M/L and dynamical friction.

Ed and Eg collisions between a giant elliptical and a small disk or irregular galaxy have received much attention since the discovery by Malin (1979), but also already by Arp (1966), that some ellipticals contain sharp-edged features indicative of dynamically cold, shell-like material. Prieur (1988, and in this volume) reviews the status of these "shell galaxies." When the small disk is a gas-rich spiral or irregular, its gas and dust settle into a warped disk within the elliptical, as, e.g., in NGC 5128 (Graham 1979) and NGC 1316 (Schweizer 1980). Although some years ago it was thought that such gaseous disks might represent equilibrium configurations in tumbling triaxial potentials, it seems to be now generally agreed upon that the majority of these disks formed by infall and have not yet reached full dynamical equilibrium.

Near-central *De collisions* between a full-size disk and a small compact object — be it an elliptical or the bulge of a spiral — form transient ring galaxies (Lynds & Toomre 1976; Theys & Spiegel 1977; Toomre 1978). Relatively little work has been done on these systems during the past decade, perhaps because the dynamical models were so convincing. Only now do observers begin to realize that ring galaxies present us with the relatively rare opportunity of studying extranuclear starbursts (Appleton and Struck-Marcell 1987).

Finally, I have always been intrigued by the isolated extragalactic H II region II Zw 40, where a *GG collision* was reported (Baldwin *et al.* 1982): when two gas-rich dwarfs collide, we may catch a glimpse of what must have been happening more frequently in the early universe. The study of this and similar mergers at the bottom of the hierarchical merging pyramid has just begun (Brinks & Klein 1988; Brinks in this volume) and holds great promise.

3. NGC 7252: Update on a Merger Remnant

The peculiar galaxy NGC 7252 has several properties that mark it as a recent merger remnant (Schweizer 1982). First, it features two long tails, one measuring 70 kpc and the other 130 kpc in projection (Fig. 2*a*, $H_0 = 50$), yet it has a single nucleus, suggesting that two disk galaxies have just completed merging. From the tail lengths and velocities, I deduced an age of $1.0^{+1.0}_{-0.5}$ Gyr since the tails started forming. Second, the main body shows delicate ripples and loops that betray a recent struggle (Fig. 2*b*), yet the mean light distribution resembles closely the $r^{1/4}$ law typical of elliptical galaxies. This suggests that the two involved disks have not only merged, but also relaxed violently. And third, the real clincher for a merger of two disks is the coexistence of *two motion*

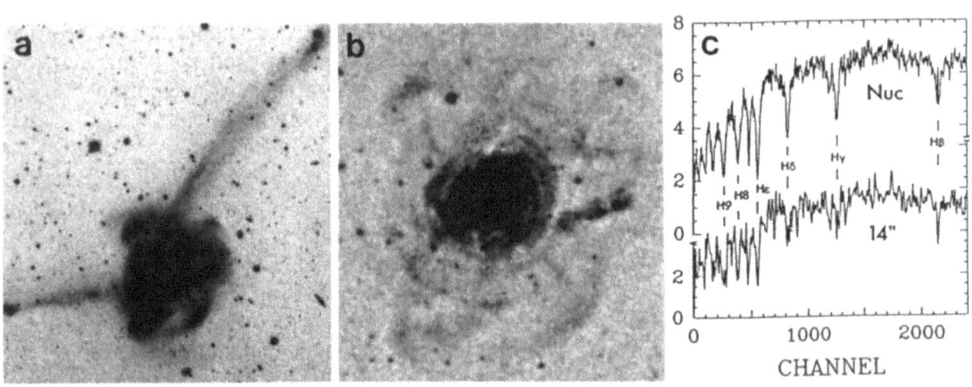

Figure 2. Merger remnant NGC 7252. (*a*) Main body with two long tails; (*b*) masked and enlarged image of main body showing ripples and loops; (*c*) spectrum of nucleus and at $r = 14''$ (7 kpc), obtained at Las Campanas 2.5-m telescope. Notice the Balmer absorption lines at both locations, corresponding to spectral type A7, and the emission-line reversals in Hβ and Hγ at the nucleus, stemming from the central ionized-gas disk.

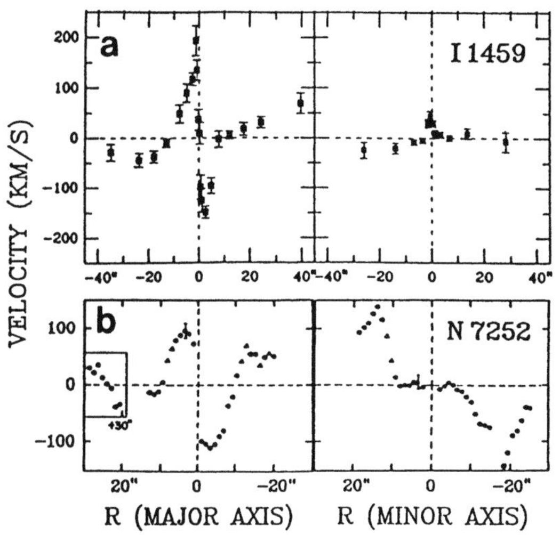

Figure 3. Comparison between kinematics of (*top*) IC 1459 (E3) and of (*bottom*) merger remnant NGC 7252. Left panels show major-axis rotation, right panels show minor-axis motions. Notice the similarities between the counterrotating core of IC 1459 and the counterrotating disk of ionized gas in NGC 7252.
(Scale: 20″ correspond to 3.3 kpc in IC 1459 and 9.4 kpc in NGC 7252 for $H_0 = 50$. IC 1459 reproduced from Franx & Illingworth 1988; NGC 7252 from Schweizer 1982.)

systems in this one galaxy. Figure 3*b* below shows radial velocities of ionized gas measured along the major and minor axes. Notice that the velocities along the major axis do not resemble a flat rotation curve at all. Rather, they oscillate between positive and negative values. Near the center, there is a bright disk of ionized gas with a regular rotation pattern that leads to the central two velocity peaks along the major axis and to a zero-rotation velocity profile along the minor axis. But beyond this central disk of $8'' = 4$ kpc radius, velocities drop sharply along the major axis, cross the zero line, and indicate that farther out the gas is counterrotating. On the eastern side, there is even a suggestion of a second velocity reversal — back to the positive sense — at $r \gtrsim 34''$.

Borne and Richstone (1990) have computed a numerical model for NGC 7252, which reproduces these observations in considerable detail. They simulated two disk galaxies with between 300 and 1000 particles each, using a fast, semi-restricted three-body algorithm. The speed of their code permitted them to survey the multidimensional parameter space of disk-disk interactions that lead to mergers. As Kirk Borne tells it, reproducing the two crossed tails was relatively easy and required only 5 runs, but reproducing the measured velocities proved much more difficult and required some 30 complete simulations. The final model with two merging disks of equal mass reproduces the oscillating velocities along the major axis, including the counterrotation, remarkably well.

Another interesting new bit of information on NGC 7252 is the recent discovery of molecular gas in it by Dupraz *et al.* (1990). Using the Swedish-ESO 15-m Submillimeter Telescope, they measured a nice CO profile indicating a molecular-gas mass of $\sim 8 \times 10^9$ $M_\odot$ (converted to $H_0 = 50$). Better yet, they have now also detected a 21-cm line signal corresponding to an about equal mass of H I. The surprise is, of course, that 1 Gyr after the collision and original starburst there is still plenty of gas left to form new stars. This fact helps explain the optical spectrum of NGC 7252 (Fig. 2*c*), which resembles the spectrum of an A7 star throughout the galaxy, and supports the view that the central disk of ionized gas is probably a settling disk (Schweizer 1982). It also suggests that *in starburst galaxies the conversion of gas to stars lasts an order of magnitude longer than is generally estimated, or at least 1 Gyr instead of the 50–100 Myr usually quoted for IRAS galaxies.* The reason, I guess, is that long after the initial collision neutral gas keeps falling from the outskirts into the center of a merger remnant, where it gets compressed and forms new molecules. NGC 7252 certainly serves as a good example that galaxy building through mergers is a protracted process!

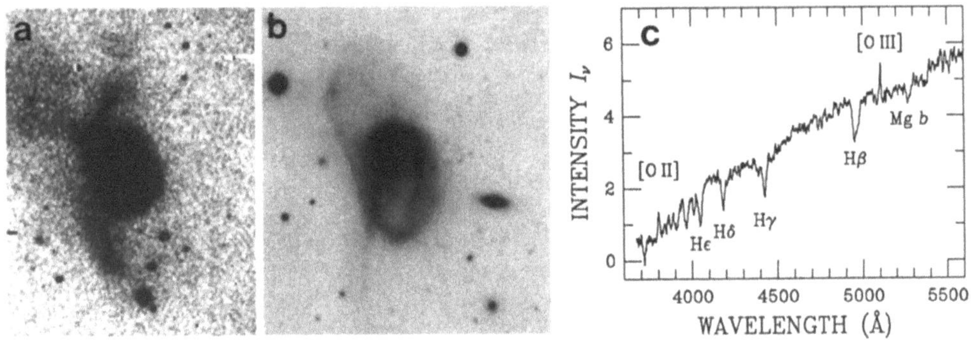

Figure 4. Two intermediate-age remnants of DD mergers with elliptical-like properties. (*a*) Mrk 231 (from Hamilton & Keel 1987); and (*b*) NGC 3921 and (*c*) its spectrum.

If giant ellipticals form mostly through DD mergers, then NGC 7252 may be regarded as a 1 Gyr old proto-elliptical and should show similar kinematic properties as elliptical galaxies. You can, therefore, imagine my delight when in early 1987 John Tonry showed me his (still unpublished) results on strangely rotating cores in a half dozen or so ellipticals, and when Franx & Illingworth (1988), Jedrzejewski & Schechter (1988), and Bender (1988) announced independent discoveries of counterrotating cores in ellipticals. Figure 3 compares major-axis and minor-axis velocities for the stars in IC 1459, an E3, and for the ionized gas in NGC 7252. This comparison suggests two results: First, oddly rotating cores in E's are not necessarily the remnants of spinning victims that fell in, the core-within-a-core idea proposed by Ostriker and suggested for NGC 5813 by Kormendy (1984). Rather, they are more likely the result of gas settling into a central disk and forming new stars within a merger remnant that may well have involved two roughly equal partners. Certainly, *NGC 7252 has enough gas left over to build itself a nice counterrotating core of a few* $10^{10}\ M_\odot$. And second, if NGC 7252 can serve as a guide, kinematic subsystems should be found not only at the centers of ellipticals, but also at larger radii. For example, even in an E with no rotation along the central minor axis, one might find velocity increases farther out similar to those observed in NGC 7252.

Finally, I wish to emphasize that NGC 7252 is not an isolated case of a merger remnant with elliptical-like properties. Several similar merger remnants have been studied recently. Figure 4 shows two of them: Mrk 231 and NGC 3921. Mrk 231 is one of the most luminous IRAS galaxies known; it, too, features two long, slender tails, odd loops in its body, an underlying Balmer absorption spectrum, and a mean $r^{1/4}$ brightness profile indicative of past violent relaxation (Hamilton & Keel 1987). And NGC 3921 is yet another bonafide merger remnant with tails, ripples, and a late A-type spectrum indicative of a starburst roughly 1 Gyr old (Schweizer 1978 and unpublished).

4. Starbursts Young and Old

The history of these A-type galactic spectra, now often dubbed "post-starburst spectra," is interesting. Zwicky (1964) knew already of several galaxies with such spectra, but was not sure how to interpret them. Sargent (1970), in his spectroscopic survey of compact galaxies, noted a half dozen objects with pure Balmer absorption spectra, but did not comment on their nature. One of the first to realize that such spectra implied *global youth* was Arp (1969), who pointed out that of six compact companions at

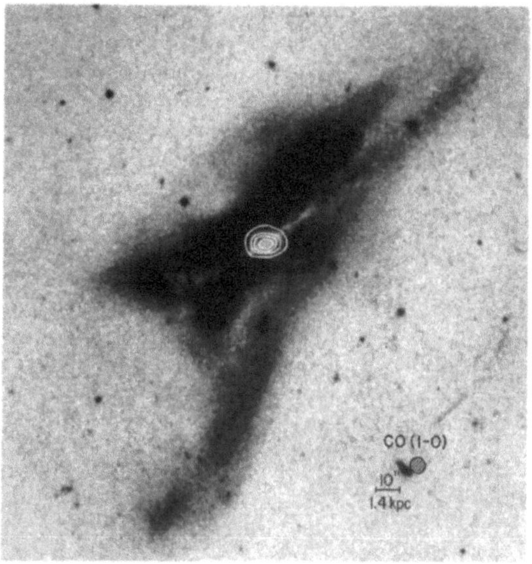

Figure 5. Map of CO emission in colliding pair of galaxies NGC 520 (from Sanders *et al.* 1988). Notice the enormous central concentration of the molecular gas, fuel for an intense burst of star formation.

the ends of spiral arms of larger disk galaxies, four showed this sign of global youth. One need not necessarily agree with Arp that this youth implies recent ejection of the companions. Yet his conclusion that the phenomenon is caused by large numbers of aging upper main-sequence stars stands. In fact, several of the compact companions studied by him appear on various lists of IRAS-luminous galaxies.

Since the early 1970's, progress has been made on two opposite fronts: First, thanks to new infrared techniques and IRAS, we have learned much about *young* starbursts, by which I mean starbursts $\lesssim$100 Myr old. And second, there has been recent progress in tracing *old* starbursts, say older than about 2 Gyr.

In 1983, the IRAS sky survey opened our eyes to the existence of a population of extremely infrared-bright galaxies that appear relatively faint at optical wavelengths, such as Arp 220 and NGC 6240. Nearly invariably, deeper optical images of such IRAS galaxies reveal distortions or tidal tails, suggesting collisions and mergers involving disk galaxies. There is now abundant evidence that these systems form new stars at rates 10–100 times higher than normal in spiral galaxies, as shown by high infrared fluxes and direct spectroscopic signatures of massive young stars (e.g., CO absorption bands at 1–3 μm, emission lines of Wolf-Rayet stars, etc.). The fuel for these *young starbursts* is provided by vast quantities of molecular gas, amounting to $4 \times 10^9 - 4 \times 10^{10}\ M_\odot$ (Young *et al.* 1984; Sanders *et al.* 1986) or 2–20 times the molecular-gas content of the entire Milky Way. From ratios between H_2, H I, and [Fe II] lines in the 1–3 μm region (e.g., Lester *et al.* 1988), it appears that most of the H_2 is excited by shocks, which is hardly surprising given that typical collision velocities between galaxies are several 100 km s^{-1}. The high densities and rapid cooling achieved in the post-shock regions are believed to form near-ideal conditions for rapid cloud fragmentation and star formation (Joseph & Wright 1985; and Joseph in this volume).

The one point about young starbursts which I wish to emphasize is the generally high central concentration of molecules and of the associated star formation. Figure 5 shows an isophotal map of CO emission in NGC 520, obtained by Sanders *et al.* (1988) with the Owens Valley millimeter-wave interferometer and superposed on an optical photograph of this colliding pair. About 1/3 of the total CO emission appears concentrated in a small central region of only 1 1/2 kpc diameter. The molecular-gas mass within this region is $2 \times 10^9 M_\odot$, and it exceeds $10^{10} M_\odot$ in another case. These enormous central concentrations of fuel for star formation solve an old puzzle, namely, how ellipticals can be remnants of disk-disk mergers, yet have central mass densities 100–1,000 times higher than the precursor disks do. Handwavingly, some of us have long argued that conversion of gas to stars might solve the problem, but here — I believe — is strong evidence. It seems likely that some of this gas will settle into a small central star-forming disk that will later be observable as a kinematically distinct subsystem in merger remnants like NGC 7252 and in the ellipticals themselves. Interestingly, the counterrotating cores of IC 1459 (Franx & Illingworth 1988) and NGC 5322 (Bender 1988 and in this volume) are both relatively cold and disk-like.

Concerning *old*, and perhaps weaker, *starbursts*, there are two interesting new developments. First, Carter *et al.* (1988) find that about 20% of "shell ellipticals" show evidence for recent star formation in the form of enhanced Balmer absorption lines. Even though it is my impression that less than half of their sample are ellipticals — many others being S0's, Sa's, and outright peculiar objects similar to NGC 7252 — what matters is that the same accretion events that lead to shell formation may occasionally also lead to central star formation (see Prieur's review in this volume).

Second, even in bona fide ellipticals there is now evidence for relatively old starbursts. Pat Seitzer and I have been conducting an imaging survey of a sample of 74 nearby E and S0 galaxies at redshifts less than 4000 km s^{-1} (see our paper in this volume). Using various masking techniques, we searched for fine structure such as ripples, strong boxiness, X-structure, and plumes, and found that about 2/3 of all E's and $\gtrsim 1/2$ of the S0's show one or several of these signatures that we have come to associate with accretions and mergers. The interesting discovery is that the presence of such signatures correlates with luminosity and spectroscopic line strengths. Figure 6*a* shows the absolute magnitude of all E + S0's plotted versus our fine structure parameter Σ, which ranges between zero for no detected fine structure and 8 for much fine structure

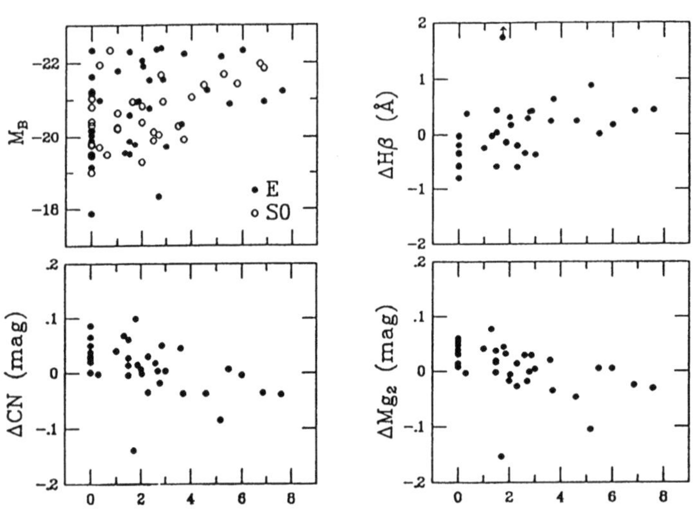

Figure 6. Correlations between luminosity, line-strength residuals, and fine-structure parameter Σ for E and S0 galaxies. (*a*) (*top left*) M_B versus Σ for 36 E's and 38 S0's; (*b*)–(*d*) residuals $\Delta H\beta$, ΔCN, and ΔMg_2 versus Σ for 35 E's. Note that $\Delta H\beta$ correlates with Σ, whereas both ΔCN and ΔMg_2 anticorrelate.

of different kinds. Note that the 12 galaxies with the most fine structure (Σ=4–8) are all brighter than $M_B \approx -20.8$. Figures 6*b*–*d* show the three line-strength residuals $\Delta H\beta$, ΔCN, and ΔMg_2 plotted versus Σ for 35 ellipticals. We obtained these residuals from line strengths measured by Faber *et al.* (1990) by calculating linear fits of $H\beta$ vs. M_B, CN vs. M_B, and Mg_2 vs. M_B for a subset of 241 E + S0 galaxies. As Figure 6 shows, *ellipticals with much fine structure tend to be luminous and to have stronger-than-average Hβ absorption and weaker-than-average CN and Mg₂ features for their luminosity.* The most likely explanation, we believe, is that these E's have experienced more recent, and perhaps also more gas-rich, mergers which have enhanced their young populations. These young populations in turn strengthen the Balmer absorption lines and dilute the CN and Mg absorption features. Hence, we may be witnessing starbursts older than those in Mrk 231, NGC 7252, and NGC 3921 ($\lesssim$2 Gyr), yet clearly younger than 10 Gyr. If this hypothesis is correct, *ellipticals with much fine structure may represent the missing population of older merger remnants that King* (1977) *asked Toomre* (1977) *to reveal.*

5. Evidence for the Delayed Formation of Bulges

There is intriguing evidence that mergers may have formed not only ellipticals, but also bulges of disk galaxies. Probably the best-known case of a disk galaxy with a blue bulge is NGC 5102. Figure 7*a* shows two spectra obtained by Gallagher *et al.* (1975) with apertures centered on the nucleus and at a radius of $8'' \approx 200$ pc in the bulge; both spectra feature strong Balmer absorption lines. These spectra suggest that NGC 5102 has just formed or at least enhanced its bulge. Yet this S0 galaxy also possesses an exponential disk of at least 3–4 scale lengths. The blue-light bulge-to-disk ratio lies between 1/4 (van Driel 1987) and ~1/2 (Yoshizawa & Wakamatsu 1975). A rough estimate suggests that the present bulge will fade 1–2 mag over the next few Gyr, thus leading to an aged $B/D \approx 1/10$. Clearly, the recent bulge building episode in NGC 5102 was not insignificant, and we are forced to conclude that some delayed bulge building is still occurring at the present time.

Simple logic supports this conclusion. Mergers between two disk galaxies of nearly equal mass lead to strong violent relaxation and, hence, to remnant ellipticals (TT; Toomre 1977; Barnes 1988). On the other hand, the infall of a small companion of $\ll$1% mass will hardly disturb a major disk galaxy. Thus, in between these two extremes, there must be a range of mass ratios for which mergers lead only to the partial destruction of the main disk and, therefore, presumably to bulge building within this disk (Quinn & Goodman 1986). I suggest that NGC 5102 is a good candidate for a recent event of this sort.

The objection is often raised that disks are very fragile (Gunn 1987) and that the infall of a small mass of a few percent will nearly disrupt a disk (Quinn 1987). Yet the example of NGC 5102 suggests that Nature may know how to build bulges within preexisting disks better than we do. Let me cite some further observational evidence in support of this statement.

First, NGC 5102 is not an isolated case. There are several other nearby disk galaxies with blue bulges, e.g., NGC 3156, IC 2035 (Caldwell 1982), VII Zw 421, and II Zw 67 (Sparke *et al.* 1980); hence, the phenomenon cannot be all that rare. Second, some polar-ring galaxies bear unmistakable evidence for a second event, namely accretion over the poles, yet possess blue bulges with strong Balmer absorption lines in addition to a well defined exponential disk. NGC 4650A, whose bulge spectrum is shown in Figure 7*b*, is a good example (Schechter *et al.* 1984; Whitmore *et al.* 1987). We have every reason to believe that these accretions involved more nearly 10% of the luminous mass than 1%. Third, ripples are known to occur not only in S0 galaxies, like NGC 4382 (Schweizer & Seitzer 1988) with its enhanced Balmer absorption lines (Bica

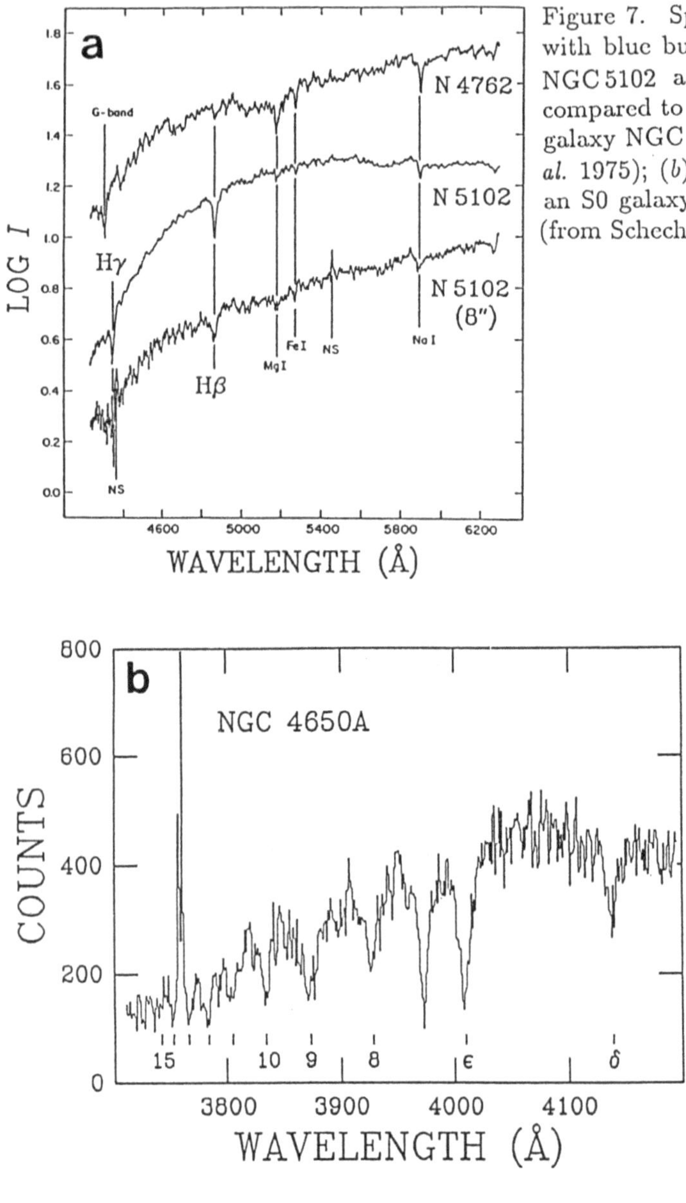

Figure 7. Spectra of two disk galaxies with blue bulges. (*a*) Spectra of NGC 5102 at center and 8″ radius, compared to spectrum of normal S0 galaxy NGC 4762 (from Gallagher *et al.* 1975); (*b*) spectrum of NGC 4650A, an S0 galaxy with a polar ring (from Schechter *et al.* 1984).

& Alloin 1987), but also in S0/Sa galaxies like NGC 474 = Arp 227, and even in Sa galaxies like the classical Seyfert NGC 5548, all of which have exponential disks, hefty bulges, and ripples that again contain $\gtrsim$5% of the total luminosity. Clearly, these disks have survived recent accretion events relatively intact.

Fourth, the fascinating galaxy IC 4767 studied by Whitmore & Bell (1988) has about 1/3 of its total luminosity each in an exponential disk, an X-structure, and a central bulge. We do not know for certain which of these three components formed first. Yet, simulations of accretions in non-spherical potentials suggest that the potential must have been flattened significantly for the X-structure to form (Hernquist & Quinn 1989). This would seem to favor the notion that there may have been a disk already at the time of accretion, and that presumably some of the most strongly perturbed disk stars made it into the bulge. Interestingly, Shaw (1987) estimates that in properly oriented galaxies fully 20% of all bulges appear boxy or peanut shaped. Should we really believe, as Gunn (1987) advocates, that all merger events forming these shapes occurred before *any* stellar disks were formed? In view of all the above evidence, I find such a scenario unnecessarily contrived.

6. Summary and Conclusions

Since Holmberg's and Zwicky's pioneering studies, our understanding of interactions between galaxies has increased dramatically. Simple gravitational models of interacting and merging galaxies now reproduce the observed bridges, tails, ripples, X-structures, and — in one case — even a counterrotating core; and modern N-body simulations of mergers between disk galaxies do, indeed, produce elliptical-like remnants. We are also beginning to understand some of the complex phenomena associated with collisions of the gaseous contents of galaxies. It seems increasingly clear that collisions and mergers of galaxies shock the gas on galactic scales and trigger starbursts that can convert up to ~10% of the total (non-dark) mass into nearly coeval stars.

The A-type absorption-line spectra of intermediate-age (10^{8-9} yr) remnants of disk-disk mergers provide evidence that galaxy-wide starbursts took place. Yet in several of their properties these systems resemble elliptical galaxies, thus suggesting that they are "young" proto-ellipticals. Merger remnants over 2 Gyr old may now have been found among ellipticals with much fine structure. These galaxies tend to have enhanced Hβ absorption lines and weak CN and Mg absorption features for their luminosity. The old objection that DD mergers could not have formed ellipticals because the latter have 10^2–10^3 times higher central densities than the precursor disks is obsolete: young merger remnants are now known to contain enormous central concentrations of molecular gas and star formation.

Specifically, optical and radio observations of the merger remnant NGC 7252 suggest that much of its remaining gas has settled into a central disk that contains $\sim 10^{10}\,M_\odot$ of mostly molecular gas and that counter-rotates. A comparison between the kinematics of this object and that of ellipticals suggests that (1) counterrotating cores in E's may be secondary disks formed during mergers from collapsed gas, rather than remnants of spinning victims, and (2) kinematic subsystems should eventually be found not just at the centers of elliptical galaxies, but also at larger radii.

There is growing morphological and spectroscopic evidence that even some bulges of disk galaxies formed through mergers. Blue bulges, ripples, and occasional X-structures all suggest that the delayed formation of at least minor ($\lesssim$10%) bulges through mass transfers and accretions of satellites is still possible in present-day disk galaxies.

The big picture emerging from studies of galaxy interactions is that the formation of galaxies was not nearly as rapid and complete as was thought two decades ago: mergers are still occurring and producing new, more massive galaxies of different morphological types. The available evidence suggests that *the position of a galaxy in Hubble's morphological sequence may depend mainly on the number and severity of merger events in its past history.* Protogalactic gas clouds that collapsed in relative isolation produced the pure disk galaxies at one end of the Hubble sequence (e.g., M 33). The most destructive collisions and mergers of such disks, namely those between disks of nearly equal mass, produced the giant ellipticals at the other end of the sequence. In between, mergers involving components of unequal mass produced galaxies with a wide range of bulge-to-disk ratios.

Amusingly, when left–right inverted Hubble's (1936) famous tuning-fork diagram of morphological types depicts our current understanding of galaxy evolution through

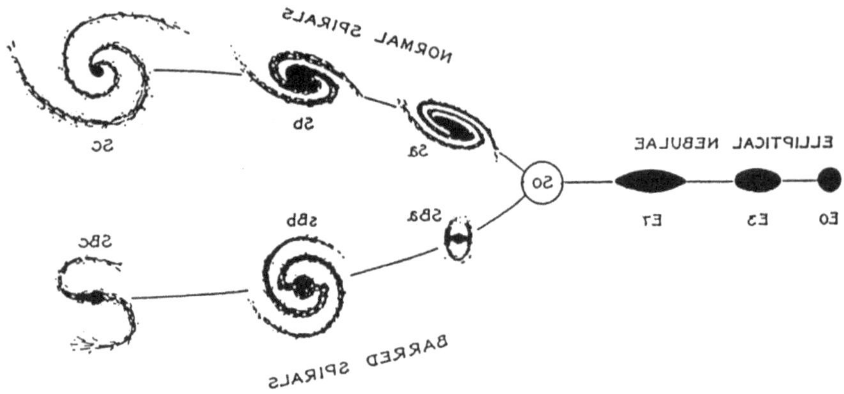

Figure 8. Hubble's tuning-fork diagram, when left–right inverted, represents schematically merging disks that form S0 and elliptical galaxies...

mergers remarkably well (Figure 8). Could it be that, like Leonardo da Vinci before him, Edwin Hubble put his most secret thoughts in mirror writing?

I thank Kirk Borne, Christophe Dupraz, and Sandra Faber for communicating their results in advance of publication, and Chip Arp, Marijn Franx, John Gallagher, Donald Hamilton, David Malin, David Sanders, and Paul Schechter for their kind permission to reproduce some figures.

References

Alladin, S.M. 1965, *Ap. J.,* **141**, 768.
Appleton, P.N., and Struck-Marcell, C. 1987, *Ap. J.,* **312**, 566.
Arp, H. 1966, *Atlas of Peculiar Galaxies* (California Institute of Technology, Pasadena).
Arp, H. 1969, *Astr. Ap.,* **3**, 418.
Baldwin, J.A., Spinrad, H., and Terlevich, R. 1982, *M.N.R.A.S.,* **198**, 535.
Barnes, J.E. 1988, *Ap. J.,* **331**, 699.
Bender, R. 1988, *Astr. Ap.,* **202**, L5.
Bica, E., and Alloin, D. 1987, *Astr. Ap. Suppl.,* **70**, 281.
Borne, K.D., Balcells, M., and Hoessel, J.G. 1988, *Ap. J.,* **333**, 567.
Borne, K.D., and Richstone, D.O. 1990, preprint.
Brinks, E., and Klein, U. 1988, *M.N.R.A.S.,* **231**, 63p.
Caldwell, C.N. 1982, Ph. D. thesis, Yale University, esp. p. 68.
Carter, D., Prieur, J.-L., Wilkinson, A., Sparks, W.B., & Malin, D.F. 1988, *M.N.R.A.S.* **235**, 813.
Dupraz, C., Casoli, F., Combes, F., and Kazès, I. 1990, preprint.
Eneev, T.M., Kozlov, N.N., and Sunyaev, R.A. 1973, *Astr. Ap.,* **22**, 41.
Faber, S.M., Burstein, D., Dalle Ore, C., and Gonzalez, J. 1990, in preparation.
Franx, M., and Illingworth, G.D. 1988, *Ap. J. (Letters),* **327**, L55.
Gallagher, J.S., Faber, S.M., and Balick, B. 1975, *Ap. J.,* **202**, 7.
Gatley, I., DePoy, D.L., and Fowler, A.M. 1988, *Science,* **242**, 1264.
Graham, J.A. 1979, *Ap. J.,* **232**, 60.
Gunn, J.E. 1987, in *Nearly Normal Galaxies*, ed. S.M. Faber (Springer, New York), p. 459.
Hamilton, D., and Keel, W.C. 1987, *Ap. J.,* **321**, 211.
Hernquist, L., and Quinn, P.J. 1989, *Ap. J.,* **342**, 1.

Holmberg, E. 1941, *Ap. J.,* **94**, 385.
Hubble, E. 1936, *The Realm of the Nebulae* (Yale Univ. Press, New Haven), esp. p. 45.
Jedrzejewski, R., and Schechter, P. 1988, *Ap. J. (Letters),* **330**, L87.
Joseph, R.D., and Wright, G.S. 1985, *M.N.R.A.S.,* **214**, 87.
King, I.R. 1977, in *The Evolution of Galaxies and Stellar Populations*, ed. B.M. Tinsley and R.B. Larson (Yale Univ. Observatory, New Haven), p. 418.
Kormendy, J. 1984, *Ap. J.,* **287**, 577.
Larson, R.B., and Tinsley, B.M. 1978, *Ap. J.,* **219**, 46.
Lester, D.F., Harvey, P.M., and Carr, J. 1988, *Ap. J.,* **329**, 641.
Lynds, R., and Toomre, A. 1976, *Ap. J.,* **209**, 382.
Malin, D.F. 1979, *Nature*, **277**, 279.
Malin, D.F., and Carter, D. 1980, *Nature*, **285**, 643.
Marcelin, M., Lecoarer, E., Boulesteix, J., Georgelin, Y., and Monnet, G. 1987, *Astr. Ap.,* **179**, 101.
Pease, F.G. 1920, *Ap. J.,* **51**, 276.
Pfleiderer, J., and Siedentopf, H. 1961, *Zeitschr. Astroph.*, **51**, 201.
Prieur, J.-L. 1988, *Ap. J.,* **326**, 596.
Quinn, P.J. 1987, in *Nearly Normal Galaxies*, ed. S.M. Faber (Springer, New York), p. 138.
Quinn, P.J., and Goodman, J. 1986, *Ap. J.,* **309**, 472.
Sanders, D.B., *et al.* 1986, *Ap. J. (Letters),* **305**, L45.
Sanders, D.B., Scoville, N.Z., Sargent, A.I., and Soifer, B.T. 1988, *Ap. J. (Letters),* **324**, L55.
Sargent, W.L.W. 1970, *Ap. J.,* **160**, 405.
Schechter, P.L., Ulrich, M.-H., and Boksenberg, A. 1984, *Ap. J.,* **277**, 526.
Schweizer, F. 1978, in *Structure and Properties of Nearby Galaxies*, ed. E.M. Berkhuijsen and R. Wielebinski (Reidel, Dordrecht), p. 279.
Schweizer, F. 1980, *Ap. J.,* **237**, 303.
Schweizer, F. 1982, *Ap. J.,* **252**, 455.
Schweizer, F. 1983, in *Internal Kinematics and Dynamics of Galaxies*, ed. E. Athanassoula (Reidel, Dordrecht), p. 319.
Schweizer, F., and Seitzer, P. 1988, *Ap. J.,* **328**, 88.
Shaw, M.A. 1987, *M.N.R.A.S.,* **229**, 691.
Sparke, L.S., Kormendy, J., and Spinrad, H. 1980, *Ap. J.,* **235**, 755.
Theys, J.C., and Spiegel, E.A. 1977, *Ap. J.,* **212**, 616.
Toomre, A., and Toomre, J. 1972, *Ap. J.,* **178**, 623 (TT).
Toomre, A. 1977, in *The Evolution of Galaxies and Stellar Populations*, ed. B.M. Tinsley and R.B. Larson (Yale Univ. Observatory, New Haven), p. 401.
Toomre, A. 1978, in *The Large Scale Structure of the Universe*, ed. M.S. Longair and J. Einasto (Reidel, Dordrecht), p. 109.
Ulrich, M.-H. 1975, *Pub. A.S.P.,* **87**, 965.
van Driel, W. 1987, Ph. D. thesis, Univ. Groningen.
Vorontsov-Velyaminov, B.A. 1959, *Atlas and Catalogue of Interacting Galaxies* (Sternberg Institute, Moscow).
Whitmore, B.C., McElroy, D.B., and Schweizer, F. 1987, *Ap. J.,* **314**, 439.
Whitmore, B.C., and Bell, M. 1988, *Ap. J.,* **324**, 741.
Wright, A.E. 1972, *M.N.R.A.S.,* **157**, 309.
Yabushita, S. 1971, *M.N.R.A.S.,* **153**, 97.
Yoshizawa, M., and Wakamatsu, K. 1975, *Astr. Ap.,* **44**, 363.
Young, J.S., Kenney, J., Lord, S.D., and Schloerb, F.P. 1984, *Ap. J. (Letters),* **287**, L65.
Zwicky, F. 1956, *Ergebnisse der Exakten Naturwissenschaften*, **29**, 344.
Zwicky, F. 1959, in *Handbuch d. Physik*, **53**, 373
Zwicky, F. 1964, *Ap. J.,* **140**, 1467.

Status of Shell Galaxies

J.-L. Prieur

European Southern Observatory, Karl-Schwarzschild-Str. 2
D-8046 Garching, Fed. Rep. of Germany

Introduction

"Shells" or "ripples" around galaxies have been known for a few decades. A set of peculiar galaxies classified as "shell galaxies" appeared in Arp's catalogue (Arp, 1966). But only recently has special attention has been drawn to them. The improvement of photographic emulsions, and the application of new photographic techniques such as "unsharp masking" and "contrast enhancement" by Malin (1977, 1981) has led to the discovery of shells and other peculiar features in a substantial fraction of early-type galaxies. The publication of Malin and Carter's (1983) catalogue of 137 shell galaxies found in the SRC/ESO Southern Sky Survey clearly established the importance of this phenomenon and showed that about 10% of ellipticals are surrounded by shells. They define this class of objects as: *"These galaxies appear at first sight to be normal ellipticals, but closer inspection reveals one or more edged-brightened structures"*.

The first photometric observations (Carter *et al.*, 1982) with IPCS and photomultipliers showed that the shells were made of stars.

Shells are faint structures and are difficult to detect since they are superimposed upon the bright background of the galaxy. The luminosity contrast of the shells is very small, and sometimes comparable to the deviations of the isophotes to perfect ellipses. Processing the images (fig 1) is generally necessary to detect the shells: for example with unsharp masking, developed by Malin (1981) to process photographically large plates, or with gradient maps (Wilkinson *et al.*, 1987a).

I will briefly present here the main results which have been published in the last four or five years. I will also refer to the statistical study of Malin and Carter's catalogue which I have been doing in collaboration with A. Wilkinson, D. Carter, W. Sparks, and D. Malin (referenced as Prieur, 1988b, and Wilkinson *et al.*, 1989a).

I. Models

The models which have been proposed to account for the properties of the shells since the beginning of the 1980's can be sorted into two classes according to the origin of the stars forming the shells: internal (formation *in situ*) or external (accretion of stars from a companion galaxy).

I-1 Internal models

Following Fabian *et al.* (1980), Williams and Christiansen (1985) proposed that shell stars have formed within a giant shell of intra-galactic gas compressed with a shock wave, and have been orbiting in phase since. A few authors have published different versions of this model: Bertschinger (1985), Umenura and Ikeuchi (1987), and Lowenstein *et al.* (1987).

All these models try to explain a rather odd geometry for this star formation, which has to be quick and massive. Then this scenario must cease to be effective rather soon after the galaxy formation, to account for the old population of stars which is observed. But the sharpness of these structures, which should last over a Hubble time (hundreds or thousands of orbits) may be a real problem. No study has been done about this, but I doubt that a family of stars formed a Hubble time ago with *approximately* the same initial conditions will remain orbiting

Dynamics and Interactions of Galaxies Editor: R. Wielen

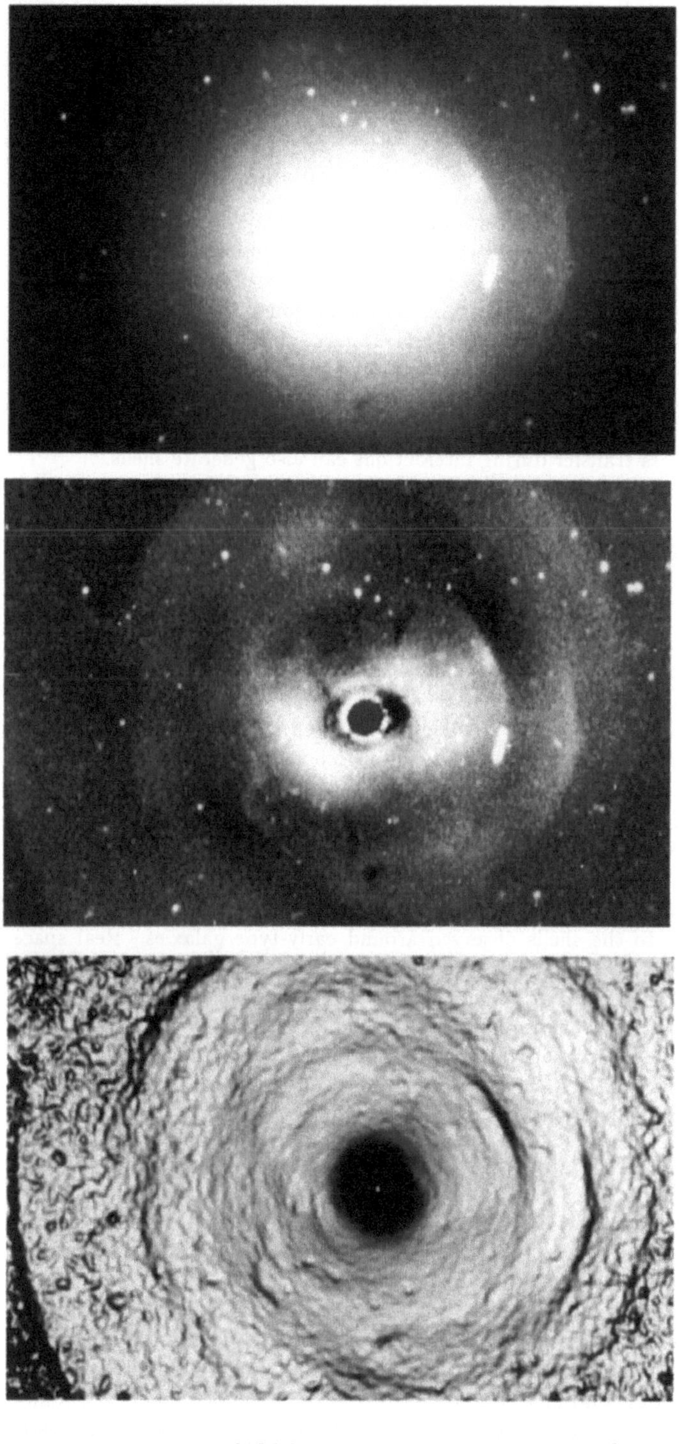

Fig. 1. CCD picture of NGC 474 in B, taken at the prime focus of CFHT. Top, image non processed; middle, a model of the galaxy has been removed; bottom, gradient map. North is on the left, East on the bottom. The bright trail on the left is probably due to an asteroid. The field is about 2'×3'.

in phase with the accuracy needed to make the edges of the shells very sharp, sometimes not even resolved (cf section III.3).

I may also mention a peculiar model by Kundt and Krause (1985) where the shells would be filaments of plasma... I think that the spectroscopy alone is sufficient to discard this hypothesis (cf section IV.1).

I-2 Merging models

The idea that shells would result from a merger was first proposed by Schweizer (1980) and based on the observation of NGC 1316. Following this, Quinn (1984) proposed a model based on wrapping the location of the companion stars in phase space and numerical simulations. When a small disk merges with a big elliptical in a radial orbit, the companion stars form density waves with sharp edges, at a location close to their apocenter. Dupraz and Combes (1986) showed that small ellipticals could also form shells, and studied the effects of non spherical potentials. Hernquist and Quinn (1988, 1989) presented a comprehensive set of simulations to explore the space of parameters. They showed that both wrapping in phase and real space can be involved in shell formation, and that mass transfer during interactions can also generate shells.

Recently Nulsen (1989) studied analytically the process of shell formation in 3-dimensions with phase wrapping ("shear shells"). According to him, the shape of shear shells materializes the invariant surfaces (orbits) in an integrable potential, and can be used to describe the potential of the galaxy.

Nevertheless there is a point which deserves some discussion. He claims that shear shells cannot be visible in a spherical potential, and that Hernquist and Quinn's (1988) simulations introduced an additional constraint for the initial conditions (no dispersion in the module of the velocity for the stars at a given radius) which made them *artificially* visible. This is very serious since this remark would invalidate the results of their paper, often quoted in the literature. I have done some simulations to check this and not found significant differences when adding dispersion in the velocity module. Nulsen (1989) also says that no shells should be visible in Keplerian potential when the stars have non radial orbits. The tests I have done seem also to contradict this. So I think that Hernquist and Quinn (1988)'s results are valid, and that something is missing in Nulsen's analytical modelling.

Numerical simulations have shown that collisions of galaxies produce very easily transient sharp-edged structures similar to the shells observed around early-type galaxies. Real space wrapping gives birth to short time lived shells with complex structures whereas phase wrapping shells produce simpler structures which can last for a few free fall times. The companion can be a spiral or a spheroid, and its mass can range from a few hundredths to a few tenths of that of the galaxy. New N-body simulations of collisions of galaxies of comparable size have also produced shells (Barnes, 1989, Salmon *et al.*, 1989), but they tend to be less "organized" than the shells created with smaller companions.

II. Morphology

To quantify things it is convenient to fit arcs of ellipses to the shells, since they are generally very close to being arcs of ellipses. (Prieur, 1988a, 1988b, Wilkinson *et al.*, 1989a). When the angular extension is large enough, ellipses with a free center can be fitted to these arcs, and in most cases the centre of the ellipses can be taken equal to the centre of the galaxy, within the error bars. We have not observed cases where the convexity is orientated towards the center of the galaxy as in some simulations generated by spatial wrapping (cf. fig 2 of Hernquist and Quinn, 1988). This does not mean that spatial wrapping should be excluded, and for instance the "spiral-like" structure very close to the center of 0422-476 (Wilkinson *et al.*, 1989b) suggests that spatial wrapping takes place there.

It should also be noted that for all the systems we know, when shells are well aligned along a certain axis, this axis is always very close to the major axis of the galaxy (Wilkinson *et al.*,

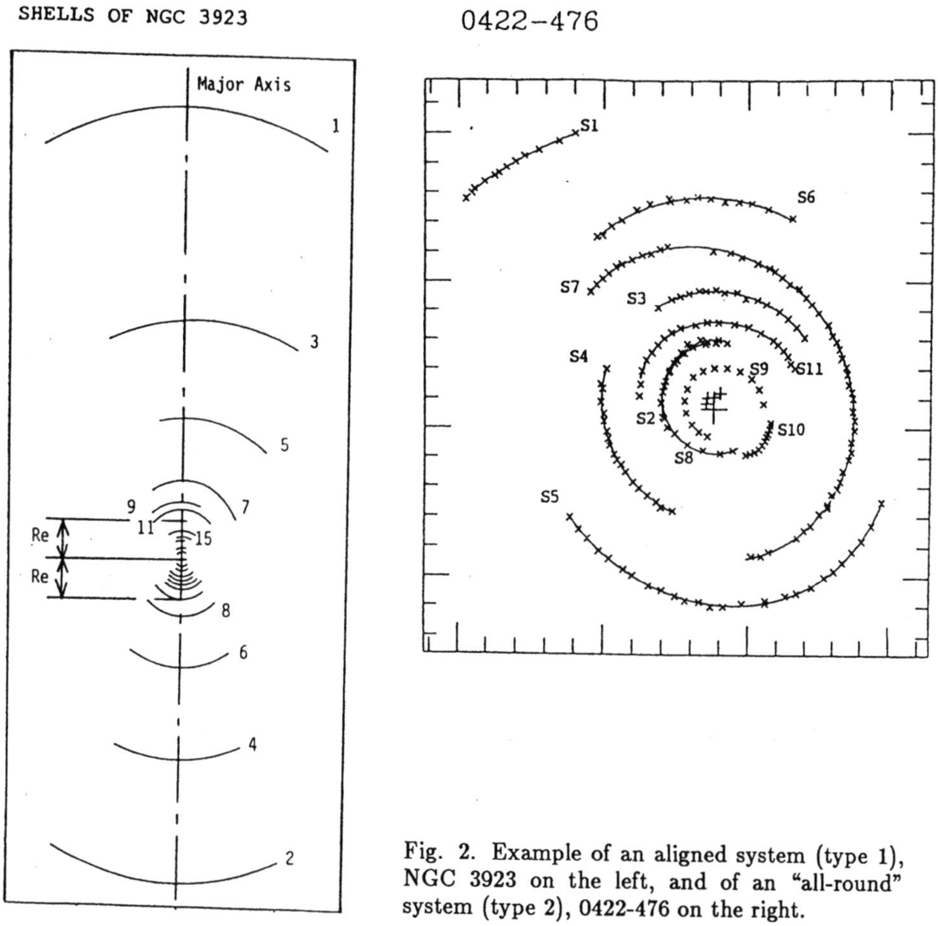

Fig. 2. Example of an aligned system (type 1), NGC 3923 on the left, and of an "all-round" system (type 2), 0422-476 on the right.

1989a). This is probably due to the constraints on the orbits of the stars in a non-spherical potential (box orbits).

II-1 Classification in 3 types

Shell systems can be classified in three types according to their morphology (fig 2) (Prieur, 1988b, Wilkinson *et al.*, 1989a): type 1 are constituted by the aligned systems (ex: NGC 3923), type 2 by the systems where the position angles of the shells are randomly distributed around the galaxy ("all-round" systems, ex: 0422-476), and type 3 class by all the remaining objects which cannot be classified as type 1 or type 2, either because the small number of shells does not allow their classification, or because shells do not appear in their "usual way", as sharp-edged concentric structures (ex: 0201-196).

It appears that these 3 types are observed in equal proportions in Malin and Carter's catalogue (Prieur, 1988b, Wilkinson *et al.*, 1989a).

II-2 Correlation type-ellipticity

A correlation has been found between the morphological types previously defined and the apparent ellipticity of the galaxy (Prieur, 1988b). Elongated galaxies are preferentially surrounded by aligned systems of shells (type 1), and round-like galaxies by type 2 shell systems (fig 3). There is no specific trend for type 3 systems.

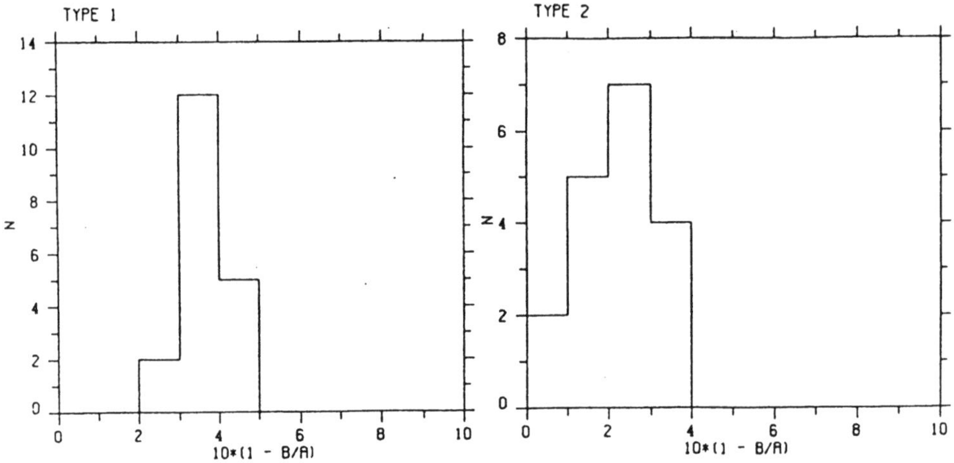

Fig. 3. Histogram of the shell galaxies versus the apparent ellipticity of the galaxy: type 1 systems on the left, type 2 on the right (Prieur, 1988b).

One could think that this is only a projection effect. Type 1 systems would appear as type 2 system when seen face on, and/or inversely for type 2 systems when seen edge on. Numerical simulations have shown that aligned systems could not appear like type 2 systems, even seen face-on (Dupraz and Combes, 1986, Hernquist and Quinn, 1989). But the question is still open to know whether "all-round" systems can look like aligned systems when seen edge-on. Hernquist and Quinn (1989) say it is possible, but Dupraz and Combes (1986) disagree.

This correlation could also be real and intrinsic to the systems in 3-dimensions, as expected by Dupraz and Combes (1986). According to them aligned systems can only form in prolate potentials and "all-round" systems in prolate potentials. This is due to a "focusing" effect of the potential on the shell geometry.

Dupraz and Combes (1986) used an elliptical potential with a constant ellipticity for the equipotentials, which could be a too crude approximation. They have changed this since, and allowed the ellipticity to become spherical at large radii, but they found the same results (Dupraz, private communication). For them the geometry in the central parts, where the deflection occurs (nearly radial orbits), is essential, and the outer parts play a negligible role.

Hernquist and Quinn (1989) disagree with this view. They admit that there is a focusing effect in prolate potentials, but that this effect disappears when the potential becomes spherical at large radii, which is a more realistic case, and argue that other parameters of the collision play also an important role. They generate aligned systems in oblate potentials (cf. their fig 5), and "all-round" systems in prolate potentials (cf. their fig 4). This would preclude the correlation that we see.

The model 3 of Hernquist and Quinn (1989) with a potential elliptical even at large radius, shows a very efficient "locking" of the shells along the major axis, very similar to the simulations of Dupraz and Combes. The correlation we observe may be an indication that the potential remains elliptical, even at large radii and therefore that there would be dark matter around these objects, and that the dark haloes would have the same shape as the luminous components...

Another reason for this correlation could be that the mass of the companion was not negligible, and that the potential of the host galaxy has been seriously modified during the merging. Radial encounters would have "streched" the host galaxy along the direction of impact as shown by recent N-body simulations (Salmon *et al.*, 1989). As encounters with a small impact parameter are more favourable for shell formation, this effect could explain why galaxies with small ellipticity are much less numerous in our histograms (fig 3), than what is expected for early-type galaxies (cf. Sandage *et al.*, 1970, de Vaucouleurs and Binney, 1981).

II-3 Radial distribution of the shells

With Quinn's (1984) assumptions of a quick and total disruption of a small companion and no significant loss of energy for the stars after this event, the radial distribution of the shells depends only on the potential of the host galaxy, and can be computed in a simple way. A series of papers have been published about the interpretation of the shell distribution of NGC 3923, NGC 3051, and NGC 1344: Quinn (1984), Dupraz *et al.* (1987), and Hernquist and Quinn (1987a, 1987b). It was found that the shell distribution of NGC 3923 (the best studied system) could not be fitted with a potential corresponding to the luminosity profile. A few attempts (Dupraz *et al.* 1987, Hernquist and Quinn, 1987b) have been done, without success, to explain this distribution by alternatives to Newton's laws: MOdified Newtonian Dynamics (MOND, Milgrom, 1983), Finite Length-scale Anti-Gravity (FLAG, Sanders, 1984), and REM (REvised MOND, Sanders, 1986). It appeared that dark matter provided a good alternative and with slightly different values for the number of shells and the radii, different authors obtained a good fit for NGC 3923 when adding a dark halo, less concentrated than the luminous component, and a mass ratio of about 40 for the dark/luminous components (Quinn, 1984, Hernquist and Quinn, 1987a, Prieur, 1988b). But even in the best fits with dark matter, the inner shells are discrepant by being too close to the center.

In the last few years, it has been established that shells can be found frequently very close to the nucleus (table 1) (Wilkinson *et al.* 1987a, Prieur 1988ab) at only a few kpcs, which implies that a dissipative process has played a role in the formation of shells.

Dupraz and Combes (1987) presented an analytical model where the companion has a high central density and is disrupted very slowly. Only the outer stars of the companion are lost in the first passage through the central regions of the galaxy, and the remaining core loses slowly its stars while orbiting in the potential of the galaxy, and being slowed down by dynamical friction. They showed that the radial distribution of shells around NGC 3923 could be explained without adding a dark halo. The disruption of the companion was unfortunately very simplified to be treated analytically.

In a recent paper, Heisler and White (1989) emphasize the importance of treating correctly the self-gravity of the companion in the collision. As already shown by Dupraz (1984), self-gravity increases the range of orbits used by the companion stars. Heisler and White find that the maximum range in radius they can obtain with reasonable collision parameters is about 10, without dynamical friction. The observations have shown that this range can be higher for some systems (0351-550, 0422-476, 1252-266, NGC 1344, and NGC 3923 for instance), and dynamical friction would have to be invoked for these objects.

Full N-body simulations should provide a definitive answer to the problem of shell distribution by answering the fundamental question: *"how is the companion disrupted and what is the role of dynamical friction?"*

Salmon *et al.* (1989) have recently done such a simulation involving 160 000 stars. The companion survived for a few oscillations and generated multiple systems of shells (one at each passage in the dense regions of the galaxy). The final range in radii was about 60, which is what we observe for NGC 3923. This shows clearly that we are not far from being able to model shell systems such as NGC 3923, in a very simple way.

One of the direct implications of the scenario involving a slow disruption of the companion (i.e. its central density is larger or comparable to that of the galaxy) is that we should observe multiple nuclei. There is already one example (4C 29.30, van Breugel *et al.*, 1986), but no real investigation to study this point has been done so far. We have not found clear evidence for multiple nuclei in the objects of Malin and Carter's catalogue we have studied up to now (Wilkinson *et al.*, 1989a).

I have fitted theoretical radial distribution of shells based on Quinn's (1984) model to 8 aligned systems, and found that it worked well for 6 systems without adding dark matter (Prieur, 1988b). A dark halo was needed only for NGC 3923 and 0351-550, which happened to be the shell systems with the lowest radii of the innermost shell (Prieur, 1988b). This may be an indication that the radial distribution is curved by the dynamical friction and not by dark haloes. This would not imply that dark matter is not present, but simply that it is not required explicitly, as thought before.

III. Photometry

III-1 Integrated luminosity

A lower limit of the total luminosity of the infalling companion can be obtained by integrating the luminosity of the shell system (table 1). According to numerical simulations, about half of the companion stars are "lost" in an average merger (Hernquist and Quinn, 1988). Observations suggest therefore that shells can be formed even when the mass ratio companion/galaxy is as high as a few tenths, which was not expected by Quinn (1984). In the last few years, observations and simulations (cf. section II-2) have been going in the same direction in showing that even major collisions where the two components are considerably modified can lead to shell formation. The detection of shells around NGC 7252 (Schweizer, 1982), was already a very good observational argument in showing this.

Table 1. Total luminosity of the shell systems as a fraction of the luminosity of the galaxy (col. 2 and 3) (references: *Fort *et al.*, 1986, †Pence, 1986, ‡Schombert and Wallin, 1987, •Prieur, 1988a, °Wilkinson *et al.* 1989b), total number of shells (col. 4), distance of the closest shell as a fraction of r_e (col. 5), morphological type as defined in II-1 (col. 6), and colours of the shells compared to that of the galaxy (col. 7).

Name (1)	Total lum. (%) (2)	Total lum. (%) (3)	N_{shells} (4)	d_{min} (r_e) (5)	Type (6)	Shell colours (7)
NGC 474	6–7‡	$> 3.3 \pm 1$	13	—	2	bluer, same col.
		$\approx$ 6–8°				
0140-658	—	$> 3. \pm 1$	6	2.7	2	same col.
		$\approx$ 3–4°				
NGC 1210	—	$> 1.2 \pm 0.5$	11	0.7	2	bluer, same col.
		$\approx$ 1.5–2.°				
0422-476	—	$> 3. \pm 0.5$	10	1.8	2	big scatter
		$\approx$ 3.–4.°				
NGC 2865	11–22*	—	7	—	2	bluer
NGC 3051	5†	—	> 5	—	1?	same col.
NGC 3923	3–8*	$> 3. \pm 1$	22	0.2	1	same col.
	> 0.5†	$\approx 5 \pm 1.5$•				
NGC 5018	9–18*	—	10	—	2	bluer

We can also have an idea on how the companion was disrupted by examining the radial distribution of the integrated light contained in each shell. For NGC 3923 (which has more than 22 shells) the four outer shells contain about half of the total luminosity. This suggests that the companion has lost most of its stars at its first passage in the central regions of the galaxy (Prieur, 1988a).

III-2 Colour indices

Shell colours are difficult to measure even in wide band photometry, and the results have generally large error bars. Only a few studies have been published on this subject so far (Fort *et al.*, 1986, Schombert and Wallin, 1987, Prieur, 1988b).

Because of the low contrast of the shells, the contribution of the galaxy has to be subtracted from the different images taken with different filters. The colour gradients which are intrinsic to the galaxy are sometimes of the same order as the difference in colours of the shells relative to the galaxy. Therefore it is not possible to subtract the galaxy in exactly the same way in different colours when the galaxy itself exhibits differences in these colours.

Fort *et al.* (1986) studied NGC 2865, NGC 3923, and NGC 5018, and found that shells can be bluer (for NGC 2865, NGC 5018), or of the same colour of the galaxy (NGC 3923). Schombert and Wallin (1987) found that the outer shells of NGC 474 had the same colour as the galaxy. I have done some measurements of the inner shells of NGC 474 (only the inner because of the small field of the CCD we used), and found a large scatter, but colours in average bluer than the galaxy.

So far we have done measurements of about 60 shells (Prieur, 1988a, Wilkinson *et al.*, 1989a) and found that the mean of the shell colours is generally close to that of the galaxy, sometimes bluer ($\Delta_{B-R} \approx 0.1$–0.2) but that the scatter varies a lot from one system to the other. Some shells can even be much redder than the galaxy ($\Delta_{B-R} \approx 0.2$) in systems with large scatter (table 1) (Prieur, 1988b).

An unexpected result has been the finding of a significant azimutal colour differences in some of the shells: $\Delta_{B-V} \approx 0.3$ for 1A and 1B of NGC 5018, and $\Delta_{B-V} \approx 0.3$ for 2A and 2B of NGC 2865 (Fort *et al.*, 1986) and even $\Delta_{B-R} \approx 0.4$ for 10A and 10B of NGC 474 (Prieur, 1988b). Schombert and Wallin (1987) have also noticed a scatter intrinsic to the shells. As I said previously, one has to be careful since the subtraction of the galaxy plays a key role in deriving the colour indices. Error bars are difficult to estimate and may be of the order of 0.2, but we still think that this effect is real, at least for NGC 474 for which we can see this effect directly on the ratio of B and R images (cf. photo 8, Prieur, 1988b).

I have done some simulations to see how the different components of the companion (bulge-disc for example) behaved in the shell formation process (Prieur, 1988b). I found that they can account for colour gradients within the shells, but that this effect fades with time. Nevertheless more realistic simulations than mine are needed to conclude about this and check if this effect can account *quantitatively* of the differences in colours that we have measured (dust also could provide a simpler explanation...).

III-3 Profiles

Shell profiles can have very sharp edges (fig 4), even sometimes unresolved, which is well explained by phase space or real space wrapping. The inner slope is much shallower but we have never observed plateau-like shells as mentioned by some observers (Pence, 1986, Schombert and Wallin, 1987). We are not sure that the profiles mentioned in these two papers have not been affected by a bad subtraction of the galaxy since this shape is only observed for *inner* shells.

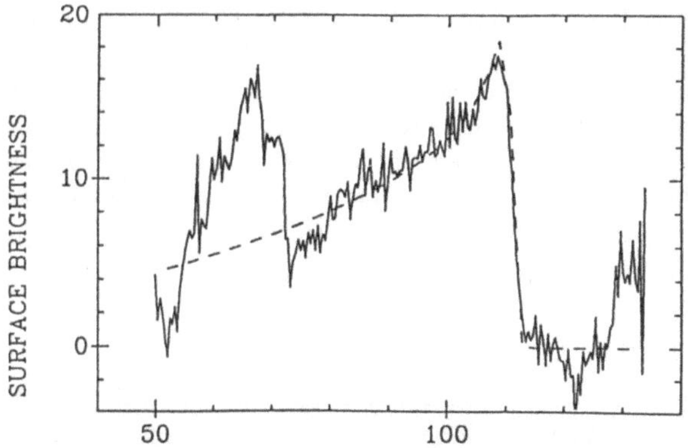

Fig. 4. Profile of shell n° 5 of NGC 474 (close to the left edge of fig 1), and fit with a radial distribution in $x^{-0.2}$. Radius in arcsec, and arbitrary units for the surface brightness.

In Quinn's (1984) model, the gradient in periods wraps the location of the companion stars in phase space around the center and thins it progressively with time, so that it can be assimilated to a thin surface. Shells appear where this surface has a fold when projected on the plane of the sky. For a shell resulting from wrapping in real space, the shell surface brightness should decrease as $x^{-1/2}$ in the inner part, whereas for a phase space wrapping, the radial density distribution would decrease as $x^{-1/2}$ and the projected surface brightness should show a *plateau*-like shape. (Hernquist and Quinn 1988, Nulsen, 1989). This property is independent of the potential in which the particles evolve, but should be valid only very close to the edge, as it is an asymptotic behaviour. This could be a very nice way of distinguishing between wrapping in phase space or in real space

We have observed around 100 shell profiles and not found clear "plateau-like" profiles. I have done some simulations and found a very good agreement with the profiles we have observed, but not any plateau-like profile, again (Prieur, 1988b). I think that even the simulations given as an example by Hernquist and Quinn (1988) to illustrate the plateau shape show clearly that the profiles do not have this shape (!) (cf. their fig 10).

IV. Spectroscopy of the shells

IV-1 Nature of the stellar population

All the spectroscopic attempts to show that the stellar population of the shells was different from that of the underlying galaxy have failed until now: Quinn (1982) with NGC 3923 and NGC 1344, and Pence (1986) with NGC 3923 and NGC 3051 (for NGC 3923, this is in agreement with the CCD photometry of NGC 3923 by Fort *et al.*, 1986). So the spectroscopy confirms the first photometric measurements in visible and infra-red (Carter *et al.*, 1982), and shows that *shells are formed from an old population of stars, very similar to that of the underlying galaxy.* It would be interesting to observe in spectroscopy some of the shells for which the wide-band photometry has shown differences in colours.

IV-2 Kinematics

Kinematic differences between two stellar subsystems formed with the same kind of population without any emission lines, and only broad absorption lines are difficult to detect. Bosma *et al.* (1985) observed NGC 1316 with long-slit spectroscopy and found a jump in the radial velocity of about 30 km/s at the location of a shell. But it was only a two-sigma detection, and new observations are needed to confirm this.

Ford *et al.* (1989) have observed the planetary nebulae of NGC 5128, and found that at the the location of shells numbered 12 and 13 by Malin *et al.* (1983), the velocity dispersion dropped to about 60 km/s and the rotation velocity to about 30 km/s. This is the first clear evidence that shells are kinematically independent from the rest of the stars forming the galaxy. In phase space wrapping, shells are density waves which propagate slowly towards the outer parts of the galaxy. We expect a lower velocity dispersion (population of stars close to their apocenter) and a small expanding velocity of typically 20–50 km/s. The observed values for NGC 5128 are therefore in good agreement with the merging predictions.

V. Study of the host galaxy

V-1 Spectroscopy: recent star formation

Carter *et al.* (1988) have found evidence of recent star formation in the center of 20 out of 100 shell galaxies of Malin and Carter's (1983) catalogue. The spectra do not go far enough to see if there are some changes with the radius, but the integrated colour indices which we have for some of them, are rather red and suggest that this star formation took place in the central regions. A recent study of 0140-658 support this conclusion: this "post star burst" (PSB) shell galaxy exhibits a strong colour gradient, bluer in the center with B-R=1.0 at r=2" and B-R=1.3 at r=8" (Wilkinson *et al.*, 1989a). For NGC 7252 also, Schweizer (1982) found evidences of a recent burst of star formation in the center, associated with an ongoing merger.

The spectra of the PSB shell galaxies are similar to the "E+A" active galaxies found in some remote clusters of galaxies (Dressler and Gunn, 1983). These observations suggest therefore that the accretion of a companion is an efficient process to reactivate star formation in elliptical galaxies, and that galaxy interactions may be the origin of the abnormally high rate of PSB galaxies in remote clusters.

V-2 Correlation with other properties

Since the publication of Malin and Carter's (1983) catalogue, many studies have been done to find a correlation between the presence of shells and other properties. Except for the star formation detection (cf previous section), all these studies have failed. No significant correlation was found with the presence of dust and gas (Sadler and Gehard, 1985, Athanassoula and Bosma, 1985), the detection by IRAS (Wilkinson *et al.*, 1987c), or the radio-emission (Wilkinson *et al.*, 1987b).

V-2 Correlation with other properties, repartition in the Hubble sequence

Schweizer and Seitzer have undertaken a survey with deep CCD frames to search for shells and other features in a sample of about 200 nearby early type galaxies. The final results have not been published yet, but they have found sharp edged structures (that they call "ripples") around 50 % of ellipticals and 30 % of lenticulars (Schweizer and Seitzer, 1989). They have also presented an estimate of the fraction of shell detection in the Hubble sequence (Schweizer and Seitzer, 1988) and found about one percent for disk galaxies. The lower rate of detection of shells in disk galaxies, can be explain by the difficulty to detect faint sharp edged structures in the background of spiral arms and gas clouds. Also, according to simulations, the domain of the collision parameters which generate shells is smaller with disks than with ellipticals (Dupraz and Combes, 1986, Hernquist and Quinn, 1989).

In a famous paper, Toomre (1977) proposed that ellipticals could have been formed with the merging of spirals, and that the observed present rate of interacting spirals could even account for the fraction of ellipticals in the universe (about 10 %). The formation of ellipticals from mergers has been a matter of (sometimes passionate) debates and I will not discuss this here. Nevertheless the estimated short lifetime of the shell systems (about 1–2 billion years) and the high rate of shell detection in ellipticals (about 10 % according to Malin and Carter, 1983) implies that nearly all the ellipticals have undergone at least one merger which has lead to shell formation (though this may be not true for cluster ellipticals where the shell detection rate is much smaller). The higher detection rate found by Schweizer and Seitzer (1989) reinforces this statement and increases the number of mergers to a few events per galaxy...

VI. Discussion

Now let us examine briefly some arguments for and against the main classes of models we mentioned in the introduction.

One of the objections against internal models that we have not mentioned in section I-1, has been the interleaving of the shells in aligned systems. When sorting the shells according to their radius, two successive shells are located in opposite sectors. This property has been observed by many authors (Malin and Carter, 1983, Quinn, 1984, Prieur, 1987,...), and is naturally explained by wrapping in phase space or real space. Most of internal models simply ignored this point ("pure coincidence..."), except Lowenstein *et al.* (1987) who associate a massive star formation with phase wrapping. If a substantial quantity of hot gas turns into stars in a short time scale, and in a small region of space, it can generate shells in the same way as a companion galaxy. But as the lifetime of wrapping systems is rather short (a few billion years at most), this star formation has to be recent, and then the photometry and spectroscopy disagree...

The main argument against the merging origin for the shells put forward by some authors, has been the very low frequency of shell galaxies in clusters. It is still not completely clear why we observe so few shell galaxies in clusters or small groups, where common sense would say that galaxies have greater chances of collisions. Only 4 % of the objects of Malin and Carter's catalogue belong to rich clusters.

One reason could be that outer shells are destroyed by close encounter with other cluster members. Since inner shells are difficult to detect by visual inspection on the sky survey, there could be such an artefact causing a lower detection of systems with inner shells only. Malin and Carter processed photographically some plates of nearby clusters with unsharp masking, but they did not find any significant result. We are waiting for the results of Schweizer and Seitzer's deep survey which is free of such a bias, to see if they come to the same conclusions.

I think that the main reason is that the collisions in clusters occur with high relative velocities, which is not favourable to mergers and shell formation (Quinn, 1984, Dupraz and Combes, 1986). Numerical simulations would be welcome to quantify this.

Conclusion

Observations have been converging in the last few years to establishing that the merger scenario was the only one which accounted for the observed properties of the shells. Nevertheless it is not as clear as a few years ago that shell stars behave like test-particles launched at the same time in the potential of the galaxy, and that they may be used for detecting dark matter around galaxies. The companion galaxy may have been rather massive (a few tenths of the mass of the galaxy), progressively disrupted, and slowed down by dynamical friction.

There is some hope that N-body simulations will soon provide a definitive answer to the problem of the companion disruption and the role of dynamical friction. I think it would be very interesting to simulate NGC 3923, for which comprehensive data about the shell system and the galaxy are available.

With numerical simulations we are doing a sort of experimental astronomy, and I think it is essential to make a big effort to dominate these simulations, and try to model analytically the results. Before any extrapolation, we must first understand what is generated by the simulations for which we can change all the parameters. The debate about the influence of the potential of the galaxy on shell morphology which has been going on for the last two years, has been a good illustration that this is not an easy task, but nevertheless essential.

The very high detection rate of shells in early type galaxies (30—50 %) suggests that nearly all the early type galaxies from the field or belonging to small groups have undergone at least one merger with a companion since the begining of the universe. This striking result emphasizes the key role of gravitational interactions for galaxy formation and evolution.

References

Arp, H.C. 1966, *Atlas of Peculiar Galaxies* (California Institute of Technology Press)

Athanassoula, E., Bosma, A. 1985, *Ann. Rev. Astron. Astrophys.* **23**, 147.

Barnes, J., 1989, this volume.

Bertschinger, E. 1985, *Astrophys. J. Suppl.* **58**, 39.

Binney, J., de Vaucouleurs, G., 1981, *Mon. Not. R. Astron. Soc.* **194**, 679.

Bosma, A., Smith, R.M., Wellington, K.J., 1985, *Mon. Not. R. Astron. Soc.* **212**, 301.

Carter, D., Prieur, J.-L., Wilkinson, A., Sparks, W.B., Malin, D.F., 1988, *Mon. Not. R. Astron. Soc.* **235**, 813.

Carter, D., Allen, D.A., Malin, D.F., 1982, *Nature* **295**, 126.

Dressler, A., Gunn, J.E., 1983, *Astrophys. J.* **270**, 7.

Dupraz, C., 1984, Thèse Université Paris VII.

Dupraz, C., Combes, F. 1986, *Astron. Astrophys.* **166**, 53.

Dupraz, C., Combes, F. 1987, *Astron. Astrophys. Letters* **185**, L1.

Fabian, A.C., Nulsen, P.E.J., Stewart, G.C. 1980, *Nature* **287**, 613.
Ford, H.C., Hui, X., Freeman, K.C., Dopita, M., Meatheringham, S., Ciardullo, R., Jacoby, G., 1989, in preparation.
Fort, B., Prieur, J.-L., Carter, D., Meatheringham, S.J., Vigroux, L. 1986, *Astrophys. J.* **306**, 110.
Heisler, J., White, S.D.M., 1989, preprint.
Hernquist, L., Quinn, P.J. 1988, *Astrophys. J.* **331**, 682.
Hernquist, L., Quinn, P.J. 1989, *Astrophys. J.* **342**, 1.
Kundt, W., Krause, M., 1985, *Astron. Astrophys.* **142**, 150.
Loewenstein, M., Fabian, A.C., Nulsen, P.E.J., 1987, *Mon. Not. R. Astron. Soc.* **229**, 129.
Pence, W.D. 1986, *Astrophys. J.* **310**, 597.
Lynden-Bell, D. 1967, *Mon. Not. R. Astron. Soc.* **136**, 101.
Malin, D.F. 1977, A.A.S. *Photo-Bulletin* **16**, 10.
Malin, D.F. 1981, A.A.S. *Photo-Bulletin* **27**, 4.
Malin, D.F., Quinn, P.J., Graham, J.A., 1983, *Astrophys. J. Letters* **272**, 15.
Milgrom, M., 1983, *Astrophys. J.* **270**, 365.
Nulsen, P.E.J., 1989, preprint.
Pence, W.D. 1986, *Astrophys. J.* **310**, 597.
Prieur, J.-L. 1988a, *Astrophys. J.* **326**, 596.
Prieur, J.-L. 1988b, Thèse Université P. Sabatier, Toulouse (France).
Quinn, P.J., 1982, Ph.D. Thesis, Australian National University, Canberra.
Quinn, P.J., 1984, *Astrophys. J.* **279**, 596.
Sadler, E.M., Gerhard, O.E., 1985, *Mon. Not. R. Astron. Soc.* **214**, 177.
Salmon, J., Quinn, P.J., Warren, M., 1989, preprint.
Sandage, A., Freeman, K.C., Stokes, N.R., 1970, *Astrophys. J.* **160**, 831.
Sanders, R.H., 1984, *Astron. Astrophys.* **136**, L21.
Sanders, R.H., 1986, *Mon. Not. R. Astron. Soc.* **223**, 539.
Schombert, J.A., Wallin, J.F., 1987, *Astron. J.* **94**, 300.
Schweizer, F. 1980, *Astrophys. J.* **237**, 303.
Schweizer, F. 1982, *Astrophys. J.* **252**, 455.
Schweizer, F., Seitzer, P., 1988, *Astrophys. J.* **328**, 88.
Schweizer, F., Seitzer, P., 1989, this volume.
Toomre, A., 1977, in *"The Evolution of Galaxies and Stellar Populations"*, Ed. B.M. Tinsley and R.B. Larson (New-Haven: Yale University Observatory), p 401.
Umenura, M., Ikeuchi, S., 1987, *Astrophys. J.* **319**, 601.
van Breugel, W.J.M., Heckman, T.M., Miley, G.K., Filippenko, A.V., 1986, *Astrophys. J.* **311**, 58.
Wilkinson, A., Sparks, W.B., Carter, D., Malin, D.F. 1987a, *"Structure and Dynamics of Elliptical Galaxies"*, IAU Symp. n° 127, Ed. T. de Zeeuw (Dortrecht Reidel), p 465.
Wilkinson, A., Browne, I.W.A., Kotanyi, C., Christiansen, W.A., Williams, R.E.,
Sparks, W.B., 1987b, *Mon. Not. R. Astron. Soc.* **224**, 895.
Wilkinson, A., Browne, I.W.A., Wolstencroft, R.O.E., 1987c, *Mon. Not. R. Astron. Soc.* **228**, 933.
Wilkinson, A., Prieur, J.-L., Carter, D., Sparks, W.B., Malin, D.F. 1989a, in preparation.
Wilkinson, A., Prieur, J.-L., Carter, D., Malin, D.F., Pence, W.D., Sparks, W.B., 1989b, this volume.
Williams, R.E., and Christiansen, W.A., 1985, *Astrophys. J.* **291**, 80.

Retrograde Streaming in SB0 Galaxies

D. Bettoni[1], *G. Fasano*[1], *and G. Galletta*[2]

[1]Osservatorio Astronomico, Vicolo Osservatorio 5, I-35122 Padova, Italy
[2]Dipartimento di Astronomica, Università di Padova, Vicolo Osservatorio 5, I-35122 Padova, Italy

1 Introduction

In recent years, two new features were discovered in galaxy kinematics: the counterrotation of gas with respect to stars and the kinematical decoupling of part of a galaxy from its main body. Both phenomena indicates the existence of retrograde motions within stellar systems. They appear in galaxies of different morphological type and size, and in spite of their empirical resemblance, it is not completely clear if they represent the same physical process and/or if they have the same origin.

During a long-term project on kinematics and photometry of barred S0 galaxies, we analysed the velocity field in a sample of 15 systems with different bar orientation with respect to the sky. In this sample, two systems (NGC 4546, Galletta, 1987) and NGC 2217 (Bettoni and Galletta, 1988) show gas rotating with projected velocities opposite with respect to that of the stellar disc. In addition, seven galaxies from our sample (Bettoni 1989) and from the literature (Kormendy 1983, Jarvis *et al.* 1988) show a typical wavy trend in their bar rotation curve, indicating retrograde *stellar* streaming in the intermediate part of the bar.

To discuss the retrograde streaming observed in S0 galaxies, it is necessary first to define the two peculiar properties that we expect to find in galaxies:

Counterrotation of gas when gas and stars stream in two planes with antiparallel spins.

Retrograde stellar motions when predominant stellar streaming, inside a defined region (e.g. the galaxy core), is opposite to that outside it.

Counterrotation of gas produces on the spectra coexisting emission and absorption lines inclined in opposite direction. While the observation of this feature in edge-on systems clearly implies opposite directions of motion, this may not be true for systems with low inclination of the principal plane with respect to the sky or when warp of the gas plane is present. In these cases projection effects can produce apparent counterrotation between gas and stars. We shall discuss these special cases later.

Retrograde stellar streaming can produce on the contrary stellar rotation curves that reverse at a some radius the direction of motion. These motions are clearly observable when the two opposing streams are spatially separated within the galaxy. This is the case for galaxies in which the core is counterrotating with respect to the main body (see Kormendy and Djorgovski 1989 for a review). But if this type of motion is mixed with many other families of stellar orbits, the rotation curve may show only a dip or an asymmetry in the line profile. These latter cases are more difficult to discover.

Dynamics and Interactions of Galaxies Editor: R. Wielen

2 Gas counterrotation

Historically, the evidence of gas counterrotation was first found in 1984 for two elliptical galaxies, NGC 5898 (Bettoni 1984, Bertola and Bettoni 1988) and NGC 7079 (Caldwell *et al.* 1984, 1986). Both authors suggest that this phenomenon could be explained if the gas has been acquired from outside, owing to the impossibility of reconciling it with the evolutionary gas⇔stars interaction by means of mass loss and star formation.

But the case of elliptical galaxies is in some way particular. In a quite drastic picture, some elliptical galaxies could be the result of a merger of two or more similarly sized object, and the gas could be the residual of one of these objects, in randomly oriented rotation. This hypothesis has been widely discussed in this Conference, and we shall not discuss it more in detail. We only note that S0's exhibit structures too complex and symmetric to be the result of a melting between galaxies.

As noted in the Introduction, gas counterrotation has been observed also in SB0's. If this gas, as seems plausible, originates from interaction with the environment, the acquired object must be small enough not to perturb the global structure of the galaxy. This should be the case with cannibalism of a dwarf magellanic spiral or an irregular galaxy. We exclude primordial intergalactic clouds as sources of the observed gas because in all the cases it is associated with a non-negligible quantity of dust, a stellar product.

The first example is NGC 4546, an almost edge-on SB0 possessing a stellar disc permeated by a gaseous disc of about the same extent but rotating in opposite direction (Galletta 1987, Bettoni *et al.* 1989b). The gaseous structure is not yet relaxed, but partially settled on the plane of the disc. The analysis of the observations again suggests that a small quantity of gas and dust, estimated as a few 10^8 $M_\odot$ (Bieging 1978, Bettoni *et al.* 1989b), has been captured by NGC 4546 within the past 10^9 years. None of the closest objects appears to be a good candidate for this gas exchange, and we conclude that the acquired galaxy has been completely absorbed by the $\sim 4 \times 10^{10}$ $M_\odot$ of stars present in NGC 4546 within 4 kpc from center. At the time of acquisition, we presume that NGC 4546 must be already gas-free, since gas and stars are crossing with relative velocities as great as 400 km s^{-1}. If some gas clouds moving like the stars were present at that epoch, we expect that collisions between clouds at a so high velocity difference would generate peculiarities (star formations, shock waves and so on) that are actually not observed.

Similar cases of almost edge-on S0's with gas counterrotation are known to exist: they are NGC 7007 (Dettmar *et al.* 1989) and NGC 1216 (Rubin 1988). All these galaxies are quite obvious cases of antiparallel spin between gas and stars.

On the contrary, the second case of gas counterrotation present in our sample is more ambiguous. NGC 2217 (Bettoni *et al.* 1989a), an almost face-on galaxy with a quite strong bar, presents in the inner regions gas clouds circulating around the bar major axis and with projected velocities opposite to that of the stars. However, in the outer regions the gas orbits are co-rotating and probably co-planar with that of the stars. The maximum velocities detected are considerable, both for stars ($V_{max} \sim 100$ km s^{-1}) and for the gas ($V_{max} \sim 150$ km s^{-1}). Since the roundness of the outer isophotes suggests an inclination $i \simeq 21°$, the deprojected stellar velocities should reach about 300 km s^{-1}, typical of a massive disc galaxy.

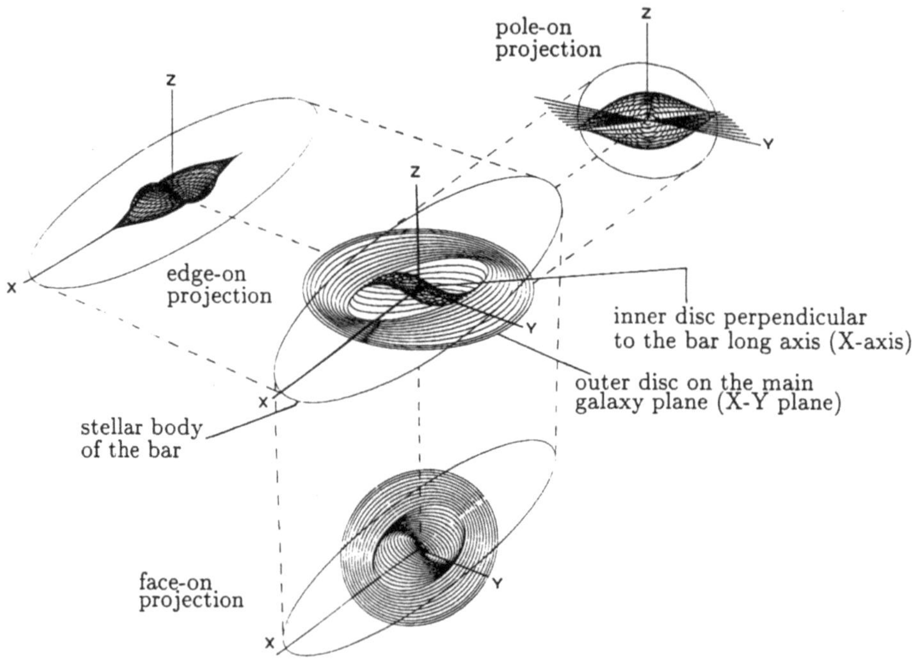

Figure 1: Model of warped and twisted gas disc for NGC 2217

On the basis of the analysis of the gas velocity curves at four different P.A. on the sky, Bettoni *et al.* (1989a) deduce that the gas kinematics can be fairly well modelled with a warped disc oriented in the inner part perpendicularly to the bar major axis, and in the outer part along the main galactic plane (Fig.1). This configuration is that expected in theoretical models of an elongated potential tumbling around its minor axis (Merrit and de Zeeuw 1983, Tohline *et al.* 1982). If we assume that the gas in NGC 2217 lies in such a configuration, the gas motions, when seen from a line-of-sight close to the rotation axis, will project on the sky as they were *retrograde* with respect to the stellar motions. The maximum rotational velocity requested by this model is 300 km s^{-1}. Since the gas in such a model rotates in a continuous way from the center to the outer edge where the motions of the gas and that of the stars share the same direction, we must conclude that the *observed* counterrotation of gas in NGC 2217 does not correspond to a true antiparallelism of the gas and stellar spins. This projected counterrotation is due only to the particular position of the line-of-sight with respect to two almost perpendicular streams. A two component model, with true counterrotation near the center and corotation outside, is more difficult to apply here. In any case, an external origin for the gas cannot be excluded at the present.

3 Retrograde stellar motions

As in the case of gas counterrotation, stellar motions retrograde with respect to the rotation of the main galaxy were first discovered in elliptical galaxies, with three objects of this type presented in the same year (Franx and Illingworth, 1988; Jedrzejewski and Schechter, 1988; Bender, 1988). The explanation of this feature was again ascribed to a merger event.

Since S0's and elliptical co-exist in similar environments, we ask ourselves if a retrograde streaming of stars, as defined in the Introduction, exists for this type of galaxy also.

In our sample, four galaxies show a rotation curve along the bar having a wavy pattern with amplitude of about 25 km s^{-1} (Bettoni 1989). This feature appears also in three galaxies whose bar rotation curve has been presented in the literature (Kormendy 1983, Jarvis *et al* 1988). The velocity curve, after a first rise from the nucleus outward, reverses its direction, reaching minimum values at about 1/3 of the bar extension and increase again in amplitude, reaching values typical of the general rotation near the end of the bar. The velocities oscillate around the systemic velocity for bars seen close to the disc minor axis, or are superimposed to the galactic rotation for the remaining systems. This phenomenon is restricted to regions near the projected bar major axis.

An analysis of the phenomenon (Bettoni 1989) shows that, despite its resemblance with the retrograde stellar streaming observed in the elliptical galaxies, it probably has a different nature. Since it is observed in all SB0's studied having intermediate inclination with respect to the sky, it is difficult to ascribe it to a merging process, whose spins have only 50% of chance to be retrograde. We are inclined to interpret it as an intrinsic feature of the bar kinematics. Among the possible interpretations, there is the presence of true reverse velocities (Pfenniger 1984) or projection effects due to mixing of orbits of different eccentricity (Sparke, private communication). Therefore, although Ellipticals and S0's share the same environment and while collision phenomena have no morphological preferences, we must conclude that retrograde stellar streaming in S0's has not yet been demonstrated.

References

Bender,R., 1988, Astron. Astrophys., **202**, *L5.*
Bertola,F., Bettoni,D., 1988, Astrophys.J., **329**,*102.*
Bettoni,D., 1984, The Messenger, **37**, *17.*
Bettoni,D., 1989, Astron. J., **97**, *79.*
Bettoni,D., Galletta, G., 1988, The Messenger, **54**, *51.*
Bettoni,D., Fasano, G., Galletta,G., 1989a, in preparation.
Bettoni,D., Galletta,G., Oosterloo, T., 1989b, in preparation.
Bieging,J.H, 1978, Astron. Astrophys., **64**, *23.*
Caldwell, N., Kirshner, R.P., Richstone, D., 1984 Bull.A.A.S. **16**, *455.*
Caldwell, N., Kirshner, R.P., Richstone, D., 1986 Astrophys.J., **305**, *136.*
Dettmar,R.J., Jullien-Dettmar, M., Barteldrees, A., 1989, preprint.
Franx,M., Illingworth,G.D., 1988, Astrophys.J. Lett., **327**, *L55,*
Merrit, D., de Zeeuw, T., 1983 Astrophys.J., **267**, *L19.*
Galletta,G., 1987, Astrophys. J., **318**, *531.*
Jarvis,B.J., Dubath,L., Martinet,R., Bacon,R. 1988, Astron. Astrophys., **79**, *513.*
Jedrzejewski,R. Schechter,P.L., 1988, Astrophys.J. Lett., **330**, *L87,*
Kormendy,J., 1983, Astrophys.J., **275**, *529.*
Kormendy,J., Djorgovskyi S., 1989, Ann. Rev. Astron. Astrophys., **27**, in press.
Pfenniger,D., 1984, Astron. Astrophys., **134**, *373,*
Rubin,V.C., 1988, in *Large scale motions in the Universe: a Vatican study week* ed.W.C.Rubin & C.V.Cohen (Princeton Univ.Press: Princeton)
Tohline,J.E., Simonson,G.F., Caldwell,N., 1982, Astrophys. J., **252**, *92.*

Gas, Dust, and Stellar Populations in the Antennae (NGC 4038/4039)*

O.-G. Richter[1,**] *and M.R. Rosa*[2,**]

[1]Space Telescope Science Institute, 3700 San Martin Drive, Baltimore, MD21218, USA

[2]ST-ECF, ESO, Karl-Schwarzschild-Str. 2, D-8046 Garching, Fed. Rep. of Germany

**Affiliated to the Astrophysics Division of the Space Science Department of ESA

Because of its spectacular optical appearance the interacting galaxy pair NGC 4038/39 has received much attention. Toomre and Toomre (1972) and, more recently, Barnes (1988) modelled a galaxy encounter that produced a system quite like the observed galaxies. Even though extensive observational material exists from the radio to the X-ray domain, the detailed geometry and dynamics of the encounter as well as most of the physical processes induced by the violent interaction have escaped satisfactory selfconsitent explanation so far.

New material is presented here consisting of UBVRI CCD photometry, Hα and [O III] CCD images obtained at the MPIA 2.2m telescope at ESO as well as spectrophotometry of a large number of emission regions secured at the ESO 1.52m telescope.

The I filter observations, least affected by internal absorption, were used to determine the true shape of the underlying old stellar population. An I image from which a scaled U band frame has been subtracted reveals two almost featureless disks. The small size of the usable field precludes a complete discussion of the disks' full extent, since much larger total extents of the luminous components of NGC 4038 and NGC 4039 are known from Schweizer's (1978) very deep photographs. Nevertheless, the population II in NGC 4038 seems to reside in a disk somewhat more inclined to the line of sight than would be indicated by the outermost blue isophotes. This impression is corroborated on a "true colour" RGB image generated from U, V, and I frames, where the individual frames were corrected in a best effort for emission line contamination in order to provide pure continuum images.

The region slightly east of a line between the two nuclei shows heavy absorption by dust. This area coincides with the strongest extended 6cm radio continuum emission found by Hummel and van der Hulst (1986). Also the $0.7 - 2.8$ keV X-ray contour map (Fabbiano and Trinchieri, 1983) seems to show a peak in that region, while the other band ($0.2 - 0.7$ keV) shows a double source roughly on opposite sides of the dust. The near-infrared image published by Gatley et al. (1988) shows no particularly striking structure in the dust region. More interestingly, the neutral hydrogen distribution mapped by van der Hulst (1979) and Mahoney et al. (1987) shows particularly high H I column densities in the dust region. All this indicates that the encounter led to a severe redistribution of the interstellar material from the two original galaxies. We determined the extinction in this region via a V-I image using the north-eastern part of NGC 4038's disk as a reference to find E(V-I) over the face of the entire disk. The resulting absorption in V has peaks at about 10 mag. Clearly, only luminous material located in front of this dust will be detectable.

Figure 1 shows (U-B) - (B-V) diagrams of NGC 4038 and 4039. The roughly 4000 points each represent the colours of all independant 1×1 arcsec^2 (140×140 pc^2) areas within the main bodies of the two galaxies. Thin lines indicate the ZAMS. The loci of integrated colours of synthetic spiral galaxies with an IMF slope of 2.1 from Searle et al. (1973) are given by a thick line. In the diagram of NGC 4038 we also show the loci of a 10^6 year and a 3×10^6 year old star cluster from the above reference with the effect of visual absorption of 1 mag indicated. It is obvious that NGC 4039 is entirely dominated by a rather old population suffering comperatively

* Based on observations collected at ESO, La Silla (Chile)

Dynamics and Interactions of Galaxies Editor: R. Wielen

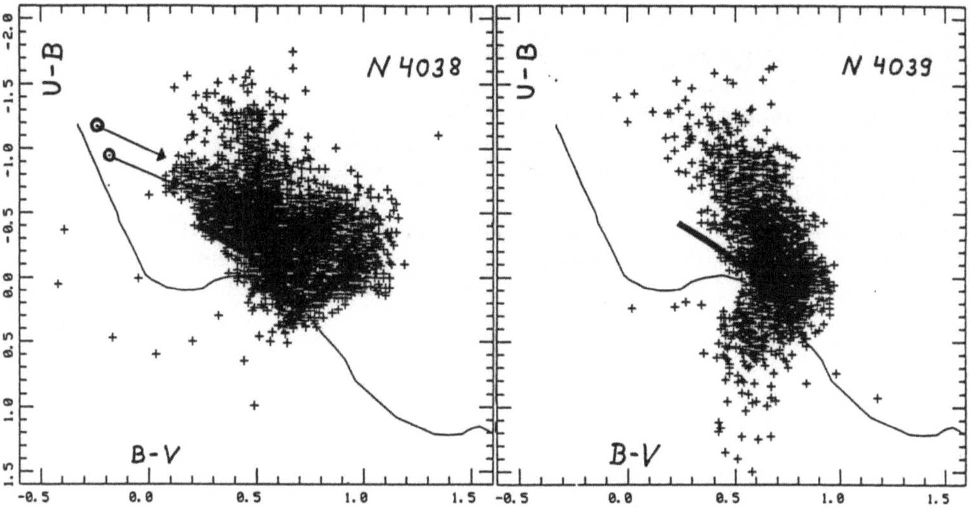

Figure 1

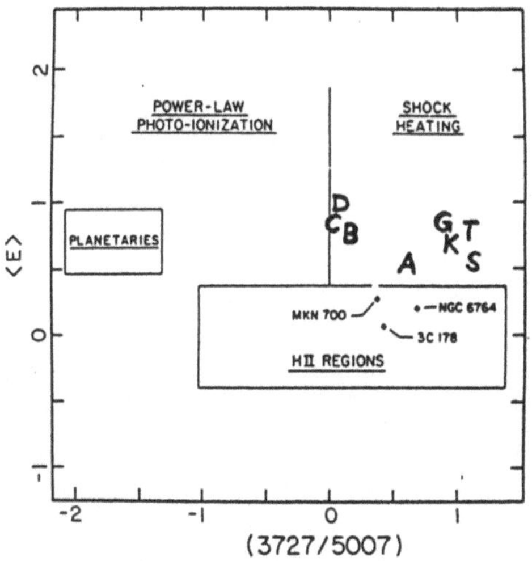

Figure 2

little obscuration. NGC 4038 consists of 3 components, (a) a population similar to that of NGC 4039, (b) a heavily reddened version thereof, and (c) severely reddened clusters of massive stars.

The enormously blue outer western rim of NGC 4038 does *not* consist of normal H II regions. The spectra show a very strong blue continuum but very little Balmer and forbidden line emission. The [O III]/Hβ ratios are exceedingly small. The location of these "emission knots" (cf. Rubin et al., 1970) in various line-ratio diagrams is far away from those of normal giant H II regions (e.g. 30 Dor, or even high metal abundance objects in M 101), and does not coincide with liner type emission or Herbig-Haro objects (see Fig. 2, template from Baldwin et al., 1981). N, O and S abundances (indicative T_e's derived from [O II]/[O III] ratios) seem to be higher than solar, provided the ionizing sources are stellar photons. Interestingly, similar emission line regions are observed in the dust lane of Cen A. Attractive explanations would be either highly evolved OB associations providing strong blue continua from B supergiants, or ionization by cloud/cloud collisions.

References

Baldwin,J.A., Phillips,M.M., Terlevich,R.: 1981, *Publ.A.S.P.* 93, 5.
Barnes,J.E.: 1988, *Astrophys.J.* 331, 699.
Fabbiano,G., Trinchieri,G.: 1983, *Astrophys.J.* 266, L5.
Gatley,I., DePoy,D.L., Fowler,A.M.: 1988, *Science* 242, 1264.
Hummel,E., van der Hulst,J.M.: 1986, *Astron.Astrophys.* 155, 151.
Mahoney,J.H., van der Hulst,J.M., Burke,B.F.: 1987, preprint.
Rubin,V.C., Ford Jr.,W.K., D'Odorico,S.: 1970, *Astrophys.J.* 160, 801.
Schweizer,F,: 1978, *I.A.U. Symposium No.* 77, 279.
Searle,L., Sargent,W.L.W., Bagnuolo,W.G.: 1973, *Astrophys.J.* 179, 427.
Toomre,A., Toomre,J.: 1972, *Astrophys.J.* 178, 623.
van der Hulst,J.M.: 1979, *Astron.Astrophys.* 71, 131.

Structure of the NGC 1023 System

V.L. Afanasiev[1], *M. Capaccioli*[2], *H. Lorenz*[3], *and G.M. Richter*[3]

[1]Special Astrophysical Observatory, SU-357140 Zelenchukskaja, USSR
[2]Institute of Astronomy, University of Padova, I-351000 Padova, Italy
[3]Central Institute for Astrophysics, DDR-1561 Potsdam, German Democratic Republic

ABSTRACT: The controversial origin of the neutral hydrogen cloud in the vicinity of the S0 galaxy NGC 1023 is re-discussed on the basis of new optical data. We assume that NGC 1023A has experienced a tidal interaction with NGC 1023. The model explains some of the pecularities of NGC 1023A, but is probably unable to account for all of the HI mass about NGC 1023.

1. Introduction

The SB0 galaxy NGC 1023 is the brightest member of the Perseus group of galaxies rich in late-type spirals and dwarf galaxies and was discovered to be one of the nearest objects of a rare class of neutral-hydrogen rich lenticular galaxies (van Woerden, 1977). Studies at increasing sensitivity and resolution led to the picture of the HI distribution given by Sancisi et al. (1984). NGC 1023 appears surrounded by 1.5 10^9 M_0 of neutral hydrogen with a rather complex density distribution and kinematics. Two clumpy ring-like cloudes of HI with different mean systemic velocities are identified, carrying almost the same amount of material and showing an overall rotation in the same sence as the stellar body, although in different planes. The "red cloud" has a strong condensation superposed on the optical image of NGC 1023A and it seems morphologically and kinematically linked to a "tail" at higher redshift.

2. NGC 1023 / NGC 1023A an interacting system

The interpretation of this picture is closely connected to the nature of HI surrounding early-type galaxies. This material is not evidently associated with regions of star formation, tends to be distributed mainly outside the optical boundaries and shows a fairly regular rotation with momenta not always aligned with those of the stellar bodies. The main question is:

accretion at the expense of intergalactic HI cloudes or dwarf galaxies

or

remnants of previous disks mainly consumed by earlier star formation.

NGC 1023A has been investigated in detail by Capaccioli et al. (1986). The basic data on NGC 1023 and NGC 1023A are summarized in their tables 4 and 5.

The classification of NGC 1023A as a Magellanic irregular or a low surface brightness late-type dwarf, consistent with the absolute magnitude and with the linear size of the object, is in contrast with the quite smooth texture of the galaxy image, with the lack of emission lines, with the colours to be far too red and with the low content of HI. The proximity of NGC 1023A to NGC 1023 as well as the provocativ morphology of the neutral hydrogen in the area of NGC 1023 (particularly the "tail") suggest a possible way to remove the discrepancies.

Following the general scenario developed by Sancisi et al. (1984) we assume that NGC 1023A has experienced a tidal interaction with the more massive galaxy NGC 1023 about 10^8 yr ago. Although insufficiently strong to disrupt the galaxy, the fly-by encounter may have caused the removal of a large fraction of the interstellar material of NGC 1023A. The consequences of that are a cut of the value of M_{HI}/L_B and a sudden reduction of the star formation activity, which may account for the observed lack of structures and of emission lines in the galaxy and for their red colours.

If we take the ratio of all the HI mass around NGC 1023 to the present luminosity of NGC 1023A, we obtain $M_{HI}/L_B=5$, a large value at the upper boundary of the distribution of dwarf galaxies in the L_B vs. M_{HI}/L_B plane. The agreement improves but not significantly if we increase the B-luminosity before the encounter even by 0.5 mag., but we have to consider the effect of the fragmentation of the stellar body during the encounter. This is difficult to quantify, but the lack of major optical debris indicates that the correction to the luminosity is likely to be small. Thus, although not totally inconsistent with the data, it does not seem likely that all of the HI around NGC 1023 has been supplied by the dwarf galaxy: This supports the pressumption first indicated by Sancisi et al. (1984) that NGC 1023A is probably the donor of just the HI "tail" and the "red cloud".

References

Capaccioli, M., Lorenz, H. and Afanasjev, V. L.: 1986, Astron. Astrophys. 169, 54.

Sancisi, R., van Worden, H., Davies, R. D., Hart, L.: 1984, Monthly Notices Roy. Astron. Soc. 210, 514.

van Woerden, H.: 1977, Topics in Interstellar Matter, ed. van Woerden, Reidel, Dortrecht, p. 261.

The Detailed Velocity Field of the Ionized Gas in the Interacting Pair of Galaxies NGC 2535-36

P. Amram, M. Marcelin, J. Boulesteix, and E. Le Coarer

Observatoire de Marseille, 2, Place Le Verrier,
F-13248 Marseille Cedex 4, France

SUMMARY. The radial velocities of the ionized gas in the double system of galaxies NGC 2535-36 (Arp 82) have been measured with a scanning Pérot-Fabry interferometer and a two-dimensional photon counting system (the CIGALE equipment). Some 3200 velocity points cover the whole galaxy and its companion with a spatial resolution of 1"5 and a velocity accuracy better than 19 km s^{-1}. The redshifts are found to be 4095 km s^{-1} ± 10 km s^{-1} for NGC 2535 and 4085 km s^{-1} ± 10 km s^{-1} for NGC 2536 (heliocentric radial velocities). Assuming a distance of 41 Mpc the masses are found to be around 3 10^{10} M⊙ for the main galaxy and 10^9 M⊙ for its companion.

VELOCITY FIELD. The grand design of the isovelocity lines of each of the two galaxies is that of normal rotating disks with differential rotation. Although the velocity field of the companion exhibits severe distorsions the major axis and minor axis remain fairly perpendicular so that there is no clear sign of expansion (nor for the main galaxy).

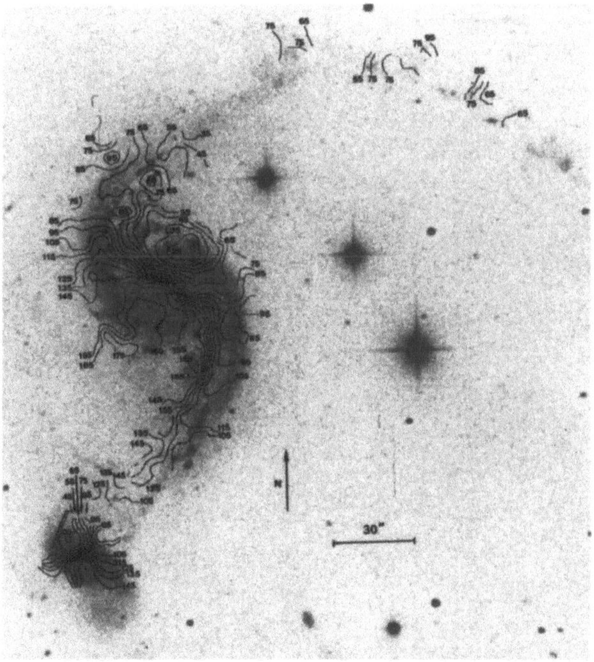

FIGURE 1. Isovelocity lines of NGC 2535-36 (minus 4000 km s^{-1}).The isoradial velocity lines have been drawn from some 3200 velocity points. They are superimposed to an Arp's photograph of Arp 82.

A remarkable result comes from the radial velocities measured in the spiral arms outside from the central disk. For both of them the radial velocity remains almost constant all along the arm, around 4075 km s^{-1} for the northern arm and around 4125 km s^{-1} for the southern one, that is to say close to the systemic velocity.

Assuming that the motions remain purely rotational it suggests that the arms are rotating slower than the central disk unless the arms are not in the same plane as this disk.

ROTATION CURVE. Figure 2, figure 3 and figure 4 show the rotation curves of NGC 2535 and its companion, NGC 2536. The parameters used for drawing these curves are the following ones:

	-NGC 2535	-NGC 2536
Systemic velocity	4095 km s^{-1} (± 10 km s^{-1})	4085 km s^{-1}(± 10 km s^{-1})
P.A. of major axis	157° (± 5°)	38° (± 5°)
Inclination (inner disk)	43° (± 10°)	49° (± 5°)

The rotation curve of the central disk of NGC 2535 is fairly symmetric (Fig. 2). A sort of a plateau around 110 km s^{-1} is rapidly reached (within 6" from the nucleus) and remains flat up to the edge of the inner disk at some 25" from the nucleus (around 20" for the North Western quadrant and almost 30" for

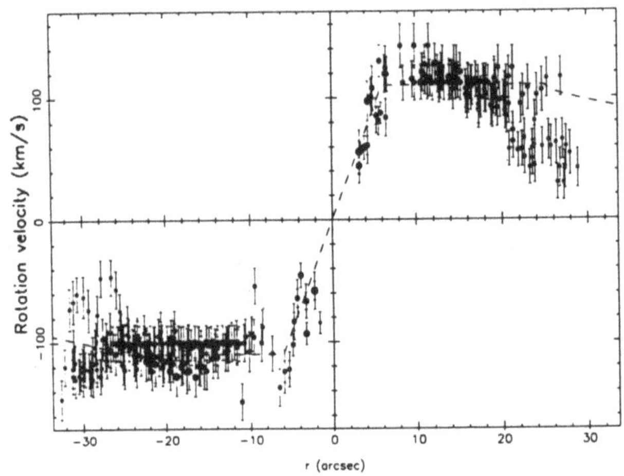

FIGURE 2. Rotation curve of the central disk of NGC 2535. The rotation velocities are computed from the radial velocity points found within 30° from the major axis (within 60° between the nucleus and a raduis of 6")

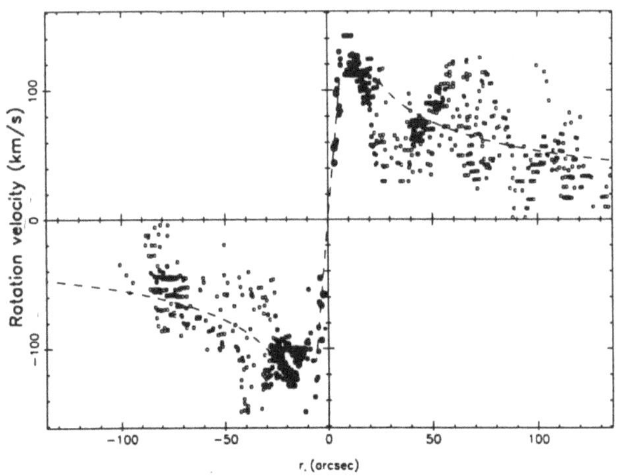

FIGURE 3. Rotation curve of NGC 2535. Same thing as Fig. 2 extended up to the outerost point where velocities have been measured.

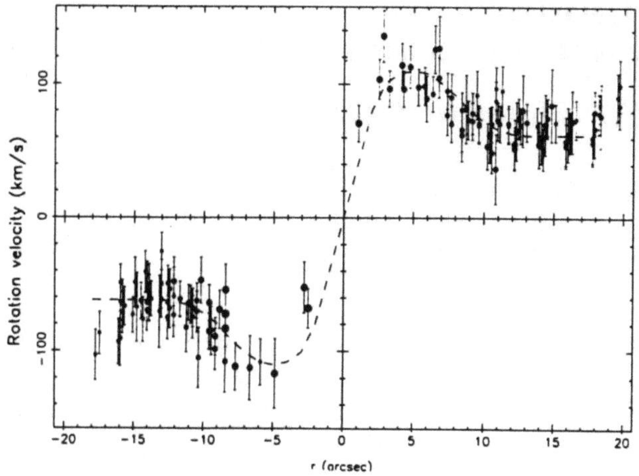

FIGURE 4. Rotation curve of NGC 2536. The rotation velocities are computed from the radial velocity points found within 50° from the major axis of the companion.

the South Eastern quadrant). Drawing the rotation curve beyond 25" (Fig. 3) shows a strong dispersion but the general tendency is a rapid decrease (even stronger than the keplerian dashed curve). Of course this rotation curve is significant providing that the observed motions occur in the same plane.

For interpreting this rotation curve we may distinguish two main hypothesis:

a) The galaxy is seen in a single plane inclined with respect to the plane of the sky. Then the rotation curve is significant and, although strongly dispersed (this dispersion may be explained by strong non circular motions being probably a consequence of the interaction, suggesting that it is in a quite advanced state) it shows an overall keplerian decrease beyond the central disk. Although some mass may have been lost because of the interaction it is highly improbable that the mass beyond 25" from the nucleus may be neglected since the spiral arms can be clearly seen up to 2.7' from the nucleus, that it is to say 6 times farther out than the central disk. This fact casts a doubt on the validity of the hypothesis.

b) The galaxy is warped beyond the central disk. Then the rotation curve of figure 3 is no more significant since it assumes that the inclination is not changing with the distance to the center. The true rotation curve may as well remain flat within the optical limits while the computed rotation curve decreases rapidly. The data (kinematics as well as photometry) in the outer parts of the galaxy are not sufficient to determine the law describing the warp unambiguously. However it is interesting to note that an inclination of the galactic disk varying progressively from 45° for the inner disk to 20° for the outer part of the galaxy (tip of the spiral ams) is sufficient to simulate a keplerian decrease of the rotation curve when computed as for figure 3 meanwhile the true rotation curve remains flat. In this case we assumed a constant P.A. for the major axis but the things may still be more complicated since the major axis may as well rotate depending on the type of warp. Anyway it let an attractive solution that the outer parts of the spiral arms are no more lying in the plane of the central disk but much closer to the plane of the sky.

SIMULATION. We are now working on models in order to simulate the morphology and the kinematics of NGC 2535-36 (collaboration with O. BIENAYME, Besançon Observatory).

Kinematics of Interacting Galaxies VV 317: Motions in the Bar of One of the Components

V.P. Arkhipova[1], *V.L. Afanasiev*[2], *A.N. Burenkov*[2], *O.K. Sil'chenko*[1,2], *and A.V. Zasov*[1]

[1]Sternberg Astronomical Institute, Moscow, USSR
[2]Special Astrophysical Observatory, Nizhnij Arkhyz, USSR

The pair of interacting galaxies VV 317 presents a divided system consisting of two galaxies of type Sc-Scd (for the photograph - see [1]). VV317*a*, m_{pg}=14.8 [2], is a spiral galaxy of type SBc seen face-on. The short broad arms emerge from the ends of the bright bar. The structure of the galaxy disk is of c-d type, it contains small bright HII-regions and a lot of dust.

The spectral observations of VV 317a have been carried out at the prime focus of the 6-meter telescope of the Special Astrophysical Observatory of the USSR Academy of Sciences in May, 1988. We have used the long-slit spectrograph UAGS with the TV two-dimensional detector [3]. For the radial velocity determination we used the H_α and [NII] emission lines. We estimate our precision of radial velocities as ±20 km/s for the best points.

The systemic velocity of the galaxy estimated from the outer parts of radial velocity curves using 3 spectrograms is equal to 4500 km/s; this value accords well with the literature data. But the behaviour of the gas velocity field in the centre of the galaxy is quite unusual.

The distribution of the gas radial velocities in the bar along the slit positioned at P.A.=355° is shown in Fig. The centre of the continuum coincides with R=0". The position angle of the bar is $P.A._b$=14°, so the slit crossed the bar at the angle ~19° from the north to the south. The H_α and [NII] emission lines are rather intense between R -7.5" and +7" corresponding to the bar inner region with the radius of 2 kpc. The characteristic feature of the radial velocity distribution at P.A.=355° is a depression near the centre of the galaxy with some compensation in the nucleus. The amplitude of the "blue" velocity deviation is 80 km/s, and the half-width of the depression is 4"≡1.2 kpc. As the galaxy is seen face-on, we may interpret this feature as gas outflow directed toward us perpendicular to the galactic disk in the central region with the radius ⩽1 kpc. Probably gas outflow away from us is hidden by the bar and disk containing a lot of dust. If it is so, it is the evidence for that we deal not with accretion but with the gas outflow from the centre of the galaxy in this case.

Gas outflows are frequent enough in Seyfert galaxies; the active nucleus stimulates the gas radial motion providing the energy

Dynamics and Interactions of Galaxies Editor: R. Wielen

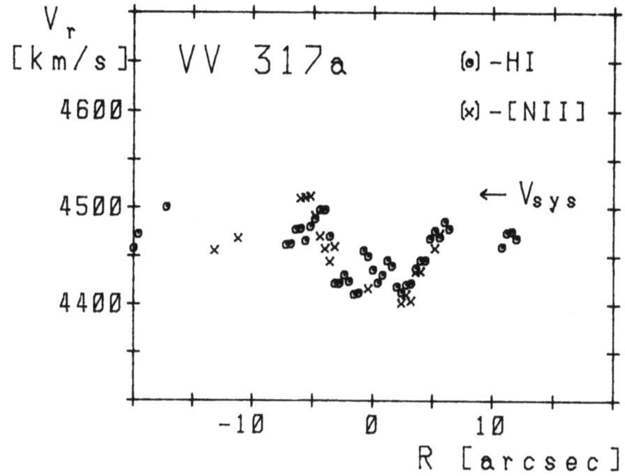

Figure. VV317a: the line-of-sight velocity distribution along the direction P.A.355° (19° to the bar): dots - Hα, crosses - [NII]λ6583. The abscissa zero is the nucleus of the galaxy

supply.The nucleus of VV317*a* shows no sign of activity; what cause may stimulate the gas outflow? A "hot" bar (with a large stellar velocity dispersion) rotates more slowly than surrounding gaseous disk. Therefore shock regions form near the edges of the bar (see, for example, [4] or [5]). Perhaps we have seen these regions in VV317*a*, they are identified by their high surface brightness on the screen of the TV-guide. There the gas of the disk loses momentum and becomes to be able to fall in the nucleus. A part of infalling gas can be fueled by the star formation, and a rest of it may outflow up and down in a cone resembling bipolar outflows of young stars. So, any dynamical simulation of the gas response in a barred galaxy potential must be three-dimensional because significant gas flows perpendicular to a galactic plane evidently exist.

References

1 B.A. Vorontsov-Velyaminov: *Atlas i katalog vzaimodejstvuyuschikh galaktik.* Part 1 (Moscow Univ. Press, Moscow 1959)
2 F. Zwicky, E. Herzog: *Catalogue of Galaxies and of Clusters of Galaxies* (Zurich 1966)
3 V.L. Afanasiev, A.N. Burenkov, A.V. Zhuravkov, N.K. Shpekina, V.G. Shtol': *SAO Report N 145* (Nizhnij Arkhyz 1987) (in Russian)
4 T. Matsuda, M. Inoue, K. Sawada, E. Shima, K.-i. Wakamatsu: *Monthly Notices of Roy. Astron. Soc.* **229**, 295 (1987)
5 T.S. Van Albada, R.H. Sanders: *Monthly Notices of Roy. Astron. Soc.* **201**, 303 (1982)

The Interacting System AM 2208-251

F. Bertola[1], *W.K. Huchtmeier*[2], *and W.W. Zeilinger*[1,*]

[1]Dipartimento di Astronomia, Università Padova, Vicolo Osservatorio 5, I-35122 Padova, Italy

[2]MPI für Radioastronomie, Auf dem Hügel 69, D-5300 Bonn, Fed. Rep. of Germany

*Affiliated to the Astrophysics Division of the Space Science Department of ESA

An interesting case of interaction between the disk galaxy AM 2208-251 and a compact elliptical-like object has been investigated. AM 2208-251 is classified as S0 (Corwin *et al.* 1985) and possesses a pronounced dust lane along its major axis. On the ESO/SRC J survey plate there is evidence of a bridge extending from AM 2208-251 towards the compact object. The companion lies on the Eastern side of the galaxy, exactly in the direction defined by the major axis of AM 2208-251. A contourplot of the digitized ESO/SRC J plate is presented in the composite of Fig. 1.

Spectroscopic observations carried out at the ESO 3.6-m telescope reveal in the light of the [O II] $\lambda\lambda$ 3727-29 Å emission a definite connection between the two galaxies (see rotation curve of Fig. 1). The spectrum of the compact galaxy exhibit a series of emission lines: [O II] ($\lambda\lambda$ 3727-29 Å), [Ne III] (λ 3868 Å) and Hδ (λ 4100 Å), while the continuum is else featureless. AM 2208-251 shows on the contrary a normal early-type galaxy spectrum.

The analysis of the radial velocities of both emission and absorption lines confirmed that we are dealing with a physical system. A distance of 99 Mpc (H_0 = 50) was deduced for the two systems. Despite its apparent roundness the companion shows a rather steep velocity gradient and its gas shows clear indications of counter-rotation with respect to AM 2208-251. It is also remarkable that the radial velocities systematically decrease between the two systems.

Observations of the HI profile at the Effelsberg 100-m radio telescope confirm the above findings (see HI velocity profile in Fig. 1). The observations reveal a disturbed double-horn profile, else typical of spiral galaxies. A HI mass of 2.6×10^{10} $M_\odot$ was derived for the system. VLA observations demonstrate nicely the interacting nature of this galaxy system. Two distinct HI-arms emerge from the disk galaxy as typical for gravitationally interacting galaxies.

The presence of a gaseous bridge between AM 2208-251 and the compact companion, in addition to the continuity in the measured the velocities of the gas indicates that the two galaxies are physically connected. Morphologically this system resembles those described as of M 51 type, where a disk galaxy is connected to a compact companion (Vorontsov-Velyaminov 1959, Arp & Madore 1987). The peculiar position of the companion, which lies on the principal plane of the AM 2208-251, suggests that also its orbit lies on that plane.

References

Arp, H.C. & Madore, B.C., 1987. *"A Catalogue of Southern Peculiar Galaxies and Associations"* Vol I & II (Cam. Univ. Press).

Corwin, H.G.Jr., de Vaucouleurs, A. & de Vaucouleurs, G. 1985. *University of Texas Monographs in Astronomy* No. 5.

Vorontsov-Velyaminov, B.A., 1959. *"Atlas and Catalog of Interacting Galaxies"* (Sternberg:Moscow).

Dynamics and Interactions of Galaxies Editor: R. Wielen

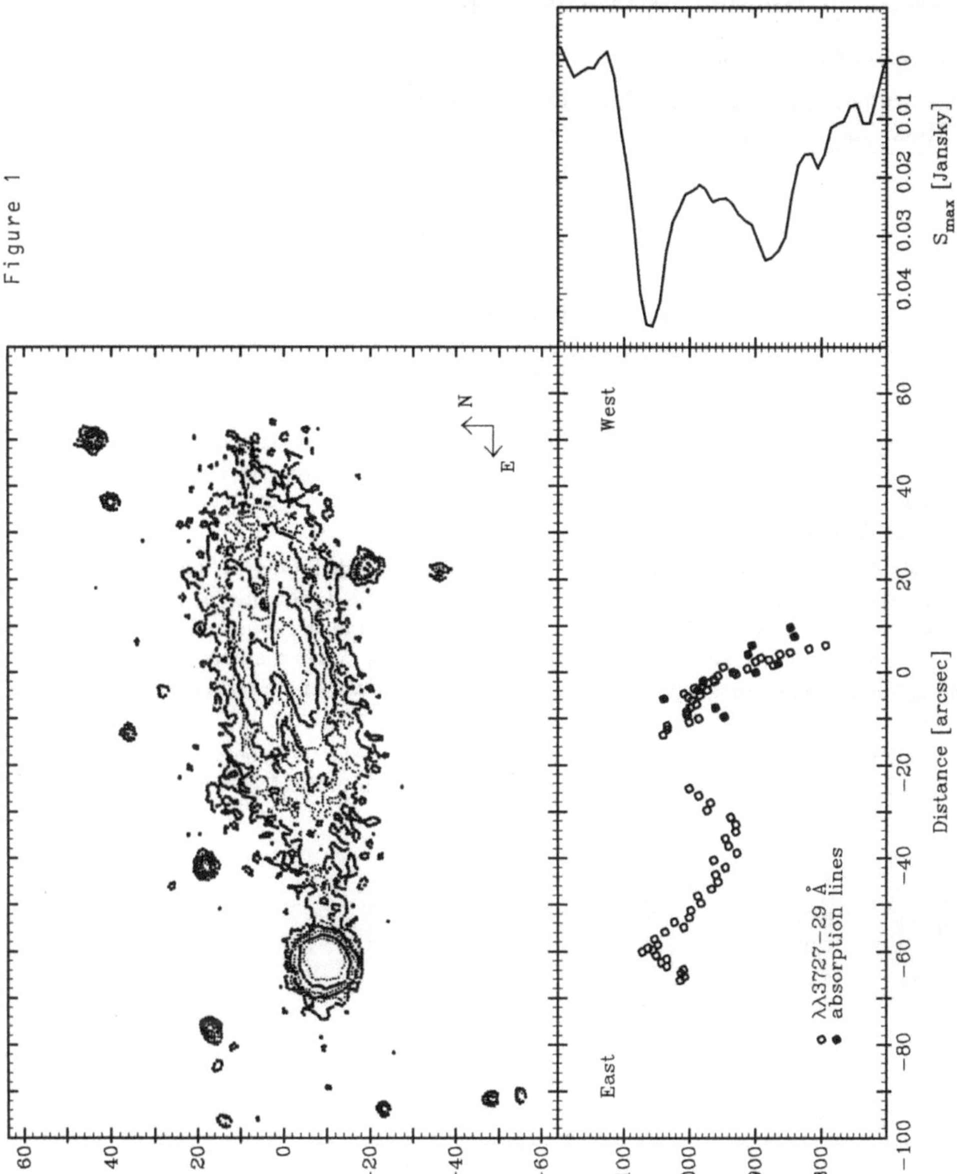
N
E
West
East
Distance [arcsec]
Velocity [km s⁻¹]
λλ3727–29 Å
absorption lines
S_{max} [Jansky]

Figure 1

UGC 9588: Another Merger in Progress?

E. Davoust and P. Poulain

Observatoire Midi-Pyrénées, 14, Avenue Eduard-Belin,
F-31400 Toulouse, France

There is increasing evidence that gravitational interactions and mergers have a role to play in the evolution of elliptical galaxies. They may explain the present morphology of some, if not all, ellipticals. They may also be at the origin of observed radio emission and nuclear activity in galaxies.

To investigate this possible role, we have undertaken an extensive photometric and kinematic survey of close pairs of elliptical galaxies. In the course of this survey, we have encountered two strongly disrupted pairs which exhibit emission lines, and which are probably in the process of merging. We report here on one of these pairs.

UGC 9588 is a double nucleus galaxy with two extended tails of luminous material, presumably of tidal origin. A series of CCD images in V and R, obtained at the 2m telescope of Observatoire du Pic du Midi, also show luminous material in the approximate, although not exact, continuation of the tails beyond the two nuclei. The outer contours (at the level V = 24 mag/arsec^2) are strongly perturbed (Figure 1). The two nuclei are separated by 5.7"(= 1.1 Kpc).

The results of the photometric study are summarized in the table below. It gives the total magnitude of the system within a radius of 45", and the magnitudes of the two nuclei, integrated through a 8" aperture.

	V	R	V - R
nucleus 1	16.94	16.18	0.76
nucleus 2	17.25	16.27	0.98
system	14.70	14.15	0.55

Nucleus 2, which is redder than nucleus 1, is presumably obscured by dust, a conspicuous feature of mergers and of Seyfert 2 nuclei. The whole system is bluer than the nuclei, suggesting the presence of a young population of stars produced in the outer regions by the interaction.

A spectrum of the galaxy was obtained with the Carelec spectrograph on the 1.93m telescope at Observatoire de Haute-Provence. The spectrum was binned in wavelengths to pixels of 210.8 km/s. It shows strong and narrow Hβ and [OIII] emission lines over an extended region, as well as a faint and narrow HeI emission line at 5875 Å and even fainter but broad [OI] lines at 6300 and 6363 Å. The strongest line, [OIII] at 5007 Å, was detected over 20"; a narrow-band CCD image of the system at that wavelength would certainly show the active regions with considerable details. The (peak) ratio of the two green [OIII] lines is approximately constant at 2.5. The (peak) ratio [OIII](5007)/Hβ is 5.0 in nucleus 1 and decreases outwards (2.0 at 10.8" East and 3.4 at 4.8" West).

This spectrum is that of a galaxy which has undergone a recent burst of star formation. HII regions excited by young stars produce the extended emission line region, Hβ is narrow, and, contrary to the case of LINER's and Seyferts, the two [OI] lines are very faint.

Dynamics and Interactions of Galaxies Editor: R. Wielen

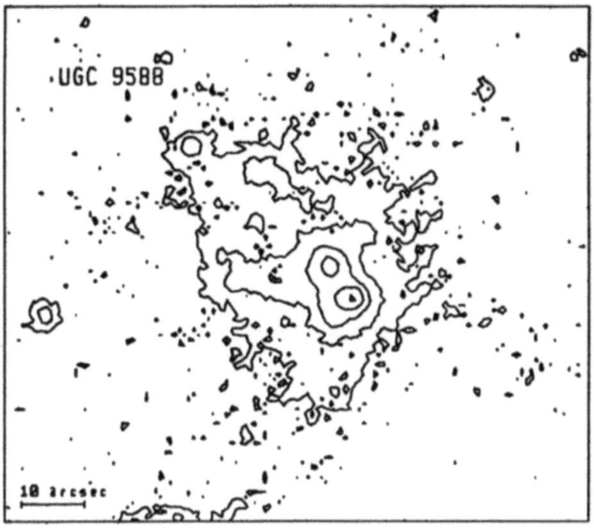

Figure 1 : *Isophotal map of a 20 min. CCD exposure of UGC 9588 in the V band. Contours are one magnitude apart, and the faintest level is at 24 mag/arcsec^2. The scale in arcsec is indicated*

The emission line spectrum of UGC 9588 was first detected by Wasilewski (1983) in an objective prism survey for low-redshift emission-line galaxies. He finds a relative flux $F(\lambda 5007)/F(H\beta) = 2.9$, an approximate redshift of z = 0.0091, and notes that the continuum is not blue.

The redshift of the emission lines on our spectrum is 3030 km/s (corrected for solar motion), which places UGC 9588 at a distance of 40 Mpc. There is no evidence from the emission lines (at the resolution of our data) for any relative motion of the nuclei, or for any internal motions in the system. This is at variance with the situation in other merger candidates; UGC 9588 might be seen face on.

Most probable mergers reported so far are Markarian galaxies. This is not surprising, since 90% of all Markarian galaxies have emission lines, and 5% of them also have double nuclei. Mergers have also been found among IRAS galaxies.

In this respect, UGC 9588 is remarkably different, since it has no strong ultraviolet nor infrared excess. No HI velocity has been reported. The rather faint absolute magnitude of the system ($M_v = -18.33$) is probably the cause for this lack of activity at other wavelengths. UGC 9588 is thus a **scaled-down** version of the above mergers.

It should be pointed out that the presence of emission lines is not a necessary consequence of mergers. NGC 3640 for example (Prugniel *et al*, 1988) does not present such features and has presumably reached a later stage of the merging process.

In conclusion, we recommend follow-up observations of this system by radioastronomers, to look for possible HI and/or CO emissions.

References

Prugniel, Ph., Nieto, J.-L., Bender, R., Davoust, E. : 1988, *Astron. Astrophys.* **204**, 61

Wasilewski, A.J. : 1983, *Astrophys. J.* **272**, 68

The Optical, Radio, and Far-Infrared Properties of the Interacting Pair NGC 2207/IC 2163

W. van Driel, K. Brink, and T. de Jong

Astronomical Institute, University of Amsterdam, Roetersstraat 15, NL-1018 WB Amsterdam, The Netherlands

In order to determine the detailed spatial correlation between the optical, radio continuum, and far-infrared morphologies of interacting galaxies we are studying the southern interacting galaxy pair NGC 2207/ IC 2163, which has a mean redshift of 2500 km/s, and a difference in redshift of 50 km/s. We will assume a distance of 25 Mpc.

NGC 2207 is classified in the RC2 as SAB(rs)bcp, and IC 2163 as SB(rs)cp. NGC 2207 has a M_B of -21.3. Their nuclei are separated by 12 kpc, and the E-W extent of the sytem is $\sim$ 30 kpc. The most obvious sign of interaction is the "tail" of IC 2163. The outer rotation curve declines faster than Keplerian, presumably indicating dynamical effects.

Optical data

Recently we made B,V,R, and Hα CCD images of three fields covering the entire system with the 1.5 m Danish telescope at ESO. Shown here is a contour plot of the blue image

An interesting feature is a quite red (B-R $\sim$ 1.8) point source in one of the brightest HII regions in NGC 2207, which coincides with the strongest radio continuum source in the system.

Radio continuum data

The other Figure shows Condon's (1983) 20 cm VLA radio continuum map. The global spectral index is 0.86.

The radio features are clearly correlated with optical structures, especially with the Hα image. The strongest radio emission occurs in the spiral arm which lies between the galaxies, and in the outermost arm on the opposite side of NGC 2207 (where the point source coincides with the abovementioned red optical point source), but these arms are not at all prominent in Hα.

This $\sim$ 3 mJy point source has a radio power comparable to that of the strongest star forming regions in M 82. Hence, it is more likely to contain a reddened cluster of newly formed stars, rather than a single supernova remnant .

Far-infrared data

The system was observed twice at 50 μm and 100 μm with the CPC detector of the IRAS satellite, which has a higher resolution ($\sim$ 90″), but a lower sensitivity, than the survey instrument (which is unable to separate both galaxies). The CPC map shows two clearly separated sources. Although IC 2163 is the smaller galaxy, it is brighter in the far-infrared, and has a 3 times higher far-IR/blue luminosity ratio than its larger neighbour.

Dynamics and Interactions of Galaxies Editor: R. Wielen

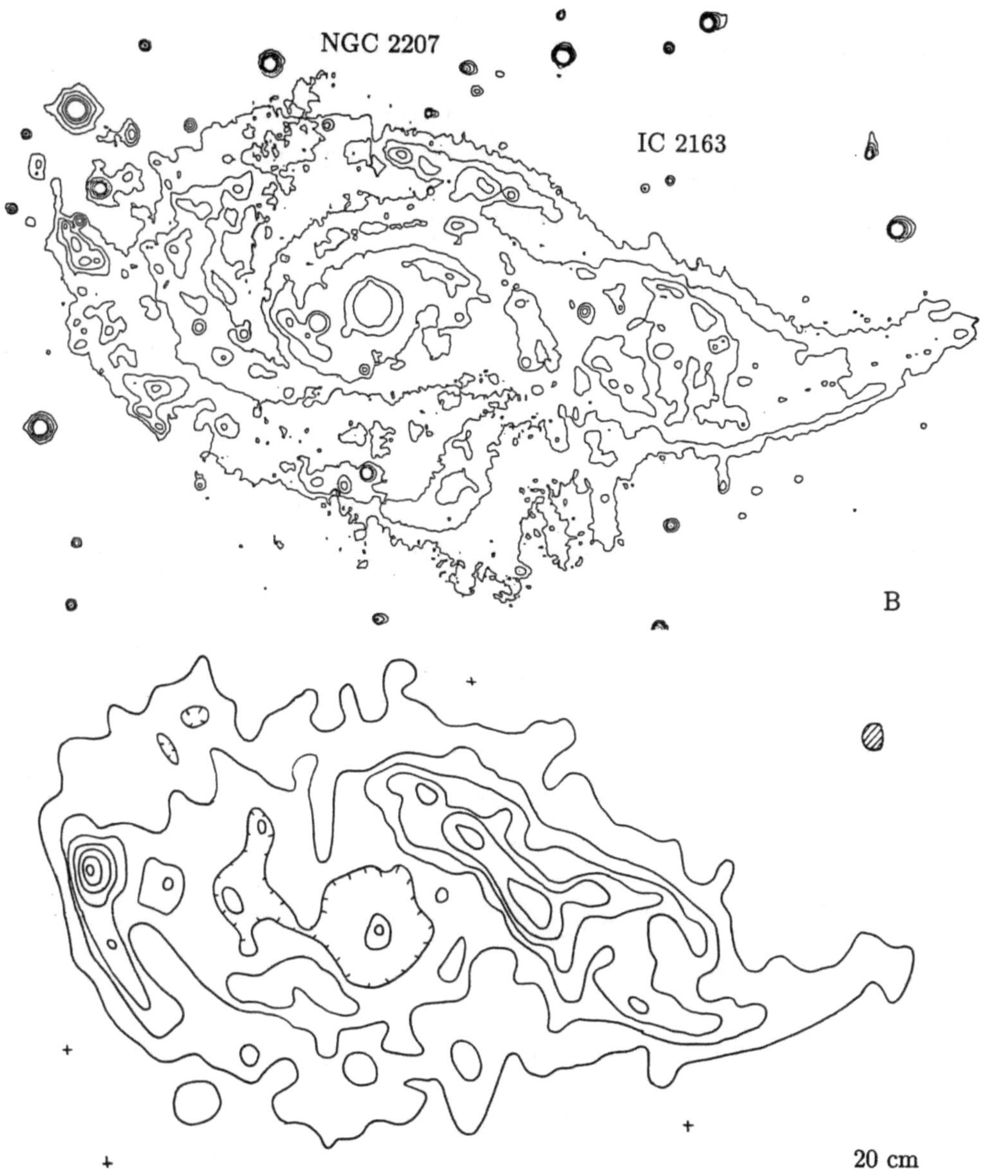

NGC 2207/IC 2163. Top: composite of three B band CCD images, contours are 20(0.75)23.5 mag/arcsec^2. Bottom: 20 cm radio continuum map (from Condon, 1983), resolution $\sim 7''$, contours 1,3,5,7,9,11,13 $\times$ 0.24 mJy/beam.

Reference: Condon, J.J.: 1983 *Astrophys. J. Suppl.* **65**, 485

0422-476: A Shell Galaxy with Non-Aligned Shells

A. Wilkinson[1], *J.L. Prieur*[2], *D. Carter*[3], *D. Malin*[4], *W.D. Pence*[5], *and W.B. Sparks*[5]

[1]University of Manchester, Manchester M13 9PL, UK
[2]ESO, Karl-Schwarzschild-Str. 2, D-8046 Garching, Fed. Rep. of Germany
[3]Royal Greenwich Observatory
[4]Schmidt Telescope, Anglo-Australian Observatory, Coonabarabran, NSW 2357, Australia
[5]Space Telescope Science Institute, 3700 San Martin Drive, Baltimore, MD 21218, USA

1. Description

0422-476, a nearly face-on S0-a galaxy, is the best example in the Malin/Carter (1983) catalogue of a shell system with shells distributed randomly "all-round" the galaxy. The shells are so bright they distort the isophotes and galaxy profile at most radii. Eleven shell arcs are distinguishable from 2.7 to 31.2 kpc radius. In the outer parts they interleave in radius, at about 11 kpc they have the same elliptical radius and within about 10 kpc of the nucleus they are diffuse and show spiral-like structure. No discordant pitch angles are seen. An extensive but faint disk (scale length 7.2 kpc) visible up to nearly 4 scale lengths underlies the outer interleaving shell region.

2. Discs in shell systems

Evidence is beginning to accumulate that all the "all-round" class contain discs.

2.1 Statistical evidence from other shell galaxies

1) The "all-round" class have types E0.5 to E3.8, and the "aligned" class E2.5 to E5. The "all rounds" are round and the "aligned" class elongated (Prieur,1988).
2) The "all-round" class have lower average velocity dispersions, ($\sim 215 \pm 35 km/sec$) typical of S0s. The "aligned" class have velocity dispersions typical of true ellipticals ($\sim 260 \pm 40 km/sec$).
3) 26% of the "all-round" galaxies have evidence of A-stars in their nuclear spectra, while only 16% have emission lines, in a sample of 19 galaxies. 43 % of the "aligned" shell

Dynamics and Interactions of Galaxies Editor: R. Wielen

galaxies have emission lines in their nuclear spectra, while only 14 % have A-stars, in a sample of 14 galaxies. This suggests that the "all-round" class have been more efficient at converting accreted material into stars, which may have happened more easily in a disc.

Shells therefore appear to have been formed round disc systems in ~ 30% of the Malin/Carter catalogue (Wilkinson et al.,1989 in preparation). This either suggests that discs are not fragile, or that mergers are an important factor in producing discs in S0s. In 0422-476 the shells lie close to or in the disc plane, and this is likely to be true for most of the systems recognised as belonging to the "all-round" class. This could be evidence either of the effective action of dynamical friction, or again that the intruder played an important part in creating the disc.

3. Shell colours

The colours reveal a perturbed N/S system (suggesting the direction of the primary encounter?) with shells rather bluer than the main galaxy. Redder E/W material has either been added during a secondary encounter or obscured by dust which is not apparent visually. However, some blue and red shells have the same elliptical radius, which is difficult to reconcile with an origin in different encounters. The colours may also vary along a single shell.

4. Origin of the shell system

Phase or spatial wrapping (eg. Hernquist and Quinn, 1987; Dupraz and Combes, 1986) is required to explain interleaving, and the small radii of the inner structures require the companion to be slowly disrupted and experience dynamical friction as it spirals to the centre. Elliptical rings could perhaps be produced by an instability in a coldish close-to-oblate triaxial galaxy or in a not-very-elliptical disc (Merritt 1989), or could occur at a disc resonance. Spiral-like structure suggests some form of density wave in the disc, and dissipative formation of that disc. Is 0422-476 a case where the beginning of true spiral structure may have been triggered in an S0 with an originally featureless disc by the accretion of a small amount of gaseous material from an interloper which left evidence of its stellar component in the outer shell system?

References

C. Dupraz, F. Combes: Astron. Astrophys. 166, 53 (1986)

L. Hernquist, P.J. Quinn: Astrophys. J. 312, 1 (1987)

D.F. Malin, D. Carter: Astrophys. J. 274, 534 (1983)

D. Merritt: Instabilities of Slowly-rotating Galaxies (1989 Florida Workshop proceedings)

J-L. Prieur: PhD. Thesis, Toulouse (1988)

Part IV

M51 and Similar Systems of Galaxies

Dynamical Status of M51

L. Hernquist

The Institute for Advanced Study, Princeton, NJ08540, USA

I. Introduction

M51 enjoys a special place in the history of observational astronomy. It was the first galaxy observed to have spiral structure, by Lord Rosse in 1845 using the 72-inch reflector at Burr Castle. Over a century later, M51 became the first galaxy outside the Local Group to be mapped extensively at radio wavelengths by Mathewson, van der Kruit, and Brouw [1]. Thereafter, it was one of the first galaxies observed at CO [2], far-infrared [3], and x-ray [4] wavelengths. Among the important recent findings are the fact that the vast majority of gas in M51 is molecular [5] and the demonstration that the blue light, CO, H-α, radio continuum, far-infrared, and x-ray emission all follow nearly the same exponential surface-brightness profile.

In spite of the many detailed observations of M51, the origin of its "grand-design" spiral structure remains enigmatic. Models based on long-lived density waves generated internally have not been conspicuously successful at reproducing the morphology of this galaxy. The recent discovery by Rots *et al.* of an extended HI tail associated with M51 supports the alternate point of view which maintains that at least some of the observed structure was externally induced. While it is by no means obvious that tides are the sole or even the most important process at play in this system, the present review focuses on the interesting possibility that nearly all the structure in M51 originated as a consequence of its tidal interaction with NGC 5195.

II. Theoretical Work on Tides

It was recognized long ago that the companion galaxy NGC 5195 may have been responsible for exciting the large, bisymmetric spiral arms in M51. Indeed, Chamberlin in 1901 [6] was one of the first to suggest that gravitational interactions might yield spirals. However, our understanding of the possible link between tidal effects and grand-design spiral structure advanced slowly until the advent of high-speed computers in the 1960's and 1970's. Toomre and Toomre ([7], hereafter TT) pioneered the application of computational methods to galactic interactions with their study of the response of test-particle disks to tidal perturbations in 1972. In particular, TT examined the kinematics of the M51/NGC 5195 system, using a restricted three-body method.

In their modeling of M51, TT were motivated by deep optical plates taken by van den Bergh [8]. A similar, but even more interesting, composite image of M51 prepared a few years later by Burkhead [9] is shown in Figure 1, where a standard photograph has been overlaid on five deeply exposed IIIaJ negatives. All frames were taken by M.S. Burkhead using the Palomar 48-inch Schmidt telescope. Note especially the abrupt, yet smooth outline of M51 to the west and the faint "streamers" from NGC 5195 pointing towards 2 and 8 o'clock. Displayed side-by-side with Burkhead's plate is a snapshot from a simulation performed by Toomre ([10], hereafter T78), which is essentially a refined version of the model presented in TT. Figure 2 displays a view projected onto the plane of the sky at a time 2×10^8 years after pericenter, assuming a mass of $10^{11} M_{\odot}$ for M51 and a pericenter distance $r_p = 25$ kpc. The primary-to-companion mass ratio is 3:1, and the plane of the disk is inclined by 22.5° with respect to the sky. TT and

Dynamics and Interactions of Galaxies Editor: R. Wielen

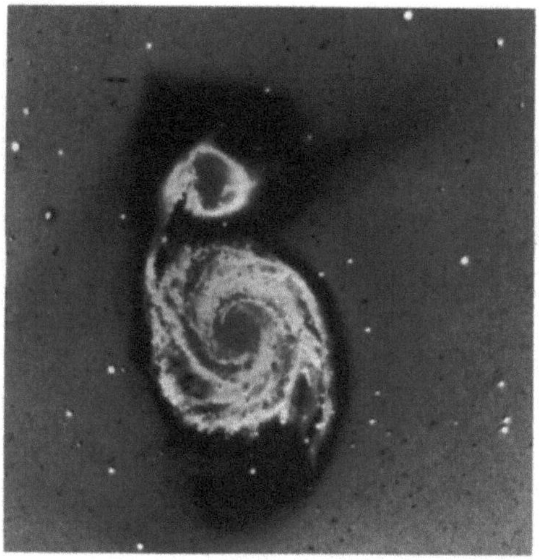

Figure 1: *Composite image of M51 and NGC 5195, by M.S. Burkhead.*

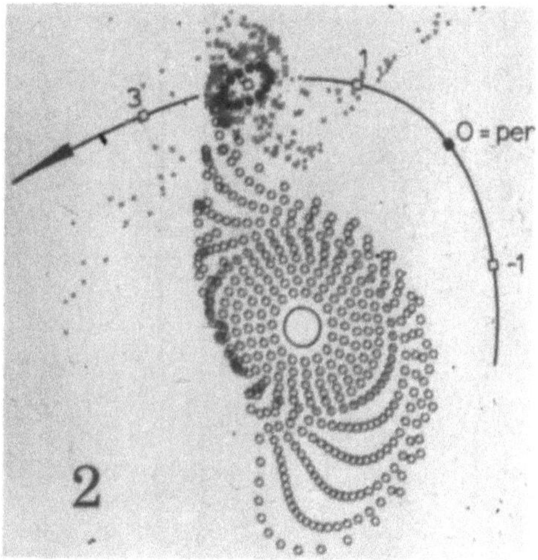

Figure 2: *Frame from the restricted three-body simulation of Toomre [10].*

T78 favored a plunging orbit to account for the short bridge arm between M51 and NGC 5195 and the rather large radial velocity separation $\Delta v = v_{5195} - v_{M51} \approx 120$ km/sec [11,12].

In Figure 2 the inclination and argument of pericenter of NGC 5195's orbit (as defined by TT's Figure 6) are $i = 75°$ and $\omega = -25°$, respectively. TT and T78's best fits to the structure in M51 are obtained relatively soon after pericenter.

In spite of their simplicity, the experiments performed by TT and T78 match the outer morphologies of M51 and NGC 5195 rather well. However, as emphasized by TT and T78, their models are totally incapable of reproducing the inner spiral structure in M51.

An early demonstration that this failure was likely due to neglect of self-gravity by TT and T78 was provided by Zang [see 13]. One of his models is shown in Figure 3, which reports an N-body simulation of a perturbed Mestel disk with a perfectly flat rotation curve. Initially, an $m = 2$ quadrupole impulse was imposed, boosting the velocities of the outermost particles by $\approx 20\%$. A strong, two-armed pattern quickly develops as a result of this impulse and propagates nearly to the center of the disk. As noted by Toomre [13] these arms result from swing-amplification of weak kinematic waves in the outer parts of the disk. Similar results have been reported recently by Sundelius *et al.* [14] and by Sundelius (this volume).

While informative, calculations like that in Figure 3 are not true models of M51 since they are not attempts to match the structure of its disk or that of NGC 5195 in detail. Moreover, in view of the recent HI observations by Rots *et al.* (this volume), it is not difficult to see that a substantial revision of the parameters adopted by TT and T78 is called for. Assuming that the long HI tail discovered by Rots *et al.* is of tidal origin, it is possible to use its length to estimate the time since pericenter. Using an impulse approximation (*e.g.* [15]) the time since pericenter can be expressed in terms of the mass of the perturber, m, the primary mass, M, the pericenter distance, r_p, the orbital eccentricity, e, and the length of the tail, L. For the values $M = 10^{11} M_\odot$, $r_p = 25$ kpc, $m/M = 1/3$, and $e = 0.8$ adopted by TT and T78, the predicted time since pericenter appropriate for the length of the tail observed by Rots *et al.* is roughly 16×10^8 years. This value is a factor of 8 larger than that in Figure 2 taken literally, or about a factor longer than if the higher exterior speeds from a massive halo had been included in the

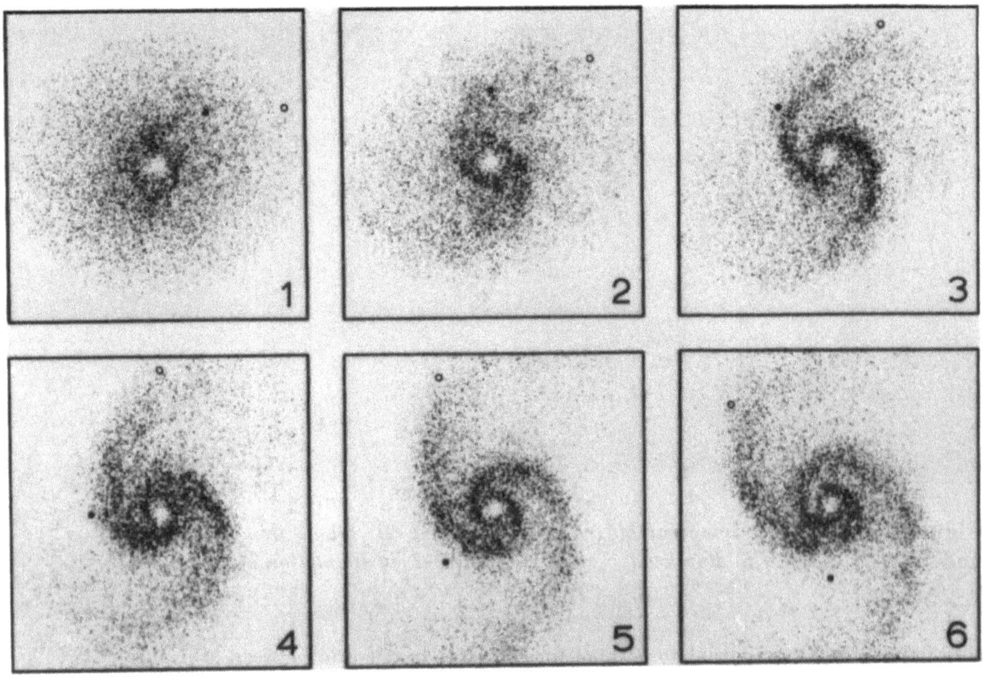

Figure 3: *Simulation of perturbed self-gravitating disk by Zang. Two "clock" particles moving in circles indicate the orbital periods.*

TT-style kinematics. As a result, new models should focus on shaper impulses, more massive primaries, and times since pericenter considerably larger than considered by TT and T78.

III. Self–consistent Models

Given the recent advances in computational techniques (*e.g.* [16,17]), Josh Barnes and I were invited by the Scientific Organizing Committee for this meeting to re–examine M51 in detail, to see whether it is indeed possible to offer a single tidal explanation for its HI distribution at large radii as well as its beautiful spiral structure in the interior. We accepted this challenge gladly, but soon discovered that a solution was not immediately forthcoming. Here indeed is a rather complex story, and what we can present at the moment is more in the nature of a progress report than any definitive work.

We use fully self–consistent methods to perform our calculations. Since the M51/NGC 5195 problem is intrinsically three–dimensional, we compute gravitational forces using a hierarchical tree algorithm which does not suffer from grid–based limitations on the global geometry or local spatial resolution [18]. In some cases we also include gasdynamical effects using a hybrid N–body/hydrodynamics code [17]. Although the distinction between gas and collisionless material is unlikely to have significant dynamical consequences for this application, we are eager to compare even the predicted distribution of gas with observations. All simulations described here were performed on the Cray Y–MP at the Pittsburgh Supercomputing Center using our existing codes optimized for vector supercomputers [19,20].

The galaxies used in our simulations are based on the bulge/disk/halo (BDH) models developed by Barnes ([16]; also this volume). Our disks have exponential surface–brightness

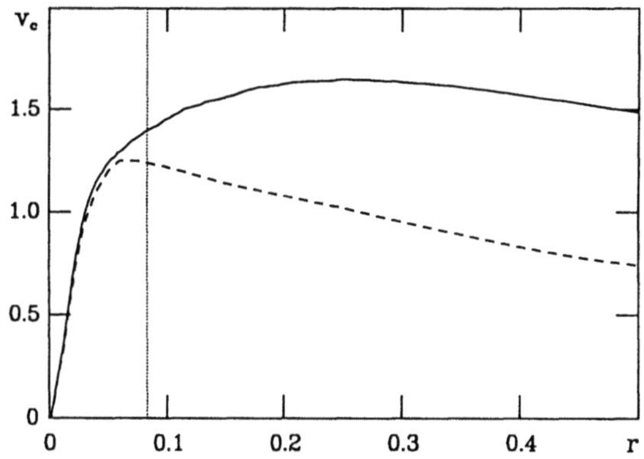

Figure 4: *Circular velocity versus radius for BDH model with mass ratio B:D:H = 1:3:16. Dashed line shows rotation curve for bulge and disk alone, while solid line includes halo component. Vertical dotted line at* $r = 1/12$ *indicates the exponential scale-length of the disk.*

profiles, and bulges and halos are currently both represented by $W_0 = 3$ King models [21]. In the calculations presented here, the B:D:H mass ratio is always 1:3:16. The rotation curve for one of these models is shown in Figure 4. Here, and in the following discussion, units are such that $G = 1$, the total mass $M = M_D + M_B + M_H = 1.25$, the exponential scale-length of the disks is $h = 1/12$, and one time unit denotes the rotation period at $3h$. Scaled to our own galaxy, the units of length, time, and mass are roughly 40 kpc, 2.5×10^8 years, and $2.2 \times 10^{11} M_\odot$.

Figure 5 shows an example of the structure which develops in an *isolated* BDH model. In this calculation 15% of the disk mass is gaseous, in accord with CO observations of M51. Note especially the rapid growth of spiral structure between $t = 0$ and $t = 0.75$. These patterns are not time-stationary, evolving on a dynamical time-scale. A Fourier decomposition into logarithmic spirals confirms one's visual impression that the spiral structure is dominated by azimuthal wave-numbers $m \approx 4 - 6$. At no time is significant $m = 2$ power present. All of these properties are consistent with the interpretation that the structure arises from swing-amplified noise [13]. The model shown in Figure 5 was evolved for five additional rotations to check its long-term stability. It suffered only minor changes in its gross properties over this time.

IV. Encounters

In our first tidal experiment, we ran a self-consistent version of the model favored by TT and T78; *i.e.* the same initial orbital parameters as in T78 and a mass ratio of 3:1. Results from this simulation are shown in Figure 6, near the time when the primary morphology most closely approximates that in the T78 model. The cross locates the center of mass of the companion which is otherwise not displayed. Pericenter is reached at $t \approx 1.0$ time units. The best match to the results of TT and T78 occurs for times $t \approx 1.5$. As can be seen by comparing Figures 2 and 6, there is some agreement between the outer morphologies of the test-particle and self-gravitating disks. However, there are also noticeable differences. For instance, when projected onto the plane of the sky, the companion in the self-consistent simulation lags behind that at equivalent times in the restricted three-body calculation shown in Figure 2. Thus the companion in Figure 1 is not correctly placed at $t = 1.5$, when the outer morphologies in the two models match most closely. At somewhat later times, $t \approx 1.75$, the companion is nearer its proper location relative to M51, but now the tidal debris is more extended than in the T78 model. Also, there is a hint of a two-armed spiral developing already at $t = 1.5$, which is completely absent in the test-particle calculations.

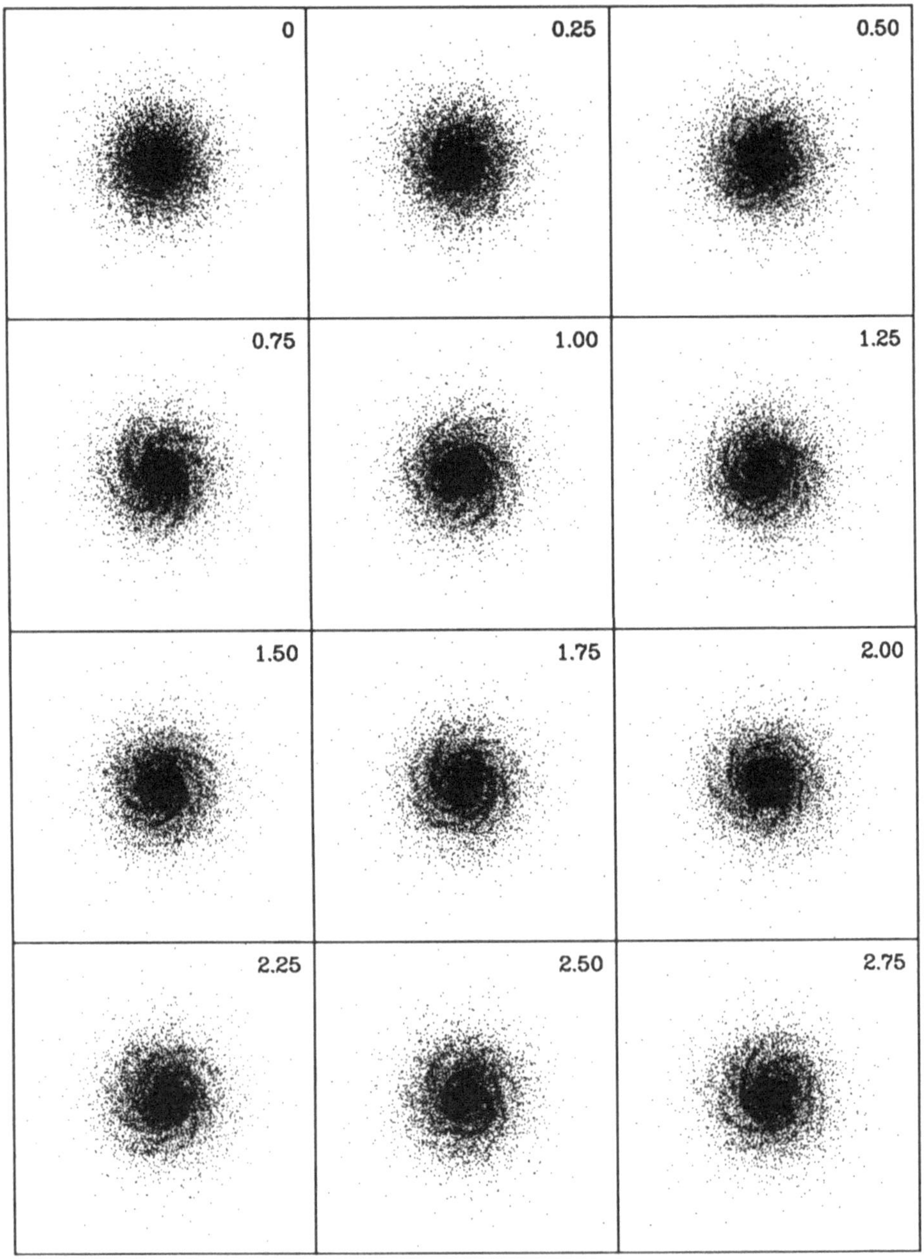

Figure 5: *Structure arising in an isolated BDH model. Time in upper right of each frame is in units of the rotation period at three exponential scale-lengths of the disk. Plotted are the* $N = 24,576$ *particles comprising the disk. Not shown are the* $N = 16,384$ *particles representing the dark matter halo or the* $N = 4096$ *particles comprising the bulge.*

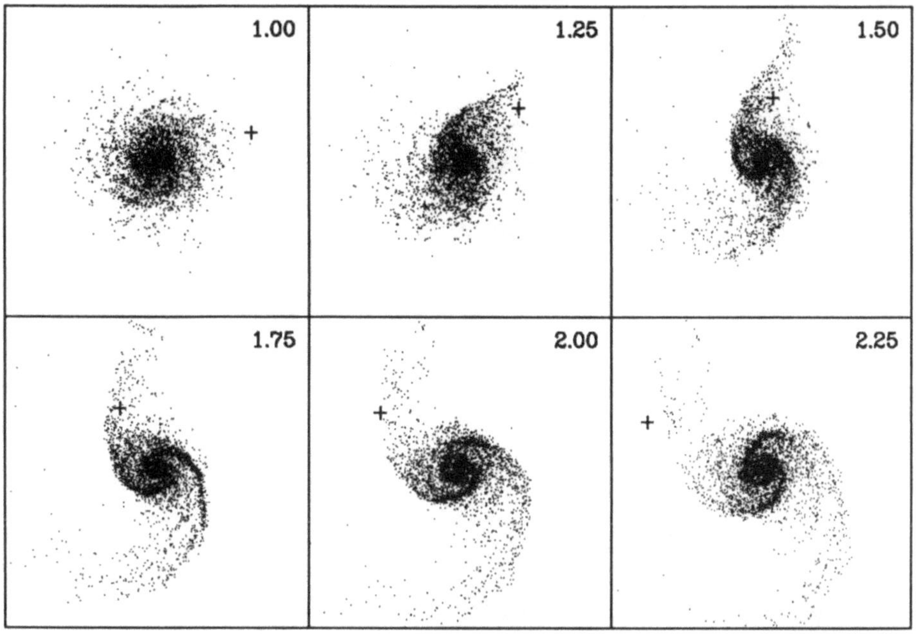

Figure 6: *Self-consistent version of the M51 model of Toomre [10]. Units of time are as in Figure 5. Plotted are the* $N = 6144$ *particles comprising the disk and the* $N = 2048$ *particles representing the bulge. The 1/3-mass companion is located by the + symbol.*

If the HI tail discovered by Rots *et al.* is of tidal origin, then the M51/NGC 5195 system is considerably further beyond pericenter than the simulation pictured in Figure 2. Eventually the disk in Figure 6 does, indeed, develop a long tail resembling the HI feature. However, at these late times the companion has already ceased receding, and it is again approaching the primary, at variance with observations. Clearly the actual orbit must be different from that inferred by TT and T78 from the data then available to them.

Encouraged by our ability to excite large-amplitude density waves tidally, we decided to conduct a modest parameter survey attempting to match the kinematic properties of M51 in detail. In view of the rather large velocity difference between M51 and NGC 5195, the actual orbit is likely along our present line-of-sight. To check the sensitivity of the response of the primary disk to the orbit of the companion, we varied the orbital inclination, i, and argument of pericenter, ω. We simplified this task by focusing exclusively on parabolic orbits. Figure 7 shows frames from three different runs at identical times. In all three models the orbital inclination is $i = -70°$ and the argument of pericenter is, from left to right, $\omega = -30°, -15°$ and $0°$.

As is apparent from Figure 7, the development of inner spiral structure and extended tidal tails depends only weakly on ω for plunging encounters. A similar conclusion obtains for the inclination, i. Thus, the orbit is constrained mainly by the relative positions and velocities of the two galaxies. This simplifies our task considerably since, in essence, it implies that the problem of matching the morphology of M51 decouples to a large extent from attempts to fit the observed properties of NGC 5195.

Next, we changed the strength of the impulse delivered to the primary during the interaction by independently varying the mass ratio and the pericenter distance. A synopsis of these results is shown in Figure 8 which displays frames from three different runs at the same time. From

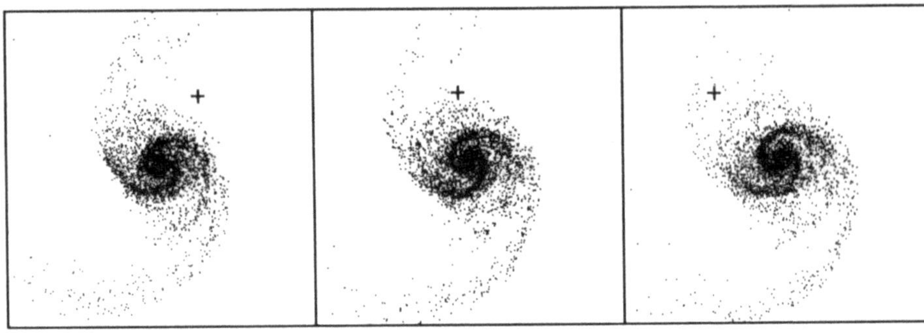

Figure 7: *Structure induced in disks by encounters with companions on orbits with* $\omega = -30°, -15°$ *and* $0°$. *Frames show responses at roughly the same time past pericenter.*

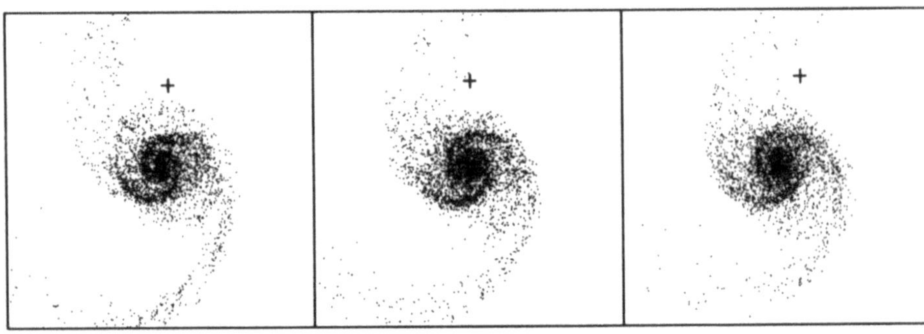

Figure 8: *Structure induced in disks by decreasingly strong impulse, from left to right. Again frames show responses at roughly the same time past pericenter.*

left to right, the pericenter distance and mass ratio are: $(r_p, M/m) = (0.5, 2), (0.6, 3), (0.7, 4)$. The same parabolic orbit was used in these three runs; specifically, $i = -70°$ and $\omega = -15°$.

Not surprisingly, sharper impulses result in more robust and longer–lived two–armed spirals, while simultaneously producing more extended and well–defined tails. The model in the right-most panel in Figure 8 fails badly on both counts. We thus decided to focus our attention on sharp–impulse events with pericenter distances and mass ratios obeying $r_p < 0.7$ and $M/m \lesssim 3$. These parameters are probably bounded from below by $r_p > 0.3 - 0.4$, to avoid "ringing" the disk (*e.g.* [22]; James, this volume), and $M/m \gtrsim 2$ [12]. Furthermore, events more violent than those pictured in Figure 8 can, in some cases, induce bar–making in the disk. Based on our preliminary results, the most promising values are $r_p \sim 0.4 - 0.5$ and $M/m \sim 2$. Note that this range in r_p corresponds to roughly $5 - 6$ exponential scale–lengths, which is $\approx 17 - 20.4(D/9.6\text{Mpc})$ kpc, where D is the distance to M51 (*e.g.* [5]). It must be emphasized, however, that these values are derived from calculations involving one particular galaxy model. If, as is argued below, it becomes desirable to improve the rotation curve of its disk, then slightly weaker collisions may again become viable.

Our best attempt to date at fitting the structure in M51 is shown in Figure 9, for a calculation with $r_p = 0.5$, *i.e.* 6 exponential scale–lengths, and $M/m = 2$. The orbit is initially parabolic with $i = -70°$ and $\omega = -45°$. This value of ω differs considerably from that used by previous workers and is motivated by our need to maintain a large relative velocity between M51 and NGC 5195 for a sufficiently long time to allow an extended tail to develop, as implied by (1).

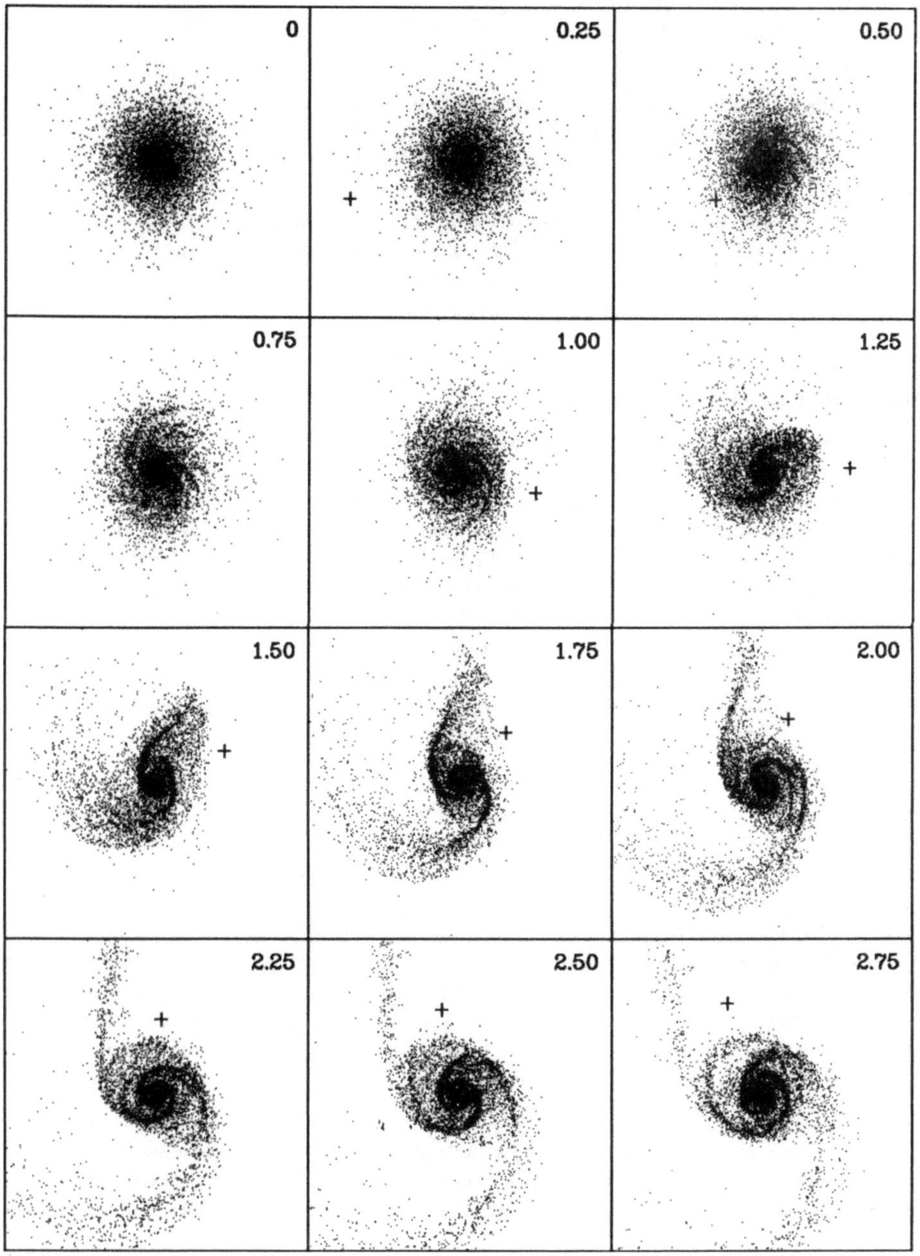

Figure 9: *Response of BDH model to an encounter with a companion half as massive as itself. The primary model and its particle number are identical to those in Figure 5. Center of mass of the companion is again located by the + symbol.*

As in Figure 5, 15% of the disk mass is gaseous. The location of the center of mass of the companion is denoted by a cross. The view in Figure 9 is projected onto the plane of the sky, which is inclined by 25° with respect to the disk plane.

At early times $t < 1.0$ the structure in the disk is quite similar to the isolated model shown in Figure 5. A two-armed spiral does not develop until well after pericenter, which occurs at $t \approx 1.0$. The best match to the spiral structure and HI tail in M51 is to be found at times between $2.0 \lesssim t < 2.5$. The velocity difference between the primary and companion is also roughly consistent with observations during this time interval. Scaled to the observed properties of M51, $\Delta v \approx 150(D/9.6\text{Mpc})^{-3/2}$ km/sec at $t = 2.0$ and $\Delta v \approx 100(D/9.6\text{Mpc})^{-3/2}$ km/sec at $t = 2.25$. In truth, the magnitude of our simulated Δv is already a bit too low when the spiral structure and tidal tail in the model most closely resemble M51.

V. Successes and Failures

Though the results presented here are encouraging, we certainly cannot claim to have unraveled all of M51's secrets. Among the positive aspects of our modeling: We have shown that strong two-armed spirals can be tidally induced in disks for parameters similar to those of the M51/NGC 5195 system, supporting the view that at least this famous grand-design spiral is probably of tidal origin. Close scrutiny of our models shows that the spiral arms are indeed density waves and not material structures, but even so it appears that they will not last for very long. We have also shown that the outer HI tail discovered by Rots *et al.* can be accounted for naturally by the collision between M51 and NGC 5195, provided that we are indeed viewing the system at a more advanced state than postulated by TT and T78. Relative to TT and T78, we currently favor a more massive companion, a smaller pericenter distance, a somewhat different orbital geometry, and, by a factor $\sim 2 - 3$, a longer wait past pericenter.

On the negative side, there is no denying that even during the most favorable time interval, the inner spiral structure in the model shown in Figure 9 is only a rough approximation to that in M51. At the time $t = 2.25$ the spiral structure in the model noticeably lags behind that of M51, in phase or position angle as viewed on the sky. The discrepancy can be reduced somewhat by waiting longer, but it then becomes impossible to match the observed relative velocities of M51 and NGC 5195. To resolve such difficulties it appears that it will be necessary to persuade the spiral arms to wrap more quickly. This can probably be achieved by modifying the shape of the rotation curve near the center of the disk. Structure will develop and revolve more rapidly if the rotation curve rises more sharply in the inner regions than in the present models. Observations do, in fact, suggest that the rotation curve of M51 remains roughly "flat" or constant to much smaller radii than that in Figure 4 [23–25], and we are eager to prepare and test some other model galaxies in which that flaw has been remedied.

We have not yet attempted to match the structure observed in the companion galaxy NGC 5195. Although we do not anticipate serious difficulty with this endeavor, given the earlier successes of TT and T78, we have yet to show that our new orbital parameters can reproduce NGC 5195's tidal streamers.

A more serious embarrassment at the moment is the presence of a northern bridge arm in our models, a feature which seems to be absent from both optical and HI observations of M51. Perhaps an even deeper, more plunging encounter than considered here can disperse the disk material which forms this bridge. Alternatively, M51 may, like the outer gas disk of M101, have been seriously off-center or asymmetric prior to its collision with NGC 5195.

It appears that self-consistent models can teach us much about the development of spiral structure during galactic encounters. Yet, judging by the difficulty of reproducing the detailed morphology of an object as well-studied as M51, it is also quite clear that much remains to be learned. To paraphrase Lord Rosse (see [26]): although many of our conjectures now seem at least plausible, a thorough understanding of galaxies like M51 remains elusive.

Acknowledgements

I thank my collaborator Josh Barnes for permitting me to quote results from our work in advance of publication. We are especially grateful to Alar Toomre for his considerable scientific input into this project. Our work is supported in part by a grant from the Pittsburgh Supercomputing Center and by New Jersey High Technology grant No. 89–240090–2.

References

1. D.S. Mathewson, P.C. van der Kruit, W.N. Brouw: *Astron. Astrophys.* **17**, 468 (1972)
2. L.J. Rickard, P. Palmer, M. Morris, B.E. Turner, B. Zuckerman: *Ap. J.* **213**, 673 (1977)
3. J. Smith: *Ap. J.* **261**, 463 (1982)
4. G.G.C. Palumbo, G. Fabbiano, C. Fransson, G. Trinchieri: *Ap. J.* **298**, 259 (1985)
5. N. Scoville, J.S. Young: *Ap. J.* **265**, 148 (1983)
6. T.C. Chamberlin: *Ap. J.* **14**, 17 (1901)
7. A. Toomre, J. Toomre: *Ap. J.* **178**, 623 (1972) [TT]
8. S. van den Bergh: *Astrophys. Letters* **4**, 117 (1969)
9. M.S. Burkhead: *Ap. J. Suppl.* **38**, 147 (1978)
10. A. Toomre: in *IAU Symposium No. 79: The Large Scale Structure of the Universe*, eds. M.S. Longair & J. Einasto (Reidel, Dordrecht 1978) pp. 109–116, [T78]
11. E.M. Burbidge G.R. Burbidge: *Ap. J.* **140**, 1445 (1964)
12. F. Schweizer: *Ap. J.* **211**, 324 (1977)
13. A. Toomre: in *The Structure and Evolution of Normal Galaxies*, eds. S.M. Fall & D. Lynden-Bell (Cambridge University Press, Cambridge 1981) pp. 111–136
14. B. Sundelius, M. Thomasson, M.J. Valtonen, G.G. Byrd: *Astron. Astrophys.* **174**, 67 (1987)
15. J. Binney, S. Tremaine: *Galactic Dynamics* (Princeton University Press 1987)
16. J. Barnes: *Ap. J.* **331**, 699 (1988)
17. L. Hernquist, N. Katz: *Ap. J. Suppl.* **70**, 419 (1989)
18. J. Barnes, P. Hut: *Nature* **324**, 446 (1986)
19. L. Hernquist: *Ap. J. Suppl.* **64**, 715 (1987)
20. J. Barnes, P. Hut: *Ap. J. Suppl.* **70**, 389 (1989)
21. I.R. King: *A. J.* **71**, 64 (1966)
22. R. Lynds, A. Toomre: *Ap. J.* **209**, 382 (1976)
23. R.B. Tully: *Ap. J. Suppl.* **27**, 415 (1974)
24. R.B. Tully: *Ap. J. Suppl.* **27**, 437 (1974)
25. R.B. Tully: *Ap. J. Suppl.* **27**, 449 (1974)
26. A. Toomre: *Ann. Rev. Astron. Astrophys.* **15**, 437 (1977)

Further Simulations of M51-Type Systems

B. Sundelius

Department of Astronomy/Astrophysics, Chalmers University of Technology, S-41296 Göteborg, Sweden

In this talk I will give a short description of some of the activities in our research-group in Göteborg. The group consists of the graduate students Stefan Engström, Maria Sundin and Magnus Thomasson, guest scientist Karl Johan Donner from Helsinki and myself.

Video movie

A numerical simulation of a disc galaxy perturbed by a passing point mass is shown in a video movie. The simulation is made using a two-dimensional polar coordinate N-body code with 60.000 fully gravitationally interacting particles. The disc is initially stable, its mass divided into 50 % cold Mestel disc and 50 % inert halo. The mass of the disturber is 22 % of the total galaxy and it passes in the disc plane on an initially parabolic direct orbit.

As the disturber approaches the galaxy, a sharply defined arm quickly forms in the disc in the direction towards the disturber. Due to the self-gravity a counter-arm appears shortly after. These first two arms are material, i.e. they contain particles that stay in the arm for a considerable time. Somewhat later, approximately half a disc rotation after the pericentre passage, a second pattern, consisting of two density wave arms, starts to grow outwards from the centre of the disc. At the same time the material arms start to disperse, but for a certain time interval both arm patterns are simultaneously present, superimposed on each other. Since there is a difference in the pitch angle between the material arms and the density wave arms, the arms in the resulting pattern will split up and give the impression of a fork in the arm.

A much observed system of interacting galaxies is the NGC 5426/27 pair. Enrique Perez at Instituto de Astrofisica Canarias has kindly supplied me with the CCD image of NGC 5427 in [OIII]5007 which is shown in Figure 1. We notice the forked arm pattern, and if we compare the image to a suitable frame from the video movie (shown in Figure 2) we will find a strong resemblance. This is even more striking if we calculate the tidal forces, as in Byrd et al. (1986). In units of the perturbed galaxy's mass and radius we find

$$\lg\,(m/r^3) = -1.18$$

Dynamics and Interactions of Galaxies Editor: R. Wielen

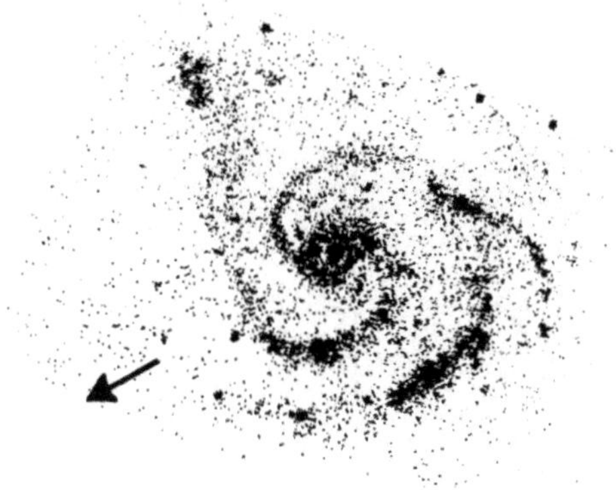

Figure 1

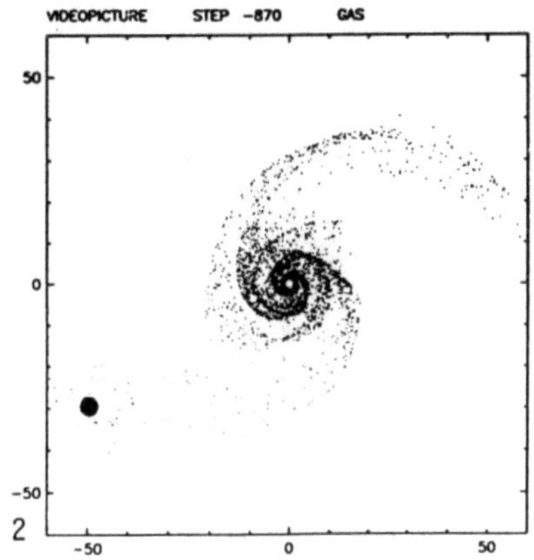

Figure 2

Figure 1. CCD image of NGC 5427. The arrow indicates the direction towards NGC 5426.

Figure 2. Step 870 in the video movie. The black dot gives the position of the disturber.

for the numerical simulation and

$$\lg\ (m/r^3) \geq -1.25$$

for the observed galaxy (using values from Blackman (1982)). Although, the resemblance is perhaps nothing more than a lucky fortuity since there remains several parameters that haven't been compared.

Kinematic explanation of sharp edges and voids

Sharp edges of spiral arms and voids between the arms are features that are often observed, both in numerical simulations and in real galaxies; see e.g. Elmegreen & Elmegreen (1989) for the possibility of sharp edges being a signature of companion-driven spirals. We will here investigate the origin of edges and voids using an impulse approximation for the perturbation, which eventually leads to a kinematic explanation.

Figure 3 shows, step by step, how the investigation is done. We start with a self-gravitating disc perturbed by a small disturber on a parabolic orbit. In the next step the disc is no longer self-gravitating and we have enhanced the outer part of it in order to see density variations more clearly. The impulse approximation now consists of the exchange of the small disturber with long duration for a more massive one that exists only during a very short time interval. This is done keeping the magnitude of the perturbation constant. The sharp edge and the void are still there.

The investigation of spirals is made more difficult by the differential rotation of the disc that draws whatever pattern into tight spirals. It is therefore easier to analyze the perturbation in a representation where this effect has been removed.

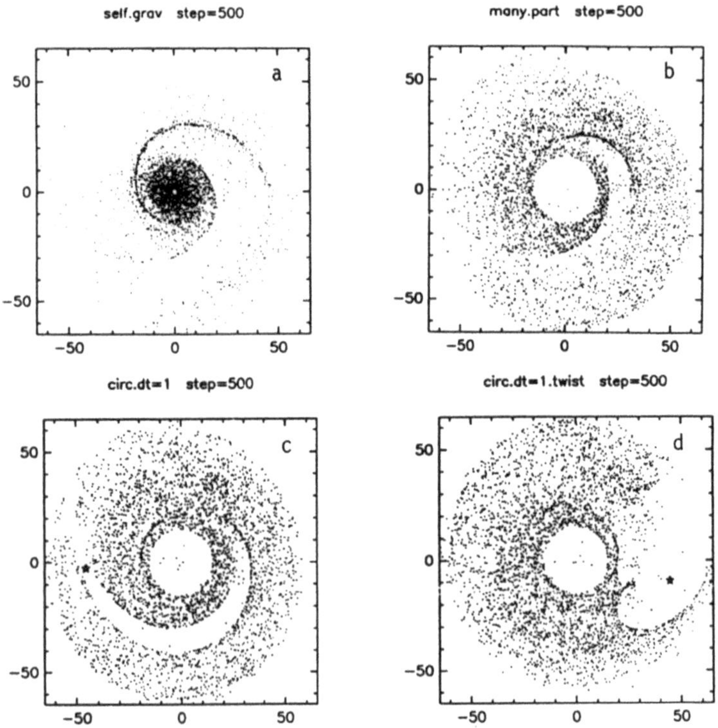

Figure 3. (a): A self-gravitating disc perturbed by a 5 % disturber. (b): Same as (a) but the disc is no longer self-gravitating and its outer part have been enhanced. (c): The impulse approximation, a short and strong perturbation. (d): The disc in (c) rewinded.

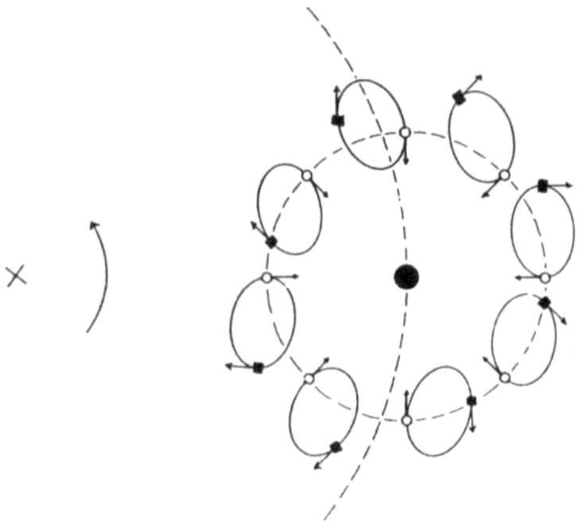

Figure 4. The epicycle approximation. Small unfilled circles: starting positions for particles; filled boxes: their positions half an epicycle later.

We produce this by "rewinding" the disc, i.e. remove the undisturbed differential rotation. This results in a disc with a crescent-shaped hole around the disturber. The size of this hole will grow with time.

We can understand the origin of this hole using the epicycle approximation (Figure 4). Particles starting on a circle around the disturber will go on epicycles with the same amplitude. Half an epicycle later they will all tend to move away from the disturber, creating the observed hole. This hole is then widened azimuthally by the deferent drift, creating sharp edges ahead of and behind the hole. Finally, the differential rotation will draw the hole and its sharp edges into the observed spiral pattern.

For a more detailed explanation of the removal of the differential rotation, see the poster paper by Donner et al. presented at this conference.

References

Blackman, C.P.: Monthly Notices Roy. Astron. Soc. 200, 407 (1982)

Byrd, G.G., Valtonen, M.J., Sundelius, B. & Valtaoja, L.: Astron. Astrophys. 166, 75 (1986)

Elmegreen, B.G. & Elmegreen, D.M.: to appear in Evolutionary Phenomena in Galaxies ed. Beckman, J. & Pagel, B. (Cambridge University Press, 1989)

Neutral Hydrogen near M51

A.H. Rots[1], *A. Bosma*[2], *J.M. van der Hulst*[3], *E. Athanassoula*[2], *and P.C. Crane*[1]

[1]N.R.A.O., P.O Box 0, Socorro, NM87801, USA*

[2]Observatoire de Marseille, 2, Place Le Verrier, F-13248 Marseille Cedex 4, France

[3]Kapteyn Astronomical Institute, Postbus 800,
NL-9700 AV Groningen, The Netherlands

1. Introduction

We observed the nearby Sc galaxy M51 (NGC 5194) and its companion NGC 5195 with the VLA in the 21-cm line to study the distribution and kinematics of the HI. M51 is a key object in the study of galaxies: its beautiful spiral structure with the clear separation of the dust lanes and associated radio continuum emission from the bright HII regions and OB-associations has led to the acceptance of a density wave picture of spiral structure (Mathewson et al. 1972), while the reproduction of the interaction between the two galaxies (and other pairs) by test particle simulations (Toomre and Toomre 1972) showed clearly that gravitational forces alone can indeed produce narrow bridges and tails. More recently (Hernquist, this volume), a link has been sought between the gravitational interaction and consequent spiral structure development by the action of the swing amplifier mechanism (Toomre 1981).

2. Observations and Results

Our observations with the VLA B and D array resulted in maps at various resolutions. The maps at low resolution demonstrate best the complicated structure of the outer gas disk of M51 and reveal the presence of a large HI tail not previously mapped with adequate sensitivity and resolution. The presence of this tail was not predicted by the simple model discussed by Toomre and Toomre (1972), and thus new and more sophisticated models are called for. Here we describe in some detail the data at 34" resolution. To obtain linear sizes we adopt a distance of 9.6 Mpc to M51, so that 1 arcmin corresponds to 2.8 kpc.

The channel maps, at intervals of 10.33 km/s, are shown in Figure 1. At first sight the bright features display the change in shape and position one expects for a differentially rotating galaxy. There are, however, several peculiar features one discovers at close inspection. A remarkable aspect is that channels close or near to the systemic velocity have an S or an H-shape rather than the familiar linear or mild U-shape expected for an axisymmetric galaxy with a flat rotation curve. Thus

* The National Radio Astronomy Observatory is operated by Associated Universities, Inc., under cooperative agreement with the National Science Foundation.

Dynamics and Interactions of Galaxies Editor: R. Wielen

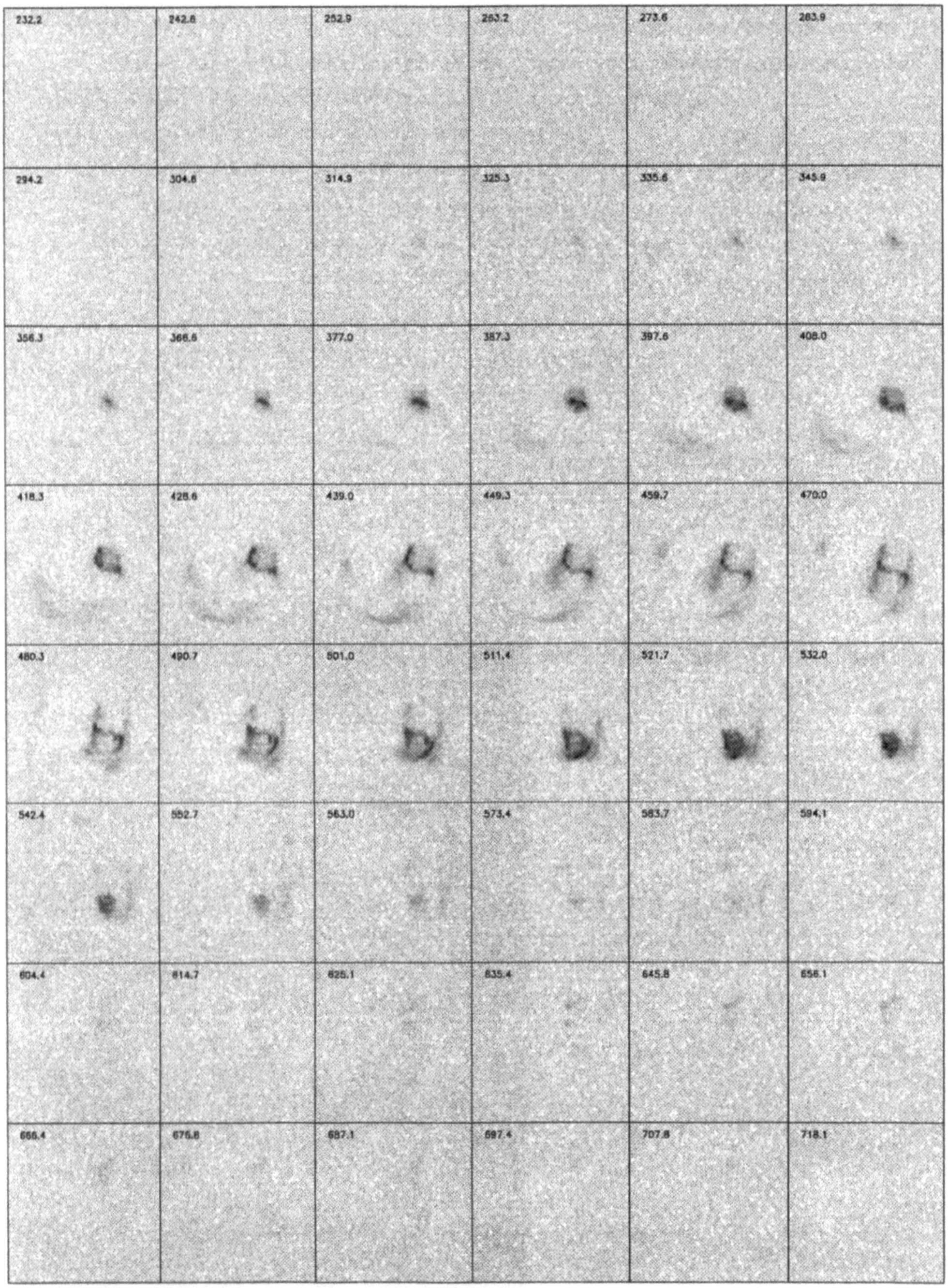

Figure 1. Grey scale representation of the channel maps at 34" resolution. The heliocentric velocity of each channel is indicated in the upper left corner.

highly non-circular and/or non-planar motions are present, which could be due in part to the interaction, as already suggested by Toomre (1979). The outlying HI becomes visible in channels between 377 and 470 km/s stretching from the south-east side of M51 toward the east and west. In the west it seems to connect to the outskirts of the galaxy itself, but in the east it reaches large distances. Faint HI features can also be seen in the north between 594 and 687 km/s.

The velocity field (Figure 3) shows indeed the differential rotation, as suggested by the channel maps. Quite clear, however, are the large velocity deviations associated with the outer spiral arms, especially in the east. The outer tail structure is not moving in harmony with the disk, but appears to have its own radial velocity field. This indicates that the tail probably has a completely different spatial orientation than the galaxy and/or follows a different motion.

The HI distribution is shown in Figure 2 on an overlay, and in Figure 4. The following description can be given :

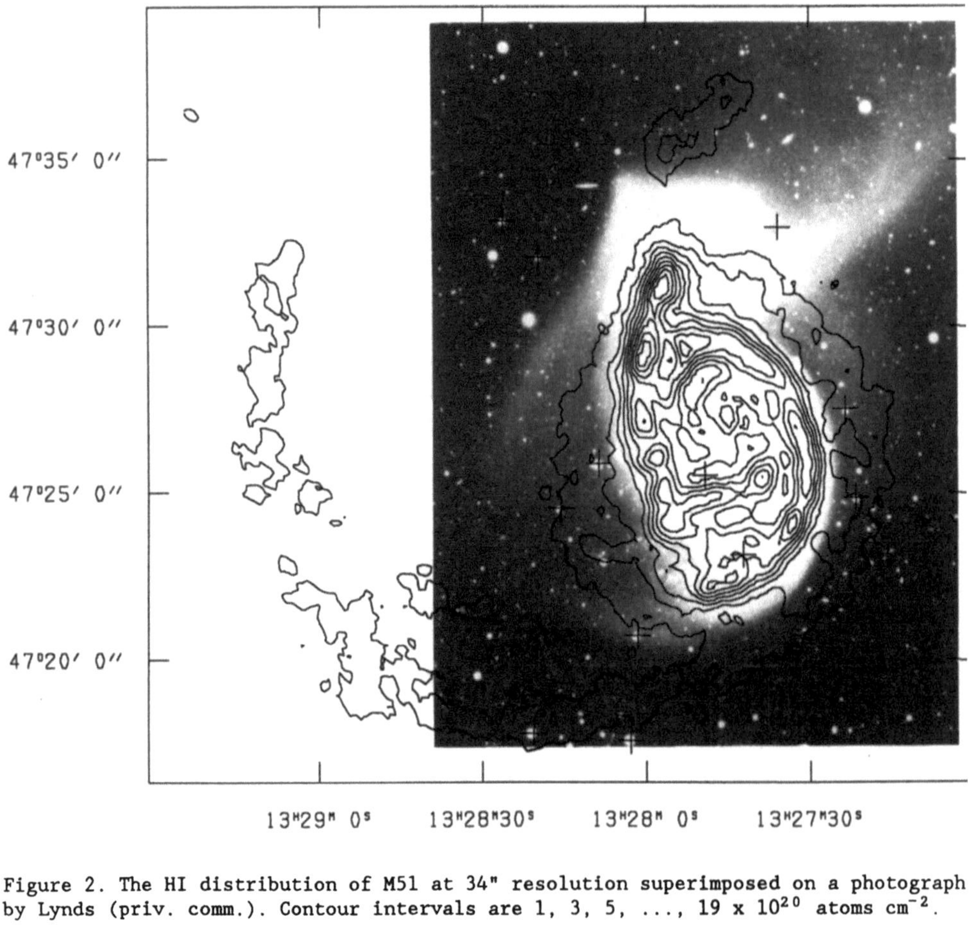

Figure 2. The HI distribution of M51 at 34" resolution superimposed on a photograph by Lynds (priv. comm.). Contour intervals are 1, 3, 5, ..., 19 x 10^{20} atoms cm^{-2}.

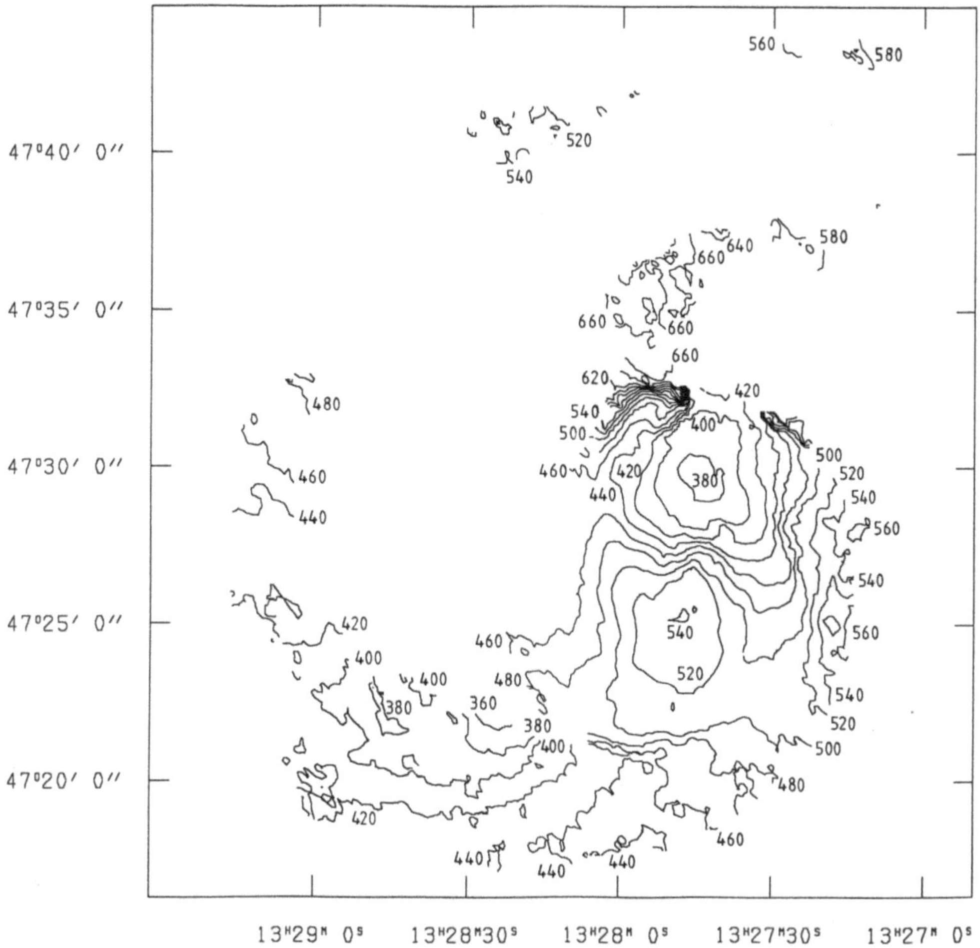

Figure 3. Intensity weighted mean velocity field of M51 at 34" resolution. The heliocentric velocities are marked at each contour in km/s.

a). In the inner parts the spiral structure is obvious in the HI. The width of the arms is wider than that of the CO (which is in general concentrated in a relatively narrow ridge in the inner part of the arms, cf. Lo et al. 1986 and Vogel et al. 1987) and similar to the Hα. In the innermost region the HI distribution shows a clear minimum, presumably where most of the gas is in molecular form. A sharp edge of the bright disk of M51 on the west side is also present in the HI distribution, but the two do not coincide completely in the southwest and southern regions.
b). At a density level of less than 3 x 10^{20} cm^{-2} there is a faint HI envelope, only partly corresponding to a faint optical envelope discernable in the deep photograph by Burkhead (1978) (cf. Fig. 2). The outer HI has a different major axis position angle from the bright HI disk, suggesting that it is either not circularly symmetric or has a different inclination. In fact, on the eastern and western parts

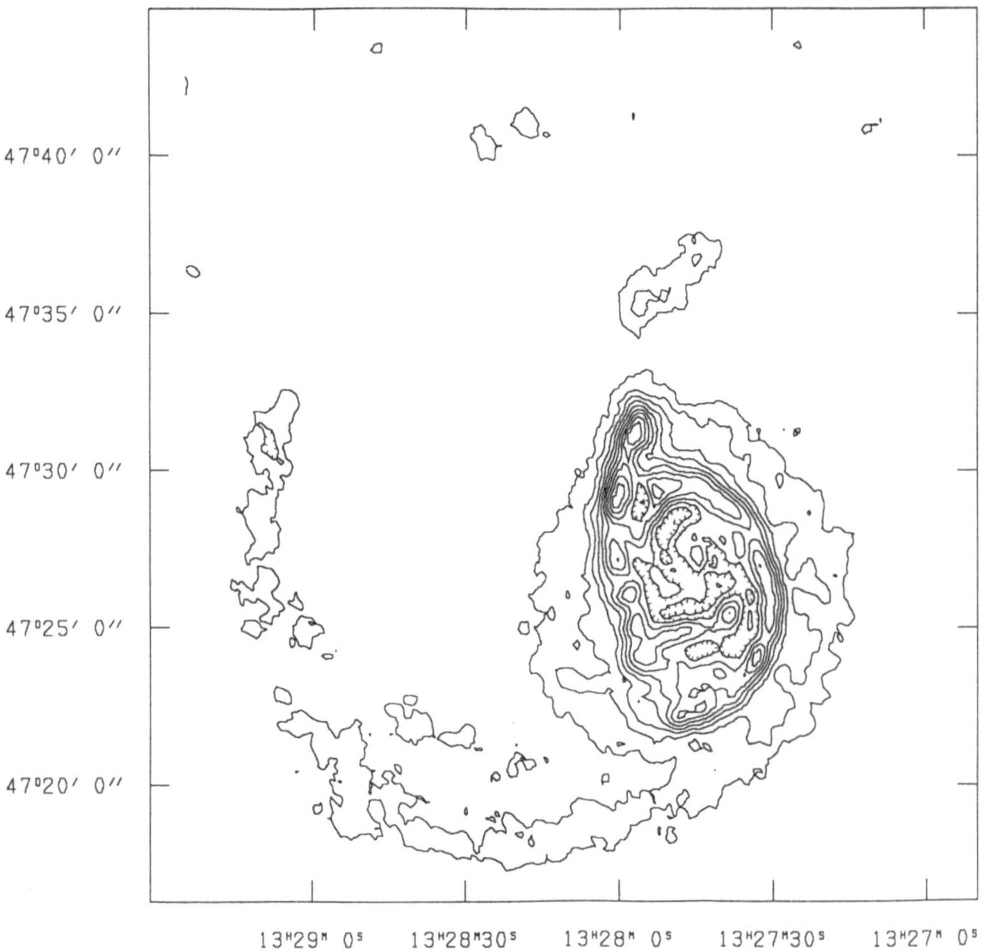

Figure 4. The HI distribution of M51 for the same area and scale as Figure 3. The contour intervals are 1, 3, 5, ..., 19 x 10^{20} atoms cm^{-2}.

it corresponds in part to the anomalous velocities in the channel maps. It is very difficult, however, to disentangle unambiguously the faint emission from the bright emission in all the channel maps.

c). The tail does not have an optical counterpart, is fairly broad and has a double structure, which can be best seen in Figure 1. It is not entirely clear in what manner the tail connects to M51 in the western part, but it seems certainly to be distinct from the bright spiral arms. The tail describes an elliptical arc, with inner projected radius of 15 kpc, and outer projected radius of 40 kpc. Its overall projected length of 90 kpc is similar to the estimated length of the Magellanic Stream, the bridge between M81 and NGC 3077 (cf. Van der Hulst 1979), or the HI plume associated with NGC 3628 (Rots 1978, Haynes et al. 1979). Its HI mass is estimated to be 5 10^8 $M_\odot$.

d). The HI around NGC 5195 and north of it is very faint and has radial velocities ranging from 510 to 700 km/s. This is quite different from the regular radial velocities of the main spiral arm running up (in projection) to the companion. However, there is no direct association of this HI with any feature in the optical image. Recent CO-data reported by Sage (1989) show a large velocity range (480 to 750 km/s), but this gas is at the center and southside of it. Note that Schweizer (1977) derives from optical emission and absorption lines a systemic velocity of about 610 km/s for NGC 5195. The predominant lack of correspondance of the outer HI and optical features suggest that each could have a different origin, and most likely most of the optical features result from the partly shattered companion NGC 5195.

3. Concluding remarks

The observations presented here show that the structure and kinematics of the HI in M51 and NGC 5195 are extremely complex. This makes it very difficult to study questions like : what is the underlying mass distribution of M51, does it have a dark matter component, and if so, how important is it. Only successful numerical simulations of the interaction can constrain the mass distribution and at the same time shed light on the origin of the beautiful spiral structure in the main disk.

References

Burkhead, M.S.: 1978, Ap. J. Suppl. 38, 147

Haynes, M.P., Giovanelli, R., Roberts, M.S.: 1979, Ap. J. 229, 83

Lo, K.Y., Ball, R., Masson, C.R., Phillips, T.G, Scott, S., Woody, D.P.: 1987, Ap. J. 317, L63

Mathewson, D.S., Van der Kruit, P.C., Brouw, W.N.: 1972, Astr. Ap. 17, 468

Rots, A.H.: 1978, A. J. 83, 219

Sage, L. J.: 1989, Ap. J. 344, 200

Schweizer, F.: 1977, Ap. J. 211, 324

Toomre, A., Toomre, J.: 1972, Ap. J. 178, 623

Toomre, A.: 1979, in "The Large Scale Structure of the Universe", I.A.U. Symp. 79, ed. M.S. Longair, J. Einasto, p. 109, Reidel (Dordrecht)

Toomre, A.: 1981, in "The Structure and Evolution of Normal Galaxies", eds. S.M. Fall and D. Lynden-Bell, Cambridge University press

Van der Hulst, J. M.: 1979, Astr. Ap. 75, 97

Vogel, S.N., Kulkarni, S.R., Scoville, N.Z.: 1988, Nature 334, 402

M51's Material Arms and Density Waves

G.G. Byrd and S. Howard

Dept. of Physics and Astronomy, University of Alabama,
Tuscaloosa, AL 35487-0324, USA

We find that the entire spiral structure in M51 is consistent with a tidal passage of its companion, NGC 5195, in a highly inclined orbit $\approx$ 70 million years ago. Fig. 1 shows dust lane arms of M51 from a photograph. Azimuths and radii of observed arms "time reversed" back to 70 million years ago are seen as two straight lines which point to the hypothesized disk crossing point of the companion to the northwest. This direction agrees with that found in Toomre and Toomre's (1972) non-self-gravitating simulations. Fig. 2 is a preliminary self-gravitating computer simulation of M51 including disk gas cloud collisional dissipation and gravitational clumping. Disk gas arms (representing HI, H_2) are shown at the present epoch. Compare to the observed dust lane arms in Fig. 1, noting the steepness of the outer arms, their symmetric kinks going inward and the size of the pattern relative to companion position. Note the long arm opposite the present companion position and the "sling shot" gas feature above the present position of the companion in the simulations. Compare to the HI observations of Rots, *et al.* elsewhere in this volume. The companion passed near the disk edge to the northwest as marked. Compare to the line direction in Fig. 1. The companion then moved beyond the disk plane to its present position seen through M51's disk. Fig. 3 shows an unperturbed disk (to a different scale) at the same time step as Fig. 2. This essential numerical test shows no large changes in structure or spontaneous spiral arms. The Fig. 2 outer arms wind up like material arms. The arm particles originated in a roughly fan-shaped region around a line from the crossing-point to the nucleus.

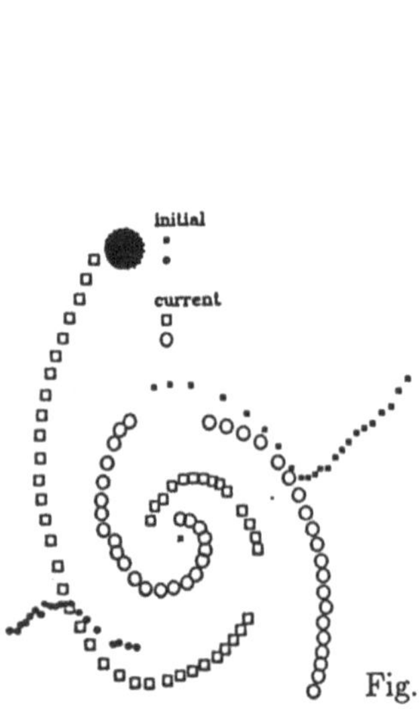

Fig. 1

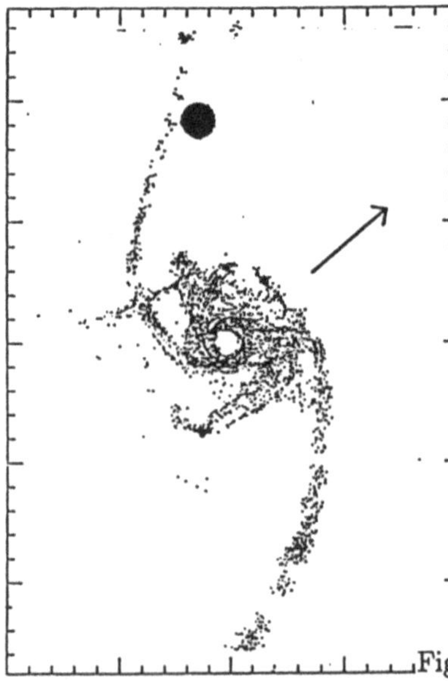

Fig. 2 - step 100

Dynamics and Interactions of Galaxies Editor: R. Wielen

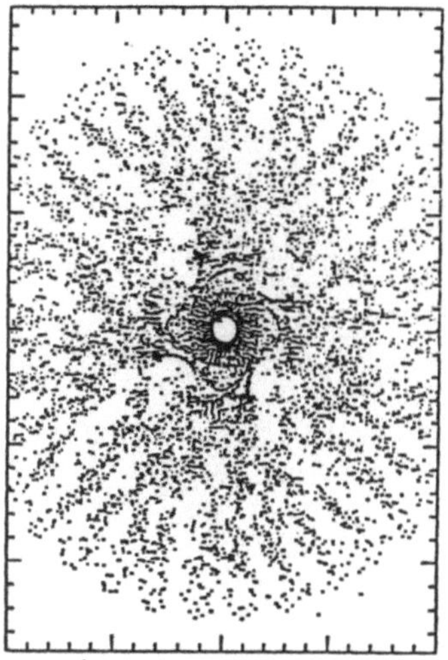

Fig. 3 - step 100

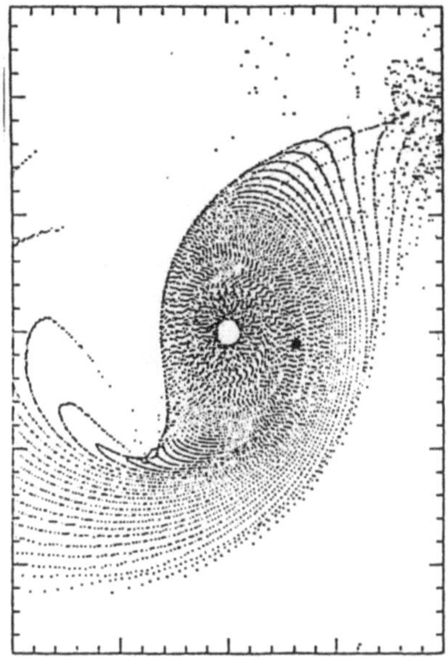

Fig. 4 - step 120

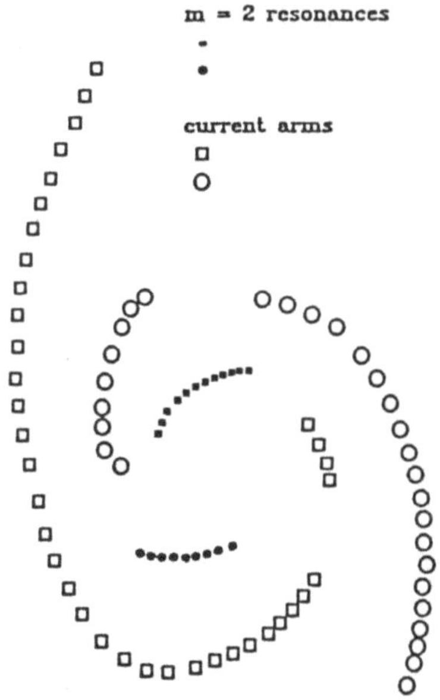

Fig. 5

Our simulations show that the companion cannot directly excite the spiral pattern in the inner part of the disk. The inner material orbits several times during the tidal perturbation so that inward and outward pushes cancel. Self-gravitation of the disk is essential with the outer tidal arms triggering inner structure. Compare the weak inner arms in Fig. 4 (which has very little self-gravity) to the stronger arms in Fig. 6b (which has self-gravity). Because the time is short since the companion crossed the disk, we propose that the inner structure is a young density wave with each part resonantly excited by corresponding parts of the tidal arms 90° ahead and behind. Fig. 5 shows the positions of corresponding portions of the observed outer tidal arms(open symbols) and the calculated density wave (filled symbols) which match

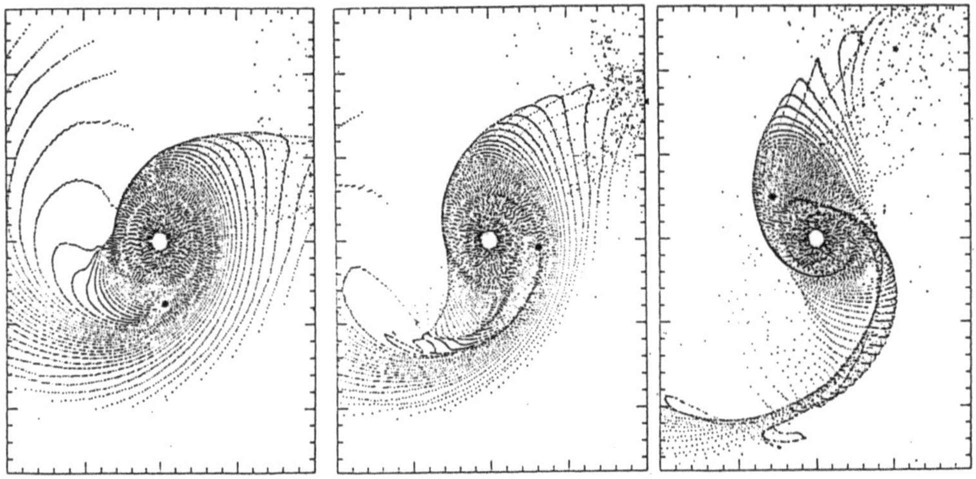

Fig. 6a - step 100 Fig. 6b - step 120 Fig. 6c - step 150

observed IR arms in M51's disk fairly well. Figs. 6a,b,c show a computer simulation of inner density wave particle motion. They demonstrate inner density wave formation with a time sequence of a perturbed self-gravitating disk (halo/disk ratio, $\mathcal{H}/\mathcal{D}$ = 1.5). Note how the marked particle passes through the inner arm in text-book density wave behavior.

The transition between *directly* excited outer arms and *indirectly* excited inner arms explains the symmetrically placed kinks in M51's arms. See Fig. 1's observed present-epoch arms and the breakdown in the time-reversal straight line at 140″. This 140″ kink radius and the fit of the density wave to observations in Fig. 5 analytically imply that M51's $\mathcal{H}/\mathcal{D}$ is between 1.5 and 4. We have tried different computer simulations and find that those with $\mathcal{H}/\mathcal{D}$ in this range best match M51. Our results apply only within the observed disk of M51.

References:

Toomre,A., Toomre,J. (1972) *Astrophys. J.* **166**, 465.

Rots,A.H., Bosma,A.,van der Hulst,J.M, Athanassoula,E., Crane, P. this volume.

Part V

Star-Formation Bursts and the Interactions of Galaxies

Infrared Studies of Interacting and Merging Galaxies

R.D. Joseph

Astrophysics Group, Blackett, Laboratory, Imperial College, London SW7, UK
Present address: Institute for Astronomy, University of Hawaii, 2680 Woodlawn Drive, Honolulu, HI96822, USA

INTRODUCTION

While the wide-ranging implications of interactions between galaxies were anticipated in the classic paper by Toomre & Toomre (1972), observational interest in interacting galaxies was confined to only a few studies in the pre-IRAS era (*e.g.* Schweizer, Larson & Tinsley, Joseph *et al.* 1984, Lonsdale *et al.* 1984, Joseph & Wright 1985, Cutri & McAlary 1985). Particularly among infrareders, interest in interactions was stimulated by the possibility that they might provide the elusive trigger for starburst activity in galaxies. Of course, IRAS provided massive evidence in support of this idea. However, there has also been a growing realisation that interactions between disc galaxies may have even more fundamental significance in extragalactic astrophysics. Interactions, and the associated intense bursts of star formation they induce, may play a dominant role in the spectral, chemical, and morphological evolution of most galaxies, and they could provide the key to the origin of other forms of activity in galaxies.

1 INFRARED ACTIVITY IN INTERACTING GALAXIES

In the post-IRAS era interactions have become a "buzz" subject, and the terms "interaction" and "starburst" are often used almost interchangeably. I should like to avoid pre-judging this question. I will therefore first define what I mean by "infrared activity" in galaxies generally, and then examine the observational evidence for infrared activity in interacting galaxies. In the following sections I will summarise the observational evidence on whether the underlying energy source for this activity is due to a starburst or to some other mechanism.

Following Balick & Heckman (1982) I take as a definition of infrared activity in galaxies an infrared luminosity which is larger than can be produced by thermonuclear reactions in a "normal" population of stars. The minimum mass-to-light ratio which can be maintained by thermonuclear reactions for a time, τ, is

$$M/L = M/(0.007\varepsilon Mc^2/\tau),$$

where ε is the fraction of the mass undergoing hydrogen burning. For $\varepsilon = 10$ % and a Hubble time $\tau = 10^{10}$ yr, M/L = 1 in solar units. Thus, $M/L_{IR} < 1$ will be our quantitative definition of infrared activity in galaxies. Such small mass/infrared-light ratios are found when the infrared continuum energy distribution longward of $\cong 2$ μm rises steeply toward longer wavelengths, as is the case for classic starburst galaxies (*cf.* Telesco 1988).

Dynamics and Interactions of Galaxies Editor: R. Wielen

One of the first attempts to systematically investigate infrared activity in interacting galaxies was the study by Joseph *et al.* (1984). They defined a sample of interacting galaxies from the Arp (1966) *Atlas*, with representatives of major morphological types, and obtained JHKL photometry for 22 pairs. Using the criterion of a K-L colour index > 0.5 as a semi-quantitative indicator of a rising infrared continuum energy distribution and therefore of infrared activity (K-L = 0.3 ± 0.1 for normal spirals), they found that:

a) there is a K-L excess for one member of an interacting pair for ≅ 85 % of the sample, and

b) never does more than one member of an interacting pair show a K-L excess.

The former point demonstrates that, if the K-L colour excess does indeed indicate infrared activity, then interactions are extremely efficient in inducing infrared activity in galaxies. The latter point shows that there is a causal connection between the interactions and the colour excess.

The second important feature of the infrared activity in interacting galaxies comes from studies at longer infrared wavelengths by Lonsdale *et al.* (1984), Cutri & McAlary (1985), and Wright *et al.* (1988). The energy distributions of these galaxies are similar to those of HII regions in the Galaxy. They are quasi-thermal in character, and can be understood as thermal emission by dust grains at a range of temperatures.

Thirdly, the interaction-induced infrared activity in galaxies tends to have larger infrared luminosity than in classic non-interacting starbursts galaxies. Wright *et al.* (1988) compiled a sample of all (≅ 40) the interacting and merging galaxies for which 10 μm photometry in a small (5 - 8 arcsec) aperture is available. Their results, shown in Table 1, demonstrate that the nuclear 10 μm luminosities of interacting galaxies are at least a factor of 3 times larger than classic starburst galaxies like NGC253 and M82.

The luminosities of the mergers are particularly noteworthy in Table 1--the merger luminosities are about an order of magnitude larger than those of non-merging interactions, and they are roughly the equal of Seyfert galaxies. As Joseph & Wright (1985) first pointed out, mergers produce the most powerful infrared galaxies known, with luminosities rivalling those of quasars and Seyferts. This pre-IRAS conclusion has been strikingly confirmed by study of a sample of the brightest galaxies in the IRAS survey (Soifer *et al.* 1987). There are ten galaxies in this sample with infrared luminosities exceeding 10^{12} $L_{\odot}$. Sanders *et al.* (1988) show that these ultra-luminous infrared galaxies all show

Table 1. Central 10 μm luminosities of several classes of galaxies.

Class of galaxy	N	Range ($L_{\odot}$)	$<L_{10\mu m}$ ($L_{\odot}$)>
Bright spirals	17	10^5-$7x10^8$	$2x10^8$
Starbursts: M82			10^9
NGC253			$6x10^8$
NGC2903			10^8
Interacting	24	$4x10^7$-$7x10^9$	$2x10^9$
Merging	9	$4x10^9$-$5x10^{10}$	$2x10^{10}$
Seyferts	50	$4x10^8$-10^{11}	$4x10^{10}$

evidence of interactions or mergers in deep CCD images. Moreover, a comparison with the quasars in the same volume of space shows that in the bolometric luminosity range $\log(L_{bol}/L_{\odot}) = 12.0 - 12.5$, these merging galaxies outnumber the quasars by a factor $\cong 3.5$. It seems that the ultra-luminous infrared galaxies produced by mergers are the dominant high luminosity objects in the local universe.

2 THE ENERGY SOURCE

The infrared radiation from galaxies is usually assumed to be thermal emission by dust grains which are heated by harder radiation from some underlying energy source. Candidates for the energy source are generally either thermonuclear energy from massive early-type stars produced in a recent burst of star formation ("starbursts") or gravitational energy released by accretion onto a black hole ("monsters"). We shall now investigate the evidence in favour of either of these two energy sources using the infrared, optical, and radio data available for the Wright *et al.* (1988) sample of interacting galaxies.

2.1 The Spatial Extent of the Mid-IR Emission

A dust grain at a distance R from a luminosity source L will reach an equilibrium temperature T at which its thermal emission equals the rate of absorption of energy from the luminosity source. For typical grain parameters this gives (*cf.* Graham *et al.* 1983)

$$R = 50(L/L_{\odot})^{1/2} T^{-5/2} \text{ pc.}$$

If there is a single central source of luminosity $\cong 10^{10}\ L_{\odot}$, *i.e.* an AGN, it can heat dust grains to temperature $\cong 300$ K, *i.e.* hot enough to that they radiate at 10 μm, only out to a distance of $\cong 3$ pc. Thus, if there is measurable spatial extent at 10 or 20 μm, it cannot be due to dust heated by a single, central source. Such spatial extent must be due to a number of distributed sources of luminosity-- precisely the description of a burst of star formation.

This argument is complicated by the recent realisation that very small grains may be present which are not in thermal equilibrium with the energy sources which heat them. Such grains could be heated sufficiently by absorption of a single ultraviolet photon to re-radiate at wavelengths of 10 μm, or even shorter (*cf.* Sellgren 1984). This possibility does not, howeve r, undercut the use of extended 10 μm emission to distinguish between starbursts and monsters. Aitken & Roche (1985) have shown that the "unidentified" features at 7.7, 8.6 & 11.3 μm, which appear to be carried by the very small grains, are strikingly absent in the mid-infrared spectra of Seyfert galaxies and quasars, presumably because they cannot survive the hard ultraviolet radiation field in the centres of these galaxies. On the other hand, Roche & Aitken (1985) also show that the spectral features associated with very small grains are common in starburst galaxies. Thus, if very small grains are responsible for any spatially extended mid-infrared emission (which we interpret as evidence for a starburst), this would appear to be likely only if the galaxy were not dominated by a monster. Thus the presence or absence of very small grains would not alter the conclusions.

For this sample of 39 galaxies, four interacting galaxies, NGC2798, NGC3227, NGC3690, and NGC5194, and five merging galaxies, NGC6240, NGC1614, NGC3256, NGC3310, and IC883, have measured spatial extent at 10 μm. Thus 25% of the sample is resolved at 10 μm. This suggests rather strongly that the 10 μm emission in these galaxies is not powered by a single quasar-like nucleus; on the other hand, starbursts rather naturally produce such spatial extent.

2.2 Optical Spectra

Another approach to determining the underlying energy source is to use diagnostic diagrams for various optical line intensity ratios, such as those developed by Baldwin, Phillips & Terlevich (1981). The required optical spectra are available for ≅ 31 of the interacting galaxies in the sample, and Wright et al. (1988) have plotted [OIII]/Hβ *vs.* [NII]/Hα line intensity ratios for these in their Fig. 4. For 21 of the interacting galaxies, the excitation is characteristic of HII regions, 4 are like Liners, and 6 are like Seyferts. Thus the excitation spectrum is characteristic of HII regions in two-thirds of the cases, and strongly supports the idea of an associated starburst.

However, the evidence for excitation characteristic of conditions associated with shocks or AGN's (*cf.* DePoy *et al.* 1987, Sanders *et al.* 1988) should not be ignored. Indeed, as we remark below, there is accumulating body of evidence associating interactions with other forms of activity in galaxies. Here we only point out that in Section 4.3 we consider the possibility that "Warmers," extreme Wolf-Rayet stars whose existence is inferred by Terlevich & Melnick (1985), account for some of the associations found between interactions and Seyfert-type activity. If there were such a stellar evolutionary phase, it would be natural to find some excitation characteristic of a harder spectrum present as the consequence of a starburst, and therefore it would not be unlikely that the ten non-HII galaxies have also experienced a starburst which has now evolved to the 'Warmer' phase a la Terlevich & Melnick.

2.3 Radio data

The characteristics of the radio emission from these galaxies provide another approach to distinguishing starbursts from monsters as the underlying energy sources in interacting galaxies. The hallmarks of radio emission from Seyfert-like nuclei are flat spectra and small, usually unresolved sizes. By contrast, steep spectra and extended nuclear radio emission would be expected for the supernovae and supernova remnants associated with a starburst.

Radio data is available for three-fourths of the sample of interacting galaxies (Wright *et al.* 1988). Virtually all have a steep spectral index, ≅ -0.7. For 80 % of the sample the nuclear radio source is resolved, and the sizes are of the order of several arcsec. These radio features are not those of "monsters," but are qualitatively consistent with the supernova activity subsequent to a starburst.

2.4 Ultraviolet spectral features

The presence of strong ultraviolet emission lines due to CIV, HeII, and CIII is one of the hallmarks of Seyfert-type activity in galaxies. By contrast, one might expect starbursts to show absorption features due to CIV and SiIV, since these features are characteristic of giant and supergiant stars. In fact, SiIV absorption is luminosity-dependent. Wallborn & Panek (1984) have shown that these absorption features are strongest in supergiants and virtually zero in dwarfs

We have devoted an entire observing shift on IUE to the most nearby example of an ultra-luminous merging galaxy, NGC3256 (Joseph *et al.* 1986). The low resolution spectrum shows that the most prominent ultraviolet features are those due to SiIV and CIV in absorption, and there are no emission features typical of Seyfert-type activity. This is strong supporting evidence, from a rather different observational perspective, that there is strong starburst activity in this copybook example of a merging galaxy.

2.5 Infrared spectral features

A number of groups have been investigating the infrared spectra of interacting galaxies in the H & K windows (Fischer *et al.* 1983, Joseph *et al.* 1984, Rieke *et al.* 1985, DePoy *et al.* 1987, Joseph *et al.* 1987). Infrared spectroscopy is an almost completely undeveloped tool for investigating physical processes in galaxies and these kinds of studies have enormous potential (*cf.* Joseph 1989).

Three results from analysis of our infrared spectra of galaxies are particularly relevant to the question under consideration here. Firstly, we (Prestwich *et al.* 1990) find the ratio of Brackett γ flux to IRAS 25 μm flux to be remarkably constant, and the same for classic starburst galaxies as for interacting and merging galaxies. Secondly, we have detected the 3.28 μm solid-state feature in virtually every one of the half-dozen or so interacting and merging galaxies we have attempted. Finally, we find deep stellar CO absorption features *suggestive* of a large supergiant population. (We are working on a programme to make more quantitative statements based on the CO feature--*cf.* Doyon et al. (1989). These features are all qualitatively suggestive of starbursts.

In summary, we have adduced observational evidence with the broad perspective provided by insights from variety of ultraviolet, optical, infrared, and radio measurements. All of these observational approaches point toward the ubiquitous presence of starbursts in these luminous IR galaxies. While it is difficult to rule out the proposal by Harwit *et al.* (1987) suggesting that the kinetic energy of the collision provides the infrared luminosity we see in ultraluminous mergers, the direct evidence for the presence of recent star formation from the ultraviolet data and the CO absorption is probably the strongest argument against this idea. These data do not exclude the concurrent presence of a monster in some or all of these galaxies. However, it is difficult to avoid the conclusion that, in the large majority of cases, starbursts are the dominant energy sources driving the large infrared luminosities we have shown to be a common feature of these galaxies. Apparently, interactions produce starbursts.

3 MODELS OF INTERACTION-INDUCED STARBURSTS

We have used a simple analytical model for starbursts first presented by Telesco & Gatley (1984), to investigate the characteristics of these interaction-induced starbursts (*cf.* Wright *et al.* 1988). In this approach one writes the initial mass function (IMF), luminosity on the main sequence, lifetime on the main sequence, and Lyman continuum luminosity of the stars in a starburst as power-law functions of the stellar mass. Since most of the mass in a starburst is in low-mass stars, the mass-to-light ratio is a sensitive function of the lower mass cutoff to the IMF, if the age of the starburst is known.

Using the infrared luminosity inside the small, 5 - 8 arcsec apertures, and the corresponding masses inside these radii, Wright *et al.* (1988) attempted to place constraints on the lower mass cutoff to the IMF. There are rotation curves available for about one quarter of their sample from which masses, and hence mass/light ratios, can be determined. Assuming a starburst age of 10^7 yr, a Miller-Scalo IMF, and that 10 % of the measured mass would have been available for star formation at the beginning of the starburst, they found that the lower mass cutoff to the IMF must be 3 - 6 $M_{\odot}$. Apparently interaction-induced starbursts do not make solar-type stars.

4 LARGER IMPLICATIONS OF INTERACTION-INDUCED STARBURSTS

On the most naive assumptions, the frequency of interaction between randomly-distributed field galaxies will be proportional to the peculiar velocity divided by the interaction mean free path. The former is proportional to (1+z), and the latter to $(1+z)^{-3}$, so we expect that the frequency of interaction would go as $(1+z)^4$. The present day frequency of interaction is estimated by various authors (*cf.* Vorontsov-Velayaminov 1977) to be 5%. This suggests that virtually every galaxy in the sky has experienced at least one interaction-induced starburst back to z = 1 or 2.

It is difficult enough to find an objective estimate of the present-day frequency of interaction between galaxies, since whether a galaxy is morphologically "disturbed" or merely "nearly normal" is a subjective judgement. It is therefore even more awkward to investigate how the frequency of interaction increases with redshift. Zepf & Koo (1989) have recently attempted to address this question by avoiding discussion of whether or not there is morphological evidence for interaction, and instead by asking how the galaxy-galaxy two-point correlation function changes with redshift. In a sample of 1000 galaxies going as faint as $m_J = 22$ they found excess pairs at small separations compared with extrapolations of correlation functions at larger correlations. They argue that the frequency of interaction is $(1 + z)^{4 \pm 2.5}$. Although this result needs confirmation, it tends to support the naive analysis given above, and underscores the potential significance of interactions for the early evolution of galaxies.

If it is true that most galaxies have experienced one or more interaction-induced starbursts, this fact has profound implications for the spectral, chemical, and morphological evolution of galaxies. Several examples of such effects follow.

4.1. Spectral evolution: the Butcher-Oemler Effect

The Butcher-Oemler effect (Butcher & Oemler, 1978), *viz.* that there are more blue galaxies in distant rich clusters of galaxies than in nearby ones, has received much attention, and seems to be confirmed (Butcher & Oemler 1984). One obvious interpretation of the effect is that a high rate of star formation would deplete the interstellar medium of gas, and choke off further star formation activity. The colours of the galaxies would then evolve toward the red, producing the observed effect. Both ram-pressure-induced star formation and star formation triggered by interactions have been proposed for the origin of the putative high star formation rates such explanations require.

Lavery & Henry (1988) have recently tested these ideas by spectroscopy and imaging of blue Butcher-Oemler cluster galaxies at $z \cong 0.2$. They find the images of ten such galaxies, with emission lines characteristic of recent star formation, to have close neighbours or other evidence of interaction. While this result raises some fundamental questions regarding the dynamics and triggering of star formation, it indicates how important interactions may be in determining the physical evolution of galaxies, even at relatively recent epochs.

4.2 Morphological effects: formation of elliptical galaxies

Toomre (1977) and others have pointed out that the violent stellar dynamical relaxation resulting from the merger of two disc galaxies will produce a light distribution much like that of an elliptical galaxy. There are several potential problems with this proposition, one of which is the fact that mergers generally look nothing like elliptical galaxies (*cf.*, however Schweizer (1982) for the case of NGC7252). This might be simply because optical images are strongly biased by hot blue stars and extinction effects.

To test this idea we have begun a programme of infrared imaging of galaxies believed to be examples of mergers (Wright *et al.* 1990). Figure 1 shows the 2.2 μm surface brightness profile of the celebrated merging galaxy Arp220, and another luminous merger, NGC2623 (Arp143). Although the optical images shows features of double or multiple nucleii, and plumes and tails, the underlying stellar distributions certainly follow a de Vaucouleurs $r^{1/4}$ light profile over a scale of at least 10 kpc. Clearly more infrared imaging will be required before this can be established as a general result, but again, the evidence is beginning to be intriguing.

Another problem with the merger scenario for the formation of ellipticals is that disc galaxies have substantial dust and gas content, whereas ellipticals do not. This difficulty may also be resolved by the *extended* starburst induced in a merger. Graham *et al.* (1984) showed that for a $M/L < 1$, there

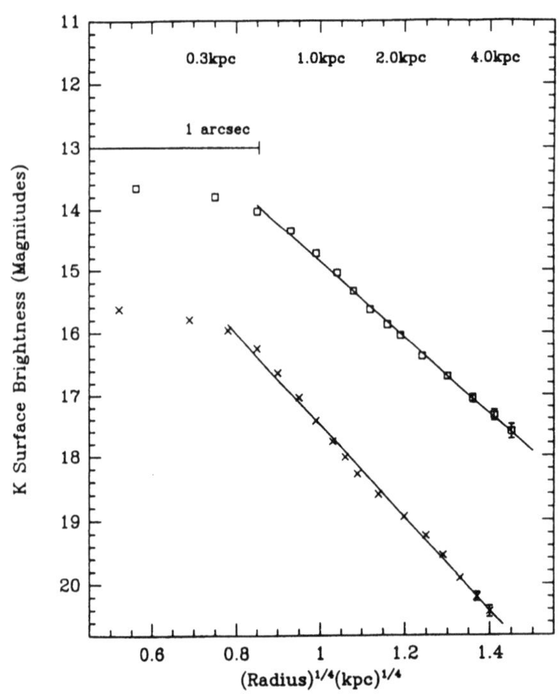

Figure 1. Surface brightness profiles at 2.2 μm of Arp220 (□)and NGC2623 (x).

is sufficient energy in the supernovae which accompany a starburst to drive a galactic wind whichsweeps the galaxy free of dust and gas. The extended starbursts we have found in mergers thus may be able to sweep a large fraction of the dust and gas out of the merger remnant, leaving it as free of dust and gas as any elliptical galaxy.

Similar processes following the mergers of proto-disc galaxies at the epoch of galaxy formation might, in fact, account for the entire Hubble sequence of spiral galaxies. The Hubble sequence is essentially a sequence in bulge/disc ratio. Norman (1988), among others, has pointed out that interactions between a spiral galaxy and gas-rich dwarfs will determine the bulge/disc ratio. The ensuing starbursts and violent relaxation would produce the bulge properties of spirals which we observe at the present time.

4.3 The relation of interactions to other forms of activity in galaxies

In their 1972 paper, Toomre & Toomre suggested that interactions might be responsible for the plethora of phenomena associated with "active" galactic nuclei. Evidence continues to grow in support of this idea. Interactions appear with exceptional frequency in other classes of active galaxies in addition to starburst galaxies.

Several studies (*cf.* Balick & Heckman 1982, Keel *et al.* 1985) have found a high occurrence of Seyfert-type activity in samples of interacting galaxies. Lilly & Longair (1984) find a ultraviolet excess associated with extended [OIII] emission lines in 3CR radio galaxies, which they interpret in terms of interaction-induced star formation. Imaging of a sample of powerful radio galaxies by Heckman *et al.* (1986) reveals that more than a quarter of them a strongly peculiar in their optical morphology, suggestive of mergers and interactions. And optical images of a sample of low redshift quasars (Smith *et al.* 1986) shows that at least half the sample are strongly disturbed or show evidence of companions and interactions. Some of these conclusions are beset with the subjective nature of "disturbance," and it is not yet clear how many "normal" galaxies will look "disturbed" at the relatively faint surface brightnesses at which several of these distant samples are detected. However, this *caveat* cannot be applied to the study of interacting (*i.e.* pairs of galaxies) galaxies investigated by Keel *et al.*, and the Seyfert activity they found quite plausibly extends to these other forms of activity. Thus, the association between interactions does seem to be real, and it is very likely indicative of something more fundamental which links interactions and AGNs.

What is the link between interaction-induced starbursts and other forms of activity in galactic nuclei? There are two novel suggestions current, both of which are discussed in more detail elsewhere in this volume. One link might be "Warmers," *i.e.* extreme Wolf-Rayet stars whose existence has been shown by Terlevich & Melnick (1985) to be quite plausible. Terlevich & Melnick show that such stars and the associated Type Ib supernovae in a dense interstellar medium can reproduce most of the features traditionally associated with AGNs. Support for this idea would come from the detection of Wolf-Rayet features in the spectra of interacting galaxies. Such features have indeed been detected in the spectrum of one ultra-luminous IRAS galaxy, IRAS01003-2238 (Armus *et al.* 1988).

The other proposal, due to Norman & Scoville (1988), is that the starburst induced by the interaction is concentrated in the central few parsec. If there is a seed black hole in the nucleus, mass loss in stellar winds accretes onto the black hole, while radiation from the black hole ionises the mass-loss envelopes of red giants in the cluster, producing the characteristic AGN spectral features. This model appears to require an enormously concentrated starburst--parsecs, rather than the kiloparsecs which characterise the data adduced in Section 2.1 above, and it would seem to falter on at least this point, if not also on some others.

In summary, the causal connection(s) between interactions and various forms of activity in galaxies, including starbursts, are not completely clear, although there are tantalising associations. Elucidation of the physics of these inter-relations continues to be one of the outstanding questions of extragalactic astrophysics.

CONCLUSIONS

It now seems clear that interactions do indeed play a major role in the evolution of galaxies. *i*) They induce starbursts with apparently high efficiency. *ii*) These starbursts have higher luminosities than "classic" starbursts in non-interacting galaxies. *iii*) Mergers of disc galaxies produce ultra-luminous infrared galaxies, with infrared luminosities up to $10^{13}\ L_{\odot}$. They may be the dominant high luminosity objects in the universe. *iv*) Interaction-induced starbursts are likely to be significant in determining the morphological, spectral, and chemical evolution of most galaxies. *v*) Interactions are clearly associated with other types of activity in galaxies beside starburst activity, and interaction-induced starbursts may provide the key needed to understand the origin of activity in galaxies.

REFERENCES

Aitken, D. K. & Roche, P. F., 1985. Mon. Not. R. astr. Soc., 213, 777.

Armus, L., Heckman, T. M. & Miley, G. K., 1988. Astrophys. J., 326, L45.

Arp, H. C., 1966. *Atlas of Peculiar Galaxies*, California Institute of Technology, Pasadena.

Baldwin, J. A., Phillips, M. & Terlevich, R., 1981. Pub. Astron. Soc. Pacific, 93, 5.

Balick, B. & Heckman, T. M., 1982. Ann. Rev. Astr. Astrophys., 20, 431.

Butcher, H. & Oemler, A., Jr., 1978. Astrophys. J., 219, 18.

Butcher, H. & Oemler, Al., Jr., 1984. Astrophys. J., 285, 426.

Cutri, R. M. & McAlary, C. W., 1985. Astrophys. J., 296, 90.

Doyon, R., Joseph, R. D. & Wright, G. S., 1989. *Infrared Spectrosopy in Astronomy*, p. 477, ed. B. H. Kaldeich, European Space Agency, ESA SP-290.

DePoy, D. 1987. *Star Formation in Galaxies*, p. 701, ed. C. J. L. Persson, NASA, Conference Publication 2466.

Fischer, J., Simon, M., Benson, J., and Solomon, P. M. 1983, Astrophys. J. (Letters), 273, L27.

Graham, J. R., Meikle, W. P. S., Selby, M. J., Allen, D. A., Evans, A., Pearce, G., Bode, M. F., Longmore, A. J. & Williams, P. M. 1983. Nature, 304, 709.

Graham, J. R., Wright, G. S., Meikle, W. P. S., Joseph, R. D. & Bode, M. F., 1984. Nature, 310, 213.

Harwit, M., Houck, J. R., Soifer, B. T. & Palumbo, G. G. C., 1987. Astrophys. J., 315, 28.

Heckman, T., Smith, E. P., Baum, S. A., van Breugel, W. J. M., Miley, G. K., Illingworth, G.D., Bothun, G. D. & Balick, B., 1986. Astrophys. J., 311, 526.

Joseph, R. D., Meikle, W. P. S., Robertson, N. A. & Wright, G. S., 1984. Mon. Not. R. astr. Soc., 209, 111.

Joseph, R. D., Wright, G. S., & Wade, R. 1984, Nature, 311, 132.

Joseph, R. D., 1986. *Light on Dark Matter,* p. 447, ed. F. P. Israel, Reidel, Dordrecht, Holland.

Joseph, R. D. & Wright, G. S., 1985. Mon. Not. R. astr. Soc., 214, 87.

Joseph, R. D., Wright, G. S. & Prestwich, A. H., 1986. *New Insights in Astrophysics,* ESA SP-263, p. 597.

Joseph, R. D., Wright, G. S., Wade, R., Graham, J. R., Gatley, I. & Prestwich, A., 1987. *Star Formation in Galaxies,* p. 421, ed. C. J. L. Persson, NASA, Conference Publication 2466.

Joseph, R. D., 1989. *Infrared Spectrosopy in Astronomy,* p. 439, ed. B. H. Kaldeich, European Space Agency, ESA SP-290.

Keel, W. C., Kennicutt, R. C., Hummel, E. & van der Hulst, J. M., 1985. Astr. J., 90, 708.

Lavery, R. J. & Henry, J. P., 1988. Astrophys. J., 330, 596.

Lilly, S. J. & Longair, M. S., 1984. Mon. Not. R. astr. Soc., 211, 833.

Lonsdale, C. J., Persson, S. E. & Matthews, K., 1984. Astrophys. J., 287, 95.

Norman, C. A., 1988. *Starbursts and Galaxy Evolution,* p. 483, eds. Thuan, T. X., Montmerle, T. & Van, J. T. T., Editions Frontieres, Gif Sur Yvette, France.

Norman, C. & Scoville, N., 1988. Astrophys. J., 332, 124.

Prestwich, A. H., Joseph, R. D. & Wright, G. S., 1990. In preparation.

Roche, P. F. & Aitken, D. K., 1985. Mon. Not. R. astr. Soc., 213, 789.

Rieke, G. H., Cutri, R., Black, J. H., Kailey, W. F., McAlary, C. W., Lebofsky, M. J., and Elston, R. 1985, Astrophys. J., 290, 116.

Sanders, D. B., Soifer, B. T., Elias, J. H., Madore, B. F., Matthews, K., Neugebauer, G. & Scoville, N. Z., 1987. Astrophys. J., 325, 74.

Schweizer, F., 1982. Astrophys. J., 252, 455.

Sellgren, K., 1984. Astrophys. J., 277, 623.

Smith, E. P., Heckman, T. M., Bothun, G. D., Romanishin, W. & Balick, B., 1986. Astrophys. J., 306, 64.

Soifer, B. T., Sanders, D. B., Madore, B. F., Neugebauer, G., Danielson, G. E., Elias, J. H., Lonsdale, C. J. & Rice, W. L., 1987. Astrophys. J., 320, 238.

Telesco, C. M. & Gatley, I., 1984. Astrophys. J., 284, 557.

Terlevich, R. & Melnick, J., 1985. Mon. Not. R. astr. Soc., 213, 841.

Toomre, A. & Toomre, J., 1972. Astrophys. J., 178, 623.

Toomre, A., 1977. *Evolution of Galaxies and Stellar Populations,* p. 401, eds. Tinsley, B. M. & Larson, R. B., Yale University Observatory, New Haven, Connecticut.

Walborn, N. R. & Panek, R. J., 1984. Astrophys. J., 280, L27.

Wright, G. S., Joseph, R. D., Robertson, N. A., James, P. A. & Meikle, W. P. S., 1988. Mon. Not. R. astr. Soc., 233, 1.

Wright, G. S., James, P. A., Joseph, R. D. & McLean, I. S., 1990. Nature, 344, 417.

Vorontsov-Velyaminov, B. A., 1977. Astr. Astrophys. Suppl., 28, 1.

Zepf, S. E. & Koo, D. C., 1989. Astrophys. J., 337, 34.

Interacting Systems in an IRAS-Selected Sample of Galaxies

U. Klaas

Max-Planck-Institut für Astronomie, Königstuhl 17,
D-6900 Heidelberg, Fed. Rep. of Germany

Abstract: In an investigation of the physical properties of 100 optically faint extragalactic IRAS sources, deep optical CCD images and long-slit spectra have been obtained with the 3.5 m telescope of the Calar Alto Observatory. These data show that interacting systems are very frequent among infrared selected galaxies and that interaction is a very effective trigger in generating infrared-luminous systems due to high starburst activity.

1. Introduction

We have performed a study of optically faint (m_B > 15 mag) previously unidentified IRAS galaxies in two regions of the sky each covering an area of 300 deg^2 (Klaas 1987). About 320 sources were selected from the IRAS Point Source Catalog (1985) whose flux criteria indicated an extragalactic nature.

2. Optical observations

We have observed 99 fields in an R-filter using the prime focus CCD camera of the 3.5 m telescope on Calar Alto Observatory, Spain. Typical exposure times were 1000 to 1500 s which resulted in surface brightness limits down to 27 mag $arcsec^{-2}$. In addition, images in a second filter (B) have been obtained for 77 fields. So far, long-slit spectroscopy has been performed in 55 fields with a total sum of 75 spectra using the Boller and Chivens Cassegrain spectrograph attached to the 3.5 m telescope.

3. Results and discussion

3.1 Proportion of peculiar objects in this optical sample

The composition of the sample of 99 fields which was observed in the optical can be regarded as more or less arbitrary, as the selection of objects depended mainly on observability during available telescope time. 11 fields containing sources which were only detected at 100 µm by IRAS have no suitable optical counterparts and are most probable due to infrared cirrus. From the remaining 88 fields 55, i.e. more than 60% contain objects with significant pecularities.

Dynamics and Interactions of Galaxies Editor: R. Wielen

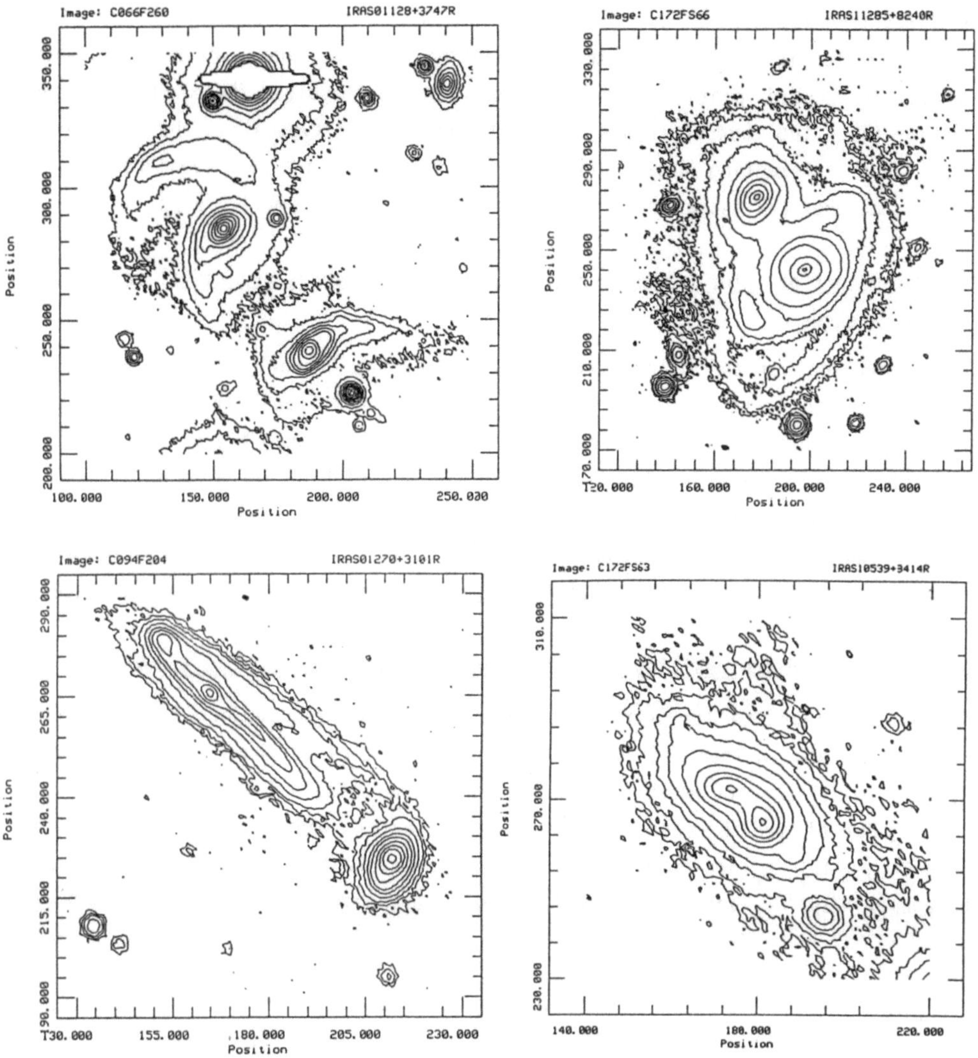

fig. 1: Examples of interacting systems identified as optical counterparts of extragalactic IRAS sources. The scales of the axes are in pixel units, each pixel corresponds to 0.5 x 0.5 $arcsec^2$ on the sky.

3.2 Importance of interactions

Most of the peculiar structures which stand out clearly only on the deep direct images indicate interactions of galaxies: pronounced tidal tails and arms, light bridges between two neighboring galaxies, common light halos or double nuclei (fig. 1). In most cases spectroscopic data are available which confirm the close physical association of the objects. 33 of the optically investigated IRAS sources are definitely classified as interacting systems because of their morphology and spectroscopic data. Thus, *interaction plays a major role in triggering infrared activity in galaxies.*

In addition to the 33 confirmed interacting systems, there are 21 cases where interaction could play a more or less important role. These objects show pecularities, too, but these are not as pronounced and typical as the features of the 'true' interacting ones. Sometimes redshift data are not available yet to verify the physical association of objects which stand in small groups.

3.3 Parameters of interaction strength: projected distances and radial velocity differences

The separations of the galaxy centers (fig. 2a) are - nearly without exception-roughly one to three galaxy diameters. This result reflects the obvious aspect that tight interactions cause stronger disturbances and hence higher activity than less tight ones. It could also mean that the stage of high infrared brightness in an interacting system is coupled to the phase of nearest encounter during the passage of two galaxies. In addition, the small radial velocity differences found for many systems (fig. 2b) suggest that these galaxies are already in a bound state.

3.4 Far-infrared luminosities of interacting and non-interacting systems

The histogram of the far-infrared luminosities of all our objects with measured redshifts (total number: 49) presented in fig. 3 clearly shows that interacting systems are - in the mean - more luminous than single objects. Below $L_{FIR} = 10^{11}\ L_{\odot}$ 35% of all objects are interacting galaxies, whereas 77% of all objects with $L_{FIR} > 10^{11}\ L_{\odot}$ are interacting. 79% of all interacting systems in this sample have $L_{FIR} > 10^{11}\ L_{\odot}$ and above $4 \times 10^{11}\ L_{\odot}$ nearly all objects seem to be interacting.

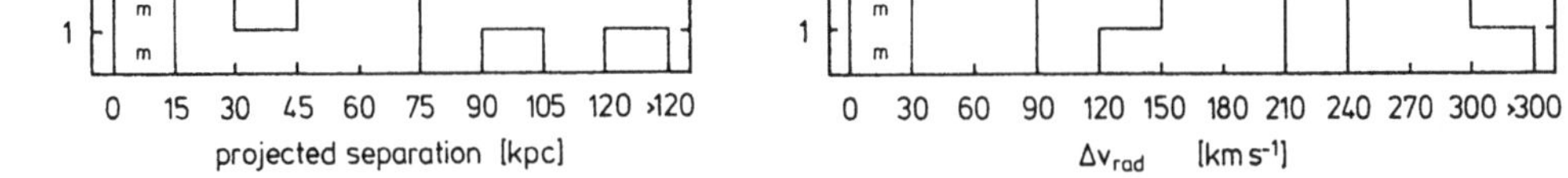

fig. 2: (a) Distribution of the projected distances between the optical galaxy centers for a total number of 29 interacting systems. $H_0 = 50$ km s^{-1} Mpc^{-1} has been used to derive the distances. (b) Distribution of the radial velocity differences for a total number of 25 interacting systems. Redshifts have been derived from the Hα lines. The objects labelled "m" are possible merger candidates where only one common body has remained.

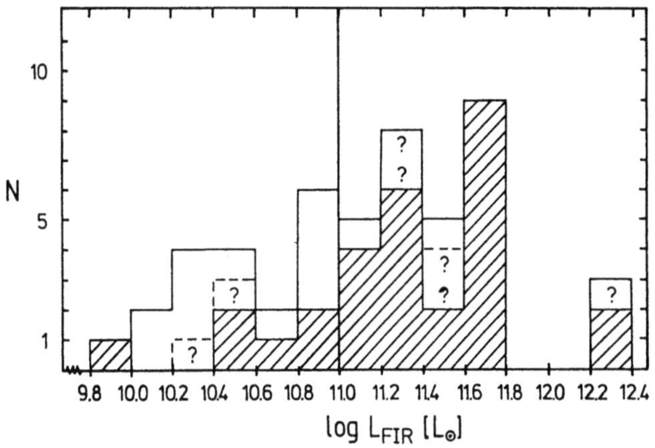

fig. 3: Far-infrared luminosity distribution for 49 interacting and non-interacting objects. Galaxies are defined as luminous in the far-infrared if $\log L_{FIR} > 11.0$. Hatched areas represent interacting systems; objects for which interactions could not be unambigously determined are indicated by a "?".

Thus, *interaction is not only a very common phenomenon in IRAS-selected galaxy samples but is also very effective in triggering infrared activity*. It should be kept in mind, however that other observers have found far lower percentages of interacting systems among their most luminous infrared galaxies (Lawrence et al. 1989) and that interactions may not be the only way to generate high infrared luminosities.

3.5 Optical line emission

In nearly all the interacting galaxies investigated in our study, optical activity has been induced. In general, all components participating in the interaction show emission line spectra. The line emission can be concentrated in the central regions of the galaxies but can also be very extended (up to several tens of kpc) and the maximum emission can originate in extra-nuclear regions. In summary, the most common spectrum in the interacting systems is of type HII-region/starburst. Thus, *interactions induce a large amount of star formation and starburst activity that shows up as high infrared luminosities from the heated interstellar dust associated with the star forming zones.*

References:

IRAS Point Source Catalog: Joint IRAS Science Working Group (1985)

Klaas, U.: in Starbursts and Galaxy Evolution, ed. by T.X. Thuan et al. (Editions Frontières, 1987) pp. 263-268

Lawrence, A., Leech, K., Rowan-Robinson, M., Jones, D. H. P., and Wall, J. V.: this conference proceedings

II Zwicky 33: Star Formation Induced by a Recent Interaction

E. Brinks

Royal Greenwich Observatory, Herstmonceux Castle, Hailsham BN27 1RP, UK

ABSTRACT: The discovery is reported of an HI cloud at the same redshift as II Zwicky 33 with an HI mass of 1.2×10^9 $M_{\odot}$. Only very weak optical emission is seen at its position in the light of Hα and in broad band B pictures whereas II Zw 33 itself shows signs of violent star formation. It is suggestive that the two clouds are tidally interacting and that this has triggered the present burst of star formation. The companion galaxy is practically a dark galaxy and might be of primordial composition. The total mass of the companion is of the same order as the mass in neutral gas and there is no need to invoke any dark matter.

INTRODUCTION

HII galaxies, also known as extra-galactic HII regions or as blue compact dwarf galaxies, are gas rich, intrinsically small and have an amorphous optical appearance. Their optical spectra indicate a low heavy element abundance. This is taken as evidence that they can not have been actively forming stars over a prolonged period of time. Here it is assumed, perhaps wrongly, that HII galaxies don't loose any material in a galactic wind as a result of the star formation. A second argument that these systems are young in terms of their star formation history is that they would have run out of fuel. These facts were already noticed by Searle and Sargent (1972) who postulated that star formation in HII galaxies must be episodic. The burst which we are witnessing can be either the very first star forming event, as perhaps in I Zwicky 18 (Kunth and Sargent 1986), or it can be that we happen to see these galaxies in an active, short lived ($\sim 10^7$ year) phase of star formation which is followed by a much longer period ($\sim 10^9$ year) of quiescence. As yet it is not clear why HII galaxies remain in a quiescent state for most of a Hubble time and what triggers them to produce a burst of star formation. In what follows I will argue that the trigger in many cases is provided by tidal interaction.

OBSERVATIONS AND RESULTS

II Zwicky 33 was observed in the line of HI with the NRAO Very Large Array (VLA) in its B-configuration on September 1, 1986. The velocity resolution after Hanning smoothing was 10.4 km s^{-1}. In total 225 minutes were spent on source. The usual calibration and reduction procedures were followed. A full account of these and similar observations of other dwarf galaxies will be given elsewhere. When inspecting the first set of maps emission was seen some 4 arcminutes to the South of II Zw 33. On larger maps this emission appeared to be a companion cloud at about the same systemic velocity, slightly smaller in size and HI content. Figure 1 shows the total HI surface

Dynamics and Interactions of Galaxies Editor: R. Wielen

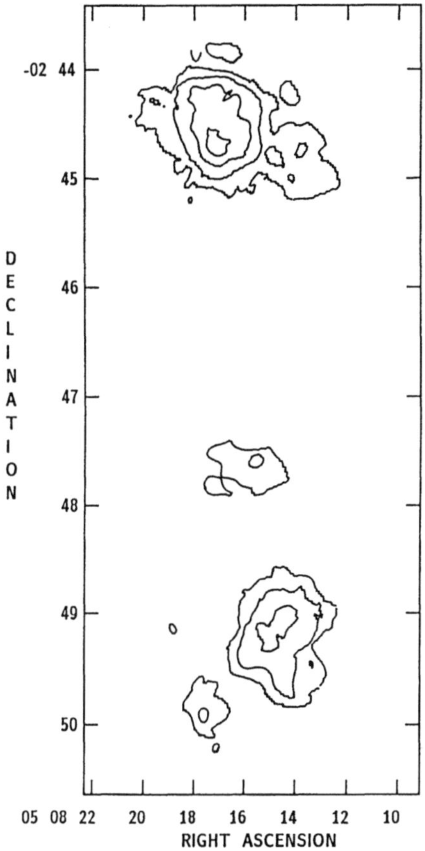

Figure 1: Total HI surface brightness map of II Zw 33 and its companion cloud located 4 arcmin to the South. The map is smoothed to a circular beam of 12 arcsec. Contour levels are drawn at 0.5, 1.5, 3.0 and 4.5 in units of 7.8×10^{20} atom cm^{-2}.

brightness map of II Zw 33 and its companion II Zw 33B. Assuming a distance for both systems of 38 Mpc (H_o = 75 km s^{-1} Mpc^{-1}) we derive HI masses of 2×10^9 $M_\odot$and 1.2×10^9 $M_\odot$, respectively. The HI mass for II Zw 33 is in excellent agreement with the value quoted by Thuan and Martin (1981) which shows that we did not miss any extended HI. The velocity field for II Zw 33 is rather disturbed. The companion shows a regular pattern which, if interpreted as solid body rotation, leads to a total mass of 1.4×10^9 $M_\odot$. No correction for inclination has been applied and the mass derived is likely to be a lower limit. This mass estimate is of the same order as the total amount of neutral gas, H and He, which was derived under the standard assumption that the gas is optically thin.

At optical wavelengths II Zw 33 is a characteristic star forming dwarf galaxy. Figure 2 shows a CCD image through a properly redshifted narrow band Hα interference filter, obtained with the 1.5-m Danish/ESO telescope on La Silla. The continuum has been removed by subtracting a frame obtained through a similar filter but with the central wavelength shifted outside the line of Hα. It shows a fairly small system, $\sim$ 5 kpc in length consisting of a string of bright knots, laid out in an S-shape, indicating recent star formation. At the position of II Zw 33B hardly any emission can be seen on sky survey plates. Stacking two on-band CCD exposures with a total exposure time of 30 minutes through the same Hα filter as was used for II Zw 33 reveals, after correcting for continuum emission, some very faint emission (see Figure 3). Equally faint emission is

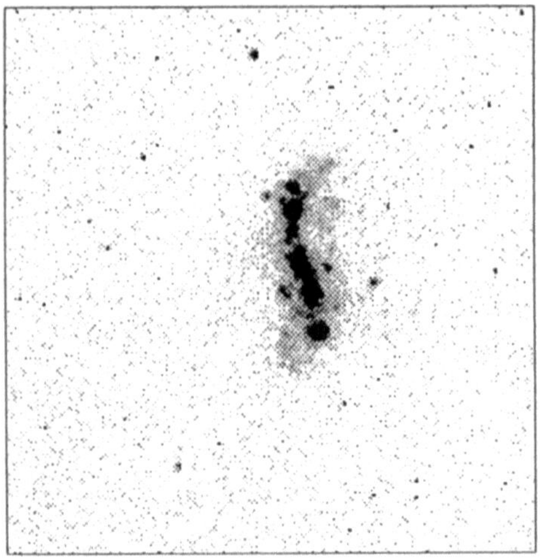

Figure 2: Narrow band CCD image in the light of Hα of II Zwicky 33. The map is at *twice* the scale of the HI map. The underlying continuum is subtracted but the data are as yet uncalibrated. The small specks are cosmic ray hits.

Figure 3: Narrow band CCD image in the light of Hα showing the faint fuzz which is associated with the companion galaxy II Zwicky 33B. Same specifications apply as for Figure 2.

visible in a broad band *B* CCD image. No emission is seen at near infrared wavelengths (Gunn *i*) indicating the absence of older stars. II Zw 33 is listed in the *IRAS* survey and in a recent study Hunter *et al.* (1989) find several infrared peaks surrounding it, but none of these peaks coincides with II Zw 33B.

DISCUSSION

The fact that II Zw 33 has a companion cloud has several interesting consequences. Firstly it provides evidence that objects which consist almost entirely of neutral gas exist in our local volume of space. It is a matter of semantics if II Zw 33B should be called an intergalactic HI cloud or a subluminous dwarf galaxy. The fact that some very weak optical emission is associated with the cloud shows that some star formation is going on, but at such a low level that surveys at optical wavelengths would have never been able to pick up this object. HI clouds which are invisible at optical wavelengths are certainly not common (Hoffman 1989). Lo and Sargent (1979) put a limit of 0.02 Mpc^{-3} for clouds more massive than 4.6×10^8 $M_\odot$ which translates to less than one such object per Mpc^2 at the distance of II Zw 33B. But it should be born in mind that they looked for signals in groups of galaxies and it is quite conceivable that groups are too hostile an environment for isolated clouds to survive.

Secondly, the presence of II Zw 33B suggests that star formation in II Zw 33 was induced by tidal interaction. The projected distance between the two galaxies is 51 kpc or only about four galaxy diameters but as nothing more is known about their space velocities it is not possible to put a firm a limit to the time of closest approach. However, evidence is accumulating that interactions are important, also for dwarf systems. For example, Brinks and Klein (1988) have shown that II Zw 40 might be composed of two interacting HI clouds of about 2×10^8 $M_\odot$ each, one of them showing induced star formation. Other examples are Michigan 160 (Axon *et al.* 1988), a very gas rich object with more than 50% of its mass in the form of HI and which is currently forming stars at a high rate, perhaps for the first time and probably triggered by its companion. Bergvall and Jörsäter state that the fuzzy gas rich component is responsible for the star burst in ESO400-G43. In addition He2-10 shows two nuclei and is clearly a merging system (Johannson 1987). Lastly Hunter *et al.* (1989) mention that they find a higher frequency of nearby blue objects in their survey of normal irregular galaxies which is consistent with them being interacting systems. Finally, II Zwicky 33B must be one of the lowest luminosity dwarf galaxy found thusfar and optically it is an almost dark system. But contrary to, for example, DDO 155 (Carignan and Freeman 1988) from a first assessment of the data there is no need to invoke huge amounts of dark matter. The fact that no older disk stars are seen nor apparently any red supergiants suggests that this cloud is of primordial composition which agrees with the idea put forward by Skillman *et al.* (1988) that the lowest luminosity systems have the lowest abundances.

The National Radio Astronomy Observatory is operated by Associated Universities, Inc., under contract with the National Science Foundation. Part of this work is based on observations obtained at the European Southern Observatory, La Silla, Chile.

REFERENCES

Axon, D.J., Staveley-Smith, L., Fosbury, R.A.E., Danziger, J., Boksenberg, A., Davies, R.D.: 1988, *Monthly Notices Roy. Astron. Soc.* **231**, 1077

Bergvall, N., Jörsäter, S.:1988, *Nature* **331**, 589

Brinks, E., Klein, U.: 1988, *Monthly Notices Roy. Astron. Soc.* **231**, 63p

Carignan, C., Freeman, K.C.: 1988, *Astrophys. J. Lett.* **332**, L33

Hoffman, G.L., Helou, G., Salpeter, E.E., Lewis, B.M.: 1989, *Astrophys. J.* **339**, 812

Hunter, D. A., Gallagher, J.S., Rice, W.L., Gillett, F.C.: 1989 *Astrophys. J.* **336**, 152

Johansson, L.: 1987 *Astron. Astrophys.* **182**, 179

Kunth, D., Sargent, W.L.W.: 1986, *Astrophys. J.* **300**, 496

Lo, K.Y., Sargent, W.L.W.: 1979, *Astrophys. J.* **227**, 756

Searle, L., Sargent, W.L.W.: 1972, *Astrophys. J.* **173**, 25

Skillman, E.D., Melnick, J., Terlevich, R., Moles, M.: 1988 *Astron. Astrophys.* **196**, 31

Thuan, T.X., Martin, G.E.: 1981, *Astrophys. J.* **247**, 823

IRAS Observations of NGC 4696 – Cooling or Evaporation Flow?

T. de Jong[1], *H.U. Nørgaard-Nielsen*[2], *H.E. Jørgensen*[3], *and L. Hansen*[3]

[1]Astronomical Institute "Anton Pannekoek", University of Amsterdam, Roetersstraat 15, NL-1018 WB Amsterdam, The Netherlands
[2]Danish Space Research Institute, Gl. Lundtoftevej 7, DK-2800 Lyngby, Denmark
[3]Copenhagen University Observatory, Østervoldgade 3, DK-1350 Copenhagen, Denmark

We have analysed 60 and 100 μm IRAS observations of NGC 4696, the central elliptical galaxy of the Centaurus cluster. Ground-based optical photometry and spectroscopic data as well as ultraviolet (IUE) and X-ray (EINSTEIN) measurements constrain the interpretation of the infrared data. The dust responsible for the infrared emission has a temperature of 24 K, a total mass of 2.5 10^6 $M_\odot$ and emits a total infrared luminosity of 1.5 10^9 $L_\odot$. It is heated by photons from the interstellar radiation field in the elliptical galaxy and by hot electrons from the X-ray gas. About 10 % of the radiating dust mass is located in a small "horse shoe" shaped dust lane in the centre of NGC 4696. We argue that the other 90 % is distributed over a much larger region (R $\simeq$ 10 kpc) in the galaxy.

Dust grains have a lifetime against destruction by sputtering in the hot X-ray emitting gas of $\sim 10^7$ yr so that they must be replenished at a rate of $\sim$0.2 $M_\odot \mathrm{yr}^{-1}$ to be consistent with the observed dust mass. We suggest that this replenishment takes place by evaporation of cool gas clouds. In line with this suggestion we propose that the 10^7 K gas observed in the central 70 kpc radius region of the Centaurus cluster is produced by evaporation and subsequent heating of cool clouds rather than by cooling of hot gas, as often suggested. The evaporating clouds are brought in by a gas-rich dwarf galaxy that has recently collided with NGC 4696.

To be able to maintain the evaporation and to balance the heat losses of the hot gas by X-ray emission and by collisions with dust grains, a source of heat is required. We suggest that the transport of heat into the core of the X-ray emitting gas by electron conduction and the release of kinetic energy of the colliding galaxy can provide the required heating.

A more elaborate version of this paper will be published elsewhere (de Jong et al. 1989). Sparks et al. (1989, see also these proceedings) have independently arrived at similar conclusions from an analysis of optical data of NGC 4696.

References

T. de Jong, H.U. Nørgaard-Nielsen, H.E. Jørgensen, L. Hansen: *Astron. Astrophys.*, in press (1989)
W.B. Sparks, F. Macchetto, D. Golombek: *Astrophys. J.*, in press (1989)

Dynamics and Interactions of Galaxies Editor: R. Wielen

IRAS Galaxies and Morphology

R.G. Clowes[1], *S.K. Leggett*[2], *A. Savage*[3], *and L.E. Campusano*[4]

[1]Royal Observatory, Blackford Hill, Edinburgh EH9 3HJ, Scotland
[2]US Naval Observatory, Flagstaff, AZ 86002, USA
[3]UK Schmidt Telescope, Anglo-Australian Observatory, Coonabarabran, NSW 2357, Australia
[4]Universidad de Chile, Casillo 36-D, Santiago de Chile, Chile

1 Introduction

There is currently a general re-appreciation of the possible importance of morphology, especially interactions and mergers, to the creation of starbursts and active nuclei in galaxies. Much of the new enthusiasm is attributable to the surprising result from IRAS that there are relatively local galaxies — the ultraluminous IRAS galaxies — that are quasar-like, at least in their luminosities in the far-infrared ($L_{IR} \sim 10^{12} L_{\odot}$).

Observations of ultraluminous galaxies have so far been few and have led to contradictory or unsatisfactory results. The difficulties arise because the galaxies are rare among IRAS galaxies in general, and samples have been one or more of very small, colour-selected and incomplete. Colour-selection implies very serious incompleteness (e.g. Low *et al.* 1988, Sanders *et al.* 1988), because of the emphasis on the 25 μm flux, which is, in fact, only rarely defined for these galaxies. However, the results that have emerged so far are these. Firstly, Sanders *et al.* (1988) claimed that interactions are the initiators of the high luminosities in at least $\sim$ 75% of cases, while Lawrence (1988) disputed this, claiming that the fraction is $\sim$ 40%. Secondly, Low *et al.* (1988) found evidence that the incidence of definite quasar activity may be almost total at the very highest luminosities, and that the space density, although very uncertain, may represent a very significant and previously unnoticed population of local quasars (perhaps $\sim$ 50% more).

For reliable work on the detailed relationship between IRAS galaxies in general — not just the ultraluminous galaxies — and optical morphology we felt that a large programme of optical identification of IRAS point sources was necessary. Our identifications, stars and galaxies, are in these areas of sky: $\sim$ 2800 deg^2 of the south galactic cap (Savage *et al.*, in preparation) — 304 deg^2 of this area have already been published by Wolstencroft *et al.* 1986 in a pilot study; $\sim$ 314 deg^2 in the north galactic cap (Clowes *et al.* in preparation); the Virgo cluster (Leggett *et al.* 1987); and the Coma, Fornax and Abell 1060 clusters (Wang *et al.* in preparation). The programme is so large that, for example, $\sim$ 50 ultraluminous galaxies are expected in a sample that is nearly complete and is free from infrared colour biases.

2 Dependences on morphology

Wolstencroft *et al.* (1986), for the pilot study in the south galactic cap, found that (i) the luminosity ratio L_{IR}/L_B correlates with the IRAS flux-density ratio $f_\nu(100\mu m)/f_\nu(60\mu m)$, with the warmest galaxies having the highest L_{IR}/L_B, and (ii) early-type spirals have higher L_{IR}/L_B than late-type spirals. Leggett *et al.* (1987), for the Virgo cluster, confirmed the correlation of L_{IR}/L_B with $f_\nu(100\mu m)/f_\nu(60\mu m)$, and, additionally, found that (iii) L_{IR}/L_B correlates with L_{IR}. High values of L_{IR}/L_B therefore select candidates for ultraluminous galaxies. The separation in L_{IR}/L_B between early- and late-type spirals is much less marked here, with values for late types remaining unchanged. This and the fact that the Virgo galaxies, being relatively local, define a sample of lower average infrared luminosity imply that (iv) the central regions of early types increasingly dominate the disk as infrared luminosity increases. A comparison of the

Dynamics and Interactions of Galaxies Editor: R. Wielen

Virgo galaxies with field galaxies of similar redshifts showed that (v) the cluster environment has no detectable influence on the infrared properties.

3 Ultraluminous and disturbed galaxies

Leggett *et al.* (1987) estimate that $L_{IR}/L_B \gtrsim 30$ implies quasar-like luminosities of $L_{IR} \sim 10^{12} L_{\odot}$. More specifically, we now estimate, incorporating data from Wolstencroft *et al.* (1988), that $\sim 60\%$ of the galaxies selected by $L_{IR}/L_B \geq 30$ have high luminosities $L_{IR} \gtrsim 5 \times 10^{11} L_{\odot}$ for $H_0 = 75$ kms^{-1}Mpc^{-1}, and that only $\sim 30\%$ of the galaxies with these luminosities are omitted by the selection. For our south and north galactic cap areas there are ~ 85 galaxies with $L_{IR}/L_B \geq 30$, and so ~ 50 of these are expected to have $L_{IR} \gtrsim 5 \times 10^{11} L_{\odot}$. 50 galaxies represent $\sim 4\%$ of the total.

About $\sim 25\%$ of these candidates for ultraluminous galaxies appear on the UK Schmidt plates to be interacting, associated with companions, or otherwise disturbed; typically, though, they are too faint to tell. But from our nearly complete identifications we find that the number of IRAS galaxies in general that are involved in interactions, mergers and other disturbances (tidal tails, jets) far exceeds the number that can be ultraluminous. So if interactions and mergers are necessary for the creation of ultraluminous galaxies then they are not immediately obvious as sufficient. Stein (1988) has, in different circumstances, reached a similar conclusion.

We will complete towards the end of 1989 an observational programme of B, I CCD imaging and optical spectroscopy at CTIO and Las Campanas in an attempt to determine (i) those morphological configurations that lead to ultraluminous galaxies and active nuclei and those that do not, and (ii) the space density of ultraluminous galaxies. At this time the first observations have just been obtained.

4 Some examples of highly disturbed galaxies

We have already obtained at CTIO (briefly, during time allocated to a different project) CCD R images and optical spectroscopy for the following three disturbed galaxies from the identifications in the south galactic cap. None of these galaxies are ultraluminous, despite being highly disturbed. $H_0 = 75$ kms^{-1}Mpc^{-1} is adopted.

IRAS 03210–4644. This galaxy has a northern arm and a long, highly curved southern tail. The main body of the galaxy has Seyfert-2 emission lines, and a redshift of 0.063. $L_{IR} = 8.60 \times 10^{10} L_{\odot}$; $L_{IR}/L_B = 5.37$. The southern object, which might be associated, has no distinctive features in its spectrum.

IRAS 03279–2856. Interacting spiral and elliptical, both having redshifts of 0.038. The spiral has a very bright nucleus and LINER emission lines. $L_{IR} = 1.99 \times 10^{10} L_{\odot}$; $L_{IR}/L_B = 0.44$.

IRAS 03372–3341. A complex galaxy, with a northerly loop. There are two knots: the northern is a foreground star, the southern has Seyfert-2 emission lines, and a redshift of 0.052. $L_{IR} = 8.91 \times 10^{10} L_{\odot}$; $L_{IR}/L_B = 2.45$.

Lawrence,A., 1988. *Comets to Cosmology* p. 217, Springer-Verlag.

Leggett,S.K., Clowes,R.G., Kalafi,M., MacGillivray,H.T., Puxley,P.J., Savage,A. & Wolstencroft,R.D., 1987. *Mon. Not. R. astr. Soc.,* **227,** 563.

Low,F.J., Huchra,J.P., Kleinmann,S.G. & Cutri,R.M., 1988. Ap. J. **327,** L41.

Sanders,D.B., Soifer,B.T., Elias,J.H., Neugebauer,G. & Matthews,K., 1988. Ap. J. **328,** L35.

Stein,W.A., 1988. Astron. J. **96,** 1861.

Wolstencroft,R.D., Savage,A., Clowes,R.G., MacGillivray,H.T., Leggett,S.K. & Kalafi,M., 1986. *Mon. Not. R. astr. Soc.,* **223,** 279.

Wolstencroft,R.D., Unger,S.W., Pedlar,A., Heasley,J.N., Parker,Q.A., Menzies,J.W., Savage,A., MacGillivray,H.T., Leggett,S.K., Wang,G. & Clowes,R.G., 1988. *Comets to Cosmology* p. 202, Springer-Verlag.

Interacting and Active IRAS Galaxies

K.J. Leech[1], A. Lawrence[1], M. Rowan-Robinson[1], J. Crawford[1], M.V. Penston[2], R.J. Terlevich[2], D.H.P. Jones[2], and J.V. Wall[2]

[1]School of Mathematical Sciences, Queen Mary College, Mile End Road, London E1 4NS, UK
[2]Royal Greenwich Observatory, Hailsham, E. Sussex BN27 1RP, UK

Aim

We decided to study a complete sample of, predominantly high IR luminosity, IRAS galaxies to determine the fractions that were interacting or active.

Sample Selection

The sample of IRAS galaxies discussed in this paper were selected from the IRAS galaxy sample of Lawrence *et al.* (1986) using the following criteria:

•$z > 0.01$ So that the image of the galaxy would fit onto our CCD chip.
•$S_{60\mu m} > 2$Jy
or $S_{60\mu m} > 1$Jy and Log($L_{60\mu m}$ / $L_{\odot}$) > 11.5.
These criteria select 61 galaxies biased towards high luminosity objects.

The control galaxies were taken from the Anglo-Australian Redshift Survey (Peterson *et al.* 1986), excluding galaxies classified as ellipticals or lenticulars, selected so that their redshift and optical magnitude distributions matched as closely as possible those for the IRAS sample.

Observations

Sixty of the IRAS galaxies and 87 control galaxies were imaged with CCD's at the Observatorio del Roque de los Muchachos (La Palma) at the f/15 Cassegrain focus of the 1M Jacobus Kapteyn Telescope (JKT) in April 1986 and at the f/3 prime focus of the 2.4M Isaac Newton Telescope (INT) in June 1987. Exposure times were between 300 and 600 seconds on the JKT and 150 seconds on the INT. Spectroscopic observations were made of 56 IRAS galaxies using the Image Photon Counting System/Intermediate Dispersion Spectrograph at the Cassegrain focus of the INT in June 1986, with exposure times of between 500 and 1000 seconds.

Classification Schemes

It is difficult to objectively classify galaxies as interacting or non-interacting. We therefore settled upon a seven-point system to classify galaxies according to how close other galaxies were from them. The system is: (0) galaxy isolated; (1) galaxy has a distant faint (within 4 optical mag) companion; (2) galaxy has a distant bright (within 2 mag) companion; (3) galaxy has a near (within 40kpc projected distance) faint companion; (4) galaxy has a near bright companion; (5) galaxy shows definite signs of interaction with another (e.g. bridges or tails); (6) merging object (two nuclei in a common envelope) or highly peculiar.

The Baldwin, Phillips and Terlevich (1981) classification scheme was used to classify the IRAS galaxies spectroscopically. Emission line luminosities and galaxy colours were also derived from the spectra.

Dynamics and Interactions of Galaxies Editor: R. Wielen

Results and conclusions

The results and conclusions from this study are presented in detail in Leech *et al.* (1989) and Lawrence *et al.* (1989). Relevant conclusions from those papers are:

•Only 46±13% of high luminosity ($L_{60\mu m} > 10^{11} L_\odot$) IRAS galaxies are in interacting or merging systems. The rest appear to be non- interacting.

•11±8% of low luminosity ($L_{60\mu m} < 10^{11} L_\odot$) IRAS galaxies are in interacting or merging systems, a fraction similar to that of the control sample.

•The majority of IRAS galaxies show H II region-like spectra. Only seven, distributed randomly in IR luminosity, showed spectra indicative of an active nucleus.

•The Hα luminosity is correlated with the $60\mu m$ luminosity. The cause of the IR luminosity is visible in the optical, and from the above evidence is mostly star formation.

The figure below shows the interaction class (from the above system), spectral classification and IR luminosities of the 56 galaxies we have spectral information for. Unless otherwise noted all galaxies show H II region-like spectra. An L next to a galaxy interaction class means that it is a Liner, an S means that it is a Seyfert.

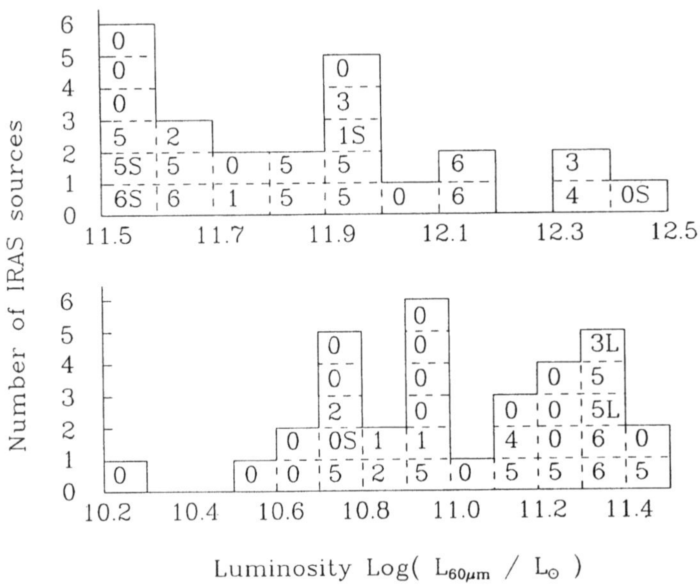

Bibliography

J.A. Baldwin, M.M. Phillips, & R. Terlevich: *Publs. astr. Soc. Pacif.*, 93, 5 (1981)

A. Lawrence, D. Walker, M. Rowan-Robinson, K.J. Leech, & M.V. Penston: *Mon. Not. R. astr. Soc.* 219, 687 (1986)

A. Lawrence, M. Rowan-Robinson, K.J. Leech, D.H.P. Jones, & J. Wall: *Mon. Not. R. astr. Soc.* accepted (1989)

K.J. Leech, M.V. Penston, R.J. Terlevich, A. Lawrence, M. Rowan-Robinson, J. & Crawford: *Mon. Not. R. astr. Soc.* accepted (1989)

B.A. Peterson, R.S. Ellis, G. Efstathiou, T. Shanks, A.J. Bean, R. Fong, & Z. Zen-Long: *Mon. Not. R. astr. Soc.* 221, 233 (1986)

Observations of an IRAS-Selected Sample of Interacting Galaxies

D. Lutz

Max-Planck-Institut für Astronomie, Königstuhl,
D-6900 Heidelberg, Fed. Rep. of Germany

Abstract: A sample of interacting and merged IRAS-galaxies has been studied with near-infrared photometry, optical CCD imaging and long slit spectroscopy. Using the observed K-L color excesses to infer the distribution of far-infrared emission, it is found that both galaxies show enhanced infrared emission in many interacting systems.

Using the IRAS database, several studies have shown that galaxy interactions can efficiently trigger enhanced infrared emission due to star formation and/or active nuclei. Data of better spatial resolution than provided by the large IRAS beam are needed to locate the source of emission. I have used ground-based near-infrared photometry and optical data to look for the infrared active regions in selected interacting systems.

From the IRAS point source catalog sources which fulfilled the following criteria were selected: High flux quality in the 25μm-100μm bands, $F_{60\mu m} \geq 5Jy$, $F_{25\mu m} \leq 0.5 * F_{60\mu m}$, $|b_{II}| \geq 15^{o}$, observable from La Silla in June. All sources were inspected on POSS prints or ESO/SERC film copies. Galactic sources, isolated galaxies and galaxies with large angular extent were discarded. The final sample comprises 19 systems, ranging from wide pairs to mergers. JHKL photometry, CCD broadband imaging and long slit spectra were obtained at the ESO/MPI 2.2m telescope.

The far-infrared properties of these systems show a clear trend with the interaction strength. FIR luminosities of more than $4 \cdot 10^{11}$ $L_{\odot}$ and L_{FIR}/L_B ratios of more than 7 occur only in the close systems, where the separation of galaxies is smaller than the galaxy diameters, but there are also close interactions showing relatively low FIR activity. The increase of FIR luminosity with interaction strength is reflected in changing FIR colors: The 60μm/100μm color gets warmer, the 12μm/25μm color gets colder with increasing interaction strength. This is easily explained in a two-component model of FIR emission, where a burst component more and more dominates a diffuse component.

The near-infrared colors can be modelled by normal galaxy colors, additional reddening by dust extinction corresponding to $A_V \approx 1...4$mag plus "hot dust" emission which contributes slightly to the K and strongly to the L flux, resulting in K-L colors larger than the normal 0.3. The amount of reddening deduced from the NIR colors is similar to that found in the spectra, indicating an origin in regions of similar optical depth. Possible contributions to the K-L excess are emission of hot dust in star forming regions, an unusually strong 3.28μm PAH feature due to a strong interstellar radiation field, dust shells around evolved stars and an active nucleus. By looking only at the K-L excess relative to normal galaxies, the standard contributions of these mechanisms

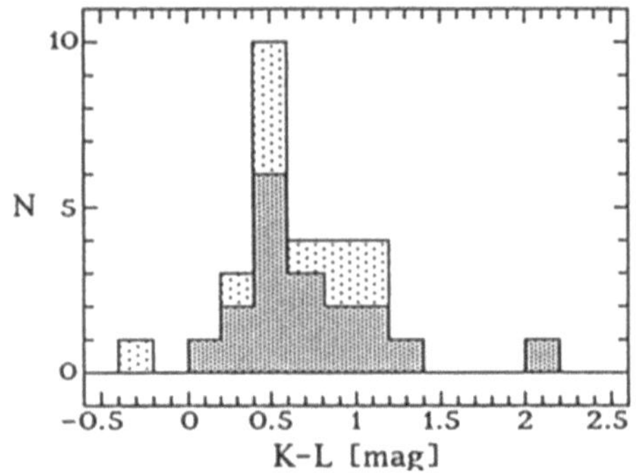

Fig.1: Histogram of K-L colors. The brighter components of interacting systems and mergers are heavily shaded. The fainter components of interacting systems are lightly shaded

are canceled. Indications for a Seyfert nucleus are found in only three of the optical spectra. The other mechanisms, related directly or indirectly to enhanced star formation, must produce the observed K-L excess. Analysis of published K,L, and 10μm data shows that the K-L excess of galaxies is correlated with the 10μm flux. If the FIR spectrum is dominated by a burst, the K-L excess must be correlated with the 60μm-100μm emission as well, allowing to locate the starburst in an interacting system.

A K-L excess, here defined as K-L ≥ 0.5, is found for two thirds of all galaxies (Fig.1). Of 11 interacting systems with both galaxies observed 1 has no K-L excess in either galaxy, 3 only in the main galaxy, 2 only in the companion, and 5 in both galaxies. For comparison, previous studies have found that only the companions are IR-active (Joseph et al. 1984) or that in most systems only the main galaxy is IR-active (Telesco et al. 1988). These different results can be explained by the different samples and observing techniques. The Joseph et al. sample is morphologically selected and its mean FIR luminosity is eight times smaller than the $10^{11.2}$ $L_{\odot}$ found here. It is therefore based on weaker interactions, which may preferentially induce sufficient tidal disturbances in the smaller galaxy only. The Telesco et al. sample is IRAS-selected and therefore presumably includes strong interactions, but the poor angular resolution of the IRAS data prevents a reliable analysis of the distribution of FIR flux in the interacting systems. The importance of this limitation is stressed by the fact that their double sources have a four times larger mean optical separation than their single sources, indicating inability to separate close double sources.

A large fraction of double IR-active systems is found only if star formation is induced with high probability during the interaction. As a consequence, the influence of the orientation of orbital parameters with respect to the disks on the efficiency of gas infall must be rather low, at least for a strong tidal disturbance. It may be noted that some of the single bursts could even be systems which will finally undergo a burst in both galaxies, but with a significant time delay.

References:

Joseph,R.D. et al.: Mon. Not. R. astr. Soc. **209**,111 (1984)
Telesco,C.M. et al.: Astrophys. Journal **329**,174 (1988)

Interactions in Extreme IRAS Galaxies

B. van den Broek, W. van Driel, and T. de Jong

Astronomical Institute, University of Amsterdam, Roetersstraat 15,
NL-1018 WB Amsterdam, The Netherlands

Introduction

From the IRAS point source catalog we have selected a sample of about 200 galaxies in a southern field of about 1000 square degrees, using strict flux density limits. For a subset of 42 galaxies with the highest $\log(L_{FIR}/L_B)$ luminosity ratio we collected optical spectra, broad band B,V,R and narrow band Hα CCD images, as well as low (10″) as high (1″) resolution VLA 6cm radio continuum maps. For 17 sources we obtained near-infrared measurements in the J,H,K bands in collaboration with I. Glass. From IPAC we recieved co-added flux densities in the 4 IRAS bands for these 42 sources. One third to one half of the 42 IRAS galaxies are interacting with other galaxies, ranging from mergers to wide-spaced interactions. In this poster we present three different types of interactions; a merger, a close- and a wide-spaced interaction.

The far-infrared radiation is produced by warm dust absorbing the UV light from young hot stars. This UV light also ionizes the gas in HII regions producing Hα light. At the end of their life these stars turn into luminous red supergiants radiating mainly in the near-infrared. When these stars become supernovae, they produce fast electrons which will emit synchroton radiation. This explains the correlations found between the far-infrared luminosity and the Hα luminosity, between the far-infrared luminosity and radio power and between the far- and near-infrared luminosities (van den Broek and de Jong, in prep.).

The interpretation of the optical images is difficult since extinction by dust plays an important role.

The individual galaxies in our sample are unresolved by the IRAS detectors. We have estimated far-infrared fluxes for the individual galaxies by distributing the IRAS flux densities over the two galaxies proportional to the observed radio flux densities. The luminosities are calculated with a Hubble constant of $H_0 = 75$ km/s/Mpc.

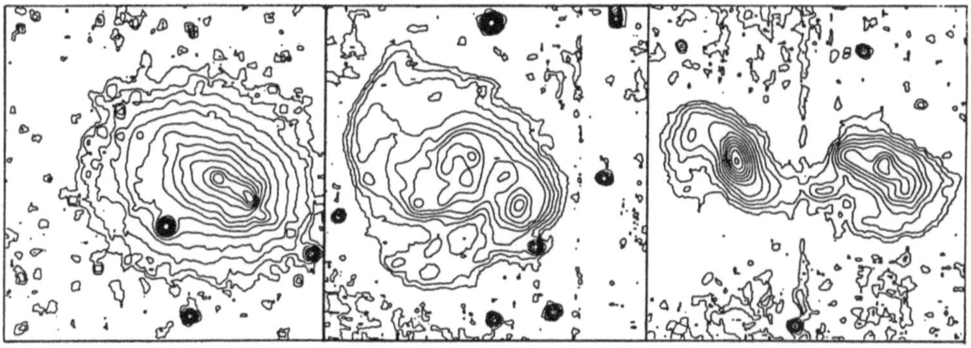

Figure 1: Contour plots (80″ by 80″) in the B band for the three galaxies; 1004-296 (left), 1304-335 (middle) and 1214-259 (right). East is to the right. The contours are separated by half a magnitude and start at 25 mag/arcsec^2.

Dynamics and Interactions of Galaxies Editor: R. Wielen

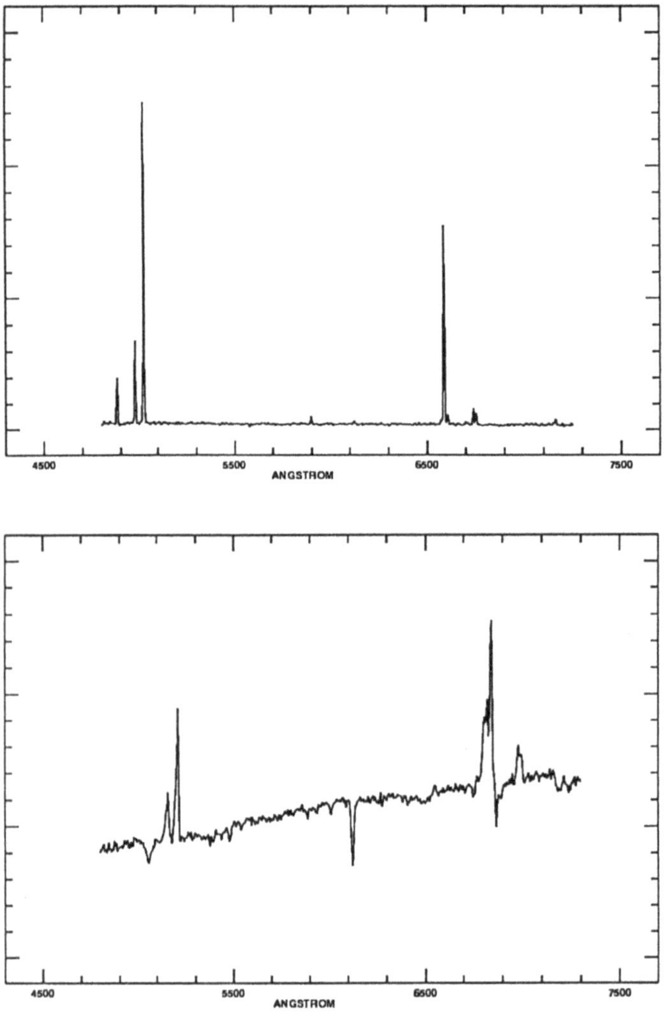

Figure 2: Spectra of 1004-296 (top) and 1214-259 (bottom)

IRAS 1004-296

This is an example of a dwarf merging system demonstrating that using $\log(L_{FIR}/L_B)$ as a selection criterion we are capable to trace such systems which are otherwise difficult to find. Its spectrum is peculiar since it shows exceptionally strong emission lines completely dominating the continuum. The ratio between the line intensities indicate an HII region-type spectrum with unusually strong O[III] lines due to highly ionized gas. We interpret the data as the result of a merging process, which enhances the starformation in two distinct regions, presumably the nuclei of the merging objects.

IRAS 1214-259

In this interacting pair the eastern galaxy has a 2 to 3 times higher radio and Hα emission than its companion. Its HII region-type spectrum indicates a high rate of starformation. The western galaxy, however, shows a complicated spectrum. Strong and broad O[III] and N[II]

Galaxy parameters

	1004-296	1304-335(NW)	1304-335(SE)	1214-259(W)	1214-259(E)
$\mathrm{Log}(L_{FIR}/L_B)$	0.31	0.86	0.96	0.52	0.99
$\mathrm{Log}(L_B/L_{\odot})$	8.81	9.97	9.57	10.11	10.06

lines indicate a Seyfert nucleus and the clear Hα and Hβ stellar absorption lines indicate a post-starburst spectrum, dominated by the light of A and F stars. The strong Na absorption is caused by high gas column density. Dust mixed with this dense gas can explain the red continuum. The Ha image shows an elongated structure pendicular to the major axis of the galaxy. This is probably a polar gas disk, a rare phenomenon, consisting of recently captured gas.

IRAS 1304-335

In this close interacting pair the north-western galaxy has bright blue HII regions to the east of the nucleus, one of which is even brighter than the nucleus. These HII regions also show up in the Hα as well as a number of HII regions on the opposite side of the nucleus. A possible explanation is that this dusty galaxy has a ring of large HII regions around its nucleus, of which the far (west) side is reddened and therefore inconspicious in the blue. Both galaxies show HII region-type spectra consistent with high starformation rates. In the south eastern galaxy the star formation is concentrated in its distinct nucleus.

The Strongly Interacting Galaxy IRAS 15320-2601

R.D. Wolstencroft[1], *S.M. Scarrott*[2], *R.F. Warren-Smith*[3], *and A. Pedlar*[4]

[1]Royal Observatory Edinburgh, Edinburgh, UK
[2]Dept. of Physics, University of Durham, UK
[3]Rutherford Appleton Laboratory, Chilton, UK
[4]Nuffield Radio Astronomy Laboratories, University of Manchester, UK

We have identified IRAS 15320-2601 with a previously uncatalogued 17th magnitude galaxy having a redshift of 20,500 km s^{-1} (Wolstencroft et al., 1988). It has a high far infrared luminosity (L_{IR} = 3.5 x $10^{11}L_{\odot}$) and is hot (F_{60}/F_{100} = 0.82) with L_{IR}/L_B = 40 suggesting that it is the site of vigorous star formation. CCD imaging (see fig 1) in (B,V,R,I) reveals clear evidence of a strong interaction, namely the presence of (1) a jet or tidal bridge to the ENE 33 kpc in length directed towards a possible companion galaxy of unknown redshift; (2) a double nucleus with components separated by 1.6 arc sec (2.1 kpc), with the line joining the nuclei lying along the axis of the bridge; (3) a faint counterjet or tidal tail extending at least 48 kpc to the WSW; and (4) a curved arm or spur emerging from the double nucleus in a direction normal to the line joining the two nuclei. Spectroscopy at the AAT in conditions of sub arc sec seeing indicates a radial velocity difference of 250 km s^{-1} between the nuclei, and emission line ratios that are typical of HII regions for both nuclei. The nucleus to the ENE is very red in B–R suggesting that it is highly obscured.

A VLA map at 6cm taken in the A/B configuration shows two sources: source 1 is essentially coincident with the double nucleus and source 2 is 3.6 arc sec to the NW in the direction of the optical spur as it emerges from the double nucleus. The two sources are unresolved at 20cm with a total flux of 12.9 mJy indicating a spectral index for the combined source of α = 0.4 ($S \propto \nu^{-\alpha}$).

The geometry of the interaction is difficult to deduce without further observations. A possible scenario is similar to that put forward by Appleton et al. (1987) to explain the ring galaxy system Arp 143. The suggested sequence is as follows: a dwarf galaxy (intruder) approaches a disk galaxy (victim) along a trajectory that is normal to the disk plane and passes through the central regions of the victim. The intruder is disrupted as it passes through the disk plane and the debris of both the stars and gas are distributed over a wide range of velocities. If a starburst is triggered as the intruder enters the disk plane an expanding shell may be seen at later stages. At the same time a ring-like density wave is produced in the disk plane. We suppose that the system is viewed from the side of the disk along which the intruder approached and that the two nuclei are viewed along almost the same line of sight. In this case we identify the bridge with the disrupted intruder, either in the form of a plume or a disturbed sheet viewed edge on, and the spur is the partial ring-like density wave in the victim plane. The nucleus of the disrupted (intruder) galaxy would be heavily reddened on this model and thus be the ENE of the two nuclei.

Dynamics and Interactions of Galaxies Editor: R. Wielen

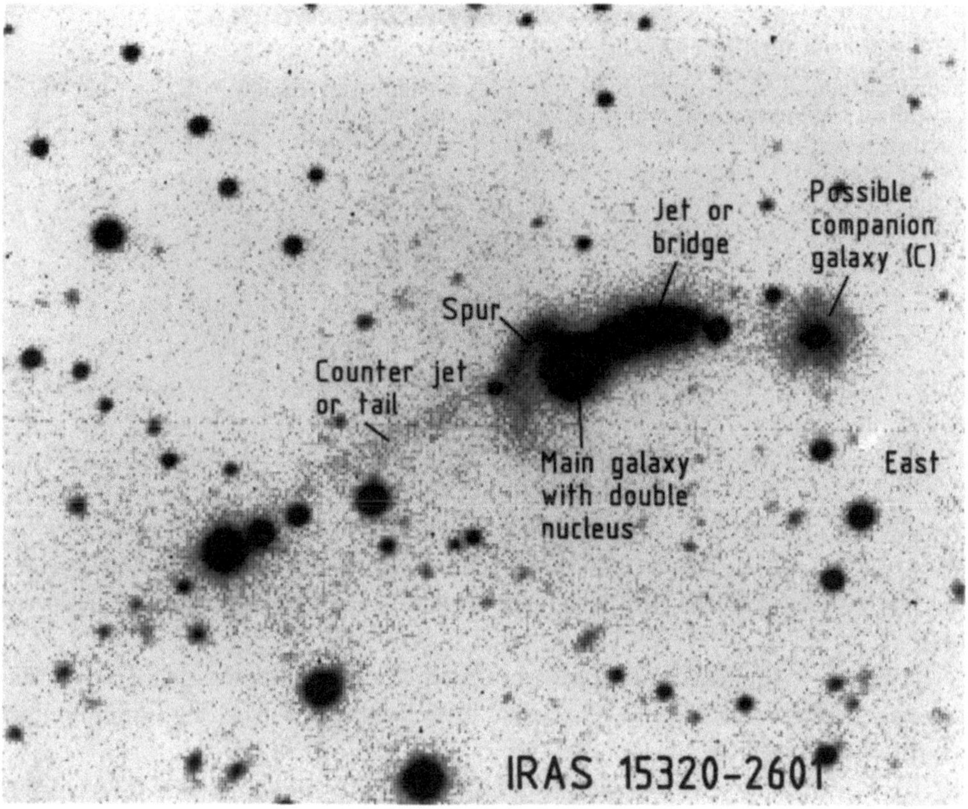

Figure 1 I band CCD image obtained using the AAT 4m telescope (0.49 arc sec = 1 pixel). North is up and East to the right.

References

R.D. Wolstencroft, S.M. Scarrott, R.F. Warren-Smith and A Pedlar: Bull. Amer. Astron. Soc. 20, 645 (1988).

P.N. Appleton, F.D. Ghigo, J.H. van Gorkom, J.M. Schombert and C. Struck-Marcell: Nature 330, 140 (1987).

Binary Starbursts in Normal and Colour-Selected IRAS Galaxies

H.M. Sopp, P. Alexander, and J.M. Riley

Cavendish Laboratory, University of Cambridge, Madingley Road,
Cambridge CB3 0HE, UK

We use IRAS colour criteria to select prototypes of IR–bright galaxies with $L_{IR} \sim 10^{11} - 10^{12} L_\odot$ and $L_{IR}/L_B \sim 10 - 100$, following Heckman, Armus and Miley 1987. This sample and the control sample (no colour criteria) have a flux limit of $S_{60\mu} \geq 2Jy$.

We made approximately 10 minute "snapshots" of all 43 galaxies at 6cm with the VLA D–array resulting in final maps with an rms noise of typically 0.07mJy/beam and a beamsize of 15″. All were detected at the 5σ level with fluxes predicted from the FIR/radio correlation implying that all the objects are star–formation dominated. Subsequent examination of the POSS prints showed that:

(i) There is a considerable excess of near radio neighbours ($< 4'$) over the expected background count (see Fig. 1), which agrees well with the number of identified companions.

(ii) The colour–selected sample is more distant and it is therefore harder to identify neighbours; the control sample has 9 out of 28 (32%) binary radio sources, and out of 9 pairs of similar magnitude galaxies < 4 galaxy diameters apart 6 (67%) are binary sources.

The discovery of so many binaries is an initially surprising result as many authors have reported that only one galaxy in each pair is active. However, on closer examination of the work of Telesco *et al.* (IRAS observations of a sample of close optical pairs) we find that of their distinguishable pairs with $S_{60\mu} \geq 2Jy$ 33% are binaries. Furthermore their IRAS data cannot separate the closest pairs so presumably the total percentage is higher. This is totally contrary to the conclusions drawn from considering the whole sample which are in general of lower IR luminosity than ours. Similarly, Lonsdale *et al.* studying the IRAS minisurvey galaxies with a flux limit of 0.5Jy found a lower level of interaction than we did.

Other surveys with similar results (finding single sources in pairs) e.g. Eales *et al.* (VLA B and C–array observations of IRAS minisurvey galaxies) and Joseph *et al.* (NIR study of strongly interacting systems) can be explained by their high resolution beams being insensitive to extended structure, so that they have overlooked discwide starbursts. This has further support from the

Dynamics and Interactions of Galaxies Editor: R. Wielen

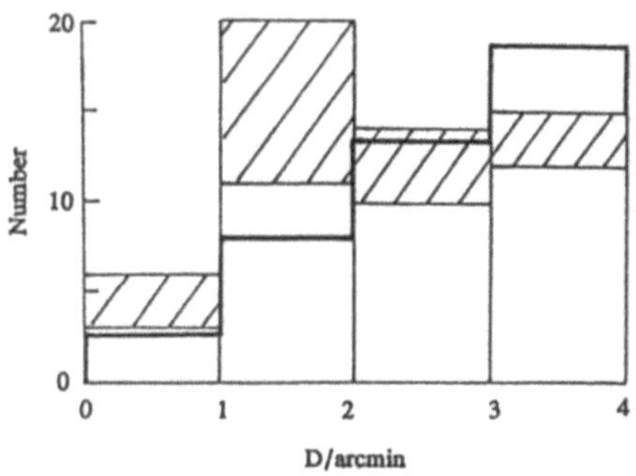

Figure 1: Radio Neighbour Statistics

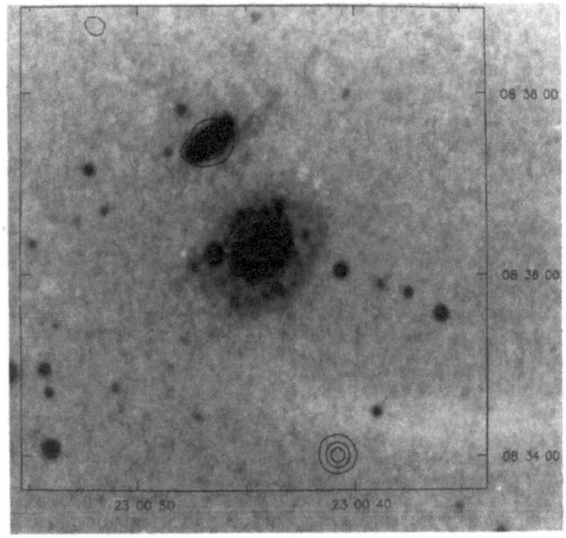

Figure 2: Example of Binary Starburst

fact that Eales *et al.*'s overall detection rate is lower than that predicted from the FIR/radio correlation assuming the sources are unresolved.

Thus it appears that:

(i) the degree of interaction found in samples increases with the flux limit

(ii) there is evidence that at least some systems may contain extended starbursts.

One scenario is that many pairs of interacting galaxies contain an active nucleus (starburst or Seyfert) in one member, which is easily detected, while the other galaxy may be undergoing a discwide starburst.

However, in at least a third of the cases one or both galaxies are inactive. One explanation of this might be an asymmetry in encounter geometry such as that discovered by Helou 1984 who found that binary galaxies occur more often with their spins opposed. Thus one will have direct spin and one retrograde spin relative to their orbit, resulting in different responses to the gravitational perturbations (Toomre & Toomre 1972, Sundelius 1989). Other effects which may be relevant to the inactivity of some galaxies include the time lag between perturbation and starburst in the disc (Bushouse 1987, Scalo & Struck–Marcell 1986) and the need for the gas density to exceed a certain threshold before star formation occurs (Kennicutt 1989).

REFERENCES

Bushouse, H.A., 1987,*Astrophys. J.*, **320**, 49.
Eales, S.A., Wynn–Williams, C.G.,& Beichman, C.A., 1988, *Astrophys. J.*, **328**, 530.

Heckman, T.M., Armus, L.,& Miley, G., 1987, *Astr. J.*, **93**, 276.
Helou, G., 1984, *Astrophys. J.*, **284**, 471.
Joseph,R.D., Meikle,W.P.S., Robertson,N.A., & Wright,G.S., 1984, *Mon. Not. R. astr. Soc.*, **209**, 111.
Kennicutt, R.C., 1989, (preprint)
Lonsdale, C.J., Persson, S.E., & Matthews, K., 1984, *Astrophys. J.*, **287**, 95.
Scalo, J.M., & Struck-Marcell, C., 1986, *Astrophys. J.*, **301**, 77.
Sundelius, B., 1989, these proceedings
Telesco, C.M., Wolstencroft, R.D., & Done, C., 1988, *Astrophys. J.*, **329**, 174.
Toomre, A., & Toomre, J., 1972, *Astrophys. J.*, **178**, 623.

Wind Dynamics in the Interacting Galaxy NGC 7714

H. Schulz[1] *and H. Hippelein*[2]

[1]Astronomisches Institut, Ruhr-Universität, Postfach 102148, D-4630 Bochum, Fed. Rep. of Germany
[2]Max-Planck-Institut für Astronomie, Königstuhl, D-6900 Heidelberg, Fed. Rep. of Germany

The interacting galaxy NGC 7714 (=Arp 284=Mkn 538) is considered to contain the "prototype star-burst galactic nucleus" by Weedman *et al.* (1981). Recently Taniguchi *et al.* (1988) found non-circular motions in its nucleus, part of which were interpreted to result from a strong bipolar wind traced by [NII] lines. The direction of this wind appeared to coincide with that of the double radio lobes detected by Weedman *et al.* This result raised our interest in this object because outflow cones or expanding bubbles related to UV-radiation beams and radio structures appear also to be found in Seyfert galaxies (Unger *et al.* 1987, Schulz 1988, 1989).

Six long-slit spectra were obtained with the Cassegrain spectrograph of the 3.5m-telescope on Calar Alto. We measured emission-line velocities of Hα, [NII]$\lambda\lambda$6548,6583 and [SII] $\lambda\lambda$6716,6731 within a rectangle of 12" x 30" centered on the nucleus of NGC 7714. The spectral resolution as indicated by the FWHM of the comparison lines is 100 km s^{-1}. Line positions and widths were measured by fitting the profiles with Gaussians.

Most line profiles were well represented by a single Gaussian. Only in the innermost 3" to 5" an enhancement of the lower blue line wings made a second Gaussian component necessary to fit the profiles. The velocity difference of the components peaked at 4" west of the nucleus with more than 150 km s^{-1}. The median blueshift of component 2 relative to the main component is 80 km s^{-1}. Having ruled out asymmetries induced by a rotating disk we conclude that outflow is the most likely explanation of the blue component. An upscaled version of superbubble-induced flow similar to that in the nearby starburst galaxy NGC 253 (Schulz 1989) could account for the blue asymmetries and the X-ray and radio luminosities given by Weedman *et al.* Our observations confirm the significance of the blue component in the nuclear region of NGC 7714 found by Taniguchi *et al.* but do not show the drastic difference between [NII] and Hα claimed by these authors. A study with higher resolution is underway to discern more clearly the direction of the flow, which does *not* appear to coincide with the radio axis.

References

Schulz, H.: 1988, *Astron. Astrophys.* **203**, 233
Schulz, H.: 1989, Radiation beams and bubbles. In Proc. ESO Workshop on Extranuclear Activity in Galaxies. Eds. R. Fosbury and E. Meurs.
Taniguchi, Y. *et al.*: 1988, *Astron. J.*
Unger, S.W. *et al.*: 1987, *Mon. Not. Roy. Astron. Soc.* **228**, 671
Weedman, D.W.: 1981, *Astrophys. J.* **248**, 105

Dynamics and Interactions of Galaxies Editor: R. Wielen

The Nuclear Gas in the Star-Burst Galaxy NGC 3310

H.-G. Grothues and Th. Schmidt-Kaler

Astronomisches Institut, Ruhr-Universität Bochum, Postfach 102148, D-4630 Bochum 1, Fed. Rep. of Germany

1. Introduction and Observations

NGC 3310 is a typical star-burst galaxy with an extended ring-shaped region of violent star formation surrounding its nucleus. A linear chain of HII-regions in the outskirts of the galaxy at about P.A.300° - interpreted as a tidal tail - and a "ripple" west of the nucleus have been regarded as clues to a merger in NGC 3310 (Balick and Heckman (1981), Schweizer (1983)). The distance of only 12 Mpc and its high surface brightness make the galaxy suitable for long slit spectroscopic investigations with high spectral and spatial resolution.

We obtained 7 photographic spectra (IIa-O with EMI image tube) and one CCD spectrum around Hα and [OIII] 5007 Å at various position angles including the major and the minor axis with the coudé spectrograph of the 2.2 m - telescope at Calar Alto/DSAZ. The spectral resolution corresponds to 34 km s^{-1} at Hα and 27 km s^{-1} at [OIII] 5007 Å. The seeing during the observation varied between 1" and 3".

2. Results and Discussion

For all emission lines with sufficiently high S/N we determined velocity (center of gravity of profile), line width (FWHM), intensity, and an asymmetry parameter A_{20} (the diference of the line centers at 20% and 50% intensities resp., in units of FWHM). Results are:

(1) The <u>heliocentric system velocity</u> of HII (1006±3) km s^{-1} seems to differ from that of HI (977±12) km s^{-1}. The center of rotation does not coincide with the centre of the bright nucleus (distance 100 pc at P.A.155°). The <u>line widths</u> reach their maxima near the center of rotation with FWHM = (200±20) km s^{-1}, while in the surrounding HII ring we find FWHM between 60 and 120 km s^{-1}.

(2) From two major axis spectra with excellent seeing the <u>rotation curve</u> could be reliably traced down to 1".5 = 100 pc. It appears

Dynamics and Interactions of Galaxies Editor: R. Wielen

to be the steepest curve ever found in a galaxy (e.g. Afanasiev et al.(1989)), with $\Delta v_{Rot}/\Delta r \approx 1000$ km s^{-1} kpc^{-1} within r=180 pc and a central mass of about $1x10^9$ $M_\odot$.

(3) The line profiles show significant <u>asymmetries</u> in two areas. Going out from the symmetric and broad profiles at the center of rotation we find exclusively blueward asymmetric profiles to the SE and redward asymmetric profiles to the NW at P.A.$(143\pm13)^\circ$.

We consider three explanations for the pecularities in NGC 3310:

(1) The spatial asymmetry distribution is interpreted as a separate, continuous line component over the nuclear region caused by an additional body (a rotating disk?) in the star-burst region. The alignment of the asymmetry distribution with the "tidal tail" then suggests to identify this body as the debris of a merging dwarf galaxy.

(2) The line asymmetry is caused by ejections from the nucleus, as proposed by Bertola and Sharp (1984), or by a strong galactic wind from the star-burst nucleus as described by McCarthy et al. (1987) for M82 and NGC 253.

(3) The asymmetry is connected to a radiation cone from the nucleus, as proposed by Schulz (1988) for NGC 4151.

3. References

Afanasiev,V.L., Sil'chenko,O.K., Zasov,A.V. (1989), Astron. Astrophys. **213**, L9

Balick,B., Heckman,T.M. (1981), Astron. Astrophys. **96**, 271

Bertola,F., Sharp,N.A. (1984), Mon. Not. R. Astron. Soc. **207**, 47

McCarthy,P.J., Heckman,T.M., van Breugel,W. (1987), Astron. J. **93**, 264

Schulz,H. (1988), Astron. Astrophys. **203**, 233

Schweizer,F. (1983), Internal kinematics and dynamics of galaxies, IAU Symp. **100**, pp.319-329

The Interacting System Mkn 305/306

M. Dietrich and W. Kollatschny

Universitätssternwarte, Geismarlandstraße 11, D-3400 Göttingen, Fed. Rep. of Germany

1. Abstract

Direct images and spectra at different slit positions of the interacting system Mkn 305/306 are discussed. Both galaxies show starburst properties due to tidal interaction. The morphology and velocity structure of Mkn 306 reveals the strongest warp of a stellar disk so far known.

2. Introduction

The galaxies Mkn 305 and Mkn 306 form a double system with 30 arcsec separation and having a common envelope at $m_B \geq 24.5$. Mkn 306 itself was classified as a double nucleus galaxy (Petrosyan et al., 1978). The optical morphology of Mkn 306 shows similarities with radio maps of warped galaxies (Bottema et al., 1987). But in the optical the warp of the stellar component is normally far weaker than in the radio for the HI-gas. The origin for the warp phenomenon is not clear yet: merging, tidal interaction or accretion of intergalactic matter (IGM).

3. Observations

Direct images of the galaxy system were taken with the Calar Alto 2.2m telescope in the B and r band (Fig. 1); low dispersion spectra (240 Å/mm) as well as high dispersion spectra (56 Å/mm) for the study of the velocity field were taken with the Calar Alto 3.5m telescope at different position angles.

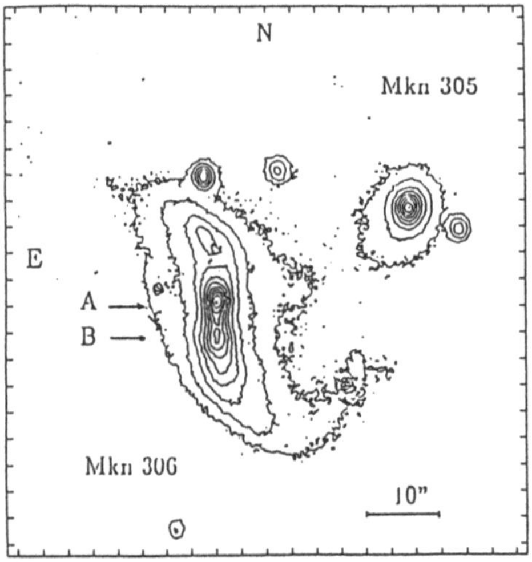

Fig. 1 — Contourplot of the interacting galaxy system

Dynamics and Interactions of Galaxies Editor: R. Wielen

4. Results

Redshifts of z=0.0195 for Mkn 306 (m_B=14.6) and z=0.0194 for Mkn 305 (m_B=16.8) were determined from the emission and absorption lines, leading to a distance of D=77.7 Mpc for the system (H_o=75 km s^{-1} Mpc^{-1}). The two central emission regions of Mkn 306 are separated by 5.6 arcsec corresponding to 2.1 kpc. The projected separation of Mkn 305 and Mkn 306 amounts to 12.4 kpc.

The spectra of Mkn 306A and B (Fig. 2) are typical for starburst activity with respect to their line intensity ratios of I([OIII]5007)/I(Hβ) vs. I([NII]6584)/I(Hα) or I([OI]6300)/I(Hα) (Veilleux and Osterbrock, 1987). The overall spectrum from the radio to the optical range shows a FIR-excess typical for dust re-emission of a strong starburst. The observed spectrum of Mkn 305 can be characterized as a post-starburst spectrum; this is very unusual considering the elliptical morphology.

The velocity field of Mkn 306 was determined parallel (Fig. 3) and perpendicular to the major axis. A comparsion of these velocity measurements shows very good correspondence with the

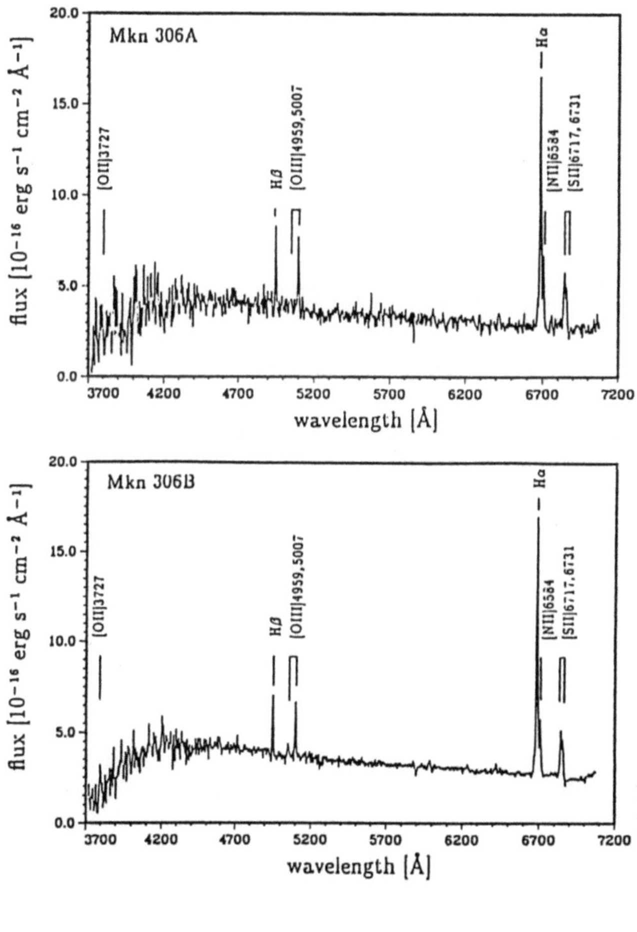

Fig. 2 — Optical spectra of Mkn 306 A, B

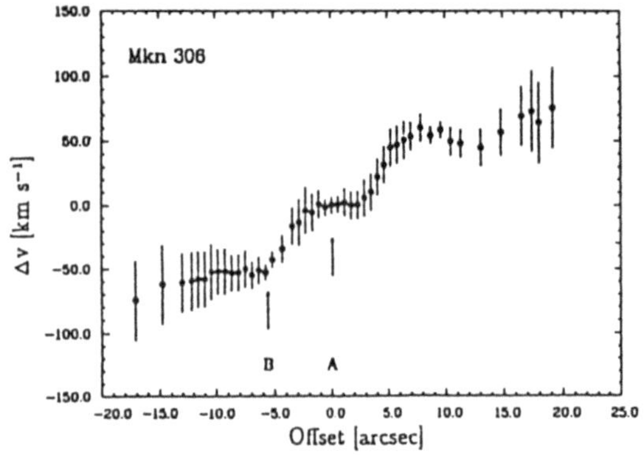

Fig. 3 — Observed velocity curve of Mkn 306 parallel to the major axis

observations and theoretical model calculations of Brinks (1985) for the warped radio velocity field of NGC 224. The warp leads to a linear velocity structure with a smaller gradient in the central region of the galaxy in comparsion to 'normal' velocity gradients; the model calculations prove the existence of the warp and the projection effect simulating the double nucleus.

The optical morphology of Mkn 306 is very similar to the radio morphology of NGC 4013 (Bottema et al., 1987). Both galaxies are shaped like an integral sign and show double nucleus structure due to the strong warp.

The possibility that the double central emission region represents the late stage of an interaction process of two different host galaxies can furthermore be excluded from the physical similarities of the two nuclei: their emission line ratios, the relative abundances, the dust content and even the Hα-luminosity and the deduced number of OB-stars.

References

Bottema, R., Shostak, G.S., van der Kruit, P.C.: Nature, **328**, 401 (1987)
Brinks, E.: PhD, Leiden, 1984
Petrosyan, A.R., Saakyan, K.A., Khachikyan, E.E.: Astrophysics, **14**, 36 (1978)
Veilleux, S., Osterbrock, D.E.: Ap.J.Suppl. **63**, 295 (1987)

Optical Structure of Blue Compact Galaxies

M. Diehl, K.J. Fricke, and H.-H. Loose

Universitätssternwarte, Geismarlandstraße 11, D-3400 Göttingen, Fed. Rep. of Germany

For only a few blue compact dwarf galaxies (BCDG's, Thuan & Martin 1981) detailed optical surface photometry exists (c.f. Kunth *et al.* 1988, Loose & Thuan 1986). Optical surface photometry is an important tool to derive the structure of the underlying galaxies in BCDG's and, together with data from other spectral ranges, to set up constraints for theoretical models (Loose & Fricke 1981) studying the mechanisms of the star formation bursts and chemical evolution of BCDG's.

We here discuss the results of detailed optical CCD-photometry obtained in the B-band of six BCDG's (Mkn 5, 36, 49, 118, 400, and 487) which have been observed using the Calar Alto 2.2m telescope. The exposure times have ranged from 1800s to 5000s. The derived surface brightness profiles have been analyzed in terms of $\mu_B(R^*)$ where $R^* = (A/\pi)^{1/2}$ and A is the area inside the isophote of brightness μ_B. In this way a standard representation of the surface brightness profiles is achieved which allows comparisons between BCDG's of different morphological shapes. For Mkn36 we also have derived the B–R color distribution.

Various functional forms $\mu_B(R^*)$ have been tried to fit the observed profiles. The form which has been found to work at best (Diehl 1988) is a 3-component model (see Fig.1):

$$\mu_B(R^*) = -2.5\log\{I_0\exp[-(R^*/a_0)^2] + I_1\exp[-(R^*/a_1)^n] + I_2\exp[-R^*/a_2]\}$$

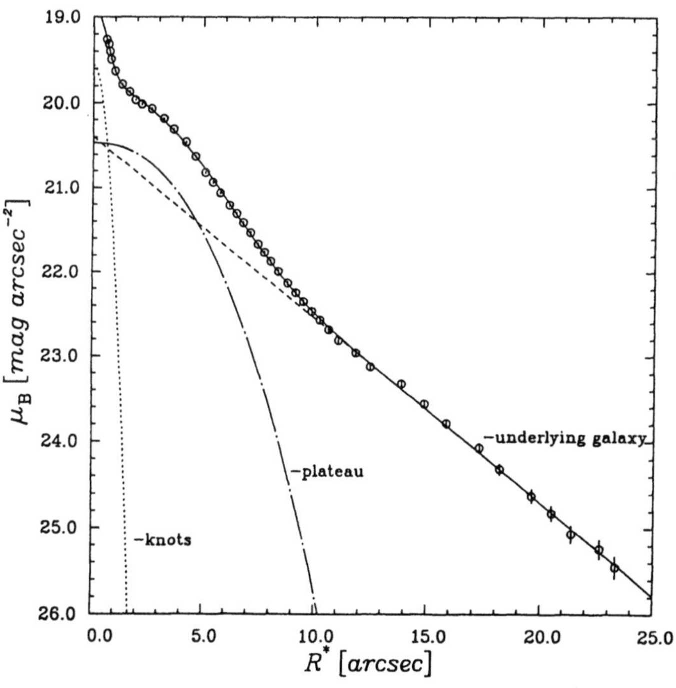

Figure 1.: Surface brightness profile of Mkn400

The first component (index '0') represents the luminosity contained in the bright knots where active star formation takes place. I_0 is the central surface brightness of this component and a_0 a measure of its spatial extend. Only the total brightness of this component is of concern here as the parameters I_0 and a_0 are strongly affected by the seeing conditions.

The second component (index '1'), designated as 'plateau', is strongest in an intermediate radius range. Besides the central luminosity I_1 and the width a_1 of the component a third parameter n enters here. However, the exponent n is correlated to the central brightness $\mu_B(I_1)$ of the plateau. This plateau component is most clearly present in BCDG's of class iE which appears to be the most abundant type (Loose & Thuan 1985). In nE–BCDG's which, in contrast to iE's, only show one central single brightness peak, this component is of low luminosity with a low value of n or is completely invisible.

The third component (index '2') is a simple exponential brightness distribution which dominates the light of the outer parts. This component is believed to represent the underlying galaxies of BCDG's. An exponential law has been shown to fit reasonably well the brightness distributions of dwarf galaxies in the Virgo cluster (Binggeli *et al.* 1984). Here it is of special advantage for it allows to extrapolate and to predict central and total luminosities of the underlying galaxies.

The widths a_1 of the plateaus and the corresponding parameters a_2 of the underlying exponentials are correlated. Expressed in radii at 25 mag arcsec^{-2}, it is found that the radii of the plateaus are ~ 0.5 of the radii of the underlying galaxies. Mkn118 seems to be a disturbed galaxy and does not fit into this scheme. The BCDG Mkn86 (iE, Loose & Thuan 1989 in prep.), however, does and was included into the sample.

If the exponential brightness distributions indeed do represent the underlying galaxies, their total luminosities should be correlated to the total masses of the BCDG's. The star formation activity, which causes the light from the two inner components, should be correlated to the total gas masses of the BCDG's. We have found tendencies in favour of these assumptions. However, the present sample is small and the total masses of the BCDG's are uncertain. More observations are needed to establish the found trends. The overall properties of the underlying galaxies, as derived here from the exponential brightness distributions, are similar to those of low surface brightness dI galaxies which have been discussed by Thuan (1985). BCDG's, therefore, might be active states of dwarf galaxies of this type.

Binggeli, B., Sandage, A., and Tarenghi, M. 1984, A.J. **89**, 64
Diehl, M. 1988, Diploma Thesis, Göttingen
Kunth, D., Maurogordato, S., and Vigroux, L. 1988, A&A **204**, 10
Loose, H.-H. and Thuan, T.X. 1986, Ap.J. **309**, 59
Loose, H.-H. and Thuan, T.X. 1985, in *Star-Forming Dwarf Galaxies* ed. D. Kunth, T.X. Thuan, and J.T.T. Van (Gif sur Yvette: Editions Frontières), p.73
Loose, H.-H. and Fricke, K.J. 1981, in *The Most Massive Stars* ed. S. D'Odorico, D. Baade, and K. Kjär, ESO Workshop, p.269
Thuan, T.X. 1985, Ap.J. **299**, 881
Thuan, T.X. and Martin, G.E. 1981, Ap.J. **247**, 823

Kinematics and Starbursts in Arp 299

H. Hippelein and K. Bernlöhr

Max-Planck-Institut für Astronomie, Königstuhl,
D-6900 Heidelberg, Fed. Rep. of Germany

Morphology

Arp 299 (Mrk 171) is an interacting pair of galaxies — NGC 3690 (west) and IC 694 (east). In the optical both galaxies appear very irregular, dominated by several bright knots with typical diameters of 2–3″ (400–600 pc at a distance of 41 Mpc). In the infrared and radio there are three dominant knots, designated as A, B, and C (Gehrz, Sramek, and Weedman 1983). While source A coincides with the center of IC 694 (see Fig. 1), sources B and C are very bright extranuclear H II regions south and north of the kinematical center of NGC 3690. The peak in the R band is 2.″5 northwest of source B and coincides with source B_2 in high resolution K band images (Eales et al. 1987). It is approximately at the kinematical center of NGC 3690.

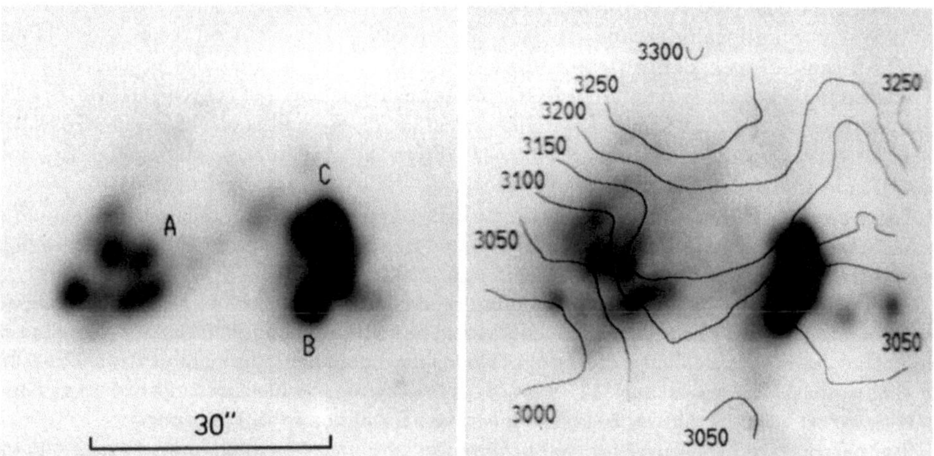

Figure 1: Left: Arp 299 in the light of Hα + [N,II] (continuum subtracted).
Right: Isovelocity curves [km s^{-1}] superimposed on an R band image.

Kinematics

High-resolution long-slit spectra at Hα (FWHM ≈ 90 km s^{-1}) were taken with a CCD camera at the 3.5 m telescope on Calar Alto. We obtained a radial velocity map with almost complete coverage of the system. In contrast to their irregular appearance NGC 3690 and IC 694 show little disturbance on the radial velocity map. NGC 3690 shows a smooth rotation curve of a spiral galaxy while IC 694 may be a barred spiral. The system velocities $cz_\odot$ are 3070 km s^{-1}

Dynamics and Interactions of Galaxies Editor: R. Wielen

(NGC 3690) and 3110 km s^{-1} (IC 694). There is no indication of a direct collision of the two galaxies (as suggested by Augarde and Lequeux 1985). The only obvious sign of interaction is the tail northwest of IC 694.

A discontinuity of the radial velocity close to the center of IC 694 as reported by Augarde et al. (1988) is not visible in our spectra. The linear part of the rotation curve of IC 694 is, however, steep and the extinction of the line emitting region is large ($A_V \approx 2.0$ from Balmer line ratios).

In the most prominent H II region (source C) the intrinisc line width is about 100 km s^{-1} (FWHM) with a fainter, broad component of 300 km s^{-1} width offset to the blue by 80 km s^{-1} and indicating turbulent motion of the gas.

Starburst(s)

Low resolution spectra covering the wavelength range 3650–7000 Å at a FWHM of 12Å show that the bright parts in Arp 299 are generally dominated by hot massive stars and strong line emission typical of star forming regions. An exception is the bright spot in the R band (source B_2) with moderately strong emission lines and Balmer lines beyond Hγ in absorption. This is the only place in the inner part of Arp 299 where an old population contributes significantly to the continuum. The colour of the continuum changes rapidly from blue 1″ north of B_2 to red 1″ south ($\Delta(B-R) = 1.6$ mag). Another exception is a bright spot 14″ west of source B where we find only hot stars with an average spectral type B2 but little emission.

Emission lines typical for shocked gas like [O II] λ 3727, [O I] $\lambda\lambda$ 6300,6363, and [N I] λ 5199 as well as [S II] $\lambda\lambda$ 6716,6731 are strong outside the very luminous emission regions, especially inbetween NGC 3690 and IC 694. Shock excitation in Arp 299 was also suggested by Fisher et al. (1983) for vibrationally excited H_2 lines. Hβ / [O III] is largest in source B which is also peculiar because of its high dust temperature seen in the mean infrared (Joy et al. 1989).

The extinction of emission lines in NGC 3690 is negligible north of source C ($H\alpha/H\beta = 2.9$) and rises from $A_V = 0.7$ in source C to $A_V \approx 2.2$ in source B_2 and south of it. In IC 694 it is about 1.2 except near source A where $A_V = 2.0$. From the near infrared colours Telesco et al. (1985) found an extinction for source A of $A_V = 4.6$.

There are no signs of an active galactic nucleus at source A. So far, the only indication of an AGN are the large ratios of radio to far infrared flux, and to Brα and Brγ fluxes as well as the spectral index of the radio source (Gehrz et al., 1983).

Although the overall rotation curve is quite orderly the disturbance of gas in both galaxies due to the interaction is large enough for shocks and line widths of about 200 km s^{-1}. Dissipation and cooling leads to the collapse of large gas clouds and consequently to giant extranuclear star forming regions. The bright knot 14″ west of source B seems to have lost most of its gas on a short time scale while in the other knots violent star formation is still going on.

The near infrared emission observed in the most prominent of these knots must be due to red super giants. We expect that the contribution of the old underlying population is small even in the near infrared. If the compact radio source in the center of IC 694 is an AGN it has a much larger extinction than the starburst region there or it must be very faint in the optical.

References

Augarde, R. and Lequeux, J. 1985, *Astr. Ap.*, **147**, 273

Augarde et al. 1988, *Astr. Ap. Suppl.*, **74**, 161

Eales et al. 1987, in *Infrared Astronomy with Arrays*, eds. C. G. Wynn-Williams, E. E. Becklin, and L. H. Good (Honolulu: University of Hawaii), p. 345

Fisher, J. et al. 1983, *Ap. J. (Letters)*, **273**, L27

Gehrz, R. D., Sramek, R. A., and Weedman, D. W. 1983, *Ap. J.*, **267**, 551

Joy, M. et al. 1989, *Ap. J.*, **339**, 100

Telesco, C. M., Decher, R., and Gatley, I. 1985, *Ap. J.*, **299**, 896

What Do We Learn from Fitting Models to Starbursts?

K. Bernlöhr

Max-Planck-Institut für Astronomie, Königstuhl,
D-6900 Heidelberg, Fed. Rep. of Germany

Introduction. Most previous starburst models try to set limits on the ages and/or lower mass cutoffs of active starbursts using mainly or exclusively the observed M/L ratio. For model fits information about the stellar population of the starburst is required. Optical spectra can give us that information provided they cover a large enough wavelength range, and the stellar continuum is not entirely dominated by stars in front of an obscured starburst.

Observations of starbursts in interacting galaxies were made with the Calar Alto 3.5m telescope. These include imaging with broad band filters and in Hα, and spectroscopy in the wavelength range 3600 to 7000 Å. Our starburst models produce total luminosities, B, V, R, and I band magnitudes, and optical spectra with stellar continuum and line emission. These models are fitted to the observations and to IRAS far-infrared data.

The models. Different sets of stellar evolutionary models by Maeder and Meynet (1988, 1989), Maeder (1976, 1981), and Matraka et al. (1981) are used for the mass range 0.85 to 60 $M_\odot$. Stars of lower masses are kept unevolved. Model stars are classified according to their spectral type (O4 to M7) and luminosity class (V, III, or I, respectively). 10^4 to 10^5 stars are summed up according to the initial mass function (IMF) and the star formation rate (SFR) of the starburst model. A synthesized starburst model spectrum is generated with flux-calibrated and dereddened spectra from the Kitt Peak library of stellar spectra (Jacoby et al., 1984). Balmer emission line fluxes are calculated for a black body spectrum of each star and case B recombination. Emission lines of He I, [N II], [O II], [O III], and [S II] are added. The ratios of all line fluxes relative to Hα flux are kept constant at values typical for starbursts. Standard interstellar extinction (foreground and stars imbedded in the dust) and a varying contribution of an independently processed underlying old population complete the model spectrum.

Models of individual interacting galaxies. The potential and the drawbacks of this method are illustrated with two pairs of interacting galaxies. These are part of a larger sample to which the model fits were applied.

Arp 284 (NGC 7714/7715): NGC 7714 is a starburst galaxy with little extinction and strong emission lines (see Fig. 1). The models indicate that the starburst started 2 to $4 \cdot 10^7$ years ago. The old stellar population contributes 40% of the light in the V band. The history of the SFR is made up of a low constant SFR of the old population from $1.3 \cdot 10^{10}$ years in the past to the beginning of the starburst, and a high constant SFR of the starburst from its beginning to now. For a solar neighbourhood IMF (Scalo 1986) and a starburst age of $3 \cdot 10^7$ years, the mass converted to stars amounts to 5% of $2 \cdot 10^{10}\,M_\odot$, the dynamical mass inside a radius of 12″ (3.3 kpc, $H_0 = 50\,\mathrm{km\,s^{-1}\,Mpc^{-1}}$). — NGC 7715 has very little extinction and very blue colours. The spectrum is dominated by late B and early A stars throughout the galaxy. There are no emission lines at all. This is either a post-starburst galaxy where star formation ceased 4 to $7 \cdot 10^7$ years ago (depending on the set of stellar evolutionary models), or a starburst galaxy with an *upper* mass cutoff of $5\,M_\odot$. The old stellar population in our models contributes only about 15% of the light in the V band. A *lower* mass cutoff of 1 to $2\,M_\odot$ is necessary if the starburst population has less than 10% of the mass of the old population.

Dynamics and Interactions of Galaxies Editor: R. Wielen

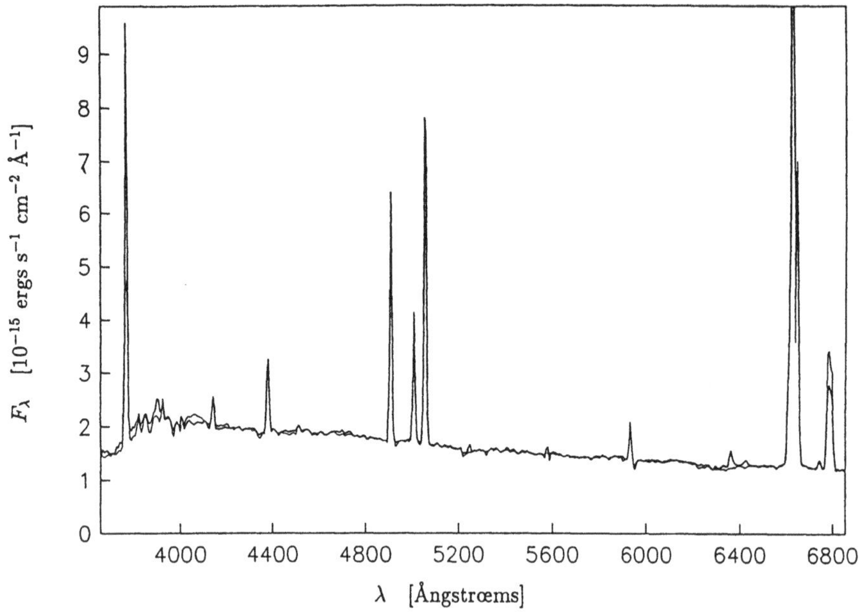

Fig. 1: Spectrum of the center of NGC 7714 (thick line) and synthesized model spectrum (thin line).

III Zw 125 (UGC 12914/12915): UGC 12914 is slightly disturbed but doesn't show a significantly enhanced SFR in its central region. However, the star formation rate in the tidal arm north of the nucleus is enhanced. Assuming a solar neighbourhood IMF our fits for this arm show that about half of the light in the V band is from an old population. A new population with decreasing SFR gives the best fit to the observed spectrum. — UGC 12915 has an extended but heavily obscured starburst. In this case the fits are very insensitive to details of the starburst population because the stellar continuum is dominated by stars in front of the starburst region.

Conclusions. The ambiguity between effects of the IMF and the SFR can not be completely resolved. The determinations of a lower mass cutoff in the IMF if starbursts predominantly form high mass stars are improved but remain indirect. The stellar compositions and ages of starbursts are estimated. In interacting pairs with an active starburst in one galaxy and a former starburst in the other we determined the time delay between the two starbursts. In the case of Arp 284 and a few other systems we have studied this time delay is of the order of $3 \cdot 10^7$ to 10^8 years.

Acknowledgements. The observations were made in collaboration with J. W. Fried.

References

Jacoby H. J., Hunter, D. A., Christian, C. A. 1984, *Ap. J. Sup.* **56**, 257
Maeder, A. 1976, *Astr. Ap.* **47**, 389
Maeder, A. 1981, *Astr. Ap.* **99**, 97
Maeder, A. 1981, *Astr. Ap.* **102**, 401
Maeder, A., and Meynet, G. 1988, *Astr. Ap. Sup.* **76**, 411
Maeder, A., and Meynet, G. 1989, *Astr. Ap.* **210**, 155
Matraka, B., Wassermann, C., and Weigert, A. 1982, *Astr. Ap.* **107**, 283
Scalo, J. M. 1986, *Fund. Cos. Phys.* **11**, 1

Near-Infrared Imaging of Interacting Galaxies

H.A. Bushouse[1] *and M.W. Werner*[2]

[1]Department of Physics and Astronomy, Northwestern University, Evanston, IL 60208, USA

[2]Space Sciences Division, NASA-Ames Research Center, MS 245-6, Moffett Field, CA 94035, USA

1. Introduction

Galaxy-galaxy interactions are known to cause dramatic changes in morphology and at the same time often induce large bursts of star formation activity. While optical images are sufficient to reveal the overall morphological peculiarities of these systems, they suffer from several problems, the greatest of which is obscuration by dust within the galaxies. Furthermore, small-scale structure seen in optical images is often dominated by the light of young and intermediate-age stars. With the advent of two-dimensional array detectors that operate in the near-infrared window, we can for the first time obtain high resolution images that escape the domination of young stars and lift the veil of dust in these galaxies. We have obtained near-infrared J and K-band images for 25 interacting galaxies at a resolution of 0.75 arcsec/pixel using the Kitt Peak National Observatory 58x62 element InSb infrared array camera. The galaxies were chosen from the larger sample compiled by Bushouse (1986), which has been studied in detail optically and in the far-infrared (Bushouse 1987; Bushouse, Lamb and Werner 1988).

2. Results

The near-infrared images have been analyzed and compared with optical R and Hα images in order to determine the degree of correlation between the optical and near-infrared morphologies of the galaxies. In general the large-scale near-infrared morphology is quite similar to the R-band morphology. Vast differences are often seen, however, in the small-scale structure. The near-infrared images often reveal features in the galaxies that are hidden in optical images due to the obscuring effect of dust. In most systems this means getting a clearer view of the near-nuclear regions of the galaxies where the effects of dust appear to be the most severe. In some cases the near-infrared images reveal the location of nuclei that are not visible at all in optical images. The inner disk and nuclear regions of the galaxies are also characterized by very smooth and regular infrared isophotes, indicating that the spatial organization of the underlying stellar population in these regions has apparently not been severely distorted by the tidal forces produced in the interaction. In general there is little correlation between the Hα and near-infrared morphologies of the galaxies that we have observed. Only the largest and brightest HII complexes appear to have infrared counterparts, presumably due to the presence of large numbers of red supergiant stars.

NGC 520, NGC 4038/9 and UGC 12915 are examples of systems in which the nuclei are either indistinct or completely invisible in optical images, but near-infrared images of these systems clearly show the locations of the nuclei. In Figure 1 we show contour plots of the R and K-band images of UGC 12915 (also featured on the conference poster) which is seen almost edge-on and has a massive dust lane running along the plane of the disk. The R-band morphology of this galaxy is quite chaotic because of the dust, while the K-band image clearly shows the nucleus of the galaxy.

By combining the near-infrared and optical images we have also been able to examine the spatial color distributions within the galaxies. Virtually all of the galaxies observed show a

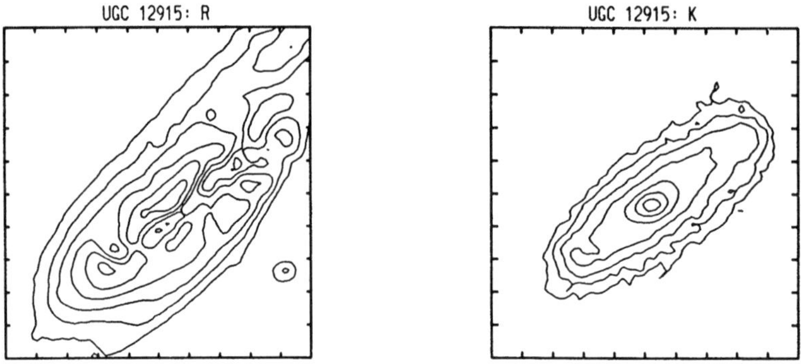

Figure 1. Contour plots of R and K-band images of UGC 12915. Contour levels are in logarithmic units of flux density.

more or less smooth radial color gradient, with the nuclei having redder J−K and R−K colors than the disk regions. The average nuclear vs. disk color differences are $\Delta(J-K)\sim0.4$ and $\Delta(R-K)\sim0.8$. A comparison of the relative scatter of the nuclear and disk colors also indicates that the disk colors are much more uniform from galaxy to galaxy than are the nuclear colors. Since these colors have been seen to correlate only very weakly with star formation activity or stellar population mix in these galaxies (Bushouse 1989) we conclude that the observed color gradients are due primarily to an increased amount of dust in the nuclear regions of the galaxies. Furthermore, the range in nuclear J−K colors is similar to that seen in the dusty regions of nearby spiral galaxies. The observed color changes imply an average increase in A_V of ~2 mag in the nuclear regions. Small-scale color changes in and around the nuclear regions of the galaxies also imply local variations in A_V from 0.5 to 2.0 mag.

3. References

Bushouse, H. A. 1986, Ph.D. thesis, University of Illinois.
———. 1987, *Ap.J.*, **320**, 49.
———. 1989, *A.J.*, submitted.
Bushouse, H. A., Lamb, S. A. and Werner, M. W. 1988, *Ap.J.*, **335**, 74.

The Role of Galactic Environment in Photometric Evolution

P. Traat

Tartu Astrophysical Observatory, SU-202444 Toravere, Estonia, USSR

Published models for photometric evolution of galaxies (e. g. Tinsley 1972, 1980; Traat and Einasto 1979; Bruzual 1983) handle the simpliest case, considering galaxies to be isolated and closed systems, not interacting with their environment and without any mass exchange between galaxies and ambient medium. In a number of cases, however, such an approach seems to be oversimplified. There are reasons to suppose that gas infall or outflow could play an important role during the early evolution of all galaxies. A well-known observational fact, supporting this point of view, is the presence in clusters of galaxies of large amounts of partially enriched gas, which has been ejected by or stripped out from cluster members. In most field galaxies the present accretion rates are thought to be negligible, but even among them a number of galaxies exists, where things can be drastically different (galaxies associated with huge gas clouds or having close neighbours).

To deal with such situations, we have developed evolutionary models for galaxies, in which the assumption of isolatedness and closedness is dropped. These models along with the classic ones have been in more detail reviewed in our recent paper (Traat 1988). As to interacting models, two most significant and obvious cases when the influence with environment must be taken into account, have been considered:

1) the tidal interaction with a neighbouring galaxy with a relatively short intense star-formation burst as its result and

2) a long-term (comparable to age of galaxies) infall or outflow of intergalactic matter into/from the galaxies.

In the first case the integrated colors of a galaxy start after the inset of extra star formation evolve very fast to blue side. The bluest colors are reached at the final moment of the burst in the case of constant SFR or, if it has a very long duration and a low intensity, some time earlier. When the SFR differs from constant, most blue colors occur after its maximum, if the burst is short, but an increase in burst length makes them first to coincide with the maximal SFR and then to precede it. When the burst is over, the colours will slowly turn redder and the original colours will be approximately (but not exactly) restored after about 1 Gy. So, in this case the path of a galaxy is an almost closed loop, its exact size and form depending on parameters of the burst.

In the case of the exchange of matter between galaxies and their environment the models reveal that depending on the rate of gas infall/outflow, all models, regardless of internal SFR fixed by initial conditions in galaxy, are capable to reproduce the whole interval of observed colours for normal galaxies. It means that the selection of a concrete model will be quite ambiguous when there are reasons for suspection that the object under consideration had been subjected to long-time interactions and mass exchange with its neighbourhood. To reconstruct the true picture of system's photometric evolution, in these occasions first the history of accretion in the system has to be postulated or restored from other data, and only after that a suitable accretional evolutionary model can be applied.

The results of these computations show very clearly, that assumptions, involved in evolutionary computations either in explicit or implicit form and used for the specification of the SFR, among them the presence or absence of interactions with the galactic environment, are able to change the evolutionary history of the photometry of galaxies most radically.

References

Bruzual, A. G. 1983, *Astrophys. J.*, 273, 105.

Tinsley, B. M. 1972, *Astron. Astrophys.*, 20, 383.

Tinsley, B. M. 1980, *Fundam. Cosm. Phys.*, 5, 287.

Traat, P., Einasto, J. 1979, *Publ. Tartu Astrophys. Obs.*, 47, 140.

Traat, P. 1988, in: The Formation and Evolution of Galaxies and Their Constituents. I, *Tartu Astrofüüs. Obs. Teated* No. 91, 23.

Bulge Morphology of Non-Interacting Starburst Galaxies

A.M. Varela[1], *J.E. Beckman*[1], *J. Cepa*[1,2], *C. Muñoz-Tuñon*[1,3], *M. Prieto*[1], *and J.M. Vilchez*[1,2]

[1]Instituto de Astrofísica de Canarias
[2]Observatorio del Roque de los Muchachos, La Palma
[3]Departamento de Física Fundamental, Universidad de la Laguna

If a starburst occurs in the circumnuclear zone of a non-interacting galaxy, a consequence is to sweep out the interstellar gas in a galactic wind (J.E. Beckman [1], C. Muñoz-Tuñón and J.E. Beckman [2]). Repeated starbursts can occur if there is a mechanism for refuelling the nuclear zone. Following Zaritsky and Lo [3], we are exploring the suggestion that departures from bulge axisymmetry can lead to this type of inflow.

We show, in NGC 4736, that the bulge indicates strong and radially variable ellipticity, evidence of star formation, and outflow, while the bulge of the non-starburst galaxy NGC 5033 gives no evidence of peculiar or variable ellipticity.

1. Observational results

We present a preliminary analysis of CCD images of the two spiral galaxies NGC 4736 and NGC 5033 obtained with the TAURUS imaging camera and GEC CCD detector at the Cassegrain focus of the 4.2 m William Herschel Telescope, at the Observatorio del Roque de los Muchachos, under excellent seeing ($\simeq$ 0.8 arcsec) and transparency conditions.

Figure 1 shows the quality of the data in the form of isophotes in B and I bands. Figure 2 shows the run of ellipticity with radius along the major-axis, and the position angles of the ellipses, while Figure 3 shows the equivalent quantities for a non-starburst spiral, NGC 5033.

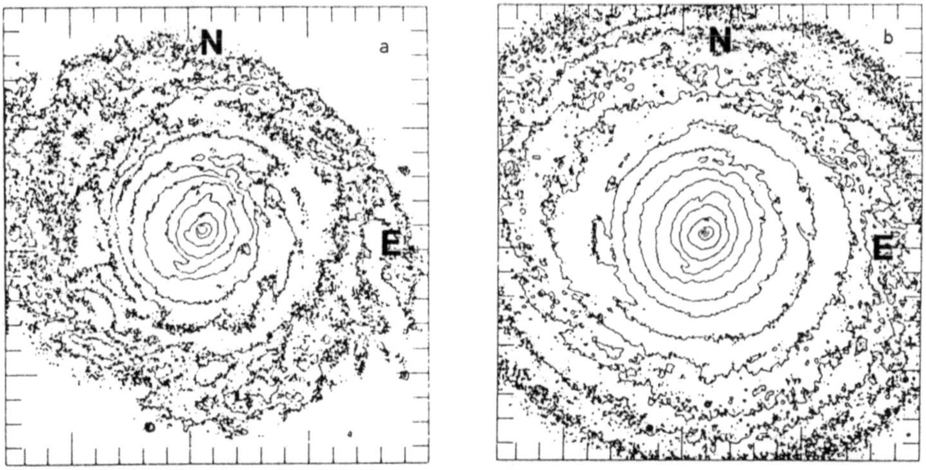

Fig. 1: Isophotes of our CCD frames of NGC 4736: (a) In the B band ; (b) In the I band. The outer contours correspond to 19.59 and 19.39 mag/arcsec^2 respectively and the spacing used is 0.5 mag

Dynamics and Interactions of Galaxies Editor: R. Wielen

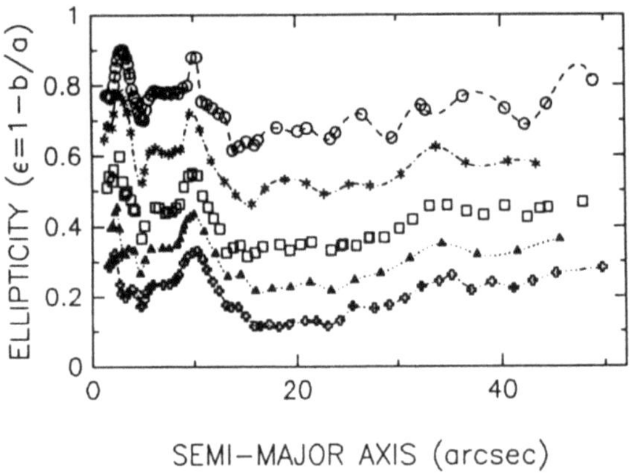

Fig. 2a: Ellipticities vs. semi-major axis for NGC 4736 (bulge + inner disc). Displacements of bands: B: .60 units; V: .45 units; R: .30 units; I: .20 units; Z: .10 units. (1 arcsec = 27 pc at the distance of the galaxy)

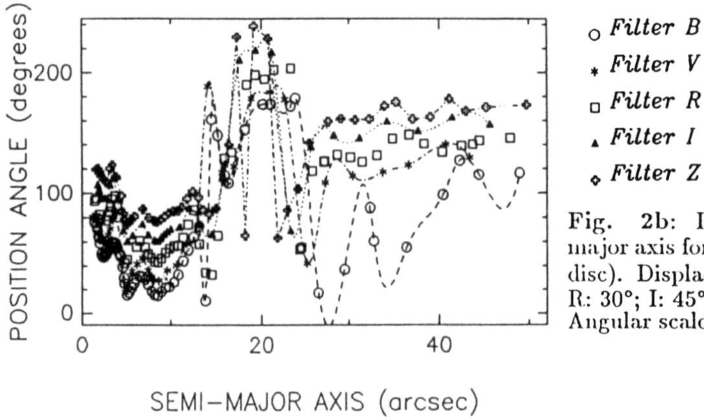

Fig. 2b: Position angles vs. semi-major axis for NGC 4736 (bulge + inner disc). Displacements of bands: V: 15°; R: 30°; I: 45°; Z: 60°; B is undisplaced. Angular scale is given in Fig. 2a

The contrast is obvious: in NGC 5033 the ellipticity runs smoothly up to the disc value, as would be expected for a projected sphere or a flattened spheroid, symmetrical to the galaxy's rotation axis, whereas in NGC 4736 the ellipticity varies substantially with radius, rising to values of up to 0.3 units and then falling sharply to a nearby face-on disc value of less than 0.1 units. The amplitudes of the variations increase from Z to V, as expected from non-axisymmetric *hot spots* of star formation, then decline slightly at B due to dust extinction. The position angles vary in phase throughout the bulge, then scatter in the near circularity of the inner disc, settling down again in the outer disc. Here again the B curve shows strong dust effects.

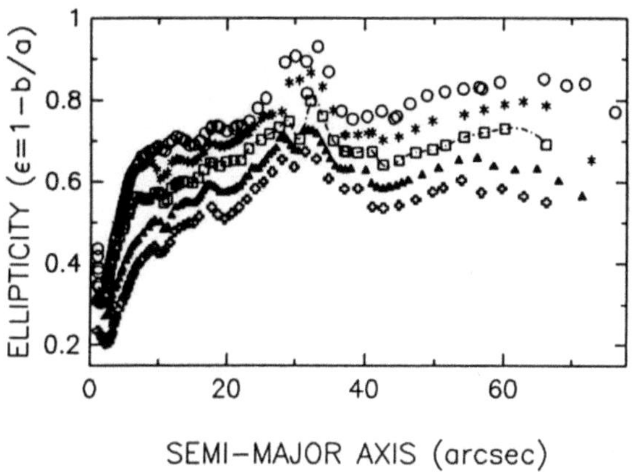

Fig. 3a: Ellipticities vs. semi-major axis for NGC 5033 in 5 bands. Displacements of bands: B: .20 units; V: .15 units; R: .10 units. I: .05 units: Z is undisplaced. (1 arcsec = 78 pc at the distance of the galaxy)

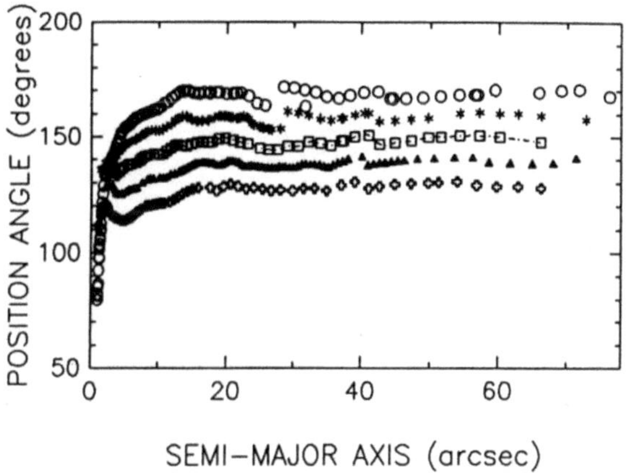

Fig. 3b: Position angles for NGC 5033 in 5 bands. Displacements of bands: B: is undisplaced: V: −10°; R: −20°; I: −30°; Z: −40°. Angular scale is given in Fig. 3a

2. Conclusions

Apart from the evident triaxiality of the bulge, we see evidence of populations of different ages, *organized* into the ovaloid structure. It will be possible to quantify these colour data to devise an age structure, which can be compared with the models needed to produce the ring of HII regions and radial outflow region seen at the edge of the bulge (Van der Kruit [4], [5]). This study forms a part of an ongoing paradigm investigation, linking circumnuclear starbursts with LINERs.

References

1 J. E. Beckman: *"Mass Outflows from Stars and Galactic Nuclei"*, in L. Bianchi and R. Gilmozze eds., Kluwer, p. 385 (1988)
2 C. Muñoz-Tuñón, J. E. Beckman: *Ap. Sp. Sci.* **147**, 173 (1988)
3 D. Zaritsky and K. Y. Lo: *Ap. J.*, **303**, 66 (1986)
4 P. C. Van der Kruit: *Ap. J.*, **188**, 3 (1974)
5 P. C. Van der Kruit: *Astr. Ap.*, **62**, 85 (1976)

Part VI

Numerical Experiments on Interacting Galaxies and Mergers

N-Body Studies of Major Mergers

J. Barnes

Canadian Institute for Theoretical Astrophysics, University of Toronto,
60 St. George Street, Toronto, Ontario M5S 1A1, Canada

Self-consistent numerical models of interacting disk galaxies are now providing detailed answers to some key dynamical questions. Here I present recent results on mergers of equal-mass disk/halo galaxies. Such collisions produce remnants with many of the properties of ordinary elliptical galaxies. In particular, the remnants exhibit de Vaucouleurs' luminosity profiles, triaxial shapes supported by anisotropic velocities, and little or no rotation. Models including gas dynamics show how a merging encounter can funnel large amounts of gas to the remnant center, fueling a starburst or an active galactic nucleus.

1. Mergers of Spherical Systems

The first self-consistent studies of merging galaxies focused on collisions between spherical, single-component systems (Toomre 1977; van Albada & van Gorkom 1977; White 1978, 1979, 1980; Roos & Norman 1979; Miller & Smith 1980, Villumsen 1982, 1983). As White (1978) pointed out, these experiments are best thought of as modeling mergers between dark halos, which presumably dominate the mass and long-term dynamics of interacting galaxies.

Perhaps most importantly, simulations of parabolic collisions, in which two equal-mass galaxies encounter each other with the velocity they would acquire falling from rest at infinity, supported earlier analytic predictions of very rapid merging (Alladin 1965). For example, in encounters with a pericentric separation R_p smaller than the half-mass radius R_h of the infalling galaxies, merging typically resulted within only one orbital period (White 1978, 1979).

The simulations also gave insight into the collective mechanisms which transfer energy from the relative motion of the galaxies to the internal degrees of freedom in each galaxy. The strongest coupling always involves those particles which orbit within their galaxy in the same direction that the two galaxies orbit each other and hence are roughly in resonance (White 1979). In a merger, the most tightly bound particles in the initial galaxies tend to become yet more tightly bound in the final remnant, while the most loosely bound particles go on to form an extended envelope and an expanding cloud of escapers. As a result, merger remnants are not homologous with their predecessors, but do tend to possess similar radial population gradients (White 1978, 1980).

Alternatively, these older experiments could be interpreted as depicting mergers between luminous galaxies. The projected surface "brightness" profiles of the resulting remnants fit reasonably well to de Vaucouleurs' law, a result which "augers well for any scheme in which elliptical galaxies are built up by ... merging" (White 1979). However, the dynamical structures of these merger remnants were inconsistent with bright elliptical galaxies: collisions with $R_p \gtrsim 0.5R_h$ typically resulted in oblate remnants supported largely by rotation (White 1979).

2. Early Studies of Disk Mergers

Closely following the studies of encounters between spherical systems, several workers tried to model collisions between disk galaxies self-consistently (Gerhard 1981; Farouki & Shapiro 1982; Quinn 1982; Negroponte & White 1983; see also White 1983). These experiments proved much

Dynamics and Interactions of Galaxies Editor: R. Wielen

more difficult, essentially because self-gravitating disks are more fragile than spherical systems. A rotating disk of stars contains a "swing amplifier" (Toomre 1981) which can boost the ambient level of fluctuations by factors of 10 or more. When a disk is modeled by only a few hundred particles, $\sqrt{N}$ fluctuations are amplified to the point where they can largely ruin the disk in only a few dynamical times. Despite this serious problem, the early simulations yielded tantalizing results concerning mergers of disk galaxies.

Except for Quinn (1982), who worked with "bare" disks stabilized by softening, these studies employed composite galaxy models, each consisting of a thin, exponential, rotating disk embedded in a spheroidal, isotropic halo. These halos were truncated at a few disk scale lengths, and disk-to-halo mass ratios of order unity were used, providing reasonably flat rotation curves and some stability against bar formation, but falling far short of the amounts of dark matter hypothesized in other contexts.

Encounters of disk/halo galaxy models corroborated the rapid merging previously seen with spherical systems. Prograde collisions, in which the spin and orbital angular momenta are parallel, produced tidal tails similar to those in restricted three-body simulations (*e.g.* Toomre & Toomre 1972) and lead to the fastest mergers (White 1979). Retrograde passages were less dramatic and noticeably slower to merge, presumably because of the lack of resonant coupling.

The merger remnants produced in these experiments had surface brightness profiles again quite similar to elliptical galaxies. Space density profiles approximating $\rho \propto r^{-3}$ were found for both the disk and halo components, although the former was more centrally condensed than the latter (Gerhard 1981). Farouki & Shapiro (1982) noted that the luminous component could become more tightly bound during a merger at the expense of the halo, and that this effect would violate the homology argument used by Ostriker (1980) to assert that merging of disk galaxies could not produce ellipticals.

The shapes and dynamical structures of these remnants were considerably more diverse than those produced in mergers of spherical systems. While oblate or oblate-triaxial shapes predominated, nearly head-on collisions were found to produce prolate triaxial remnants with little or no figure rotation (White 1983). Prograde encounters typically resulted in flattened, rapidly rotating systems, while mergers of inclined disks tended to produce rounder remnants. By colliding a remnant with an additional disk, Gerhard (1983) was even able to produce a stable system with a prolate center and an oblate envelope. But to obtain remnants consistent with the observed $v/\sigma - \varepsilon$ distribution of elliptical galaxies, it proved necessary to restrict encounters to $R_p \lesssim 0.75 R_h$ (Negroponte & White 1983).

Mergers including an approximation to the dynamics of the interstellar medium were presented by Negroponte & White (1983), using "sticky particles" which dissipate some of their kinetic energy when they collide. In these simulations typically $\sim 40\%$ of the gas particles collected at the center of the merger remnant. This outcome was particularly interesting in light of the nuclear gas disk found in NGC 7252 (Schweizer 1982).

3. A Merger Survey

The swing-amplified $\sqrt{N}$ noise present in a cool, shearing disk of particles can best be suppressed by increasing the value of N. Direct-sum N-body codes were nonetheless employed for most of the work described above, simply because the advantages of a completely general-purpose method outweighed the $O(N^2)$ cost in CPU time. Hierarchical N-body codes (*e.g.* Barnes & Hut 1989 and references therein) offer the same generality at a cost of only $O(N \log N)$. Used in conjunction with the faster computing hardware now available, these codes have permitted N to be increased by two orders of magnitude. Some early results from such experiments were presented by Barnes (1988); a preliminary report on a more ambitious study appears below.

3.1. Bulge/Disk/Halo Galaxy Models

As Negroponte & White (1983) wryly observed, "both care and uncomfortable compromise are required to construct relatively stable galaxy models for use in merger simulations". The models employed in the present simulations have three components: a central bulge, an exponential disk, and an extended halo. All three components are represented by N-body systems with thousands of particles each.

Setting up the initial data for such bulge/disk/halo models is a complex process (Barnes 1988). The bulge and halo are nested $W_0 = 3$ King (1966) models, scaled to have the same velocity dispersion. The disk, with an $e^{-\alpha r}$ radial profile and a $\mathrm{sech}^2(z/z_0)$ vertical profile with z_0 independent of r, is imposed adiabatically on the combined spheroid. Particle velocities in the disk are then set by fixing the ratio of radial-to-vertical velocity dispersion $\sigma_r/\sigma_z = 2.0$, typical of the solar neighborhood, and the Jeans equations (Binney & Tremaine 1987, § 4.2) are used to correct the rotation curve for pressure support.

In the model galaxies used for most of these calculations, the mass ratio of the bulge, disk, and halo components is 1:3:16, implying a luminous-to-dark ratio of 1:4. In arbitrary units with $G = 1$, this model has a total mass of 1.25 and a binding energy of -1.38. The disk has an inverse exponential scale of $\alpha = 12.0$ and a vertical scale of $z_0 = 0.005$, implying a value of ~ 1.4 for Toomre's Q parameter. Scaling one of these disks to the Milky Way, the units of length, time, and mass are roughly 40 kpc, 250 Myr, and 2.2×10^{11} $M_\odot$, respectively.

To check the stability of these models, tests were run using a modified hierarchical code (Barnes 1989) with accuracy parameter $\theta = 0.7$, time-step $\Delta t = 1/128$, and force softening parameter $\epsilon = 0.015$. As the models evolve, transient spiral patterns appear in the disk component; these heat the disk, increasing the radial and vertical dispersions, and causing individual particles to drift in binding energy. Measured heating rates scale roughly as $N^{-1/2}$, consistent with swing-amplified $\sqrt{N}$ noise. In a test with $N = 32768$ particles (4096, 12288, and 16384 in the bulge, disk, and halo, respectively), the mean radial dispersion σ_r grew by $\sim 30\%$ over 4 rotation periods, while the median change in the binding energy of individual disk particles was $\lesssim 10\%$.

3.2. Initial Conditions for Merger Simulations

This work is limited to parabolic encounters, in keeping with the idea that most interacting galaxies are on very eccentric orbits and are only now experiencing their first collision (Toomre & Toomre 1972). Closeness of approach is parametrized by the pericentric separation R_p which the two galaxies would reach if they maintained their initially Keplerian orbits. Most of the encounters studied so far have had $R_p = 0.2$ or 0.4. For comparison, the overall half-mass radius of our standard galaxy model is $R_h \simeq 0.25$.

Given the orbits, four angles are needed to specify the orientations of the two incoming disks. Barnes (1988) studied inclined symmetric encounters, in which both disks had the same relative orientations. The present work focuses on asymmetric encounters, attempting to span the space of orientations in a sparse but uniform manner. Table 1 lists inclinations i_1, i_2 and pericentric arguments ω_1, ω_2 for four pairs of disks; definitions of the angles i and ω are given in fig. 6 of Toomre & Toomre (1972). Also tabulated are results obtained from the $R_p = 0.4$ encounters, to be further discussed below. Most of these simulations employ $N = 32768$ particles; a few were rerun with twice this number.

Table 1. Parabolic, $R_p = 0.4$ encounters. Angles i_1, ω_1 and i_2, ω_2 specify the orientations of the two disks; t_m is the time from first encounter to final merger; $c:b:a$ are the axial ratios of the central ellipsoid; λ_c is an estimate of the dimensionless spin parameter.

No.	i_1	ω_1	i_2	ω_2	t_m	$c:b:a$	λ_c
1	0		71	30	3.5	5.1:8.7:10	0.13
2	109	-90	71	-90	4.3	7.1:7.4:10	0.10
3	109	-30	71	-30	4.5	5.8:7.5:10	0.10
4	109	30	180		5.0	4.6:7.4:10	0.04

4. Survey Results

Fig. 1 shows an intermediate stage of encounter 2, one time unit after first approach. This view is edge-on to the orbital plane, which is vertically oriented on the page; the line of sight parallels the separation vector at pericenter. The two disks spin in opposite directions. The upper one, having undergone a slightly prograde encounter, sports a well-defined tidal tail and a truncated bridge, while the lower one, which experienced the same passage as slightly retrograde, exhibits a broad fan of debris extending to the left. Such gross tidal features are reasonably well modeled by restricted three-body simulations. However, the detailed response of the galaxies, which includes the S-shaped bar in the upper disk and the nicely-wound spiral pattern in the lower one, can only be reproduced in fully self-consistent calculations.

4.1. Orbital Decay and Transfer of Angular Momentum

On general principles it seems likely that any parabolic encounter between disk/halo galaxies eventually leads to a merger, because the initial orbit is marginally bound, and hence any net

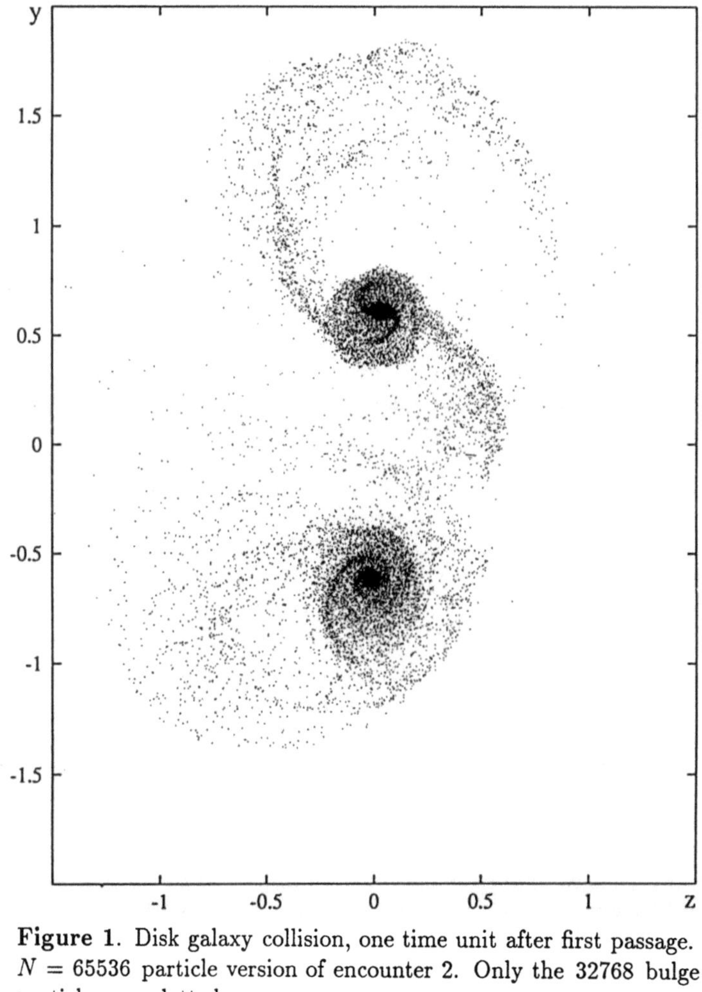

Figure 1. Disk galaxy collision, one time unit after first passage. This is a $N = 65536$ particle version of encounter 2. Only the 32768 bulge and disk particles are plotted.

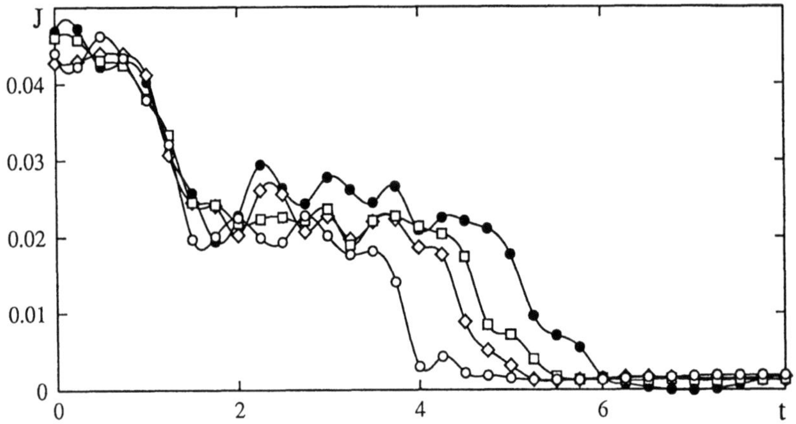

Figure 2. Orbital decay in the $R_p = 0.4$ encounters, showing the net angular momentum J_B of the bulges as a function of time. First passage occurs at time $t = 1$. The open circles, diamonds, squares, and filled circles refer to encounters 1, 2, 3, and 4, respectively.

work done in creating tidal features will trap the two galaxies into a elliptical relative orbit. Successive passages will then shrink this orbit until the two merge. In the close encounters best studied to date, merging actually takes place quite quickly. For example, the galaxies shown in fig. 1 merge at $t_m \simeq 4.3$ time units, or nominally just 10^9 years, after their first passage.

This process of orbital decay may be monitored by computing the net angular momentum J_B contained in the two galactic bulges, evaluated in their mutual center-of-mass frame. Fig. 2 shows J_B as a function of time for the four $R_p = 0.4$ parabolic encounters of table 1. The first passage occurs at time $t = 1$; shortly thereafter, all four pairs suffer a dramatic loss of angular momentum. Closer examination reveals that this angular momentum is transferred to the interpenetrating halos, which temporarily "bunch up" behind the luminous galaxies and so retard them gravitationally by dynamical friction. Subsequent passages follow a similar pattern and soon reduce the net angular momentum of the bulges to a mere few percent of their initial values.

Table 1 and fig. 2 also illustrate the effect of disk orientation. Encounter 1, in which both disks are prograde, merges only $t_m \simeq 3.5$ time units after first passage, while encounter 4, in which both spins are reversed, merges at $t_m \simeq 5.0$. This "spin-orbit" coupling may not seem large, but it is impressive given that the disks are only 15% of the total mass.

4.2. Merger Remnants: Profiles and Characteristic Scales

Galaxies typically merge from the *outside in*; as their relative orbit decays, an interacting pair experience a series of increasingly close encounters. Each successive passage tidally dislodges mass from the outskirts of the galaxies, and most of this mass becomes bound to the combined system. When merging is complete, the remnant retains considerable memory of its progenitors; the final binding energy of a particle is correlated with its initial binding energy (White 1979; Quinn, this volume).

One consequence of this orderly merging process is shown in fig. 3, which plots cumulative mass profiles, $M(< r)$, for the bulge, disk, total luminous (bulge + disk), and halo components of a typical merger remnant. At small radii the remnant is dominated by particles belonging to the bulges of the original galaxies, which finally merged with each other inside an already-coalesced halo. At large radii, on the other hand, halo particles predominate. Disk particles are distributed over a range of radii; those from the inner third or so of the initial disks are almost

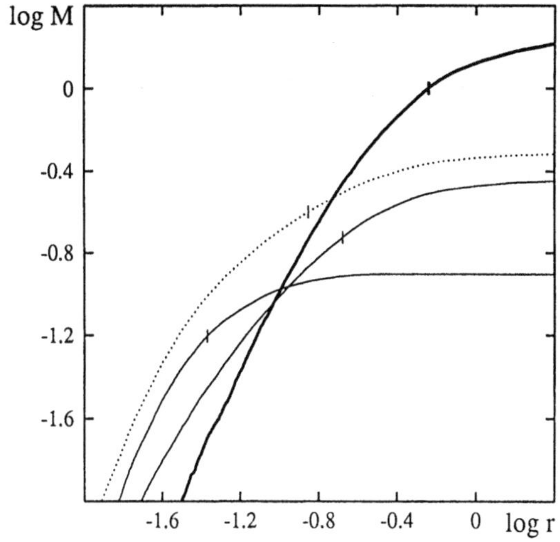

Figure 3. Cumulative mass profiles of a merger remnant. In increasing asymptotic value at large r, the four curves refer to the bulge, disk, total luminous (dotted) and dark halo (heavy) components, respectively. Short vertical lines mark half-mass radii.

as centrally concentrated as the bulges, while some particles from the edges of the disks escape to infinity.

Outside-in merging also has important consequences for the characteristic length scales and velocity dispersions of the merger remnants. The overall properties of a remnant may be predicted by a simple energy argument (Hausman & Ostriker 1978; White 1983): whenever two identical systems of mass M_i and binding energy E_i merge from a parabolic orbit, the final system must have mass $M_f \simeq 2M_i$ and binding energy $E_f \simeq 2E_i$, since only trivial amounts of mass and binding energy escape to infinity. The characteristic (*e.g.* half-mass) radius, velocity dispersion, and surface density of the remnant follow at once: $R_f \simeq 2R_i$, $v_f \simeq v_i$, and $\Sigma_f \simeq 0.5\Sigma_i$. It is important to remember that these global relations hold *only* for the *entire* remnant. By contrast, very different results are obtained when attention is limited to a distinct subcomponent. As we have seen, in collisions between galaxies with large amounts of dark matter, the central luminous components are trapped within a common dark halo, and eventually merge from a rather tightly-bound orbit. Consequently, the luminous parts of the final remnant are more compact and tightly bound than would be expected from energy conservation alone. In the simulations described here, the final half-mass radius of the total luminous component is $R_f \simeq 1.3R_i$. For the bulge component alone, the final radius is $R_f \simeq 1.4R_i$, and the velocity dispersion is $v_f \simeq 1.4v_i$.

The merging of two equal-mass galaxies is sufficiently tumultuous to produce a significant degree of violent relaxation. Here as in the collapse experiments of van Albada (1982) and McGlynn (1984), such violence results in an object with a de Vaucouleurs-law profile. Fig. 4 shows the projected surface density of luminous particles in a typical merger remnant, plotted with $-2.5\log\Sigma$ as the ordinate and $r^{1/4}$ as the abscissa. Fitting a straight line to these points, an effective radius $r_e \simeq 0.1$ is obtained; this small effective radius and correspondingly high surface brightness are consequences of the tight binding of the luminous remnant. Within a radius $r \simeq 0.02$ the luminosity profile seems to fall below the straight line. It is not coincidental that this radius is comparable to the softening length $\epsilon = 0.015$ used in these experiments. A more physical interpretation, however, may be given following May & van Albada (1984); a de Vaucouleurs law extending to $r \ll 0.02$ would require a higher phase-space density in

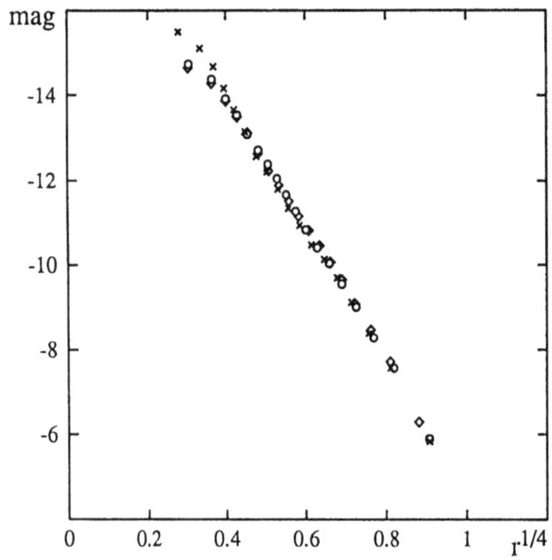

Figure 4. "Surface brightness" profile of a merger remnant. The three different plotting symbols refer to orthogonal projections.

the remnant than existed in the initial galaxies. Such an increase is precluded by Liouville's theorem, irrespective of ϵ. Mergers between galaxies with higher peak phase-space densities will presumably produce remnants with de Vaucouleurs' profiles extending to smaller r.

4.3. Merger Remnants: Structure and Dynamics

The recurrent pattern of orbital decay described above also has significant consequences for the shapes of merger remnants. Stars flung out in the earlier stages of the encounter tend to be distributed towards the orbital plane, while those which hang on to the very end wind up in an elongated central ellipsoid formed by the final, nearly head-on collision of the bulges. Hence, the remnants of bulge/disk/halo mergers tend to be basically prolate at small radii and oblate at large radii.

To describe the "average" shape of these merger remnants, I focus on a well-defined central population: the particles which made up the bulges and the inner third of the initial disks. This population, which comprises half of the total luminous mass, has a distribution function cutting off well below escape energy. Thanks to this cut-off, the moment of inertia **I** is not unduly biased by outlying particles. Over an interval of 2 time units following the merger, the eigenvalues $a^2 > b^2 > c^2$ of **I** are stable, and the eigenvectors of **I** and the kinetic energy tensor **T** are nearly parallel, showing that the central population is already close to equilibrium. Table 1 lists axial ratios for the $R_p = 0.4$ encounters. Remnant shape depends on disk orientation; note for example that the prograde encounter 1 produces the most oblate remnant.

Rotation plays only a small role in the dynamics of these central ellipsoids. To quantify this, table 1 gives the dimensionless spin parameter λ_c estimated by treating the central population as self-gravitating. The resulting values are small, $\lambda_c \sim 0.1$ as compared to the value of ~ 0.4 expected for a self-gravitating disk, and not particularly correlated with the actual flattening. Figure rotation is likewise insignificant; typical pattern speeds are only of order 0.2 radians per central dynamical time. Thus remnant shapes cannot be due to rotation, but must arise from velocity anisotropy. Analysis of individual particle trajectories is consistent with this point, and suggests that the central ellipsoids of the remnants are built largely from box orbits.

5. Mergers with Gas Dynamics

Stellar-dynamical merger simulations can only tell part of the story; many interacting galaxies have gaseous components amounting to $\sim 10\%$ of their total luminous mass. While the gas makes only a modest contribution to the net gravitational field, it behaves very differently from the stellar components of merging galaxies. Being "colder" than the stars in a disk galaxy, the gas is more responsive to tidal perturbations. Unlike stars and dark matter, gas flows cannot interpenetrate; violent shocks may develop in close encounters. These shocks can dissipate a significant fraction of the kinetic energy of the gas, with the result that some gas will pool at the center of a merger remnant. In addition, both the shocks and the central pool of gas may be susceptible to rapid star-formation, possibly triggering a "starburst".

The best way to simulate this complex and volatile situation is far from clear. Negroponte & White (1983) and several other researchers have modeled the interstellar medium of disk galaxies using sticky particles, capturing the clumpy nature of this gas without attempting to satisfy its equations in detail. An alternative approach is "smoothed particle" hydrodynamics or SPH (*e.g.* Monaghan 1985; Hernquist & Katz 1989, and references therein), in which particles representing fluid elements move in response to local pressure forces as well as gravity. This method is potentially much more faithful to the hydrodynamics and thermodynamics of real gas flows, provided they are reasonably smooth. Whichever method, however, is used, any realistic treatment of gas dynamics in interacting galaxies must await better models for star formation and the back-reactions of supernovae and stellar winds on the gas.

Using an SPH method, Lars Hernquist and I have recently simulated mergers between disk galaxies including gas dynamics. The interstellar medium of each disk, comprising 10% of the net luminous mass, is represented by 4096 particles; thus a single gas particle weighs $\sim 10^6\ M_\odot$, and structures down to a few times the mass of a giant molecular cloud can in principle be resolved. The gas is given a standard solar-abundance cooling curve, cut off below 10^4 K to prevent violent thermal instabilities from fragmenting the gas into cold dense clumps. In nature such clumps would doubtless be reheated by star formation and related effects, which are not included here. Nonetheless, the models may help to identify likely sites of star formation.

Up to first encounter the gas behaves much like the rest of the disk. After the encounter, however, significant differences develop between the gas and stars. Some tidal arms may be swept clean of gas by interpenetrating passages, but most of the gas becomes concentrated in narrow shocks within the broader tidal features traced out by the stars. In the outflung tidal tails the gas may break up into clumps reminiscent of the distributions observed in "long-tailed" galaxies (Schweizer 1978). Quite another fate awaits the gas near the centers of the galaxies. Much of this gas participates in strong bars and other non-axisymmetric features, losing kinetic energy in shocks and angular momentum via gravitational coupling to the collisionless components. As a result, gas begins to collect near the centers of the two galaxies. This process is accelerated during the final encounter and merger; in one case, $\sim 60\%$ of the gas wound up in a compact blob at the center of the merger remnant. Given the differences in method, this outcome is encouragingly similar to the results obtained by Negroponte & White (1983).

6. Conclusions

The numerical experiments described above are of course limited to a very small subset of all galactic encounters: specifically, rather close collisions of two and only two identical disk galaxies, with each galaxy containing a relatively massive and non-rotating bulge and surrounded by a much more massive and sharply truncated halo. Nonetheless, this work supports the idea that most of the strongly interacting galaxies we observe today will merge soon – within a few dynamical times, or roughly 1 Gyr. Rapid production of merger remnants could have been a serious embarrassment had the remnants turned out to be highly unusual specimens. But the results from N-body simulations, such as above, indicate that the mergers of typical disk/halo galaxies produce remnants with most of the properties of ordinary ellipticals. Hence instead of littering the universe with bizarre relics, merging appears to be a natural way to account for a rather common type of galaxy (Toomre 1977).

Transformation of two or more disk/halo galaxies into a single elliptical involves several different effects. Dynamical friction, arising as halos interpenetrate, promotes rapid orbital decay and transfers angular momentum from the luminous galaxies to the dark matter. When galaxies finally merge, they do so within a common dark halo, creating a radially segregated remnant with luminous material predominating at small radii. Violent relaxation of the existing stars sets up a luminosity profile closely following de Vaucouleurs' law; Liouville's theorem, however, limits their maximum phase-space density, imposing a finite core on this profile.

The luminous components of merger remnants are significantly more tightly bound than predicted by simple homology arguments; instead of remaining constant, the central velocity dispersion climbs by a factor of ~ 1.4 during a merger. This effect, due to energy transfer from the luminous components to the dark matter, appears capable of removing the rather modest but systematic disparities otherwise expected between merger remnants and typical elliptical galaxies. In connection with this point, it is also interesting to note that recent merger remnants seem to follow the same $L - \sigma$ relation as ordinary ellipticals (Lake & Dressler 1986).

Most of the orbital angular momentum originally residing in the luminous material is lost to the dark matter during a merger. Even for relatively wide encounters, the final collision between the central parts of the galaxies is nearly head-on. This welcome but still poorly understood phenomenon may account not only for the slowly-rotating prolate ellipsoids typically produced in merger experiments, but also perhaps for the surprising prevalence of tightly-bound shells in elliptical galaxies (see Prieur in this volume). At the same time, the orientation of the initial disks can have a significant effect on the shape and dynamics ot the remnant. Hence mergers between disk galaxies with sufficiently extensive halos seem to reproduce the diversity of structure and dynamics of real elliptical galaxies, while avoiding the overproduction of oblate, rapidly rotating remnants bemoaned in some previous studies (*e.g.* Negroponte & White 1983).

Merger simulations including gas dynamics indicate that a large fraction of the gas can be funneled to the center of the remnant (Hernquist 1989). The fate of this gas is uncertain, but it seems likely to fuel a rapid burst of star formation. Unlike the stars, the gas evades Liouville's theorem and thus can dramatically boost the maximum phase-space density of the remnant. If the gas settles into a small spinning disk it may give rise to a counter-rotating core or other kinematically distinct component (Schweizer 1982 and this volume). Rapid reprocessing of gas by high-mass stars may help steepen metallicity gradients. Finally, and yet more speculatively, a massive influx of gas may be able to fuel or even form a central black hole, providing a mechanism by which mergers can provoke nuclear activity.

I thank Lars Hernquist, Alar Toomre, and Scott Tremaine for comments on the manuscript. Numerical calculations were run at the JvNC and Pittsburgh Supercomputer centers.

REFERENCES

Alladin, S. 1965, *Ap. J.*, **141**, 768.
Barnes, J. 1988, *Ap. J.*, **331**, 699.
Barnes, J. 1989, *J. Comp. Phys.*, in press.
Barnes, J. & Hut, P. 1989, *Ap. J. Sup.*, **70**, 389.
Binney, J. & Tremaine, S. 1987, *Galactic Dynamics* (Princeton: Princeton University Press).
Farouki, R. & Shapiro, S. 1982, *Ap. J.*, **259**, 103.
Gerhard, O. 1981, *M. N. R. A. S.*, **197**, 179.
Hausman, M. & Ostriker, J. 1978, *Ap. J.*, **224**, 320.
Hernquist, L. 1989, *Nature*, **340**, 687.
Hernquist, L. & Katz, N. 1989, *Ap. J. Sup.*, **70**, 419.
King, I. 1966, *Astr. J.*, **76**, 64.
Lake, G. & Dressler, A. 1986, *Ap. J.*, **310**, 605.
May, A. & van Albada, T. 1984, *M. N. R. A. S.*, **209**, 15.

McGlynn, T. 1984, *Ap. J.* **281**, 13.
Miller, R. & Smith, B. 1980, *Ap. J.,* **235**, 421.
Monaghan, J. 1985, *Comp. Phys. Rep.,* **3**, 71.
Negroponte, J., and White, S. D. M. 1983, *M. N. R. A. S.,* **205**, 1009.
Ostriker, J. 1980, *Comments on Astrophysics* **8**, 177.
Quinn, P. 1982, *PhD thesis*, Australian National University.
Roos, N. & Norman, C. 1979, *Astr. Ap.,* **76**, 75.
Schweizer, F. 1978, in *Structure and Properties of Nearby Galaxies*, eds. E. M. Berkhuijsen and R. Wielebinski (Dordrecht: D. Reidel), p. 279.
Schweizer, F. 1982, *Ap. J.,* **252**, 455.
Toomre, A. 1977, in *The Evolution of Galaxies and Stellar Populations*, eds. B. M. Tinsley and R. B. Larson (New Haven: Yale Univ. Obs), p. 401.
Toomre, A. 1981, in *The Structure and Evolution of Normal Galaxies*, eds. S. M. Fall and D. Lynden-Bell (Cambridge: Cambridge Univ. Press.), p. 111.
Toomre, A. & Toomre, J. 1972, *Ap. J.,* **179**, 623.
van Albada, T. 1982, *M. N. R. A. S.,* **201**, 939.
van Albada, T. & van Gorkom, J. 1977, *Astr. Ap.,* **54**, 121.
Villumsen, J. 1982, *M. N. R. A. S.,* **199**, 493.
Villumsen, J. 1983, *M. N. R. A. S.,* **204**, 219.
White, S. 1978, *M. N. R. A. S.,* **184**, 185.
White, S. 1979, *M. N. R. A. S.,* **189**, 831.
White, S. 1980, *M. N. R. A. S.,* **191**, 1P.
White, S. 1983, in *Internal Kinematics and Dynamics of Galaxies*, ed. E. Athanassoula (Dordrecht: D. Reidel), p. 337.

Tidal Phenomena in Interacting Galaxies

K.D. Borne

Space Telescope Science Institute, 3700 San Martin Drive, Baltimore, MD21218, USA

We present evidence for the action of tidal stripping and tidal friction in interacting binary galaxies. This evidence comes from the morphological distortions seen in pairs of elliptical galaxies, and from the kinematic peculiarities seen both in elliptical galaxy rotation curves and in spiral galaxy HI 21cm profiles. For several pairs of interacting ellipticals the observed strengths of such tidal disturbances have been matched with appropriate interaction simulations, thereby allowing the binary mass and orbital parameters to be derived.

1. INTRODUCTION

The proliferation of numerical models demonstrating the action of dynamical friction in pairs of interacting galaxies has not been matched by an equivalent set of observations lending support to the friction hypothesis. Tidal friction has been shown theoretically to result in the merger of two interacting galaxies, but obviously it is not possible to observe this process directly in the real universe. One can demonstrate however that the simple gravitational physics that leads to tidal friction and merging also reproduces the morphological and kinematic disturbances observed in colliding galaxies. This then strengthens the validity both of the gravitational interaction simulations and of their results (i.e., friction and merging).

In confirming the validity of the gravitational interaction hypothesis for tidal distortions, one actually finds numerical simulations that match specific colliding galaxy observations. From these matching models one can derive the masses and orbital parameters of the observed pairs since the tidal distortions probe the binary gravitational potential in such systems.

2. OBSERVATIONS OF TIDAL PHENOMENA

Several pairs of interacting elliptical galaxies have been observed both photometrically and spectroscopically (Borne and Hoessel 1988; Borne, Balcells, and Hoessel 1988; Balcells, Borne, and Hoessel 1989). Most of the pairs were selected from the catalog of Karachentsev (1972) on the basis of photometric evidence of interaction. That the form of the morphological disturbances are of a certain type supports the idea that momentum is being transferred from the binary orbit into the motions of individual stars, thus giving evidence for the action of tidal friction. The form of the disturbances is usually like this: one side of each galaxy is distended while the opposite side is compressed; each galaxy in a pair is similarly disturbed, except that

Dynamics and Interactions of Galaxies Editor: R. Wielen

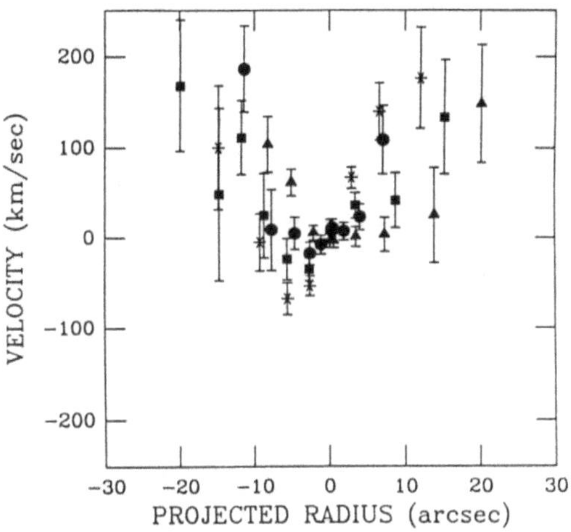

FIG. 1: U-shaped rotation profiles for four tidally-disturbed ellipticals. The zero values of projected radius and velocity correspond to the central brightness peak in the given galaxy. Velocities are plotted for NGC 2673 (circles), NGC 7236 (squares), NGC 1588 (triangles), and IC 413 (asterisks).

the parity of the tides is reversed. Such broad one-sided distensions are precisely what we expect in a hot stellar system. By contrast, the observed response of a cold disk to a collision is usually a long and relatively narrow tidal tail.

Long-slit spectroscopic observations of these interacting ellipticals have revealed some peculiar motions along the line connecting the galaxy centers. In many cases, the rotation profile is U-shaped, indicative of some form of ablation (Borne and Hoessel 1988). A selection of these profiles is shown in Figure 1. Positions and velocities are measured relative to the central brightness peak in the given galaxy. Note that both sides of each galaxy are moving in the same sense, away from the observer; that none of these show the motion toward the observer is surely a statistical effect. It is very unlikely that the upturn in velocity is due to contamination from the companion galaxy since the raw spectral data show clearly that the contribution from the other galaxy is falling rapidly and, for all practical purposes, goes to zero at a radius significantly beyond the last-measured point in these profiles.

The non-steady dynamical configuration indicated by a U-shaped rotation profile is consistent with momentum transfer from the binary galaxy system to the individual stars, and hence with the action of tidal friction and tidal stripping. Detailed physical arguments justifying this conclusion are presented by Borne and Hoessel (1988). In short, stars that are on trajectories that move in phase with the collision (i.e., on prograde orbits) couple strongly with the perturbing galaxy, drawing energy and momentum from its orbit; stars on retrograde orbits respond very weakly to the perturbation and therefore contribute very little to the observed tidal disturbances. The resonant coupling of the prograde stars has these consequences: stars that are on the leading edge of their parent galaxy's motion will be pulled toward the perturber and will accelerate away from their parent, whereas stars on the trailing edge will be dragged by the perturber and will have their motion around the parent decelerated. In each case, the stars will appear to lag behind the motion of the parent, causing a U-shaped rotation profile and asymmetric tidal lobes of the type observed. Prograde tidal coupling therefore drives the tidal stripping and friction that ultimately leads to a merger of the galaxy pair. These physical processes are made manifest in the U-shaped rotation profiles and in the one-sided tidal lobes.

The line-of-sight stellar velocity distribution in a galaxy with a U-shaped rotation profile must be asymmetric since the mean stellar velocity is displaced relative to the galaxy's central velocity. Such asymmetric velocity distributions have been measured in other samples of galaxy pairs: Dickel and Rood (1980) and Sulentic and Arp (1983) each present HI 21cm profiles for spiral galaxies found in pairs. Unlike the symmetric profiles seen in relatively isolated spirals, spirals having close companions show asymmetric profiles indicative of the tidal coupling (i.e., momentum transfer) described above. This provides further evidence of tidal friction in action.

3. MODELS OF INTERACTING ELLIPTICALS

Observations of several pairs of distorted ellipticals have been analyzed in conjunction with detailed numerical simulations in order to measure (with very few model assumptions) the masses of the ellipticals directly. Borne (1984; 1988*a*) describes the technique, while Borne (1988*b*), Borne, Balcells, and Hoessel (1988), and Balcells, Borne, and Hoessel (1989) present the results. The algorithm is powerful in that it is very fast, can be run interactively, allows for a large volume of the model parameter to be explored in a reasonable length of time, and uses a physically realistic interaction model. Many different simulations can be run using a variety of input parameters until a good match to the data is found. Figure 2 shows the observed and best-fit model surface brightness distributions for two specific pairs of interacting ellipticals.

Comparing the strengths of the morphological and kinematic tidal disturbances in interaction models with those detected observationally allows us to actually "weigh" interacting galaxy systems. For several pairs, we have found a wide range of galaxy masses, but a very limited range of M/L_B ($\approx$ 6-10 solar; $H_0 = 60$ km s^{-1} Mpc^{-1}). The following table presents the mass and orbit results derived from the analysis of four such systems.

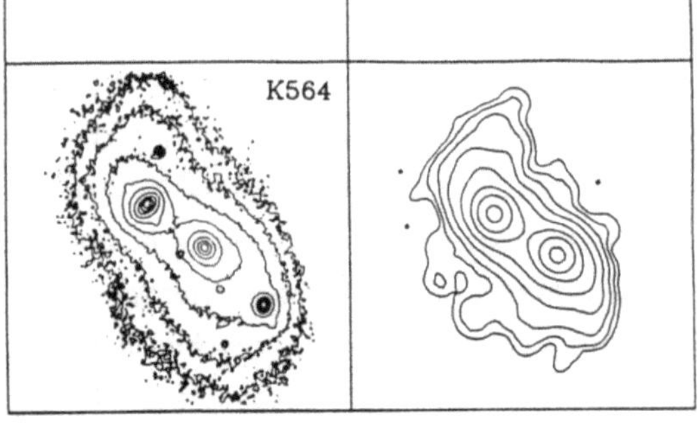

FIG. 2: Surface density maps for two pairs of interacting ellipticals. The true images are on the left, and the best-fit simulations are on the right, each of which used 4000 particles. Note how the asymmetric mass distributions in the model galaxies match the observed photometric distortions, especially at large radii. Since the motion of each model galaxy is found to be in the direction of its compressed isophotes, galaxy light distributions showing similar one-sided tidal distortions thus reveal the sense of binary motion in interacting pairs.

Table: MODEL RESULTS FOR INTERACTING ELLIPTICALS

Galaxy Pair	Initial Orbit	Final Orbit	Time since pericenter	M/L_B (solar)
NGC 1587/88	UNBOUND	UNBOUND	30×10^6 yr	7 ± 2
NGC 2672/73	BOUND	BOUND	26×10^6 yr	7 ± 2
NGC 4782/83	UNBOUND	BOUND	25×10^6 yr	10 ± 2
NGC 7236/37	UNBOUND	UNBOUND	65×10^6 yr	6 ± 2

4. CONCLUSIONS

We have confirmed the action of tidal stripping and tidal friction in interacting binary galaxies, and we have used simple numerical simulations to measure the masses of these systems.

The evidence for tidal friction includes the tidal disturbances that are observed both in photometric imaging data and in multi-position spectroscopic data. Numerical simulations of interacting binary galaxies are able to reproduce the observed tidal effects. Since the simple gravitational physics used in the interaction simulations can explain all of the observed phenomena and since that same physics also produces tidal friction and merging, then friction must be active in the observed systems. Binary orbital energy and angular momentum are being transferred to the very stars that contribute to the observed tidal disturbances. Furthermore, that many of these stars are being tidally stripped is supported by the same simulations, in which tidal stripping of such "disturbed" stars is "observed" over the course of the integration.

The kinematic and morphological distortions observed in several pairs of interacting ellipticals have been used to derive their binary mass and orbital parameters. By matching the spectroscopic and imaging data with appropriate interaction simulations, we find $M/L_B \approx$ 6-10 $M_\odot/L_\odot$ and a mixture of bound and unbound orbits among the pairs in our sample.

REFERENCES

1 Balcells, M., Borne, K. D., and Hoessel, J. G. 1989, *Astrophys. J.*, **336**, 655.
2 Borne, K. D. 1984, *Astrophys. J.*, **287**, 503.
3 Borne, K. D. 1988*a*, *Astrophys. J.*, **330**, 38.
4 Borne, K. D. 1988*b*, *Astrophys. J.*, **330**, 61.
5 Borne, K. D., Balcells, M., and Hoessel, J. G. 1988, *Astrophys. J.*, **333**, 567.
6 Borne, K. D., and Hoessel, J. G. 1988, *Astrophys. J.*, **330**, 51.
7 Dickel, J. R., and Rood, H. J. 1980, *Astron. J.*, **85**, 1003.
8 Karachentsev, I. D. 1972, *Comm. Spec. Astrophys. Obs. USSR*, **7**, 3.
9 Sulentic, J. W., and Arp, H. 1983, *Astron. J.*, **88**, 489.

Self-consistent Simulations of Ring Galaxies

P.N. Appleton[1] *and R.A. James*[2]

[1]Department of Physics, Iowa State University, IO, USA

[2]Department of Astronomy, University of Manchester, Manchester M13 9PL, UK

Introduction

We have run a series of simulations to explore the collision model for the formation of ring galaxies (Lynds and Toomre, 1976, Toomre, 1978). It is possible to choose collision parameters to reproduce various characteristics of observed systems. We present choices which (i) lead to double ring systems in which the primary and secondary rings are in different planes and (ii) generate spokes connecting the nucleus to the primary ring, sometimes crossing it and joining up with external spiral features. The latter are frequently tightly wound so that part of the ring appears to be double. Some models resemble the Lindsay-Shapley ring AM 0644-741. (In referrring to observed systems we use the notation of Arp and Madore, 1987. The galaxies mentioned were selected from those listed by Few and Madore, 1986).

Lynds and Toomre argue that the formation of rings requires the intruder to penetrate the target disc within the central 15 percent of its radius. Its orbit can be inclined to the disc normal at up to 45 degrees without destroying the rings. We may speculate that the presence of a massive halo leads to a relaxation of these limits. Huang and Stewart (1988) found such a relaxation and were able to produce ring structures with an intruder crossing the disc at an angle of 80 degrees to the disc normal. Unfortunately they assumed a rectilinear path for the intruder relative to the disc, a poor assumption for all but the most hyperbolic encounters. The results below support the speculation with ring structures arising from almost normal penetration at radii up to 40 percent of the disc radius. For more central encounters we have obtained good ring structures with penetration at an angle of 60 degrees to the disc normal. So far this is the largest angle used in our simulations and we cannot rule out the production of rings from more extreme cases.

Details of the model

Our target galaxies comprise an almost spherical halo together with a disc containing either 0.15 or 0.3 of the galaxy's mass. The intruders are initially spherical and have one third of the mass of the target in all the examples presented here. All three components of the system are modelled self-consistently using a particle-mesh code. The mesh used to calculate potentials contains 33x33x65 points and the initial line joining the centres of the two galaxies is parallel to the long dimension. With a scale of 1.333 kpc per mesh cell we can place the initial centres of the galaxies at a separation of 37.333 kpc and still have almost all the model particles on the mesh at the moment of projection.

The discs contain 15000 particles with either 35000 or 85000 in the halo, depending on the halo mass fraction. All particles have equal mass and the number in the intruder is chosen to give the correct mass ratio. The initial disc is tilted so that the intruder penetrates it at a controllable angle. The various components of the system are relaxed separately to give reasonably quiet initial conditions at projection. Most encounters are mildly hyperbolic. The initial relative velocity consists of a parabolic component along the line of centres, together with

Dynamics and Interactions of Galaxies Editor: R. Wielen

a perpendicular component to displace the penetration point from the centre of the target galaxy. This component is set parallel to the tilt axis to give prograde or contragrade encounters or at right angles to give encounters with the leading or the trailing part of the disc. Our simulations to date are confined to these four possibilities.

Results

Table 1 lists the simulations reported in this paper. In all cases penetration of the disc by the intruder takes place approximately 160 million years after projection. In those figures relating to simulations the time unit is a million years, measured from the projection epoch. Where given, the first two integers printed below the time are (in degrees) the co-latitude of the observer, measured from the initial line of centres, and the longitude measured from the initial mean symmetry plane. The third integer identifies the stellar component and is unity for all the discs shown here. The edges of the mesh are shown in projection and serve to frame the pictures. The horizontal distance between mesh edges is 40 kpc in all cases. With the exception of figure 3 all the figures are on the same scale.

Figures 1 and 2 show the separation of the disc into components in distinct planes 160 million years after penetration. Figure 1 shows the view in a direction parallel to the original disc plane and figure 2 a view parallel to the plane containing the central part of the disc. The same separation into two components is visible, but less obvious, in run 86. Figure 3 shows the view along the initial line of centres for this run and is on a scale which permits us to see both the original (or primary) ring formed by the encounter and an inner secondary ring with complex internal structure. Although the resolution in this feature is less than we would

Table 1 - model parameters

Run no.	Disc mass fraction	Encounter type	Tilt angle (degrees)	Penetration radius (kpc)
48	0.30:1	central	60	0.0
52	0.30:1	contragrade	45	5.6
86	0.15:1	central	45	0.0
100	0.15:1	prograde	45	2.3

Figure 1

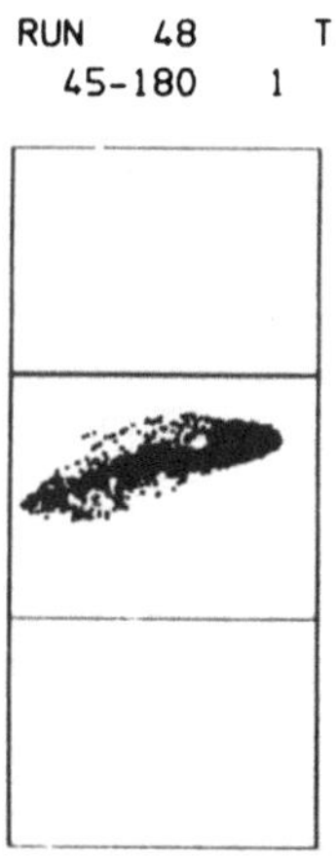

Figure 2

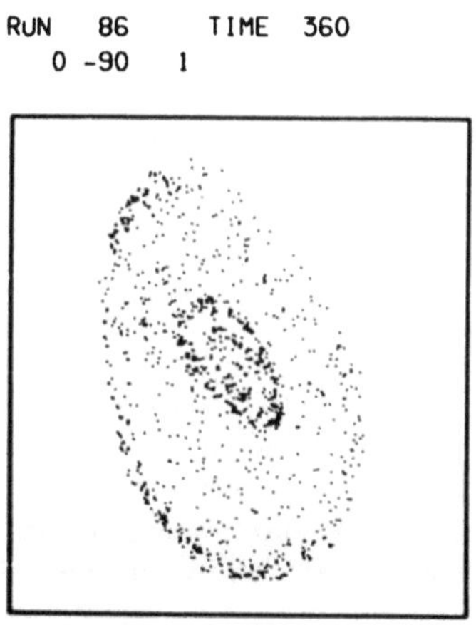

Figure 3

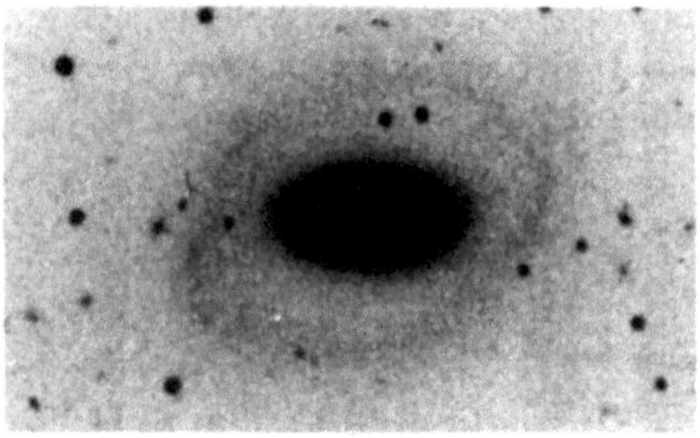

Figure 4(a)

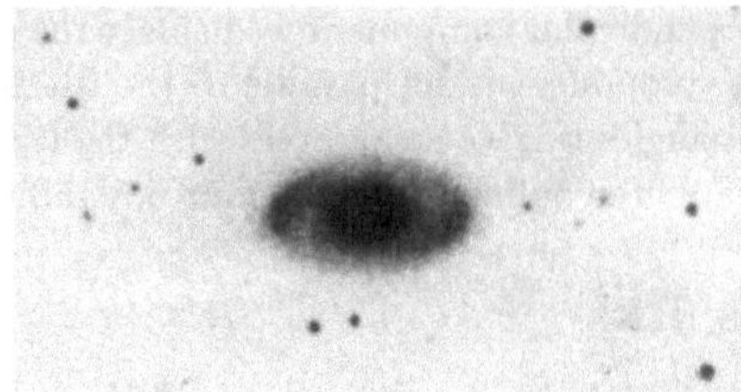

Figure 4(b) AM 2036-734

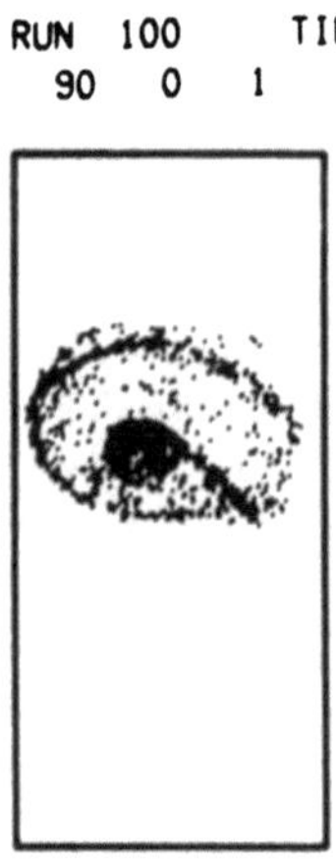

Figure 5

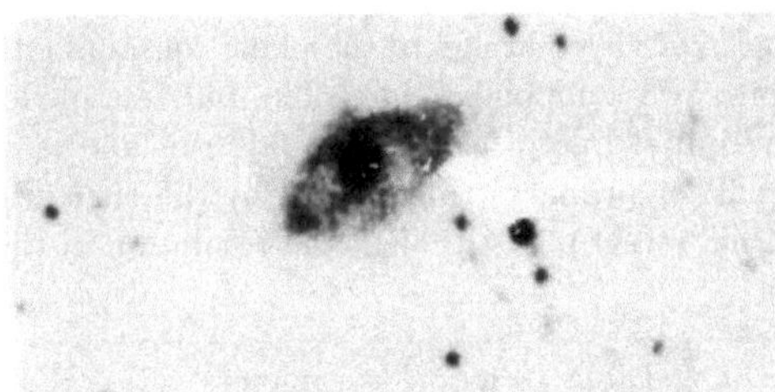

Figure 6
AM0438-661

wish, it does appear to be genuine. Such rings appear in many of our models and that shown is not an isolated accident. We would note in particular the misalignment of the major axes of the primary and secondary rings. It is instructive to compare these pictures with figures 4a, 4b, showing the southern ring galaxy AM 2036-734. These figures were printed at different densities from the same negative to show both the inner and the faint outer rings. This system has a conspicuous misalignment between the major axes of the primary and secondary rings. As it is more symmetrical than the models of run 86 and there is no obvious companion to produce the rings, we cannot argue that run 86 is in any sense a model for AM 2036-734. We do however suggest that the separation of the disc into components in different planes offers a possible mechanism for producing misaligned rings and that further exploration may lead to an acceptable model.

Figure 5, run 100, shows the effect of a prograde encounter approximately 200 million years after penetration. The strong spoke visible in this model resembles the spoke in the observed galaxy AM 0438- 661 (figure 6). We ran a parallel simulation departing from an identical initial system, but with the initial systemic velocity component perpendicular to the line of centres reversed to give a contragrade encounter. The spoke did not appear in the comparison simulation and the interaction between the galaxies was considerably weakened. The spoke clearly arises from the weak resonance between the orbital motion of the intruder and that of individual disc stars. We must however remark that although the stellar response of the disc can form spokes, the motion of the gas in ring galaxies may differ considerably from that of the stars, especially near the centre where hydrodynamic effects are most dramatic (Appleton and Struck-Marcell, 1987, Struck-Marcell and Appleton, 1987). There seems little prospect of reproducing the rich internal structure of the Cartwheel by purely stellar dynamical models.

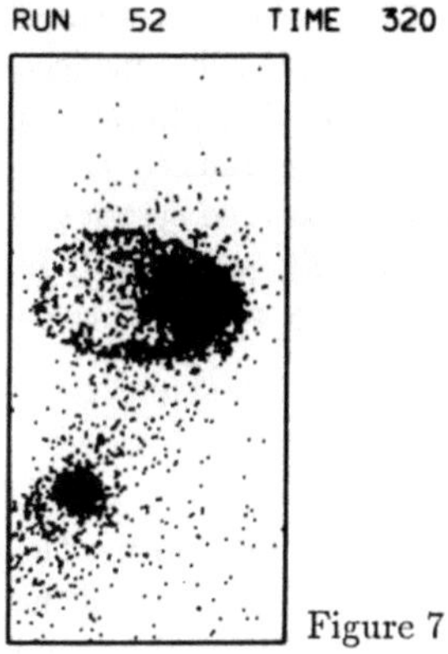

Figure 7

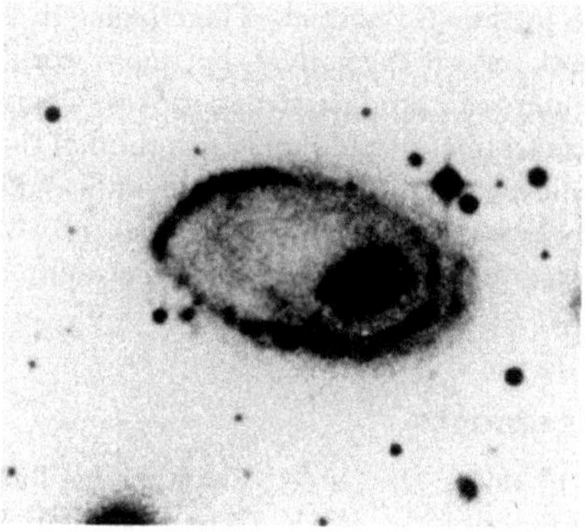

Figure 8
AM 0644-741

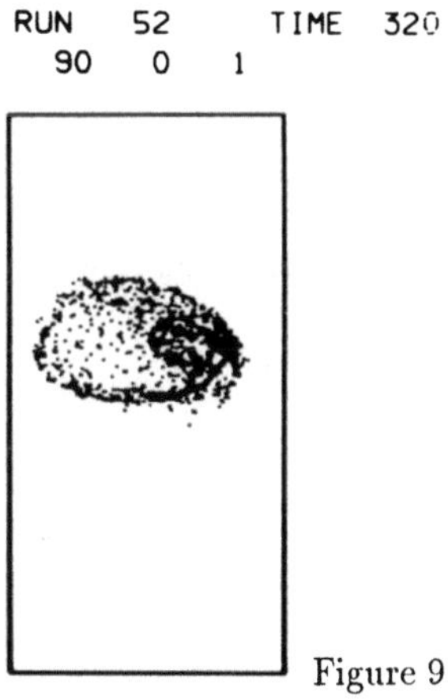

Figure 9

The strong interaction characteristic of prograde encounters causes considerable damage to the disc and penetration at large radii tends to produce one-armed spiral rather than ring features. Contragrade encounters can produce convincing ring structures with larger penetration radii. Figure 7 shows the halo, intruder and disc for run 52 at a time 160 million years after penetration. Figure 8 shows the Lindsay-Shapley ring (AM 0644- 741). To assist in comparing with the simulation we have rotated this picture anti-clockwise through approximately 90 degrees. The two rings are similar in shape, and the nuclei have similar offsets from the ring centres. In figure 8 a candidate for the intruder is just visible at the edge of the picture in almost exactly the same position as the intruder in figure 7. It is tempting to extend the comparison to the tightly wrapped spiral feature which gives an apparent doubling of the ring of AM 0644-741 near the nucleus. Figure 9, the disc component of figure 7, shows a similar feature but unfortunately it winds in the wrong direction. It is difficult to assess the significance of the feature in the simulation. It lasts less than 40 million years and the ring is severely warped so that its appearance is likely to depend sensitively on viewing angle.

Conclusions

Observed ring galaxies show individual features such as mis-aligned primary and secondary rings, spokes and partial doubling of the primary ring. This doubling is probably due to the presence of tightly wound spiral features. Simulations can produce similar features if the orbital parameters are chosen correctly. Interesting features seem to arise when the intruder's orbit crosses the disc at a large angle to the disc normal. There seems to be a possibility of producing mis-aligned rings from almost central encounters, but more work is needed to generate plausible models. Prograde encounters can produce strong spokes. Contragrade encounters permit the formation of rings from penetrations at radii up to one third of the disc radius and produce transient spiral features and doubled rings, although the identification of such features in observed systems is dubious. Offsets between the nucleus and the centre of the ring arise naturally, particularly in the case of contragrade encounters where penetration of the disc at fairly large radii still leads to ring formation.

Acknowledgements

We would like to thank Dr. M. A. Shaw for his illuminating comments on this work. The advice and assistance of Mr. W. V. Garner in obtaining the photographs required were of the greatest value and we have gained enormously from his persistence and ingenuity. We would like to thank the SERC for support and computing facilities under grants SGC/14116 and GRE/28727. Figures 4, 6 and 8 were taken from the J plates of the ESO-SERC Southern Sky Survey and we gratefully acknowledge our debt in this respect.

References

Appleton, P. N. and Struck-Marcell, C., 1987, Astrophys. J., 312, 566.
Arp, H. C. and Madore, B. F., 1987, "A Catalogue of Southern Peculiar Galaxies and Associations", (Cambridge University Press).
Few, J. M. A. and Madore, B. F., 1986, Mon. Not. Roy. Astron. Soc., 222, 673.
Huang, S-n and Stewart, P., 1988, Astron. Astrophys., 197, 14.
Lynds, R. and Toomre, A., 1976, Astrophys. J. 209, 382.
Toomre, A., 1978, in "The Large Scale Structure of the Universe", ed. M. S. Longair and J. Einasto, (Reidel, Dordrecht) p 109.

Gas Dynamics in a Tidal Interaction: Formation of Rings

F. Combes, C. Dupraz, and M. Gerin

Ecole Normale Supérieure, Paris, and
Observatoire de Meudon, F-92195 Meudon, France

1. Introduction

Some ring galaxies, of which the Cartwheel is the archetype, form during the near-central head-on intrusion of a small companion in a disk galaxy (Lynds and Toomre, 1976; Theys and Spiegel 1976). Such rings are of essentially stellar nature. On the contrary, many spiral galaxies exhibit ring-like features that mostly consist of gas and young stars and do not involve the old stellar population. Among the latter, we shall only consider objects with nuclear rings, located at the turnover of the rotation curve, usually within the inner kiloparsec. In particular, we shall leave apart the galaxies which have annular distributions of gas, at radii around 4-5 kpc like our Galaxy, M31, and M81, since they may have a different origin.

The nuclear rings are very active sites of star formation. They are conspicuous in Hα in the so-called "hot-spot" galaxies of Sersic (1973). The ring-like morphology is also detected in radio-continuum with the VLA (Hummel *et al.*, 1987; Garcia-Baretto and Pismis, 1985). As for the molecular gas, an increasing number of high-resolution CO observations (e.g., of M82, NGC 1068, NGC 1097, and NGC 4736) reveal that it is concentrated in the rings. Arsenault (1989) has recently performed a statistical study of 170 normal and 97 starburst galaxies, and finds that the presence of a nuclear ring is well correlated with an enhanced star-formation activity.

Various mechanisms have been proposed to account for the existence of these nuclear rings. It has for instance been suggested (Icke, 1979; Silchenko and Lipunov, 1987; Lesch *et al.*, 1989) that viscous torques could dissipate the energy of the gas component in redistributing its angular momentum, thus driving it inwards towards the center. In order to explain the formation of a nuclear ring, it is necessary that the gas stops before it has reached the center, which implies that the viscous torques must cancel in the central region. This only happens if the inner rotation curve is sufficiently linear, as the viscous torques are proportional to the gradient $d\Omega/dr$ of differential rotation. However the gas present in the central parts is not affected by viscous torques and is not depleted inside the ring.

Another possible mechanism consists of the action of gravitational torques exerted by a non-axisymmetric potential on a gaseous spiral pattern. It has indeed already been shown that the bi-symmetric potential Φ produced by a stellar bar ($\Phi \propto \cos 2\theta$) can make the gas accumulate in inner or outer rings through this mechanism (Schwarz, 1984; Combes and Gerin, 1985). Such a bi-symmetric potential is also found in the tidal interaction of the galaxy with a companion. Therefore, nuclear rings may also be expected to form in tidally interacting galaxies.

2. Simulations

A galaxy is represented by stellar and gaseous components, and both are self-gravitating. We compute their self-gravity by a 2D N-body code, based on the Fast Fourier Transforms method. The FFT are calculated on a 256x256 grid, but the useful space is restricted to 128x128 cells, to avoid images. The cell has size of 0.25 kpc. There are 4 10^4 particles representing the stellar component, and between 10^3 and 2 10^3 clouds.

The gas clouds ensemble is simulated according to the code developped by Combes & Gerin (1985). The clouds are spread about a mass spectrum, from 10^4 to 10^7 $M_{\odot}$; the

latter represents the Giant Molecular Clouds (GMC). Small clouds grow through collisions. The life-time of GMC against dispersion by star-formation is 4 10^7 yrs. Kinetic energy is re-injected in the gas component when a GMC is dispersed. The cell (spatial resolution) of the collision grid is 130pc (240x240). The mass fraction in the gas is M_{gas}/M_{tot} = 3.6%.
We present here two simulations, run with the same initial galaxy: the first is a control run, with the galaxy isolated; in the second, an equal-mass companion turns around the galaxy in a bound orbit, which pericenter is 20 kpc, the apocenter 30kpc and the period 8 10^8 yrs. The initial galaxy is composed of a spherical massive bulge, containing 63% of the total mass, represented by a Plummer potential of radial-scale 1.5 kpc. The stellar disk is a Toomre disk, of scale 3kpc, with an initial velocity dispersion such that Q=1 at each radius. The gaseous disk is exponential with characteristic scale of 5 kpc.

3. What forms the ring?

The two simulations are presented in Fig. 1. The evolution of the stellar and gaseous components are displayed separately. Let us first describe the evolution of the isolated galaxy. After the apparition of a transient spiral pattern, the stellar component settles into a bar-like feature. Then, a spiral structure develops in the gas outside the stellar bar, roughly between the corotation (CR) and outer Lindblad resonance (OLR) of the bar. This spiral has the same pattern speed as the bar itself, about the maximum of the $\Omega-\kappa/2$ curve. The bar exerts a gravitational torque on this spiral, which is negative inside CR. This explains that the gas inside CR is falling towards the inner Lindblad resonance (ILR), where it accumulates in a nuclear ring (Combes 1987) in a time scale of a few 10^8 years. For the same reason, the gas outside CR is subject to a positive gravitational torque and slowly moves outwards. After a lapse of several 10^9 years, the ring finally collapses under the action of either enhanced collision rate or dynamical friction of giant clouds due to the high central density.

In the case of a colliding galaxy, the transient spiral is enhanced by the tides (bridge and counterarm). At each passage, the companion generates a bi-symmetric density perturbation, and a two-armed spiral wave, which pattern speed is about the angular velocity of the companion at pericenter (Fig. 2). However, a bar emerges in the stellar disk in about the same time scale as for the isolated galaxy (Fig. 3). Moreover, it must be stressed that the angular velocity of the bar is about the same as above, i.e., the maximum of the $\Omega-\kappa/2$ curve, independently of the orbital angular velocity of the companion. Inside the bar, the behaviour of the gas is the same as if the galaxy were isolated, which results in the formation of a nuclear ring at the bar ILR. Outside the bar, the tidal potential due to the companion dominates over the bi-symmetric $\cos 2\theta$ term due to the bar. As bar and companion have different angular velocities, the bar has no secular action on the gaseous spiral wave: the gas only undergoes the negative gravitational torque exerted by the companion, since the whole disk is inside the companion CR.

This is reflected in the final radial gaseous distribution (Fig. 4): in an isolated galaxy, gas outside CR is driven away towards the OLR, while in the interacting galaxy, the whole gas is driven inwards. Accordingly, the nuclear ring that forms at the bar ILR is much more contrasted if the galaxy is colliding.

4. Discussion: the role of viscous torques

In the above picture, a minimum dissipation is required, such that the gas stays in a spiral structure, but how important are the viscous torques ? At least in the isolated galaxy, they must be negligible with respect to gravitational torques, since the gas is driven outwards outside the bar CR.
We have estimated the viscous torques, by computing the net balance between loss and gain of angular momentum in cloud collisions, for a given initial momentum per unit mass. The result is plotted in fig. 5, where it can be seen that the viscous torques are two orders of magnitude lower than the gravitational torques, computed from the total gravitational force. To test the relevance of this estimation, we also simulated a viscous

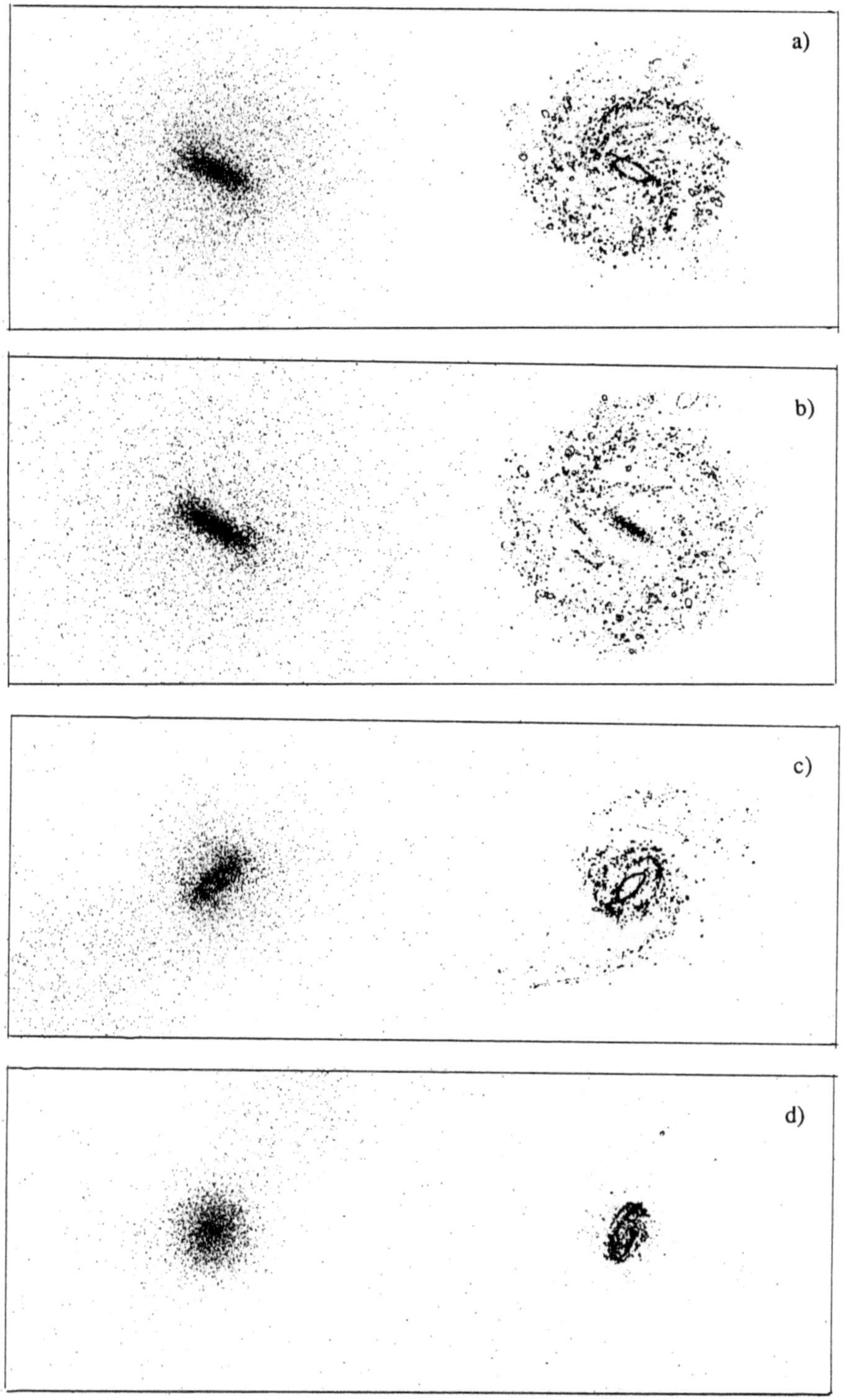

Fig1: a)3 10^9 yrs and b) 9 10^9 yrs plots for an isolated galaxy: left = stars, right= gas c) and d) same for a tidally-perturbed galaxy (companion in a bound orbit).

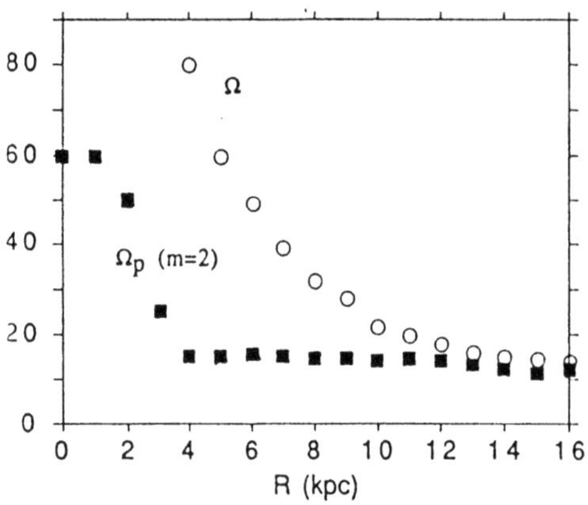

Fig2: Pattern speed of the spiral wave generated by the companion passage at pericenter.

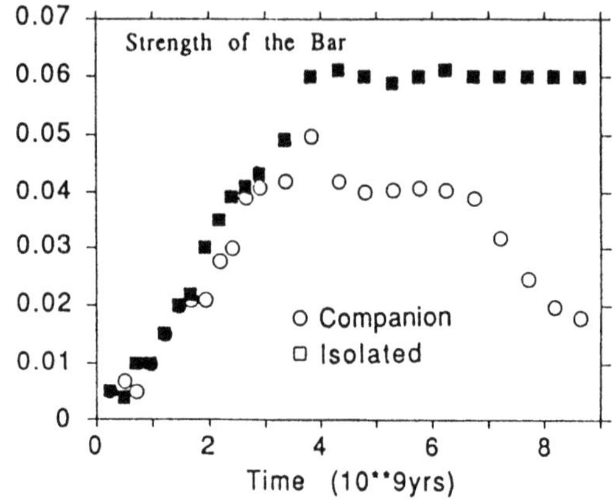

Fig3: Strength of the bar versus time for the two runs: the repeated passages of the companion destroys the bar.

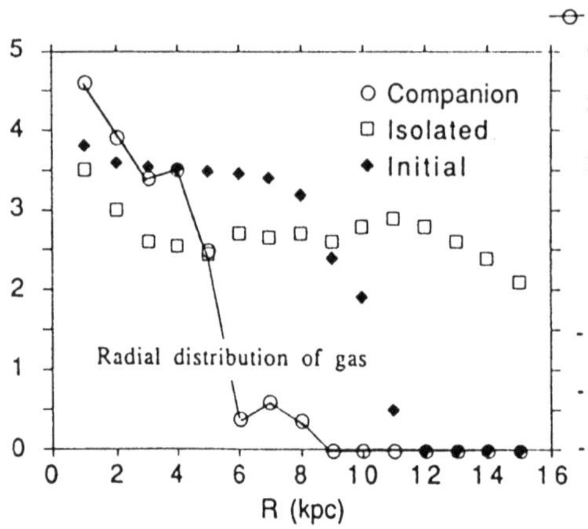

Fig4: Initial and final gaseous distributions for the two runs.

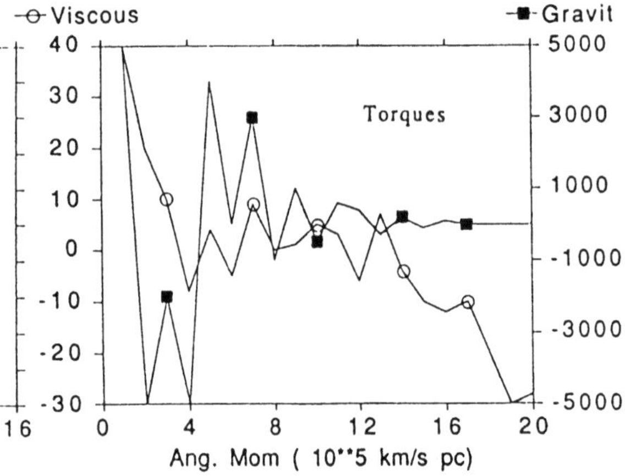

Fig5: Estimation of viscous and gravitational torques, in the case of the perturbed galaxy.

isolated galaxy, where the cloud masses were all identical: collisions were inelastic with a restitution factor of 0.5, and there was no more energy re-injected in the interstellar medium by star formation. In that run, the whole gas mass fell into the center in 2 10^9 yrs, without ring formation, and the viscous torques estimated as above were three times larger than the gravitational torques.

That the viscous torques are negligible can also be seen on test-runs with no self-gravity included (no gravitational torques): the initial gas density remains similar to the initial until a Hubble time. The energy re-injected in the GMC dispersion compensates exactly the energy loss in collisions. This is an important factor often ignored in the viscosity estimations: part of the nuclear energy produced in stars is transformed in kinetic energy of the gas, via supernovae explosions, HII regions expansions, stellar winds, etc. This neutralises the viscous energy losses, and indirectly the effects of viscous torques.

5. Conclusion

Nuclear rings form by gravitational torques. In the case of a tidally-perturbed galaxy, two waves at different pattern-speeds develop. All the gas is gathered towards the center. In contrast, for an isolated galaxy, gas outside corotation is driven outwards. To produce a central starburst or to feed the nucleus, mergers are therefore very efficient. Bars are also, but not as much, since they can gather only gas inside their corotation.

References

Arsenault R. (1989) *Astron. and Astrophys.* **217**, 66

Combes F., Gerin M. (1985) *Astron. and Astrophys.* **150**, 327

Combes F.(1987)*in* "Galactic and Extragalactic Star Formation", ed R. Pudritz and M. Fich

Garcia-Barretto J.A. and P. Pismis (1985) IAU **115**, 626

Hummel E., van der Hulst J.M., Keel W.C.(1987) *Astron. and Astrophys.* **172**, 32

Icke V. (1979) *Astron. and Astrophys.* **78**, 21

Lesch H. et al. (1989) poster, *these Proceedings.*

Lynds R., Toomre A. (1976) *Astrophys. J.* **209**, 382

Schwarz M.P. (1984) *Monthly Notices of Roy. Ast. Soc.* **209**, 93

Sersic P. (1973) *PASP* **79**, 152

Silchenko O.K., and Lipunov V.M. (1987) *Astrofizika* **26**, 443

Theys J.C. Spiegel E.A.(1976) *Astrophys. J.* **208**, 650

Merger Origin for Counterrotating Cores in Elliptical Galaxies

M. Balcells[1] *and P.J. Quinn*[2]

[1]Washburn Observatory, University of Wisconsin-Madison, and
Kapteyn Lab, Postbus 800, NL-9700 AV Groningen, The Netherlands
[2]Space Telescope Science Institute, 3700 San Martin Drive, Baltimore, MD21218, USA
Mt. Stromlo and Siding Spring Observatories, The Australian National University,
Woden Post Office, ACT 2606, Canberra, Australia

Using numerical simulations of the merger between a high- and a low-luminosity elliptical galaxy we show that this type of merger can result in an elliptical with a counterrotating core. Only merger orbits retrograde with respect to the primary spin generate counterrotating cores, because the initial secondary spin does not survive the merger even though the secondary core is not disrupted during the merger. The shape and the rotation axis of the primary are severely modified during the merger. The remnants of the simulations show a similarities with elliptical galaxies with peculiar core kinematics.

Introduction

Kinematically distinct nuclei in elliptical galaxies have been regarded as evidence for accretion in ellipticals (Franx and Illingworth 1988; Jedrzejewski and Schechter 1988; Bender 1988), and they may be one of the only merger signatures that survive beyond the violent relaxation phase. Understanding the mechanism for their formation may provide clues on a process that might be common in the life of an elliptical galaxy. We have explored the conjecture by Kormendy (1984) that these systems represent the undisrupted core of a smaller, denser elliptical that has merged with a large elliptical galaxy. The study is based on numerical simulations using the TREECODE of Hernquist (1987), involving a high number of particles (~50,000). Details are given in Balcells and Quinn (1990).

Results

For the mass-ratios explored (5:1 and 10:1), counterrotating cores can arise from this type of merger. Two conditions have to be met. First, a merger orbit *retrograde* with respect to the initial primary spin is necessary, because the secondary spin does not survive the merger; tidal torquing alignes the spin of the secondary core with the orbital angular momentum. Second, the orbital angular momentum $\mathbf{J}_{\mathrm{orb}}$ has to be small compared to the total primary spin; otherwise, the absorption of $\mathbf{J}_{\mathrm{orb}}$ by the primary particles reverses the rotation sense in the main body of the remnant, inhibiting counterrotation. At the core, not only the secondary material but also the primary material counterrotates, and the velocity distribution function is fairly symmetric. A typical amplitude for the counterrotation is 40 km/s for a galaxy with $M = 5 \cdot 10^{11} M_0$ and $r_{1/2} = 10$ kpc, thus amplitudes can reach 100 km/s for giant ellipticals. Counterrotation can be seen with slits placed along any photometric axis, major, minor or intermediate. A typical rotation curve is shown in Figure 1.

Approximately half of the secondary mass remains self-bound throughout the merger, and it dominates the central surface density of the remnant. The remnant velocity dispersion profile shows a dip at the center, corresponding to the light being dominated by a less massive system.

For the mass ratios explored, the effects on the surface density profile are subtle. For $M1/M2 = 10$, a hump is seen in the surface density profile at the radius of the counterrotating core. In all cases the effective radius of the galaxy increases, and the core radius decreases during the merger. Isophote twists

Dynamics and Interactions of Galaxies Editor: R. Wielen

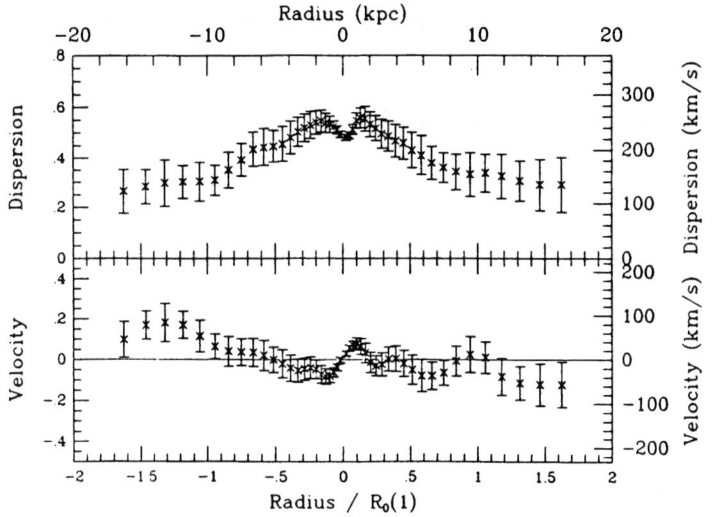

FIGURE 1: Rotation curve and velocity dispersion profile for the remnant of a retrograde merger with M_1/M_2=10. The right and top scales refer to a galaxy with $M=5\cdot10^{11}\,M_0$ and $r_{1/2}$=10 kpc Counterrotation peaks at 1kpc from the center with a maximum of 40 km/s.

are seen at all radii outside the region of the counterrotating core. The orientation of the primary major axis and its rotation may change by more than 90° during the merger.

A disk of stripped secondary material is formed around the counterrotating core. This disk also counterrotates, and a smooth transition occurs from the self-gravitating core to the inner part of this disk.

Discussion

Our results indicate that core counterrotation can result from a merger involving purely stellar, non-dissipative dynamics. Dynamical friction is effective in sending the secondary core to the remnant center.

While some systems exist with properties which differ from the predictions of our experiments, e.g. the high rotation of IC 1459 (Franx and Illingworth 1988), and the asymmetric line profiles shown by Bender (this conference), each of the properties described in the previous section have been observed in some elliptical galaxies with peculiar core kinematics (cf. the rotation curves in Jedrzejewski and Schechter 1988 and Bender 1988). This suggests that some counterrotating cores have formed in largely dissipationless merger events.

Acknowledgements

We thank Lars Hernquist for making his TREECODE available to us, and the SDSC and PSC for allocations of CRAY time to this project. Support to MB was provided by a subcontract to NASA grant NAS7-918.

References

Balcells, M. and Quinn, P. J. *Ap. J.,* submitted (1990)
Bender, R.: *Astr. Ap.*, **202**, L5 (1988)
Franx, M., and Illingworth, G. D.: *Ap. J. (Letters),* **327**, L55 (1988)
Hernquist, L.: *Ap. J., Suppl.,* **64**, 715 (1987)
Jedrzejewski, R., and Schechter, P. L.: *Ap. J. (Letters),* **330**, L87 (1988)
Kormendy, J.: *Ap. J.* **287**, 577 (1984)

A Simplified Numerical Treatment of Encounters Between Galaxies

R. Bien, B. Fuchs, and R. Wielen

Astronomisches Rechen-Institut, Mönchhofstraße 12–14,
D-6900 Heidelberg, Fed. Rep. of Germany

In the following, we discuss the present status of our research on the dynamics of interacting galaxies.

Our basic tool is a three-dimensional Cartesian particle-mesh code. Poisson's equation is solved via Fast Fourier Transform, as usual. A well-known problem with such codes is that in many cases the potentials of centrally condensed galaxies cannot be represented satisfactorily. The situation gets worse when interactions are simulated. For instance, an acceptable resolution of the cores and - after the encounter - a reasonably wide spatial separation of the partners are two conflicting conditions.

To avoid these difficulties, we introduce additional grids in regions where the mass is highly concentrated. Here, we describe our technique by means of a simple model: A Plummer sphere ('target galaxy') collides with a rigid Plummer potential ('perturber') with the same total mass and Plummer radius (R_{Pl}). The moving target galaxy is treated in *three* 32^3 grids which we re-construct for every new time step.

- Grid no. 1 contains the 'core', i. e. all particles inside the radius $R_C = 2R_{Pl} = 2$ cell lengths.
- Grid no. 2 contains the outer region, i. e. all particles outside the radius R_C. This 'halo' is truncated at $8R_{Pl} = 8$ cell length, a radius containing 98% of the mass of the whole Plummer sphere.
- Grid no. 3 also contains the core, but now R_C corresponds to 12 cell length, i. e. the resolution of the core is considerably increased.

The three grids yield potentials Φ_1, Φ_2, Φ_3, which - after a re-scaling of Φ_3 - can be 'glued together' to form the total potential.

If a particle is

- outside R_C, we evaluate $\Phi_1 + \Phi_2$ at the position of the particle,
- if inside R_C, we take $\Phi_2 + \Phi_3$.

The acceleration is then known and is used to push the particle via the leapfrog-scheme. The position of the perturber is calculated at each time step depending on the orbit of the target galaxy.

This simplified treatment is fast, and therefore a survey is possible for various values of the impact parameter and the initial velocity. The results agree very well with direct N-body calculations as long as the encounter is not too disruptive. Since the target galaxy can be modelled by a large number of particles (e. g. 50000), 'noise' is largely reduced.

The generalization to a self-consistent treatment of both galaxies is straightforward. Each galaxy needs *five* grids. The first three grids play the same role as before. Nos. 4 and 5 have the same length scales and are used to represent the whole region where the two galaxies move. Thus they are chosen as coarse as possible.

Dynamics and Interactions of Galaxies Editor: R. Wielen

A full account of our technique will be given elsewhere. For demonstration only, the figure shows the logarithmic surface density of two isotropic de Vaucouleurs spheres, a few crossing times after the closest approach. The projection is on to the orbital plane, the unit of the length scale is 1 de Vaucouleurs radius. A shift of both cores is clearly visible (cf. L. A. Aguilar and S. D. M. White, 1986, Astrophys. J. **307**, 97).

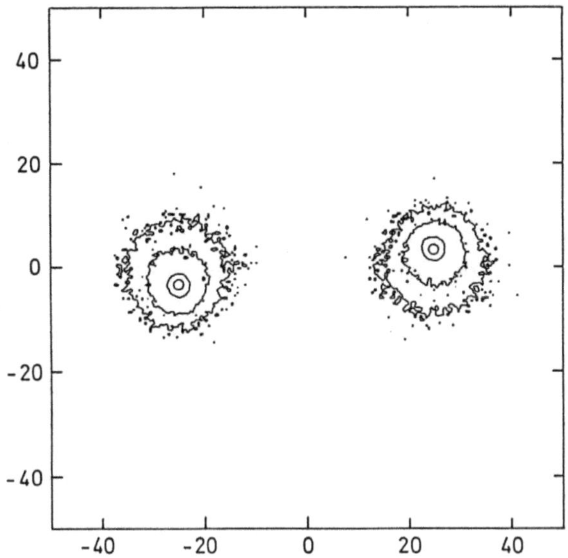

Numerical Simulation of Encounters of Galaxies

P. Londrillo[1] *and A. Messina*[2]

[1]Osservatorio Astronomico, via Zamboni 33, I-40126 Bologna, Italy
[2]Dipartimento di Astronomia, via Zamboni 33, I-40126 Bologna, Italy

Introduction

Numerical simulation of collisionless gravitating systems requires large dynamic range in three–dimensional space resolution to treat structures which have an highly inhomogeneous structure, with field gradients covering quite different length scales, up to infinity.

To simulate galaxies we introduced a code (Londrillo and Messina, 1989) which, preserving the advantages of the surface-harmonic expansion as used in the van Albada code (1982), introduces a mapping which permits the extension of spectral techniques to the radial coordinate. The code allows high accuracy combined with very short computation time.

This basic version of the code has been employed to treat the merging with multicenters, extending to three dimensions the technique suggested by van Albada and van Gorkom (1977).

In this way, the basic formulation of the code allows the accurate computation of field quantities for all the structures present. In the following, we present the main characteristic of the basic version of the code and some preliminary results.

The basic code

In spherical coordinates, (r, θ, ϕ), the unbounded r domain is mapped into a finite x domain:

$$r(x) = L \tan^{\alpha}(x/2), \qquad 0 \leq x < \pi. \tag{1}$$

L gives the radial distance corresponding to half the mapped interval and therefore allows the optimal allocation of the physical object on the working space x. The second parameter α, instead, controls the resolving power of the mapping at small and large distances.

A polynomial expansion $(z = cos\theta, \alpha = 2)$ is introduced in the form:

$$\Psi(r, z, \phi) = \sum_{n=1}^{N-1} \sum_{l=0}^{M-1} \sum_{m=-l}^{l} \tilde{\Psi}_{n,l,m} cos(nx) P_l^{(m)}(z) e^{im\phi}, \tag{2}$$

where $P_l^{(m)}(z)$ are the associated Legendre polynomials. The Poisson equation reduces to:

$$\sum_{k=n-2}^{k=n+2} R_{n,k} \tilde{\Psi}_{k,l,m} + l(l+1)\tilde{\Psi}_{n,l,m} = -\tilde{D}_{n,l,m}, \tag{3}$$

$$l = 0, ..M-1, \qquad -l \leq m \leq l, \qquad n = 0, ..N-1.$$

where $\tilde{D}_{n,l,m}$ and $R_{n,k}$ are the spectral components of $4\pi G\rho r^2$ and of the radial part of the Laplace operator, respectively. The solution is obtained by direct inversion since all the submatrices

$$T_l(n, n') = R(n, n') + l(l+1)\delta_{n,n'}, \qquad n = 1, N-1,$$

are non–singular.

Dynamics and Interactions of Galaxies Editor: R. Wielen

Preliminary results

We have studied the interaction: a) of two identical Plummer polytropes, and b) of two cold spherical objects with initial density proportional to $1/r$. In both the models, the centers of the two groups (each containing $2 \cdot 10^4$ particles) are initially at a distance about three times the average radius of the distribution. In the two cases, the conservation of momenta and energy is well satisfied (*cf.* Fig. 1 for the merging of two polytropes).

Fig. 2 shows, at left, the contour levels in a meridian plane at the times indicated (in units of crossing time) for the merging b). The computation was performed with a grid (64, 24, 48) in (r, θ, ϕ), in $\simeq 50$ min CPU time on a CRAY X-MP 48. At right, Fig. 2 shows the brightness radial distribution for the same simulation. Notwithstanding the system has not yet reached equilibrium, the fit to the de Vacouleur law (full curve) is better than any one obtained with a dissipationless collapse, starting with a virial ratio $10^{-4} < 2T/|W| < 10^{-1}$ (Londrillo, Messina & Stiavelli, 1989).

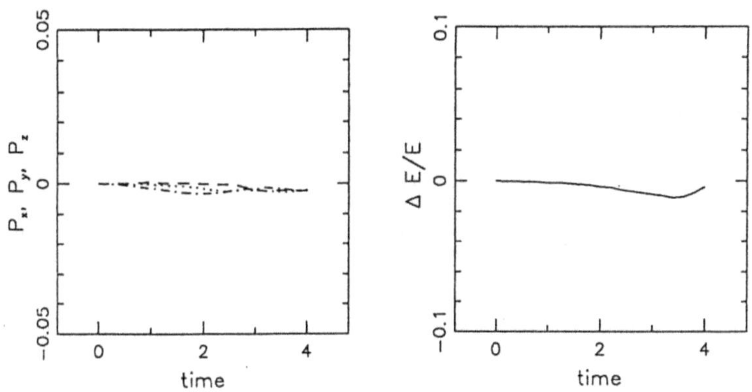

Fig. 1

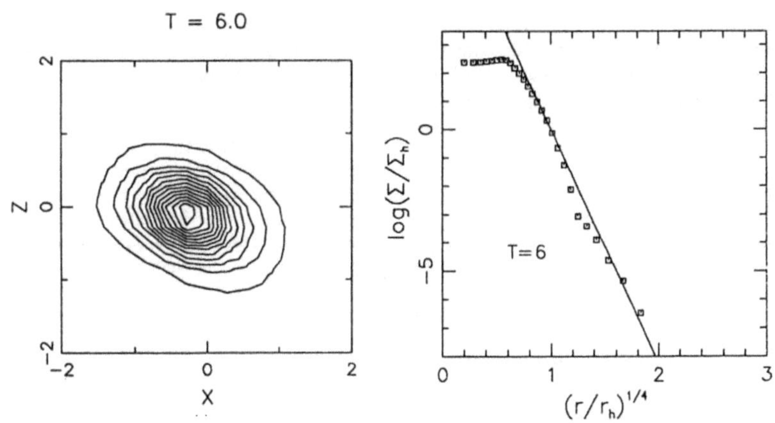

Fig. 2

References

van Albada, T.S., 1982. *Mon. Not. R. astr. Soc.* **201**, 939
van Albada, T.S. and van Gorkom, 1977, *Astron. Astrophys.*, **54**, 121
Londrillo, P. and Messina, A., 1989. *Mon. Not. R. astr. Soc.*, to be published
Londrillo, P., Messina, A. and Stiavelli, M., 1989, in preparation

Using Parallel Computers for Very Large N-Body Simulations: Shell Formation Using 180 K Particles

J. Salmon[1], *P.J. Quinn*[2], *and M. Warren*[1]

[1]Caltech Concurrent Computing Project

[2]Mt. Stromlo and Siding Spring Observatories, The Australian National University, Woden Post Office, ACT 2606, Canberra, Australia

We have used a parallel version of the Barnes-Hut C treecode running on a parallel supercomputer (a 512 node NCUBE) at the Caltech Concurrent Computing Project to follow the evolution of systems with very large numbers of particles (N>180,000). Many important problems in astrophysics demand large N and in particular we have studied the fully selfconsistent formation of stellar shells around an elliptical parent galaxy via the merger with another elliptical of one tenth its mass. The final system shows a spread in shell radii similar to that in observed shell galaxies. Major changes in the structure and dynamics of the primary have resulted from the merger.

Large N problems : the motivation

Traditionally, N-body simulations of collisionless systems have been run with the largest number of particles possible given the limitations on machine time and memory. This push to large N was motivated by the need to control the collisionality of the simulation over the desired number of dynamical times. The arrival of methods with computational times that scaled like $N Log N$ instead of N^2 (PM, PPPM, Treecodes) meant that for a fixed computing resource,10,000 - 100,000 particles could be evolved with high dynamical resolution and in runs of only a few tens of hours on machines like the CRAY 1. In trying to achieve very large N, most of the NlogN codes run up against serious memory constraints on machines like the CRAY X-MP . In an attempt to address this memory problem we have developed the Barnes-Hut treecode on a parallel architecture in which individual nodes have substantial amounts of memory. The 512 node NCUBE at the Caltech Concurrent Computing Project has 512K of memory per node giving a total machine memory of 256 MgB which is more than ten times the typical maximum process size on the X-MP.

Tree Codes on Parallel Computers

The basic problem in designing a parallel algorithm is the decision of "who gets what data, when?" Given an appropriate organization of the data, termed a "decomposition", one is usually able to carry over the basic sequential algorithm with little or no change. That is, each processor behaves as though it is performing the original sequential algorithm on a subset of the complete data set.

The simplest decomposition to consider is one in which each processor is assigned all the bodies in a region of space. The processor boundaries are chosen to assign approximately equal work loads to the processors. Given a decomposition of the particles which balances the load, it is tempting to think we can construct the tree on each processor, merge them into a collective tree which is copied to each processor and proceed with the force evaluation. Unfortunately, this leads to severe problems with the available memory. One is limited to treating problems whose entire tree fits into the 512 kbytes available to each processor. Thus, this approach provides extremely good utilization of the parallel processor when the total number of bodies is less than a few thousand, but it doesn't work at all for large numbers of bodies.

At this point, it is crucial to realize that by the very nature of the Barnes-Hut(1986) algorithm, it is not necessary to store the entire tree in each processor. Since each processor is concerned with computing forces only within a restricted region of space, it will need to access the most detailed levels of the tree only in a restricted region. Distant regions of space may be accounted for with much less data.

Thus, the parallel force calculation algorithm proceeds as follows :

(a) Determine processor boundaries by recursive bisection of the estimated work.
(b) Build the tree for the local particles.
(c) Communicate to obtain only those details of other processors' trees which may be necessary for force computations.
(d) For each local particle descend the tree and compute the force.

The performance of the parallel algorithm may be measured several ways. Of primary interest is the overall performance. Counting all of the overheads, load imbalances, i/o operations, etc., the system computes the force on our 180,000 bodies in approximately 200 seconds (θ = 0.8,monopole). In addition, it is of interest how this figure scales with the number of processors. Obviously, we were not able to evaluate the forces on a 180,000 body

Dynamics and Interactions of Galaxies Editor: R. Wielen

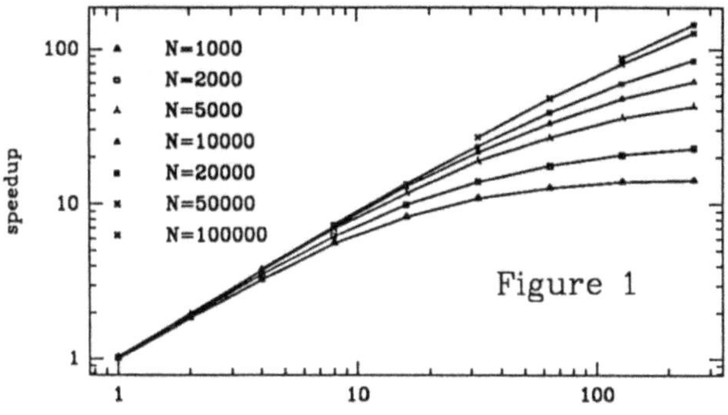

system on a single processor with 512 kbytes of memory. Nevertheless, extrapolation from a series of runs with fewer particles indicates that we obtained a speedup of 300 on 512 processors. Figure 1 shows the speedup resulting from the use of parallel processing as a function of the number of processors, as well as the number of bodies. It is typical of such plots that for a fixed problem size, the speedup begins to level off and the addition of more processors does little to reduce the overall running time. On the other hand, if the problem size is allowed to grow with the number of processors, the speedup continues to increase up to very large numbers of processors. Thus, large parallel computers are generally well matched to large problems, but are a poor match to small problems.

Shell Formation

Since the early 1980s, stellar shells around ellipticals have been recognized as the product of a collision between a large elliptical and a compact and/or dynamically cold companion galaxy. Models of shell formation by Quinn (1984), Dupraz and Combes (1986) and Hernquist and Quinn (1988) have all used rigid potential models for the primary and hence have ignored important selfconsistent effects like dynamical friction and tidal stripping. As a result several important features of shells galaxies, like the small radii of the innermost shells, have remained unexplained. In order to address important dynamical effects such as the orbit decay and progressive tidal distruction of the companion as well as the selfconsistent effects on the primary, we have undertaken a study of shell formation using the treecode on the NCUBE.

Figure 2 shows the distribution of companion particles near the end of the simulation. The core of the companion stayed intact for five pericentre passages and was destroyed in the final pass. The following is a summary of the main points of the simulation with regard to shell production :

- O Shell making particles are removed from the companion after each close encounter with the core of the primary. As a consequence the final shell distribution is complex, consisting of several overlapping shell systems.
- O Since the companion core continues to lose orbital energy over the course of the simulation, the final shells are confined to smaller radii than the shells that resulted from earlier pericentre passages . The innermost shells

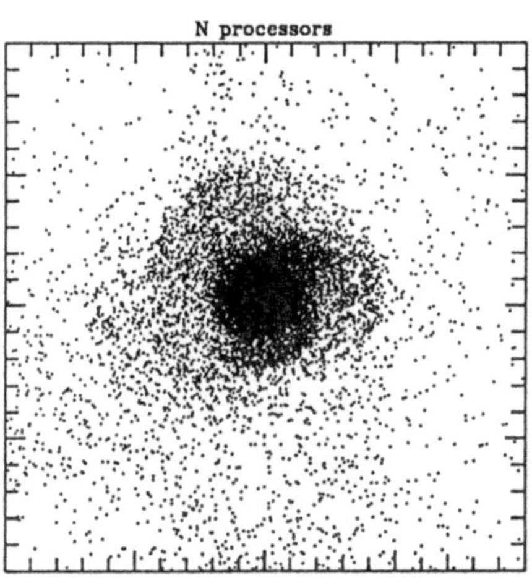

Figure 2 : The distribution of companion particles at the end of the simulation seen from about the original orbit plane. The box is 15 half-light radii of the initial primary on an edge.

are within 0.3 effective radii and the outer shells extend to more than 20 effective radii which gives a shell radius range similar to many observed shell galaxies.

O Because of the complex nature of the multiple shell system, it is very difficult in this case to apply the techniques outlined by Hernquist and Quinn (1987) to probe the underlying potential using the distribution of shell radii.

Conclusion

Many problems in galaxy formation and evolution will require the use of very large numbers of particles in N-body simulations. The model of shell formation we ran has addressed a number of outstanding issues in shell formation; in particular the radius of the inner most shells. The simulation has also given us valuable new insights into the effects shell making mergers have on the primary galaxies. These effects may explain many generic features of ellipticals (photometric and kinematic axis alignment, rotation curve shape, see Quinn and Zurek, this volume). The simulation also teaches us about the complex process of tidal disruption and the dynamics of mergers. By comparing this simulation with others on the formation of counterrotating cores by Balcells and Quinn (1989) and the cosmological simulations of Quinn, Salmon and Zurek(1986), we can start to see common processes emerge that may tell us a great deal about the way mergers work in general and how they may have influenced the properties of galaxies along the Hubble sequence.

Acknowledgements

We would like to thank the Caltech Concurrent Computing Project for providing travel funds and CPU time on the NCUBE. This work was supported in part by Department of Energy, Applied Mathematical Sciences grant DE-FG03-85ER25009.

References

Balcells, M. and Quinn, P.J., 1989, preprint.
Barnes,J. and Hut,P., 1986, *Nature*, **324**, 446.
Dupraz, C. and Combes, F., 1986, *Astr. & Ap.*, **166**, 53.
Hernquist, L. and Quinn, P.J., 1987, *Ap.J.*, **312**, 1.
Hernquist, L. and Quinn, P.J., 1988, *Ap.J.*, **331**, 682.
Quinn, P.J., 1984, *Ap.J.*, **279**, 596.
Quinn, P.J. and Zurek, W.H., 1988, *Ap.J.*, **331**, 1.
Quinn, P.J. and Zurek, W.H., 1989, preprint.
Quinn, P.J., Salmon, J.K. and Zurek, W.H., 1986, *Nature*, **322**, 392.

N-Body Integration on a Transputer Array

W.L. Sweatman

Department of Mathematics, University of Edingburgh,
Mayfield Road, Edinburgh, EH9 3JZ, UK

1. ABSTRACT: This paper describes an N-body program written to run on a transputer array. The computer environment and algorithms used are outlined, and then results are given relating efficiency to the size of the array.

2. COMPUTER ENVIRONMENT

The Edinburgh Concurrent Supercomputer is an array of several hundred transputers. Each transputer contains a processor (rated at about 1 Mflop), memory and communications. Transputers can be connected together to form large "concurrent" systems. The program runs on such a system.

3. ALGORITHM

The program divides into two sections: the master(1 transputer) controls the system, initialises the calculation, and collects the results; the slaves (s transputers) perform the bulk of the calculations. Essentially the algorithm uses an individual variable timestep for each body; however, to spread the load across the transputer array, s bodies are updated at once. In each cycle of computation the master finds the next s bodies to be updated; the earliest update time is used for all s bodies. Each slave updates one of the s bodies, and then communicates the new data to the other transputers.

The algorithm is a multistep integrator similar to that of NBODY1 (Aarseth 1985), but of one order lower. The individual timestep computed for each body is chosen to keep the error in energy to within a preset limit.

4. RESULTS

The results here were obtained from a set of N identical masses whose initial positions and velocities were generated from a Plummer model.

Empirically, for most of the results, the processing time per update is given to within 10% by the formula

$$\frac{1}{8000}\cdot\frac{N}{s} + \frac{7}{4000}\cdot s^{1/4}$$

Dynamics and Interactions of Galaxies Editor: R. Wielen

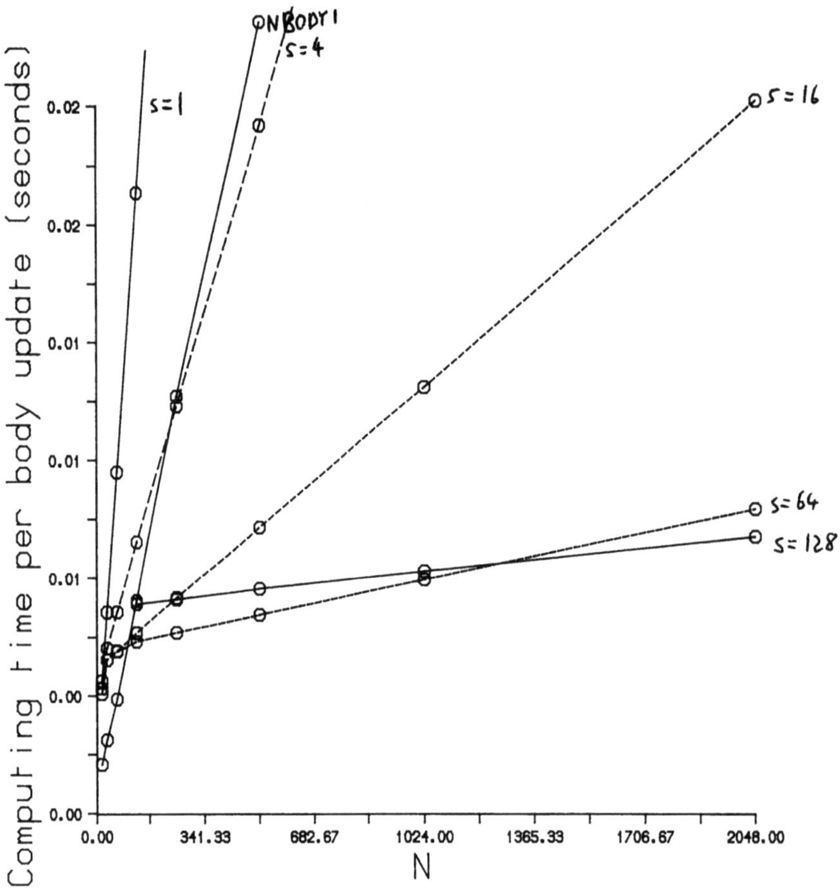

Fig. 1

The two contributions are mainly due to computation and communication, respectively. Fig. 1 shows the linear relation between N and processing time per body update, for various numbers of slaves(s). Also shown is Aarseth's NBODY1 running on a 2 Mflop scalar machine.

Using larger s the program takes less time per body update, but it also forces some updates sooner than are required by the timestep criterion. Fig. 2 shows how steps taken relate to N, for various s. As N increases the disadvantage of simultaneously updating s bodies gradually disappears.

Figs. 3, 4 plot, against N, the processing time to update the system a given fraction of the crossing time, again for various s. They show that as N is varied the optimal value of s takes different values.

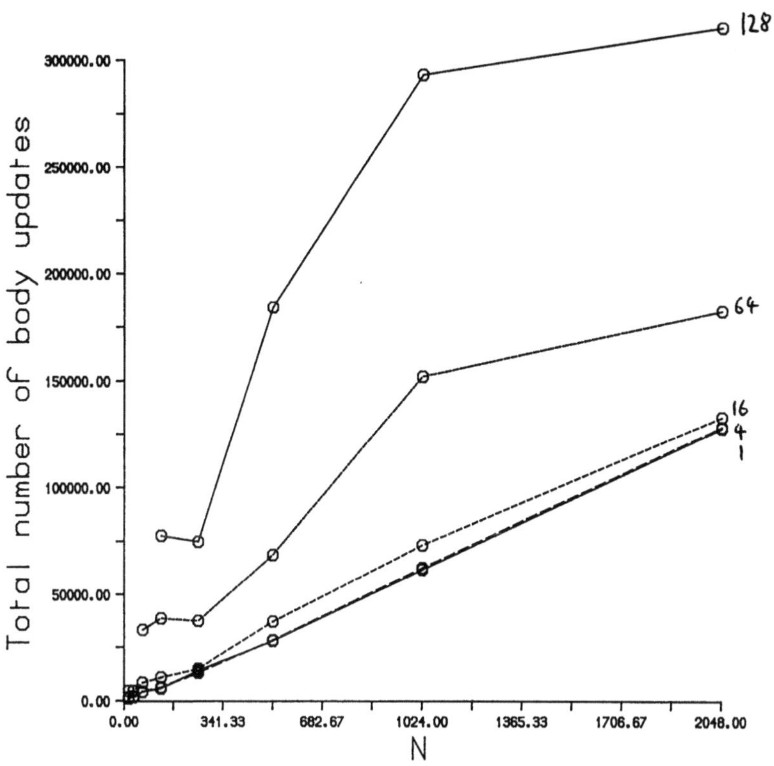
128
64
16
4
1
300000.00
250000.00
200000.00
150000.00
100000.00
50000.00
0.00
Total number of body updates
0.00
341.33
682.67
1024.00
1365.33
1706.67
2048.00
N

Fig. 2

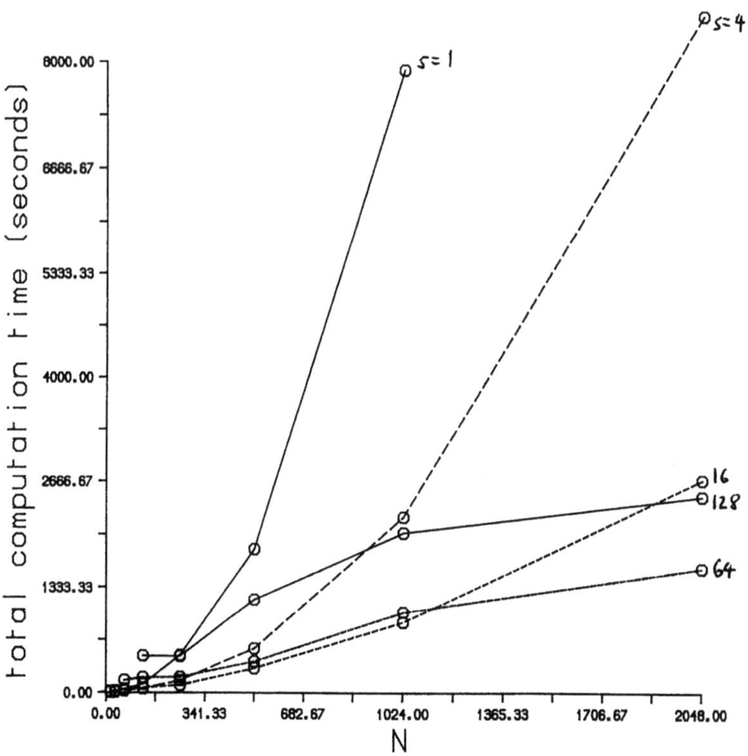
s=1
s=4
16
128
64
8000.00
6666.67
5333.33
4000.00
2666.67
1333.33
0.00
total computation time (seconds)
0.00
341.33
682.67
1024.00
1365.33
1706.67
2048.00
N

Fig. 3

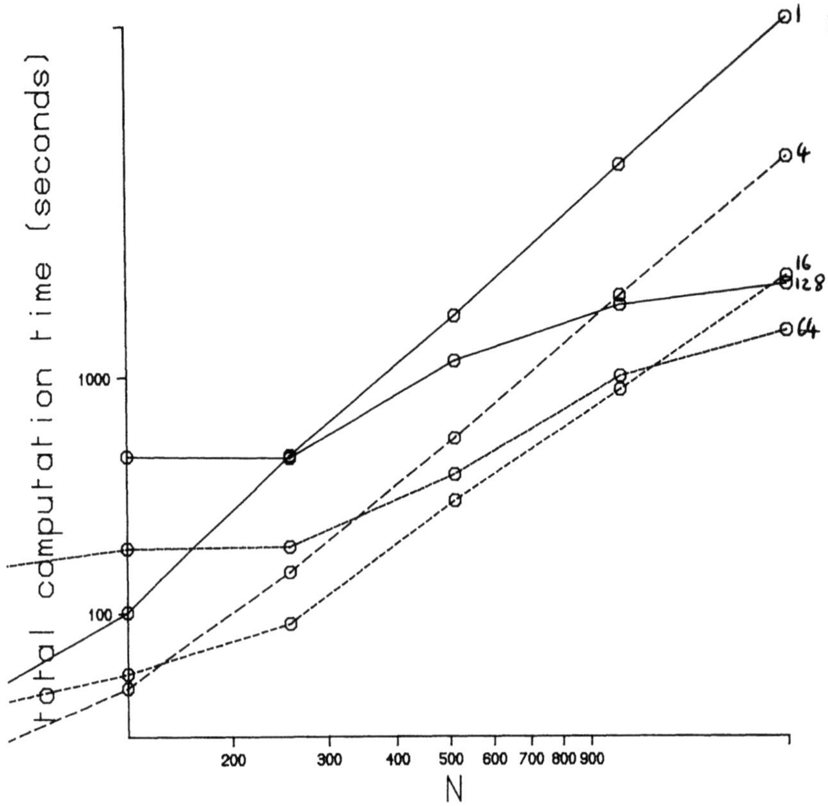

Fig. 4

5. REFERENCE

1. S.J.Aarseth: in Multiple Time Scales (eds. J.U.Brackbill and B.I.Cohen, Academic Press, 1985) pp. 377-418.

Models of Colliding Galaxies: Kinetic Energy and Density Enhancements

R.A. Gerber[1,2], *S.A. Lamb*[1,2], *R.H. Miller*[3], *and B.F. Smith*[2]

[1]University of Illinois, Urbana-Champaign, IL, USA
[2]Space Sciences Division, NASA-Ames Research Center, MS 245-6, Moffett Field, CA94035, USA
[3]University of Chicago

1 Introduction

Collisions and close encounters between galaxies have a profound effect upon the galaxies involved. For example, morphological changes can be pronounced. A very obvious effect of interactions on galaxies is a mean increase in their far infrared luminosities as detected by IRAS (1). This observed excess in infrared luminosity has often been invoked as evidence of increased star formation in these systems (2).

Theoretical calculations show that a galaxy initially experiences a roughly homologous infall as a result of a close passage (3,4) with the stars gaining kinetic energy and increasing their space density (5). The interstellar medium will gain kinetic energy along with the stars during this infall stage, resulting in an increase in cloud collisions which may lead to enhanced rates of star formation.

Here we show the results of using N-body models to obtain information which will be useful in probing some aspects of the relationship between galaxy collisions and star formation. The models allow us to identify regions of increased density in the systems and to calculate the increase in kinetic energy which is available to trigger star formation. The calculations model the stellar component of galaxies, but it is anticipated that the gas will follow the stars during the initial infall stage that we are concerned with here.

2 The Models and Results

The experimental collisions we use are those described in (3,5). These are the results of fully self consistent N-body calculations in which the self-gravity of each galaxy is included. The models are produced using a particle-mesh technique and contain approximately 100,000 particles. The collisions have varying values of orbital angular momentum and two different values of initial orbital energy corresponding to a parabolic sequence, with orbital kinetic energy equal to potential energy, and a hyperbolic sequence; with a difference in initial orbital kinetic energy between the two series of about a factor of two. All encounters are interpenetrating, ranging from almost head on to almost grazing.

We analyze collision experiments of spherical galaxies, for both equal mass encounters and galaxies with a mass ratio of 2:1. We also investigate collisions between equal mass spherical galaxies with cold, thin, rotating disks embedded in them. In these the disks contribute only 1 percent of the total mass. Each spherical galaxy is the stellar dynamical analog of an n=3 polytrope and the disks have a surface density that falls off as R^{-1}. In the following discussion, we give time units in terms of a typical stellar crossing time in an unperturbed galaxy, t ,and distances in terms of the unperturbed galaxy half-mass radius, l.

A Equal mass galaxies. The total kinetic energy and the density peak somewhat after close approach. For example, the rms velocity in a head on parabolic collision increases 42 percent over the starting value, reaching its maximum 0.12 t after close approach. The relative increase in the rms velocity decreases with increasing impact parameter, and very little effect is noticed for the encounter with our largest impact parameter (2.6 l.). We find that the parabolic collisions lead to larger peak rms velocity than do the hyperbolic encounters for a given separation at close approach, although the difference between the two series is only on the order of a few percent.

The galaxies reach their most contracted state up to $0.25t$ after close approach and, in the case of the almost head on parabolic collision, the central density of the galaxy's own particles has doubled by this time. Generally, when one takes into account the contributions of the particles of both galaxies the density can be as high as 2.5 times the initial value.

Dynamics and Interactions of Galaxies Editor: R. Wielen

In the disk-halo experiments we find that face on collisions produce a larger increase in disk particle kinetic energy than those in which the disks hit edge on. We find that the disk particles pile up into spiral and strong linear features soon after closest approach and the disks contract, following the overall gravitational potential produced by the halo.

B Unequal mass galaxies. The collision affects both galaxies in the unequal mass pairs, but it has a more profound effect on the less massive of the two. For example, the kinetic energy of the less massive one more than triples during the head on parabolic collision. There is a time lag between the kinetic energy peaks of the two galaxies, with the larger galaxy reacting more slowly and reaching its energy peak up to .25 *t* after the smaller galaxy, where t is defined as the stellar crossing time in the unperturbed larger galaxy. We find that the rms velocity of the larger galaxy increases more than that of the smaller galaxy. A major feature of the unequal mass experiments is the destruction of the smaller galaxy except at our larger impact parameters.

3 Implications for Star Formation

We expect that any gas present in real galaxies will follow the stars during the initial infall stage of the collision and this will lead to an increase in cloud-cloud and cloud-intercloud collisions. Interpenetration of the galaxies may also produce very high energy collisions in the interstellar material and together these effects may result in either disruption of the star forming regions or enhancement in the rate of star formation due to an increase in the number of star forming centers and possibly to an increase in the size or activity of these regions. Our calculations give us an approximation to the density and kinetic energy increases of interstellar material during the early infall stages and allow us to estimate timescales for possible increases in the star formation rate. We note that for a head-on, parabolic collision between equal mass spherical galaxies of mass 5×10^{10} solar masses and radius 16 kpc, the kinetic energy peaks 8 million years after minimum separation and the central density of the galaxies' own particles reaches its maximum another 8 million years later. Thus one might expect to observe a significant increase in the star formation rate about 10^7 years after the first passage of such galaxies. The timescale for merger of these galaxies (about 10^9 years) is not the relevant one for the initiation of star formation.

Our finding that the larger and smaller galaxies in an interacting system react dynamically on different time scales suggests that star formation may be initiated at different times in such a pair. For example, if the larger galaxy has a mass of 6.7×10^{10} solar masses and a radius of 16 kpc and the smaller galaxy has a mass of 3.4×10^{10} solar masses and a radius of 11 kpc, the larger may have a time lag of up to 16 million years in its kinetic energy peak. This may explain why some pairs of interacting disk galaxies, which are otherwise similar, exhibit very different current star formation characteristics (6).

Some interacting systems, such as UGC 12914/5, are observed to have large HII regions far away from either nucleus. These systems may be explained as the result of an interpenetrating collision of two disk galaxies in which a high density was obtained part way between the two nuclei. If the angular momentum and kinetic energy roughly cancel then a somewhat detached star forming region may result. We see evidence of this possibility in collisions with impact parameters around 0.9 to 1.3 *l*. (from 4 kpc to 6 kpc for a galaxy of 5×10^{10} solar masses).

Interacting counter rotating disks may help reduce the total angular momentum of the gas in interacting systems and thus allow the gas to spiral in towards the nuclei and accumulate there. Strong density enhancements are found in the disk particles during collision, and these are often in the form of strong, if temporary, linear (bar-like) features. The existence of such features in real systems appears correlated with high rates of star formation .

References

1 Bushouse, H.A., Lamb, S.A., and Werner, M.W., 1988, *Ap.J.*, **325,** 74, and references within.
2 Bergvall, N. and Johansson, L., 1985, *Astron. Astrophys.*, **149,** 475.
3 Miller, R.H. and Smith, B.F., 1980, *Ap.J.*, **235,** 421.
4 Miller, R.H., 1989, "Stellar Dynamics", in *Numerical Methods in Astrophysics,* (ed. P.R. Woodward), to be published.
5 Gerber, R.A., Lamb, S.A., Miller, R.H., and Smith, B.F., 1989, in preparation.
6 Bushouse, H.A., 1987, *Ap.J.*, **320,** 49, and references within.

Remnants of Strong Tidal Interactions

T.A. McGlynn

Space Telescope Science Institute, 3700 San Martin Drive, Baltimore, MD21218, USA

1. Introduction. This paper reported on n-body calculations studying tidal interactions which stripped roughly half the mass of a stellar system. The mass ratio between primary and secondary was large, $m_p/m_s = 10^2 - 10^6$. Such strong interactions may be important in the early development of globular cluster systems (Elson *et al.* 1987), in understanding the origin of Population II stars (Fall and Rees 1977), tidal stripping of dwarf companion galaxies (*e.g.*, Murai and Fujimoto 1985), in interactions between galaxies in groups and clusters (Aguilar and White 1985) and in interactions between galaxies and clusters (Miller 1986; Fitchett *et al.* 1990).

2. Encounter Models. Nine models were started as strongly concentrated King models and were dropped through point-mass and King external potentials. The potentials were evaluated using the tree code algorithm developed by Barnes and Hut (1984). This code does not put any restrictions on the geometry or density profiles of the developing systems. Eight of the systems had remnants which survived encounters which stripped about half the mass of the original system. However, except in the most centrally concentrated King models, the cores were puffed up and the systems would likely not survive a series of shocks. For most of the systems the density within the half-mass radius of the remnant was close to the density within the half-mass radius of the initial system.

In such strong interactions, the energy integral appears to dominate the dynamics of the system. In tidal encounters, a characteristic $\rho \propto r^{-4}$ density profile and flat distribution in energy, $N(E)$, is observed. These distributions are expected if the tidal interactions essentially scatter the energies of the particles in the system. Measures of the importance of angular momentum in the final systems indicate that it is of secondary importance although there were indications of both a preference for radial orbits and an imparted rotation in some of the remnants. The energies of particles stripped from the systems could be understood by simple kinematic descriptions of the stripping process where particles are instantaneously removed from the influence of the secondary mass.

3. N(E) and $\rho(r)$. $N(E)$ and $\rho(r)$ are closely related in systems where the distribution function is a function of energy alone and where the potential of the system can be described as a power law. If $\Phi(r) = \alpha r^n$ and $N(E) = \gamma E^m$ then the density profile of the system is $\rho(r) = \beta r^{nm+n-3}$. In the centers of systems where the potential is finite this implies $mn = 1$ while at large radii $\rho(r) = \beta r^{-m-4}$. For the flat energy distributions here we expect $m = 0$ hence $\rho \propto r^{-4}$ (also see Jaffe 1987).

4. Steady Tides. Three models were run with a system in a circular orbit in a King external potential. In constant tides, stripping does not appear to affect the central properties of the system, nor does it appear to show strong preferences in the angular momentum of particles stripped. Remnants are reasonably appoximated by simply removing all particles with a binding energy less than some nominal value. The final tidal radius is larger than classical tidal radius (*e.g.*, Keenan and Innanen 1975). Current observations of the globular-like clusters in the Magellenic clouds (e.g. Elson *et al.* 1987) indicate that they likely exceed their tidal radii and will have a large fraction of mass stripped from them in their next few orbits, but this process should not affect the central properties of the clusters according to the results of these simulations.

5. Comparison with Observations. A sample of galaxy profiles (Pelletier *et al.*1989) shows a variety of outer profiles at large radii with a significant fraction showing R^{-3} surface density profiles expected if strong tidal interactions took place in their development. There is a tendency for the r^{-3} profiles to be seen in the lowest mass galaxies which is where the tidal effects are expected to be most extreme.

The complete analysis of these results may be found in McGlynn (1990).

References.

Aguilar, L., and White, S.D.M., 1985, *Ap.J.* **295**, 374.

Barnes, J., and Hut, P., 1986, *Nature* **324**, 446.

Elson, R.A.W., Fall, S.M., and Freeman, K.C, 1987, *Ap.J.* **323**, 54.

Fall, S.M., and Rees, M, J., 1977, *MNRAS* **181**, 37P.

Fitchett, M.J., McGlynn, T.A., Quinn, P.J., and Zepf, S.E., 1990, preprint.

Jaffe, W, 1987 in *IAU Symposium 127, Dynamics of Elliptical Galaxies*, ed. T.de Zeeuw (Dordrecht: Reidel), 511.

Keenan, D.W., and Innanen, K.A., 1975, *A.J.* **80**, 290.

McGlynn, T.A., 1990, to be published in *Ap.J.*

Miller, R, 1986, *Astr. and Ap.* **167**, 41.

Murai, T., and Fujimoto, M., 1986, *Ap.and Sp.Sci.* **119**, 169.

Pelletier, R.F, Davies, R.L, Illingworth, G.D., Davis, L., and Cawson, M., 1989, preprint.

Headon-Collisions Between Plummer Models

P. Teuben[1] *and P. Hut*[2]

[1]Astronomy Program, University of Maryland, College Park, MD 20742, USA
[2]The Institute for Advanced Study, Princeton, NJ 08540, USA

We present a framework for a detailed analysis of encounters between stellar systems, on two levels: an interaction analysis and a scattering analysis. The interaction analysis covers a closeup investigation of the details of the time-dependent deformations of the systems during an encounter. The scattering analysis describes the encounter as a transformation between the asymptotic states of the systems in the early past and the far future. Such gravitational scattering experiments can be described in a way similar to laboratory experiments in other branches of physics, such as nuclear, atomic and molecular physics.

Our interaction analysis compares different ways of measuring physical variables such as density (spatial, projected as well as phase space density), and of classifying and describing the different parts of the system, including the clouds of escaping stars produced during an encounter. We also discuss different ways of following the trajectories of individual star systems and star clumps. From a practical point of view, we discuss strategies for automatizing the setup, execution and analysis of simulations of encounters between galaxies. We present real-life examples of command sequences, in the form of shell scripts, which orchestrate such experiments.

Our scattering analysis is focused on the relation between observable physical variables before and after the encounter on the one hand, and their dependence on the choice of non-physical parameters and numerical methods characterizing the experiments on the other hand.

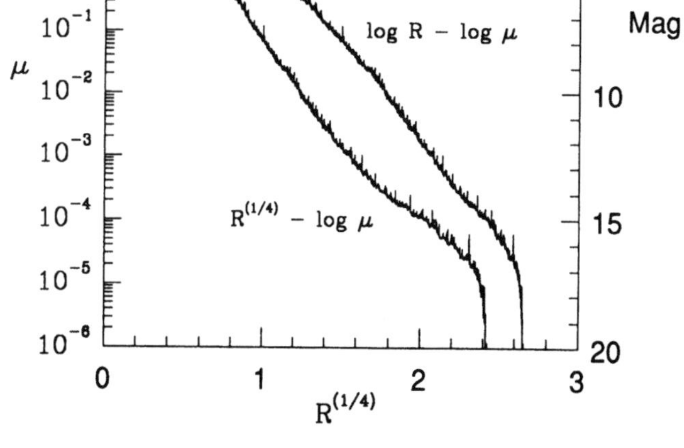

We start with a detailed investigation of isolated spherical galaxy models, realized as a finite set of mass points. Already on this level there are interesting relations between the deviations from the ideal mathematical models from which they are constructed, both for individual observable variables and their correlations. Knowledge of these deviations and their time dependence then allows us to discriminate between these one-galaxy errors, and the additional two-galaxy errors introduced during the simulations of an encounter. We illustrate our discussion with the results of a large variety of test runs. The detailed understanding of the process of gravitational scattering simulations, should anable us to put error bars on the physical measurements presented in subsequent analysis.

In the current work we present a detailed analysis of the simplest type of encounters between galaxies, in which two spherically symmetric, isotropic galaxies undergo a head-on encounter. Among our results are the dependence of energy loss and mass loss as a function of encounter velocity, as well as an accurate determination of the criterion for merging. A qualitatively new result is the determination of the shape of the cloud of escaping particles, both in the merging and non-merging cases. In addition, we present a detailed analysis of the luminosity profiles and three-dimensional shape of the galaxies after an encounter, and of their orbit structure.

In the accompanying figure the projected surface density of a merger product is shown. The obvious bump at a radius of about 20 units is due to the relatively early time of this snapshot, and will smoothen out in time.

Sharp Edges Due to Close Interactions

K.J. Donner, S. Engström, B. Sundelius, and M. Sundin

Department of Astronomy/Astrophysics, Chalmers University of Technology, S-41296 Göteborg, Sweden

Sharp edges of spiral arms, found in simulations and observations of galaxies in interaction, are explained kinematically using the impulse approximation.

We investigate the kinematics of a galactic disc perturbed by a disturber on an unbound orbit. Such encounters can be expected to be common if galaxies are initially formed in weakly bound, low angular momentum systems. Here we shall be concerned with smaller disturbing masses (less than 10 % of the main galaxy) as larger disturbances will tend to disrupt the disc. We present results for the case of a 5 % disturber passing through the outer parts of a Kuzmin disc.

The impulse approximation consists in calculating the perturbation due to a passing mass, assuming that each star remains stationary during the encounter. The case the approximation mimicks is really that of an extreme hyperbolic passage through the disc. This calculation connects to the parabolic encounter by using the same magnitude of injected momentum. Comparisons with simulations, self-gravitating and not, are given in Sundelius' talk in these proceedings. They show that the feature of sharp edges is a general one, at least when considering unbound orbits. The thus perturbed stellar orbits can be approximated by epicyclic oscillations about a deferent on a circular orbit. The resulting motion is:

$$R = R_0 + \Delta R + a\cdot\cos(\kappa t + \beta) \qquad (1)$$

$$\phi = \phi_0 + (\Omega + \Delta\Omega)\cdot t + (2\Omega/\kappa)\cdot(a/R_0')\cdot\sin(\kappa t + \beta). \qquad (2)$$

R is radial coordinate, ΔR is the change in radius due to injected angular momentum. a, κ and β are epicyclic amplitude, frequency and phase respectively. t is the time after the disturbance, ϕ is polar angular coordinate, Ω is angular velocity of the original circular orbit, $\Delta\Omega$ change in Ω due to angular momentum transfer. The amplitude, a, is proportional to the disturbance $M\cdot\Delta t$, where M is the mass of the disturber and Δt is the effective duration of the encounter.

In a differentially rotating disc, every disturbance will wind up as a spiral. To overcome this visually confusing fact one can try to remove the differential rotation. The equations above suggest two ways of doing this:

(1) in what can be called the epicycle representation, the stars can be plotted keeping the position of their deferents stationary at their perturbed position (subtract $(\Omega + \Delta\Omega)\cdot t$ from (2)).

Dynamics and Interactions of Galaxies Editor: R. Wielen

(2) in a Lagrangian representation the positions are plotted keeping the unperturbed deferent stationary (subtract $\Omega \cdot t$ from (2)).

Working in such a representation of the perturbed stars orbits, it can be seen that particles on circles around the disturber will move on epicycles as in Figure 4 in Sundelius' talk, these proceedings. Stars closer to the disturber will receive a stronger impulse and overtake those starting farther away. A hole thus created will map on to a hole in the ordinary, differentially rotated representation. The edges of the hole will become the sharp edges of the spiral arms. We search for the folds in this mapping by calculating the Jacobian of the transformation:

$$(R_0, \phi_0) \rightarrow (R, \phi).$$

This can be done analytically in a local approximation around the disturber, valid for small disturbances. One then finds that the size of the hole scales with the disturbance as $(M \cdot \Delta t)^{1/3}$, a result verified by simulations.

Part VII

Elliptical Galaxies

New Clues to Structure and Formation of Elliptical Galaxies

R. Bender

Landessternwarte, Königstuhl, D-6900 Heidelberg, Fed. Rep. of Germany

1. Introduction

This review confines to two topics which gave important new clues to the structure and formation of elliptical galaxies in the past two years: isophotal shapes and core kinematics.

Excellent reviews have appeared recently (*e.g.* Nieto 1989, Davies 1989, Kormendy and Djorgovski 1989) and I refer to these for a more comprehensive account to all previous work. Nevertheless, I will list some properties of early-type galaxies which, according to my opinion, are among the most important ones:

- Both ellipticals and bulges show surface-brightness profiles rather well described by an $r^{1/4}$-profile (de Vaucouleurs 1948, 1953). Although systematic deviations from the $r^{1/4}$-law are frequently present (*e.g.* Kormendy 1977, Capaccioli, 1985) and show a dependence on galaxy luminosity (*e.g.* Schombert 1986, 1987), the more or less universal applicability of this law to ellipsoidal stellar systems gives an important clue to common underlying principles of their formation.
- Luminous ellipticals are frequently flattened due to anisotropic velocity dispersions (Bertola and Capaccioli 1975, Illingworth 1977), while ellipticals of lower luminosity ($M_T > -20.5$ mag, $H_o = 50$ km/s/Mpc) are in general supported by rotation (Davies et al. 1983). The latter objects resemble the bulges of disk–galaxies in their kinematics (*e.g.* Kormendy and Illingworth 1982, but also Whitmore et al. 1984, Gerhard et al. 1989). This result is pointing to more dissipation in lower luminosity ellipticals and bulges and also may indicate the increasing influence of merging with increasing luminosity of the objects.
- The anisotropy of luminous ellipticals implies that these objects can be triaxial (see Binney 1982 and references therein). This is partially supported by the presence of isophotal twists, but the conclusion is nonunique because isophotal twists can be due to interactions (see *e.g.* Kormendy 1982). Also, statistical tests based on *e.g.* major- and minor-axis kinematics (Binney 1985) appeared to be not very conclusive (see *e.g.* Davies 1989, Nieto 1989); the reason for that can most likely be found in the non-homogeneity of the class of ellipticals.
- The global optical parameters of elliptical galaxies (blue luminosity L_B, half-light radius r_e, mean surface brightness $< \mu >$, velocity dispersions σ, and colors) are intimately related ('fundamental plane', Djorgovski and Davis 1987, and references therein). The importance of these parameter relations is based on the fact that they provide a powerful distance indicater (see also Dressler et al. 1987). Whether they also allow insight in the formation process of these objects or are independent from it and just represent relations valid for all kinds of relaxed stellar systems is still a matter of debate (Nieto et al. 1989).
- A result often not considered sufficiently by optical astronomers is the detection of huge X–ray halos around luminous elliptical galaxies (*e.g.* Forman et al. 1985, Fabbiano and Trinchieri 1985, Canizares et al. 1987 and references therein). Although there exists a correlation between optical and X–ray luminosity ($L_X \approx L_B^{1.7\pm0.3}$), there is a large scatter about this relation. As stressed by Canizares et al. (1987), the investigation of the origin of this scatter and its relation to a further parameter is crucial for a better understanding of the status

Dynamics and Interactions of Galaxies Editor: R. Wielen

and evolution of the X–ray halos and the galaxies themselves (here, isophotal shape analysis can give important new clues, see below). Furthermore, the mere presence of the halos most likely indicates large amounts of dark matter around X–ray luminous galaxies (Forman et al. 1985, Fabian et al. 1986)

- Colors and color gradients as well as metallicity indices are intimately linked to population and metallicity gradients in ellipticals and bulges. Therefore, they can in principle give an important contribution to the understanding of the structure of these objects. However, because of the difficulty to distinguish between metallicity changes and population gradients and also because of the large observational problems (Capaccioli 1987), the study of colors and metallicities was up to now unable to give further insight in the structure and formation of early–type galaxies (for a review, see Kormendy and Djorgovski 1989). This situation will hopefully improve in the future by the larger fields of new CCDs and the use of focal reducers.

In summary, valuable new insight in the structure of elliptical galaxies has been obtained in the last fifteen years, but also revealed many new aspects showing the complexity of these objects. It has become a common opinion that beyond the very global correlations all ellipticals are individuals reflecting very individual histories.

In this paper, I will argue that the last statement is only partially true. Especially the investigation of isophote shapes and their relation to other parameters allows to sub-classify ellipticals according to physical properties. Furthermore, both isophotal shape analysis and the detection of peculiar core kinematics brought us a step further in the long standing question of the frequency of recent mergers among luminous elliptical galaxies (Toomre 1977).

2. Kinematically decoupled cores in elliptical galaxies

Observations in recent years have shown that the cores of elliptical galaxies are not just the isotropic central regions of violently relaxed stellar systems, but in general have a nontrivial structure. Photometric evidence for this fact was obtained mainly by Lauer (1985a) and Kormendy (1987) after the pioneering work of Schweizer (1979) (see Kormendy and Djorgovski 1989 for a review).

Kinematic evidence for peculiar core properties in ellipticals was first obtained by Efstathiou et al. (1982) for NGC 5813. Besides a strong increase in rotation velocity in the core, this galaxy shows a core-within-a-core surface-brightness profile supporting additionally the existence of distinct subcomponents. Subsequently, Kormendy (1984) interpreted these observations as the result of a merging between a compact low-luminosity and a high-luminosity elliptical.

Our knowledge on kinematical sub-components in the cores of ellipticals increased immensely in the last two years. Striking cases are the counter-rotating cores in IC 1459 (Franx and Illingworth 1988), NGC 3608 (Jedrzejewski and Schechter, 1988) and NGC 5322 (Bender 1988b). Even more astonishing is the detection of cores rotating at skew angles: NGC 4365 (Bender 1987, 1988b) and NGC 4406 (Bender 1987,1988b, Franx et al. 1989). While counter-rotation between core and main body can be a stable phenomenon in an axisymmetric galaxy, inclinations around 90° between the angular momenta of core and main body indicate a triaxial shape or an even not yet completely static configuration (cf. Smith et al. 1989, Miller and Smith 1989). The most strange case for skew core kinematics is NGC 4365 (Fig. 1): the core rotates about the photometric minor axis of the galaxy, while the main body's rotation axis lies near the photometric major axis of NGC 4365. Unfortunately, because of the small core size, little is known about the photometric core structure of NGC 4365 up to now.

A galaxy where we have more detailed knowledge on the photometric and kinematic structure is NGC 5322 (Bender et al. 1989b). Here, signicificant constraints on the core structrure can be derived:

NGC 5322 exhibits a main body rotating along its major axis with 30 km/s and core counter-rotating with an amplitude of $\approx$ 70km/s (Fig.2). No rotation along the minor axis is found,

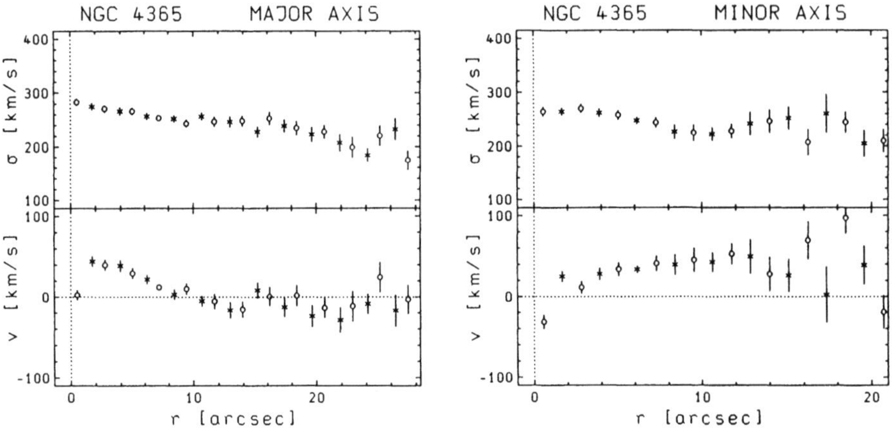

Figure 1: *Rotation velocity v and velocity dispersion σ of NGC 4365 along major and minor axis. The curves were folded around the center with different symbols referring to different sides.*

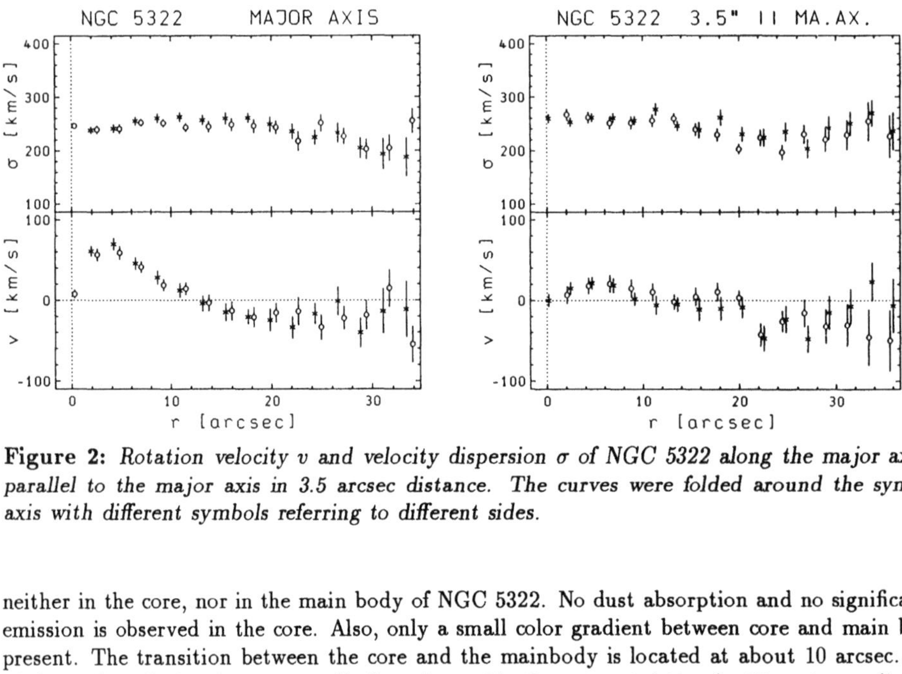

Figure 2: *Rotation velocity v and velocity dispersion σ of NGC 5322 along the major axis and parallel to the major axis in 3.5 arcsec distance. The curves were folded around the symmetry axis with different symbols referring to different sides.*

neither in the core, nor in the main body of NGC 5322. No dust absorption and no significant gas emission is observed in the core. Also, only a small color gradient between core and main body is present. The transition between the core and the mainbody is located at about 10 arcsec. Inside of this radius the isophotes are well aligned, outside they start twisting by 3° up to a radius of 50 arcsec (Bender et al. 1987). The isophotal shape (see Sect. 3) changes dramatically between the core and the outer parts. While the main body of NGC 5322 appears to be boxy, the core shows pointed isophotes from 8 arcsec inwards down to a radius of 2 arcsec (Fig.3). The fact that the pointedness is visible even at such small radii under moderate seeing conditions (the images of NGC 5322 were obtained with a seeing of ≈ 1.6 arcsec) suggests the existence of a flat ($b/a < 0.3$) and bright additional component in the core. It is plausible to assume that this component is responsible for the counter-rotation. Indeed, a spectrogram taken in 3.5 arcsec distance to the center and parallel to the major axis shows (Fig.2) that the amplitude of the counter-rotation decreases very fast perpendicular to the major axis.

Further constraints can be obtained by the analysis of the line-of-sight velocity distributions (broadening functions) in the core. The broadening functions shown in the following were obtained

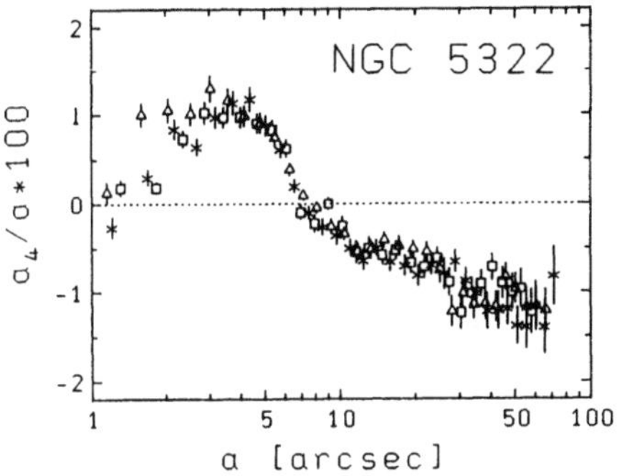

Figure 3: *The isophotal shape parameter a_4/a as a function of the semi-major axis length a of NGC 5322. Negative a_4/a indicate boxy isophotes, positive a_4/a pointed isophotes.*

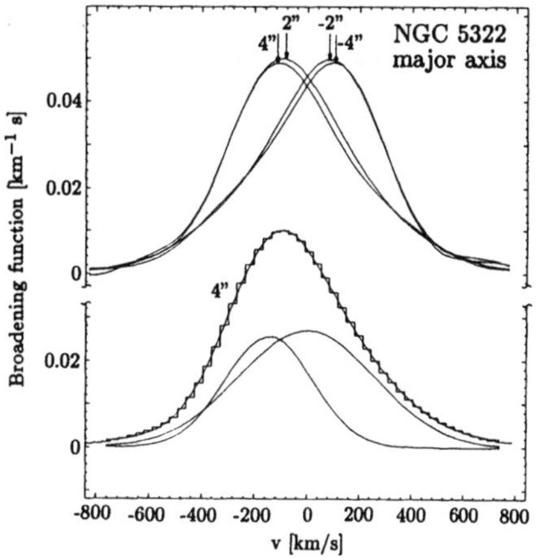

Figure 4: top: *The line-of-sight velocity distributions (broadening functions) of NGC 5322 on the projected major axis at ±2 arcsec and ±4 arcsec distance to the center.* **bottom:** *The broadening function at 4 arcsec (shown as a step function) fitted by the superposition of two Gaussians (full lines).*

using a new method to analyse the kinematics of elliptical galaxies (Bender 1989). Fig. 4 shows the broadening functions at ±2arcsec and ±4arcsec. It is evident that the wings of the broadening functions of different sides from the center match very well, the counter-rotation is just due to a shift of the peaks. A perfect fit to the broadening function at 4 arcsec can be obtained by the superposition of two Gaussians (Fig. 4), one component having no rotation and a velocity dispersion of 260 km/s, the other one rotating with 145 km/s and having a velocity dispersion of 165 km/s. The fast-rotating component contributes nearly 40 % to the total light on the major axis at 4 arcsec, this is roughly the same value one calculates from the simulation of the a_4/a profile by the superposition of an ellipsoid and a disk-like exponential component convolved with the seeing (a more detailed discussion will be given in Bender et al. 1989b).

NGC 5322 also shows a core-within-a-core surface-brightness profile similar to NGC 5813 (however, the effect is not as pronounced, see Surma and Bender, poster contribution to this conference). The central light excess in NGC 5322 is consistent with being due to the contribution of a flat counter-rotating component.

Finally, it is a very suggestive fact that the radio jet of NGC 5322 (Feretti et al. 1984) is perpendicular to the presumed 'disk' in the core.

Conclusively it can be stated that the core of NGC 5322 most likely consists of two kinematically decoupled components. One of them counter-rotates rapidly with respect to the main body of the galaxy and contributes about one third to the light of the core. The structure of NGC 5322 closely resembles the one inferred for IC 1459 by Illingworth and Franx (1989).

After this exemplifying discussion of the properties of NGC 5322, it is interesting to look for common photometric properties of all ellipticals with peculiar core kinematics. For the sake of completeness I first list all objects where we have firm or weaker indications for decoupled cores:
well established cases: NGC 3608, NGC 4365, NGC 4406, NGC 4494, NGC 5322, IC 1459, NGC 5813, NGC 7626
possible cases: NGC 1395, NGC 1439, NGC 1700, NGC 4278, NGC 4472, NGC 5982, NGC 7619.
The references are Franx and Illingworth (1988), Jedrzejewski and Schechter (1988), Bender (1988b), Franx et al. (1989), Wagner (1989), Bender and Möllenhoff (1989).

The objects listed cover a rather broad range in luminosity from $L_B \approx 10^{10} L_\odot$ (H_o = 50km/s/Mpc) to the most luminous ellipticals.

The radius where the kinematics of the core shows up is in all cases at least a factor 2 larger than the photometric core radius (defined as the radius where the surface brightness has dropped to half its central value). Therefore, photometric core radii can in general not provide a useful estimate of the extent of the physical core.

The main bodies of NGC 3608, NGC 4365, NGC 4406, NGC 4472, NGC 5322, and NGC 5982 show boxy isophotes. Irregular isophote shapes (indicated by deviations not axisymmetric to major and minor axis, see Sect. 3) and/or dust are present in NGC 1395, NGC 1439, NGC 4278, IC 1459, NGC 5813, and NGC 7626. NGC 4494 and NGC 7619 show very weak pointedness at the significance level, and only NGC 1700 has $a_4/a \approx 1\%$. Obviously, peculiar core kinematics is much more frequent in boxy and irregular ellipticals than in ellipticals with definitely pointed isophotes, although the frequency of boxy and irregular E's among luminous E's amounts to only $\approx 50\%$.

Another interesting aspect concerns the correlation with environment: there is none! Ellipticals with peculiar core kinematics occur in all kinds of environments, ranging from dominant cluster galaxies (NGC 7626) and ellipticals in clusters (NGC 4365, NGC 4406) to objects in small groups (NGC 5322, IC 1459).

Which explanations can be found for the origin of the decoupled core kinematics? Of course, merging is the most obvious way. Several scenarios can be imagined:

Accretion of a gas–rich companion: Gas settles in the core, subsequent star–formation leads to a stellar counter-rotating disk. The core-rotation reflects the orbital angular momentum. Early stages could be similar to the dust-lane ellipticals with counter-rotating gas (Bertola & Bettoni 1988, Galletta, this conference, Zeilinger, this conference).

Merging of a luminous elliptical with a low–luminosity compact elliptical which settles in the core of the merger, —> core-within-a-core (Kormendy 1984, Balcells & Quinn 1989). Compact ellipticals are mostly observed in the vicinity of bright galaxies (Nieto and Prugniel 1987, and Nieto, this conference).

Merging of two spirals. Scenario advocated by Schweizer (this conference) and also suggested by the observations of obvious mergers (Schweizer 1982) and of IRAS galaxies (this conference).

And the combination of all above possibilities:

Multiple merging in compact groups: Barnes (this conference) showed that the dynamical evolution of a compact group most likely results in an elliptical. In fact, NGC 5322 and IC 1459 are members of small groups with crossing times of $0.09H_o^{-1}$ and $0.04H_o^{-1}$ (Geller and Huchra 1982).

Weak counter-rotation can possibly be explained by less violent processes like:

'Primordial' misalignment: Barnes & Efstathiou (1987) found misaligned angular momenta between core and main body in their N-body simulations of collapsing proto-galaxies.

Tidal torques: may probably reverse the outer parts of a lower luminosity companion of a massive galaxy (May et al. 1985), as may have happened in the case of NGC 3608. However streaming in combination with counter figure rotation is difficult to achieve (Vietri 1986).

Bar motions: Especially low amplitude counter–streaming could be due to motions in a bar (e.g. Bettoni 1989). This may apply to NGC 4494, which shows a rotation curve similar to M31 (however, on much larger scales).

Numerical simulations and the investigation of the stability of the observed phenomena are important to constrain the possible formation scenarios and also may give hints to the shape of the central potential. The extensive N-body simulations of the merging between a high luminosity and a low luminosity elliptical performed by *e.g.* Balcells and Quinn (1989), Smith et al. (1989) are already able to account for many aspects of the above described phenomena. Both further modelling and the detailed decomposition of the different core components as shown in the preliminary analysis of NGC 5322 will hopefully give a much more profound insight in the structure of these objects in the future.

This section should not end without some sceptical remarks. Although, merging is very attractive to explain decoupled core kinematics, some fundamental problems of this scenario still deserve a more thorough analysis: the high central phase space densities of ellipticals compared to spirals (Carlberg 1986), the globular cluster frequency and further difficulties summarized *e.g.* by Tremaine 1981, Gunn 1987 and v.d. Bergh, this conference.

3. Isophote shapes

Already in 1951 Evans recognized that isophotes of elliptical galaxies may deviate significantly from pure ellipses:
'... *isophotal maps often show irregularities in the outer parts of the nebulae. Examples are the rift and tongue in the north following quadrant of NGC 1549, the southern patch of NGC 1380 and so on. It was thought first that they merely represent errors of measurement at the limit of faintness which could be reached, but an examination of the material suggest that these structures are represented on the plates and may have some basis in reality.*'
In 1975 Wilson noted that an adequate description of these deviations can be obtained by a Fourier series expansion in azimuth and that the dominant contribution should be due to the $\cos(4\phi)$–term.

The systematic investigation of the isophotal shapes of elliptical galaxies started only a few years ago (Carter 1979, 1987, Lauer 1985b, Jedrzejewski 1987, Jedrzejewski et al. 1987, Michard and Simien 1988, Bender and Möllenhoff 1987, Bender et al. 1987, 1988).

About 80% of the galaxies classified as ellipticals show systematic deviations from elliptical isophotes at the present detection limit ($\approx 0.5\%$ of the isophotal diameter). For the majority of the galaxies the $\cos(4\phi)$–term was indeed found to be the most significant one.

In the following we will designate the deviations by the scale-independent a_4/a parameter, *i.e.* by the $\cos(4\phi)$–term normalized to the semi-major axis a of the corresponding isophote. Because of the self-similarity of the isophotal shapes in most objects, this normalization allows to characterize most objects by a single a_4/a, what is crucial for a statistical investigation (for a description of how to obtain characteristic a_4/a, see Bender et al. 1989a). Isophotes exhibiting positive a_4/a have a pointed appearence, while negative a_4/a is caused by a box-like shape.

Pointed isophotes in ellipticals are usually interpreted to indicate the presence of weak stellar disks with high inclination against the line-of-sight. This interpretation is supported by the global properties (see below) of these objects, now frequently called 'disky' ellipticals. However, the pointedness may not always be due to a real disk, but may at least sometimes be caused by specific phase-space distributions (see e.g. the density distributions for certain Stäckel models, de Zeeuw et al. 1986).

In contrast to the rather straight-forward interpretation of pointed isophotes, boxy isophotes may be due to a variety of orbit populations (see Combes and Sanders 1981, Binney and Petrou 1985, May et al. 1985, Petrou 1987, Rowley 1988), but up to now our knowledge about the origin of boxiness in specific objects is still very limited. Constraints can only be derived from a detailed investigation and modelling of the kinematical properties of boxy ellipticals (see below).

Some objects can not be characterized sufficiently by a single a_4/a value, because other Fourier-coefficients dominate over a_4/a. Strong odd Fourier–terms (3ϕ, 5ϕ terms) indicate a generally irregular isophote shape, the presence of a $\sin(4\phi)$-term points to a disk inclined against the major axis of the bulge or a light excess only along one diagonal of the isophotes (causing a kind of 'skew boxiness' see Prugniel et al. 1988). The ellipticals with irregular isophote shapes show in general similar properties to the boxy ellipticals, so the discussion of boxy E's presented below also applies *cum grano salis* to irregular E's. The number of elliptical galaxies with irregular isophote shapes amounts to less than 1/3 of all objects observed so far (there exist indications that irregular objects are more frequent in dense environments, like *e.g.* in the central part of the Fornax cluster).

Besides the three principal types of isophotal shapes (boxy, pointed, irregular) some ellipticals show a radial transition from inner pointed to outer boxy isophotes and vice versa. Since this

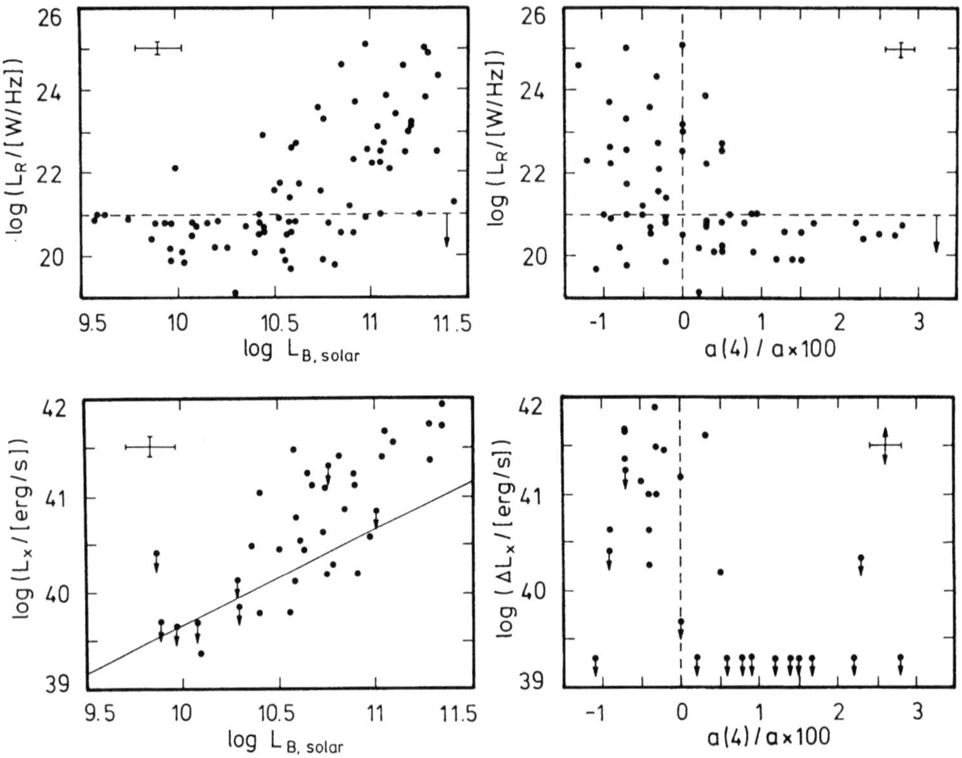

Figure 5: *The radio and X-ray properties of elliptical galaxies against blue luminosity L_B and isophote shape a_4/a. In the ΔL_X against a_4/a diagram the contribution of the discrete sources to the total X-ray emission (indicated by the line in the L_X against L_B plot) is subtracted according to Canizares et al. (1987). L_R gives the total radio luminosity at 1.4 GHz, values below* 10^{21} *W/Hz are nearly all upper limits.*

behaviour is rather rare and only leads to minor refinements of the principal results described below, I will not discuss this point further but refer to Nieto and Bender (1989).

The relevance of isophote shapes for our understanding of the structure of elliptical galaxies became clear by the statistical analysis of a large sample of objects.

Only weak correlations between isophotal shapes and the global optical parameters of luminous ellipticals are found (Bender et al. 1989a). There is a tendency that boxy ellipticals have on average slightly higher luminosity than disky objects. Furthermore, it is indicated that higher ellipticity implies stronger deviations, *i.e.* the largest a_4/a values occur in 'edge-on' systems (Bender et al. 1989a).

On the other hand, clear correlations between a_4/a and radio, X-ray, and kinematical properties are found. Therefore, isophotal shape analysis appeared to be a very simple and direct method to identify physically distinct objects.

The radio and X-ray properties of disky and boxy ellipticals are illustrated in Fig. 5 (taken from Bender et al. 1989a). It is obvious that a_4/a is a crucial second parameter (besides luminosity) for both radio emission and the presence of gaseous X-ray halos. The radio and X-ray data also show that the different kinematics of disky and boxy E's cannot be explained simply by different inclinations of the same type of objects.

The kinematical properties of luminous elliptical galaxies are presented in Fig. 6 (the data were compiled from Schechter and Gunn 1979, Davies et al. 1983, Davies and Illingworth 1983,

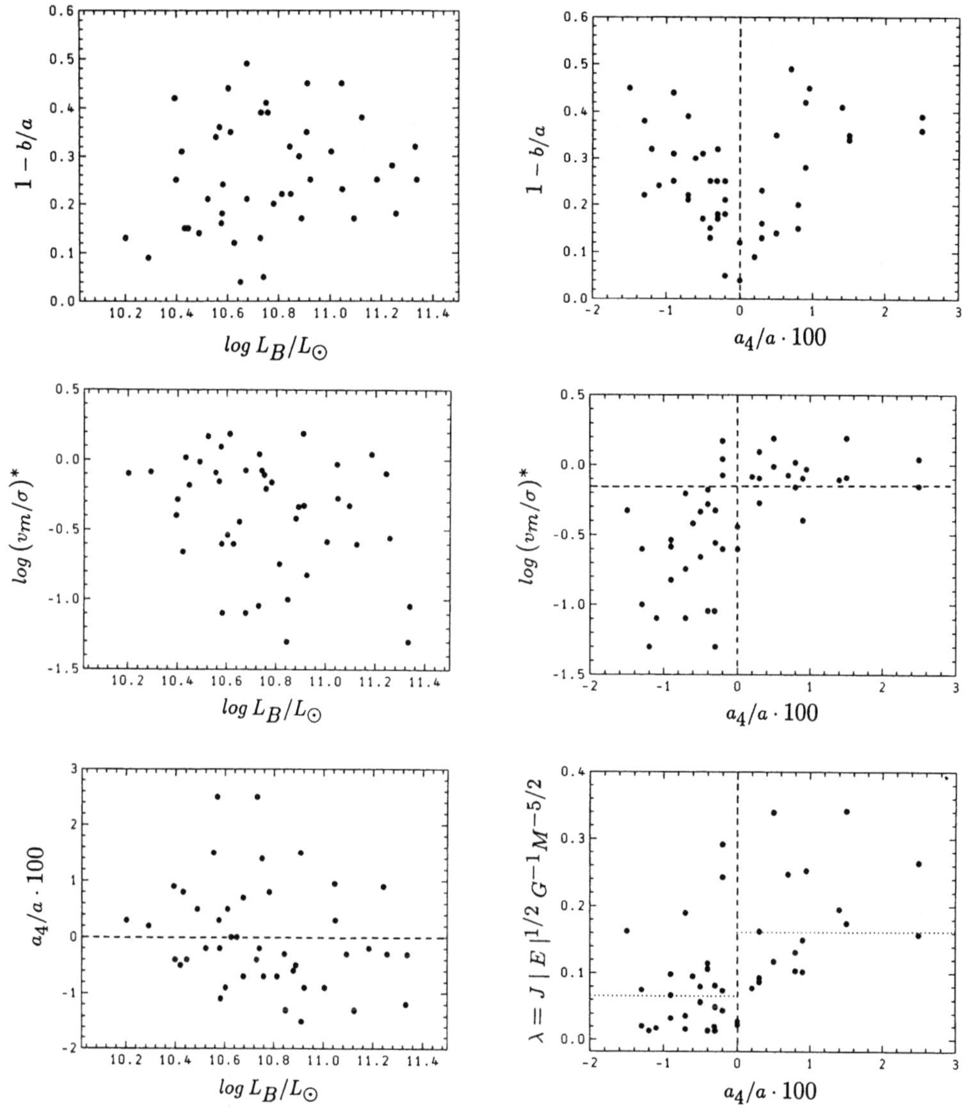

Figure 6: *The relations between morphological and kinematical parameters for luminous elliptical galaxies with available kinematics and isophotal shape parameter a_4/a (L_B = blue luminosity, $1 - b/a$ = ellipticity, $(v_m/\sigma)^*$ = velocity anisotropy, λ = dimensionless spin parameter, and a_4/a = isophotal shape parameter). The dotted lines in the λ-a_4/a diagram indicate the median λ values.*

Bender 1987, Jedrzejewski and Schechter 1989, Franx et al. 1989, Bender and Möllenhoff 1989). In order to avoid confusion with the properties of low luminosity ellipticals, the figures are confined to objects with $L_B > 10^{10.2} L_\odot$ (for the properties of low luminosity ellipticals, see Nieto, this conference). While no relation between luminosity L_B, ellipticity $1 - b/a$ and anisotropy $(v_m/\sigma)^*$ is observed, there is a clear dependence of $(v_m/\sigma)^*$ on the isophotal shape parameter a_4/a (cf. Carter 1987, Nieto et al. 1988, Bender 1988a). Fig. 6 also shows the dimensionless spin parameter λ against a_4/a (the λ values were derived according to Davies et al. 1983; random orientations were assumed for the boxy objects, edge-on orientation for the disky ellipticals). The mean λ value of boxy E's is roughly a factor 2.5 lower than that of the disky E's.

According to these plots and the further papers referenced in the text, we can summarize the properties of disky and boxy elliptical galaxies in the following way:

- **Ellipticals with pointed isophotes** resemble in their kinematical, radio and X-ray properties S0-galaxies. They are in general flattened by rotation, are much weaker radio emitters than ellipticals on average, and their X-ray emission can fully be accounted for by a standard population of discrete sources (no hot gaseous halo is required). Because of this similarity, it is likely that pointed isophotes indeed indicate the presence of a disk and that disky ellipticals form a continuous sequence in disk-to-bulge ratio with S0-galaxies. This interpretation is supported by the detailed surface photometry of a few of these objects (Capaccioli 1987, Capaccioli et al. 1987). However, the final proof that we really observe disks comparable to those in S0-galaxies has to be given by more detailed future observations. Taking inclination effects into account, a rough estimate of the frequency of disky ellipticals among luminous E's yields values between 20 % and 50 %.

- **Ellipticals with boxy isophotes** owe their shape to various degrees of velocity anisotropies. They are frequently radioloud and their X-ray emission is much higher than expected from their number of discrete sources, consequently, hot gaseous coronae must be present. These properties show that radio activity is favored by anisotropy (triaxiality) and the presence of X-ray halos, a result, which is also expected from theoretical arguments (*e.g.* Norman and Silk 1983). Boxy E's seem to have higher mass-to-light ratios than disky E's (Bender et al. 1989a). That is not only indicated by the presence of X-ray halos but also follows from a simple estimate of their masses using central velocity dispersions and half-light radii (the latter estimate, however, has to be taken with some care because of the large variety of velocity anisotropies in boxy E's).

  About 1/3 of the nearby bright *boxy* ellipticals shows decoupled core kinematics demonstrating the influence of merging or accretion on their evolution. Other signs of interaction are frequently present as well (Nieto 1989). These properties indicate that boxy E's may represent the final state of radio-luminous galaxies which were found to suffer interactions frequently (Heckman et al. 1986, Smith and Heckman 1989).

  Boxy E's also have λ values consistent with their formation in a hierarchical merging scenario ($< \lambda_{boxy} > \approx 0.06$, in accordance with *e.g.* Efstathiou and Jones, 1979). Of course, that does not imply that boxy E's must have been formed by *recent* mergers. It is also possible that the merging process took place earlier in their evolution.

The above splitting into two groups was done mainly to clarify the different properties of disky and boxy objects. It does not necessarily follow that there indeed exist two clearly distinct groups of objects, although this seems to be likely. It cannot be excluded that a continuous transition between the two types of objects exists. Anyhow, an important consequence of this result is that any statistical investigation or test applied to the class of ellipticals as a whole (*e.g.* the kinematical shape tests of Binney 1985 or Levison 1987) has to be handled with care, since the result can be severely influenced by the confusion of different types of objects. The same argument applies to distance indicator relations as well, since disky and boxy ellipticals seem to have different M/L.

4. Conclusions

In this paper I reviewed the most recent progress in our understanding of the structure and formation of elliptical galaxies. I showed that we have evidence that the class of luminous elliptical galaxies contains objects with strikingly different physical properties and most likely also different evolutionary histories.

The analysis of isophote shapes allows a grouping of these objects according to their physical properties.

The extreme representatives in this isophotal shape order are on the one hand ellipticals containing stellar disks and being in their structure very similar to S0-galaxies and on the other hand objects most likely being merger remnants. Therefore, the *early-type* Hubble sequence seems to be partly determined by the degree of merging or interaction the galaxies experienced. This merging could have happened during the early collapse phase of the galaxies as well as more recently.

However, the merging rate alone is unlikely to determine the Hubble type completely. Certainly other protogalactic parameters also play an important role, like *e.g.* the specific angular momentum (Fall 1983) or the properties of the dark halos. Especially the hypothesis that the halo density may be of relevance (Zurek et al. 1989) is consistent with the finding that the halo concentration of spiral galaxies increases with decreasing disk-to-bulge ratio (Athanassoula et al. 1988). This trend could be imagined to continue down to very low disk-to-bulge ratio, *i.e.* to disky ellipticals and finally to boxy ellipticals. In this scenario the high M/L of the X-ray luminous boxy E's could be explained by a higher concentration of dark matter.

Acknowledgements. I am grateful to Massimo Capaccioli, Roger Davies, George Djorgovski, Ortwin Gerhard, John Kormendy, Claus Möllenhoff, Jean-Luc Nieto, Peter Quinn, Francois Schweizer, Peter Surma and Stefan Wagner and many others for discussions, suggestions and comments. I also acknowledge support by the Deutsche Forschungsgemeinschaft (SFB 328).

References

Athanassoula, E., Bosma, A., Papaioannou, S.: 1987, *Astron. Astrophys.* **179**, 23
Balcells, M., Quinn, P.J.: 1989, *Astrophys. J.* submitted
Barnes, J., Efstathiou, G.: 1987, *Astrophys. J.* **319**, 575
Bender, R.: 1987, Ph.D. thesis, University of Heidelberg 1987
Bender, R., Möllenhoff, C.: 1987, *Astron. Astrophys.* **177**, 71
Bender, R., Döbereiner, S., Möllenhoff, C.: 1987, *Astron. Astrophys. Letters* **177**, L53
Bender, R.: 1988a, *Astron. Astrophys. Letters* **193**, L7
Bender, R.: 1988b, *Astron. Astrophys. Letters* **202**, L5
Bender, R., Döbereiner, S., Möllenhoff, C.: 1988, *Astron. Astrophys.* Suppl. Series **74**, 385
Bender, R., Surma, P., Döbereiner, S., Möllenhoff, C., Madejsky, R.: 1989a, *Astron. Astrophys.* **217**, 35
Bender, R.: 1989, *Astron. Astrophys.* in press
Bender, R., Möllenhoff, C.: 1989, in preparation
Bender, R., Möllenhoff, C., Surma, P.: 1989b, in preparation
Bertola, F., Capaccioli, M.: 1975, *Astrophys. J.* **200**, 439
Bertola, F., Bettoni, D.: 1988, *Astrophys. J.* **329**, 102
Bettoni, D.: 1989, *Astron. J.* **97**, 79
Binney, J.: 1982, *Ann. Rev. Astron. Astrophys.* **20**, 399
Binney, J.: 1985, *Monthly Notices Roy. Astron. Soc.* **212**, 767
Binney, J.J., Petrou, M.: 1985, *Monthly Notices Roy. Astron. Soc.* **214**, 449
Canizares, C.R., Fabbiano, G., Trinchieri, G.: 1987, *Astrophys. J.* **312**, 503
Capaccioli, M.: 1985, in *New Aspects of Galaxy Photometry*, ed, J.-L. Nieto, Springer Verlag, Heidelberg
Capaccioli, M., Held, E.V., Nieto, J.–L.: 1987, *Astron. J.* **94**, 1519
Capaccioli, M.: 1987, in *Structure and Dynamics of Elliptical Galaxies*, Ed. T. de Zeeuw, p.47, Reidel Dordrecht
Carter, D.: 1978, *Monthly Notices Roy. Astron. Soc.* **182**, 797
Carter, D.: 1987, *Astrophys. J.* **312**, 514
Carlberg, R.: 1986, *Astrophys. J.* **310**, 593
Combes, F., Sanders, R.H.: 1981, *Astron. Astrophys.* **96**, 164
Davies, R.L., Illingworth, G.D.: 1983, *Astrophys. J.* **266**, 516

Davies, R.L.: 1989, in 'Les Mondes des Galaxies', eds L. Bottinelli and H.G. Corwin, Springer Verlag, New York
Davies, R.L. Efstathiou, G., Fall, S.M., Illingworth, G.D., Schechter, P.: 1983, *Astrophys. J.* **266**, 41
Djorgovski, S., Davis, M.: 1987, *Astrophys. J.* **313**, 59
Dressler, A., Lynden-Bell, D., Burstein, D., Davies, R.L., Faber, S.M., Terlevich, R.J., Wegner, G.: 1987, *Astrophys. J.* **313**, 42
Efstathiou, G., Jones, B.J.T.: 1979, *Monthly Notices Roy. Astron. Soc.* **186**, 133
Efstathiou, G., Ellis, R.S., Carter, D.: 1982, *Monthly Notices Roy. Astron. Soc.* **201**, 975
Evans, D.S.: 1951, *Monthly Notices Roy. Astron. Soc.* **111**, 526
Fabian, A.C., Thomas, P.A., Fall, S.M., White III, R.E.: 1986 *Monthly Notices Roy. Astron. Soc.* **221**, 1049
Fabbiano, G., Trinchieri, G.: 1985, *Astrophys. J.* **296**, 430
Fall, S.M: 1983, in *Internal Kinematics and Dynamics of Galaxies, IAU Symp. 100*, p.391, ed. E. Athanassoula, Reidel Dordrecht
Forman, W., Jones, C., Tucker, W.: 1985, *Astrophys. J.* **293**,102
Franx, M., Illingworth, G.: 1988, *Astrophys. J. Letters* **327**, L55
Franx, M., Illingworth, Heckman, T.M.: 1989, preprint
Geller, M.J., Huchra, J.P.: 1983, *Astrophys. J. Suppl. Series* **52**, 61
Gerhard, O.E., Vietri, M., Kent, S.M.: 1989, preprint
Gunn, J.E.: 1987, in *Nearly Normal Galaxies, From the Planck Time to the Present*, ed. S.M. Faber, Springer Verlag, New York
Heckman, T.M., Smith, E.P., Baum, S.A., van Breugel, W.J.M., Miley, G.K., Illingworth, G.D., Bothun, G.D., Balick, B.: 1986, *Astrophys. J.* **311**, 526
Illingworth, G.: 1977, *Astrophys. J. Letters* **218**, L43
Illingworth, G., Franx, M.: 1989, in *Dynamics of Dense Stellar Systems*, ed. D. Merrit, Cambridge University Press
Jedrzejewski, R.I.: 1987 *Monthly Notices Roy. Astron. Soc.* **226**, 747
Jedrzejewski, R.I., Davies, R.L., Illingworth, G.: 1987 *Astron. J.* **94**, 1508
Jedrzejewski, R.I., Schechter, P.L.: 1988, *Astrophys. J. Letters* **330**, L87
Jedrzejewski, R.I., Schechter, P.L.: 1989, *Astron. J.* **98**, 147
Kormendy, J.: 1977, *Astrophys. J.* **218**, 333
Kormendy, J.: 1982, in 'Morpholgy and Dynamics of Galaxies', eds.Martinet, L., Major, M., Saas Fee Lectures
Kormendy, J.: 1984, *Astrophys. J.* **287**, 577
Kormendy, J.: 1987, in *Structure and Dynamics of Elliptical Galaxies, IAU Symp. 127*, ed. T. d. Zeeuw, p.17, Reidel Dordrecht
Kormendy, J., Illingworth, G.: 1982, *Astrophys. J.* **256**, 460
Kormendy, J., Djorgovski, S.: 1989, *Ann. Rev. Astron. Astrophys.* in press
Lauer, T.R.: 1985a, *Astrophys. J.* **292**, 104
Lauer, T.R.: 1985b, *Monthly Notices Roy. Astron. Soc.* **216**, 429
Levison, H.F.: 1987, *Astrophys. J. Letters* **320**, L93
May, A., van Albada, T.S., Norman, C.A.: 1985, *Monthly Notices Roy. Astron. Soc.* **214**, 131
Michard, R., Simien, F.: 1988, *Astron. Astrophys. Suppl. Series* **74**, 25
Miller, R.H., Smith, B.F.: 1989, *Astrophys. J.* submitted
Nieto, J.-L., Prugniel, P.: 1987, in *Structure and Dynamics of Elliptical Galaxies, IAU Symp. 127*, ed. T. d. Zeeuw, p.99, Reidel Dordrecht
Nieto, J.-L.: 1989, 2[da] Reunion de Astronomia Extragalactica, Academia Nacional de Ciencias de Cordoba, in press
Nieto, J.–L., Bender, R.: 1988, *Astron. Astrophys.* **215**, 266
Nieto, J.-L., Bender, R., Davoust, E., Prugniel, P.: 1989, *Astron. Astrophys.* submitted
Nieto, J.–L., Capaccioli, M., Held, E.V.: 1988, *Astron. Astrophys. Letters* **195**, L1
Norman, C., Silk, J.: 1983, *Astrophys. J.* **266**, 502

Petrou, M.: 1987, *Monthly Notices Roy. Astron. Soc.* **226**, 111

Prugniel, P., Nieto, J.-L., Bender, R., Davoust, E.: 1988, *Astron. Astrophys.* **204**, 61

Rowley, G.: 1988, *Astrophys. J.* **331**, 124

Schechter, P.L., Gunn, J.E.: 1979, *Astrophys. J.* **229**, 472

Schombert, J.M.: 1986, *Astrophys. J. Suppl. Series* **60**, 603

Schombert, J.M.: 1987, *Astrophys. J. Suppl. Series* **64**, 643

Schweizer, F.: 1979, **233**, 23

Schweizer, F.: 1982, **252**, 455

Smith, E.P., Heckman, T.M.: 1989, *Astrophys. J.* **341**, 658

Smith, B.F., Miller, R.H., Roelofs, G.R.: 1989, Poster presented at the AAS meeting, Boston

Toomre, A.: 1977, in *The Evolution of Galaxies and Stellar Populations*, eds. B.M. Tinsley and R.B. Larson, p.401, Yale University Observatory

Tremaine, S.: 1981, in *The Structure and Evolution of Normal Galaxies*, eds. S.M. Fall and D. Lynden-Bell, p.67, Cambridge University Press

de Vaucouleurs, G.: 1948, *Ann. Astrophys.* **11**, 247

de Vaucouleurs, G.: 1953, *Monthly Notices Roy. Astron. Soc.* **113**, 134

Vietri, M.: 1986, *Astrophys. J.* **306**, 48

Wagner, S.: 1989, in preparation

Wilson, C.P.: 1975, *Astron. J.* **80**, 175

Whitmore, B.C., Rubin, V.C., Ford, W.K.: 1984, *Astrophys. J.* **287**, 66

de Zeeuw, T., Peletier, R., Franx, M.: 1986, *Monthly Notices Roy. Astron. Soc.* **221**, 1001

Zurek, W.H., Quinn, P.J., Salmon, J.K.: 1988, *Astrophys. J.* **330**, 519

Two-Dimensional Velocity Maps of Cores of Elliptical Galaxies

S.J. Wagner

Landessternwarte, Königstuhl, D-6900 Heidelberg, Fed. Rep. of Germany

1. Introduction

Recently several elliptical galaxies were found to show decoupled core kinematics. In these objects the angular momenta of cores and outer parts are misaligned. Most of these discoveries resulted from kinematic investigations based on spectra extracted along the projected major and minor axes. It is therefore not possible to determine whether the angular momentum vectors are indeed antiparallel (as suggested e.g. in IC 1459, Franx and Illingworth, 1988) or orthogonal (as found e.g. in the case of NGC 5982, Wagner et al., 1988) or whether misalignments occur at different angles as well. In the course of mapping the velocity fields of several 'bonna fide' elliptical galaxies, the nuclear regions were covered in detail. The twist and distortions in the isodispersion and isovelocity maps of the central parts were studied. Evidence for kinematical misalignments by small angles was found. Such cases cannot be identified from major axis and minor axis kinematics alone and hence may occur much more often than suspected at present. Twists of the rotational velocities are expected for allmost all viewing angles in triaxial galaxies. Triaxiality alone however cannot account for distortions which do not show point-reflection symmetry.

2. Observations and Reductions

The sample of galaxies studied in this project is distance limited with the only other selection criterion being the position in the sky. No selection according to isophotal behaviour was introduced. The seven objects which were mapped include galaxies with boxy and pointed isophotes, different amounts of isophotal twists and different ellipticities. All of them were observed along 4 to 8 different position angles. In most cases all cuts include the nuclei. This has the advantage that the core regions are well covered and all profiles could be checked for consistency, requiring similar results in the centre. Observations and reduction procedures are described in Wagner, 1989 (in preparation).

The rather large binning along the slit, chosen to achieve large radial coverage is a disadvantage as fas ar the analysis of the nuclear profiles is concerned. To illustrate the coverage Fig. 1 shows the slit positions used in the mapping of NGC 5982. Fig. 1b is a blow-up of the central parts showing the true slitwidths. The dotted sections one of the slits indicate the pixel size. The reconstruction of the exact slit positions is important for deriving maps. To check these positions, light-profiles were extracted along rectangles placed on direct images taken in the V band. They were folded with the seeing derived during the campaign from spectra of template stars, rebinned to the spatial resolution of the galaxy spectra and compared to the profiles extracted from the longslit spectrograms by integrating along dispersion. The exact slit positions turned out to be very sensitive to small displacements allowing *a posteriori* a high-precision determination of the true slit positions.

Dynamics and Interactions of Galaxies Editor: R. Wielen

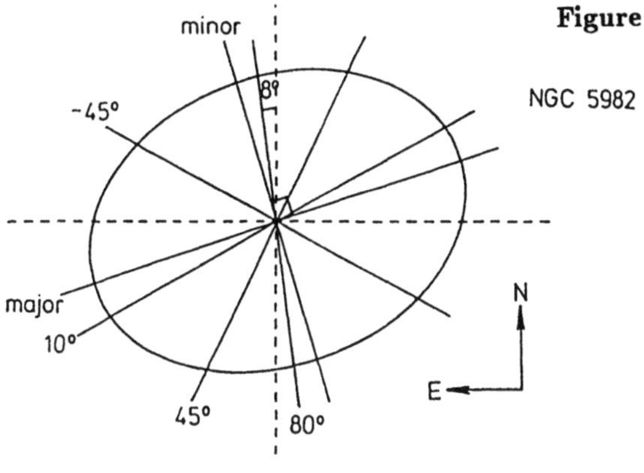

Figure 1a

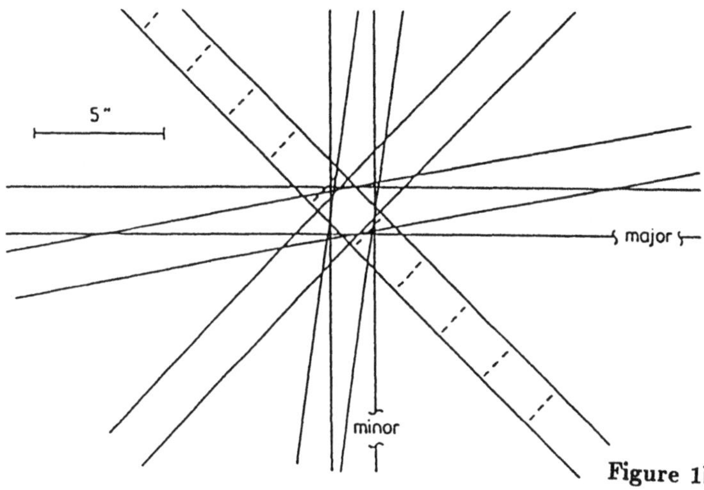

Figure 1b

3. Kinematical maps

Figs. 2 and 3 show the rotation and dispersion curves of NGC 5982 along major and minor axes. A hump in velocities along the minor-axis rotation curve is clearly visible. By combining all six longslit spectra it is possible to derive maps of the velocity field and the velocity dispersions. The contours are displayed in Fig. 4. For clearity only the centre is shown. Parts at larger radii have been omitted. The rotation seen along the minor axis can be traced in all spectra taken at skewed angles as well. Obviously the projected velocity field is not axisymmetric. The hump, which is clearly evident in the minor-axis rotation curve, is not an indication of an overall minor-axis rotation. As suggested by Wagner et al., 1988, the distorted velocity fields indicate a kinematically decoupled nuclear component. In many cases the velocity fields of arbitrarily projected, triaxial galaxies show similar distortions (e.g. Statler, 1987). Due to the large amount of free parameters in modelling one particular object, it is difficult to rule out such an explanation in a specific case. In general the twists in the velocity-dispersion contours of projected triaxial galaxies are accompanied by twists of the isophotes. This is not observed in NGC 5982, whose position angle remains constant within an accuracy of 1° out to 30" and little, if any, $\leq 2°$ change further out. In addition, the twist shown in Fig 4a requires a very contrived geometry it be explained by triaxiality.

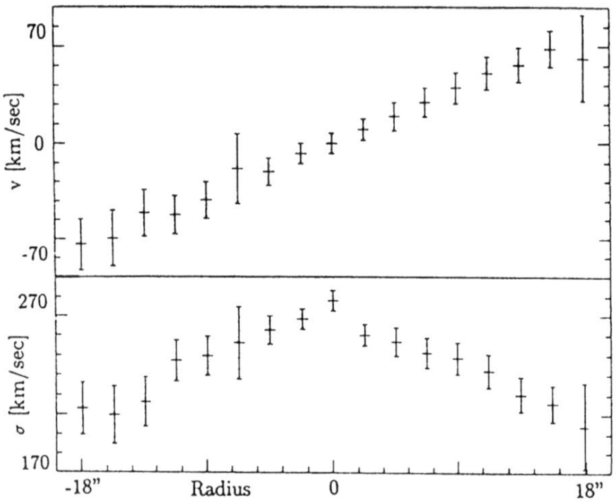

Figure 2
Major-axis kinematics

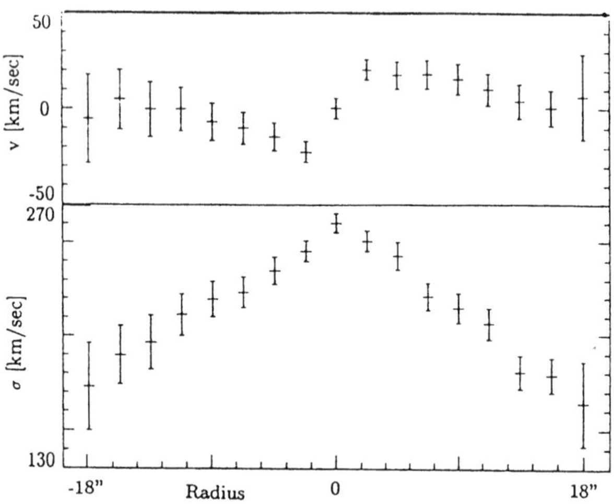

Figure 3
Minor-axis kinematics

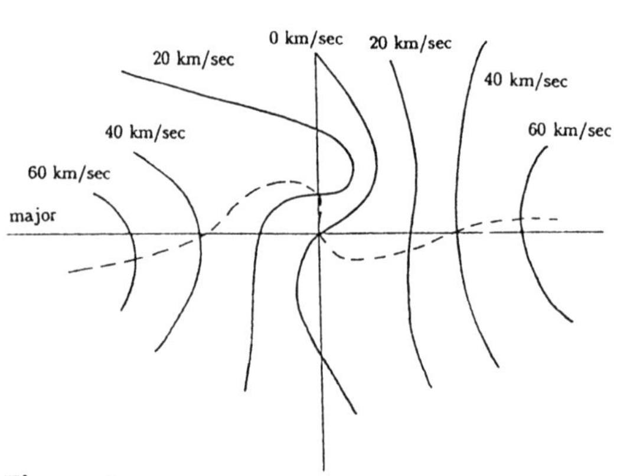

Figure 4a
Velocity map

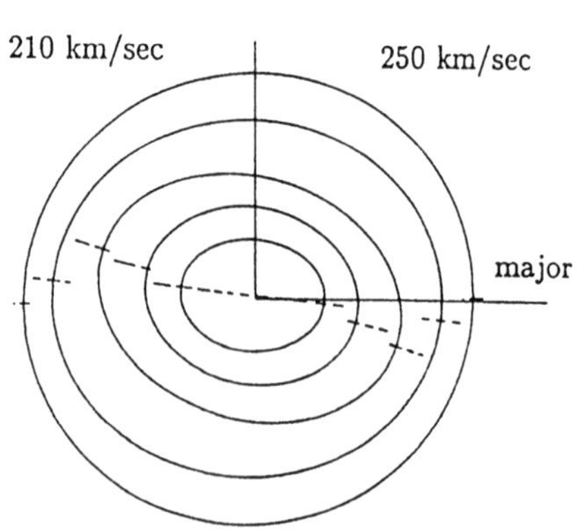

Figure 4b
Velocity-dispersion map

4. Statistics

In the course of this project 7 elliptical galaxies were covered. Individual maps will be published by Wagner, 1989 (in preparation). The kinematical information was examined for twists of the iso-velocity and iso-dispersion contours as well as for deviations of point-reflection symmetry. While the former effects could be caused by projection effects of triaxial galaxies, the latter results imply strong - and recent - perturbations or merging events. Since the coverage of the objects is different, 'twists' in the context mentioned above were defined very conservative to require a change in position angle in excess of 30° for the major axis of the (ellipsoidal) contours of equal velocity dispersion or of the steepest gradient in the isovelocity maps respectively. These two lines are indicated in Fig. 3 (dot-dashed). Deviations from point-reflection symmetry were defined to require two or more adjacent points along two or more adjacent rays to deviate from the corresponding (reflected) points by an amount of 3 σ (error bars) or more.

A second test on all galaxies checked the probability to discover any of the kinematical anomalies from major-axis and minor-axis data alone (i.e. non-monotonic variations of $v_{major}(r)$ / $v_{minor}(r)$, $\sigma_{major}(r)$ /$\sigma_{minor}(r)$ or differences of the two sides in the folded kinematical profiles along the projected principal axes).

Table 1 lists the occurence of twists and asymmetries and indicates whether any of these features can be derived from longslit-spectroscopy along the projected principal axes. All of the galaxies had been classified as ellipticals and none of them shows strong morphological perturbations. Indications of kinematical peculiarities had been found in NGC 5982 by Wagner et al., 1988 in NGC 7626 by Jedrezejewski and Schechter, 1988 and in NGC 1395 by Franx, 1988. A detailed investigation of the isophotes of NGC 7626 exhibits an anomoulus behaviour in the whole radial range (Bender, 1987). Seitzer and Schweitzer (this volume) discovered several distinct structures in this object.

Out of the seven galaxies studied, only one does not show any deviations from a co-axial (with respect to photometry) and symmetric velocity field. Perfectly regular kinematics seems to be an exception rather than the rule. This assumption is strengthened by the observation that such perturbations cannot generally be discovered from major axis spectra alone. NGC 7626 is the only case in which peculiarities are immediately obvious from the kinematics derived along one of the projected principal axes. In the two further cases of NGC 5982 and NGC 1395 the combination of major and minor axis spectra reveals kinematical peculiarities. In view of the fact that only few galaxies have been observed along other axes with sufficient S/N and sufficient radial coverage to detect twists (e.g. Binney et al., 1989), it is possible that a majority of elliptical galaxies do show distortions of their velocity fields of some kind. Although this conclusion is based on a small number of galaxies, it is emphasised again that this sample is not biased against any kinematical or morphological selection criterion. In particular the galaxies do not show significant isophotal twists or any significant signal in fourier coefficients higher than the fourth cosine. Furthermore the galaxies are not members of dense groups or clusters. They even live in a somewhat sparser environment than the average elliptical galaxy. In case the kinematical features have to be attributed to interactions, such perturbations seem to be persistent and are likely to

Table 1

galaxy	velocity twists	dispersion twists	velocity asymmetry	dispersion asymmetry	twists in maps only	asymmetry in maps only
NGC 1395	+	–	–	–	+	
NGC 1399	+	+	–	+	+	+
NGC 5557	–	–	+	–		+
NGC 5982	+	–	+	–	–	–
NGC 6482	–	–	–	–		
NGC 7619	+	–	+	–	+	+
NGC 7626	+	+	+	+	–	+

be rather common among elliptical galaxies. In case the twists do not require external influences and are simply the result of a projected velocity field of a triaxial galaxy, we conclude that twisted velocity fields are not associated with similar twists in the isophotes.

5. Broadening functions

A superposition of two or more components with different kinematical behaviour will cause the resulting broadening function (i.e. the luminosity-weighted distribution of the velocities along the line-of sight) to deviate from a symmetric single-peaked shape. These deviations are much larger than the asymmetries caused by line-of-sight integration in a single component with shallow gradients of the projected velocities along the line-of-sight. A detailed investigation of the observed broadening functions allows a straight forward kinematical decomposition. In the case of NGC 4594 it was shown by Wagner et al., 1989 that the rapidly rotating nuclear disc, discovered by Jarvis and Dubath, 1988 and Kormendy, 1988 can be studied directly as an individual component from the correlation-functions (i.e. the con volutions of the velocity distributions with the instrumental profiles). Filter techniques (Franx and Illingworth, 1988 and Bender, 1989) can be used to improve the S/N and thus the ability to measure the asymmetries in a quantitative way.

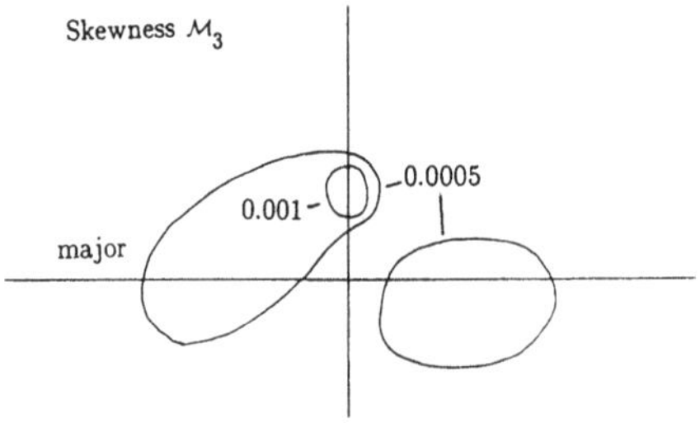

Figure 5

The large "apertures" used in the longslit-spectroscopy described above, induced integration over a large projected area and do not permit a detailed analysis of the broadening functions. Nevertheless it is possible not only to study the spatial distribution of the first and second moment (being closely related to the velocities and velocity dispersions but also the skewness (third moment) of the broadening functions. Isocontours of the map depicting the third moment are reproduced in Fig. 5. Future investigations of the higher moments of the broadening functions should increase the information which can be extracted from spectroscopic observations of sufficiently high S/N. They may also help to clearify the issue whether the distortions of the velocity fields are due to triaxiality of perturbations introduced from outside.

References

Bender, R.: 1987, Ph.D. thesis, University of Heidelberg
Bender, R.: 1989, preprint
Binney, J., Davies, R. and Illingworth, G.: 1989, preprint
Franx, M.: 1988, Ph.D. thesis, University of Leiden
Franx, M. and Illingworth, G.: 1988, *Astrophys.J.* **327**, L 55
Jarvis, B. and Dubath, ?.: 1988, *Astron. Astrophys.* **201**, L 31
Jedrzejewski, R. and Schechter, P.: 1988, *Astrophys.J.* **330**, L 87
Kormendy, J.: 1988, *Astrophys.J.* **335**, 40
Statler, T.: 1987, *Astrophys.J.* 321, 113
Tonry, J.L.,and Davis, M.: 1979, *Astron.J.***84**, 1511
Wagner, S.J., Bender, R. and Möllenhoff, C.: 1988, *Astron. Astrophys.* **195**, L 5
Wagner, S.J., Dettmar, R.-J. and Bender, R.: 1988, *Astron. Astrophys.* **215**, 243

Counter-Rotation in Dust-Lane Ellipticals

F. Bertola[1], *D. Bettoni*[2], *L.M. Buson*[2], *and W.W. Zeilinger*[1,*]

[1]Dipartimento di Astronomica, Università di Padova, Vicolo Osservatorio 5, I-35122 Padova, Italy
[2]Osservatorio Astronomico, Vicolo Osservatorio 5, I-35122 Padova, Italy
*Affiliated to the Astrophysics Division of the Space Science Department of ESA

The class of dust-lane ellipticals is characterized by the presence of a lane of absorbing matter which crosses an elliptical-like galaxy body. The dust lane consists of gas and dust and appears ring-like or as a narrow disk seen edge-on. Bertola & Galletta (1978) pointed out the case of minor-axis dust-lane ellipticals, NGC 5128 (Cen A) being a prototype. Successively in a survey carried out by Hawarden *et al.* (1981) also galaxies having dust lanes along their major axis or along intermediate axes have been identified. A list of "bona fide" dust-lane ellipticals has been presented by Bertola (1987). Thirty objects are found to have minor-axis dust lanes, while on the contrary only nine major-axis dust-lane ellipticals were identified. In addition seven galaxies having a dust lane along an intermediate axis were listed.

In the recent years this type of galaxy has been subject of several detailed studies. In Table 1 an overview is given on the available photometric and kinematic data on the galaxies listed by Bertola (1987). In several cases the dust lane shows a considerable degree of warping in the outer part. This peculiar morphology can be interpreted as due to the dust lane being produced by a second event in the history of the galaxy as suggested by Tubbs (1980) and Simonson (1982). The dust lane can be considered as material acquired from outside which is now in the process of settling in an equilibrium configuration in the potential of the elliptical galaxy. The orientation of the angular momentum vectors of gas and stars depends on the geometry of the potential of the elliptical galaxy and on the impact parameters. Since the most general shape of elliptical galaxies is a triaxial ellipsoid and the infalling mass is very small compared to the total mass, there are two possible equilibrium configurations. These are the planes defined either by major and intermediate axis or by minor and intermediate axis. Consequently, depending on the infall angle, both minor-axis and major-axis dust lanes should be observable. In the first case the orthogonal angular momenta of gas and stars are expected while in the latter case one should expect a roughly equal distribution of objects having either co- or counter-rotating gas.

As a matter of fact, in all galaxies but Anon 0632-629, Anon 0641-412 and IC 4320, of the sample in Table 1 ionized gas was detected. In the class of minor-axis dust-lane ellipticals the angular momentum vectors of gas and stars are found to be orthogonal. A first evidence for the expected counter-rotation was found in the E0 galaxy NGC 5898 (Bertola & Bettoni 1988). In this galaxy it was concluded on basis of geometrical considerations that the projected rotation axis of the stars coincides with the axis of the dust ring and that therefore the two angular momenta are intrinsically anti-parallel. The definite proof was found in the limited sample of four major-axis dust-lane ellipticals, which have to be considered to be the best representants of this class (Bertola, Buson & Zeilinger 1988). Two cases of counter-rotation (Anon 1029-459 and NGC 3528) and two galaxies (NGC 4370 and NGC 5745) with co-rotating gas were found. The rotation curves of gas (full circles) and stars (open circles) for the five galaxies are presented in Figure 1. In addition, Anon 1029-459 and NGC 4370 show a warped dust lane, which again may be an indication of an ongoing settling process. On the basis of the above evidence it must be concluded that the observed decoupling of the kinematics of gas and stars in both minor-axis and major-axis dust-lane ellipticals gives strong support to the idea that gas and dust are acquired from outside, or in any case, due to a second event.

About the progenitor of the infalling material very little is known. The presence of dust indicates that it is very likely already processed material and not primordial. Since the major part of these objects are affiliated to groups, a small gas-rich spiral would be a good candidate. A similar case of second event is noticed in some ellipticals having gaseous disks but not dust. Both major-axis disks (e.g. NGC 7097 - Caldwell, Kirshner & Richstone 1986) and minor-axis disks (e.g. NGC 5077 - Bertola *et al.* 1989) are found.

Dynamics and Interactions of Galaxies Editor: R. Wielen

Table 1: Data on Bona-Fide Dust-Lane Ellipticals

object	m_B	cz	M_B	ε	V_{max}	σ_o	V_m/σ_o	dust	L_*,L_{gas}	ref.
A 0151-498	13.12	6170	-22.34	.33	50	257	0.19	minor	⊥	10
NGC 1052	11.53	1471	-20.81	.31	96	245	0.39	minor	⊥	7
NGC 1947	11.49	1160	-20.34	.10	57	150	0.38	minor	⊥	5
A 0609-331		11300		.50	160	265	0.60	minor	⊥	8
A 0632-629		8550		.43	145	215	0.67	minor	?	8
A 0641-412		10900		.14	100	160	0.63	minor	?	8
A 1029-459	12.67	2764	-21.04	.30	210	260	0.81	major	↑↓	3,4
NGC 3528	12.38	3635	-21.90	.35	120	270	0.44	major	↑↓	3,4
NGC 4370	10.06	750	-20.82	.47	60	120	0.50	major	↑↑	3,4
NGC 4374	10.80	1000	-20.71	.20	68	290	0.23	skew	?	5
NGC 4589	12.00	1824	-20.81	.19	100	215	0.47	minor	⊥	9
NGC 5128	7.96	497	-22.60	.23	80	140	0.57	minor	⊥	1,12
NGC 5266	12.27	2880	-21.53	.29	140	200	0.70	minor	⊥	6,8,11
IC 4320	13.76	6670	-21.87	.00	40	270	0.15	minor	?	5
NGC 5363	11.08	1121	-20.67	.44	135	199	0.67	minor	⊥	10
NGC 5745	14.02	7020	-21.72	.47	220	160	1.38	major	↑↑	3,4
NGC 5898	12.60	2160	-20.60	.00	80	230	0.35	E0	↑↓	2
NGC 7070A	13.40	2382	-19.99	.60	30	100	0.30	skew	⊥	10

References to Table 1:

1 Bertola, F., Galletta, G. & Zeilinger, W.W., 1985. *Ap.J.*, **292**, L51.
2 Bertola, F. & Bettoni, D., 1988. *Ap.J.*, **329**, 102.
3 Bertola, F., Galletta, G., Kotanyi, C. & Zeilinger, W.W., 1988. *M.N.R.A.S.*, **234**, 733.
4 Bertola, F., Buson, L.M. & Zeilinger, W.W., 1988. *Nature*, **335**, 705.
5 Bertola, F., Galletta, G. & Zeilinger, W.W., 1989. *in preparation*.
6 Caldwell, N., 1984. *Ap.J.*, **278**, 96.
7 Davies, R.L. & Illingworth, G.D., 1986. *Ap.J.*, **302**, 234.
8 Möllenhoff, C. & Marenbach, G., 1986. *Astr. Astrophys.*, **154**, 219.
9 Möllenhoff, C. & Bender, R., 1989. *Astr. Astrophys.*, **214**, 61.
10 Sharples, R.M., Carter, D., Hawarden, T.G. & Longmore, A.J., 1983. *M.N.R.A.S.*, **202**, 37.
11 Varnas, S., Bertola, F., Galletta, G., Freeman, K. & Carter, D., 1987. *Ap.J.*, **313**, 69.
12 Wilkinson, A., Sharples, R.M., Fosbury, R. & Wallace, P.T., 1986. M.N.R.A.S., **218**, 297.

There is no reason to believe that these phenomena of acquisition are limited only to the class of elliptical galaxies. On the contrary these features may be used to identify the same acquisition processes also in other type of galaxies. A morphological evidence of an accretion event could be the presence of a warped dust lane, which is noted not only in dust-lane ellipticals like NGC 5128 but also in some S0 galaxies (e.g. NGC 5866, Sandage 1961).

In spiral galaxies, however, accretion events do not seem to produce pronounced effects because of the pre-existing gas disk. Polar gas will be swept up by the gas in the disk and forced to settle in the plane of the galaxy. Gas falling directly into the disk is not likely to produce the effect of counter-rotation, because the pre-existing gas disk is much more massive. S0 galaxies are therefore the only possible field of study, because of their relatively low intrinsic gas content.

From the morphological point of view there is a striking similarity between the class of minor-axis dust-lane ellipticals and polar-ring S0s which are characterized by an extended ring of gas and stars along the minor axis of the disk galaxy. As counter-part for the major-axis dust-lane ellipticals the class of S0s with extended emission lines may be considered. The S0 galaxies contained in the "Revised Shapley-Ames Catalog of Bright Galaxies" (Sandage & Tammann 1987) represent a homogenous and complete

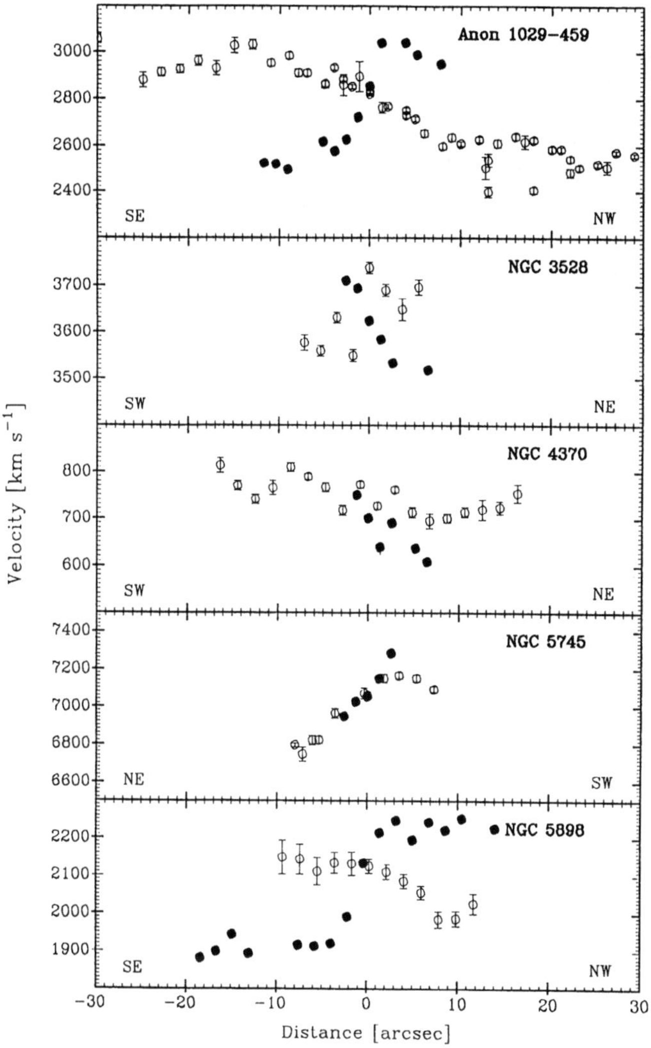

sample. On the basis of the available literature (e.g. Bettoni & Buson 1987) 12 galaxies are found to have extended emission lines. In this sample eight galaxies have been studied kinematically sofar. It results that one polar-ring S0 is found (NGC 2685), three galaxies with counter-rotating gas (NGC 2217 - Bettoni, Fasano & Galletta 1989, NGC 4546 - Galletta 1987 and NGC 7007 - Dettmar 1989) and four galaxies with co-rotating gas (NGC 2768, NGC 5846, NGC 5866 and NGC 6868 - Bertola, Buson & Zeilinger 1989). These numbers are consistent with the accretion picture, similar to the one for dust-lane ellipticals.

These facts induce now a more general question: Is all the gas present in S0 galaxies of external origin as suggested for the neutral hydrogen in S0 galaxies (Wardle & Knapp 1986)? This hypothesis can be verified by the detection of counter-rotation in 50% of the S0s with emission lines in the complete sample selected from the "Revised Shapley-Ames Catalog of Bright Galaxies".

References

Bertola, F. 1987. in *"Structure and Dynamics of Elliptical Galaxies"*, IAU Symp. 127, p. 135, ed. T de Zeeuw (Reidel:Dordrecht).

Bertola, F. & Galletta, G., 1978. *Astrophys. J.*, **226**, L115.

Bertola, F. & Bettoni, D., 1988. *Astrophys. J.*, **329**, 102.

Bertola, F., Bettoni, D., Danziger, I.J., Sadler, E.M., Sparke, L.S. & de Zeeuw, T., 1989. *preprint.*

Bertola, F., Buson, L.M. & Zeilinger, W.W., 1988. *Nature*, **335**, 705.

Bertola, F., Buson, L.M. & Zeilinger, W.W., 1989. *in preparation.*

Bettoni, D. & Buson, L.M., 1987. *Astr. Astrophys. Suppl.*, **67**, 341.

Bettoni, D., Fasano, G. & Galletta, G., 1989. *in preparation.*

Caldwell, Kirshner & Richstone, 1986. *Astrophys.J.*, **305**, 136.

Galletta, G., 1987. *Astrophys. J.*, **318**, 531.

Dettmar, R.-J., 1989. *private communication.*

Hawarden, T.G., Elson, R.A.W., Longmore, A.J., Tritton, S.B. & Corwin, H.G.Jr., 1981. *Mon. Not. R. astr. Soc.*, **196**, 747.

Sandage, A.R., 1961. "The Hubble Atlas of Galaxies" (Carnegie:Washington).

Sandage, A.R. & Tammann, G., 1987. *"Revised Shapley-Ames Catalog of Bright Galaxies"* (Carnegie:Washington).

Simonson, G.F., 1982. *Ph.D. thesis* Yale University.

Tubbs, A.D., 1980. *Astrophys. J.*, **241**, 969.

Wardle, M. & Knapp, G.R., 1986. *Astron. J.*, **91**, 23.

Is the Kinematic Diversity in Ellipticals Reflected in Their Color Gradients?

M. Franx[1] *and G. Illingworth*[2]

[1]Harvard University, Center for Astrophysics, 60 Garden Street, Cambridge, MA 02138, USA
[2]Lick Observatory, University of California, Santa Cruz, CA 95064, USA

Recent observations of the kinematics of elliptical galaxies have shown a variety of outstanding features: a significant fraction of galaxies have kinematically-distinct cores, while some even appear to rotate around their long axes. These discoveries lead one to ask whether the stellar populations in ellipticals differ in any way that can be related to the dynamical properties of the galaxies. Such relationships could provide further clues about the formation process in early-type galaxies. Remarkably enough, the colors and color gradients of ellipticals are very regular. Even more surprising, the local color within an elliptical appears to be related to the local escape velocity. The most straightforward interpretation is that the mean metallicity is a function of the local escape velocity. Such a relation could result from the star formation and metallicity enrichment being regulated by interplay between the input of energy from supernovae, and the temperature and pressure of the hot X-ray gas in ellipticals. Since color gradients may not be strongly affected by subsequent merging and evolution, we may have a diagnostic of early events during formation.

1. INTRODUCTION

The apparent simplicity of elliptical galaxies belies a wealth of fascinating complexity. The early discoveries of slow rotation, shells, X-ray halo's, and major and minor axis gas disks have been followed more recently by the detection of minor-axis rotators and kinematically-distinct stellar cores. These latest results show that elliptical galaxies have a large variety of dynamical structures. How did this arise? Have galaxies with such structures formed in quite different ways? Are there differences in the stellar populations that are related to the dynamical features?

We will address some of these questions with an example of the complex kinematics in ellipticals in §2, and with a discussion of the observed colors and color gradients in §3.

2. STELLAR KINEMATICS

Amongst the galaxies with remarkable kinematics, NGC 4406 is probably the most outstanding. It has recently been observed by three different groups (Wagner, Bender and Möllenhoff 1987; Franx, Illingworth and Heckman 1989; Jedrzejewski and Schechter 1989), with consistent results. We show the kinematics in Figure 1, where the velocity dispersion (σ) and rotation velocity V in km s^{-1} have been folded about the nucleus and plotted against radius in arcsec for the major and minor axis. Along the major axis, the radial velocity peaks very near to the core, with a maximum of about 70 km s^{-1}, and then drops to essentially zero velocity at larger radii. Along the minor axis, the radial velocity rises slowly with radius, reaching a velocity of 30 $\pm$ 10 km s^{-1}. Hence the angular momenta of the inner and the outer parts are projected orthogonal to each other, and the angular momentum of the outer parts projects along the apparent major axis of the galaxy!

The intrinsic orientations of the angular momenta are not straightfoward to determine from the kinematic data, mainly because the intrinsic short and long axes for a triaxial figure do not

Dynamics and Interactions of Galaxies Editor: R. Wielen

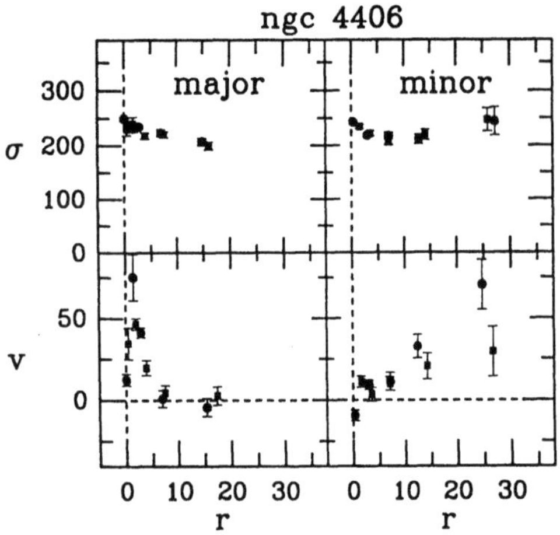

Fig. 1. – See text.

necessarily project onto the apparent minor and major axes. However, in the case of NGC 4406 the core rotation gives an extra constraint. In a non-rotating triaxial potential three orbital families contribute to the streaming: two produce streaming around the intrinsic long axis, and one produces streaming around the short axis.

In a first attempt to model NGC 4406 it is reasonable to assume that the misalignment of the angular momentum of the core and the outer part is produced because one of the systems rotates around the intrinsic short axis, and one rotates around the intrinsic long axis. Let us assume first that the outer parts rotate around the intrinsic short axis. The figure is projected is such a way that the intrinsic short axis projects along the apparent major axis. In that case we are viewing the system in the plane of the intrinsic short and long axis. This implies that the angular momentum of the core would project along the same line as the angular momentum of the outer parts. This is clearly not the case. Thus, the 90° misalignment could not be produced with the above assumption. So we have to conclude that the outer parts are rotating around the intrinsic long axis, and the core around the intrinsic short axis. NGC 4406 may not be the only system which rotates around its long axis; several other galaxies have their angular momenta projected along the apparent major axis (e.g., NGC 4261, Davies and Birkinshaw 1986). However, only a detailed statistical analysis of a sufficiantly large sample can tell what fraction of such minor-axis rotators are due to projection effects (Franx, Illingworth and de Zeeuw 1989).

Kinematically-distinct cores are a common feature of elliptical galaxies; they appear in ≈30% of ellipticals. Ellipticals with such cores are listed by Illingworth and Franx (1989) and Bender (1990). It is not clear how these systems formed. While the cores are characterised as "kinematically-distinct", it is unlikely that we are seeing a "core-within-a-core" structure. The velocity dispersion near the center of many of these systems remains high, and analysis of the line profiles in the counter-rotating core of IC 1459 by Franx and Illingworth (1988) showed that they were asymmetric. These asymmetries can be modelled by a superposition of a cold, rapidly-rotating component and a hot, non-rotating component, with the cold component contributing some 25% of the light within a few arcseconds. Hence, in such a model the rapidly-rotating component contributes only a small fraction of the light, and presumably the density, at any point – it can be thought of as a disk-like structure rotating within the potential of the core of the main galaxy. Such systems may form by the merger or acquisition of a satellite galaxy. Alternatively, they could be the end product of the infall of gas and subsequent star formation near the center.

The formation of the galaxies which rotate around their long axis is even more uncertain. It seems unlikely that such galaxies form in a dissapative scenario, where the angular momentum

of the galaxy and its shape are relatively tightly coupled. It is plausible that mergers of equal-mass systems could produce long-axis rotators, but it remains to be demonstrated that they can be produced in mergers at even the small fraction (< 10% ?) required (e.g., Barnes 1990).

Whatever the formation scenario of these systems, it seems likely that they involve accretion of material, or even "full-fledged" mergers. One might therefor speculate that such galaxies differ in their stellar populations from "ordinary ellipticals". In the next section we discuss recent observations of colors and color gradients in ellipticals which address this question.

3. COLOR GRADIENTS IN ELLIPTICAL GALAXIES

It is well known that mean color and absolute magnitude of elliptical galaxies correlate, in the sense that bright galaxies are relatively red (e.g., Sandage and Visvanathan 1978). In addition, within galaxies the color generally changes from the center outward, in the sense that the centers are redder (e.g., de Vaucouleurs and de Vaucouleurs 1972, and references in Franx, Illingworth and Heckman 1989). The uniformity of the color characteristics within and between galaxies is striking. We have illustrated this in Fig. 2, where we plot the $B - R$ and $U - R$ color profiles against each other for the ellipticals observed by Franx, Illingworth and Heckman (1989) (different symbols for each galaxy profile). The gradients follow the theoretical locii from Arimoto and Yoshii (1987) for 15 Gyr models with two different metallicities (overlapping solid lines) and a 10 Gyr model (dashed line). The sensitivity to wider ranges of metallicity and age needs to be explored, but these data suggest that a simple model, a single parameter perhaps, may govern the color properties of ellipticals.

Before discussing such a picture, we should note that it has become customary to view these color differences as metallicity differences. The available color- and line-strength gradient data are consistent with such an assumption, but further and more detailed observations, particularly in the UV and the IR, plus more extensive modelling are needed before age differences can be excluded. The uniformity of the color properties of ellipticals, as we shall see, argues for a global property such as metallicity as the driver for the observed colors. For the following discussion we shall assume that metallicity is the determining parameter.

It is generally thought that metallicity gradients are caused by dissipation during star formation (see, e.g., models by Larson 1975, Carlberg 1984a, b). The gas is enriched by the ejecta from stars, and the dissipation causes the more metal-rich gas to flow towards the center. However, other processes might also cause or contribute to such gradients. At the end of the period of star formation, a strong wind may develop, carrying away the residual gas. If the star formation was more efficient near the center, then the mean metallicity in the inner regions is likely to be higher than the metallicity in the outer parts. The yield of the star formation may even have been different in the inner parts compared to the outer parts, resulting in a gradient. While the merging of subclumps is now being viewed as an important process during the development of ellipticals, it is unlikely that the gradients in the progenitor clumps were destroyed (White 1979), though they could have been weakened in successive mergers.

The comments above suggest that a multitude of processes may have contributed to metallicity gradients. Some of these processes suggest tests that may help us discriminate between the processes. In the dissipative model one might expect a correlation between rotation and color gradients, since strong dissipation may have enhanced both the color gradient and the rotation of the galaxy. Such a correlation does not seem to be present in the current data. If merging played an important role one might expect that luminous galaxies would have smaller gradients than lower luminosity galaxies. The data do not show such a correlation. A larger sample of color gradient data, collected from the literature by Kormendy and Djorgovski (1989), also does not show any striking correlations with color gradients. Even more surprising is the fact that galaxies with kinematical features in their core, such as in NGC 4406, seem to have normal color gradients! The process that formed these features is either a normal part of (elliptical) galaxy

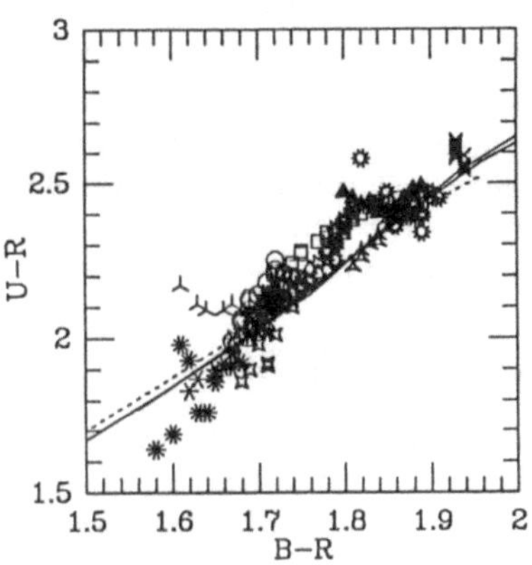

Fig. 2. – See text.

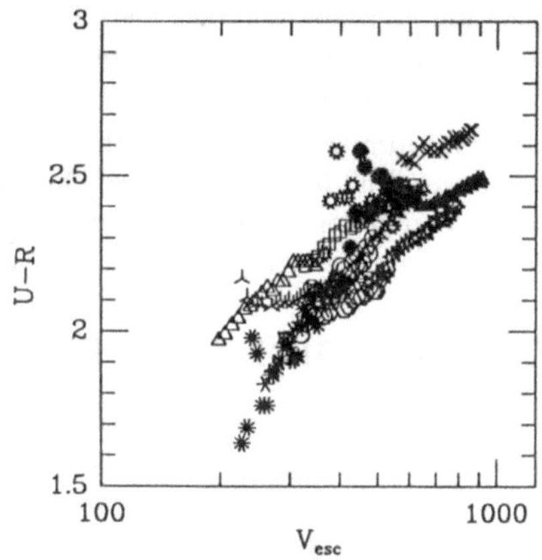

Fig. 3. – See text.

formation, or the features and there associated material are either indistinguishable from the the usual population in ellipticals or do not contribute significantly to the total light.

Given these difficulties in understanding the implications of color gradients one might take a simple approach towards the data by asking: *is there a single relation of the form "local color = f(structural and dynamical parameters)" that explains both the color gradients within galaxies and the color differences between galaxies.* It is clear that such an approach is simplistic, and has to be used with care, but it may give some valuable insight into the physical processes at work. The straightforward color versus density or surface brightness relations are not convincing. Taking a more sophisticated view, and building upon the idea that the color-lumininosity relation is produced by a relation between binding energy and mean metallicity (e.g., Larson 1974), we have derived local color as a function of local binding energy. We have expressed the binding energy as the local escape velocity. The resulting relation for the galaxies in the sample of Franx, Illingworth and Heckman is shown in Figure 3, where we plot the color gradient data of Figure 2, *with no zero point shifts applied*, i.e., as calibrated, as a function of escape velocity. The escape velocity was calculated assuming an $r^{1/4}$ profile for all the galaxies. The measured central velocity dispersion was used to calculate the escape velocity at an effective radius, where we took the effective radius and velocity dispersion from Faber *et al.* (1989).

The relation in Figure 3 is surprisingly tight. The useful range in $U - R$ is about 0.7 mag, while the spread about the relation is approximately 0.1 mag. While a larger sample is needed to verify this result, it is of some interest to think about how such a relation may arise and its implications.

The important factors are (1) the $U - R$ color gradients of the galaxies in the sample are roughly constant at -0.23 mag per decade in radius, (2) the uniformity of the surface brightness profiles and structure of elliptical galaxies, with only small variations in the shape of the velocity dispersion profile (typically slightly negative logarithmic gradient), (3) the good correlation between the mean color and the velocity dispersion (e.g., Wirth and Gallagher 1984). Together these give a qualitative rationale for the existence of the relation seen in Figure 3. This will be explored in more detail by Franx and Illingworth (1990).

The escape velocities shown in Figure 3 are probably under-estimated because we did not include dark matter. However, the inclusion of dark matter does not necessarily change the result in a significant way. Franx and Illingworth (1990) obtain a similar relation for models which do incorporate dark matter; they assume that the density profiles of halos are similar, and that the total M/L is constant.

What are the physical processes that would result in a relation between escape velocity and mean metallicity? A purely local process could be regulation of star formation by a hot (X-ray) galactic medium. Such a hot X-ray gas component has a pressure that is a function of the local escape velocity. A plausible scenario is one in which star formation proceeds until the cold gas is heated to a temperature and pressure higher than the X-ray component, and hence escapes from the locus of star formation (and from the galaxy?). For regions where the pressure of the X-ray gas is lower (lower escape velocity), star formation will be stopped after fewer generations of stars, and hence the self-enrichment will be lower. This simple-minded scenario needs to be explored more quantitatively, since it ignores the effects of cooling and conduction. Another possible explanation might be that the IMF and the yield of the star formation is a function of the local pressure of the hot gas. More global mechanisms involving, for example, the coupling of dissipation and metal enrichment, are harder to envisage, but should not be excluded given our limited understanding of galaxy formation.

SUMMARY

The contrast between the uniformity of the structural properties of ellipticals, particularly the uniformity of their color gradients and the existence of a global color-luminosity relation, and the diversity of their kinematic properties is striking. If larger samples show that the color relations noted here, particularly that of local color with escape velocity, are generic to ellipticals, theories of the formation of such galaxies will have to encompass their structural uniformity and kinematic diversity.

REFERENCES

Arimoto, N., and Yoshii, Y. 1987, *Astr. Ap.*, **173**, 23.
Barnes, J. E. 1990, in preparation.
Bender, R. 1990, in preparation.
Carlberg, R. G. 1984a *Ap. J.*, **286**, 403.
Carlberg, R. G. 1984b *Ap. J.*, **286**, 416.
Davies, R. L., and Birkinshaw, M. 1986, *Ap. J. (Letters)*, **303**, L45.
de Vaucouleurs, G. and de Vaucouleurs, A. 1972, *Mem. R.A.S.*, **77**, 1.
Faber, S. M., Wegner, G., Burstein, D., Davies, R. L., Dressler, A., Lynden-Bell, D., and Terlevich, R. J. 1989, *Ap. J. Suppl.*, **69**, 763.
Franx, M., and Illingworth, G. D. 1988, *Ap. J. (Letters)*, **327**, L55.
Franx, M., Illingworth, G. D., and Heckman, T. M. 1989, *A. J.*, **98**, 538.
Franx, M., Illingworth, G. D., and de Zeeuw, P. T., 1990, in preparation.
Franx, M., Illingworth, G. D. 1990, in preparation.
Illingworth, G. D. and Franx, M. 1989, in *Dynamics of Dense Stellar Systems*, ed. D. Merritt, Cambridge University Press., p 13.
Kormendy, J., and Djorgovski, S. 1989, *Ann. Rev. Astron. Astrophys.*, **27**, in press.
Larson, R. 1974, *M.N.R.A.S.*, **169**, 229.
Larson, R. 1975, *M.N.R.A.S.*, **173**, 671.
Jedrzejewski, R. I., and Schechter, P. L. 1989, *A. J.*, **98**, 147.
Sandage, A., and Visvanathan, N. 1978, *Ap. J.*, **223**, 707.
Wagner, S. J., Bender, R. and Möllenhoff, C. 1988, *Astr. Ap. (Letters)*, **195**, L5.
White, S. D. M. 1979, *M.N.R.A.S.*, **189**, 831.
Wirth, A., and Gallagher, J. F. 1984, *Ap. J.*, **282**, 85.

Low-Mass Ellipticals: Properties and Formation Processes

J.-L. Nieto

Observatoire de Toulouse, 14 Avenue Edouard Belin, F-31400 Toulouse, France

1. Introduction

The title of this talk seems to imply that some properties of E's correlate with mass. Indeed, the internal kinematics, the radio emission, the nature of the emission-line spectrum and the isophotal class are such properties. However, low-mass E's, as a family, still have to be defined. They should be those whose properties show a *continuous* transition with normal E's. Two types of objects, morphologically different, are candidates to be the low-mass end of the E luminosity function : compact M32-type E's (CE's), of high surface brightness, or dwarf ellipticals (dE's, the faintest of them being called dwarf spheroidals), of low surface brightness. Before choosing between the two, we have to ask ourselves if the two low-mass families are well separated from each other and if one of them really is well separated from the E family. For this purpose, we also need to understand the origin of compactness in M32-type systems. If we inevitably call for more observations, the present data however allow us to suggest some clues about this question.

2. The low-mass end of the E luminosity function : the problem

In their extensive photometric study of the Virgo cluster, Binggeli *et al.* (1984) presented an (SB, M_B) diagram for a complete sample of E and E-like objects. (M_B is the absolute magnitude in the B band; SB is the mean surface brightness within the effective isophote throughout this paper). This diagram shows a smooth transition between dE's and normal E's, which suggests that dE's are the natural continuation of the E luminosity function toward low masses. CE's would not be excluded from the E family but their extreme densities would result from strong tidal stripping from their massive neighbor, following Faber (1973).

At variance with this view, three main lines of arguments were presented by Wirth and Gallagher (1984) to suggest that CE's are genuine E's (compactness being *intrinsic* to the formation of this type of E's), while dE's would form by another process : *i)* CE's, like normal E's, follow in a first approximation an $r^{1/4}$ profile while dE's follow instead an exponential one, *ii)* compactness is not related to environmental effects (isolated CE's do exist) *iii)* CE's follow the extrapolation toward low masses of the (color-luminosity-velocity dispersion) relations followed by E's. These suggestions were later supported by Kormendy (1985), who found strong discontinuities in the core properties between normal E's (including M32) and dE's.

These considerations deserve a few comments :

(1) The (SB, M_B) diagram built by Binggeli *et al.* presents a complete sample of E-like galaxies. In the contrary, Wirth and Gallagher and Kormendy tend to compare objects having extreme properties, namely an extreme compacity, with objects of very low surface brightness. This is true in particular for the core properties that are strongly affected by observational biases (distance effects). This may explain the observed discontinuities.

(2) The spectroscopic counterpart of this study (a $\log\sigma, M_B$ relation) is still untouched, because velocity dispersions are difficult to obtain for low surface brightness objects.

Dynamics and Interactions of Galaxies Editor: R. Wielen

From these comments, it is logical to work in two directions : (1) to look for objects presenting characteristics (notably *SB*) *intermediate* between those of normal E's, CE's and dE's, and (2) to investigate in detail their kinematical properties. These two aspects will naturally lead us to check the location of all these systems with respect to the *Fundamental Plane* defined for all E's, including CE's (Djorgovski and Davis, 1987).

3. Recent observational results

a) Photometry

A large number of closely interacting low-mass galaxies, compact candidates, has been investigated (Prugniel *et al.*, 1987, 1989; Prugniel, 1989) : *all* turned out to be of intermediate *SB* only. In the large sample of *isolated* objects classified compact by Nilson (1973), those ($\simeq 20\%$) having a low luminosity ($M_B > -18$) hardly reach an intermediate *SB* (Davoust *et al.*, in preparation). Figure 1 is a compilation of these results : the separation between the different families is not as strong as found earlier. Objects as compact as the three prototypes, M32, NGC 4486B and NGC 5846A, are certainly very difficult to detect at large distances, but they are not rare. In addition, intermediate-*SB* cases have profiles intermediate between $r^{1/4}$ and exponential (Prugniel *et al.*, 1989; Prugniel, 1989).

b) Kinematics

The detailed kinematics of 14 low-mass E and E-like objects (CE's, closely interacting E's and one dE, N) covering a range of 5 mag in *SB* has recently been obtained (Bender, Nieto, Prugniel, in preparation). It appears (Figure 2) that CE's and the interacting galaxies are *all* flattened by rotation while those not interacting (numbering 7) have all types of behaviors; among these, 5 of them tend to show no rotation along the major axis and must be anisotropic systems. (This contradicts the long-standing correlation between mass and kinematical behavior found by Davies *et al.*, 1983). In addition, the ($\log\sigma$, M_B) diagram confirms previous results showing that interacting objects are hotter than non-interacting ones of same mass.

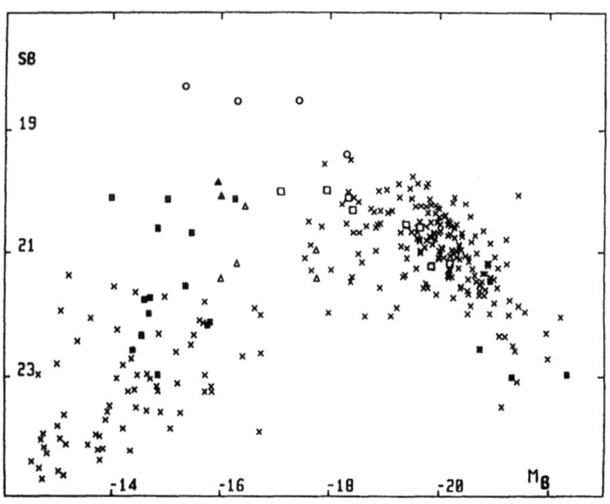

Figure 1 : The blue mean surface brightness within the effective isophote, *SB*, versus M_B for different types of E-like galaxies. Crosses : normal E's in the upper right of the diagram and dE's in the lower left; black squares : compact candidates in the Virgo cluster (Binggeli *et al.*, 1984); open circles : classical CE's, open triangles : isolated low-mass E's of intermediate *SB* classified as compact by Nilson (1973); open squares : E's interacting with a massive neighbor; full triangles : NGC 1097A and NGC 1510. Although quite incomplete, such a diagram shows no clear separation between the different families.

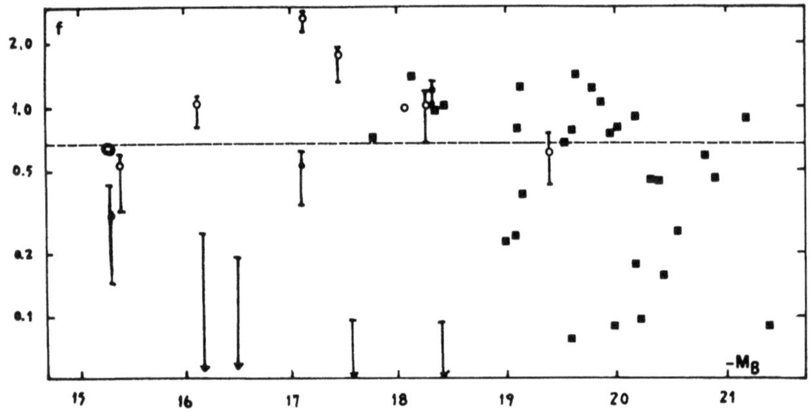

Figure 2 : Anisotropy factor versus M_B for E's and E-like galaxies, adapted from Bender (1988). Galaxies below the dashed line are likely to be anisotropic systems. The brightest galaxies (full squares) are from Bender. Interacting galaxies (including M32 and NGC 4478) are plotted with open circles, noninteracting ones with full circles (for 4 of them, only upper limits can be given). The distance scale is consistent with H_0=100km/s/Mpc.

This sample of 14 objects is certainly small and probably affected by some (known or unknown) observational biases : for instance, all interacting objects have been chosen of high SB. However, no strong discontinuity is apparent in Figure 2, nor in the (logσ, M_B) diagram that would suggest two classes of low-mass E's.

c) The Fundamental Plane

All low-mass (cE's as well as dE's) systems are in or close to the Fundamental Plane. Because this Plane is a robust consequence of the Virial Theorem applied to hot relaxed stellar systems (including globular clusters), it is unable to provide by itself any evidence for discontinuities between these low-mass systems (see Nieto *et al.*, 1989).

4. Discussion : the formation of (low-mass) E's

a) The origin of compactness

These observational data provide a firm ground for proposing a possible scenario for the origin of compactness (Nieto *et al.*, in preparation). The fast rotation of CE's and strongly interacting objects can result either from (recent) torques induced by the interaction or a high angular momentum acquired at an early stage of formation of the galaxy. In the second hypothesis, progenitors of CE's would be bulges of S0-like objects (or disk-E's or early-type spirals) having a low mass (since the mass loss rate is likely to be small during the interaction; see Nieto and Prugniel, 1987). At this intermediate stage of merging, the disk has been destroyed first (or maybe never formed in this rich environment), but the remaining $r^{1/4}$ bulge has been hot enough to resist any kinematical changes due to the interaction.

Now, to reproduce high stellar densities as those observed in CE's, we may consider the loss of a small fraction of mass that would carry away a large amount of angular momentum (Prugniel *et al.*, 1989), but this may be difficult to concile with the fast rotation of these objects. On the other hand, dissipation certainly is required (Carlberg, 1986). We note that all CE's are interacting with massive *gas-rich* galaxies : M87 and NGC 5846, both surrounded by X-ray haloes, and M31. Therefore, we may imagine either a continuous accretion of gas swept by the companion during its journey through the halo of the massive galaxy or a periodic one each time M32 crosses the disk of M31. Dissipation would then take place in the central region of the companion so that small amounts of star formation, probably unobservable or extremely difficult to observe would be responsible for the high central stellar density. (A strong dissipation at an early stage of interaction produced by a sudden accretion of gaseous material from the gas-rich neighbor is not excluded). We indeed observe present star formation in low-mass objects near gas-rich galaxies (*e.g.* the case of NGC 1510; see Eichendorf and Nieto, 1984).

Compactness would then be due to accretion (plus dissipation) of small amounts of gaseous material by the bulge of an early-type system, whose disk has not survived the interaction. Mass loss appears as an unimportant process in itself, but may contribute to compactness by carrying away part of the angular momentum. This scenario and the above observational results imply that isolated genuinely compact E's do not exist (at variance with Wirth and Gallagher's suggestion).

b) The low-mass end of the E luminosity function

The end of the luminosity function should be represented by objects with properties showing a continuous transition with normal E's. This suggests a continuity in the formation processes as well. Recent data have indicated that normal E's are divided into two classes of objects with different historical pasts, namely disk-E's, *e.g.* galaxies having kept their original structure, and merger products, *e.g.* objects showing obvious signatures of recent mergings or accretions (see Nieto and Bender, 1989, and Bender, this workshop). Except for a few specific cases, these two classes correlate with luminosity (Nieto and Bender, unpublished poster) : merger products are $\simeq$ 1 mag higher than disk-E's (which in turn are $\simeq$ 1.5 mag brighter than S0's). We thus expect the low-mass end of the E family to be made of galaxies whose structure results mainly from pure dissipative collapse. Clearly, these galaxies, if they exist, cannot be CE's (nor closely interacting low-mass E's), which result, as shown above, from a quite specific process. We have found no clear separation between normal E's, low-mass E's located in different environments, and dE's : are therefore dE's the low-mass members of the E family? If these objects form by dissipative collapse, is their anisotropy a signature of a recent radial infall (Binney, 1976)?

In conclusion, it results from these data and speculative remarks that solving the question of the low-mass end of the E family requires avoiding the observational bias towards the highest SB objects only and concentrating on objects more difficult to observe, namely dE's. It is only when a larger set of data is obtained that we may be on a firmer ground.

References

Bender, R., 1988, *Astron. Astrophys.* **193**, 7

Binney, J., 1976, *Comm. in Astrophys.*, **8**, 27

Binggeli, B., Sandage, A., Tarenghi, M., 1984, *Astron. J.* **89**, 64

Carlberg, R.G., 1986, *Astrophys. J.* **310**, 593

Davies, R.L. *et al.*, 1983, *Astrophys. J.* **266**, 41

Djorgovski, S., Davis, M., 1987, *Astrophys. J.* **313**, 59

Eichendorf, W., Nieto, J.-L., 1984, *Astron. Astrophys.* **132**, 342

Faber, S.M., 1973, *Astrophys. J.* **179**, 423

Kormendy, J., 1985, *Astrophys. J.* **295**, 73

Nieto, J.-L., Bender, R., 1989, *Astron. Astrophys.* **215**, 266

Nieto, J.-L., Bender, R., Davoust, E., Prugniel, P., 1989, *Astron. Astrophys.* **000**, 000

Nieto, J.-L., Prugniel, P., 1987, *Astron. Astrophys.* **186**, 30

Nilson, P., 1973, *Uppsala General Catalogue of Galaxies*, Uppsala 1973
Prugniel, P., 1989, Thèse, Université Paul Sabatier, Toulouse
Prugniel, P., Nieto, J.-L., Simien, F., 1987, *Astron. Astrophys.* **173**, 49
Prugniel, P., Davoust, E. Nieto, J.-L., 1989, *Astron. Astrophys.* **000**, 000
Wirth, A., Gallagher, J.S., 1984, *Astrophys. J.* **282**, 85

Settling of Gas Disks in Elliptical Galaxies

T. de Zeeuw

Theoretical Astrophysics, California Institute of Technology,
Pasadena, CA91125, USA

Introduction

Many elliptical galaxies are known to have regular dust lanes and/or disks of cold gas (Ebneter & Balick 1985; Sadler & Gerhard 1985; Bertola 1987). These are usually aligned with the photometric axes of the galaxy, and are often warped. In many cases the gas appears to be kinematically decoupled from the stars, and it is therefore generally believed that this gas was accreted, most likely by capture of a small gas–rich disk galaxy (e.g., Athanassoula & Bosma 1985). The infalling gas is cold, and it will smear out into an inclined ring in a few orbital periods. Differential precession between neighboring orbits then causes cloud–cloud collisions, which are dissipative, and slowly change the orientation of the ring. The expected result is that the gas eventually settles on a set of simple non-selfintersecting closed orbits, which occur at most in a few preferred orientations, namely those where the differential precession vanishes (Tohline, Simonson & Caldwell 1982; Steiman–Cameron & Durisen 1982). Once the gas has reached a preferred plane, its subsequent evolution is expected to be very slow.

Many details of the infall scenario are only partially understood. What is the time scale for settling of the accreted material? Are any of the observed gas disks settled? How do the settling times and the observed properties of the gas disks depend on the structure and shape of the galaxy in which it resides? Is the infalling material responsible for the formation of new stars, and hence can it give rise to distinct substructure in the central regions, such as counter–rotating cores (Illingworth & Franx 1989)? How much mass, if any, is delivered to the nucleus as food for the central monster (Gunn 1979)?

The fact that many elliptical galaxies have acquired gas from the outside implies that galaxy mergers may be rather common, with most galaxies having undergone at least one such event. However, an accurate determination of the merger rate from the observed fraction of elliptical galaxies with gas disks requires detailed answers to the first three questions mentioned above. Here we briefly review the current status of our understanding of these issues, and in particular consider the implications of the fact that most elliptical galaxies have triaxial shapes (Binney 1985; Franx 1988),

Preferred Planes and Closed Orbits

In an axisymmetric galaxy, cold gas can settle only on the circular orbits in the equatorial plane, unless self–gravity of the gas is important (Sparke 1986). If the galaxy is oblate, one will observe the gas disk always along the apparent major axis. If it is prolate, it will appear elongated along the minor axis of the galaxy. Hence in this situation a skewed or warped gas disk is not settled, and such a configuration would be expected to occur infrequently if the settling time is short compared to the lifetime of the disk. If all elliptical galaxies were axisymmetric, it would be trivial to deduce the intrinsic shapes of those with settled gas disks.

In a triaxial galaxy, there are various possible morphologies for a settled gas disk (Steiman–Cameron & Durisen 1982; Merritt & de Zeeuw 1983). If the figure of the galaxy is non–rotating, then the gas will settle on roughly elliptic closed orbits which all lie either in the plane perpendicular to the long axis, or in the plane perpendicular to the short axis of the galaxy. When projected on the sky, there will generally be a significant misalignment between the kinematic axes of the gas and the photometric axes of the stellar light, and the gas disk may also appear warped (de Zeeuw & Franx 1989). This results from the fact that for a triaxial light distribution the projected principal axes generally do not coincide with the major and minor axis of the projected surface brightness distribution (Stark 1977).

If the figure of the galaxy tumbles, e.g., about the short axis, the gas may settle either on roughly elliptic orbits in the equatorial plane or on the family of anomalous orbits, which do not lie in any of the principal planes, but form a stable configuration which is intrinsically warped (Heisler, Merritt & Schwarzschild 1982; Magnenat 1982). The observed properties of such gas disks have not been modeled in any detail, but it is evident that a wide variety of morphologies and velocity fields can be obtained.

In triaxial systems with finite core radii, the elliptic closed orbits become more elongated near the center, so that there will be increasing shear between neighboring orbits. This will induce substantial turbulence in the gas, which is likely to be accompanied by shocks (cf. van Albada & Sanders 1982), and will result in a steady supply of gas to the nucleus. It follows that the gas cannot settle here, and that the closed orbit description of the gas motions must break down in the central regions of a triaxial galaxy with a core.

Settling Times

The time scale τ_s for settling of the infalling gas is expected to be a few times the differential precession time scale, which is of the order of 10 or 20 orbital periods. This can be quite long, especially at large radii. Early analytic estimates of τ_s by Tohline, Simonson & Caldwell (1982) have recently been rediscussed by Steiman–Cameron & Durisen (1988). These authors use a semi–analytic approach in which the cloud–cloud interactions are described via an effective viscosity. They find that in this appoximation τ_s is somewhat larger than thought earlier, and is likely to be of the order of 2–3 times 10^9 years at an effective radius.

Habe & Ikeuchi (1985, 1988) have used smooth particle hydrodynamics (SPH) to study numerically the evolution of an inclined circular ring in a fixed triaxial potential, which was chosen to be logarithmic and therefore describes a galaxy with a massive halo. They considered both nonrotating and rotating figures. In their simulations the gas usually manages to find the preferred planes, and subsequently stays confined to a ring of simple closed orbits. The dissipation in these simulations is due not only to cloud–cloud collisions, but also due to shocks. The settling time scales are of the order of 1–2.5 times 10^9 years. Most interestingly, in these simulations a nuclear disk forms with about 10% of the gas mass. Varnas et al. (1987) performed similar SPH simulations, and found that figure rotation decreases the amount of gas deposited in the center. This can be understood qualitatively by noting that all the models studied possess cores. In the nonrotating case the closed orbits become linear in the core, so that it is not surprising that the gas destined for the core actually ends up in the nucleus, and a ring remains. In the rotating case the orbits reach a limiting elongation.

These simulations confirm that a substantial fraction of an infalling cold gas disk can indeed settle in a preferred plane of a triaxial galaxy in a time that is shorter than the Hubble time. They furthermore show that the closed orbit description for the settled gas is useful. However, it is evident that much further work is necessary. More realistic initial conditions should be considered, as well as a larger variety of shapes of the host galaxy. The mass in the gas disk should be varied, so as to examine the effects of self–gravity (cf. Sparke 1986). This will also require relaxing the condition that the gravitational potential of the host galaxy is fixed, which will allow the galaxy to react to the infalling material in a consistent fashion. The combination of an accurate smooth particle hydrodynamics code with an N–body tree code for following the stellar dynamics (Hernquist & Katz 1989) has made it feasible to explore the gas settling in the required detail. Work along these lines is in progress.

Gas Disks as Diagnostics of Galaxy Structure

We can now ask the further question whether the observed morphology and kinematics of the gas can be used as a diagnostic of the galaxy structure, not only of its shape, but also of the density profile, so that a comparison with the observed light distribution will give the mass–to–light ratio M/L as a function of radius, which might help answer the question whether elliptical galaxies, like spirals, contain substantial amounts of dark matter.

To date, only a few elliptical galaxies with prominent gas disks have been studied in detail. The warped dustlanes in Centaurus A and NGC 5266 can be modeled by the gas having settled on the anomalous orbit family, but the morphology requires tumbling of the galaxy figure opposite to the stellar streaming (van Albada, Kotanyi & Schwarzschild 1982; Varnas et al. 1987). This is not impossible (Vietri 1989), but is not easy to arrange over a substantial part of the galaxy. No detailed comparisons of the gas velocity fields with the model predictions have been attempted. For Centaurus A this would be a very worthwhile exercise, since two–dimensional velocity information is now available not only from Hα measurements (Bland, Taylor & Atherton 1987), but also from HI and CO line emission (van Gorkom et al. 1989; Phillips, Sanders & Sargent 1988).

Recently, the kinematics of the gas in the bulge of the galaxy NGC4845 have been studied in detail (Bertola, Rubin & Zeilinger 1989; Gerhard, Vietri & Kent, 1989). The emission line velocities measured along four different position angles can be accurately fitted with a simple model in which the gas has settled on the elliptic closed orbits that exist in the equatorial plane of a triaxial bulge, which has a shape that is consistent with the observed isophote twist. The gas orbits in the best fitting model turn out to be quite elongated, and have axis ratios as low as 1/3. Although in this case there is no need to invoke infalling gas, the fact that the model and the observations match so well provides circumstantial evidence that cold gas can stay for a long time on elongated orbits. Furthermore, this study shows that the two–dimensional velocity field of a gas disk can usefully constrain the intrinsic shape of the host galaxy.

A similar investigation has recently been done for the E3 galaxy NGC 5077 (Bertola et al. 1989). This galaxy has a disk of ionized gas along its apparent minor axis, and gas velocities have been measured along seven position angles out to a about an effective radius. A simple model with the gas on circular orbits in a spherical potential fits the observed gas kinematics, but would imply a sharply declining M/L in the central regions, similar to what has been reported for the E4 galaxy NGC 7097 by Caldwell, Kirshner & Richstone (1986). However, it turns out that constant M/L triaxial mass models with non–rotating figures and the gas in either of the two preferred planes can be constructed which are consistent with the morphology and kinematics of both the gas and the stars. In addition, a constant M/L oblate model with the gas in a warped polar ring of the kind discussed by Sparke (1986) can be fitted to the data. This shows that it is important to use the proper geometry when deriving M/L from gas kinematics, and, more specifically, that there is no reason to suspect that M/L varies strongly over the luminous part of elliptical galaxies.

Concluding Remarks

It is likely that in many cases the gas disks seen in elliptical galaxies have an external origin. However, the details of the settling process are incompletely understood at present, although there seems to be a consensus that the settling time scales are long. We obviously need realistic simulations of the whole process, with a detailed analysis not only of the morphology but also of the kinematics of the gas. Only then will it be possible to estimate the rate of merging with any confidence.

At the same time it should be pointed out that even the current, incomplete and simple models cannot be discriminated observationally. We need high–accuracy two–dimensional velocity information for a number of elliptical galaxies with gas. This can give valuable insight in the settling process, but also in the structure of the underlying galaxy, and may reveal the presence of dark matter in these systems.

It is a pleasure to thank Elaine Sadler, Lars Hernquist, and Ewine van Dishoeck for enlightening conversations. This research was supported by NSF Grants AST 86–2199 and 88–15132.

References

Athanassoula, E.O., & Bosma, A., 1985. *Ann. Rev. Astr. Astrophys.*, **23**, 147.
Bertola, F. 1987, in *IAU Symposium No. 127, Structure and Dynamics of Elliptical Galaxies*, ed. P.T. de Zeeuw (Dordrecht: Reidel), p. 135.
Bertola, F., Rubin, V., & Zeilinger, W., 1989. *Ap. J. (Letters)*, in press.
Bertola, F., Bettoni, D., Danziger, I.J., Sadler, E.M., Sparke, L.S., & de Zeeuw, P.T., 1989. *Ap. J.*, submitted.
Binney, J.J., 1985. *M.N.R.A.S.*, **212**, 769.
Bland, J., Taylor, K., & Atherton, P.D., 1987. *M.N.R.A.S.*, **228**, 595.
Caldwell, N., Kirshner, R.P. & Richstone, D.O. 1986. *Ap. J.*, **305**, 136.
de Zeeuw, P.T., & Franx, M., 1989. *Ap. J.*, **343**, 617.
Ebneter, K., & Balick, B. 1985, *A. J.*, **90**, 183.
Franx, M., 1988. *PhD Thesis*, Leiden University.
Gerhard, O.E., Vietri, M., & Kent, S.M., 1989. *Ap. J. (Letters)*, in press.
Gunn, J.E., 1979. In *Active Galactic Nuclei*, ed. C. Hazard & S. Mitton (Cambridge: Cambridge University Press), p. 213.
Habe, A., & Ikeuchi, S., 1985. *Ap. J.*, **289**, 540.
Habe, A., & Ikeuchi, S., 1985. *Ap. J.*, **326**, 84.
Heisler, J., Merritt, D.R., & Schwarzschild, M., 1982. *Ap. J.*, **258**, 490.
Hernquist, L., & Katz, N., 1989. *Ap. J. Suppl.*, **70**, 419.
Illingworth, G.D., & Franx, M., 1989. In *Dynamics of Dense Stellar Systems*, ed. D.R. Merritt (Cambridge University Press), p. 13.
Magnenat, P., 1982. *Astr. Ap.*, **108**, 89.
Merritt, D.R., & de Zeeuw, P.T. 1983, *Ap. J. (Letters)*, **267**, L23.
Phillips, T.G., Sanders, D.B., & Sargent, A.I., 1988. Conference on *Submillimeter and Millimeterwave Astronomy*, Kona, Hawaii.
Sadler, E.M., & Gerhard, O.E., 1985. *M.N.R.A.S.*, **214**, 177.
Sparke, L.S. 1986, *M.N.R.A.S.*, **219**, 657.
Stark, A.A., 1977. *Ap. J.*, **213**, 368.
Steiman–Cameron, T.Y., & Durisen, R.H. 1982, *Ap. J. (Letters)*, **263**, L63.
Steiman–Cameron, T.Y., & Durisen, R.H. 1988, *Ap. J.*, **325**, 26.
Tohline, J.E., Simonson, G.F., & Caldwell, N. 1982, *Ap. J.*, **252**, 92.
van Albada, T.S., Kotanyi, C.G., & Schwarzschild, M. 1982, *M.N.R.A.S.*, **198**, 303.
van Albada, T.S., & Sanders, R.H., 1982. *M.N.R.A.S.*, **201**, 303.
van Gorkom, J.H., van der Hulst, J.M., Haschick, A.D., & Tubbs, A., 1989. Preprint.
Varnas, S.R., Bertola, F., Galletta, G., Freeman, K.C., & Carter, D., 1987. *Ap. J.*, **313**, 69.
Vietri, M., 1989. In preparation.

Structure and Colour Gradients in Elliptical Galaxies

R. Peletier[1], *R.L. Davies*[2], *and G. Illingworth*[3]

[1]Kapteyn Laboratorium, Postbus 800, NL-9700 AV Groningen, The Netherlands
[2]Department of Astrophysics, Nuclear Physics Laboratory,
Keble Road, Oxford, OX1 3RH, UK
[3]Lick Observatory, University of California, Santa Cruz, CA 95064, USA

Abstract: We have obtained colour profiles in $U-R$ and $B-R$ for a sample of 39 nearby, bright elliptical galaxies. We find that colour gradients in $U-R$ and $B-R$ correlate well with each other, and are consistent with being caused by metallicity changes. There is no correlation of colour gradients with luminosity, however, the lowest luminosity galaxies in the sample ($M_B > -20$) do not show any colour gradients.

In order to study the intrinsic properties of elliptical galaxies and the mechanisms operating during their formation and evolution, the stellar populations, and especially stellar population changes across the galaxy, contain important information. We have studied the $U-R$ and $B-R$ behaviour for a large sample of ellipticals by analyzing combined CCD frames from two telescopes, to have a large field of view, and good sampling near the core. Our results agree well with the current literature. For details see Peletier *et al.* (1989).

We observe that the colours of elliptical galaxies do not vary much as a function of radius. However, gradients in $B-R$ correlate well with those in $U-R$ (Fig. 1). Most of the scatter here is still caused by the observations. The ratio between the gradients in $U-R$ and $B-R$

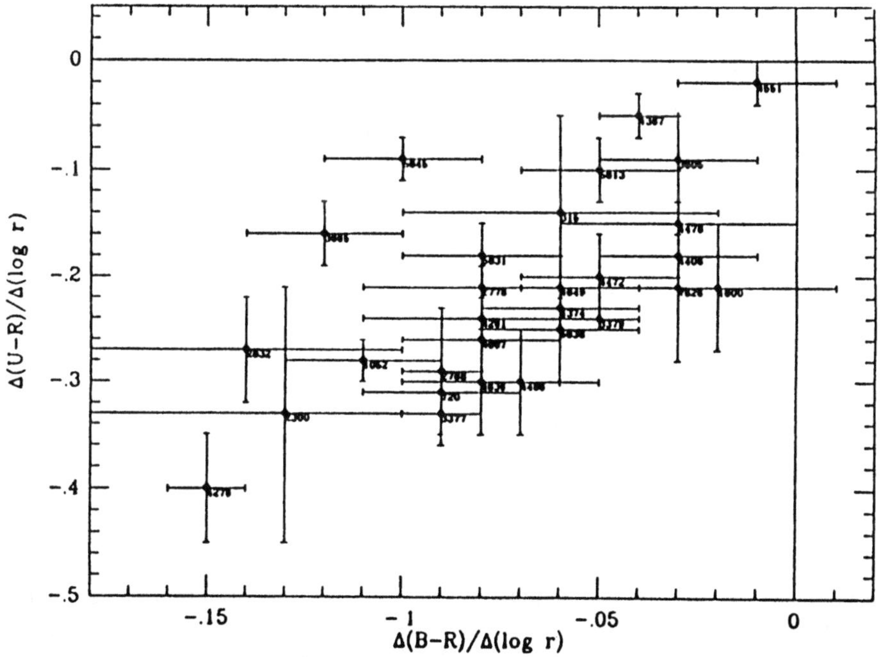

Figure 1: Logarithmic colour gradients in $U-R$ vs. $B-R$.

Dynamics and Interactions of Galaxies Editor: R. Wielen

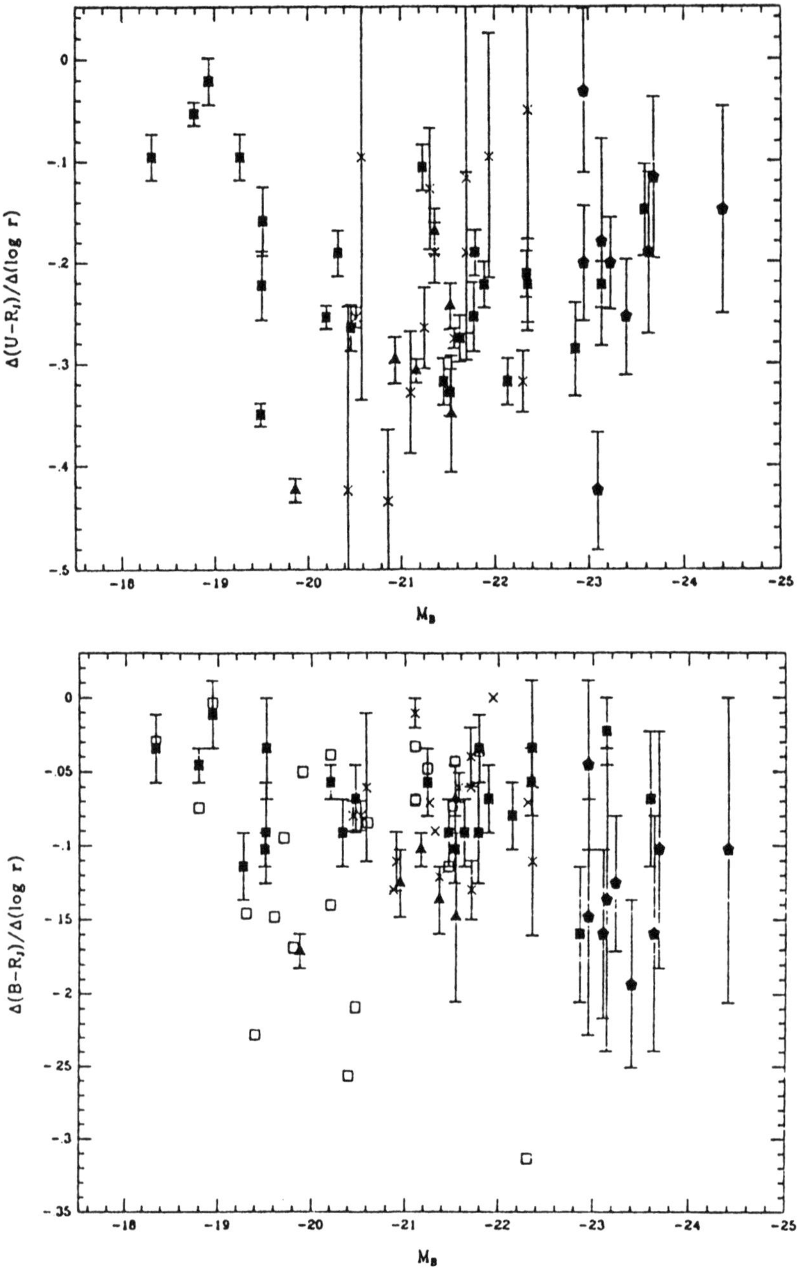

Figure 2: Logarithmic colour gradients in $U-R$ and $B-R$ vs. total blue luminosity

agrees very well with the ratio expected if colour changes are caused purely by metallicity variations.

Colour gradients can provide very important constraints on the formation process. The luminosity dependence of colour gradients is a good discriminator against various models of galaxy formation. If galaxies were formed purely by dissipation, the most massive galaxies would have the largest colour gradients (Carlberg, 1984). Mergers on the other hand cause initial colour gradients to decrease. If elliptical galaxies were formed by mergers, one would expect colour gradients to diminish with increasing luminosity, since the number of mergers undergone would be expected to increase with luminosity. In Fig. 2 we have plotted our gradients vs. luminosity. Also included are data from Franx, Illingworth & Heckman (1989) (crosses) and Vader *et al.* (1988) (open squares). From Fig. 2 we conclude the following:

- There is a large, intrinsic scatter in the amplitudes of colour gradients at all luminosities, with the possible exception of the highest and the lowest.
- There is no significant relation between colour gradient and luminosity, nor is there a relation between colour gradient and $(v/\sigma)^*$.
- The lowest luminosity galaxies ($M_B > -20$) have small or zero gradients. They have boxy isophotes and are rotationally flattened, and are the companions to giant ellipticals. It appears that this group may be physically distinct from the other galaxies, and that their evolution may have been strongly influenced by their massive neighbours.

No relation has been found between colour gradient and boxyness. As far as we can tell, colour gradients are independent of galaxy shape.

References

Carlberg, R.G., 1984, *Astrophys. J.*, **286**, 403

Franx, M., Illingworth, G.D. and Heckman, T.M., 1989, *Astron. J.*, **98**, 538

Peletier, R.F., Davies, R.L., Illingworth, G.D., Davis, L. and Cawson, M., 1989, *Astron. J.*, , in press

Vader, J.P., Vigroux, L., Lachièze-Rey, M. and Souviron, J., 1988, *Astron. Astrophys.*, **203**, 217

A Survey for Fine Structure in E+S0 Galaxies

P. Seitzer[1] *and F. Schweizer*[2]

[1]Space Telescope Science Institute, 3700 San Martin Drive, Baltimore, MD 21218, USA

[2]Department of Terrestrial Magnetism, Carnegie Institution of Washington, Washington, DC 20015, USA

ABSTRACT. We present the results of an imaging survey aimed at finding and cataloguing various kinds of fine structure in 74 nearby E + S0 galaxies. More than half of the objects feature at least one of the following: ripples ("shells"), strongly boxy isophotes, X-structure, and plumes, thus suggesting a multi-component origin (mergers, accretions) for these galaxies. The presence of fine structure correlates with spectral line strengths measured by Faber *et al.* (1990): ellipticals with ripples, plumes, etc., tend to have stronger-than-average Hβ lines and weaker-than-average CN and Mg_2 indices for their luminosity, suggesting that — compared to their featureless counterparts — they may have experienced more recent star formation.

1. Introduction

There is mounting evidence that many E and S0 galaxies show signatures of past mergers: kinematic subsystems, sharp-edged ripples, faint disks in E's, X-structures, boxy isophotes, and luminous plumes. Models explaining ripples and X-structures as the remains of small accreted galaxies (Quinn 1984; Dupraz & Combes 1986; Hernquist & Quinn 1988, 1989) have found wide-spread acceptance. In order to study the frequency and strength of various types of fine structure, we have been conducting an imaging survey of a well-defined sample of 74 E + S0 galaxies with a distance cutoff at $v_0 =$ 4000 km s^{-1}. The discovery of ripples in a number of bona fide disk galaxies has been described elsewhere (Schweizer & Seitzer 1988). Here, we discuss the statistics of fine structures found in our sample and report the discovery of correlations between the presence of fine structure and spectroscopic line strengths.

2. Observations and Reduction

The sample galaxies were selected from the RC2 and RSA according to criteria given in Schweizer and Seitzer (1988). A typical galaxy image obtained with the CCD camera of the 0.9-m telescope at Kitt Peak consists of a series of consecutive exposures totaling 40–60 min in the R passband. Typical S/N values achieved on sky are 55/□″ for an RCA chip and 130/□″ for a Tektronix chip.

All galaxy images were cleaned, calibrated, and masked with various derivative images obtained by convolution with $\sigma = 5''$ and $10''$ Gaussians. Extensive visual evaluation of all unmasked and masked images and of contour plots led to a catalog of fine structures detected in each galaxy. Figure 1 shows some examples of observed structures.

3. Results

Fine structure exists in a surprisingly large fraction of the observed galaxies. Fully 56% of the E's and 32% of the S0's feature ripples. In about 10% of the sample we also find X-structure, and one third of the galaxies feature obviously boxy isophotes ($\delta_4 \gtrsim 0.02$) at some, often outer, radius. Nearly 70% of E's and 53% of S0's possess at least one

Dynamics and Interactions of Galaxies Editor: R. Wielen

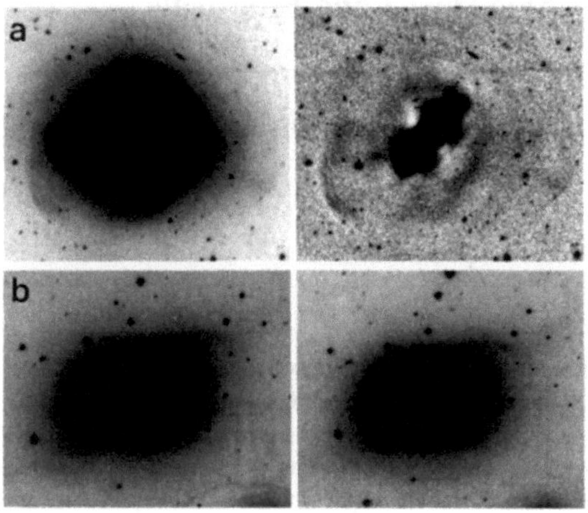

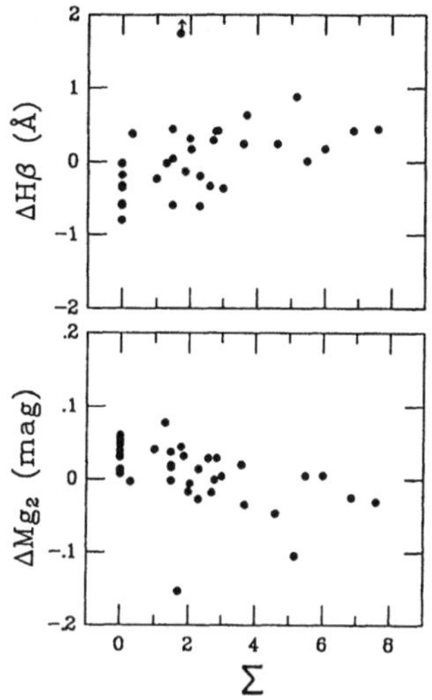

Figure 1. (*left*) Examples of fine structure. (*a*) NGC 3610 (E5) direct and masked; changes from a thin disk plus X-structure near the center to boxy with ripples at $r \gtrsim 55''$. (*b*) NGC 1700 (direct, 2 CCD's) features two broad arms or tails that produce boxy isophotes and an isophotal twist; this E3 galaxy may be the remnant of two merged disks.

Figure 2. (*right*) Correlations between line-strength residuals and fine-structure parameter Σ for 35 ellipticals: (*top*) $\Delta H\beta$ versus Σ, and (*bottom*) ΔMg_2 versus Σ.

of four types of fine structure thought to be merger signatures: ripples, boxiness, X-structure, and plumes.

Two main conclusions from this survey are: (1) There is morphological evidence for past mergers in at least half of all nearby E and S0 galaxies. (2) If ripple lifetimes are $\lesssim$2 Gyr, as model simulations suggest, then an average E experiences at least 3–4 accretion events per Hubble time and an average S0 at least 2–3 such events; alternatively, if ripples keep forming for a prolonged period after a major merger due to material falling back, then they may trace single events that formed the present-day E + S0 galaxies.

The amount of fine structure seems to correlate with spectral line strength and luminosity. Figure 2 shows line-strength residuals $\Delta H\beta$ and ΔMg_2 plotted versus the fine-structure parameter $\Sigma = S + \log(1+n) + B + X + P$, where S, B, X, P are our visual estimates of ripple strength, boxiness, X-structure, and plumes, and n is the number of ripples. We obtained these residuals from line strengths measured by Faber *et al.* (1990) by making linear fits of $H\beta$ vs. M_B and Mg_2 vs. M_B for a subset of 241 E + S0 galaxies. The plots show that ellipticals with much fine structure tend to have stronger-than-average $H\beta$ lines and weaker-than-average Mg_2 (and, not shown here, CN) indices for their luminosity. A similar plot of M_B vs. Σ shows that the galaxies with the most fine structure ($\Sigma = 4$–8) are all brighter than $M_B \approx -20.8$. These various correlations suggest that ellipticals with much fine structure have experienced more recent star formation than their featureless counterparts.

References

Dupraz, C., and Combes, F. 1986, *Astron. Astrophys.*, **166**, 53.
Faber,S.M., Burstein, D., Dalle Ore, C, and Gonzalez, J. 1990, in preparation.
Hernquist, L., and Quinn, P.J. 1988, 1989, *Astrophys. J.*, **331**, 682, and **342**, 1.
Quinn, P.J. 1984. *Astrophys. J.*, **279**, 596.
Schweizer, F., and Seitzer, P. 1988. *Astrophys. J.*, **328**, 88.

The Various Kinematics of Minor-Axis Dust-Lane Ellipticals

C. Möllenhoff and R. Bender

Landessternwarte, Königstuhl, D-6900 Heidelberg, Fed. Rep. of Germany

Minor-axis dust-lane ellipticals are otherwise 'normal' elliptical galaxies displaying a fast rotating belt of dust and gas along their minor axis. It is generally accepted that the dust lanes originate from infall events of gaseous material into the elliptical galaxies. This is especially supported by the discrepancy between the angular momenta of the stellar and gaseous components. Two well investigated minor-axis dust-lane ellipticals, NGC 5128 = Cen A (Bertola et al, 1985; Wilkinson et al, 1986), and NGC 5266 (Möllenhoff and Marenbach, 1986, Varnas et al, 1987) rotate along their major axes, orthogonally to the dust lanes.

A very interesting object of this class is NGC 4589 (Möllenhoff, Bender 1989): While the dust/gas rotates rapidly along the minor axis, the stellar velocity field exhibits rotation along both the major and the minor axis, the axis of zero rotation lies $\approx 45°$ in between. NGC 4589 is most easily explained as the remnant of a merging event of two galaxies, the skew stellar velocity field reflecting the rotational and orbital kinematics of the parent objects.

Here we present new morphological (CCD camera of the Landessternwarte, Calar Alto 1.23m telescope) and kinematical results (B&C spectrograph with CCD, Calar Alto 3.5m telescope) of four further minor axis dust-lane ellipticals: NGC 5354, NGC 5846, NGC 5485, and NGC 6703. In spite of their similar morphology new permutations of different orientations between stellar and gas kinematics occur. This is a further support for the infall or merger origin of these dust-lane ellipticals.

NGC 5354 has a prominent minor-axis dust-lane. There are no signs of any physical connection to the optical neighbour NGC 5353 (Fig. 1). No stellar rotation could be detected, neither along the minor nor along the major axis.

NGC 5846 is a nearly round galaxy ($\epsilon = 0.05$) in interaction with the southern companion NGC 5846A (Nieto, Prugniel 1987). The dust lane of NGC 5846 is rather weak and of patchy structure, with a fast rotation of $\pm$ 200 km/sec along P.A. = 70°. No stellar rotation could be detected.

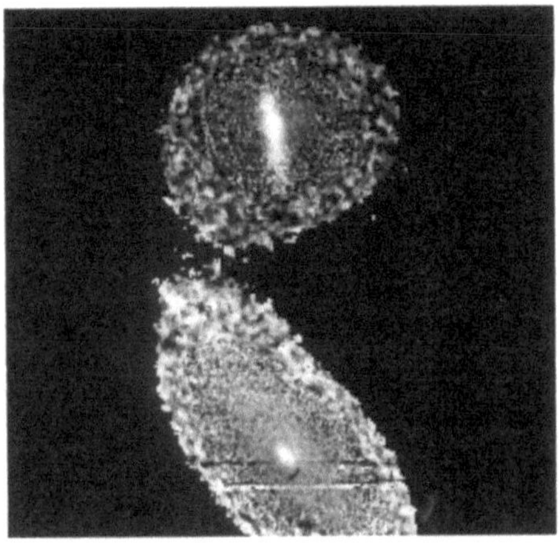

Fig.1: The V - I color index image shows the strong dust lane of NGC 5354. The southern NGC 5353 is an optical companion

Dynamics and Interactions of Galaxies Editor: R. Wielen

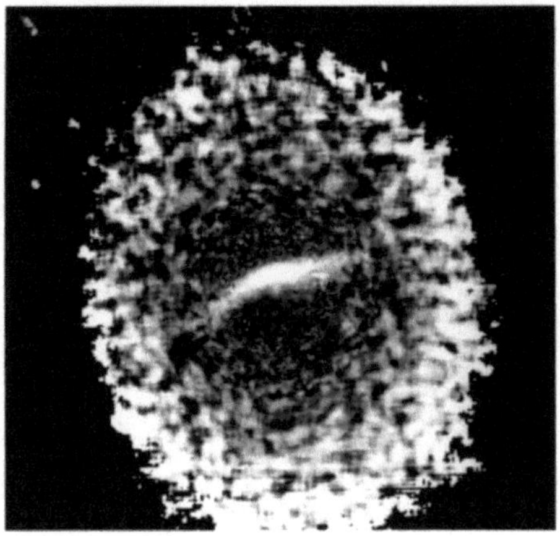

Fig.2: The V - I color index image of NGC 5485 shows the prominent dust lane along the minor axis

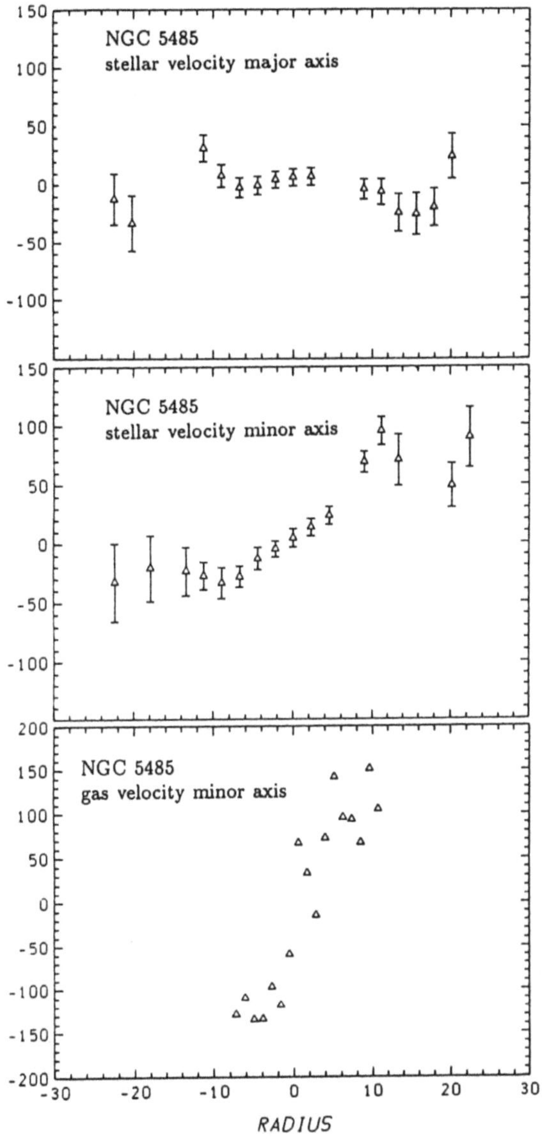

Fig.3: The stars in NGC 5485 do not rotate along the major but along the minor axis, however with a smaller amplitude than the gas

NGC 5485 exhibits an exceptional strong minor axis dust lane, slightly inclined to the line-of-sight (Fig. 2). The dust and gas rotate fast along the minor axis with an amplitude of ± 150 km/sec. The stars show rotation along the minor axis as well (± 50 km/sec) but not along the major axis (Fig. 3). This gives strong arguments for NGC 5485 to be of triaxial or prolate (spindle-like) shape.

NGC 6702: An image of the dust lane of this object was shown in Davoust et al (1987). The gas of the dust lane exhibits fast rotation along the minor axis with an amplitude of more than ± 200 km/sec. The stellar component shows no rotation along the major axis and indications for a weak rotation along the minor axis (Fig. 4).

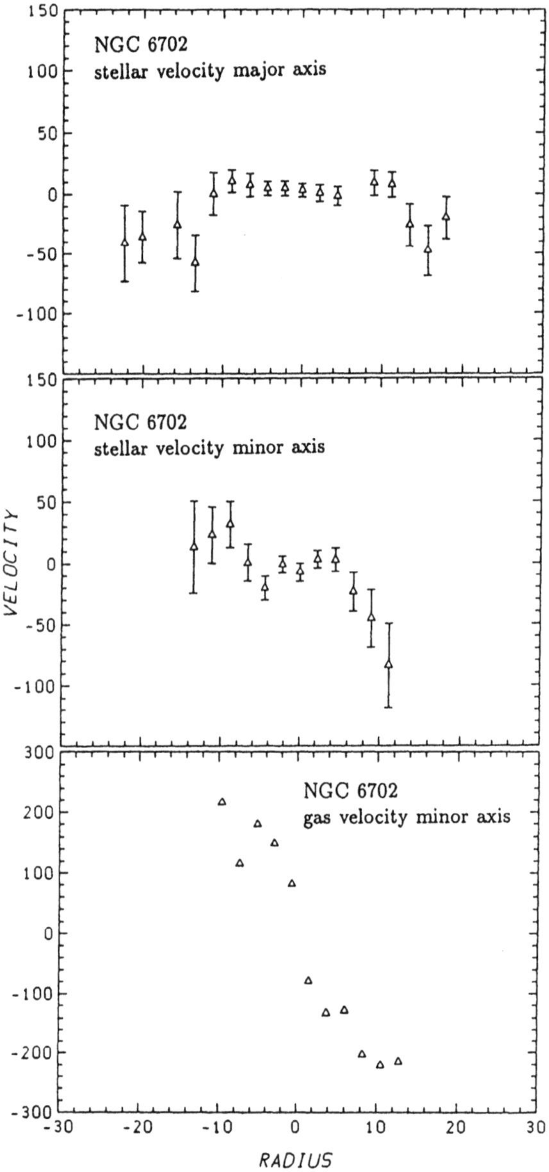

Fig.4: While the gas in NGC 6702 rotates fast along the minor axis, there is no stellar rotation along the major and marginal rotation along the minor axis

References:

Bertola, F., Galletta, G., Zeilinger, W.W.: 1985, Astrophys. J. **292**, L51
Davoust, M., Capaccioli, M., Lelièvre, G., Nieto, J.-L.: 1987, IAU Symp. No. 127, p.407
Möllenhoff, C., Bender, R.: 1989, Astron. Astrophys. **214**, 61
Möllenhoff, C., Marenbach, G.: 1986, Astron. Astrophys. **154**, 219
Nieto, J.-L., Prugniel, P.: 1987, Astron. Astrophys. **186**, 30 Varnas, S.R., Bertola, F., Galletta, G., Freeman, K.C., Darter, D.: 1987, Astrop. J. **313**, 69
Wilkinson, A., Sharples, R.M., Fosbury, R.A.E., Wallace, P.T.: 1986, MNRAS **218**, 297

Rummaging Through Elliptical Galaxies

J.L. Tonry

Massachusetts Institute ot Technology, Department of Physics,
Cambridge, MA 02139, USA

High signal-to-noise CCD observations of 13 Virgo galaxies are presented. Smooth galaxy models have been subtracted from the original data, allowing inspection of the contents of these galaxies. Essentially all globular clusters are visible between 1 and 10 kpc radius.

An image of each galaxy is shown, before and after subtraction of the smooth model. The parameters ellipse fits to isophotes of these models is plotted, as a function of radius, along with the degree of 4θ (boxy) deviation. Globular clusters and other objects are identified, and their density is plotted against radius for each galaxy.

There is a great variability in both amplitude and in shape (boxy versus pointy) of the deviation from pure elliptical isophotes. This variability occurs between different galaxies, and even within the same galaxy. The globular cluster populations are generally more extended than the stars in these galaxies, and there is a moderate amount of variability in the specific frequency with which they occur.

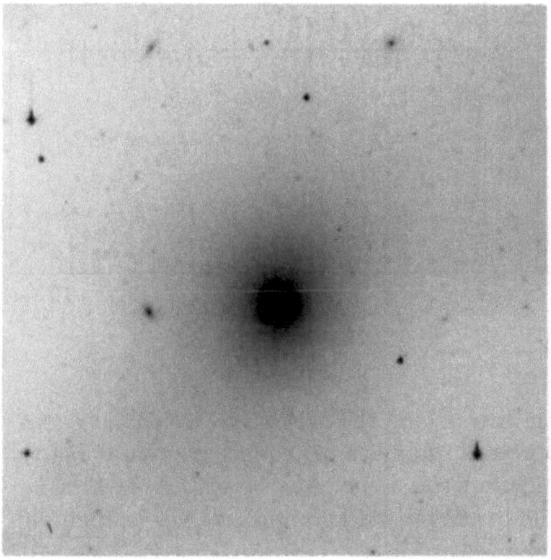

A CCD image of N4636 and its surroundings. The image is about 4 arcmin = 20 kpc on a side, each pixel is 0.29 arcsec and the seeing is typically 0.8 arcsec fwhm. The exposures are 2000 sec in the R band on the KPNO 4-m telescope, yielding $1e^-$ from an object with m_R = 34 (M_R = +2.8 at Virgo, assumed to be 17 Mpc distant) and $2 \times 10^5 e^-$ arcsec^{-2} from the sky.

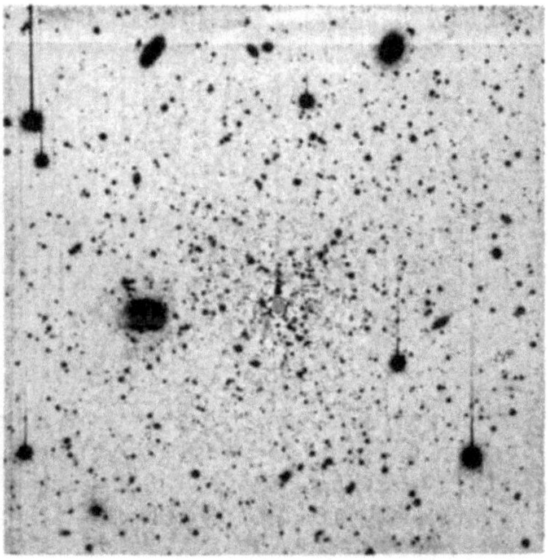

The image of N4636 after a smooth model of the galaxy has been subtracted. Black corresponds to about 500 e^-, and objects whose apparent magnitude is m_R = 25.6 (M_R = −5.5 at Virgo) can be detected at 5σ confidence (flux = 2500 e^-, peak = 300 e^-). The Poisson statistics of the number of giant stars in the galaxy found in each pixel causes the mottling seen near the center.

Dynamics and Interactions of Galaxies Editor: R. Wielen

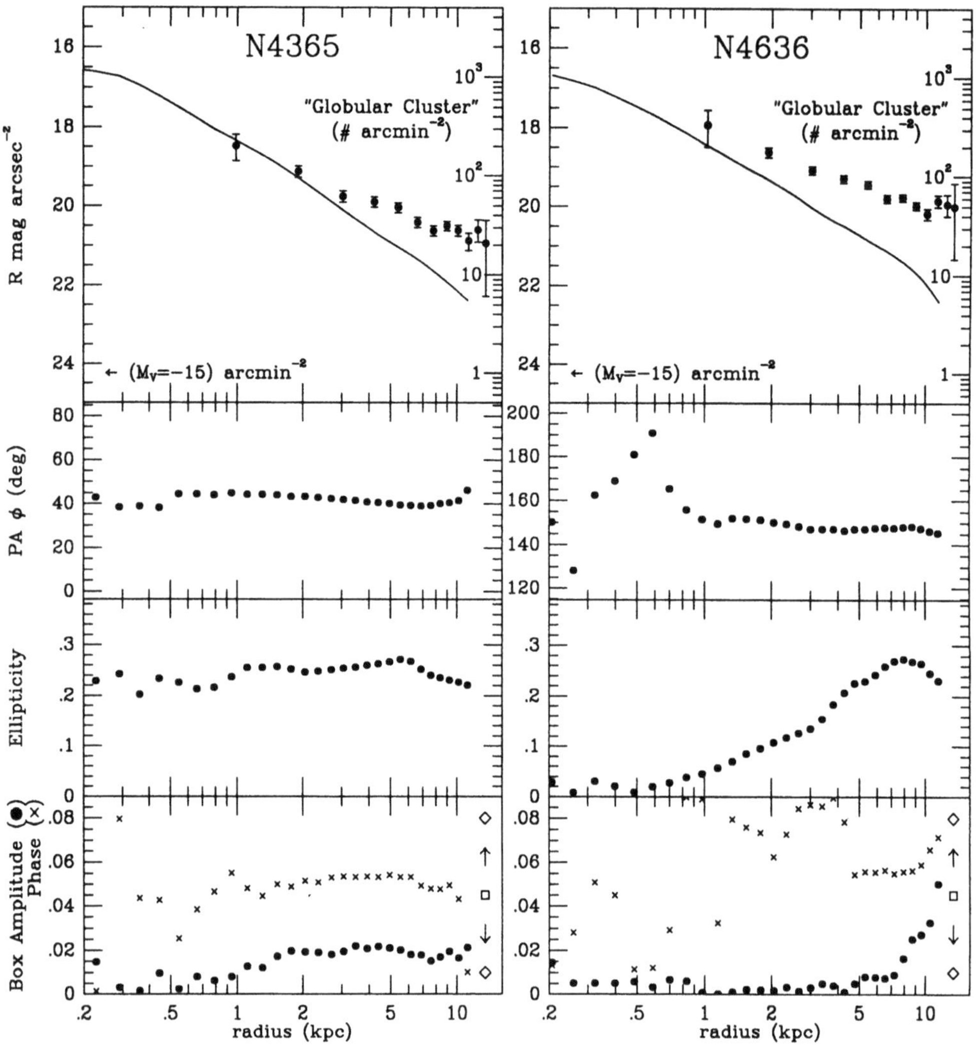

(1) The upper plots show the galaxy surface brightness along the major axis as a solid curve, and points giving the density of all other sources (in number per square arcmin at right). The two scales are aligned with a factor of 1 point source per $M_V = -15$. Note that for this completeness limit (roughly R=23.5), the density of background galaxies is about 10 per square arcmin.

(2) The second plots show the galaxy position angle (E from N) as a function of radius.

(3) The third plots show the galaxy ellipticity $(1 - b/a)$ as a function of radius.

(4) The lower plots display the boxiness of the isophotes. The solid dots give the amplitude of a $\cos 4\theta$ isophote component relative to the overall isophote level, and the crosses show the phase angle of the boxiness with respect to the major axis. As indicated by the label at the right, the phase angle ranges from 0° (pointy) to 45° (boxy) to 90° (pointy isophotes).

Photometrical and Kinematical Sub-Systems in Elliptical Galaxies

P. Surma and R. Bender

Landessternwarte, Königstuhl, D-6900 Heidelberg, Fed. Rep. of Germany

Problem, Observation and Reduction Techniques: Knowledge of the light distribution in elliptical galaxies is interesting for questions about their formation and evolution. Prevailing limitations of brightness profiles are central seeing and resolution problems and -essential for the outer regions- difficulties in subtracting the sky-brightness properly.

An especially elaborated R-band-photometry was obtained using the GEC P8602/A CCD-camera and the focal reducer of the Landessternwarte at the 1.23m telescope at Calar Alto (1.56 arcsec/pixel, 15x10 arcmin field). In order to prevent our measurement of sky-brightness from being 'polluted' by galaxy light from the outer regions, 'empty' sky frames about 20-30 arcmin north and south of the galaxies were taken immediately before and after each galaxy exposure. The subtraction of mean sky-frames (free of stars and presumably also free of residual galaxy light) yields galaxy frames with a zero background, which was tested to be flat to the order of 0.1% of the initial sky-level. By this new, very precise method profiles down to levels 5 mag fainter than sky and out to radii of several arcmin are routinely observed, with overall accuracies of about 0.1 mag. A comparison with the extensive study of NGC 3379 by de Vaucouleurs and Capaccioli (dVC: 1979 ApJS 40,699; CdV: 1983 ApJS 52,465) was performed, showing coincidences within some 0.01 mag, at radii ranging from 6 to 240 arcsec and only a 1.6% deviation from their measurement of $r_e = 58.69$ arcsec. The two-component structure of NGC 3379's light profile is exactly re-found.

Results and Discussion: The brightness profiles of a major part of elliptical galaxies show a two-component structure similar to NGC 3379, i.e. the faint outer region is well-fitted by a shallow $r^{1/4}$ -law over a wide range of radii from $\lessapprox 10$ to >200 arcsec. Centrally superimposed on this main body a steeper 'core-region' of amplitudes up to 0.8 mag is found. In all cases of ellipticals showing kinematical sub-systems (Jedrzejewski,Schechter:1988, ApJ 330,L87; Bender: 1988, A&A 202,L5) we observe distinct photometric components at corresponding radii (see Fig.1 for an example).

Calculations in CdV show, that seeing effects can only introduce up to 0.1 mag of similarly shaped excess light. Independent tests without focal reducer revealed

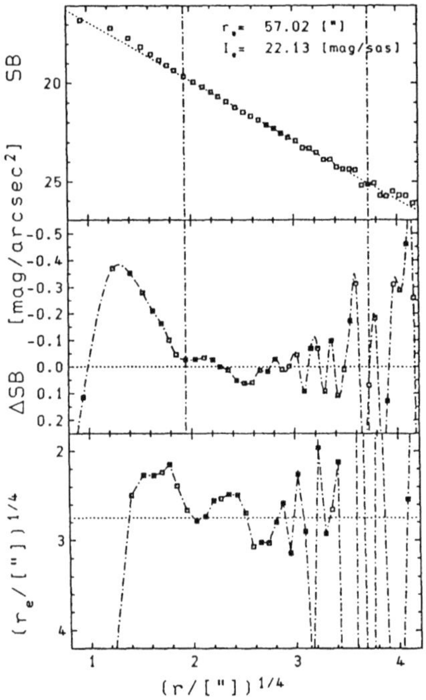

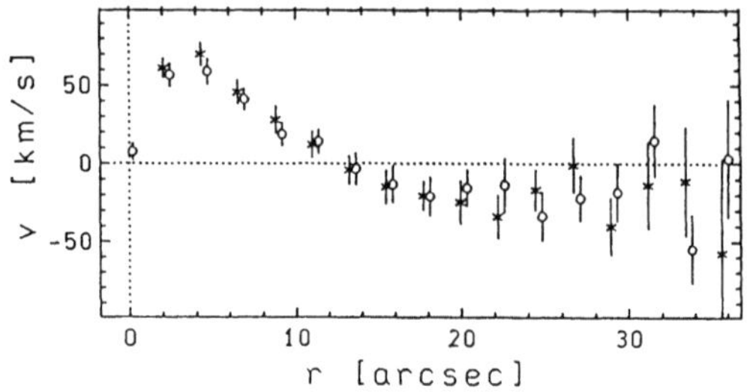

Fig.1 NGC 5322 - major axis profiles

left (top to bottom) plot versus $r^{1/4}$: (1) surface brightness with $r^{1/4}$ -law (vertical lines indicate fit area); (2) residuals from $r^{1/4}$ -fit; (3) local slope of SB-profile expressed in terms of $r_e^{1/4}$ (r_e =57" from fit indicated), steep core-region, but shallow outer region; note the end of the photometrical core-region at $r^{1/4} \approx 1.9$, i.e. r≈13"

right: rotation velocity versus linear radius, counter-rotating core within r≈12"

only minor changes in the profile of the core-region (apart from the very center), while resolution changed by a factor of 3 down to 1.2 arcsec FWHM. Above that the close radial correspondence with kinematically outstanding features can hardly be understood as an artifact due to seeing.

In this context the elliptical galaxy NGC 5813 which was discussed to be the pathological case of a merger remnant (Kormendy: 1984, ApJ 287,577) merely seems to be one object (out of many) fitting into this pattern of quite 'normal' two-component galaxy structure, since photometrically this behaviour is also exhibited by kinematically unspectacular objects like NGC 4472, 4486, 4621, 4636 and 4649. No correlation between this photometrical phenomenon and dominance of neighboured galaxies is found, as it has been proposed by Kormendy 1977 (ApJ 218,333) in terms of interaction-triggered 'tidal halos'. Thus profiles reaching faint brightness levels near 25 mag seem to be vital for a physically reasonable characterization of the overall light distribution; both for recognition of photometrical sub-systems and for deriving accurate profile parameters (e.g. r_e and I_e), the latter being a severe problem with photometries of too limited extent. Therefore it may be advisable to be cautious about inconsistencies in parameter fitting, because this may influence many related topics, e.g. minimal manifold discussions or distance determination of ellipticals (Djorgovski, Davis: 1987, ApJ 313,59).

A detailed description and discussion of this project will be published in A&A (Surma, Seifert, Bender: in preparation).

Further Evidence of Continuity Between Elliptical and Disk Galaxies

M. Capaccioli[1], *N. Caon*[2], *and R. Rampazzo*[3]

[1]Osservatorio Astronomico, Vicolo Osservatorio 5, I-35122 Padova, Italy
[2]Dipartimento di Astronomica, Università di Padova, Vicolo Osservatorio 5, I-35122 Padova, Italy
[3]Osservatorio Astronomico di Brera, Milano, Italy

The analysis of the deviations of galaxy isophotes from pure ellipses has greatly contributed to the growing idea that ellipticals constitute a composite, non-homogeneous family, in which 'boxy' and 'pointed' E's represent independent physical classes (Bender *et al.* 1989, and references therein).

Since pointed isophotes in S0 galaxies indicate the presence of a disk, we are motivated to inquire if the disky E's and the S0 galaxies may be interpreted in terms of a continuous structural/physical sequence characterized by the variation of some morphological/genetic parameter (for instance, the bulge–to–disk ratio as tracer of the average specific angular momentum).

An attack on the problem of a link between disky E's and lenticulars requires a comparative analysis of the isophotal properties of a large, homogeneous and unbiased sample of E and S0 galaxies (*cf.* Bender *et al.* 1989) We report here on the study of a small but statistically significant sample of early–type galaxies, all members of the Virgo cluster (Caon *et al.* 1989).

The 33 galaxies in our sample are all E's and non–barred S0's (classified according to Binggeli *et al.* 1985), which are brighter than $B_T = 14$ mag and which are contained within the fields of two deep UK Schmidt plates (centers at Dec.= $11°30'$, and R.A.= 12^h22^m and R.A.= 12^h36^m respectively). The photometric mapping for the inner regions ($\mu_B \lesssim 22.5$) rests on CCD material collected with the 2.2 m MPI–ESO and 1.5 m Danish telescopes at La Silla. In the outer regions ($22.5 \lesssim \mu_B \lesssim 28$) we used UK Schmidt photographic material. The reduction procedure is the same as in Capaccioli *et al.* (1988). In particular, the modelling of the isophotes with ellipses (and the investigation on the occurrence of disk's signatures) has been made using the fitting procedure implemented in the Interactive Numerical Mapping Package (*cf.* Capaccioli 1989) and applied by Capaccioli *et al.* (1987, 1988, 1989; see also Capaccioli and Vietri 1989).

Figure 1 shows a striking similarity for the E and the S0 samples: in both cases $\epsilon_{max} \simeq 0.3$ represents a well defined boundary value for the presence/absence of the disky phenomenology. The existence of such a boundary is easy to understand for the S0 sample, because the morphological classification is informative about the 3–D structure. When the inclination i of the principal plane is $\lesssim 45°$, the apparent ellipticities of disk and bulge become very similar, although the intrinsic flattening of these two components are different. In other words, the appearance of the *lemon shape* effect in the isophotes of S0's depends upon the inclination. Fig. 1 suggests that this is also the case for the E group. In other words, the geometrical behaviour of the isophotes of most if not all the E's of our sample is accounted for by the assumption that, alike S0 galaxies, these objects consist of the superposition of two components, an oblate (or moderately triaxial, with $b/a > 0.5 > c/a$) spheroid and a thin, very likely faint disk. As already stressed elsewhere (Capaccioli 1987), in this scenario the disk may have no dynamical significance; it is only a feature used to establish an analogy between a known structure (S0's bulges) and an unknown one.

The applicability of our intrinsic–shape analysis based on isophotal fitting is limited the suggestion that a large fraction of E's are shape–wise akin to lenticulars. It is not even possible to establish which fraction, since our sample seems biased toward intrinsically faint objects.

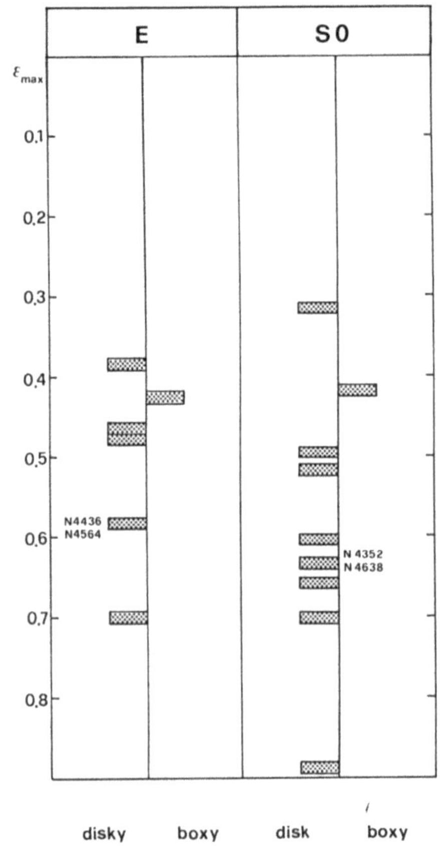

Figure 1. Histogram of E and S0 galaxies in our sample (classification according to Binggeli *et al.* 1985) showing disky and boxy isophotes. The objects indicated in the figure share the same ϵ_{max}.

References

Bender, R., Capaccioli, M., Macchetto, F., and Nieto, J.-L., 1989: *The Messenger*, **55**, 6.

Binggeli, B., Sandage, A., and Tammann, G.A., 1985: *Astron. J.*, **90**, 1687.

Caon, N., Capaccioli, M. and Rampazzo, R. 1989: *in preparation.*

Capaccioli, M., 1987: IAU Symp. No. 127 *Structure and Dynamics of Elliptical Galaxies*, (Princeton), ed.T.P. de Zeeuw, *Reidel*: Dordrecht, p. 47.

Capaccioli, M., 1989: in '*Proc. Second Summer School on Extragalactic Astronomy*', (Cordoba: Argentina), *in press.*

Capaccioli, M., and Vietri, M., 1989: in '*Origin and Structure of Galaxies*', Proc. of the Yellow Mountain Summer School on Relativistic Astrophysics, (China), p. 150.

Capaccioli, M., Held, E.V., and Nieto, J-L., 1987: *Astron. J.*, **94**, 1519.

Capaccioli, M., Piotto, G., and Rampazzo, R., 1988: *Astron. J.*, **96**, 487.

Capaccioli, M., Held, E.V., and Lorenz, H., 1989: *in preparation.*

Kinematics of the Interacting Elliptical Galaxy Pair NGC 4782/4783*

R. Madejsky and R. Bender

Landessternwarte Königstuhl, D-6900 Heidelberg, Fed. Rep. of Germany

Abstract The peculiar morphology of the elliptical galaxy pair NGC 4782/4783 reflects a stage of strong tidal interaction. Both galaxy centers are displaced with respect to the centers of the outer isophotes. Long-slit spectroscopy reveals disturbed kinematical properties. The velocity dispersion increases radially from the centers of both galaxies towards the centers of the outer isophotes to maximum values larger than 400 kms^{-1}. It is supposed that dynamical friction strongly decelerates both galaxies on their orbits. In the central parts the crossing-time of the stars is much less than the encounter timescale. Therefore, the reaction of the cores is similar to the behaviour of a solid body while the sticky envelopes are subject to differential tidal forces.

Kinematical properties We present kinematical data for both galaxies in the interacting galaxy pair NGC 4782/4783. With the B&C-spectrograph at the 3.6m telescope of the European Southern Observatory, ESO, Chile, we obtained for each slit position two spectra in two 90 min. exposures. The slit positions (p.a. $= 110^o$) were chosen perpendicular to the line connecting both nuclei since there are located the major morphological disturbances (see e.g. Madejsky, 1989). In addition, any mutual contamination from the superposition of the luminosity profiles is minimized. The northern galaxy NGC 4783 has a central radial velocity of 3900 kms^{-1}. The radial velocity is constant toward the east part, but increases by an amount of 80 kms^{-1} towards the west part of the galaxy. The southern galaxy NGC 4782 has a central radial velocity of 4580 kms^{-1}, this implies a minimum velocity difference of 680 kms^{-1} between both galaxies. The radial velocity begins to rise 10 $arcsec$ east of the center of NGC 4782 by an amount of 130 kms^{-1}, this is like in NGC 4783 toward the position of the supposed original center of the galaxies. Apart from this characteristic the radial velocity curves are very regular and no rotation seems to be present (see upper panels of Fig. 1 and Fig. 2). The velocity dispersion in NGC 4782 (southern component, slit position p.a. $= 110^o$) is shown in the lower panel of Fig. 1. The central velocity dispersion of 315 kms^{-1} increases rapidly radially and reaches a maximum value of 430 kms^{-1} at a position 14 $arcsec$ east of the center. In the opposite direction the velocity dispersion increases only marginally and decreases below the central value at 12 $arcsec$. The northern galaxy NGC 4783 has a central velocity dispersion of 260 kms^{-1} which increases rapidly in both directions. A maximum value of 410 kms^{-1} is reached at 12 $arcsec$ on both sides of the center.

Discussion The observations presented in Figs. 1 and 2 represent direct evidence for strongly disturbed kinematical properties. During the high-velocity encounter of NGC 4782 and NGC 4783 with a radial velocity difference of more than 680 kms^{-1} the gravitational potential is varying rapidly with time. The innermost stars are not affected very much by the changing potential if their orbital period or crossing-time t_{cr} is much shorter than the perturbation, i.e. the encounter timescale T_{enc}. The radius r of such a core can be defined by the condition $t_{cr} \approx r/\sigma \ll T_{enc} \approx b/\Delta v$ where σ is the velocity dispersion at radius r, b is the impact parameter and Δv is the velocity difference between both galaxies at closest approach. For the cores the perturbing force is averaged over many orbital periods (Lauer, 1988). Therefore, compared to the outer parts, the structure of the core is not changed considerably during the encounter. For the outermost stars with orbital periods much larger than the encounter timescale the encounter can be roughly described by the impulsive approximation (Aguilar and White, 1986). There, sig-

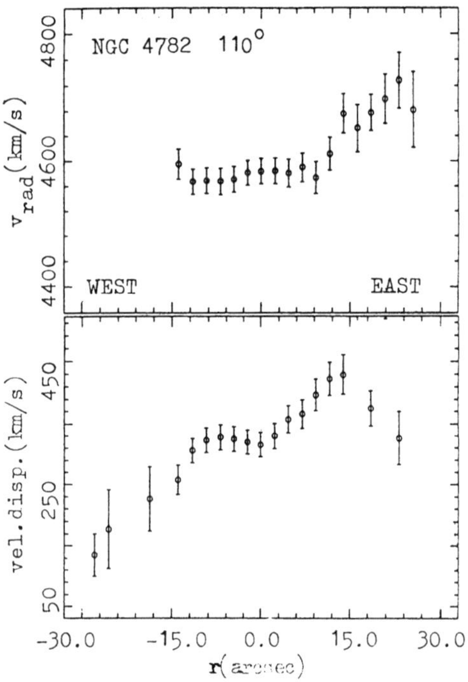

Fig. 1: The radial velocity of the southern component NGC 4782 at position angle p.a.=110° is displayed in the upper panel. The radial velocity increases by about 80 kms^{-1} towards the west part which coincides with the original position of the core. The velocity dispersion with a central value of 315 kms^{-1} increases radially by more than 100 kms^{-1} and reaches the maximum value of 430 kms^{-1} at a distance 14 *arcsec* from the center.

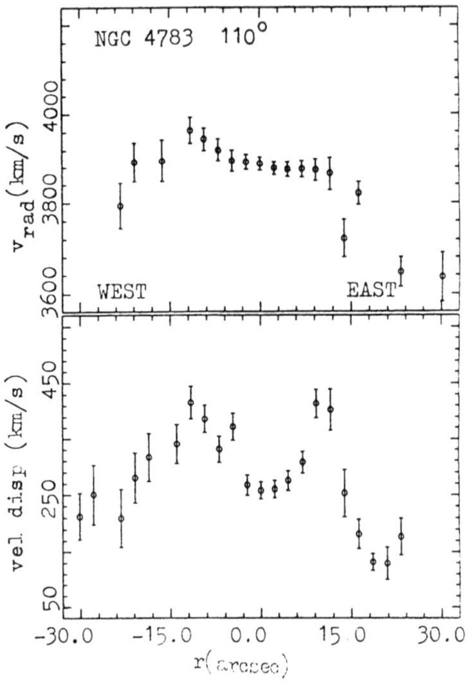

Fig. 2: The velocity dispersion of the northern component NGC 4783 at position angle p.a.=110° is displayed. The velocity dispersion increases radially from a central value of 260 kms^{-1} to a maximum value of more than 410 kms^{-1} at a distance 12 *arcsec* on both sides from the center. This very high velocity dispersion outside the center cannot reflect an equilibrium state and indicates the ongoing interaction.

nificant energy transfer from the orbital motion into the halo can be expected. This differentiated interaction leads to the effect that the cores appear to be displaced from their original positions.

References

Aguilar, L.A., White, S.D.M., 1986, Astrophys. J. **307**, 97
Lauer, T., 1988, Astrophys. J. **325**, 49
Madejsky, R., 1989, Astrophys. Space Sci., in press

The Kinetic Energy and Angular Momentum in Elliptical Galaxies

G. Busarello and G. Longo

Osservatorio Astronomico di Capodimonte, via Moiariello 16, I-80131 Napoli, Italy

A method to evaluate from the observed rotation curves (RC) and velocity dispersion profiles (VDP) of elliptical galaxies, the angular momentum, the total mass, and the two terms of the kinetic energy due to the rotational and to the random motions is proposed.

The method

In order to understand the true physical status of the elliptical galaxies, it is needed to evaluate the contribution of the rotational and of the residual motions to the total kinetic energy of the system. We propose a procedure to derive the two terms of the kinetic energy from the observed velocity dispersion profile and the rotation curve. In the following discussion it will be assumed that E galaxies are spheroidal systems and that their rotation axes are oriented perpendicularly to the line of sight. This hypothesis does not affect the result by more than the cosine of the inclination angle. The spatial luminosity distribution given by Mellier and Mathez (1987) was adopted. The deprojection of the observed VDP and RC is performed by fitting the observed quantities to a suitable analytic expression, and then by solving the corresponding Abel integral equation.The kinetic energies per unit mass are then computed assuming spheroidal symmetry for the $\sigma(\bar{r})$ and cylindrical simmetry for the $V(\bar{r})$. To obtain the total kinetic energy due to random motions, hypotheses on the anysotropy of the velocity field need to be made, but the true value is not far from the one obtained by assuming isotropy (Busarello et al, 1988, Busarello and Longo, 1989). The total mass can, therefore, be derived from the virial theorem :

$$2T = -U = U^{*} f \frac{GM_T}{r_e}$$

where $U^{*} = 0.33611$ depends only on the radial structure of the density distribution (Mellier and Mathez, 1987) and f is a factor which depends only on the ellipticity $\varepsilon = 1 - c$ of the density distribution (Busarello et al., 1988). As shown in table 1, the contribution to the total mass of the rotational kinetic energy is not negligible. It is necessary to point out that, while the kinetic energies are expressed in unit of luminous mass, the potential energy, as derived from the virial theorem, gives total mass, and therefore, the values for M/r_e shown in table 1 need to be corrected for the luminous-to-total mass ratio. The total angular momentum per unit mass is computed from the spatial rotation curve, assuming cylindrical symmetry for the rotation. Owing to the dependence on r, the computation of angular momentum is more sensitive to the extrapolation of the spatial RC to the outer regions. The estimates of the error are in the range between ~ 10 and ~ 30 percent (Busarello and Longo, 1989). The method was applied to the original data for the 10 faint ellipticals in the sample by Davies et al. (1983). Table 1 gives, in column 4 and 5, the V/σ values by us and by Davies et al. (1983), respectively. Column 6 and 7 give the values of the ratios M/r_e and J/r_e.

The systematic difference between the two sets of data is due to the fact that Davies and collaborators used a value of the velocity dispersion obtained at the center or from an average over the central regions, i.e. in a region of the galaxies where the relative importance of random

Dynamics and Interactions of Galaxies Editor: R. Wielen

Table 1

ident.	V	σ	(V/σ)	$(V/\sigma)_D$	M/r_e $(\times 10^{23})$	J/r_e
N2778	89.92	87.72	1.03	0.70	0.151	73.02
N3605	48.67	51.31	0.95	0.58	0.520	49.15
N3818	96.15	82.98	1.16	0.75	0.147	60.40
N3904	50.47	110.70	0.46	0.31	0.189	56.65
N4387	52.53	55.84	0.94	0.56	0.607	63.43
N4478	52.79	77.23	0.68	0.44	0.962	56.29
N4551	35.68	61.96	0.58	0.33	0.602	34.77
N4742	78.55	36.78	2.14	1.24	0.505	41.34
N5638	87.22	69.93	1.25	0.41	0.102	96.70
N5845	125.28	106.13	1.18	0.65	0.245	11.68

motions is higher, leading, therefore, to an overestimate of the the relative importance of T_σ over T_V. The application of the above procedure to all the galaxies in the catalogue of kinematical properties of early type systems (Busarello, Longo and Di Martino, 1989) is in progress.

The authors thank M.Capaccioli for very helpful suggestions.

References

Busarello G., Filippi S., Ruffini R., 1988, Astron. and Ap., **197**, 91

Busarello G., Di Martino F., Longo G., 1989, Astron. and Ap., in press

Busarello G., Longo G., 1989, in preparation

Davies R.L., Efsathiou G., Fall S.M., Illingworth G., 1983, Ap. J., **266**, 41

Mellier Y., Mathez G., 1987 Astron. and Ap., **175**, 1

Tidal Changes in Energy and Angular Momentum of a Spheroidal Galaxy

G. Som Sunder[1], *R.K. Kochhar*[1], *and S.M. Alladin*[2]

[1]Indian Institute of Astrophysics, Bangalore 560034, India

[2]Department of Astronomy, Osmania University, Hyderabad 500007, India

Consider an oblate spheroidal stellar system, mass M, semiaxes $a_1 = a_2 > a_3$ being tidally perturbed by a mass point M' moving in a straight line with a large velocity $\underline{v}$ and pericentric distance $\underline{p}$ Fig.1. The change in the velocity u of an individual star at $\underline{x}$, to first order in x/p, is (Knobloch 1978)

$$\Delta u_i = \eta \, \Sigma \, \alpha_{ip} x_p \,, \quad \text{with} \quad \alpha_{ij} = 2\hat{p}_i \hat{p}_j + \hat{\nu}_i \hat{\nu}_j - \delta_{ij} \,,$$

$$\eta = \frac{2GM'}{p^2 v} \,, \quad \hat{p} = \underline{p}/p, \; \hat{\nu} = \underline{v}/v. \tag{1}$$

1. Energy

Under the impulse approximation, the change in the total energy tensor is given by the change in the kinetic energy tensor (cf. Som Sunder & Kochhar 1986)

$$\Delta K_{ii} = \frac{1}{2} \eta^2 \, I_{11} \, (1 - \hat{\nu}_i^2 - y \, \alpha_{i3}^2) \,, \tag{2}$$

where y is the asphericity : $a_3^2 = a_1^2 = a \;(1-y)$. The scalar change in the energy is

$$\Delta K = \eta^2 \, I_{11} \, [1 - \frac{1}{2} y \, (1 - \cos^2 \psi \sin^2 \theta)] \,. \tag{3}$$

For y = 0, this reduces to the well known Spitzer (1958) result. The effect of asphericity is zero for $\psi = 0$, $\theta = 90°$, i.e., when the perturber is moving perpendicular to the equatorial plane. Averaging over all angles we get

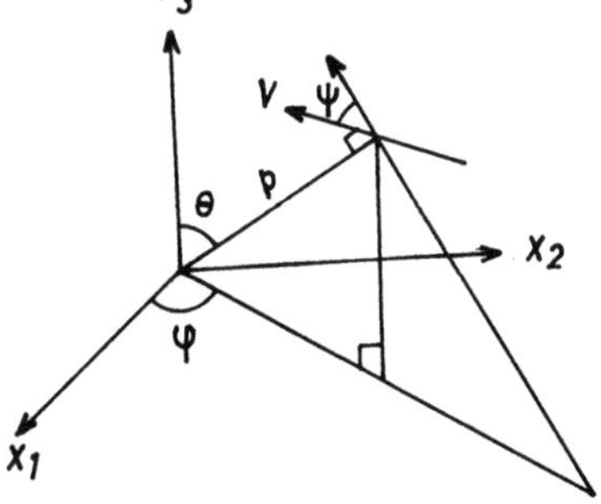

Fig.1. The geometry of the problem

$$< \Delta K > = \eta^2 I_{11} (1 - \frac{1}{3} y) . \qquad (4)$$

Thus other things being equal, energy transferred is lower for a disc (y = 1) than for a sphere (y = 0), and lower for a sphere than for a spindle-shaped galaxy (y= -2).

The fractional increase in the total kinetic energy K is given by

$$\frac{\Delta K}{|E|} = \frac{8}{3} (1 - \frac{1}{4} e^2) \frac{e}{\sin^{-1} e} \cdot \frac{GM'}{pv^2} \frac{M'}{M} (\frac{a_1}{p})^3 . \qquad (5)$$

$\frac{\Delta K}{|E|}$ falls off with increasing e.

2. Angular momentum

The change in the angular momentum of an ellipsoidal test galaxy is given by

$$\Delta J_i = \eta \alpha_{jk} (I_{kk} - I_{jj}) \qquad i,j,k \text{ cyclic} \qquad (6)$$

Note that ΔJ_i arises because of the presence of quadrupole moment. If any two axes of the ellipsoidal test galaxy are equal, the tidally transferred angular momentum does not have any component in the perpendicular direction. For an oblate spheroidal galaxy, the tidal angular momentum vector lies in the equatorial plane:

$$\Delta J_1 = - \eta \alpha_{23} (I_{11} - I_{33}); \qquad \Delta J_2 = +\eta \alpha_{13} (I_{11} - I_{33}); \qquad J_3 = 0. \qquad (7)$$

Averaging over angles,

$$< \Delta J_1^2 >^{1/2} = < \Delta J_2^2 >^{1/2} = \frac{1}{\sqrt{5}} \eta I_{11} y. \qquad (8)$$

This is similar to Peebles' (1969) result. ΔJ increase with increasing departure from sphericity.

The fractional increase in the angular momentum due to tidal interaction is

$$\frac{\Delta J}{J^2} = \frac{2}{3} \frac{e^4 (1-e^2)^{1/2}}{\Omega^2} \frac{GM'}{pv^2} \frac{M'}{M} (\frac{a_1}{p})^3 , \qquad (9)$$

where Ω is the angular velocity of the Mclaurin spheroid in units of $(\pi G \rho)^{1/2}$:

$$\Omega^2 = \frac{2(1-e^2)}{e^3} [(3 - 2e^2) \sin^{-1} e - 3e (1-e^2)^{1/2}] \qquad (10)$$

A detailed paper is under preparation.

References

E,Knobloch : Astrophys. J. Suppl. 38, 353 (1978).

P.J.E.Peebles : Astrophys. J. 155, 393 (1969).

G.Sum Sunder,G., R.K.Kochhar : Mon. Not. R. astr. Soc. 221, 553 (1986).

L.Spitzer : Astrophys. J. 127, 17 (1958).

Effects of Gravitational Interactions on the Structure of Elliptical Galaxies

Ph. Prugniel

ESO, Karl-Schwarzschild-Str. 2, D-8046 Garching, Fed. Rep. of Germany

The structure and the dynamics of elliptical galaxies are undoubtedly affected by gravitational interactions; encounters and merging. However, it is in general difficult to get an idea of the history of individual galaxies. Exceptions are the few well known cases of merger remnants, but these galaxies are still on their way toward relaxation. Old mergers, like NGC 3640 (Prugniel et al., 1988) and NGC 6776 (Sansom et al., 1988) and some "boxy" ellipticals are probably closer to the "normal" equilibrium state of ellipticals.

Pairs of elliptical galaxies, as they provide cases of ongoing interactions, allow the study of the response of these systems to the gravitational perturbation (e.g. Borne, 1984). We have conducted a detailed study of such pairs based on surface photometry and kinematic analyses (Prugniel et al., 1989). This poster gathers updated results from this study with data taken from the literature and draws some general conclusions for the structure of the manifold of elliptical galaxies.

The main difficulty in the study of pairs is the lack of information concerning their real separation. Even a selection of pairs with low red-shift differences cannot avoid projection effects within a cluster. For our purpose, in order to avoid this problem as much as possible, we have selected pairs on the following criteria:

- Low red-shift difference
- Close and relatively isolated
- Small and compact companion of a large galaxy
- Morphological signs of interaction (see Davoust and Prugniel, 1989)

The last condition is the best evidence for interaction and the first is a minimal requirement.

Results

It is now well established that ellipticals form a two-dimensional family in the three-dimensional space: $m_e - B_T - \log\sigma$ (e.g. Djorgovski and Davis, 1987). Actually, the galaxies are constrained to this plane (the fundamental plane: FP) by the virial theorem (Djorgovski et al., 1988). Thus, strong perturbations should lead, before virialization, to departures from this plane (note, however, that galaxy-to-galaxy variation of mass/luminosity and effects of rotation also broaden the FP). The distribution of the galaxies within the FP contains certainly information about the formation (amount of dissipation) and evolution processes. In this respect, beside chemical enrichment, gravitational encounters are likely to be a major factor. Indeed, we have shown (Prugniel et al., 1989) that paired galaxies tend to delineate a narrow sequence in the FP, corresponding to relatively high velocity dispersion and surface brightness.

Figure 1 shows the FP "edge on". It appears clearly that the paired ellipticals belong to the FP defined by a reference sample of non-interacting galaxies. There is no evidence for a broadening of the FP for the paired ellipticals, in other words, the fundamental plane is robust against the interactions.

Dynamics and Interactions of Galaxies Editor: R. Wielen

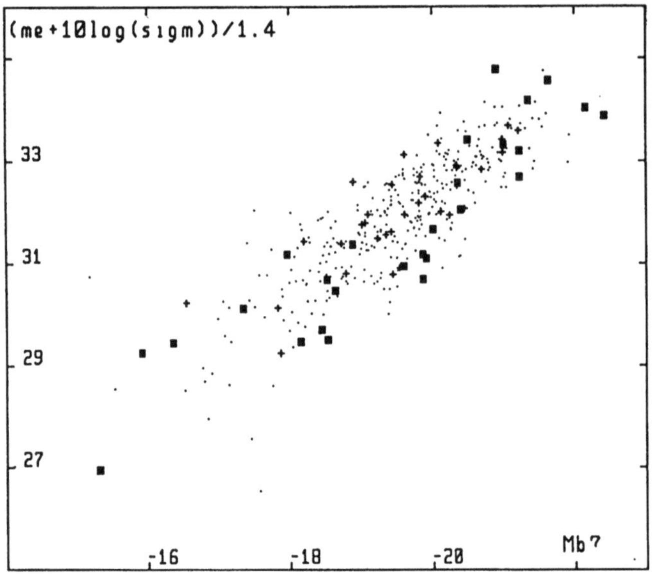

Figure 1: Fundamental plane "edge on"

The small dots stand for galaxies from a sample of non-interacting galaxies. Squares are for the paired galaxies from Prugniel et al. (1989) plus NGC 1587/8, NGC 4467, NGC 4782/3 and NGC 7236/7. The crosses are for Turner's pairs (White et al., 1987, MNRAS **203**, 701). *me* is the mean surface brightness within the effective aperture, *sigm* is the central velocity dispersion and *Mb* the absolute B-magnitude. The distances are computed from the velocity flow model from Burstein et al. (1989, in *Large Scale Structure and Motions in the Universe*). The distance modulus of Virgo is 30.60.

Conclusion

What does this robustness of the FP mean? There are two alternatives, first: The gravitational interaction has only a minor effect on the global equilibrium of these paired galaxies, or second: The perturbation that arises at the peri-galactic passage virializes within a small fraction of the orbital period.

It is possible to check these alternatives. For a typical pair (e.g. NGC 741/2 or NGC 2693/4) the ratio of the dynamical time at the effective radius to the orbital time is of the order of 0.1 for the small galaxy and of 0.25 for the large one. Thus, it seems likely that about 1/4 of the main members of the observed pairs are still unrelaxed, while small galaxies are in general observed in a state closer to equilibrium. We can also evaluate the strength of the tidal perturbation. Typically, for the small companions of large galaxies, the potential energy due to the tidal field is a few tenths of the gravitational energy, while for the large galaxies it is only between 0.01 and 0.1. Due to the non-linearity of the self-graviting response, we expect strong departure from the FP mainly for the small galaxies.

Finally, the first alternative applies to large galaxies (weakly perturbated), while the second applies to small galaxies (with short virialization times). The net effect on observable quantities, which is expected to be of the order of a few tenths of the kinetic energy, is not more important than the galaxy-to-galaxy variation of rotational/random kinetic energy or than observational uncertainties.

The main observable effect of the gravitational encounters appears to be a long term evolution that affects the distribution of the galaxies in the fundamental plane.

References

Borne, K.D.: 1984, *Astrophys. J.* **287**, 503

Davoust, E., Prugniel, Ph.: 1988, *Astron. Astrophys. Letters* **201**, L30

Djorgovski, S., Davis, M.: 1987, *Astrophys. J.* **313**, 59

Djorgovski, S., de Carvalho, R., Han, M.-S.:1988, in *The extragalactic Distance Sacale*, Ed. van den Bergh, S. and Pritchet, J., page 329

Prugniel, Ph., Nieto, J.-L., Bender, R., Davoust, E.: 1988, *Astron. Astrophys.* **204**, 61

Prugniel, Ph., Davoust, E., Nieto, J.-L.: 1989, *Astron. Astrophys.* **000**, 000

Sansom, A.E., Reid, I.N., Boisson, C.: 1988, *Mon. Not. Roy. Ast. Soc.* **234**, 247

Part VIII

Infall

Gas-Hungry Sc Spirals

A. Toomre

Massachusetts Institute of Technology, 2-371, Cambridge, MA 02139, USA

1. Introduction

The various interactions between galaxies that have been the principal theme of this conference would be much too infrequent if they had to rely solely on the accidental passages of truly independent travelers in the vastness of space. Fortunately, we know now that there is a lot more to interacting galaxies than this charming old notion of "grand ships passing in the night". Our ships seem to collide or suffer alarmingly close scrapes at their observed sizable rates mainly because they never quite managed to escape from each other amidst the great flying-apart known as the Hubble flow. Their own gravities and especially those from any associated dark matter seem to have been more than enough to overcome or "close" this flow in the vicinities referred to as pairs, groups or clusters — and in the likely graveyards that are elliptical in shape, slow in rotation, and de Vaucouleurs' in profile. In short, it has been recognized more and more in recent years that our story of interacting galaxies must be primarily one of *fallback*, with greatly increased collision risks that would worry even Lloyd's.

Viewed like that, the observed recent interactions between galaxies do not differ vastly from the fallback that must have formed the individual galaxies in the first place. At issue are only the content and timing: Scarcely anyone — certainly not the present writer — would dispute that at very early times the Hubble flow can have "rained" back only gas and/or the dark matter, whereas nowadays this continuing precipitation clearly includes entire galaxies, sometimes leading to spectacular fireworks. And yet, is the original era of gas infall really over?

Over the past two decades a number of authors including Oort (1969, 1970), Larson (1972, 1976), Gott (1975) and perhaps most consistently Gunn (see 1982, 1987 for two good summaries) have already answered, or have claimed to answer, this nagging gaseous question with emphatic negatives. Of course, to this day it remains frustratingly hard — despite even the "high-velocity clouds" discovered in the 1960s, though still at what distance? — to detect the stuff actually arriving, either here or elsewhere. Indeed there have been several painstaking searches such as the one conducted by Lo & Sargent (1979), which in essence located no other promising reservoirs whatsoever, at least of neutral atomic hydrogen. Nonetheless, on various grounds of theory or interpretation, Gunn and the others persuaded even myself long ago that we must remain alert to the possibility that the disks of present-day galaxies may well continue to accrete late-arriving virgin gas at rates that are by no means trifling and that this "afterbirth", as Oort termed it, may actually tend to double their surface densities (especially at the larger disk radii) in something like the next Hubble time. In other words, galactic disks may still be growing quite significantly before our very eyes.

One reason why I have felt extra sympathetic to this theme of slow disk build-up has been my lingering suspicion (cf. Toomre 1977, 1985) that even the bulges or spheroids of galaxies like our own arose primarily from early mergers of lesser though more compact galaxies which tried to remain separate but could not. In such speculations, I have tended very much to agree with Gunn — and for once to disagree a bit with Schweizer — that existing disks consisting

Dynamics and Interactions of Galaxies Editor: R. Wielen

already of stars are fragile entities which cannot reasonably have survived intact even during any moderately-massive central mergers, at least not without suffering a great deal of heating and thickening. Hence my willingness to believe that any bulge- or spheroid-makings via early mergers involving, say, the first quarter or half of the baryonic mass destined to end up in a given galaxy would still have needed to be followed by much subsequent disk-building from yet later fallback of gas.

A second and more popular indirect reason for believing that the gas infall continues has been our old puzzle of the warps of galaxies, as already Larson used to stress and as Binney reviews nicely in this volume. But perhaps even more adamant, once one stops to think about it, is a third logically awkward truth — that ragged spiral structures, such as we find in the Sc galaxies, are still very abundant in the sky. Despite a number of alarms voiced in recent years especially by Sellwood, this last little time-bomb seems to have been ticking away vastly underappreciated by most astronomers. To repeat that alarm, the rest of this review will now dwell exclusively on the surprisingly severe demands made by the survival of the Sc spirals.

2. Swirling Hotch-Potch

In the literature on spiral dynamics, the Sc's and other multi-armed spirals have long tended to take a back seat to the more urgent challenges of "grand designs" like M51 or M81 or NGC 1300. For years, this relative neglect was quite understandable, along with the widespread presumption that success with density waves and the like in the grander arena might eventually explain even the Sc's as a minor intellectual spinoff. However, that is not how things turned out — at least in my opinion.

Ironically, it became clear only about a decade ago (cf. Toomre 1981, hereafter T81) that the basic engine behind even the grandest spiral structures was in essence nothing more than the so-called "swing amplification" which Goldreich & Lynden-Bell (1965, hereafter GLB) and Julian & Toomre (1966, hereafter JT) had jointly discovered in a much more local setting, whilst trying to make sense of just the *ragged* spirals! By a curious blindness, none of us four authors appreciated until much later how relevant this type of amplification was apt to be even for the global modes, tidal spirals and/or the bar-making. But at least for the Sc's we were not nearly so diffident in those old papers: Already GLB pursued the theme that the "swirling hotch-potch of pieces of spiral arms" in such galaxies results from this variant of the Jeans instability affecting mainly their interstellar gas, whereas JT went on to point out that with such help the gravity from any massive orbiting gas aggregate (which we would nowadays call a giant molecular cloud) will evoke a strong two-sided *wake* — looking much like a segment of a spiral arm — in the stellar medium that shears past it in a differentially rotating disk.

Likewise encouraging for the Sc's were the results of early N-body experiments with disks. For instance, though they are rightly celebrated for having first demonstrated the bar-making, the papers of Miller, Prendergast & Quirk (1970) and of Hohl (1971) also displayed plenty of ragged spiral features when either an inelastic "gas" component was included, or when the random motions in an already featureless disk of stars were gradually cooled. Figures 2 and 3 from the first paper and Figure 17 from the second were particularly striking in this regard. Better still, the ease of generating such transient spiral features in one's computer kept on being rediscovered or reaffirmed every few years since then: The latest and perhaps also the prettiest example of that genre appears in this volume as Figure 5 of the paper by Hernquist, who included it admittedly as only a control experiment.

There was, however, one major weakness to all these demonstrations. The gravitational forces involved in such constantly-changing spirals have the unwelcome side effect that they tend to increase the mean epicyclic motions of any stars or other undamped particles which take part in that chaotic dance. Typically, in just a few revolutions these would-be participants become simply too "hot" to contribute appreciably to any more fun. In other words, the vigor of these apparent "spiral instabilities" was quickly recognized to doom any hopes for their longevity — unless one adds some form of cooling or damping of the random motions, such as might be natural in a gas but surely not in a disk of stars, despite Hohl's very instructive exercises.

In retrospect, these severe heating worries had been foreshadowed by the theoretical work of Julian (1967), who demonstrated soon after we had discovered the wakes that such cooperative effects — especially when JT's stability index $Q \lesssim 1.5$ — must greatly enhance the growth of stellar random motions due to encounters with "cloud complexes" such as envisaged originally by Spitzer & Schwarzschild (1953). And yet, though Julian had been my student, I confess that for years afterwards I kept on hoping that his analyses had exaggerated those rates of heating! I agreed that he had ruled out any long-term survival of values of Q close to unity, but had Julian erred in suggesting that this index would typically tend to rise to about *twice* that level? For instance, despite adopting quite modest parameters, had he still overestimated the masses or abundance of the gaseous clumps? This seemed a valid escape route as long as it remained unclear what, if anything, such postulated clumps had to do with the ragged spirals displayed so abundantly by the galaxies and N-body experimenters alike. Today, however, such detachment would be unwise. As I am about to explain, those gas clumps and the transient spirals and hence also the heating seem very much to be parts of just a single story.

3. Sellwood and Carlberg

This old but by-then-rather-dormant story received a big boost from the work of Sellwood & Carlberg (1984, hereafter SC; see also Carlberg 1987, Sellwood 1986, 1989) in the mid-1980s. Already for several years before that, Sellwood had been alternately impressed and annoyed by the stochastic features that kept cluttering up his extensive N-body studies of disk models of galaxies. Indeed those features, like unwanted guests, had proved to be a nuisance as he sought to track down down and measure some genuinely unstable global modes of such disks. They had put him to the considerable trouble of having to devise various "quiet start" experiments in order to accomplish the latter goals.

In their joint work, SC essentially decided to turn this nuisance to good advantage. They not only conjectured anew that "most visible spirals are short-lived features organized by gravitational forces", and that any continued accretion of gas from outside was bound to be helpful. Much more important, they also demonstrated explicitly that a fairly modest but steady drizzle of new particles into initially circular orbits in one rather plausible disk/halo model of a galaxy kept an ever-changing and multi-armed spiral structure in business for at least ten or twenty revolutions as opposed to the usual few. These constantly-added computer particles were meant by SC to imitate either the newly-arrived gas itself or else stars which had recently been born from it; both would plausibly begin "cool" in the sense of random motions. In their impressive Figure 5, SC began with a composite model in which 20,000 disk particles provided only 30% of the central force field and a fixed halo supplied the rest. One revolution later they had gradually increased this active number to 21 K, and ten revolutions later to 30 K, meaning a fractional increase by one-half.

This rate of increase of the disk mass was hardly trivial, especially when one considers that SC chose to count their revolutions at a location well inside the half-mass radius of their disk. On the other hand, apart from that deposition of fresh particles into circling orbits, their experiment deliberately omitted all explicit forms of damping including that which any resident gas would doubtless experience, or the relatively small extra amount that would occur as some mass was returned to interstellar space during late phases of stellar evolution. Moreover, it seemed plausible that even a halving (or more) of the infall rate might keep a weaker spiral structure going proportionately longer, or that the real infall could be somewhat episodic, thereby perhaps reinvigorating the real Sc spirals once in a while. In short, even though they did not yet clinch their own case, the authors SC must have surprised quite a few of their readers that such an explanation of galaxies Sc was suddenly already so palatable.

The thing more troubling for me, as a fellow theorist, was that SC seemed to have no clear idea of the dynamical origins of their recurrent spiral arms — despite various sensible comments about the inevitable Poisson noise from randomly-placed particles, and also despite a whole section devoted to the swing amplification of wavelike disturbances. Indeed even the words "spiral instabilities" in the title of their paper struck me as a likely misnomer, for reasons that should become clearer shortly. To be sure, I did not (and still do not) find it implausible that their model galaxy might in addition have suffered from subtler large-scale instabilities resulting from "some slight feedback", as Carlberg (1987) put it in his review. But their interesting paper reminded me above all that noone had yet done a good N-body job of checking even locally what the swing-amplified "particle noise" should really look and act like — and I undertook to remedy that oversight.

4. Spiral Chaos in an Orbiting Patch

What follows is only a glimpse at a long series of N-body experiments in a drastically "local" or Hill approximation (cf. Spitzer & Schwarzschild 1953, GLB, or JT) that I have been conducting since the mid-1980s, initially in response to SC. This work began on my personal computer — literally at home — but it soon spilled over, nearly always in a low-priority or background mode, onto a series of increasingly powerful workstations available in our department. Just as important, much of the analysis and other head-scratching on this surprisingly rich project soon came to be shared with my old friend Agris Kalnajs in Australia; to this date, he and I have still never written a joint paper on anything, but on this topic we soon intend to.

Hill's (1878) approximation originated in lunar theory where, for the purpose of analysing the mutual interplay of the Earth and the Moon under the distant influence of the Sun, it made sense to imagine the latter yet more distant by a large factor, but also yet more massive by the same factor cubed. In the limit of huge distance (but vanishing eccentricity) this ploy nicely *linearized* the tidal force from the Sun. Alternatively one can pretend, in that subject as well as ours, that both the radial separation $x \equiv r - r_0$ and the tangential separation $y \equiv r_0(\theta - \Omega_0 t)$ of a particle from some convenient reference point, which strictly circles with the normal orbital frequency $\Omega_0 \equiv \Omega(r_0)$, are and will remain tiny compared with the radius r_0 itself. Either way, one obtains the neat approximate equations of motion

$$\ddot{x} - 2\Omega_0\dot{y} = 4\Omega_0 A_0 x + f_x \qquad (1)$$

$$\ddot{y} + 2\Omega_0\dot{x} = \qquad f_y \ , \qquad (2)$$

where

$$A_0 \equiv -\tfrac{1}{2}R_0(d\Omega/dR)_0 \qquad (3)$$

is the familiar Oort constant of differential rotation, the term $4\Omega_0 A_0 x$ gives the tidal force, and f_x and f_y refer to any additional forces of local origin. The old celestial mechanics of course demanded that $A_0 = \frac{3}{4}\Omega_0$, but herewith let us insist instead that $A_0 = \frac{1}{2}\Omega_0$ in our simplified patch of a galaxy, meaning that $V(R) \equiv R\Omega(R) \simeq$ const.

Like the linearized analyses of GLB and JT, my N-body experiments presumed that *each* of the many particles is subject to exactly the same idealized equations (1) and (2), with not even the angular speed Ω_0 permitted to vary from one particle to the next. In both examples below, the only forces between particles were taken to be gravitational, although softened at close range via the usual law $F_r(r) = -GMr/(r^2 + a^2)^{3/2}$. As an unusual extra step, these interparticle attractions were here imagined also to taper linearly to zero outwards from a separation r equal to four times the softening length a . The latter rule, adopted to reduce the burden of the force computations, meant that these particles lost all direct "gravitational" contact at separations exceeding $6.19a$! Such a strategy might well horrify anyone who remembers gravity to be a long-range force, but I attest that in the present context it was (a) surprisingly harmless — as demonstrated especially via several other choices of that taper distance — and (b) easily allowed for in all subsequent comparisons with theory.

To appreciate the particle masses M and the various lengths and speeds actually used in these experiments, we must digress briefly to the old critical length

$$\lambda_{crit} = \frac{4\pi^2 G\mu_0}{\kappa_0^2} \tag{4}$$

and the minimum radial rms speed

$$\sigma_{u,min} = \frac{3.358 G\mu_0}{\kappa_0} \tag{5}$$

related to axisymmetric Jeans instabilities in a smooth and very thin layer of star-like particles of surface density μ_0 of mass per unit area. Here $\kappa_0 = \sqrt{2}\Omega_0$ is the usual epicyclic frequency. Recall that equations (4) and (5) make no explicit allowance for any additional effects of finite thickness or other forms of softening. Also recall (from T81 or SC) that at least the wavelike swing amplification tends to be strongest at unwrapped or azimuthal wavelengths $\lambda_y \equiv X\lambda_{crit}$ near $X = 2$, whenever $V =$ const as here.

In these terms, it is easy to state now that the first of the two experiments below used a mean number density $n = 100$, reckoned in particles per λ_{crit}^2; it assigned an equal mass $M = \mu_0\lambda_{crit}^2/100$ to each of its particles. The second experiment used a slightly larger density, $n = 150$. The first experiment began absolutely cold, with its randomly-sprinkled particles ascribed only the shear-flow components $\dot{x} = 0$, $\dot{y} = -2A_0x$ appropriate to each vicinity, whereas in the second case a Schwarzschild distribution of random velocities with a rms radial speed $\sigma_u = \sigma_{u,min}$ was added as well. One could alternatively describe these starting "temperatures" as $Q = 0$ and $Q = 1$, but if so, these very low values of Q would have nothing like their usual alarming significance. The reason is that the softening length a was here set deliberately to the rather large value $0.2\lambda_{crit}$, as opposed to

$$a_{min} = \frac{\lambda_{crit}}{2\pi e} \simeq 0.0585\lambda_{crit} \tag{6}$$

which would have sufficed to kill the Jeans instability even if the random motions had been nil.

This very "soft" choice of a helped not only to ensure that the swing amplification would remain fairly sedate here, but also keep these particles — especially after some damping had been added to their equations of motion — from congealing into massive bound aggregates very reminiscent of moons, planets, giant molecular clouds, or even tightly-bound star clusters the likes of which we rarely see any longer in the present-day disks! The gist of this aside is that, among other things, these local-patch experiments with sufficiently sharp but lossy particles turned out to exhibit a fascinating tendency towards what I tend to call "moon-making" and subsequent lane-clearing. But we must leave those topics now, and return to the ragged spirals. The large value of a will help us to do just that — right after we touch briefly on the last key ingredient to these experiments.

That last question obviously was: How should one represent with only a finite number of particles a sheet that was meant to be infinite in two directions? Equally obviously, the answer had to lie in double periodicity of some sort — but owing to the basic shear flow one could here not permit any single particle to have replicas that remained fixed relative to itself at "radii" x

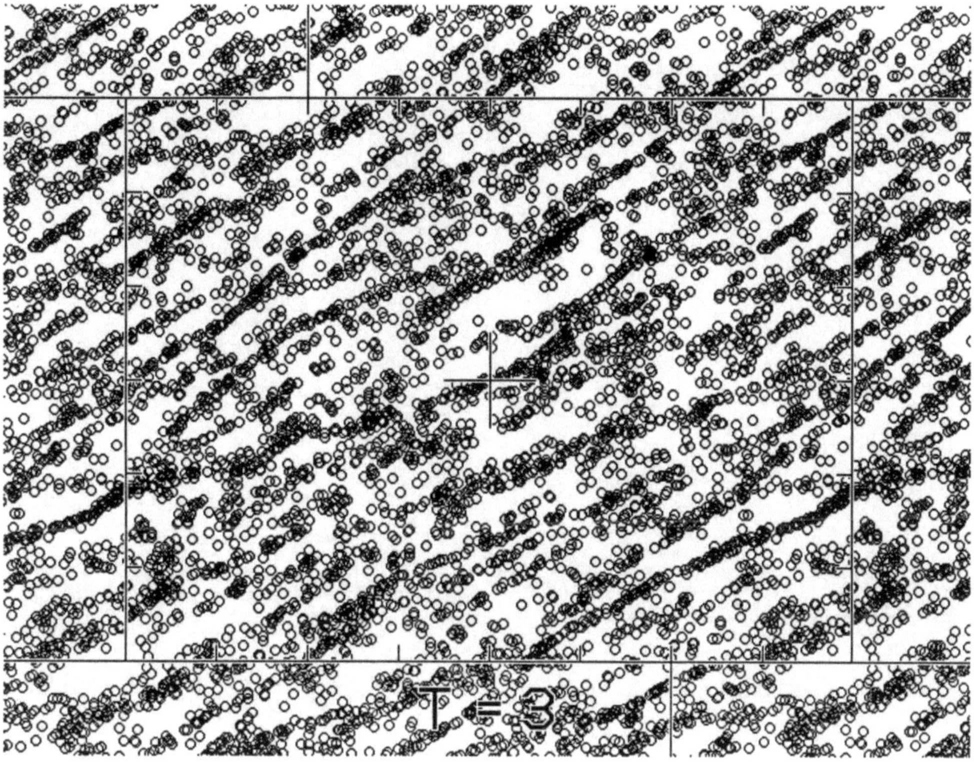

Figure 1: Locations of particles from the $n = 100$ experiment at the dimensionless time $T \equiv \Omega_0 t = 3$. Here the "radius" x increases upward, whereas the "longitude" y increases to the left. Note the staggered locations of the replicas at this instant; they had begun vertically aligned at $T = 0$, and were so situated again at $T = 4/3$ and $8/3$.

other than its own. Instead, the whole lattice of image points needed to *shear* with time, as if through a continuous sequence of parallelograms. Alternatively, any given rectangular region or "unit cell" of height h and width w in the x and y directions needed to be replicated by others sliding relative to itself with speed $\dot{y} = -2A_0 h$ immediately above, and with speed $\dot{y} = +2A_0 h$ immediately below. It turned out that this little trick had been used effectively by numerical experimenters in molecular dynamics already since the early 1970s. Closer to home, it was also used — and well described as "sliding bricks" — by Wisdom & Tremaine (1988) in their recent local simulations of planetary rings, which omitted any self-gravity but included collisions between various ring particles imagined to be identical spheres. Back in our own bailiwick, it was essential to make these bricks, cells or parallelograms quite sizable compared to the expected dominant wavelengths of order $2\lambda_{crit}$. Hence our present two examples employed $h = 6\lambda_{crit}$ and $w = 8\lambda_{crit}$, and a total of either $N = 4800$ or 7200 independent particles.

Despite this sizable particle number, an early outcome such as pictured in Figure 1 requires these days only a modest overnight job of computation, even on a personal computer. Though I am old enough to remain slightly awestruck just by that, I suspect most readers will be more impressed by the tilted striations in Figure 1 which evidently developed in a great hurry. Remember that here was a case where 4800 cold particles had been sprinkled at random, like Poisson noise, only as recently as $t = 0$ over the $6 \times 8\,\lambda_{crit}$ unit cell. Yet slightly less than half a revolution period later their configuration already exhibits this much structure! Perhaps indeed

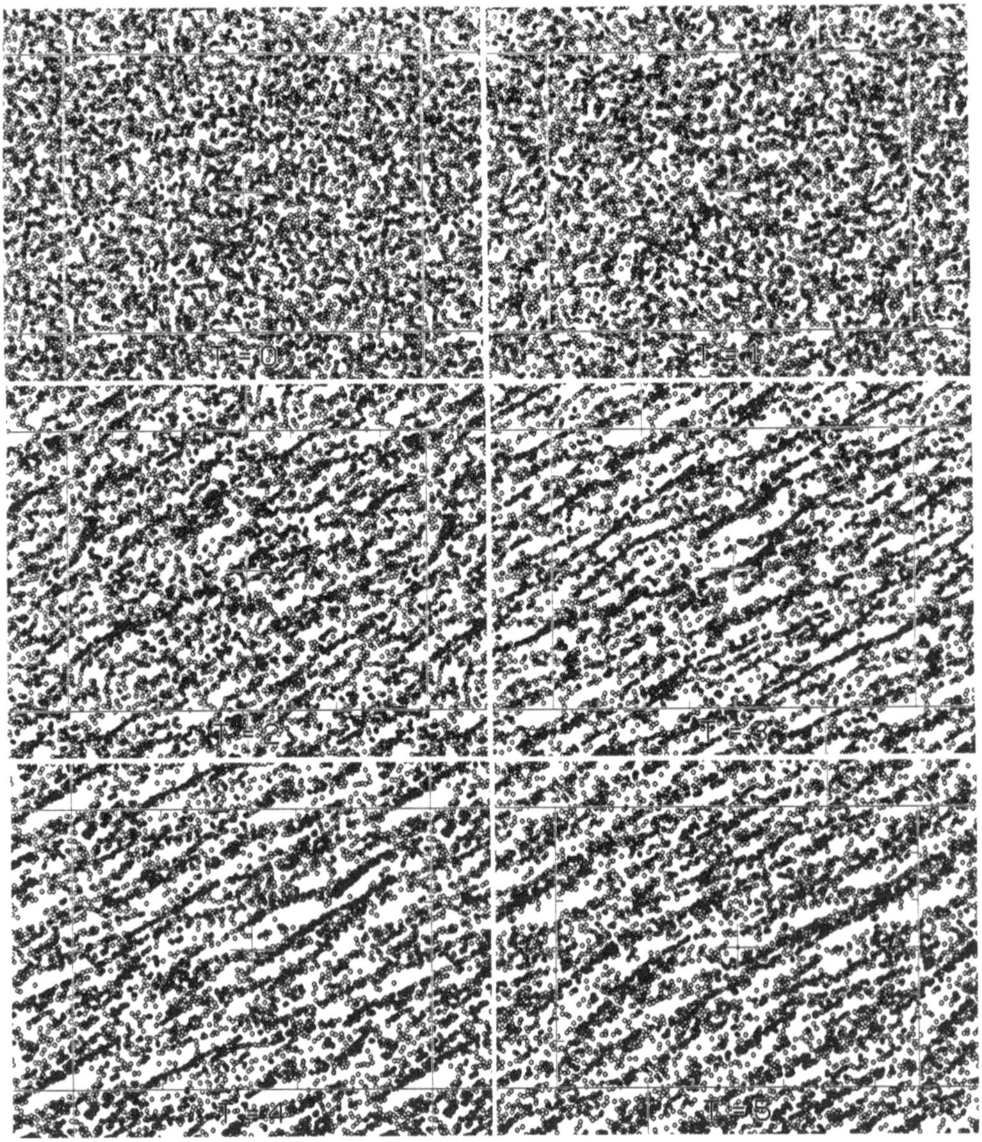

Figure 2: Time-development of the streaks in the $n = 100$ experiment, from $T = 0$ through $T = 5$. Individual particles of course travel freely through the walls of the sliding $6 \times 8\, \lambda_{crit}$ rectangles, and also closely past each other despite the plotted circles.

that fact is more impressive than any raw speed of computation. This rapid development is shown anew in Figure 2, in a more miniature but frame-by-frame fashion.

The rapid evolution evident in Figures 1 and 2 results once again from swing amplification, even though our "gravity" had been softened and and also truncated to an unusual degree. It is essential to understand at once, though, just what is — and what is not — being amplified in this instance. Clearly we had no incoming waves here to speak of, since this example started perfectly cold. Nor can there yet have been time for any appreciable feedback of density waves, even if one could spot a plausible reflector. Thus it would be logically silly to try and explain the present streak patterns as any superpositions of swing-amplified *free* waves or wavelets such as GLB focused upon in their seminal paper, or as T81 rediscussed in the context of global modes and/or major tidal responses where the wave input could hardly be doubted. Instead, one must here think much more of constantly-*forced* wavelets such as JT stressed in their own paper, and as they incorporated already into their integral equation (21) via the "imposed" transform D_{imp}, though then mostly with just a single imposed mass in mind.

I stress this distinction because, unlike some recent authors, I have long viewed this free-vs-forced emphasis as a more telling difference between GLB and JT than the usual first impression that GLB dealt with a gas and JT with stars, correct though that is. Of course the two emphases are closely related, but one should not presume — as SC did in a little blunder, later freely conceded — that the amplification factors available in the forced case are exactly or even nearly the same as the free-wave "maximum growth factors" such as T81 displayed in his Figure 7. For accurate comparisons with theory, all such factors need to be recalculated. Yet, for a cold initial-value problem such as shown in Figure 2, the specific transient behavior can be pursued even via GLB's differential equations, provided one adds the forcing term or "loading" implied in turn by each Fourier component of a given initially random distribution of particles. Such tests Kalnajs and I have by now pursued almost ad nauseam. The results leave very little doubt that Figures 1 and 2 in essence portray 4800 JT-style density wakes simultaneously and rapidly in the making. Or to put it another way, what we observe here, apart from their many replicas, are 4800 aggressors and just as many "victim" particles.

Most of these cautions should apply also to other remarkably quick N-body arousals of spiral structure such as pictured in Hernquist's Figure 5 in this volume, and in various earlier papers. Though Hernquist himself wrote nothing amiss, I wish other authors would stop calling such phenomena "spiral instabilities"! As far as I can tell, such loose talk really refers mostly just to vigorous spiral *responses*, given the nearly steady $\sqrt{N}$ noise that is inevitable whenever N particles have been sprinkled at random.

So much for the early stages of a typical undamped experiment. By contrast, Figure 3 on the left summarizes the successive outcomes obtained over a ten times longer interval from an initially "warm" though random start in an experiment where a small but vital damping term $-C_D\dot{x}$ had been included in the force f_x governing the motion of each particle via equation (1). The value of this admittedly artificial damping coefficient was here fixed at $C_D = \Omega_0/15$, meaning that the epicyclic motions of these particles would have decreased by a factor e in a time interval $\Delta T \equiv \Omega_0 \Delta t = 30$, or over about 5 revolutions, without any interparticle jostling. Given that jostling, however, the rms random motions in the sequence shown in Figure 3 remained near their initial level throughout. Obviously the main message here, as from past full-disk experiments, is that this kaleidoscope of chaotic arm features continues to churn essentially forever in the presence of adequate damping. What we observe even here, though, is in principle no "recurrent instability". It is instead just a constantly-sliding superposition of the wakes of N discrete particles, which in this case number 7200.

5. Lessons and Impressions

If pressed, I could discourse at much greater length about various subtleties — such as a gradual development of weak but tightly-pitched radial "grooves" in the locations of the guiding centers, which results in a mild overheating, or the Goldreich-Ward instability recovered from too much tangential damping — that have long distracted Kalnajs and myself in these shearing-sheet experiments, especially when conducted with yet larger boxes, or with larger number densities of particles (like $n = 600$ or even 1200), or with a sharper gravity, or just much farther in time. But here it is time instead to wrap up this discussion by asking what, if anything, such single-specie calculations portend for the real Sc disks which, as we all know, consist of one component that is doubtless very lossy and another that plainly is not.

A fine place to begin is by imagining that only 1/4 or 1/9 of the discrete mass points had been retained unchanged in the above experiments, and that the other 3/4 or 8/9 had been been pulverized into millions of pieces before being sprinkled back into the $t = 0$ opening frames. In that case, given the $\sqrt{N}$ logic, our input roughness would either have been halved or divided by three statistically. Given also the swing amplifier, we would then expect to obtain patterns with essentially the same linear scales but only one-half or one-third of the former fractional amplitudes. Sad to say, almost noone would be impressed by such pictures, given the failings of the human eyes and visual cortex, which at least in my own case do not seem to grasp easily such milder variations of density from mere dot plots!

To be sure, one could then boast that the rate of heating, being quadratic in the amplitudes, had become only 1/4 or 1/9 as embarrassing as it used to be. But that would be only a Pyrrhic victory because no editors would then wish to publish our "spiral" photographs. Of course, there would be at least two remedies to repair the latter flaw: (a) We could rearrange the mass from our surviving fraction of the original particles into chunks four or nine times as massive, but now only 1/16 or 1/81 as numerous as the ones we had to begin with. This would recover the original statistical roughness. (b) We could alternatively beg the swing amplifier to enhance the typical inputs by an extra factor of two or three — which is quite reasonable, given the excessively severe gravity softening that I chose to employ in the examples above. Either way, or in suitable combination, we would then obtain pictures about as photogenic as they used to be, and and happiness would reign again ... until we remembered the fresh "spot of difficulty" with the heating of the make-believe stars.

The good news about the heating of the stars in such hybrid situations is that we would not especially need to know either the gravitational roughness of the gas, or the strength of the swing amplifier, both of which remain very opaque in reality. What matters much more is the *product* of those individually dubious quantities — and it is this which we can, at least in principle, read off with much greater confidence than either factor alone from the observed strength and separation of the multiple arms seen in any given Sb or Sc galaxy. In other words, one can undertake to add enough gas lumps and combined swing amplification to a JT-style local simulation until the spiral intensity — determined mainly by that product — looks about right, and then use fresh versions of Julian's (1967) formulas to read off the implied rate of heating. This approach would be vaguely akin to such modern analyses as those of Lacey (1984) or Fuchs and Wielen (1987), who studied the heating of stars in the solar neighborhood primarily via as-yet-unspecified diffusion coefficients, and who only afterwards adjusted the rates so found to the actual data summarized for instance by Wielen (1977). The main difference is that with these explicit wakes we would probably be closer to the principal source of this diffusion. And surely this agenda, very little of which has yet been completed, ought to be superior to the weird on-and-off spiral arms of fixed (!) planform which Carlberg & Sellwood (1985) felt compelled to contribute to our recent literature on the heating.

The bad news about this heating is that in most cases it would still remain very considerable. One could be more technical, but it should suffice in these closing remarks to recall only that the damping rate in Figure 3 needed to be strong enough to kill the existing epicyclic motions effectively in just the next five revolutions: With the decreasing efficiency of wake heating at the higher Qs, this suggests that it might take another ten revolutions before the trailing features shown there fade away to oblivion. But this remains a relatively short time in the life of a galaxy. And once the stellar Q has risen well above 2, it is easy to estimate from the forced GLB or JT equations that such material will have grown so non-compliant that it might as well be rigid for the purpose of helping with the wakes. The same point was made very nicely by

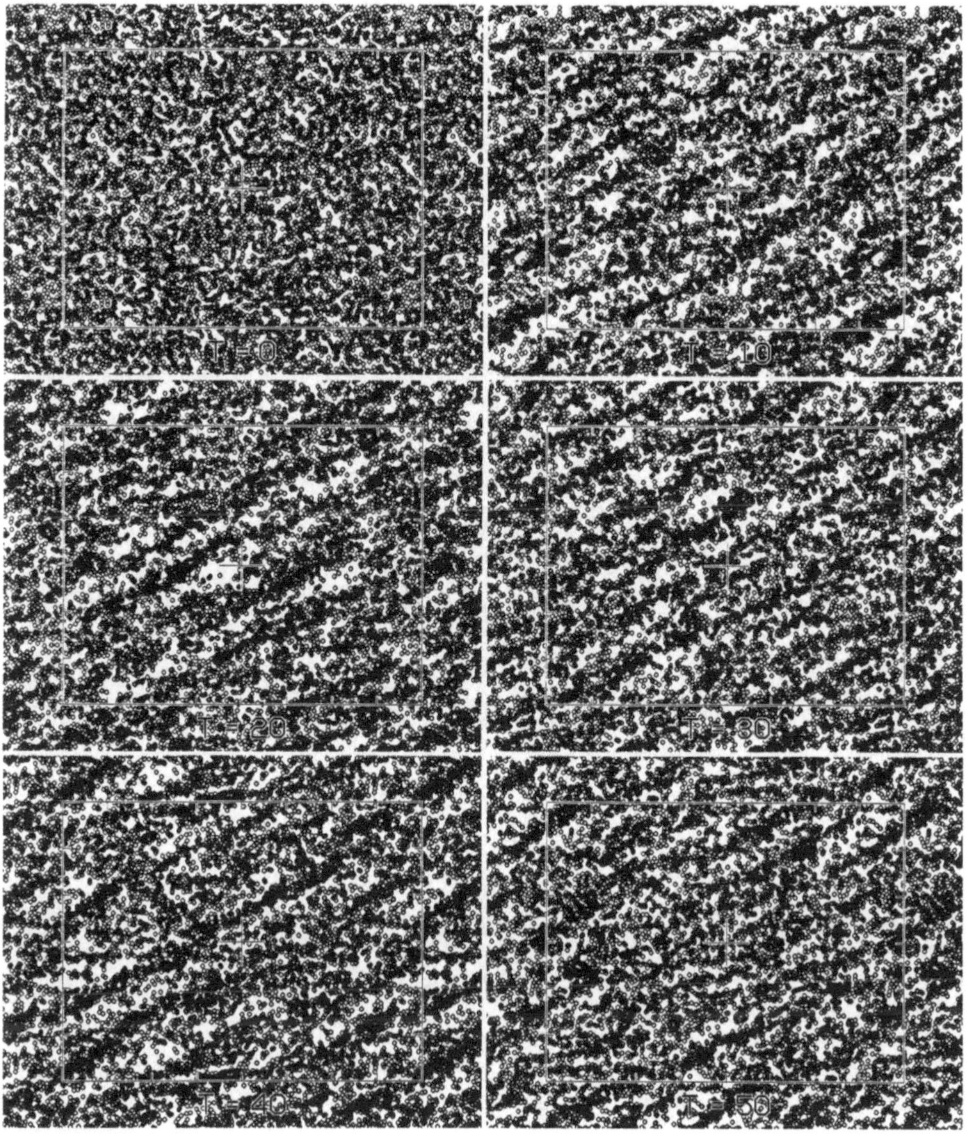

Figure 3: Six snapshots from the $n = 150$ experiment, taken at equal intervals from $T = 0$ to $T = 50$. This run included the small radial damping described in the text, chosen to maintain the epicyclic motions near their initially-assigned rms value $\sigma_{u,min}$.

Carlberg & Freedman (1985; see esp. their Figure 4) in their full-disk study including both stars and a "dissipative gas component": Once the stars tune out, the spiral structure that remains must be dominated by the self-gravity of the gas, and its linear scale can be comparable only to the much-shrunken value of λ_{crit} appropriate to that subcomponent alone.

To add some pungency to these last few comments, I have in Figure 4 taken the rash step of plotting back into logarithmic polar coordinates the particle locations that would have been seen at time $T = 80$ if we had continued Figure 3 to yet later times. There was nothing very

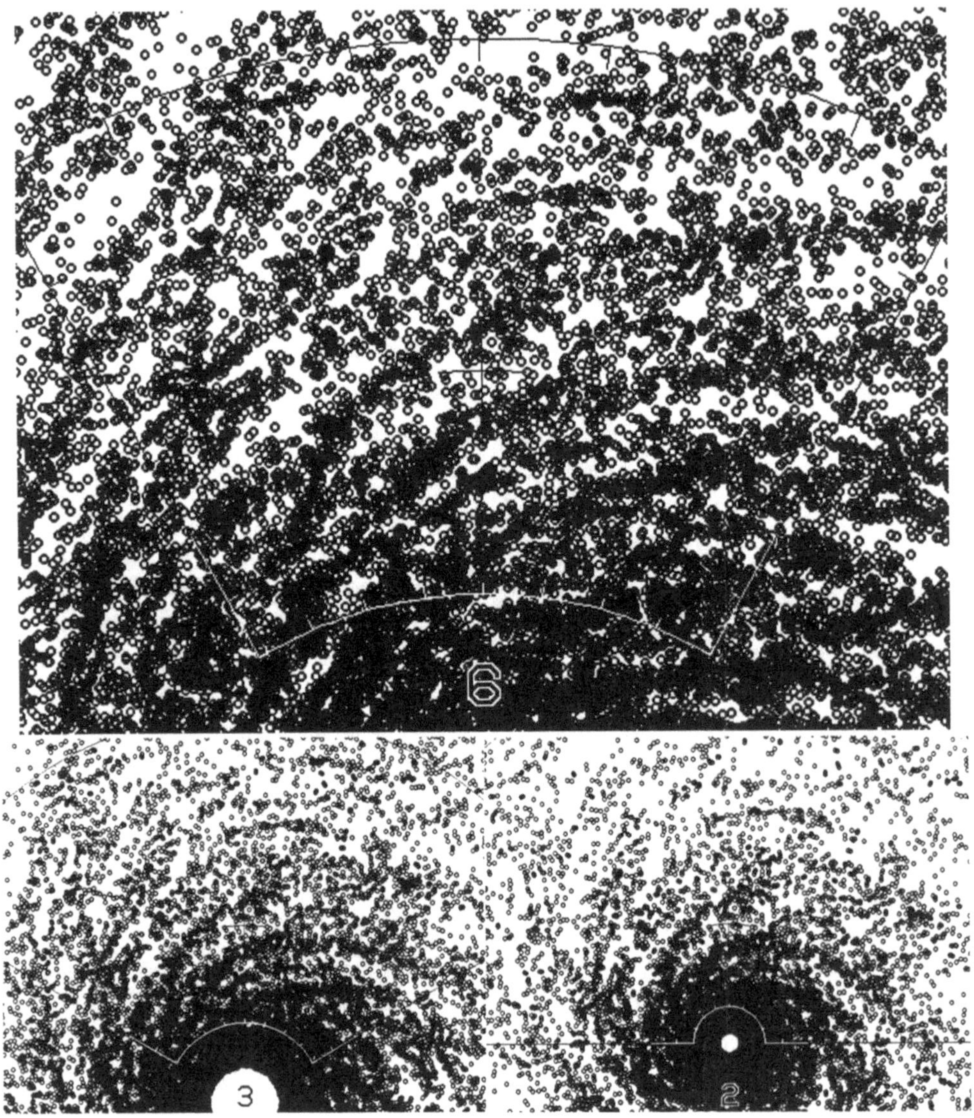

Figure 4: The $n = 150$ experiment at $T = 80$, in logarithmic polar coordinates. The big plot at the top supposes that it would have taken 6 of our unit cells to encircle the galaxy in question; the smaller plots assume that it would have required only 3 and 2.

special about that choice of time: Except for the accident of two distinct arms situated right in the middle of that frame, several of the other times (esp. $T = 40$) would have done about as well. None of this display is strictly valid in dynamical terms, and even the length scales here would make sense only if we could be sure that λ_{crit} is proportional to the galactocentric distance R. Still, when we recall that the projected density of interstellar gas in our neighborhood of the Galaxy is often quoted roughly as $\mu_g = 5-10\,\mathrm{M}_\odot/\,\mathrm{pc}^2$, we can infer from from equation (4) that the length scale $\lambda_{crit,g}$ relevant to our gas layer alone is apt to equal $1.0 - 1.5$ kpc. Adopting the lower value, and also $R_0 = 8+$ kpc for our galactocentric distance, means therefore that the uppermost diagram in Figure 4 is apt to give a slightly conservative estimate of the *spacing* of spiral arm segments that we ought to observe in our vicinity. Yet we know that our observed interarm spacing is roughly twice as great. Furthermore, it is commonly alleged also that ours is only an Sb galaxy, whereas the Sc's typically have twice as large a fraction of their disk mass in gas. But again even the lower right panel in Figure 4 still looks only like an excessively multi-armed version of an Sc. One way or another, it really seems as if we typically need at least to double the known amounts of reasonably active mass in the disks of Sc galaxies — beyond the gas amounts that they are known to possess, and which in any case ought also to clump for their own reasons to provide the necessary roughness — in order that such galaxies appear about as few-armed as they often are. And from where can they have obtained and still retain such relatively cool additional disk material if not from recent infall?

References

Carlberg, R.G. 1987, in *Nearly Normal Galaxies*, ed. S.M. Faber (Berlin: Springer), p. 129.

Carlberg, R.G. & Freedman, W.L. 1985, *Astrophys. J.*, **298**, 486.

Carlberg, R.G. & Sellwood, J.A. 1985, *Astrophys. J.*, **292**, 79.

Fuchs, B. & Wielen, R. 1987, in *The Galaxy*, ed. G. Gilmore & B. Carswell (Dordrecht: Reidel), p. 375.

Goldreich, P. & Lynden-Bell, D. 1965, *Mon. Not. Roy. Astron. Soc.*, **130**, 125 (GLB).

Gott, J.R. 1975, *Astrophys. J.*, **201**, 296.

Gunn, J.E. 1982, in *Astrophysical Cosmology*, ed. H.A. Brück, G.V. Coyne & M.S. Longair (Vatican City: Pontificia Academia Scientarium), p. 233.

Gunn, J.E. 1987, in *The Galaxy*, ed. G. Gilmore & B. Carswell (Dordrecht: Reidel), p. 413.

Hill, G.W. 1878, *Am. J. Math.*, **1**, 5.

Hohl, F. 1971, *Astrophys. J.*, **168**, 343.

Julian, W.H. 1967, *Astrophys. J.*, **148**, 175.

Julian, W.H. & Toomre, A. 1966, *Astrophys. J.*, **146**, 810 (JT).

Lacey, C.G. 1984, *Mon. Not. Roy. Astron. Soc.*, **208**, 687.

Larson, R.H. 1972, *Nature*, **236**, 21.

Larson, R.H. 1976, in *Galaxies*, Sixth Advanced Course at Saas-Fee, ed. L. Martinet & M. Mayor (Geneva Observatory), p. 67.

Lo, K.Y. & Sargent, W.L.W. 1979, *Astrophys. J.*, **227**, 756.

Miller, R.H., Prendergast, K.H. & Quirk, W.J. 1970, *Astrophys. J.*, **161**, 903.

Oort, J.H. 1969, *Nature*, **224**, 1158.

Oort, J.H. 1970, *Science*, **170**, 1363.

Sellwood, J.A. 1986, in *The Use of Supercomputers in Stellar Dynamics*, ed. P. Hut & S. McMillan (New York: Springer), p. 5.

Sellwood, J.A. 1989, in *Dynamics of Astrophysical Disks*, ed. J.A. Sellwood (Cambridge Univ. Press), p. 155.

Sellwood, J.A. & Carlberg, R.G. 1984, *Astrophys. J.*, **282**, 61 (SC).

Spitzer, L., Jr. & Schwarzschild, M. 1953, *Astrophys. J.*, **118**, 106.

Toomre, A. 1977, in *The Evolution of Galaxies and Stellar Populations*, ed. B. M. Tinsley & R. B. Larson (New Haven: Yale Univ. Observatory), p. 420.

Toomre, A. 1981, in *The Structure and Evolution of Normal Galaxies*, ed. S.M. Fall and D. Lynden-Bell (Cambridge University Press), p. 111 (T81).

Toomre, A. 1985, in *IAU Symposium No. 113, Dynamics of Star Clusters*, ed. J. Goodman & P. Hut (Dordrecht: Reidel), p. 500.

Wielen, R. 1977, *Astron. Astrophys.*, **60**, 263.

Wisdom, J. & Tremaine, S. 1988, *Astron. J.*, **95**, 925.

Observed Gas Infall into Galaxies

R. Sancisi, A. Broeils, J. Kamphuis, and T. van der Hulst

Kapteyn Astronomical Institute, University of Groningen,
Postbus 800, NL-9700 AV Groningen, The Netherlands

Abstract. Recent observations of the nearby spiral galaxy M 101 have revealed the presence of a large complex of HI clouds moving at high velocity perpendicular to the disk of that galaxy. Capture and infall of gas-rich, dwarf irregular galaxies may explain the M 101 results. Observations of a number of systems with gas-rich companions, such as NGC 3359, or with associated, peculiar HI structures suggest that recurrent, episodic gas infall into galaxies may be taking place.

1. Observations

HI features with peculiar density distributions and kinematics are seen associated with a number of isolated galaxies. They are presented here as possible, *circumstantial* evidence for gas infall into galaxies. M 101 and NGC 3359 are perhaps the two most striking examples and are briefly described here to illustrate the phenomenon.

1.1 M 101 and the high-velocity gas

Recent, sensitive 21-cm observations of the giant nearby spiral galaxy M 101 with the Westerbork Synthesis Radio Telescope have confirmed the existence of high-velocity HI gas clouds and show that the two massive features reported by Van der Hulst and Sancisi (1988) are part of a very extended and clumpy gas complex or sheet covering almost one half of the optical image of M 101 (Fig. 1). The size of the complex is approximately 15 × 25 arcmin (30 × 50 kpc at the assumed distance of 7.2 Mpc), its mass is 2 10^8 $M_\odot$, ~1% of the total HI mass of M 101. Its radial velocities are up to ~150 km/s higher than the "local" rotating gas of the disk. In the directions of the main HI high-velocity clumps, the HI layer shows holes or troughs of 2 kpc or more suggesting that the gas has been removed, perhaps blown away, from the disk. Since this gas is seen redshifted with respect to the gas in the disk, it may be located on the near side of M 101 and falling down toward its disk; alternatively it may be on the far side and moving away. The presence of holes and troughs seems to favour the second possibility. The age of the complex, as estimated from its linear sizes and velocities and also from the width of the troughs and gas random motions, is of order 10^8 year.

Various explanations have been considered. The hypothesis of a "local" disk origin for the high-velocity complex, i.e. supernova explosions or stellar winds, seems unlikely because of the too large masses, velocities and kinetic energies (10^{54-55} erg) and also the velocity one-sidedness. The possibility of a collision with some external object like an intergalactic gas cloud or a dwarf companion seems more attractive. Such a mechanism has been discussed by Tenorio-Tagle et al. (1987) for the HVCs in our galaxy in relation to Heiles' HI supershells. In this collision picture the high-velocity complex observed in M 101 would consist of gas removed from the disk of M 101 and probably also of new gas brought in by the infalling object. The latter could have been a gas-rich dwarf like DDO 154 (Krumm and Burstein 1984) or the companion of NGC 3359 (Fig. 2). It seems natural that the high-velocity gas complex will spread out and break

Dynamics and Interactions of Galaxies Editor: R. Wielen

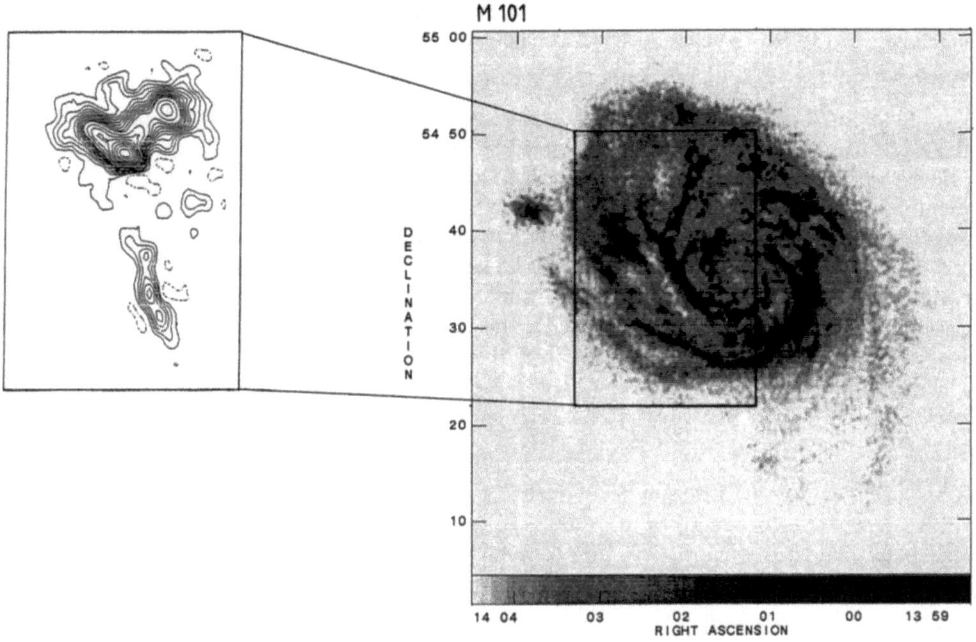

Figure 1. *Total HI map of M 101 (HPBW 23″ × 29″) and enlarged map of column densities of the high-velocity HI complex (HPBW 80″ × 80″). The latter has been obtained by integrating over the heliocentric radial velocity range 343–417 km/s. The contour levels are* −7.5, 7.5 *to* 75 10^{18} cm^{-2} *in steps of* 7.5 10^{18} *followed by* 9 *to* 18 10^{19} cm^{-2} *in steps of* 1.5 10^{19}.

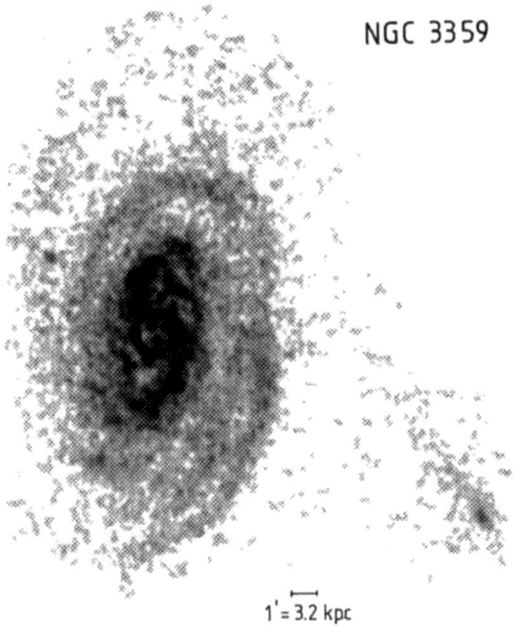

Figure 2. *Grey-scale map of the HI distribution of NGC 3359. The column densities range from* 5 10^{19} cm^{-2} *to* 2.5 10^{21} cm^{-2}. *The resolution is* 30″ × 30″. *Visible on the right are the* 1 10^8 $M_\odot$ *HI companion and bridge.*

up into smaller clouds which will eventually rain down toward the disk of M 101. This may be the analogue to the HVCs of our galaxy. The end result for M 101 will be the accretion of new material to its disk.

In addition to the high velocity gas complex there are other, possibly related, peculiarities in the HI density distribution and kinematics in M 101, which also need to be explained. These are the well-known lopsidedness, the outer spiral features and the large deviations from circular motion observed in the HI layer, in particular the remarkable, strong velocity-gradient about 8 arcmin to the south-west of the centre, where two bright HI arms seem to cross (see Fig. 1). The radial velocity increase across these two arms is almost 50 km/s. There is a remarkable symmetry between the locations of this velocity jump south-west of the centre and of the high-velocity features on the north-east described above. Furthermore, they seem to be connected by a wiggle in the velocity field running like an arc between these two locations (Sancisi, 1989a). Computer modelling is required to test the hypothesis that the encounter with a dwarf companion, as advocated above for the high velocity phenomenon, might also explain this peculiar kinematics.

1.2 NGC 3359 and its HI companion

The 21-cm line observations of the spiral galaxy NGC 3359 show an extended HI layer and a HI companion (Ball 1986, Broeils 1989). Broeils' observations (Fig. 2) indicate the presence of a weak, straight HI bridge between NGC 3359 and the companion. The latter has a head-tail structure suggesting tidal stretching and disruption. Its radial velocity is within 40 km/s of the systemic velocity of NGC 3359. Its total HI mass is about 1 10^8 $M_\odot$, ~2% of the HI mass and only 0.1% of the total mass of NGC 3359. On the Palomar Sky Survey prints there seems to be a very faint object in its direction, but it is not clear whether it is a dwarf galaxy and whether it is associated with the HI blob at all. In any case the object's M_H/L ratio must be higher than about 8. These observations seem to suggest that a gas-rich satellite is being captured and that new, fresh gas is being acquired by the disk of NGC 3359.

2. Concluding Remarks

The two cases of M 101 and NGC 3359 presented here seem to provide circumstantial evidence for infall of gas into spiral galaxies. There are several other examples of peculiar distributions and kinematics of HI around galaxies (cf. Sancisi 1989b), which might also be interpreted as a result of mergers with low mass companions and possibly gas enrichment of the host galaxy. Among these are the late type spiral galaxies NGC 3067, 1961, and 5635, the SO galaxies NGC 1023 and 4694 and the elliptical galaxy NGC 4472. Some of these may be more controversial as far as the gas infall is concerned. Their irregular HI structures are, however, reminiscent of the features often found associated with interacting systems and do suggest recent tidal interaction and advanced merging. A gas accretion by galaxies as proposed here to explain the 21-cm line observations has been discussed by Silk and Norman (1979).

In the cases of M 101 and NGC 3359 and also in most of the others, the masses involved in the hypothetical infall are negligible with respect to those of the main galaxy. Usually the observed HI features are only a few times 10^8 $M_\odot$. Such infall, therefore, would have no catastrophic effects on the galactic disks, although it may cause some transient effects such as asymmetries in the gas distribution, local density and kinematics perturbations, effects on the spiral structure, star formation etc. It may, however, play an important role in the building up and re-supplying of disks of spiral galaxies. The cases reported so far are only a few, but they may represent the presently observable, high-mass tail of the distribution pointing at short-lived events taking place in a much larger number of systems. Perhaps, instead of a gradual and homogeneous gas infall, one should regard such episodic gas accretion of typically a few 10^8 $M_\odot$ gas clumps every 10^{8-9} year over the Hubble time as capable of accounting, at least partly, for the present gas contents of spiral galaxies.

References

Ball, R. 1986, Astrophys. J. *307*, 453.

Broeils, A.H. 1989, in preparation.

van der Hulst, J.M. and Sancisi, R. 1988, Astron. J. *95*, 1354.

Krumm, N. and Burstein, D. 1984, Astron. J. *89*, 1319.

Sancisi, R. 1989a, In *"Dynamics of Astrophysical Discs"*, ed. J. Sellwood, Cambridge University Press.

Sancisi, R. 1989b, In *"Windows On Galaxies"*, Proceedings Erice Workshop, eds. G. Fabbiano, J. Gallagher and A. Renzini, Kluwer.

Silk, J. and Norman, C.A. 1979, Astrophys. J. *234*, 86.

Tenorio-Tagle, G., Franco, J., Bodenheimer, P. and Rozyczka, M. 1987, Astron. Astrophys. *179*, 219.

HI Absorption in Radio Ellipticals: More Evidence for Infall

J.H. van Gorkom[1], *G.R. Knapp*[2], *R.D. Ekers*[3], *D.D. Ekers*[3], *R.A. Laing*[4], *and K.S. Polk*[5]

[1]Columbia University, 538 W. 120th Str., New York, NY 10027, USA
[2]Princeton University Observatory, Princeton, NJ 08544, USA
[3]CSIRO Australia Telescope, P.O. Box 76, Epping NSW 2122, Australia
[4]Royal Greenwich Observatory, Herstmonceux Castle, Hailsham East Sussex BN27 1RP, UK
[5]Astronomy Department, University of Washington, Seattle, WA 98195, USA

A complete sample of 29 radio galaxies with compact cores has been observed with the VLA in the HI line to an almost uniform optical depth sensitivity of $\tau = 0.01$ at a velocity resolution of 40 km s^{-1}. HI absorption was detected in four galaxies, NGC 3894, NGC 5363, 3C236 and 4C31.04, bringing to eight the number of radio ellipticals in which HI absorption has been detected. In all cases, absorption is seen redshifted with respect to the galaxy's systemic velocity, or close to the systemic velocity or both. No blueshifted absorption components are seen. These observations demonstrate the infall of HI to the inner regions of radio elliptical galaxies. The estimated infall rates (a few hundredth of a solar mass per year) are more than adequate to fuel the nuclear activity.

1. Introduction

Why some ellipticals are giant radio galaxies and others have no radio emission down to very low limits is still not understood. It was first suggested by Shlovskii (1962) that accretion of intergalactic gas might have something to do with the nuclear radio activity of some ellipticals. Gallagher et al (1977) after detecting HI in NGC 4278 suggested that it might be the presence of cold gas that enables a galaxy to turn on in the radio, subsequently a theoretical scenario was put forward by Gunn (1979), who explained nuclear activity on all scales by the accretion of cold gas onto a central black hole. Observational evidence for this scenario could be found in the alignment of the rotation axis of the cold gas (as derived from the orientation of dust lanes) and the radio axis in elliptical galaxies (Kotanyi and Ekers, 1979). Statistical evidence for a correlation between the presence of cold gas and the likelyhood of an elliptical to be a radio galaxy was presented by Knapp and Wardle (1988) comparing HI properties with radio activity and by Walsh et al (1989) by looking at IRAS detection rates. In all of the work, it has yet to be shown that active galactic nuclei are actually accreting gas.

In this paper we present the results of an HI absorption survey, obtained with high enough angular resolution that gas motions along a narrow line of sight toward the nuclei of these radio galaxies can be unambigously determined. Combined with already existing high quality data, these results show for the first time that gas is indeed falling toward the center of the galaxies. A more extensive account of the survey is given by van Gorkom et al (1989).

Dynamics and Interactions of Galaxies Editor: R. Wielen

2. The sample, observations and data reduction

The galaxies were selected from the radio continuum surveys of optically bright galaxies and from optically identified radio galaxies found in radio surveys. Sources were selected with total continuum flux density brighter than 100 mJy, and a redshift $z < .13$, the redshift limit of the VLA. Continuum snapshots were made of the selected sources (49) with a resolution of 1 arcsec, to determine the position, current core flux density and structure at the arcsec scale for each of these sources. All sources with unresolved cores stronger than 350mJy were observed in HI with an angular resolution of 1 - 4 arcsec, covering a velocity range of 1300 km $s^{-1}$and with a final velocity resolution of 40 to 45 km s^{-1}.

3. Results

HI absorption was detected against four sources, 0116+319 (4C31.04), 1003+351 (3C236), 1146+596 (NGC 3894) and 1353+055 (NGC 5363). The line profiles are shown in Fig. 1. The vertical arrow indicates the systemic velocity of these galaxies and it can be

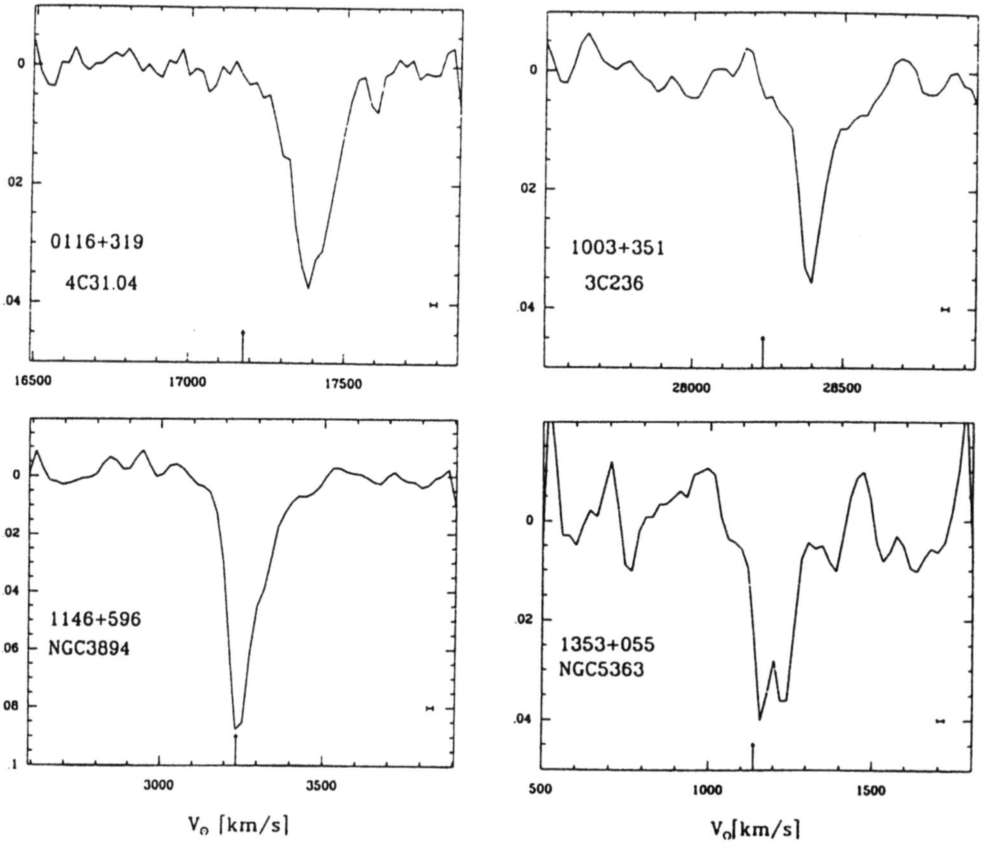

Figure 1. HI profiles towards the cenral core sources. The ordinate is optical depth. The abscissa is heliocentric radial velocity in km/s. The velocity derived from the optical redshifts is indicated by an arrow. The velocity resolution is shown by the horizontal bar in each plot.

Table 1. Absorption components toward the nuclei of radio ellipticals

Galaxy	redshift	infall velocity (km/s)	core flux (Jy)	optical depth
4C31.04	0.059	+212	2.6	0.037
3C236	0.0988	+162	2.3	0.032
NGC 3894	0.01085	+ 17	0.5	0.08
NGC 5363	0.0038	+ 22	0.15	0.04
		+ 96		0.04
NGC 315	0.01653	+400	0.4	0.5
NGC 1052	0.00492	+ 13	0.9	0.008
		+ 50		0.013
		+173		0.021
NGC 5128	0.00175	+ 3	2.9	1.28
		+ 26		0.66
		+ 45		0.34
NGC 1275	0.0174	+ 100	7.5	0.004
		+2713		0.095
		+2719		0.011
		+2725		0.014
		+2731		0.006
		+2738		0.009

seen that in all cases HI absorption occurs at velocities which are redshifted with respect to the systemic velocity. There are a few other radio galaxies toward which HI absorption has been detected and which satisfy our criteria of having a core < 1 arcsec and having been observed with an interferometer. The latter criteria are essential to unambiguously determine the radial motions of the gas. These other galaxies are NGC 315, NGC 1052, NGC 1275 and NGC 5128. In Table 1 we summarize the characteristics of the absorption features in all the above mentioned galaxies.

4. Discussion

A striking result of these observations is that in no case blueshifted absorption is seen. This seems to prove beyond doubt that infall is occurring in these galaxies. The alternative option, that we are seeing gas on elliptical orbits should give an equal number of red and blue shifted components. Current statistics is that 25 % of our complete sample of radio galaxies get detected in HI absorption, and all have a redshifted component. This makes the elliptical orbit hypothesis quite unlikely. Of course the reliability of some of the optical redshifts can be questioned. The more nearby galaxies however have redshifts based on optical absorption lines and on HI emission. Since the nearby radio galaxies all show the same phenomenon, an HI disk rotating regularly, and very close to the core, redshifted HI in absorption (eg NGC 1052, van Gorkom et al, 1986 and NGC 5128, van Gorkom et al, 1989), we believe that an extrapolation of these phenomena to higher z is justified.

The infall velocities range from 17 to 400 km s^{-1}. These velocities, although rather high, can be produced as gas loses its angular momentum in a galactic potential. Gunn (1979) showed that gas can lose its angular momentum by frictional drag by pressure-supported hot gas, which is now known to exist from X-ray data. The proposed scenario, that the gas disk breaks up due to Kelvin-Helmholz instability at the disk-hot gas interface in clouds of about 40 psec, which then gradually lose and angular momentum and spiral into the center seems perfectly consistent with the observations. In several nearby radio galaxies gas disks have been observed with central holes in HI (NGC 1052, 4278 and 5128 for example). The disks rotate regularly, so we know that the gas must have been around for at least a few rotation periods (10^8 to 10^9 years) and thus it is not the interaction (in which the gas was captured) by itself that moves the gas toward the center, it is a process that occurs more slowly later on. Either the proposed mechanism by Gunn could do it, or alternatively any deviation from axial symmetry in the galaxy's potential closer to the center could make it fall in as well.

Direct observational evidence that these clouds are actually close to the center is sparse, but there is some. For two galaxies (NGC 315 and NGC 5128) there is a core jet structure in the continuum, inside the optical body of the galaxy. In neither case do we see absorption at the redshifted velocity against the extended structure, we only see it against the core. Absorption at the systemic velocity on the other hand can be seen against the core and the jet in NGC 5128 (van der Hulst et al, 1983), as would be expected if the gas at the systemic velocity arises in the disk further out, while the gas at the redshifted velocity is close to the center. In NGC 1052 (van Gorkom et al, 1986) we see gas at non circular velocities (both redshifted absorption and blueshifted emission) in the central beam (4 kpc) only.

The accretion rates we can derive from the observations depend critically on the assumed distance (r) from the center (it goes as r^2). Assuming that the clouds are on average at a distance of a few hundred parsec, we derive typical accretion rates of a few hundredths of a solar mass per year, which is much more than needed to fuel even the most powerful of the radio sources in our sample (3C236).

References

Gallagher, J. S., Knapp, G. R., Faber, S. M., and Balick, B. (1977). Astrophys. J. **215**, 463

Gunn, J. E. (1979). In *Active Galactic Nuclei*, edited by C. Hazard and S. Mitton (Cambridge University, Cambridge), p213

Knapp, G. R. and Wardle, M. J., 1989, in preparation

Kotanyi, C. G. and Ekers, R. D. (1979). Astron. Astrophys. **73**, L1

van der Hulst, J. M., Golisch, W. F., and Haschick, A. D. (1983). Astrophys. J. Lett. **312**, L 37

van Gorkom, J. H., Knapp, G. R., Ekers, R. D., Ekers, D. D., Laing, R. A., and Polk, K. S. (1989). Astron. J. **97**, 708

van Gorkom, J. H., Knapp, G. R., Raimond, E., Faber, S. M., and Gallagher, J. S. (1986) Astron. J. **91**, 791

van Gorkom, J. H., van der Hulst, J. M., Haschick, A. D., and Tubbs, A. D. (1989). Astron. J., submitted

Walsh, D. E. P., Knapp. G. R., Wrobel, J. M., and Kim, D. W. (1989) Astrophys. J. **337**, 209

Shlovskii. I. S. (1963). **6**, 465

Gaseous Infall and Selfpropagating Starformation

J.V. Feitzinger[1,2], *E. Harfst*[1], *and M. Perschke*[1]

[1]Astronomical Institute, Ruhr-University Bochum, P.O. Box 102148, D-4630 Bochum, Fed. Rep. of Germany

[2]Observatory of the City of Bochum, Castroper Str. 67, D-4630 Bochum, Fed. Rep. of Germany

Models of stochastic selfpropagating starformation are suitable to simulate the effect of gaseous infall in galaxies. Our numerical simulations incooperate stochastic molecular cloud and star formation, three gas phases and detailed gas redistribution in neighbouring cells and perpendicular to the disk around star forming sites. Density wave phenomena, magnetic fields and external gaseous infall can be taken into account. The cell size is 200 pc. The standard model uses a flat rotation curve with v_{max} = 200 km/s.

The probability of starformation depends on three points:

gas and number density of the molecular clouds in the disk; the number of shockwaves reaching the individual starforming cells as function of time; the refractory time scales of the interstellar medium.

The model simulates the infall process by increasing the gas content of the cells in the infall region. This implies that the probability of molecular cloud and starformation is increased. The selfpropagating starformation is triggered by stars at location $\bar{r}$ at time t which resided at $\bar{r} - d\bar{r}$ at time t − dt. The starformation rate will depend on $\rho(\bar{r} - d\bar{r}, t - dt)$. Therefore, if an enhancement of cool gas is produced by infall processes, a local burst of starformation starts. In general, the morphological appearance of a normal galaxy can change drastically. Fig. 1 shows a normal galaxy, Fig. 2 gives the same galaxy in a burst mode triggered by gas infall.

Dynamics and Interactions of Galaxies Editor: R. Wielen

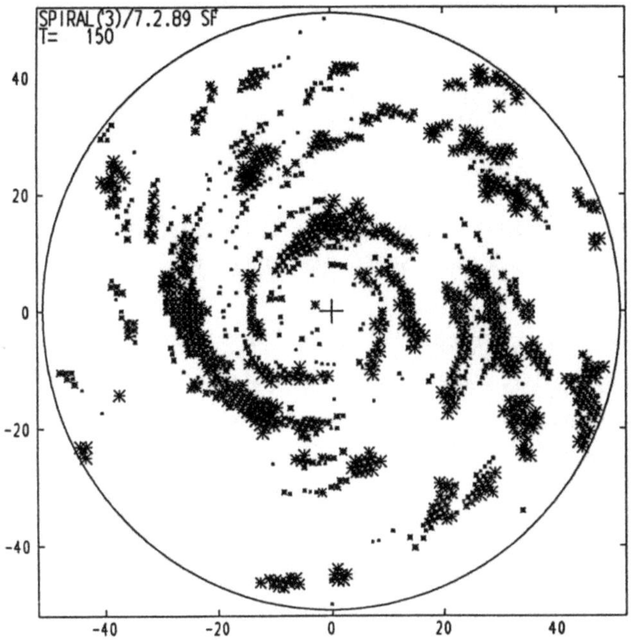

Fig. 1 Normal galaxy, Sbc type, flocculent. Pop I stars are shown; smallest dots represent an age of 10^{9} years; largest dots are the new formed stars.

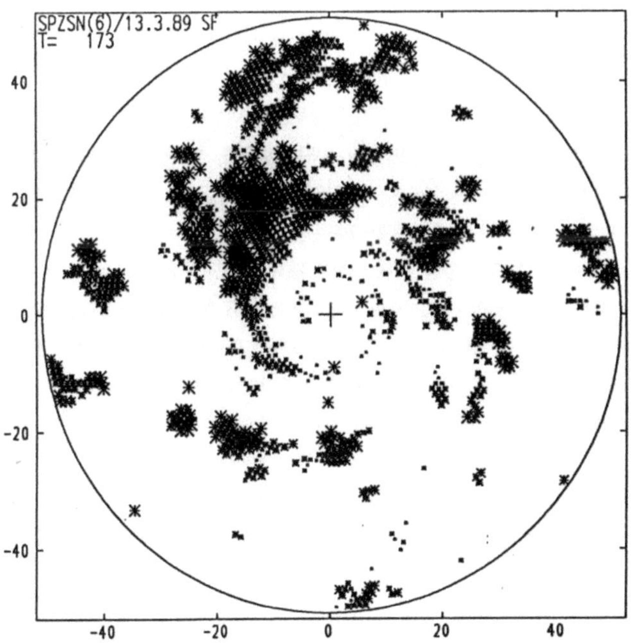

Fig. 2 Normal galaxy (as in Fig. 1) in a burst mode.

Th. Neukirch, J. V. Feitzinger, 1988, MN 235, 1343

J. V. Feitzinger et al., 1988, Vistas 30, 243

Gas and Dust in NGC 4696 – Implications for Cooling Flows

F.D. Macchetto and W.B. Sparks*

Space Telescope Science Institute, 3700 San Martin Drive,
Baltimore, MD21218, USA**
*Affiliated with the Astrophysics Division, Space Science Department,
European Space Agency (ESA)
**Operated by the Association of Universities for Research in Astronomy, Inc.,
under contract with the National Aeronautics and Space Administration

NGC4696 is the radio elliptical galaxy at the centre of the Centaurus cluster of galaxies. It is an excellent candidate for a galaxy hosting a cooling flow (Fabian *et al.* 1982, Canizares & Markert 1988). It contains extended optical line emission, but also a dust lane coincident with these filaments. In this and other 'cooling-flow' galaxies spatially extended optical emission lines have been taken as evidence that the hot X–ray gas is cooling through 10^4K and that we are seeing the thermally unstable condensations. Since the emission-line filaments are dusty however, and since that dust is apparently quite normal, such an explanation for the filament origin may be untenable.

Quantitative imaging data have been used to investigate the filaments (Sparks *et al.* 1989). A normal wavelength dependence of extinction was found, suggesting an infall (accretion) origin for this material. Since many other emission-line systems are dusty, and dust sputtering times are short in hot gas, we suggest the infall pheomenon is widely occurring. There are other good reasons to believe infall is a frequent phenomenon, such as the counter rotating stellar cores and high angular momentum gas discs found in some elliptical galaxies, and it seems likely that both the dust and emission line gas are associated with the infall. Equally, though, there are good reasons to believe the X–ray emission *is* linked to the optical line emission (Hu *et al.* 1985, for example). We suggest that effective heat conduction provides the missing link between the two processes.

If heat conduction *is* effective in the X–ray gas and infall events occur, then (i) there is a natural explanation for the correlation between optical emission and X–ray cooling time (Hu *et al.* 1985) since the line emitting gas drains energy from the hot halo and shortens the apparent 'cooling time', (ii) the presence of X–ray line emission at intermediate temperature may be accounted for since the temperature of the interface region must span the range from both extremes, and there is sufficient energy loss to give the lower temperature X–ray line emission, and (iii) the emission-line excitation mechanism may also have the same origin, either directly through collisional heating or indirectly from the absorption of UV radiation from the interface.

Heat Conduction

If a cold region is next to a hot region, the maximum rate at which energy can be transported across the interface by thermal electrons is (Cowie & McKee 1977) $Q_{\rm sat} \sim n_e v_{th} \times$

Dynamics and Interactions of Galaxies Editor: R. Wielen

$\frac{1}{2}m_e v_{th}^2 \times A$ where A is the surface area estimated to be $\sim$50 kpc^2. There is good energy balance between the emission of the dust lane ($\approx 2 \times 10^{43}$ erg s^{-1}) and the energy available for direct heat transfer ($\approx 3 \times 10^{43}$ erg s^{-1}). It is possible that much of the heat inflow may be absorbed by evaporative processes. However even a modest conversion efficiency could suffice to power the cooler gas. If conduction is effective, irrespective of the emission-line excitation mechanism, the energy removed from the X-ray gas by its interaction with the dust lane can exceed the total losses from X-ray emission and may be crucial in determining the properties of the hot gas. The "heat capacity" $\approx$ power radiated $\times$ timescale, which if $\lesssim$ dynamical timescale (reasonable given the morphology) implies the capacity to cool $\gtrsim 10^9 M_\odot$ and more than sufficient to give rise to the lower temperature X-ray line emission.

Can tangled magnetic fields suppress conduction? This remains an unresolved issue. Smooth rotation measure structure in Virgo (Laing & Bridle 1987) may imply ordered magnetic fields while Rosner & Tucker (1989) argue that even if tangled, conduction is not dramatically suppressed.

Conclusion

We favour an external accretion origin for the gas and dust in NGC4696 and propose that heat conduction links the optical and X-ray properties of this and other 'cooling-flow' galaxies. Heat conduction plus infall leads to the symptoms of a cooling-flow without requiring the existence of one (i.e. no large mass outfall), since in the absence of the cool infalling gas, X-ray radiative losses would be balanced by heat transfer.

References:

Canizares C.R. & Markert T.H., 1988, *in Cooling Flows in Clusters and Galaxies, ed. A.C. Fabian.*

Cowie, L.L. & McKee, C.F., 1977, *Ap. J.*, **211**,135.

Fabian, A.C., Atherton, P.D., Taylor, K. & Nulsen, P.E.J, 1982, *M. N. R. A. S.*, **201**,17P.

Hu, E.M., Cowie, L.L. & Wang, Z., 1985, *Ap. J. Suppl.*, **59**,447.

Laing, R.A., & Bridle, A.H., 1987, *M. N. R. A. S.*, **228**,557.

Rosner, R. & Tucker, W.H., 1989, *Ap. J.*, **338**,761.

Sparks, W.B., Macchetto, F.D. & Golombek, D., 1989, *Ap. J.*, in press. Oct 1.

Part IX

Disk Galaxies

Age-Dependence and Distribution of the Velocities of Disk Stars and Implications of a Possible Disruption of Globular Clusters by Massive Black Holes

R. Wielen and B. Fuchs

Astronomisches Rechen-Institut, Mönchhofstraße 12–14,
D-6900 Heidelberg, Fed. Rep. of Germany

1. General Considerations Concerning the Dynamical Evolution of Galactic Disks

The systematic increase in the dispersions of the peculiar velocities of disk stars as function of age provides the best direct observational evidence for a dynamical evolution of galalactic disks. This age-velocity-dispersion relation is particularly well observed for stars in the solar neighbourhood (Wielen 1974, 1977, 1982). To the increase of the velocity component W perpendicular to the galactic plane corresponds an increase of the vertical scale height of the disk stars as function of age. This age-thickness relation is well known for our Galaxy, and is also observed in external edge-on galaxies.

The age-velocity-dispersion-relation of disk stars cannot be explained solely by a decrease of the turbulence of the interstellar gas out of which the stars have been formed at various epochs during the evolution of the disk: The observed velocity dispersions (cf. Fig.1) decrease rather dramatically for young stars formed at epochs close to the present epoch. The shape of the age-velocity-dispersion relation would imply an undue preference of our present epoch of disk evolution, if the velocities of the stars observed today would simply reflect the gas velocity at the time of formation of the stars (Wielen 1977). The velocity dispersions of halo objects and of stars in a 'thick disk', however, are probably related to the particular conditions at their formation during the early phases of galactic evolution.

The observed shape of the age-velocity-dispersion relation can be understood without giving any preference to our present epoch of galactic evolution, if the disk stars are supposed to be accelerated in a stochastic way by an irregular component of the galactic gravitational field (Wielen 1977). According to this conjecture, which was originally proposed by Spitzer and Schwarzschild (1951, 1953), the increasing velocity dispersion of disk stars with increasing age would be due to the accumulative effect of numerous gravitational perturbations of the stellar orbits after the formation of the stars.

The various mechanisms for the gravitational perturbation of stellar orbits have been discussed in our previous reviews (Fuchs and Wielen 1987, Wielen and Fuchs 1989). We consider (1) 'stochastic heating' by an irregular gravitational field, (2) 'adiabatic heating or cooling' by slow changes of the regular gravitational field, in particular perpendicular to the

Dynamics and Interactions of Galaxies Editor: R. Wielen

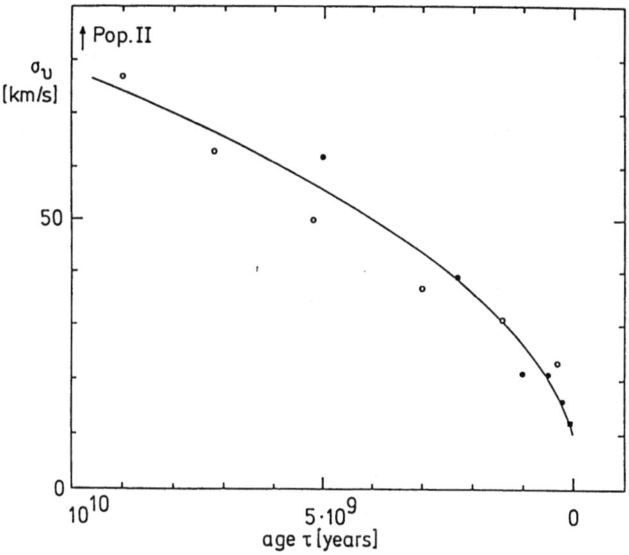

Figure 1. Age-velocity-dispersion relation of nearby stars. Observations are indicated by symbols. The theoretical σ-τ-relation assuming a constant diffusion coefficient is shown as the full line.

galactic plane, and (3) 'deflections' of stellar orbits by irregularities of the galactic gravitational field, which may be important for the axial ratio of the velocity ellipsoid, e.g. σ_U/σ_W, even if they do not contribute significantly to the 'heating' of the galactic disk.

Since the work of Spitzer and Schwarzschild (1951, 1953) a number of various physical processes have been studied as possible sources of the stochastic heating of galactic disks. Encounters among the field stars themselves are known to be completely inefficient (e.g. Chandrasekhar 1960) so that Spitzer and Schwarzschild envisaged encounters of stars with massive interstellar gas clouds. Today, giant molecular clouds (GMCs), or even more massive complexes of GMCs, are known to be efficient scatterers of disk stars. Encounters between stars and GMCs have been investigated in great detail by Icke (1982), Lacey (1984), Villumsen (1983, 1985), and recently by Binney and Lacey (1988). It appears, however, that at solar distances from the galactic center GMCs are neither abundant nor massive enough to explain the actually observed velocity dispersions of nearby disk stars. Even assuming a higher abundance and higher masses of GMCs at earlier stages of galactic evolution does not explain the rapid increase of the velocity dispersion with age during the last 10^9 years.

An interesting possibility has been put forward by Lacey and Ostriker (1985) and Ipser and Semenzato (1985). They propose that the dark matter of the galactic coronae as deduced from the flat rotation curves in the outer parts of disk galaxies consists of massive black holes. Carr and Lacey (1987) envisage compact clusters of less massive black holes. Such objects with masses of about 10^6 solar masses whould accelerate disk stars in a significant way, when they penetrate the galactic disk with typical velocities of halo objects of a few 100 km/s, and could well account for the observed velocity dispersions. Of particular interest is the very good agreement between the observed and the theoretically expected shape of the age-velocity-dispersion relation predicting an increase in the velocity dispersion σ with age τ as $\sigma \propto \tau^{1/2}$.

A further promising mechanism for the stochastic heating of disk stars are fluctuations in the gravitational field caused by instabilities in the galactic disk, such as Jeans instabilities, density wavelets, or transient spiral arms (Carlberg and Sellwood 1985).

2. Modelling the Velocity Distribution of Nearby Stars

As discussed above the basic physical mechanism for the heating of the stellar galactic disk cannot be identified with certainty at present. We have therefore adopted a phenomenological description to model the dynamical evolution of the galactic disk quantitatively (Wielen 1977, Fuchs and Wielen 1985). In this approach the stochastic heating of the disk is described by a diffusion of stellar orbits. The basic physical parameter of this diffusion process, the diffusion coefficient in velocity space, is empirically determind from the observed age-dependence of the velocity dispersion of disk stars in the solar neighbourhood (Wielen 1977).

We have experimented with various forms of the diffusion coefficient. A constant diffusion coefficient which depends neither on time nor on the stellar velocity would be ideal to explain both the age-velocity-dispersion relation as well as the present velocity distribution of nearby disk stars. Such a constant diffusion coefficient would model the perturbations due to the coronal massive black holes mentioned above. Their masses and number density are not expected to vary with time, and the relative velocity between a star and a massive black hole would be almost independent of the peculiar velocity of the star, measured relative to the local circular velocity, since the massive black hole, being a halo object, has typically a much higher velocity than a star in the disk.

In constrast to such a constant diffusion coefficient, a diffusion coefficient describing the perturbative effects by giant molecular clouds would depend on the peculiar velocity of a disk star. Since the typical peculiar velocities of GMC's are rather small, the relative velocity between a GMC and a star is determined predominantly by the peculiar velocity of the star. The corresponding diffusion coefficient is by definition a time-integral over the statistical average of the squared stochastic velocity jumps experienced by a star and is thus proportional to the length of a typical encounter or roughly inverse proportional to the relative velocity between a star and a perturber. However, a diffusion coefficient with a $1/\upsilon$-characteristic does not reproduce the observed increase of σ with τ (Wielen 1977): The predicted shape of the σ-τ relation would be $\sigma \propto \tau^{1/3}$ in contrast to the observed behaviour, approximately $\sigma \propto \tau^{1/2}$. In order to fit the observed σ-τ relation one has to assume a time-dependence of the diffusion coefficient in the sence that GMCs were more frequent in the past.

The formulation of the theory of the diffusion of stellar orbits has been developed along these lines in our earlier papers cited above. Relevant equations, numerical constants and formal results are summarized by Fuchs and Wielen (1987) and Wielen and Fuchs (1989).

The stellar distribution function is computed from a Fokker-Planck equation into which the model diffusion coefficients described above have been introduced. In the case of a constant diffusion coefficient we find an approximate analytic solution in the form of a Schwarzschild distribution function with velocity dispersions growing like $\sigma_{U,V,W} \propto \tau^{1/2}$. The constants of proportionality are given by the diffusion coefficient times certain factors which come from epicyclic theory. In Fig. 1 we show a comparison of the theoretically predicted velocity distributions with observations of nearby stars. As the most suitable observational sample of nearby disk stars, we use 317 McCormick K and M dwarfs in the Gliese Catalogue, since these stars are free from kinematical selection effects, because they have been detected on objective-prism surveys. The observed data are shown as histograms for the spatial velocity components in the radial (U), tangential (V), and the vertical (W) directions, respectively.

In order to derive a theoretical velocity distribution comparable to the observations, the velocity distributions - each a Schwarzschild distribution - of many stellar generations of varying age, born at a steady rate during the galactic evolution, have been integrated. In Fig. 2 a constant star formation rate over 10^{10} years has been assumed. Once that the σ-τ relation is calibrated and the star formation rate is adopted, there is no further free parameter available to adjust the theoretical results to the observations, so that the good agreement is an independent confirmation of the underlying physical ideas. There is an asymmetry in the observed V-velocity distribution related to the well-known asymmetric drift, which we have not modelled here but which we discuss in some detail in Wielen and Fuchs (1989).

In the case of the velocity-time-dependent diffusion coefficient no simple analytic solutions of the Fokker-Planck equation are known. We have performed a series of Monte-Carlo simulations following the paths of stars experiencing sudden random velocity changes according to a velocity-time dependent diffusion coefficient through phase space. The principal difference of the resulting velocity distributions to distributions corresponding to a constant diffusion coefficient is the development of a plateau or even a small dip at low velocities. This is due to the $1/\upsilon$-characteristic of the diffusion coefficient which depletes regions of low velocities in phase space. As in the case of the constant diffusion coefficient

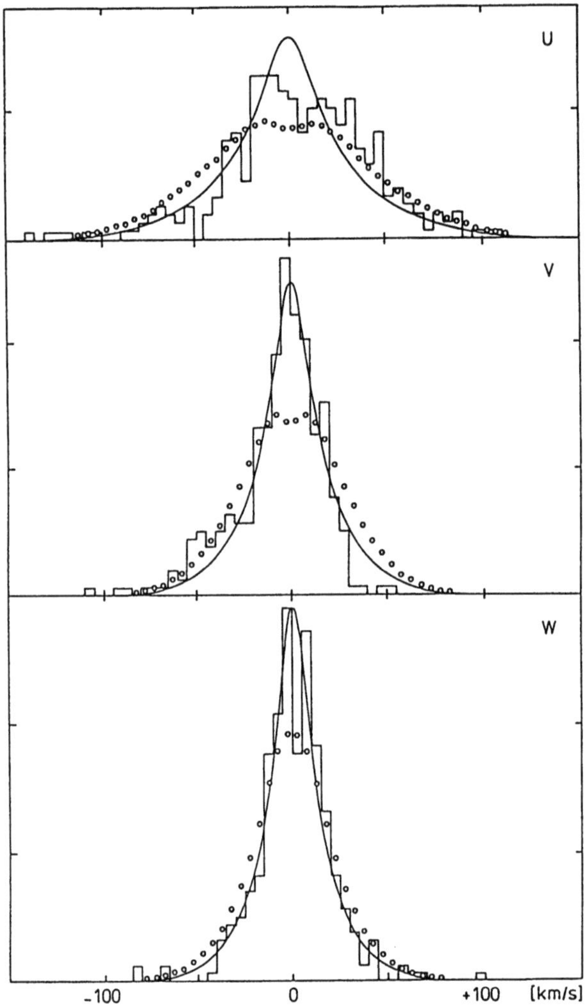

Figure 2. Velocity distributions of nearby stars at the galactic midplane. Observations of McCormick K and M dwarfs are given as histograms. Theoretical predictions for a constant diffusion coefficient are shown as full lines and predictions for a velocity-time-dep ndent diffusion coefficient as open circles, respectively.

the velocity distributions have been integrated over many generations of stars in order to compare them with the observed distribution of the McCormick K and M dwarfs. Again a constant star formation rate was assumed. As may be seen from Fig. 2 the resulting velocity distributions are broader than the velocity distributions corresponding to a constant diffusion coefficient and fit less well to the observed velocity distribution, although the σ-τ relation has been calibrated to fit exactly the observed velocity-dispersion-age relation. We thus conclude that GMCs are probably not the main cause for the stochastic heating of the galactic disk. GMCs, however, may be still very effective in perturbing stellar orbits as far as the adjustment of the axial ratio of the velocity ellipsoid of disk stars is concerned, even if they do not heat the disk.

3. Massive Black Holes in the Dark Coronae of Galaxies

As discussed in Section 1, the existence of massive black holes in the dark coronae of galaxies would allow a very good explanation of both the increase of the peculiar velocities of stars in galactic disks with time and the observed distribution of these peculiar velocities in the solar neighbourhood. If one assumes that all the dark matter of our Galaxy is in the form of massive black holes, the mean mass of such a black hole, m_n, can be derived from the observed diffusion coefficient D_0 for nearby stars and the local mass density of the dark coronal matter, ρ_{an}, which can be estimated from galactic mass models based on flat rotation curves. The observed increase of the velocities of disk stars with age in the solar neighbourhood requires that

$$m_n \, \rho_{an} = 2 \cdot 10^4 \, m_0^2 \, pc^{-3} \tag{1}$$

(Lacey 1984). Together with the local mass density of the invisible galactic corona of ρ_{an} = 0.006 m_0/pc^3, taken from the galactic mass model of Caldwell and Ostriker (1981), we derive for the mean mass of an individual massive black hole

$$m_n = 3 \cdot 10^6 \, m_0 \tag{2}$$

(Wielen 1985). It is quite remarkable that this value of m_n is nearly identical to the mass of a possible central black hole in our Galaxy, namely 1 to $3 \cdot 10^6$ m_0, derived by Genzel (1989) from completely different observations and arguments. This would support a scenario in which one of the many coronal massive black holes has settled down, perhaps due to dynamical friction, at the galactic center. More than two of these massive black holes cannot reside close to the center, because the gravitational sling-shot effect, operating between three or more massive black holes, would expell the other black holes from this region.

In elliptical galaxies, it is more difficult to confirm the existence of dark matter than in disk galaxies with flat rotation curves. Beside polar gas rings, the most direct evidence for dark matter in extended halos of elliptical galaxies stems from the interpretation of X-ray observations, e.g. for M87 by Fabricant and Gorenstein (1983). We postulate that this dark matter in elliptical galaxies is also made up by massive black holes. The simplest assumption is then to assume furthermore that the mean mass of a coronal black hole does not vary from galaxy to galaxy, and to use the value of $3 \cdot 10^6$ m_0, derived for our Galaxy, also for other galaxies.

4. Disruption of Globular Clusters by Massive Black Holes

The existence of massive black holes orbiting in the dark corona of a galaxy, envoked for speeding-up the peculiar velocities of disk stars during their fast passage through the galactic disk, would have severe implications for the stability of globular clusters. This has been pointed out by Wielen (1987, 1988) and confirmed by Ostriker, Binney and Saha (1989).

General expressions for the time of disruption of a star cluster due to encounters with black holes have been given by Wielen (1985). We have to consider two cases: Either the cluster can be disrupted already by one passage of a black hole, or many passages are required for dissolving the cluster. In the first case, we derive for the mean lifetime of the cluster

$$T_{n,1} = (3/(32\alpha))\pi^{-1/2} (Gm_c/r_c^3)^{1/2}/(G\rho_{an}) . \qquad (3)$$

In the second case, the mean lifetime is given by

$$T_{n,0} = (9/512\alpha^2)\, G^{-1} m_n^{-1}\, \rho_{an}^{-1}\, v\, m_c r_c^{-1} . \qquad (4)$$

m_c is the total initial mass of the cluster, r_c the median radius of the cluster, α the ratio between the root-mean-squared cluster radius and r_c (we set α=1), m_n the mass of a passing black hole, ν_n the number density in space of the black holes, $\rho_{an} = m_n \nu_n$ the overall mass density of the black holes, v the typical relative velocity between the cluster and a black hole, and G the gravitational constant. $T_{n,1}$ is applicable if the critical impact parameter

$$p_1 = 4(\alpha/3)^{1/2}(G/\pi)^{1/4}\, m_n^{1/2}\, v^{-1/2}(m_c/r_c^3)^{-1/4} . \qquad (5)$$

is larger than the cluster radius r_c. If p_1 is smaller than r_c, we have to use $T_{n,0}$. It is remarkable and helpful for applications that at least $T_{n,1}$ neither depends on the mass m_n of a black hole nor on the relative velocity v, but only on the overall mass densitiy ρ_{an} of the black holes.

The actual lifetime of a cluster can deviate strongly from the mean expected lifetime. The randomness of close encounters introduce a strong dispersion into the individual lifetimes of clusters. The corresponding statistical survivale rate ν of clusters as a function of age τ has been given by Wielen (1988, Fig.5) and Kuhrau (1989).

The results on the disruption of clusters by passing massive black holes have been applied to globular cluster in our Galaxy (Wielen 1988) and in the giant elliptical galaxy M87 (Wielen 1987). For our Galaxy, it was found that the initial number of globular clusters should have been higher by an order of magnitude than we do observe now. In other words, most of the originally existing globular clusters should have been disrupted within $15 \cdot 10^9$ years. Their stars are now (a small fraction of the) halo field stars.

For M87, the effect of the disruption of globular clusters by massive black holes is able to explain the relative scarcity of globular clusters in the inner parts of M87 relative to the integrated light of the field stars. The basic idea is that due to the increase of the density of the black holes towards the center of M87, most of the globular clusters within a region of about 50 kpc around the center of M87 are disrupted within a Hubble time, while the outer distribution of globular clusters remains unchanged and proportional to that of the integrated light. Ostriker et al. (1989) have essentially confirmed this result, although they propose that either not all the dark matter in M87 is in the form of black holes or that the mean mass of a black hole is significantly smaller in M87 than in our Galaxy.

5. Implications of the Disruption of Globular Clusters by Massive Black Hole for Merged Galaxies

Both disruption times $T_{n,1}$ and $T_{n,0}$ are inversely proportional to the mean density ρ_{an} of massive black holes

$$T_n \propto (1/\rho_{an}) \quad . \tag{6}$$

Now. in most cases of the meraina of two aalaxies, the mean densitv ρ_m of the merged galaxy is significantly smaller than the mean initial density ρ_i of the two galaxies before merging. This overall dilution of galaxies in most cases of merging can be easily derived from energy conservation and the virial theorem. For example, in the case of two identical galaxies of initial mass m_i and initial virial radius r_i which are falling towards each other from rest at infinity (i.e. with zero orbital energy), the resulting merged galaxy has a total mass of $m_m = 2m_i$ and a virial radius of $r_m = 2\ r_i$, since the total energy of the merger, $E_m = -Gm_m^2/(4r_m)$, is for zero orbital energy equal to the sum of the energies $E_i = -Gm_i^2/(4r_i)$) of each of the initial galaxies. The corresponding mean density ($\rho \propto m/r3$) is in this case decreased by a factor of 4 with respect to the initial galaxies, i.e. $\rho_m = \rho_i/4$.

The overall decrease in density by merging of two (or more) galaxies affects also the density of the dark coronal matter. If the number density of massive black holes, and hence also their overall mass density ρ_{an}, is significantly decreased by merging, then the lifetimes of the globular clusters are correspondingly increased, according to Eq. 6. This means that the globular clusters are <u>more slowly</u> destroyed in the merger than they were disrupted in the original galaxies. The result would be that the specific frequency, i.e. the number of globular cluster per unit mass of their parent galaxy, is expected to be <u>higher</u> in the merger than in the original galaxies.

The effect would be noticible only in galaxies or in those parts of galaxies in which the lifetimes of the globular clusters are smaller than the Hubble time. Furthermore, the density distribution in the merger is in general not homologeous to the density distributions of the original galaxies. In some numerical simulations, it is found that the central density in the merger is somewhat higher than expected from homolgy. This would reduce the effect of slower disruption of globular clusters in mergers. Nevertheless, we still expect a significantly higher specific frequency of globular clusters in a merger than in the original galaxies.

If the scenario outlined above is essentially correct, then this would weaken an argument given by van den Bergh (1986 and this conference) against the formation of elliptical galaxies by merging spirals. Van den Bergh argues that the spirals have a lower specific globular cluster frequency than ellipticals (also true if measured per unit mass instead of unit luminosity), and that merging should not change the specific globular cluster frequency. Now, our arguments show that the decrease in density due to merging would slow-down the disruption of globular clusters by massive coronal black holes. Hence we would expect in this case a higher specific frequency of globular clusters in the merger (elliptical galaxy) than in the original galaxies (spirals), just as it is observed.

References

Binney, J., Lacey, C.G.: 1988, Mon. Not. R. Astron. Soc. 230, 597.

Carlberg, R.G., Sellwood, J.A.: 1985, Astrophys. J. 292, 79.

Carr, B.J., Lacey, C.G.: 1987, Astrophys. J. 316, 23.

Chandrasekhar, S.: 1960, Principles of Stellar Dynamics, Dover Publ., New York.

Fabricant, D., Gorenstein, P.: 1983, Astrophys. J. 267, 535.

Fuchs, B., Wielen, R.: 1987, in The Galaxy, Eds. G. Gilmore and B. Carswell, D. Reidel Publ. Comp., Dordrecht, p. 375.

Icke, V.: 1982, Astrophys. J. 254, 517.

Ipser, J.R., Semenzato, R.: 1985, Astron. Astrophys. 149, 408.

Kuhrau, H.-J.: 1989, Diplomarbeit, Techn. Univ. Berlin.

Lacey, C.G.: 1984, Mon. Not. R. Astron. Soc. 208, 687.

Lacey, C.G., Ostriker, J.P.: 1985, Astrophys. J. 299, 633.

Ostriker, J.P., Binney, J., Saha, P.: 1989, Mon. Not. R. Astron. Soc. 241, 849.

Spitzer, L., Schwarzschild, M.: 1951, Astrophys. J. 114, 385.

Spitzer, L., Schwarzschild, M.: 1953, Astrophys. J. 118, 106.

van den Bergh, S.: 1986, Bull. Astron. Soc. India 14, 8.

Villumsen, J.V.: 1983, Astrophys. J. 274, 632.

Villumsen, J.V.: 1985, Astrophys. J. 290, 75.

Wielen, R.: 1974, Highlights of Astronomy 3, 395.

Wielen, R.: 1977, Astron. Astrophys. 60, 263.

Wielen, R.: 1982, in Kinematics and Dynamics of the Galaxy, Chapter 8.4 of Landolt-Börnstein, Group VI, Vol. 2 Subvol. 2c, Eds. H.H. Voigt and K. Schaifers, Springer-Verlag, Berlin, p. 208.

Wielen, R.: 1985, in IAU Symposium No. 113, Dynamics of Star Clusters, Eds. J. Goodman and P. Hut, D. Reidel Publ. Comp., Dordrecht, p. 449.

Wielen, R.: 1987, in Proceedings of the Tenth Europ. Regional Astron. Meeting of the IAU, Vol. 4, Ed. J. Palous, Publ. Astron. Inst. Czech. Acad. Sci., No. 69, p. 157.

Wielen, R.: 1988, in IAU Symposium No. 126, The Harlow Shapley Symposium on Globular Cluster Systems in Galaxies, Eds. J.E. Grindlay and A.G.D. Philip, Kluwer Acad. Publ., Dordrecht, p. 393.

Wielen, R., Fuchs, B.: 1985, in IAU Symposium No. 106, The Milky Way Galaxy, Eds. H. van Woerden, R.J. Allen, and W.B. Burton, D. Reidel Publ. Comp., Dordrecht, p. 481.

Wielen,R., Fuchs, B.: 1988, in The Outer Galaxy, Proceedings of a symposium held in honor of Frank J. Kerr, Eds. L. Blitz and F.J. Lockman, Lecture Notes in Physics, Vol. 306, Springer-Verlag, Berlin, p. 100.

Wielen, R., Fuchs, B.: 1989, in Evolutionary Phenomena in Galaxies, Eds. J.E. Beckman and B.E.J. Page, Cambridge Univ. Press, Cambridge, p. 244.

What's with Warps?

J. Binney

Department of Theoretical Physics, University of Oxford,
Oxford, OX1 3RH, UK

Most galactic disks are warped in a way that would be hard to understand if they were not self-gravitating. A massive but realistically fuzzy-edged disk cannot sustain a coherent warp when it is isolated. A sustained warp is, however, possible if the disk is embedded in moderately flattened halo. Such a warp arises because disk and halo have misaligned angular momentum vectors. Cosmic infall onto galaxies will reorient galactic angular momentum vectors every Hubble time or so. However, an indefinitely enduring warp can be excited only by a sudden, discrete infall event rather than by steady cosmic infall. Steady infall will bend a disk at points of low surface density and this may manifest itself as a warp. Such a warp will, however, disappear as soon as infall ceases.

1 Introduction

Warps are common and one of the longest standing puzzles in galactic dynamics. We still cannot claim to know why so many galaxies are warped. But I think one can now make a strong case (a) that warps are a manifestation of dark matter, and (b) that warps arise because galaxies are continuously growing concerns.

Briggs (this meeting) has reviewed the phenomenology of HI warps. From a sample of 12 galaxies for which convincing ring models can be constructed he concludes: (i) the most impressive warps occur in nearly isolated galaxies; (ii) warps start near R_{25}; (iii) interior to a transition radius $R_{\rm trans} \lesssim R_{\rm Holmberg}$ the line of nodes is straight while outside $R_{\rm trans}$ the line of nodes becomes a leading spiral that turns through $\lesssim 150°$; (iv) one can always choose an equatorial plane for one's coordinate system such that the line of nodes is straight at both small and large radii, with an abrupt change of orientation around $R_{\rm Holmberg}$.

Early theoretical work on warped galaxies, especially the classic paper by Hunter & Toomre (1969), was concerned with the dynamics of isolated disks. It is now generally supposed that even late-type galaxies, which appear to consist of a disk alone, are embedded in a dynamically dominant dark halo. This is fortunate since it is in some ways easier to understand the dynamics of a massive disk sitting in an externally generated gravitational field than to get to grips with the dynamics of an isolated disk. I therefore start by outlining the dynamics of a disk in an externally generated potential. I shall generally assume that the latter is a flattened axisymmetric object and shall make no attempt to distinguish the contributions to it from spheroid and dark halo; for brevity I shall refer simply to the "halo potential". The component of any angular momentum vector that is perpendicular to the halo potential's symmetry axis I shall refer to as "skew L". After reviewing the dynamics of embedded disks, I consider the consequences for disk dynamics of accretion by (a) the halo and (b) the disk.

Dynamics and Interactions of Galaxies Editor: R. Wielen

2 Dynamics of Warps

If we put a star on a circular orbit in a spherical potential, it will orbit in the fixed plane whose normal is parallel to the star's angular momentum vector **L**. If we now squash the potential along the direction $\hat{\mathbf{z}}$ that is inclined at angle θ to **L**, the orbital plane and its normal **L** will precess about $\hat{\mathbf{z}}$; the line of nodes in which the orbital plane cuts the plane perpendicular to $\hat{\mathbf{z}}$ rotates in the opposite sense to the sense of the star's circulation. The frequency of this precession is $\Omega_p = \nu - \Omega$, where ν is the frequency at which the star oscillates parallel to $\hat{\mathbf{z}}$ and Ω is the circular frequency. For a system with a flat rotation curve, $\Omega = v_c/R$ is inversely proportional to the radius r of the star's orbit, and $\nu > \Omega$ is also roughly proportional to $1/R$. Hence if we form an inclined disk of test particles in a flattened potential by simultaneously launching stars onto orbits of varying radii but initially parallel angular momentum vectors $\mathbf{L}(R)$, we shall find that each ring of our inclined disk precesses backwards at a different rate, with the result that the initially flat disk steadily winds up into an object with corrugations that form a leading spiral. The If $R_{\min}$ and $R_{\max}$ are the inner and outer radii of our disk, the time-scale for the disk to wind into a spiral is $N_{\text{wrap}} = \Omega(R_{\min})/[\Omega_p(R_{\min}) - \Omega_p(R_{\max})]$ galactic years at $R_{\min}$.

To see how this time-scale works out numerically, let the galaxy consist of an exponential disk of mass M_d and scale-length R_d plus spherical components that have mass $M_h(r)$ interior to spherical radius r. Then expanding the disk's potential up to quadrupole order in spherical harmonics one finds

$$\begin{aligned} \left.\frac{\Omega_p}{\Omega}\right|_R &= \frac{\nu - \Omega}{\Omega} = \frac{\nu^2 - \Omega^2}{(\nu + \Omega)\Omega} \\ &\simeq \frac{\nu^2 - \Omega^2}{2\Omega^2} \simeq \frac{4R_d^2/R^2}{1 + M_h/M_d} \qquad (R \gg R_d). \end{aligned} \tag{1}$$

Near the Sun ($R \simeq 3R_d$, $M_h \lesssim M_d$) we have $\Omega_p \simeq \Omega/4$ and $N_{\text{wrap}} \simeq 4$, while in the more extended HI disks, where $R \simeq 10R_d$ and $M_h \gtrsim 2M_d$, we have $N_{\text{wrap}} \gtrsim 100$. Thus while naive theory predicts that the warp in our Galaxy will wind up in a Gyr or so, warps far out in the HI disks of galaxies might take a hundred times as long to wind up significantly. Tubbs and Sanders (1979) argued in this way that when heavy halos are taken into account we need not be concerned for the longevity of most HI warps.

However, even spherical heavy halos do not stabilize the warps that are so often seen at the edges of optical disks ($R \lesssim 4R_d$), and we must bear in mind that it is unlikely that halos are precisely spherical. Let the halo's ellipticity be $\epsilon \equiv 1 - b/a \ll 1$. Then the halo potential will be approximately of the form $\Phi_h(R^2 + z^2/q^2)$, where the axis ratio $q \simeq 1 - 0.35\epsilon$ and the precession frequency due to the halo alone will satisfy $\Omega_p/\Omega \simeq 0.35\epsilon$. If halos resemble elliptical galaxies in shape, $\langle\epsilon\rangle \simeq 0.3$, so $N_{\text{wrap}} \lesssim 10$ and most warps will wrap up within ten local rotation times.

So far no account has been taken of the self-gravity of the disk. Could self-gravity stop warps winding up? A simple physical argument suggests that this may indeed be the case.

First we break the disk up into many thin circular rings of material and persuade ourselves that each ring will behave dynamically as if it were a rigid wire.[1] Then we

[1] This is a non-trivial step but a valid one – for a discusson see Appendix A of Ostriker & Binney (1989).

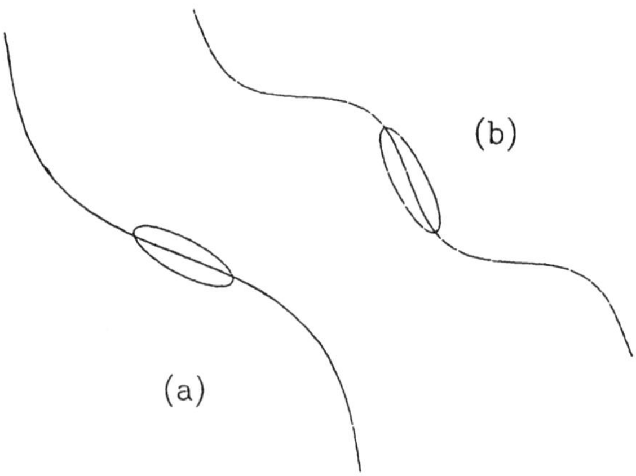

Fig. 1. A warped disk may be broken down into a series of flat rings. In a warp mode all rings have a common line of nodes and precess about a vertical axis at the same rate. Gravitational torques between the rings acting along the line of nodes hasten the precession of some rings and slow that of others. When the disk curves upwards as in (a), inner rings are slowed and outer rings are hastened. Sparke & Casertano call this a Type I warp. In the configuration shown at (b), both inner and outer rings are hastened while intermediate rings are slowed. This is a Type II warp. The flattened dark halo is not visible in these pictures, but its major axis is horizontal.

argue that warps usually wrap up because in the gravitational field of the rest of the galaxy tilted inner rings precess backwards faster than do tilted outer rings. So we can construct a disk in which all rings precess together if we can slow the precession of the inner rings and hasten that of the outer rings. Now if we think of each ring as a gyro we see that as it precesses its angular momentum vector slowly rotates around the $\hat{\mathbf{z}}$ axis in response to the gravitational torque to which it is subject. So we can speed the precession of the outer rings and slow that of the inner rings by applying torques between rings. Figure 1 illustrates how this principle establishes the curvature of a disk with a stationary warp. Notice that all torques act along the common line of nodes of the rings. In the usual case in which the precession of the inner rings has to be slowed and that of the outer rings hastened, a uniformly precessing disk is obtained when the inclination of rings increases steadily from the centre outwards. In the terminology of Sparke & Casertano (1988) this is a Type I warp. In certain halo/spheroid potentials the single-ring precession frequency Ω_p peaks well away from the centre and the precession of the central rings may be hastened by decreasing the ring inclinations away from the centre. Sparke & Casertano refer to this as a Type II warp.

In an inertial frame, the height $z(R, \phi, t)$ above the equatorial plane of the halo potential of a disk that has been arranged so that all rings precess in unison is a harmonic function of time $z = \Re\left[z_0(R, \phi)e^{i\omega t}\right]$. Thus this special configuration may be described as a normal mode and sought as the solution of an eigenvalue equation. Hunter & Toomre (1969) investigated the warp modes of isolated disks while Sparke (1984) and Sparke & Casertano (1988) investigated the warp modes of disks embedded in various types of rigid halo.

An isolated disk always has one discrete mode – the trivial mode at zero frequency that corresponds to a steady inclination of the entire disk at a common angle with respect to the coordinate system. Clearly this is not in itself interesting and an isolated disk of realistic structure generally has no other discrete mode. However, when we embed the disk in a slightly flattened halo, this trivial mode at zero frequency may go over into a non-trivial discrete mode in which the whole disk precesses as a single curved unit in the halo potential. In this configuration the disk's net angular momentum vector is inclined to the halo's symmetry axis. Sparke & Casertano (1988) find that this "modified tilt mode" exists provided (a) the halo is not too flattened, and (b) the disk does not extend too far in radius.

It is easy to understand the physical origin of these requirements. The flatter the halo and the further the disk spreads in radius the greater are the torques which must be transmitted through the body of the disk if the natural precession rates of centre and periphery are to be reconciled. If the precession rates are large and the disk extends far in radius, the disk's self-gravity may be unequal to the task laid on it, the chain snaps and centre and periphery precess on their own timescales. The maximum halo mass and ellipticity that are compatible with a discrete modified warp mode depend on the extent of the disk and the halo's core radius. Consider, for example, a disk which is embedded in a halo with $M_d(6R_d) = 2M_d$ and is exponential to $5R_d$ before tapering to a density cutoff at $R = 6R_d$. This disk has a discrete modified warp mode for halo eccentricity $\epsilon < 0.45$ when the halo's core radius is $r_c = 2R_d$ and for $\epsilon < 0.63$ when $r_c = 3.5R_d$. Thus realistically extended disks in plausible halos do admit discrete warp modes.

In addition to determining the conditions that must be fulfilled if a disk embedded in a flattened halo is to have a modified tilt mode, Sparke & Casertano showed that the condition for an embedded disk to have any other discrete warp mode is the same as the condition derived by Hunter & Toomre (1969) for an isolated disk to have *any* non-trivial discrete warp mode. This condition is that the edge of the disk must be abrupt. Hunter & Toomre argued from the theory of running bending waves in disks that this condition can be expressed quantitatively as the requirement that the integral $\int_0^\infty \mathrm{d}R/\Sigma$ converges, where $\Sigma(R)$ is the disk's surface density.

The theory of bending modes parallels the WKBJ theory of spiral structure. One first uses the WKBJ approximations[2] to study the propagation of tightly wound bending waves. Then one imagines constructing a normal mode by superposing two running waves. A trailing wave propagates in towards the inner vertical resonance at which the wave's radial wavenumber k vanishes. At the resonance, radius $R_{\min}$, the wave reflects into an outgoing leading wave, which is transformed back into an ingoing trailing wave on reflection by the edge at $R_{\max}$. The frequency of the mode is determined by the condition that the total phase change $\Delta\psi \equiv \int_{R_{\min}}^{R_{\max}} |k|\,\mathrm{d}R$ between these turning points shall be some multiple β_n of π, where $\beta_n = n+\beta_0$ is made up of an integer and an offset.[3]

The analogue for $m = 1$ bending waves of the Lin-Shu-Kalnajs dispersion relation for spiral waves is (Hunter 1969)

$$(\Omega - \omega)^2 = \nu^2 + 2\pi G\Sigma|k|. \tag{2}$$

Here $\omega < 0$ is the wave's frequency, Ω and ν are the circular and vertical frequencies due to everything except nearby disk material (which in a thin disk contributes largely to the net vertical frequency) and Σ is the disk's surface density. The frequency-determining condition is therefore

2 It is too often forgotten that two logically independent approximations are being made here: (a) that the gravitational field of the perturbation is locally generated, and (b) that the density, velocity dispersion etc in the disk change little over a wavelength.

3 Practitioners of semi-classical quantization would call β_0 the Maslov index.

$$\Delta\psi = \int_{R_{\min}}^{R_{\max}} \frac{(\Omega-\omega)^2-\nu^2}{2\pi G\Sigma}\,\mathrm{d}R = (n+\beta_0)\pi. \tag{3}$$

Near $R_{\max}$ the numerator is but a slowly varying function of R, while the denominator vanishes at $R_{\max}$. Suppose

$$\Sigma \propto (R_{\max}-R)^{\alpha} \quad \text{at} \quad R \simeq R_{\max}. \tag{4}$$

Then when $\alpha \geq 1$, $\Delta\psi$ diverges and it becomes impossible to satisfy the quantization condition (3) for finite n. Following Hunter & Toomre (1969) people usually explain this state of affairs by saying that the eigenfrequency spectrum has "become continuous". I find this misleading. What really happens when $\alpha < 1$ is that the outgoing leading wave fails to be reflected off the now fuzzy edge just as sound is not reflected off the exponential horn of an ear trumpet and ocean waves do not reflect off a beach. Rather the bulk of the wave propagates into regions of very low density, where its amplitude becomes large and dissipative processes thermalize its energy. Thus when the integral in (3) diverges, the only normal modes are those low-n modes for which the WKBJ treatment is invalid. Sparke & Casertano (1988) describe a few such modes.

In summary, studies of the normal modes of disks have established that reasonably compact disks in squashed halos are capable of precessing as coherent warped entities, but warping disturbances in extremely large disks wind up steadily and end by heating the outer disk vertically. Some embedded disks support more than the basic "modified tilt mode", but most real galaxy disks are too fuzzy-edged to support more than one warp mode.

3 Exciting a warp

Let us assume that many galaxies are structured such that they do admit a discrete warp mode. Two questions arise: (a) do warps have the correct morphology to be interpreted as excited warp modes? and (b) how might warp modes have become excited?

Sparke and Casertano (1988) show that the warps in NGC's 4565 and 4013 can be satisfactorily fitted with type I and type II warp modes respectively. So warp modes can be morphologically similar to observed warps. This supports the conjecture of Dekel & Shlosman (1983) and of Toomre (1983) that warps arise because the optical galaxy is misaligned with its embedding halo.

How did this unfortunate misalignment come about?

Primordial warps?

Consider first the possibility that the optical galaxy has been misaligned with its halo from birth. Suppose we feed a drizzle of gas into an initially empty, non-spherical heavy halo (White & Rees 1978). As each gas cloud swirls round in the halo it exchanges angular momentum with both the halo and other elements of gas. After one precession time in the halo potential the *mean* angular momentum of all gas of given $|\mathbf{L}|$ will be pretty nearly parallel to the halo's symmetry axis. Thus a self-gravitating disk which forms over more than one halo precession period will lie in one of the halo's fundamental planes. It seems to be an open and interesting question whether a gas cloud which dissipates rapidly can settle to a warped configuration rather than a principal plane. However, I suspect the interest of this question is largely academic since there are clear indications that disks have formed very slowly rather than within a halo precession time. Consequently, it is unlikely that any warp is primordial.

3.1 Realignment of galactic angular-momentum vectors

Let us then assume that the disk formed flat in a principal plane of its halo and ask how it might subsequently have become warped. Since in a warped configuration different rings have non-parallel angular momentum vectors, our question is tantamount to investigating how the disk and/or halo angular momenta became misaligned.

Gunn & Gott (1972) early emphasized the importance of prolonged cosmological infall for the evolution of galaxies and galaxy clusters. Ryden & Gunn (1987) and Ryden (1988) have presented a detailed study of such accretion within the context of cold-dark-matter cosmogony. They find that the mass M of material that has fallen in on a local maximum in the cosmic density should now be growing as $M \propto t^2$. Furthermore, (a) the ***specific*** angular momentum of the infalling material grows as $t^{2/3}$, and (b) the direction of the angular momentum vector of material that falls in after $t = t_0$ is only weakly correlated with that of material that fell in before time $t = \frac{2}{3}t_0$, where t_0 is any large time. Since the half of the galaxy's substance at time $t = t_0$ that fell in during the interval $\frac{2}{3} < t/t_0 < 1$ contains 85% of the galaxy's angular momentum at t_0, it follows that in the cold-dark-matter picture the direction of every galaxy's angular momentum vector is radically reoriented each Hubble time or so. Studies of the acquisition of angular momentum by protogalaxies in numerical simulations of the expanding Universe are in general agreement with these conclusions (Efstathiou, this meeting).

Several empirically based arguments lead to the conclusion that real galaxies do in fact continue to accrete large quantities of mass and angular momentum. These include:

(i) Evidence that spiral galaxies such as our own are locking up interstellar gas in low-mass stars so fast that existing stocks of material will be exhausted within a Hubble time (Larson *et al*, 1980; Sandage, 1986). Since it is improbable that all spirals are soon to degenerate into lenticulars, it follows that most spirals must be accreting gas at a rate $\gtrsim 0.1\,\mathrm{M_\odot\,yr^{-1}}$, and dark matter significantly faster than this.

(ii) The relative paucity within the Galactic disk of unevolved low-metallicity stars arises naturally if the Galaxy is accreting gas (e.g. Tinsley 1980).

(iii) As stellar disks age they are heated by spiral and other irregularities (Sellwood & Carlberg 1984). If this heating is not balanced by cooling, the stellar velocity dispersion in the disk will rise and the disk's spiral structure tighten. Hence the disks of Sb and later-type galaxies must be cooled. The only credible cooling mechanism is the formation of substantial quantities of fresh low-velocity stars from gas. This argues for a substantial star-formation rate sustained by prolonged infall of gas (Toomre, this meeting).

(iv) Galaxies are known to cannibalize their smaller neighbours (Tremaine & Lauer, this meeting). Over a Hubble time cannibalism is expected to increase the luminosity of a typical giant galaxy by $\sim 10\%$, mainly at the expense of gas-rich dwarfs. Evidence that our own Galaxy is active in this field includes the Magellanic Stream and the presence of young A-stars kiloparsecs above the plane (Rogers *et al*, 1981; Lance 1988). Each parcel of visible material accreted by a galaxy is probably accompanied by a much larger quantity of dark matter, and all the accreted material is expected to have high specific angular momentum about the galactic centre.

All these arguments add to a fairly convincing case for the continuing accretion by galaxies such as the Milky Way of dynamically significant quantities of mass and angular momentum. What evidence do we have regarding the speed with which the ***direction*** of the currently accreting angular momentum changes?

(i) Many attempts have been made to find correlations between the major-axis directions of galaxies (Kindl 1987 and references therein). The only alignments that have been convincingly demonstrated are those between brightest cluster galaxies and their surrounding clusters rather than straight alignments or anti-alignments of the spin axes of neighbouring galaxies. The absence of such alignments clearly indicates that the length-scale over which the angular momentum vectors of parcels of cosmic material become decorrelated is smaller than the typical separation of giant galaxies. It may well be very much smaller.

(ii) Polar rings (Sparke, this meeting) have been convincingly interpreted as the debris of cannibalized dwarfs that were accreted from near polar orbits and hence contributed angular momentum almost perpendicular to the cannibal's original angular momentum vector. The Magellanic stream around our own Galaxy is almost certainly a polar ring in the making. Thus this is no exotic phenomenon.

(iii) An impressive number of early-type galaxies have nuclei that with spins that are not parallel to those of the galaxy at large (Bender, this meeting).

Thus on the one hand the essence of a warp is misaligned angular momenta and on the other hand we expect the angular momentum of the currently accreting material to be constantly shifting. Let us therefore ask how a flat disk galaxy responds to accretion of skew **L**. It is helpful to distinguish steady accretion of a drizzle of material from episodic acquisition of discrete lumps of stuff. The first possibility may be further divided into the accretion of dark matter and of gas.

Steady off-axis accretion of dark matter

We expect dark matter and its associated angular momentum to comprise most of what galaxies are currently accreting. Material with skew **L** will distort the halo potential from simple axisymmetry and cause it to evolve gradually. Slow evolution of the potential experienced by the rapidly spinning inner galaxy will cause all inner orbits to evolve adiabatically. In particular, all closed orbits will stay closed, and the amplitude of every initially unexcited mode will remain zero. Thus a thin disk will remain thin and if the potential were to settle to a new axisymmetric equilibrium, a flat disk would remain flat.

Thus it as improbable that warp modes were excited by steady growth of the embedding halos as that they are primordial.

However, a self-gravitating disk in a slewing potential cannot be planar; reorientation of the inner disk requires transport through the disk of a component angular momentum perpendicular to the current disk axis. Gravity accomplishes this task by warping the disk into a configuration in which the (straight) line of nodes runs parallel to the direction of the new component of **L**. The curvature induced thus in the disk is largest where the disk's surface density, and therefore vertical cohesion, is smallest. Ostriker & Binney (1989) have suggested that tipped nuclear disk in the central kiloparsec of the Milky way that has been deduced from several studies of nuclear gas may arise in this way. Sadly, this hypothesis is called into question by the recently completed ^{12}CO Bell Labs survey of the galactic centre, which does not confirm the existence of a tipped nuclear disk (Bally & Stark, private communication). Nonetheless, the outer warp in the Milky Way's disk may well be caused by slewing of the sort studied by Ostriker & Binney. To check this conjecture we need to know more than we now do about the way in which infalling dark matter distorts the potential of the Galaxy at large radii.

Steady off-axis accretion of gas

Polar rings clearly demonstrate that from time to time galaxies acquire gas with skew **L**. Are warps mere transient phenomena that arise as this material is integrated into the outer disk? In this picture the warp must form a leading spiral and have a lifetime that is essentially the naive wind-up time discussed in §2. Briggs' observation that the line of nodes near R_{Holmberg} forms a leading spiral suggests that beyond R_{Holmberg} the disk may be in just such a transient condition. We then have a prediction of the rate at which such a galaxy is accreting gas; if the disk beyond R_{Holmberg} weighs M_H and the wind-up time is t_H, then the accretion rate is $\dot{M} \approx M_H/t_H$. For example, Begeman (1989) reports that for NGC 2841 $M_H \simeq 6.5 \times 10^9\,\text{M}_\odot$ so taking $t_H \simeq 20\pi R_{\text{Holmberg}}/v_c \simeq 3.3\,\text{Gyr}$ we have $\dot{M} \simeq 2\,\text{M}_\odot\,\text{yr}^{-1}$, which is not implausibly large.

Cannibalism

Consider the fate of a satellite galaxy that falls into the non-spherical halo of a large disk galaxy. The satellite sinks slowly through the squashed halo, precessing its **L** as it goes. When it reaches the edge of the disk (which we imagine to lie in the halo's equatorial plane), **L** has a certain orientation. If the satellite is pulled apart over an interval longer than $2\pi/\Omega_p$, the mean skew **L** of the pieces will be zero and the disk's modified tilt mode will not be excited when the pieces are subsequently digested by the disk – the physics here is very similar to that governing the White–Rees picture of the gradual formation of a disk galaxy at the centre of a preexisting dark halo. If on the other hand, the satellite is disrupted in less than $2\pi/\Omega_p$, it carries its instantaneous skew **L** intact into the disk. The disk now wiggles and wobbles and differentially precesses until the only relative motions between disk particles that have not damped out are circular motions. At the end of this phase the disk may well be in a normal mode configuration, i.e. one of those magic shapes in which torques between the rings of the disk nicely balance the surpluses and deficits of torque provided by Φ_h to ensure that every ring precesses at the same rate.

It is not obvious that a normal mode will be the endpoint of the disk's postprandrial relaxation, but it is likely. One of Hernquist's simulations of a satellite falling into a galaxy with a spherical halo beautifully illustrates the excitation of the (now-trivial) tilt mode. This strongly suggests that the halo could be at least slightly flattened without killing the excitation of the tilt mode.

Where does the energy end up that is goes into running bending waves when a disk eats a satellite? The gaseous portion of the disk can radiate its share of this energy, but the stellar disk cannot unless it can somehow transfer the energy to either gas or the heavy halo. Hence the stellar disk probably responds to the excitation of running bending waves by thickening. It is likely that this is the origin of at least some thick disks. Deposition by bending waves of abnormally large quantities of energy at the peripheral radii where they break may help to generate the rather sharp edges on optical disks that have been reported by van der Kruit & Searle (1982).

4 Conclusions

While our understanding of warps is still incomplete I feel the last few years have seen substantial progress towards the resolution of this old puzzle. The following are some of the points that have emerged most clearly:

(i) Warps arise because the net angular momentum vector of the relaxed portion of each galaxy is constantly changing direction. Word of the current fashion in **L** takes time to reach the fuddy-duddies in the centre, and the resulting misunderstandings manifest themselves as warps.

(ii) It is likely that the most important single misunderstanding is that between the halo, which is in touch with the very latest developments at the edge of the relaxed world, and the optical disk.

(iii) Realistic galactic disks in plausible flattened halos admit discrete normal modes, which, if excited, would allow the optical disk to retain a realistic warped configuration indefinitely.

The following conclusions strike me as very likely but not certain:

(i) Warps are probably not primordial since a disk that grows gradually in a flattened halo will lie in a fundamental plane of the halo.

(ii) If warps are actually excited discrete modes, they can be excited only by sudden events such as the swallowing within less than a halo precession period of a good-sized companion.

(iii) A galaxy that has never gulped down a good-sized companion can still be warped provided its halo is reorienting its angular-momentum vector sufficiently rapidly. A warp induced by the steady reorientation of the net angular momentum differs from an excited warp mode in that it will flatten out as soon as the reorientation ceases, while a normal mode will remain excited for ever. The warps of some galaxies might be superpositions of an excited normal mode and distortion by steady slewing of **L**.

Our understanding would be materially furthered by investigation of the following questions.

(i) All existing investigations of the normal modes of galaxies embedded in flattened halos treat the halo potential as a given, rigid object. In reality the central halo will be more tightly coupled to the optical disk than to the outlying halo, and must be treated as a dynamical entity on a par with the disk itself. Including the dynamics of the halo will modify the type I warps of Sparke & Casertano (1988) more than their type II warps.

(ii) Briggs' interesting observation that one can always find a reference plane in which the line of nodes of each galaxy consists of two straight lines suggests that disks which are too extended to support a discrete warp mode of the entire disk might naturally break into concentric annuli. Within each annulus self-gravity would be able to enforce a common precession rate, while different annuli would precess at different rates since the whole disk would be too extended to support a global warp mode.

(iii) Much more work needs to be done on the distortions that are induced in an optical disk when the embedding halo absorbs a substantial measure of skew **L**. It is clear that in these circumstances the disk must bend most noticeably in regions of low surface density such as the zone in the Milky Way between $R \simeq 2\,\mathrm{kpc}$ and $R \simeq 4\,\mathrm{kpc}$. But the response of the outer disk to the evolution of the halo cannot be predicted until we know how the halo responds to the inflow of skew **L**.

It is a pleasure to report that the old warp puzzle is still flourishing and shows promise of flowering into a powerful means of probing the long-term evolution of both galaxies and those mysterious dark halos.

My thanks are due to Eve Ostriker and Linda Sparke for commenting on a draft of this contribution.

References

Begeman, K., 1989. Astron. Astrophys., 00, 00.
Dekel, A. & Shlosman, I. 1983. In Internal Kinematics and Dynamics of Galaxies, IAU Symp. 100 ed. E. Athanassoula (Dordrecht: Reidel) p. 187.
Gunn, J. E. & Gott, J. R. 1972. Astrophys. J., 176, 1.
Hunter, C., 1969. Astrophys. J., 157, 183.
Hunter, C. & Toomre, A., 1969. Astrophys. J., 155, 747.
Kindl, E. 1987. Astron. J., 93, 1024.
Lance, C. M. 1988. Astrophys. J., 334, 927.
Larson, R. B., Tinsley, B. M. & Caldwell, C. N. 1980. Astrophys. J., 237, 692.
Ostriker, E. C. & Binney, J. J., 1989. Mon. Not. Roy. Astron. Soc., 237, 785.
Rogers, A., Harding, P. & Sadler, E. 1981. Astrophys. J., 244, 912.
Ryden, B. S. 1988. Astrophys. J., 329, 589.
Ryden, B. S. & Gunn, J. E. 1987. Astrophys. J., 318, 15.
Sandage, A. 1986. Astron. Astrophys., 161, 89.
Sellwood, J. A. & Carlberg, R. G. 1984. Astrophys. J., 282, 61.
Sparke, L. S. 1984. Astrophys. J., 280, 117.
Sparke, L. S. & Casertano, S. 1988. Mon. Not. Roy. Astron. Soc., 234, 873.
Tinsley, B. M. 1980. Fund. Cosmic Physics, 5, 287.
Toomre, A. 1983. In Internal Kinematics and Dynamics of Galaxies, IAU Symp. 100 ed. E. Athanassoula (Dordrecht: Reidel) p. 177.
Tubbs, A., D. & Sanders, R. H., 1979. Astrophys. J., 230, 736.
van der Kruit, P. C. & Searle, L. 1982. Astron. Astrophys., 110, 61.
White, S. D. M. & Rees, M. J. 1978. Mon. Not. Roy. Astron. Soc., 183, 341.

Polar Rings in Disk Galaxies

L.S. Sparke

Washburn Observatory, University of Wisconsin, Madison, WI53705, USA

A number of disk galaxies, usually S0s, show an outer ring of gas, dust and stars at a large angle to the major axis of the central galaxy. In some cases the inner galaxy is known to rotate rapidly about its short axis, consistent with the interpretation that it is a disk seen nearly edge-on; the ring, of course, rotates about its own, nearly orthogonal, minor axis. It has been estimated (Schweizer *et al.* 1983; *AJ* **88**, 909) that 'a few percent' of S0 galaxies show polar rings; they are neither common nor extremely rare. The rings are gas-rich (*e.g.* van Gorkom *et al.* 1987; *Ap J* **314**, 457), some containing several billion solar masses of neutral hydrogen; one, in NGC 3998, is seen *only* in HI (Knapp *et al.* 1985; *A Ap* **142**, 1), with no stellar ring visible. Polar rings around S0 galaxies seem to differ from gas disks in ellipticals in that the stars and ionized gas (as well as the HI) can extend far beyond the body of the galaxy.

Because of its orientation, it is fairly clear that the ring material has been acquired from outside the galaxy; the details of this accretion are more obscure. Subsequently, the ring evolves as a result of **differential precession, inflow** and **self-gravity**.

Differential precession occurs because the rings are not exactly polar and so do not lie in a symmetry plane of the inner galaxy. The galactic gravitational field has a quadrupole moment, which causes particle orbits to precess about the pole, at a rate which depends on the radius and orbital inclination. Simple estimates (*e.g.* Schechter *et al.* 1984 *Ap J* **277,** 531) suggest that differential precession will cause the ring to become noticeably twisted in much less than a Hubble time. Differential precession of an inclined gas disk can explain much of the observed ring structure in NGC 3718. This galaxy, classified SBa (peculiar), was mapped in neutral hydrogen by Schwarz (1985; *A Ap* **142,** 273); he found a very complex gas distribution, which could be modelled as a warped and twisted disk. The gas orbits are obviously quite inclined to the optical disk, which appears near to face-on, while the HI disk passes from about 75° inclination, through edge-on, to i=65° on the other side. At the same time, the major axis of the gas ring twists on the plane of the sky (see Figure 1 below).

To model this, I assumed that the potential of the galaxy was given by a combination of a thin Kuzmin disk (1956 *Astron Zh* **33,** 27; Toomre 1962 *Ap J* **138,** 385) and a spherical massive halo in which the rotation speed rises asymptotically to a constant value V_H outside a central core:

$$\rho_H(r) = \frac{\rho_o}{1 + r^2/r_H^2}.$$

The scale length of the Kuzmin disk was taken as $1'$, a quarter of the quoted optical diameter, and the parameters of the halo chosen such as to give a nearly flat rotation curve with the measured amplitude of $250 km/s$ and the largest reasonable disk mass ($M_D = 10^{11} M_\odot$ for $H_o = 75 km/s/Mpc$, so that $M/L_B = 8$; $V_H = 300 km/s$, and $r_H = 200''$). In this potential, the precession of an orbit with any given tilt to the pole could be found. The calculation was started with all the gas orbits aligned at 10° to the pole of the galaxy (representing the plane of the orbit from which the disk material was presumably captured), and precession was allowed to occur over several orbital times. The development of the gas ring is shown in Figure 1; at

Dynamics and Interactions of Galaxies Editor: R. Wielen

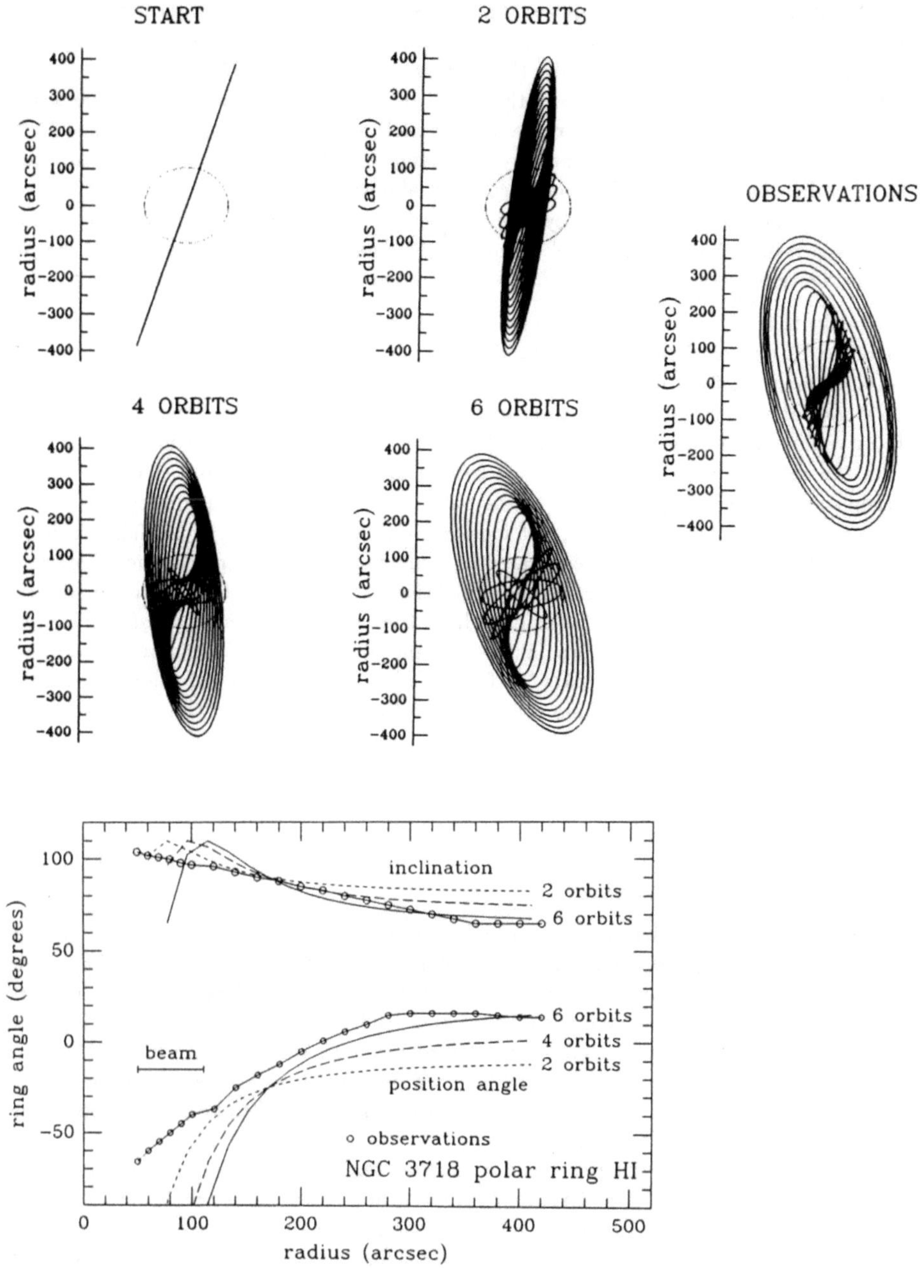

Figure 1: a precessing model for the polar ring in NGC 3718. Top left: the model, projected to look like the observed ring; the dotted ellipse shows the plane of the S0 disk. Time is measured in units of the orbital period at 250″ radius. Top right: the ring structure fitted by Schwarz (1985) to his HI observations. Bottom: inclination (upper curves) and position angle of the major axis (lower) of the ring as a function of radius. The observations (circles) are compared with the model at various times; the beam size of the HI observations is shown on the left of the figure.

each step, the model is projected to look as much as possible like the observed galaxy, with the ring orbit at 160″ seen edge-on; this requires that the S0 disk is about 30° to face-on.

The time has been measured in units of one orbit at 250″ radius, or about 6×10^8 years; after 6 orbits (3.5×10^9 years), the model (left) looks quite plausibly like the observed disk (right). The panel below shows how good the match is; starting from a nearly-planar disk, differential precession seems adequate to explain the structure of this ring, at least outside 150″ radius. Schwarz found that, in a coordinate system defined by the plane of the *innermost gas orbits*, the gas disk has a shape which is warped by almost 90° but is nearly untwisted (to within about 15°) about the pole of that system. According to the present model, this beautiful 'restricted' warp results from the gas disk being nearly polar as seen from the *disk of the S0 galaxy* – coincidentally we observe it at a time when the twisting about the pole is close to 90°.

The model fits badly in the central regions; it is hard to tell just how badly, because the beam size of the HI observations (indicated in the figure) is comparable to the range over which there is disagreement. **Inflow** may be at work here. If material is drifting slowly inwards, then gas which is now at say 100″ from the center has not been there throughout the lifetime of the ring. At any radius in the ring, a 'precession age' can be computed by comparing the observed twist about the galaxy pole with the predicted rate of differential precession. The ring orbits at 100″ radius are twisted by 62° relative to the outermost ring gas, which corresponds in the model to an age of about 1.2×10^9 years, or 5 local orbital periods. If the gas has fallen in from larger radii, the infall timescale can be estimated as being somewhat longer than the 'precession age', because the precession rate increases rapidly closer to the center. The density in the HI disk decreases sharply inward of 200″; if the gas within that radius is falling towards the center, then this argument suggests that it has taken a couple of billion years to move inwards to 100″, implying an inward drift at $2 - 3km/s$. If the model is changed by reducing the mass of the galaxy disk while increasing the halo mass to keep the rotation speed the same, all these timescales become longer: for a disk with $M/L_B = 4$, the implied inflow speed is only about $1km/s$. This inflow rate is still ten times that expected if the dominant source of viscosity is cloud-cloud collisions (Steiman-Cameron and Durisen 1988; *Ap J* **325,** 26), but it is probably too small to be excluded on the basis of Schwarz's HI observations. Observations with higher resolution (G. van Moorsel, unpublished data) are required to map the inner parts of the ring and to test whether inflow is occurring.

But not all polar rings suffer such strong differential precession: the ring in UGC 7576 is seen nearly edge-on and appears extremely flat, so that it cannot be twisted by more than a few tenths of a radian. The ring is about 12° from polar (Whitmore 1984 *AJ* **89,** 618), but appears smooth and relaxed, with little ionized gas (Mould *et al.* 1982; *Ap J* **260,** L37), suggesting that it is fairly old. Neutral hydrogen measurements (Schechter *et al.* 1984 *MNRAS* **208,** 111) show that the ring contains $5 \times 10^9 M_\odot$ of gas, probably in a wide ring – the outer edge is around twice the radius of the inner edge – so that differential precession is expected to be severe. Something is obviously acting to stabilize this ring.

The **self-gravity** of a polar ring can hold it together and prevent twisting, if enough mass is present in the ring (Sparke 1986; *MNRAS* **219,** 657); in this case, the whole ring precesses uniformly about the pole of the S0 galaxy without change of shape. A stable self-gravitating model for the ring in UGC 7576 is shown in the second figure. Following Schechter *et al.*, the ring is taken to have constant density in the range $26'' < r < 49''$. The galaxy is modelled in the same way as for NGC 3718, with a Kuzmin disk of scale length 3.8″ (as indicated from I-band photometry; Whitmore, private communication) and mass $6 \times 10^{10} M_\odot$; the halo parameters were $V_H = 265km/s$, $r_H = 10''$. Again, these were chosen to make the rotation curve nearly flat at the largest measured value of $250km/s$ out to the edge of the ring, while putting as much mass in the disk as is reasonably possible ($M/L_B = 7.5$). The polar ring was given a mass equal to that measured in gas, $7 \times 10^9 M_\odot$(corrected to $H_o = 75km/s/Mpc$, and multiplied by a factor of 1.4 to allow for the helium content). The model is displayed first with the galaxy disk

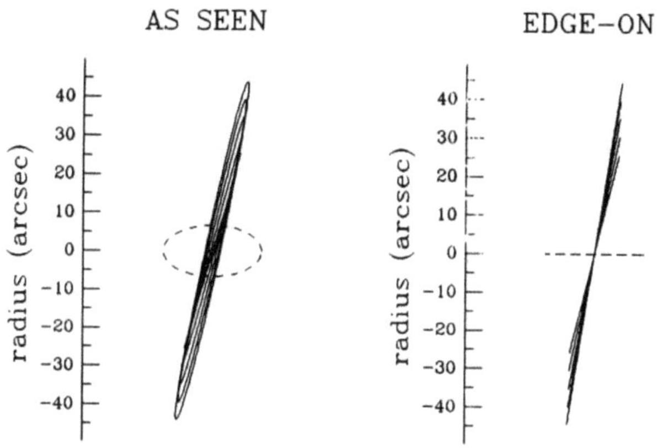

Figure 2: a self-gravitating equilibrium model for the polar ring in UGC 7576; the dotted line shows the plane of the S0 disk. The model is displayed first in the observed orientation, then with the disk edge-on to show the structure better.

tipped by 30° away from edge-on as indicated by photometry (Whitmore 1984), and then with the galaxy edge-on to show the structure more clearly. The model exhibits no twisting about the galaxy pole, and a poleward warp is evident. This warping would be less if the ring mass were larger; if the ring mass is too small, no stable solutions are found. For a given ring mass, the curvature is most critically dependent on the radial distribution of ring matter. Again, HI observations at high resolution would provide the best test of this model.

Even if self-gravity does not lock a polar ring into a steady state, it can still cause the characteristic warp up towards the pole. The ring in NGC 4650A is twisted, which shows that it is evolving in time, but it also bends towards the pole at the outer edge. Sparke *et al.* (1988; in IAU Colloquium 96, 'The Few-Body Problem', p409) show that a time-dependent model similar to that described for NGC 3718, but including the self-gravity of the ring gas, can reproduce the main features of the observations. The ring is not massive enough to hold itself together against differential precession, but self-gravity transfers angular momentum within the ring as it twists, so that the outer gas is thrown into a more nearly polar orbit. The model indicates that the ring is 3-5 orbits old, so that the material forming it may have been accreted a few billion years ago.

Eventually, an unstable ring like this will break up into a number of sub-rings which precess independently; the outer ring material is pushed into an orbit closer to the pole than the original ring plane, while inner gas sinks towards the equator of the S0 galaxy (Sparke 1986). Thus a single accreted gas cloud might give rise to a multiple ring system, such as that seen in ESO 747-G26 (Schweizer *et al.* 1983). However, dissipative processes act more rapidly at smaller radii and lower inclinations to remove ring gas; if the inner material disappears completely, the remaining ring will be thinner and in a more nearly polar orbit than that on which the gas was originally accreted. This might resolve a puzzle pointed out by Schweizer *et al.* : why so many polar rings are so nearly polar, when the initial orbits of the captured material must have been much more uniform in angle.

The dynamical modelling described above shows that much of the structure of polar rings can be described in terms of a self-gravitating ring precessing about an oblate galaxy. The effects of differential precession and self-gravity can be calculated straightforwardly; inflow and settling depend on the (unknown) effective viscosity in the gas, and are more difficult. But it seems that these mysterious rings of light suspended high above the galactic disk are shaped by the 'everyday' forces which we know must be acting in the system.

Triaxial Bulges in Spiral Galaxies

O.E. Gerhard[1] *and M. Vietri*[2]

[1]Astronomisches Rechen-Institut, Mönchhofstr. 12–14, D-6900 Heidelberg, Fed. Rep. of Germany
[2]Osservatorio Astrofisico di Arcetri, Firenze, Italy

1. It is possible to determine the axial ratios of a triaxial stellar system from two-dimensional gas-kinematic data, if the gas velocity field is sufficiently regular.
2. Several apparently 'normal' spiral galaxies show evidence for triaxial inner bulges (M31, The Galaxy, NGC 4845, NGC 2784). In NGC 4845 there can be little doubt.
3. If triaxial bulges are common, this raises interesting questions concerning their dynamics, history, and interaction with the disks of their host galaxies.

1. M31

Observational evidence: The 'Lindblad (1956) twist'. The apparent major axes of disk and bulge on the sky are not aligned. $P.A._{bulge} = 48°$ from CCD photometry of Kent(1983), $P.A._{disk} = 38°$, e.g., Light *et al.* 1974. The twist angle $\psi = 10°$.

The most natural explanation for the Lindblad twist is by projection of a triaxial bulge, cf. Stark (1977). Assuming that the minor axes of disk and bulge coincide (cf. edge-on systems) one can deproject the triaxial ellipsoid with one free parameter. For M31 most solutions are prolate-triaxial (Gerhard 1986). A similar result is found for NGC 2784 (Bertola *et al.* 1988).

2. The Galaxy

Observational evidence: 'Forbidden' velocities in the l-v- diagram. The envelope ('rotation curve') is shown in Clemens (1985). The main point is the disagreement of two scales: (i) The rotation curve peaks at $\simeq$ 500pc. (ii) from K-band photometry (Matsumoto *et al.* 1982) the bulge luminosity profile turns over at $\simeq$ 800pc. Any axisymmetric bulge model with this profile has a rotation curve that peaks slightly *outside* 800 pc. If the K-band light indeed measures the old stars, M/L of the inner $\simeq$ 1kpc is roughly constant, and if the gas velocities near the peak are correct, a consistent axisymmetric model cannot be constructed.

With a prolate-triaxial bulge one can reconcile the above two scales. This assumes that the gas streamlines are well-approximated by closed orbits in the bulge equatorial plane. The bulge must be oriented such that we measure higher velocities closer to the centre as compared to the axisymmetric case, implying that we see the bulge nearly along its intermediate axis. The data can be well-fitted for near-prolate triaxial bulge models with axial ratios around 10:6:6 (cf. Fig. 3 of Gerhard & Vietri 1986). For the M/L of the bulge we find $\simeq 0.8$ in K, or $\simeq 7$ in B.

3. NGC 4845, the best case

Observational evidence: The emission line gas velocity field (Bertola, Rubin & Zeilinger 1989; BRZ) is very regular and symmetrical about the centre; however, the gas motions are far from circular. This can already be seen from comparing the two nearly equal observed slit rotation curves along P.A.=78° and P.A.=98° without reference to any model: The measured velocities along P.A.=98° are up to $\gtrsim 50\,\mathrm{km\,s^{-1}}$ higher than consistent with an axisymmetric velocity field (Gerhard, Vietri & Kent 1989; GVK)

We have fitted the observed rotation curves with a model in which the gas moves in the equatorial plane of a triaxial bulge. The bulge is modelled as a non-rotating triaxial ellipsoid with $r^{\frac{1}{4}}$-law density profile (r_e = 45", Kent 1988). At 1 kpc figure rotation is unlikely to be

Dynamics and Interactions of Galaxies Editor: R. Wielen

important. The velocity field was constructed from numerically integrated closed orbits. The best-fitting model (see Fig. 1 in GVK) gives an excellent fit to the data. Its parameters are b/a = 0.74, c/a = 0.60, M/L = 7. The formal statistical error box for the axial ratios is b/a = 0.74 ± 0.06, c/a = 0.60 ± 0.06 at a level of $> 5\sigma$. The high significance reflects the regularity of the velocity field and the stringent constraints on the model set by 60 data points. It does not, however, include systematic errors.

A *photometric* analysis similar to that above for M31 shows that the elongated velocity field is not caused by a bar or spiral arms in the disk instead of by the bulge, and that other systematic errors do not strongly modify the above axial ratios. Photometry and kinematics agree to well within 1σ fof b/a and within 2σ for c/a, with a possible small systematic effect overestimating c/a by $\simeq 0.1$. This shows that it is the triaxiality of the *bulge* of NGC 4845 that we see in the observed velocity field.

4. Implications and further work

Genesis: Clearly, whether or not bulges of spiral galaxies are triaxial does not directly connect with their formation process. However, (i) The bulge of NGC 4845 is noticably box/X-shaped (Kent, priv. comm.). This is may hint towards past interactions. (ii) If (some) bulges are triaxial similar to large ellipticals, this might also reflect a continuity in the way the formation / evolution processes operated.

Statistics: It would be worthwhile to investigate whether few / a class of / all nuclear bulges in spiral galaxies are triaxial. All it needs is CCD photometry of the centers of face-on galaxies. If necessary, one may use HI-velocity widths to confirm that the systems are indeed face-on. It would also be interesting to see whether correlations of triaxiality with nuclear activity exist, as one might expect, or with some disk properties.

Dynamics: Bulges are traditionally taken to be oblate isotropic rotators. For NGC 4845 this clearly cannot be true; thus the stellar kinematics may not be a sensitive indicator. It is almost unnecessary to say that dynamical models for (slowly rotating) triaxial systems would help. A second aspect about which little is known is the possible interaction of the bulge quadrupole moment with the disk of the spiral galaxy.

Elliptical galaxies: The method of analysis presented here may also work for elliptical galaxies where one has no photometric constraints, provided a) the gas velocity field is sufficiently regular, and b) one does not by accident view the system in a direction too nearly along axes or in principal planes.

References

Bertola, F., Vietri, M. & Zeilinger, W., 1988. *ESO Messenger* 52,24.

Bertola, F., Rubin, V.C. & Zeilinger, W., 1989. *Astrophys. J. (Letters)* , *submitted.*

Clemens, D.P., 1985. *Astrophys. J.* 295, 422.

Gerhard, O.E., 1986. *Mon. Not. R. astr. Soc.* 216, 373.

Gerhard, O.E. & Vietri, M., 1986. *Mon. Not. R. astr. Soc.* 223, 377.

Gerhard, O.E., Vietri, M., Kent, S.M., 1989. *Astrophys. J. (Letters)* , submitted.

Kent, S.M., 1983. *Astrophys. J.* 266, 562.

Kent, S.M., 1988. *Astron. J.* 96, 514.

Light, E.S., Danielson, R.E. & Schwarzschild, M., 1974. *Astrophys. J.* 194, 257.

Lindblad, B., 1956. *Stockholms Obs.Ann.* , 19, no.2

Matsumoto, T., *et al.* , 1982. In: The Galactic Center, AIP conference no. 83, ed. G.R. Riedler & R.D. Blandford.

Stark,A.A., 1977. *Astrophys. J.* 213, 368.

The Influence of Discs on the Dynamics of "Box/Peanut" Bulges

M. Shaw[1], *R.-J. Dettmar*[2], *and A. Barteldrees*[3]

[1]Department of Astronomy, University of Manchester, Manchester, M13 9PL, UK
[2]Lowell Observatory, 1400 W. Mars Hill Road, Flagstaff, AZ 86001, USA
and Radioastronomisches Institut der Universität Bonn, Auf dem Hügel 71,
D-5300 Bonn 1, Fed. Rep. of Germany
[3]Max-Planck-Institut für Radioastronomie, Auf dem Hügel 69,
D-5300 Bonn 1, Fed. Rep. of Germany

Owing to the inability of standard fitting functions to fully account for the observed 2D luminosity distributions within the non–thin disc regions of galaxies displaying "box" or "peanut" bulge morphology (Shaw *et al.*, in preparation), we have sought to identify which of a range of readily measurable photometric parameters best describe these isophotal shapes. Principal component analyses (PCA's) have been undertaken with a view to identifying the minimal range of observational parameters required to fully describe this morphology, and thereby to define the optimal constraints to be placed on theoretical simulations of such systems.

Samples used

- Those 23 large (major axes $\gtrsim$ 5 arcmins), normal, edge–on (i $\gtrsim$ 80°) spirals and lenticulars in Shaw 1987 – the S87 sample;

- An additional 28 objects comprising 7 galaxies in literature sources; 6 using measurements taken from the SERC/J or POSS/B sky surveys; and 15 from our own, as yet unpublished, CCD data – collectively termed the "literature" sample.

Procedure

- We have undertaken principal component analyses for 24 combinations of the seven readily measurable photometric parameters ("variables") – namely several measures of bulge and disc (D_{25}) dimension, the opening angle of the box/peanut to the major axis (Θ), the type, bulge:total (B/T) luminosity ratio and the bulge concentration index. Too few reliable kinematic measures exist to be included in the analyses. For each parameter combination, both pairwise (missing variables allowed) and casewise (missing variables not allowed) analyses have been performed. The excellent agreement found for all combinations indicates that the limitations of small number statistics are minimal.

- All variables have been scaled to zero mean and unit variance, thus minimising the influence of those variables whose values are numerically larger than the rest. Also all scale parameters have been converted to linear measures, and only absolute (kpc) dimensions are included in the analyses.

Dynamics and Interactions of Galaxies Editor: R. Wielen

- Three independant criteria of "eigenvalue significance" have been adopted: either only eigenvalues > 1.0 or ranked eigenvalues whose cumulative index accounts for more than 75 % of the total observed variance are considered significant, or we have applied the Sphericity Test (Kendall *et al.* 1983). The equivalent results obtained using each criterion confirm the high statistical validity of the derived conclusions.

Results

1. The derived parameter space is strictly two dimensional regardless of the eigenvalue significance criterion imposed – the two principal eigenvalues accounting for between 76 and 82 % of the total observed variance. Thus, within those tested here, **no** single parameter exists with which one could undertake to fully characterise box/peanut bulge morphology.

2. Optimal representations of each principal dimension are provided by a measure of the scale–size of the bulge component and B/T. These conclusions are found in all samples, and the derived correlation coefficients between each of those parameters which best represent each principal eigenvector are all significant at > 98 %.

3. Contrary to the results of Whitmore (1984), D_{25} does **not** provide an adequate representation of either principal dimension. An independant PCA of the 78 non–box/peanuts contained in Shaw (1987) shows this to result from the intrinsic nature of the bulge components contained therein rather than our exclusive use of edge–on galaxies. Thus, discs appear an unimportant element in any optimal characterisation of box/peanut bulges. This concurs with Shaw (1987), who found evidence of such morphology in a wide range of Hubble types, and provides further confirmation that the presence of a disc is **not** an important factor in the dynamics of box/peanut bulges.

4. There is no statistically significant correlation between D_{25} and Θ. If a merger mechanism is invoked to account for this morphology, and if Θ reflects the orbits preferentially populated by such infalling satellites, the absence of any correlation with the size of the disc suggests that the orbits inferred from the present photometric properties of these isophotal distortions are **not** those followed by such neighbours, possibly owing to a global change in the properties of the bulge density distribution since formation.

5. Existing (observed) correlations between disc and bar parameters are currently too ill-defined to allow useful constraints to be placed on the importance, or indeed even presence, of a bar.

A complete presentation of this work can be found in Shaw *et al.* (1989, in preparation).

References

Kendall, M., Stuart, A. & Ord, J. K., 1983. "*Advanced Theory of Statistics*", vol. 3, pp 328 – 331, Griffin.
Shaw, M. A., 1987. *Mon. Not. R. astr. Soc.*, **229**, 691.
Whitmore, B. C., 1984. *Astrophys. J.*, **278**, 61.

Ring Formation in Viscous Differentially Rotating Galactic Disks

H. Lesch[1], *P.L. Biermann*[2], *H.P. Reuter*[2], *and R. Wielebinski*[2]

[1]Landessternwarte Königstuhl, D-6900 Heidelberg, Fed. Rep. of Germany
[2]Max-Planck-Institut für Radioastronomie, Auf dem Hügel 69, D-5300 Bonn, Fed. Rep. of Germany

Ringlike structures (in H_α, CO and radio continuum) in central galactic regions have been observed in a number of objects. They seem to be a basic feature of galactic structure (for references see Dickman et al., 1988). These observations can be summarized as follows:

The typical mass of these rings is $10^8 - 10^9\, M_\odot$.

The gas distribution is smoothly decreasing from the ring into the innermost galactic regions.

The rings appear always at the turnover radius where the galactic rotation turns from rigid to differential rotation.

It is a well-known fact that angular momentum is transferred outwards in viscous, differentially rotating disks, which leads to an inward massflow (Lüst, 1952; Icke, 1979).

It is plausible that mass accumulation occurs somewhere near that radius where the rotation curve begins to tend appreciably towards constant Ω . Here Ω denotes the angular velocity of the gas (Ω is constant in the rigid rotating regions and decreases with increasing radius in the differentially rotating disk). This behaviour follows from the diffusion equation for the surface density in the disk σ

$$\frac{\partial\sigma}{\partial t} = -\frac{\alpha_T < v_T^2 >}{R}\frac{\partial}{\partial R}\left(\frac{\partial}{\partial R}\left[\frac{\sigma R^3}{\Omega}\frac{d\Omega}{dR}\right] / \frac{dj}{dR}\right). \tag{1}$$

According to Eq. (1) mass accretion in a viscous differentially rotating disk occurs down to that radius where the differential rotation turns into solid body rotation.

From Eq.(1) one can define an accretion timescale (Duschl, 1988):

$$t_{ACCR}(R) = \frac{R^2\Omega}{\alpha_T < v_T^2 >}\frac{\mid \frac{d}{dR}(R^2\Omega)\mid}{\mid R^2\frac{d}{dR}(\Omega)\mid}. \tag{2}$$

In order to build up a gaseous ring the accretion time must increase strongly at the turnover radius. We consider whether observed rotation laws of spiral galaxies lead to such large accretion timescales.

Observations (Rubin et al., 1982;) indicate that most disk galaxies have rotation curves which approach solid body rotation inside and constant velocity outside. The law

$$\Omega(R) = \Omega_0/[1 + (R/R_{BULGE})^n]^{1/n} \quad (3),$$

presents a general family of such a behaviour, where n describes the sharpness of the turnover.
In order to produce a sharp transition at the inner boundary of the ring, we clearly require a dramatic increase in t_{ACCR} this condition requires $n > 2$ (Lesch et al., 1989)

The formation of a ring then puts a constraint on the gravitational potential which fixes the rotation law. *In the following we in fact assume that* $n > 2$. The disk matter accumulates approximately at the radius where $\frac{d\Omega}{dR} \to 0$ in a typical timescale

Dynamics and Interactions of Galaxies Editor: R. Wielen

$$t_{ACCR}(R) = \frac{\Omega_C\, R^2_{BULGE}}{\alpha_T < v_T^2 >} = \frac{v_{ROT}\, R_{BULGE}}{\alpha_T < v_T^2 >} =$$

$$3 \cdot 10^8\, yr\, \alpha_T^{-1} \left[\frac{v_{ROT}}{10 v_T}\right] \left[\frac{R_{BULGE}}{300pc}\right] \left[\frac{v_T}{10\, km\, s^{-1}}\right]^{-1}. \qquad (4)$$

Eq. (4) can be rewritten as a mass flow with a "typical" gas surface density σ:

$$\frac{dM}{dt} = \frac{2\pi R^2 \sigma}{t_{ACCR}} = 2\pi R^2_{BULGE}\sigma \frac{\alpha_T\ < v_T^2 >}{v_{ROT}\, R_{BULGE}}$$

$$= \frac{0.03\, M_\odot}{yr} \left[\frac{R_{BULGE}}{300\, pc}\right] \left[\frac{\sigma}{30\, M_\odot\, pc^{-2}}\right] \left[\frac{v_T}{10\, km\, s^{-1}}\right] \left[\frac{v_{ROT}}{200\, km\, s^{-1}}\right]^{-1}. \qquad (5)$$

After 10 galactic rotation periods ($\sim 3 10^9\, years$), this flux would build up a ring of about $10^8\, M_\odot$.

Summarizing the model of ring formation by viscous accretion explains the mass of the rings and their position in a galaxy in a very simple and plausible manner.

It follows that, unless rings selfdestruct easily by starbursts, all early Hubble type galaxies should show a maximum of the surface density of cool gas in the form of a ring at the location of the turnover of the rotation law.

It has been shown by Combes and Gerin (1985), that the excitation of a density wave in a galaxy leads also to the formation of ringlike structures in galaxies. This model is discussed in detail by Combes (1989).

References

Combes, F.: these proceedings (1989)

Combes, F., Gerin, A.: *Astron. Astrophys.* **150**, 327 (1985)

Dickman, R., Snell, R. and Young, I. (eds.): **Molecular Clouds in the Milky Way and External Galaxies** (Springer, Heidelberg 1988)

Duschl, W.: *Astron. Astrophys.* **194**, 33 (1988)

Icke, V.: *Astron. Astrophys.* **78**, 21 (1979)

Lesch, H., Biermann, P.L., Crusius, A., Reuter, H.P., Dahlem, M., Barteldrees, A.., Wielebinski, R.: *Mon. Not. Roy. Astr. Soc.* (in press) (1989)

Lüst, R.: *Zeitschr. f. Naturf.* **71**, 87 (1952)

Rubin, V.C., Ford, W.K., Thonnard, N.: *Astrophys. J.* **261**, 439 (1982)

Cut-Off Radii of Galatic Disks

A. Barteldrees[1] *and R.-J. Dettmar*[2]

[1]Max-Planck-Institut für Radioastronomie, Auf dem Hügel 69, D-5300 Bonn, Fed. Rep. of Germany
[2]Radioastronomisches Institut der Universität Bonn, Auf dem Hügel 71, D-5300 Bonn, Fed. Rep. of Germany

1. Introduction

The radial dependence of the surface brightness and therefore the luminous mass density is proposed to follow the exponential Freeman-law but shows in well studied cases a steeper decline at galactocentric distances of typically 4 exponential scale-lengthes h. This distance will be called the cut-off radius R_{max}. The exponential light distribution might be an effect of the angular momentum distribution of the protogalaxy (Freeman 1970; Gunn 1987; van der Kruit 1987) and therefore the cut-off gives additional information on the formation process for the thin stellar disk. Especially a constant ratio R_{max}/h and the independence of this ratio on the morphological type is considered to be an indication of the similar initial conditions for all disk galaxies and for a collapse with detailed conservation of angular momentum from an uniformly rotating sphere (van der Kruit 1987, 1988).

Here we describe first results of CCD surface photometry of several edge-on galaxies. These observations confirm the existence of cut-off radii for galactic disks. Moreover we observed several galaxies with a cut-off radius $R_{max} < 2$ exponential scale-lengthes.

2. Observation and Reduction

Our CCD surface photometry of edge-on galaxies results from several observing runs at the Calar Alto 1.2 m telescope and the ESO/MPI 2.2 m telescope on La Silla in 1987. We have obtained images in at least one colour of g, r or i in the Thuan-Gunn system. The radial extent of the observed galaxies is about 2-3 arcmin due to the field sizes obtainable with a CCD-detector at these telescopes.

After correcting the images (flat-field) the frames have been normalized in units of the night sky brightness. 20 galaxies observed in the r-band have been choosen to obtain their cut-off radii and scale-lengthes. We have made sure that for the choosen galaxies the determination of the disk scale-length is not influenced by the light of the bulge component. The selected galaxies cover the morphological types with $-2<T<9$ (or S0 to Sd) according to the classification by Lauberts (1982). We have determined R_{max} and h for each quadrant defined by major and minor axis and calculated the mean values of both parmeters.

3. Results

Our CCD photometry confirms the presence of cut-off radii of galactic disks for a larger sample of galaxies. For most of the galaxies studied here the edge-on seen disks start to drop off much steeper than the inner exponential at levels of 1/10 of the night sky. The influence of the assumed night sky level on the occurance of cut-off radii was checked by analysing the minor axis profiles of the studied galaxies. They do not show any kind of cut-off's down to a level of 1% of the night sky.

We have measured the distance independent ratio R_{max}/h because we know only a few distances for the galaxies of our sample. To proof that neither R_{max} nor h are a function of

Dynamics and Interactions of Galaxies Editor: R. Wielen

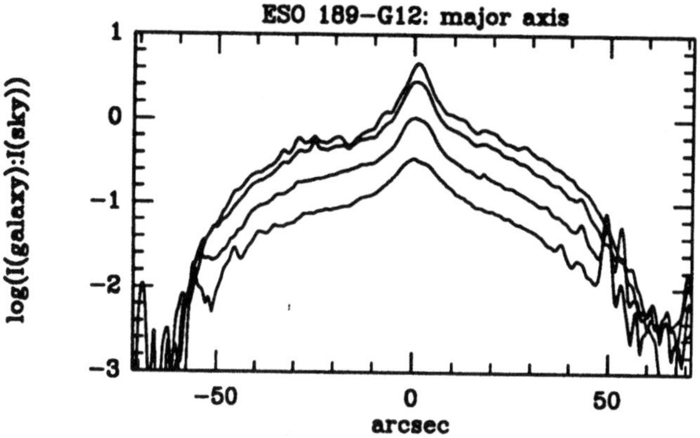

the morphological typ we compute the ratios R_{max}/R and h/R. R is the radius of each galaxy at a brightness of $\log(I_{gal}/I_{sky})$=-0.5. We could not detect any dependence of these ratios on morphological typ for our sample of 20 edge-on galaxies.

The CCD photometry confirms the disk parameters for NGC 5170 (R_{max}=25.6 kpc and h=8.6 kpc) obtained by Bottema et al. (1987) from photographic work. As an example the figure shows several cuts parallel to the major axis of the southern edge-on galaxy ESO189-G12 seperated by 1.36 arcsec starting at the major axis. The profiles show an approximatly exponential behaviour of the disk out to a radius of r=30". At 1/10 of night sky brightness a steeper decline of the cuts can be recognized.

The ratio R_{max}/h of this galaxy has a value of 1.8. The mean ratio of R_{max}/h of the sample is 2.8±1. This is a significantly lower value than R_{max}/h=4.5±1 in the van der Kruit and Searle (1982) sample. The galaxies of their sample show cut-off radii between 3.5 and 6.0 exponential scale-lenghtes. None of the galaxies studied by them has a cut-off at $R_{max}/h < 2$.

This difference might be an effect of the sample selection. Limited by the field sizes our main selection criterion was the morphological typ. Van der Kruit an Searle have studied galaxies with much larger angular extent, mainly members of the Virgo Cluster. The discrepancy between the mean values of both samples therefore might reflect a dependence of the cut-off radius or the scale-length on the environment of a galaxy, which has to be studied in further investigations. The existence of galaxies with cut-off radii less than 2 scale-lenghtes gives new constraints that models of disk formation and stability has to explain.

Acknowledgement: We wish to thank the director of the Max-Planck-Institut für Astronomie at Heidelberg for the generous allocation of observing time.

1 R. Bottema, P.C. van der Kruit, K.C. Freeman: Astron. Astrophys. 178, 77 (1987)
2 A. Lauberts: The ESO-Uppsala Survey of the ESO (B) Atlas, ESO (1982)
3 J.E. Gunn: Astrophysical Cosmology, Pont. Acad. Sci. Vatican (1982) pp. 233
4 K.C. Freeman: Astrophys. J. 160, 811 (1970)
5 J. Kormendy: Morphology and Dynamics of Galaxies, 12th Advanced Course SSAA (1982) pp. 113
6 P.C. van der Kruit, L. Searle: Astron. Astrophys. 95, 105 (1981)
7 P.C. van der Kruit, L. Searle: Astron. Astrophys. 95, 116 (1981)
8 P.C. van der Kruit, L. Searle: Astron. Astrophys. 110, 61 (1982)
9 P.C. van der Kruit: Astron. Astrophys. 173, 59 (1987)
10 P.C. van der Kruit: Astron. Astrophys. 192, 117 (1988)

Nonlinear Density Waves in Self-Gravitating Gas Sheets

B. Fuchs

Astronomisches Rechen-Institut, Mönchhofstr. 12–14,
D-6900 Heidelberg, Fed. Rep. of Germany

The dynamical evolution of spiral density waves of finite amplitude in a thin, self-gravitating, rotating gas sheet is investigated. The analysis proceeds through three stages. First the basic model is described. Next a linear stability analysis of the sheet, subjected to spiral perturbations, is carried out. The evolution of the wave amplitudes is then followed from initial exponential growth into the regime of finite amplitudes. The result is a non-deterministic *oscillation* of the perturbed surface density. This may be related to recurrent spiral instabilities of stellar disks as studied by Toomre (1989 and these proceedings) and also observed in numerical simulations of the global dynamical evolution of stellar disks (Sellwood and Athanassoula, 1986).

1 Basic Model and Basic Equations

The basic model is chosen so that it describes the local dynamics of a patch of a differentially rotating thin gas disk according to Goldreich and and Lynden-Bell (1965). The set of basic equations (see e. g. Goldreich and Tremaine, 1978) is given by the equation of continuity, Euler's equations of motion-including centrifugal and Coriolis forces-, and Poisson's equation. The disk is assumed to be self-gravitating, but is imbedded in a rigid halo potential. The unperturbed surface density and pressure of the disk are taken to be homogeneous and isotropic, respectively, and the unperturbed velocity field is a parallel shear flow. The parameters are so adjusted, that the disk is stable against axisymmetric disturbances.

2 Linear Stability Analysis

The basic equations are linearized with respect to the amplitudes of surface density and velocity perturbations of the background model. As is well known, the linearized equations may be decomposed into normal modes with respect to time as well as to the spatial coordinate in the direction of rotation (y). The velocity components may be eliminated from the equations, leading to a second order differential equation for the surface density, $\sigma_p \propto \exp(i\omega t + ik_y y)$,

$$\partial_{xx}\sigma_p + \mathcal{A}(x)\partial_x\sigma_p + \mathcal{B}(x)\sigma_p = 0, \tag{1}$$

where x indicates the (radial) spatial coordinate perpendicular to the direction of rotation. As will be shown elsewhere in detail, eqn.(1) may be approximately solved in the region near corotation at $x = 0$ by

$$\sigma_p = \sigma_{p0}\exp(ik_x x + \delta x^2). \tag{2}$$

For obvious physical reasons only solutions with $\Re(\delta) < 0$ are considered. The resulting dispersion relation ω(k) has two principal branches. One branch shows exponential growth for a certain region in k-space, representing rigidly rotating, open, trailing waves. The second branch gives damped solutions for the corresponding wave numbers.

Dynamics and Interactions of Galaxies Editor: R. Wielen

3 Nonlinear Amplitude Regime

In the nonlinear amplitude regime nonlinear terms of the basic equations have to be taken into account. In the following Fourier-components are 'lumped' together to single modes,

$$\sigma_p, \mathrm{v}_p \approx \xi(t) \exp(ik_{m,x}x + ik_{m,y}y + \delta x^2) + cc. \tag{3}$$

The time evolution of the wave amplitude in the nonlinear regime is then found to be governed by the equation

$$\partial_{tt}\xi = \lambda^2\xi - (1+\lambda^2)|\xi|^2\xi, \tag{4}$$

where the first term on the rhs describes exponential growth, λ being the linear growth rate, whereas the second term indicates nonlinear self- suppresion of modes. $\xi(t)$ is obtained from eqn.(4) as one of the Jacobian elliptic functions with *oscillating* character. The surface density perturbation oscillates about a mean value (scaled by the backgroud density) of $\lambda/\sqrt{1+\lambda^2}$. The final amplitude as well as the oscillation period, however, depend on the initial fluctuations, so that the actual evolution of the wave amplitude is not deterministic.

4 References

Goldreich, P. and Lynden-Bell, D. 1965, *Mon. Not. R. astr. Soc.*, **130**, 125.
Goldreich, P. and Tremaine, S. 1978, *Astrophys. J.*, **222**, 850.
Sellwood, J.A. and Athanassoula, E. 1986, *Mon. Not. R. astr. Soc.*, **221**, 195.
Toomre, A. 1989, in *Dynamics of Astrophysical Disks*, J. Sellwood (ed.), (in press).

Deterministic Self-Propagating Star Formation and Gas Consumption Rates in Galaxies

J. Palouš

Astronomical Institute of the Czechoslovak Academy of Sciences,
Budečská 6, 12023 Prague 2, Czechoslovakia

We discuss the gas consumption rates in galaxies with a simple model of galactic evolution, which has been introduced elsewhere / Palouš, 1989 /. In this model, the differentially rotating galactic disk is represented by three interconnected components: diffuse gas, clouds and stars. The basic mechanism is a self-regulating cycle between star formation, supershells and molecular clouds / Tenorio-Tagle and Bodenheimer, 1988 /: star-forming clouds are disrupted as a consequence of the energy injection from massive stars formed in their interiors. The expanding supershells are also created. They accumulate the ambient medium and form after certain time and at certain place in the disk the new molecular clouds. The star formation is in this model a propagating process deterministic in nature: the direction and speed of propagation are detemined by the rotation curve.

The evolution of supershells expanding in differentially rotating disks is described by Tenorio-Tagle and Palouš /1987/ and by Palouš et al. /1989/. The rate of superstructure evolution, which is the most time consuming part of the cycle, is proportional to $t_{ref} = \pi/\alpha\,\kappa\,10^9$ years, where $\alpha = 0.75\left[\sigma_g / M_o pc^{-2}\right]^{0.1}$, σ_g is the gas surface density in the galactic disk, and $\kappa^2 = 4\,\omega^2\left[1 + \frac{R}{2}\,\frac{d\omega}{dR}\right]$, ω is the angular rotation velocity in km $s^{-1} kpc^{-1}$ and R is the galactocentric distance in kpc. We can say: more rotation means shorter t_{ref} or shorter time scale of superstructure evolution.

In the present model, the molecular gas in clouds partly recycles: about 90% goes back into the diffuse component. The rest does not. It is locked up in long-living low mass stars or in the remnants after the evolution of high mass stars. The total mass in the disk of a galaxy is kept constant. We disregard any halo-disk interaction or infall, as well as the chemical evolution within the disk. The inclusion of these processes can improve our simple model.

In the beginning of the simulation, all the mass is in the diffuse component. The number of molecular clouds grows rapidly, which introduces an initial burst of star formation and high gas consumption rate. Subsequent decrease in the density of diffuse component

Dynamics and Interactions of Galaxies Editor: R. Wielen

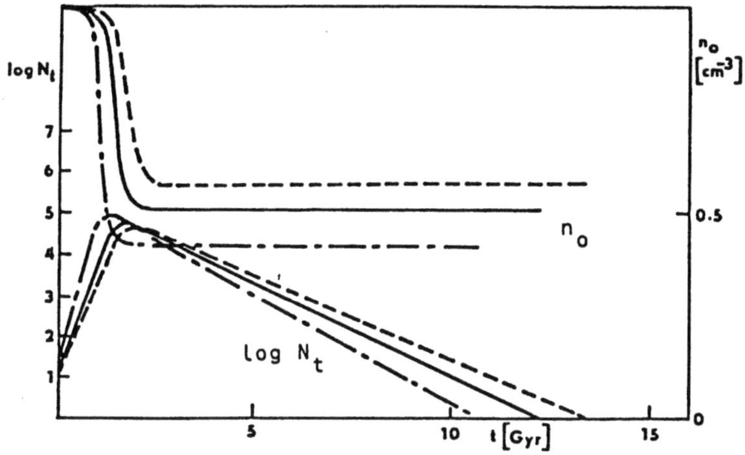

Figure 1.
The time evolution of the number of star-forming molecular clouds N_t and of the average density of the diffuse gas component n_o in a galactic disk
— — — — — — Sa, ———————— Sb, - - - - - - - - Sc

reduces the number of star-forming clouds, slows down the star formation and lowers the gas consumption rate.

Sa galaxies / see Fig. 1./, which rotate at higher rate than Sc galaxies / Rubin et al., 1985 /, suffer, in the beginning of their evolution, a steeper and larger increase in the number of star-forming clouds and experience a more violent burst of star formation than Sc galaxies. This is a direct consequence of shorter t_{ref}. Finally, at later times, the diffuse gas is much more exhausted in Sa's, yielding less clouds and lower gas consumption rate in Sa's than in Sc's.

References

J.Palouš: 1989, in Structure and Dynamics of the Interstellar Medium, eds G.Tenorio-Tagle, M.Moles and J.Melnick, Springer-Verlag, in press

J.Palouš, J.Franco, and G.Tenorio-Tagle: 1989, Astron. Astrophys., in press

V.C.Rubin, D.Burstein, W.K.Ford, Jr., and N.Thonnard: 1985, Astrophys. J. 289, 81

G.Tenorio-Tagle, and P.Bodenheimer: 1988, Ann.Rev.Astron.Astrophys. 26, 146

G.Tenorio-Tagle, and J.Palouš: 1987, Astron. Astrophys. 186, 287.

Discovery of the Large-Scale Vortices in the Gaseous Disk Component of the Spiral Galaxy Markarian 1040

V.L. Afanasiev[1], *J. Boulesteix*[2], *F. Bonnarel*[2], *S. Dodonov*[1], *and V. Vlasjuk*[1]

[1]Special Astrophysical Observatory, SU-357140 Zelenchukskaja, USSR
[2]Observatoire de Marseille, 2 Place Le Verrier, F-13248 Marseille Cedex 4, France

The velocity field of the nearby Seyfert galaxy Markarian 1040 = NGC 931 has been investigated using the CIGALE instrument [1] at the prime focus of the 6-m Telescope. The H_α- image of the galaxy shows a well developed two-armed spiral pattern (Fig.1a) in the extended (size ≈ 55 kpc, assuming H_o=50 km/s·Mpc) gaseous disc. In the velocity field two local velocity maxima at a distance of 20″ from the center are evident relative to the systemic velocity of 4930 ± 5 km/s (Fig.1b) which are located nonsymmetrically to the line of nodes. This results in a decrease of the rotation velocity as a function of radius two times faster than keplerian in the interval between 6 and 12 kpc at the Markarian 1040 rotation curve along the line of nodes (Fig.2). The local maximum at 1 kpc is caused by the compact bulge.The analysis of the velocity field shows strong deviations from an azimuthal symmetry in the mentioned distance interval. Velocities are subtracted from the observed radial velocity field according to an azimuthally symmetric rotation curve defined for each distance interval by the amplitude of the sinus-like dependence on the position angle. It has been found that the rotation curve determined in such a way is in agreement with the rotation curve of an exponential disc with a scale length of 7.5 kpc derived from the galaxy photometry in the continuum.

The residual radial velocity field (Fig.3) transformed to the plane of the galaxy indicates the presence of two rotating

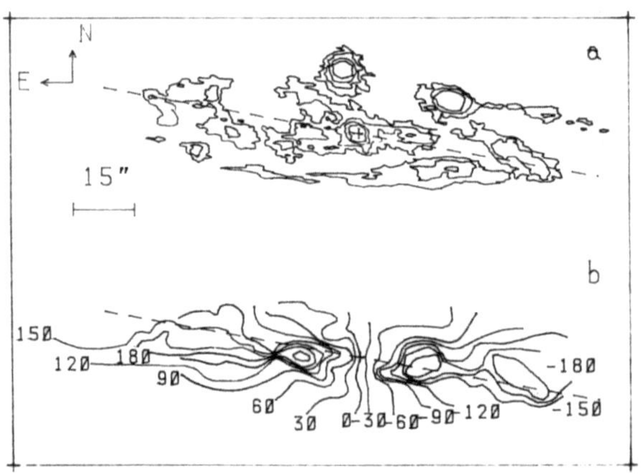

Figure 1
a) The H_α- map of Mkn 1040 obtained with the CIGALE. The size scale is 1px=0.75″; the seeing was ≈2″. The line of nodes is marked by a dashed line
b) The velocity field of Mkn 1040. The precision of measurement is ±10 km/s. The scanning step of the interferometre is 0.42 A = 19 km/s

Dynamics and Interactions of Galaxies Editor: R. Wielen

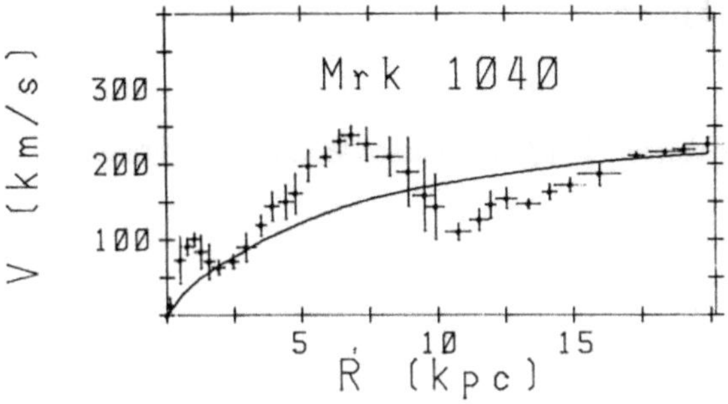

Figure 2
The rotation curve of Mkn 1040 obtained with the slit position along the line of nodes and transformed to the plane of the galaxy. The solid line marked the azimuthally-symmetric component of the rotation curve

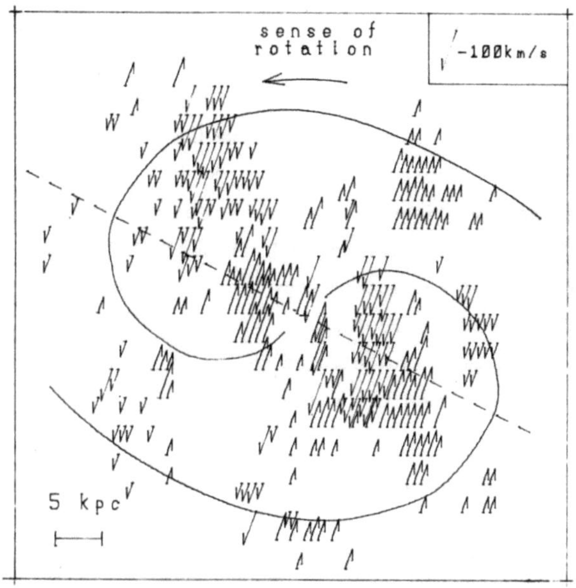

Figure 3
The residual velocity field of Mkn 1040, transformed to the plane of the galaxy, in the frame rotating with the galaxy. The dashed line indicates the line-of-nodes position. Adopted values are: the inclination $i=70^{\circ}$, the line-of-nodes P.A.$=10^{\circ}$. The upper part of the figure contains the nearest side of the galaxy

anticyclonic regions in the disc of the galaxy located between the spiral arms. The existence of global vortices located between spiral arms in galactic discs on the circle of corotation has been predicted from the hydrodynamic theory of the spiral structure [2] and from laboratory simulations in shallow water conditions [3]. The parameters of the detected vortices (distance from the center 9 kpc, size ≈ 4 kpc and linear rotation velocity ≈ 60 km/s) and of the spiral structure (Ω_p= 13 km/s·kpc, σ_{gas} ≈ 50 km/s) are in good agreement to the predictions of the hydrodynamic theory.

REFERENCES

1 J. Boulesteix, Y. Georgelin, M. Marcelin, G. Monnet: *Instrumentation in Astronomy V* (SPIE,London 1984)
2 A.M.Fridman : *Uspechy Fizich.Nauk (in Russian)*, 125, 352 (1978)
3 M.B. Nezlin, V.L. Polyachenko, E.N. Snezhkin, A.S. Trubnikov, A.M. Fridman: *Sov.Astron.Letters* 12, 213 (1986)

The Frequency of Warped HI Discs

A. Bosma and E. Athanassoula

Observatoire de Marseille, 2 Place Le Verrier, F-13248 Marseille Cedex 4, France

We have recently made a small survey of ten edge-on galaxies, using the VLA in D-configuration (resolution 40") in the 21-cm line. The purpose was to find non-warped galaxies extended in HI, suitable for a detailed study of the vertical equilibrium of their stellar and gas disks. Such a study might help to determine the shape of the dark halo component. As a byproduct, we have found several more examples of warped HI-discs. In Table 1 we collect the data on edge-on galaxies either observed by us, or already published elsewhere. For small warp amplitudes resolution effects become important : e.g. for NGC 100 our 40" data does not indicate a warp, but recent mapping at 4" resolution show a warp of small amplitude, as well as the expected flaring of the HI disc.

From the table, it can be seen that roughly two-third of the galaxies are warped. About 35 - 40 % of these warps are asymmetric, either in position or in

Table 1. Frequency of warped HI discs of edge-on galaxies

Galaxy	Warp	Symmetry	Reference
NGC 100	no	yes	This work
NGC 1560	yes	yes	,,
NGC 3044	yes	no	,,
NGC 4010	yes	yes	,,
NGC 4157	yes	no	,,
NGC 4183	yes	yes	,,
NGC 4217	no	yes	,,
NGC 4517	slight	yes	,,
UGC 9242	no	yes	,,
UGC 10288	yes	no	,,
NGC 891	slight	no	Sancisi & Allen (1979), Rupen (1989)
NGC 4013	yes	yes	Bottema et al. (1987a)
NGC 4244	yes	yes	Sancisi (1976)
NGC 4565	yes	no	Sancisi (1976), Rupen (1989)
NGC 5023	slight	yes	Bottema et al. (1986)
NGC 5084	slight	yes	Gottesman & Hawarden (1986)
NGC 5170	no	yes	Bottema et al. (1987b)
NGC 5907	yes	no	Sancisi (1976)
NGC 7814	yes	no	Van der Kruit & Searle (1982)

Dynamics and Interactions of Galaxies Editor: R. Wielen

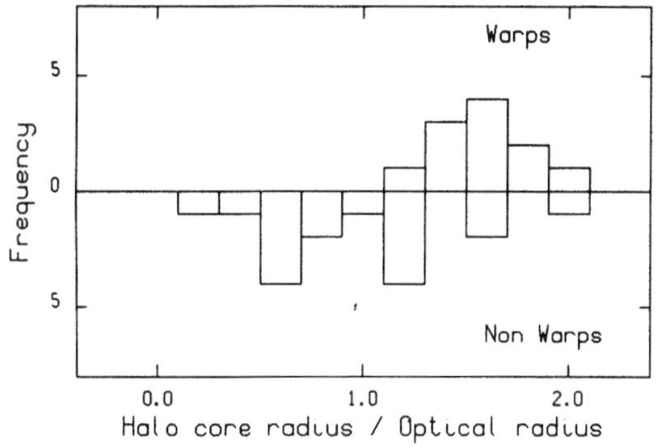

Figure 1. Histogram of the number of warped and non-warped galaxies as function of the ratio of halo core radius to optical radius.

velocity, or both. NGC 4157 is a clear case where both asymmetries occur. To stay in line with the topic of this conference, we suggest that its warped parts were more recently accreted.

Is it possible to predict whether a warp should be expected ? We plotted the frequency of warps in non edge-on galaxies as a function of the ratio of halo core radius to optical radius for a sample of galaxies for which we previously determined composite disk/halo mass models. Only solutions in which the disk mass is maximized subject to the constraint that m=1 instabilities cannot occur ('no m=1' models) are considered (cf. Athanassoula et al., 1987), and the warps are now inferred from the velocity field of a galaxy by fitting a tilted ring model. The frequency of warps in this sample is about 40%, and if both samples are combined about 55%. Figure 1 indicates that warps occur only in galaxies for which the halo core radius is larger than the optical radius. Such an effect seems in agreement with the theory of warp modes in disk galaxies embedded in oblate halos reported by Sparke and Casertano (1988). However, in their models the line of nodes of the inner disc and the outer warped parts is the same, contrary to the observations of some galaxies (cf. Briggs, this volume).

References :

Athanassoula, E., Bosma, A., Papaioannou, S.: 1987, Astr. Ap. 179, 23

Bottema, R., Shostak, G.S., Van der Kruit, P.C.: 1986, Astr. Ap. 167, 34

Bottema, R., Shostak, G.S., Van der Kruit, P.C.: 1987a, Nature 328, 401

Bottema, R., Van der Kruit, P.C., Freeman, K.C.: 1987b, Astr. Ap. 178, 77

Gottesman, S.T., Hawarden,T.G.: 1986, M.N.R.A.S. 219, 759

Rupen, M.: 1989, Ph. D. thesis, Princeton University

Sancisi, R.: 1976, Astr. Ap. 53, 159

Sancisi, R., Allen, R.J.: 1979, Astr. Ap. 74, 73

Sparke, L.S., Casertano, S.: 1988, M.N.R.A.S. 234, 873

Van der Kruit, P.C., Searle, L.T.: 1982, Astr. Ap. 110, 79

Rules of Behavior for Warped Galaxies

F.H. Briggs

Department of Physics and Astronomy, University of Pittsburgh, USA, and
Kapteyn Astronomical Institute, Postbus 800,
NL-9700 AV Groningen, The Netherlands

Introduction

A sample of galaxies is now available for which HI 21cm line observations allow the development of detailed kinematic models based on concentric, circular rings with adjustable inclinations and orbital velocity (see references 1-9). By examining these warped systems in a variety of reference frames, clear empirically-determined "rules" for the behavior of galactic warps have emerged. A complete description of the galaxy sample and the methods of analysis will be reported elsewhere [10].

The galaxies in the sample have the common trait that HI can be observed far outside the optical disk of the galaxy, typically to 2-3 times the Holmberg radius, R_{Ho}. The most extended galaxies in this sample are also the most isolated in space, a fact that argues against tidal interactions among galaxies being the drivers of warps.

The Rules

Empirically-determined rules for behavior of warped galacties are summarized below:

1. Warping within R_{Ho} appears consistent with a common (i.e. straight) line of the nodes, *LON*, measured in the plane defined by the innermost galaxy. Typically, a galaxy is planar within R_{25}, but warping becomes detectable within R_{Ho}. On the other hand, NGC 3718 is warped by nearly 90° within R_{25}[8].
2. Warps change character at a transition radius near R_{Ho}.
3. For radii larger than R_{Ho}, the *LON* measured in the plane of the inner galaxy advances in the direction of galaxy rotation for successively larger radii. Thus, the nodes lie along leading spirals in this frame of reference.
4. Surprisingly, the galaxy kinematics uniquely specify a new reference frame in which there is a common *LON* for orbits within the transition radius and also a *differently oriented* straight *LON* for the gas outside the transition radius. The new reference plane is typically inclined by less than 10° to the plane of the central galaxy disk. Within this reference frame, the *LON* of the outer system lags behind the *LON* of the inner system, relative to the direction of galaxy rotation. The physical significance of the new reference system is not clear.

Conclusions

The lack of a common *LON* throughout the entire warped disk argues against models that rely on normal bending modes to maintain warp coherence at all radii[11-14]. However, the emergent picture is consistent with galaxy models that have the *LON* regressing most rapidly for orbits that are in closest proximity to the flat, stellar disk. Thus, the apparent longevity of warps may serve to place constraints on the asphericity of the dark halo [15] or may point to a continuing acquisition of angular momentum from a "cosmic drizzle" [14] created by a smooth reversal of the Hubble flow in the vicinity of massive, isolated galaxies.

Dynamics and Interactions of Galaxies Editor: R. Wielen

Acknowledgements

During my visit to the Kapteyn Astronomical Institute, many colleagues have made essential contributions to the findings summarized here, especially K. Begeman, S. Casertano, J. Kamphuis, P. Sackett, R. Sancisi, and L. Sparke. This work has been supported by grants from the Nederlandse Organisatie voor Wetenschappelijk Onderzoek and the U.S. National Science Foundation AST 87-15070 and AST 88-06737.

References

1) Begeman, K. 1987, Ph.D. Thesis, University of Groningen.
2) Bosma, A. 1978, Ph.D. Thesis, University of Groningen.
3) Briggs, F.H. 1986, *Ap.J.*, **300**, 613.
4) Briggs, F.H., Matsik, S., and Kamphuis, J. 1989, in preparation.
5) van der Kruit, P.C., and Shostak, G.S. 1984, *Astron. Ap.*, **134**, 258.
6) Rogstad, D.H., Wright, M.C.H., and Lockhart, I.A. 1976, *Ap. J.*, **204**, 703.
7) Rogstad, D.H., Lockhart, I.A., and Wright, M.C.H. 1974, *Ap. J.*, **193**, 309.
8) Schwarz, U.J. 1985, *Astron. Ap.*, **142**, 273.
9) Shostak, G.S., and van der Kruit, P.C. 1984, *Astron. Ap.*, **132**, 20.
10) Briggs, F.H. 1989, submitted to *The Astrophysical Journal.*
11) Dekel, A., and Shlosman, I. 1983, In: *Internal Kinematics and Dynamics of Galaxies, IAU Symp. No. 100*, ed. Athanassoula, E., Reidel, Dordrecht, The Netherlands.
12) Toomre, A. 1983, In: *Internal Kinematics and Dynamics of Galaxies, IAU Symp. No. 100*, ed. Athanassoula, E., Reidel, Dordrecht, The Netherlands.
13) Sparke, L.S., and Casertano, S. 1988, *M.N.R.A.S*, **234**, 873.
14) Binney, J. 1989, this conference.
15) Tubbs, A.D., and Sanders, R.H. 1979, *Ap. J.*, **230**, 736.

Part X

Other Problems of Galactic Dynamics

Can Dumbbell Galaxies Be Equilibrium Systems?

H.-W. Rix and S.D.M. White
Steward Observatory, University of Arizona, Tucson, AZ85721, USA

Introduction

There exists a large body of work on the interactions of close galaxy pairs. However, most of the theoretical work has concentrated on violent interactions between the components which lead to rapid evolution (mergers, tidal tails,...). In this work we attempt an equilibrium description of binary systems to investigate the circumstances under which close binary galaxies can be long-lived phenomena (Rix and White,1989).

The observational motivation for this is twofold: Firstly, there exist (apparently) closely interacting pairs of field galaxies within a common envelope (dumbbells), which do not show signs of rapid evolution, e.g. NGC750/751 (see Figure 1). Secondly, binary giant ellipticals inside a common envelope are claimed to be more abundant in Abell clusters than can be

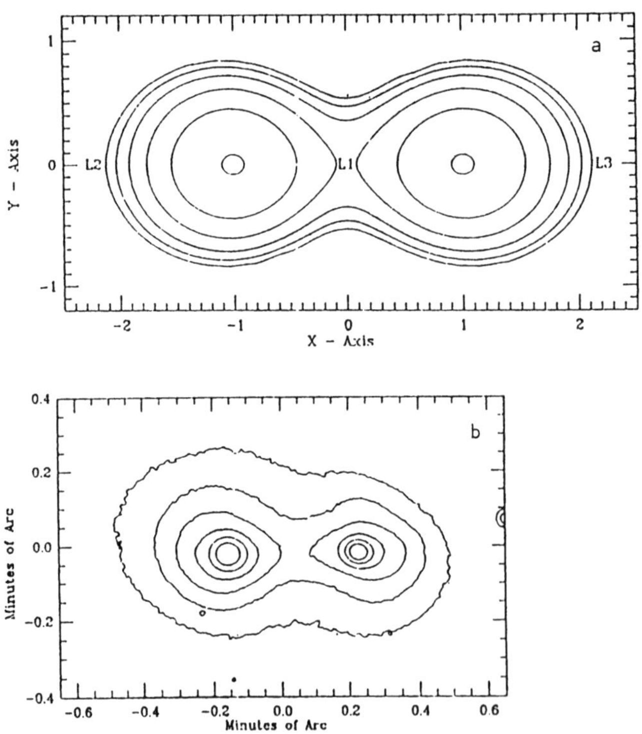

Figure 1 (a): Basic dumbbell geometry as illustrated by this contour map of the surface mass density (projected into the orbital plane) for one of our equilibrium models. L1 through L3 denote the corresponding Lagrangian points. **(b):** Isophote map of the dumbbell galaxy NGC750/751.

Dynamics and Interactions of Galaxies Editor: R. Wielen

understood in terms of standard dynamical friction estimates and line-of-sight superpositions (Rood, 1988).

By constructing strongly idealized, but self-consistent, models we address the question:

Do stable equilibria exist for the dumbbell geometry ?

Equilibria

Simple model: equal components, circular orbits, distribution functions depending on the Jacobian Energy E_J only. Choosing

$$f(E_J) \propto (E_J^0 - E_J)^{n-\frac{3}{2}}$$

(for $E_J < E_J^0$), we use the distribution function cut-off E_J^0 to regulate the extent of the common envelope and the polytropic index n to control the central concentration of each binary component.

A self-consistent solution is obtained by iteration starting from two spheres in contact.

Result: Equilibria exist with common envelopes of various extents. The common envelope cannot extend beyond the outer Lagrangian point $L3$ (see Figure 1 for an example).

Test for Stability

We employed a Barnes/Hut N-body code (using 6000 particles) to test the stability of our equilibrium models by following their time evolution for about five orbital periods.

Result: Dumbbell galaxies with common envelopes are stable on dynamical time scales, unless their envelopes extend to the outer Lagrangian point $L3$. If the envelope extends to $L3$, mass loss and angular momentum loss lead to merging of the dumbbell components on a dynamical timescale.

Are there Equilibria with Envelopes beyond L3?

Most dumbbell systems are observed to have envelopes extending beyond the outer Lagrangian points. As mentioned above, for $f = f(E_J)$ no such equilibria exist. Alternative models must involve additional integrals of motion (IOM). To look for additional integrals of motion, we investigated the orbital structure of the dumbbell potential.

Results: Two additional IOM exist for all orbits that remain localized within one half of the dumbbell. Orbits in the common envelope are largely chaotic and have no additional IOM.

It seems likely that no equilibria with envelopes extending beyond L3 exist. However, this result may not preclude the existence of quasi-equilibria with evolution timescales long compared to t_{dyn}.

Conclusions

1) We have constructed the the first fully self-consistent equilibrium models of binary galaxies with a common envelope (dumbbell galaxies). We choose distribution functions which depend only on the Jacobian energy E_J. For such models the common envelope cannot extend beyond the outer Lagrangian point $L3$.

2) Based on N-body experiments we argue that these models are stable on dynamical timescales unless their envelopes extend to $L3$.

3) Since many of the observed dumbbell systems have envelopes extending beyond $L3$, we have investigated the existence of alternative models, involving additional IOM. In an analysis of the orbital structure of the dumbbell potential we find no additional IOM for orbits in the common envelope. This appeares to rule out the existence of equilibria with envelopes extending beyond $L3$. However, quasi-equilibrium configurations with long evolution time scales may not be excluded.

References

Rood, H., 1988, PASP **100**, 1071

Rix, H.-W. and White, S.D.M., 1989, MNRAS *in press*

Dark Matter in Dumbbell Galaxies

S. Casertano and E.A. Valentijn

Kapteyn Astronomical Institute, Postbus 800,
NL-9700 AV Groningen, The Netherlands

A morphologically well-defined sample of dumbbell systems, with measured projected separations and velocity differences, provides evidence of the presence of dark matter in the center of rich clusters, on scales of $\sim 50\,\mathrm{kpc}$.

We have defined a sample of 44 dumbbell pairs, consisting of two galaxies of nearly equal luminosity, with projected separation of order of their optical size, at the center of rich clusters. For all pairs in this sample, projected separations Δr and velocity differences Δv are now available; many of them are new measurements. Our aim is to derive constraints on the mass distribution in these systems from the intrinsic velocity difference as a function of true distance, which provides the statistical equivalent of a rotation curve $v(r)$.

In order to derive this velocity curve from the projected quantities, one must correct for both projection effects and orbital eccentricity. In general, this cannot be done uniquely with only 44 objects. However, for dumbbells the observed relationship is very distinctive, in that the projected velocity Δv correlates positively with the projected separation Δr (Fig. 1). A positive correlation is hard to obtain: it requires both *circular orbits* and a *non-Keplerian potential.*

If orbits were significantly eccentric, large projected separations would be probably associated with components near the apocenter, where velocities are lower. We illustrate this in Fig. 2, where we show projected separations and velocities for a Monte Carlo simulated catalog of particles distributed at random with a flat velocity curve $v(r) =$ constant and either circular (Fig. 2a) or slightly eccentric orbits (Fig. 2b). The mean eccentricity in Fig. 2b is quite small, only 0.4 (corresponding to an axis ratio of 0.92), but is already sufficient to introduce a significant negative correlation in the projected quantities. Circular orbits indicate that dumbbell components are very likely physically associated.

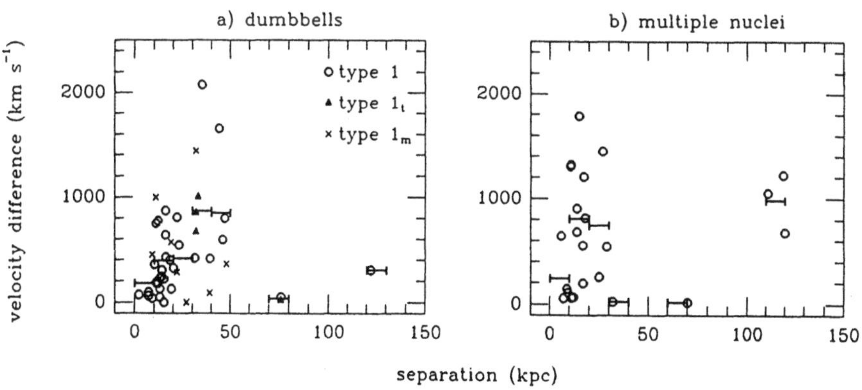

Figure 1. Projected velocity difference vs. apparent separation for our sample of dumbbells and for a comparison sample of multiple nuclei. The dumbbells show a weak, but significant positive correlation between the two quantities. No correlation is apparent for the multiple nuclei systems

Dynamics and Interactions of Galaxies Editor: R. Wielen

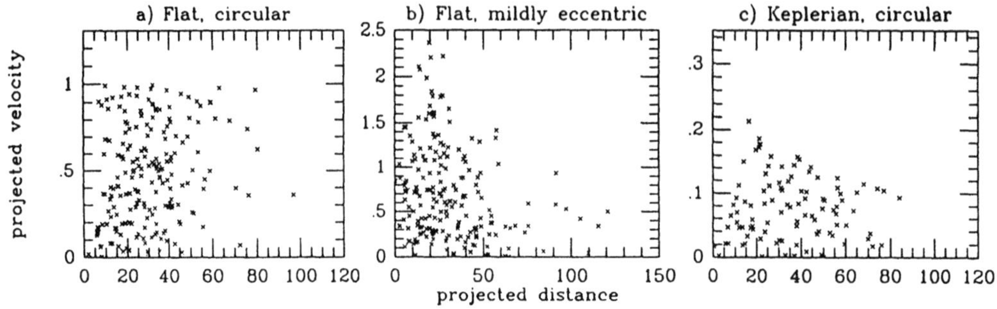

Figure 2. Projected velocity difference vs. apparent separation for simulated samples. **2a:** Flat velocity curve ($v(r) = 1$), circular orbits. The correlation between the projected quantities is negligible. **2b:** Flat velocity curve, mildly eccentric orbits. The eccentricity is distributed as $f(e) \propto e$, up to a maximum eccentricity of $e = 0.6$. Even this mild eccentricity induces a strong negative correlation between projected separation and projected distance. **2c:** Keplerian velocity curve $v(r) = r^{-1/2}$, circular orbits. Also in this case a negative correlation can be seen

A declining velocity curve, as would result from constant mass-to-light ratio, would also result in a negative correlation. Systems with a larger projected separation are likely to have larger intrinsic distances; if the intrinsic relative velocity $v(r)$ decreases, the projected velocity difference is also likely to be smaller. Projection effects are insufficient to offset this correlation, as shown in Fig. 2c.

These conclusions have been quantified by performing a large number of Monte Carlo simulations. We generated fake catalogs with a combination of flat, falling and rising rotation curves, and circular or mildly eccentric orbits. The combination that best reproduces the observed correlation is that of circular orbits and a rising rotation curve, although a flat rotation curve cannot be excluded at the 2-σ level. On the other hand, a Keplerian velocity curve is ruled out at the 99% level. The mean orbital eccentricity is likely to be small (mean axis ratio > 0.9 for flat or mildly rising rotation curve; 99% confidence), although we cannot rule out the combination of a linearly rising velocity curve $v(r) \propto r$ and isotropic velocity dispersion.*

For a more detailed description of our results, see reference [2].

Conclusions

Dumbbells are pairs of physically related galaxies moving in circular orbits around each other. Their velocity curve is not Keplerian (99% confidence) and probably rising (90% confidence). Dark matter is present on a scale of about 50 kpc; the implied M/L ratios are of about 60.

This dark matter cannot be explained by a change in the laws of gravity like MOND [1], since the typical accelerations are of order of $6 \cdot 10^{-8}\,\mathrm{cm\,s^{-2}}$, thus well in the Newtonian regime if the critical MOND acceleration is of order of that required by galactic rotation curves.

References

1) M. Milgrom, *Astrophys. J.* **270**, 371 (1983).
2) E. A. Valentijn and S. Casertano, *Astr. Astrophys.* **206**, 27 (1988).

* We thank Scott Tremaine for pointing out this possibility to us

Shape of the Dark Halo of Polar Ring Galaxy NGC 4650A

P.D. Sackett[1] *and L.S. Sparke*[2]

[1]Department of Physics and Astornomy, University of Pittsburgh, USA and
Kapteyn Astronomical Institute, Postbus 800,
NL-9700 AV Groningen, The Netherlands
[2]Washburn Observatory, University of Wisconsin, Madison, WI53705, USA

Polar ring galaxies, which have an optical ring outside and perpendicular to the plane of their central disks, provide a unique opportunity to probe the gravitational potential of a galaxy in two perpendicular planes. We place limits on the flattening of the dark matter halo of NGC 4650A by fitting the rotation curves of its central disk and polar ring with models consisting of a bulge, disk, axisymmetric dark halo and massive polar ring. Our results indicate that the data are consistent with a range of halo flattening from spherical to quite possibly as flat as E6.

NGC 4650A lies in a chain of five galaxies in the Centaurus Cluster. Its polar ring is aligned perpendicularly to the central S0 disk and is viewed nearly edge-on. The blue color of the ring and its knotty HII regions suggest recent star formation. The surface brightness profile along the major axis of the central S0 indicates that the disk is exponential with a small, bulge-like component in its center [1]. Assuming H_o = 75 km/s/Mpc, NGC 4650A is at a distance of about 35 Mpc. Its disk has a scale length of about 5″ (850 pc) and a lumonsity of not quite 3×10^9 $L_\odot$. A substantial amount of neutral hydrogen ($\geq 6.4 \times 10^9$ $M_\odot$, helium corrected) is associated with the polar ring, extending to more than twice its optical radius, and exhibiting an essentially flat rotation velocity of $\sim$ 100 km/s at large radii ($\sim$ 120″).

We construct a physical model for NGC 4650A consisting of: a spherically-symmetric Plummer bulge; a thick exponential disk; an axisymmetric dark halo, with a radial density profile of constant axial ratio and constrained to yield asymptotically flat rotation curves; and a thin, polar ring with a surface density roughly matching that indicated by HI observations. The luminosities and scale lengths of the bulge and disk are dictated by photometry. The free parameters of the model are then: the mass-to-light ratio (M/L) of the disk; M/L of the bulge; the asymptotic rotation speed, scale length, and flattening of the dark halo; and the mass of the polar ring. Closed orbits are computed in the combined gravitational field of these dynamical elements. Since the polar rotation curve is the more reliable of the two, model parameters are varied until the over-the-pole speeds of the closed polar orbits give the best fit to the observed polar rotation curve. We estimate the allowed range of halo flattening by demanding that the closed orbits in the disk plane simultaneously fit the equatorial rotation curve as well.

In the figure below, the upper panels show model equatorial rotation curves superimposed on stellar absorption-line data [1] taken along the major axis of the central disk. The data have been corrected by us for inclination and asymmetric drift. Open circles indicate unreliable data (due to ring interference, anomalously high velocity dispersion, or difficult sky subtraction). The lower panels show model polar rotation curves superimposed on emission-line data taken along the major axis of the polar ring [1 (circles), 3 (stars)]. Both sets of polar data have been corrected by us for inclination and warping of the ring.

The model curves in the left panels represent fits to the equatorial and polar rotation curves of NGC 4650A using only a dark halo component. Three representative halo flattenings are shown:

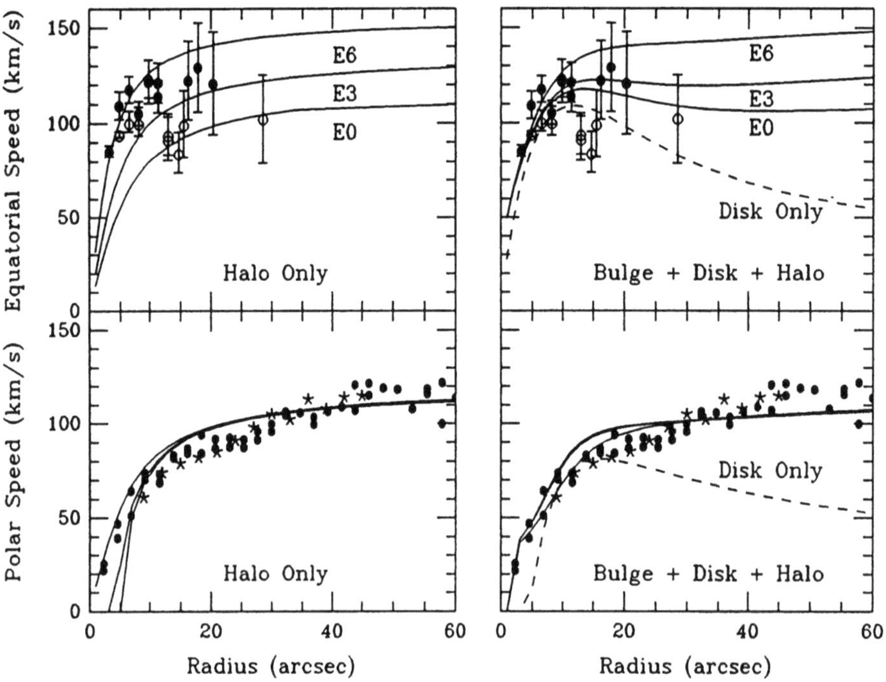

E0 (spherical halo), E3, and E6. The halo parameters are adjusted so that all halo flattenings produce similar, reasonable fits to the polar curve; the resulting equatorial curves are then quite different. Although halo flattenings of E4 to E5 best reproduce the disk velocities, none of these halo-only models provide acceptable equatorial fits. The panels on the right show the effect of adding mass to the luminous components of the galaxy by increasing the M/L of the disk and bulge to maximum values consistent with both rotation curves (disk: $M/L = 2.5\ M_\odot/L_\odot$, bulge: $M/L = 0.5\ M_\odot/L_\odot$). For comparison, the dashed line shows the model rotation curves for the disk alone. The discrepancy between the dashed line and the polar rotation data indicates quite clearly that the mass in the luminous component of NGC 4650A is not sufficient to support the high rotation velocities observed at large distance along the ring. The addition of massive disk and bulge components improves the equatorial fits for all halo flattenings, halos flattened by E3 to E4 do slightly better than the rest.

The substantial mass of the polar ring (at least equal to that of the central S0) is probably reponsible for the characteristic "wiggle" in the polar rotation curve between 20″ and 50″. The effect of adding a massive ring to the models is to provide a better fit to this feature and to admit slightly more flattened halos. This effect is currently under study.

[1] Whitmore, B.C., McElroy, D.B. & Schweizer, F., 1987. *Astrophys. J.*, *314*, 439.
[2] van Gorkom, J.H., Schechter, P.L. & Kristian, J., 1987. *Astrophys. J.*, *314*, 457.
[3] Nicholson, R.A., Taylor, K., Sparks, W.B. & Bland, J., 1987. In *Structure & Dynamics of Elliptical Galaxies, IAU Symposium # 127*, p. 415, ed T. de Zeeuw, D. Reidel Publishing Company.

Self-consistent Models of Spherical Galaxies with Dark Matter

R.P. Saglia, G. Bertin, and M. Stiavelli
Scuola Normale Superiore, I-56126 Pisa, Italy

A well-known result of the analysis of extended HI rotation curves of spiral galaxies is that these objects have an M/L ratio that increases with the galactocentric distance. For the case of elliptical galaxies, for an approximately constant M/L ratio, the rotation curve of a disk of gas, if present in the galaxy, would show a keplerian decrease beyond the half luminosity radius R_e . In addition, the stellar velocity dispersion profile projected along the line of sight $\sigma_P(R)$ would also be rapidly decreasing with the distance (for a wide class of models where pressure anisotropy is the one suggested by numerical experiments of galaxy formation (van Albada 1982) and mean motions of the stars are dynamically insignificant). Therefore evidence for the presence of a dark component in elliptical galaxies, distributed differently from the luminous matter, could be obtained if new kinematical data on a larger radial extent will show that σ_P is only slowly decreasing (Bertin et al. 1989), thereby confirming the indications deriving from the observations of the few ellipticals that possess a gaseous disk (see the flat HI rotation curves measured by Raimond et al. 1981 and by Schweizer, van Gorkom, and Seitzer 1989).

A survey of two-component models of spherical galaxies has been carried out to ascertain to what extent and under which conditions kinematical profiles can be changed, while the density distribution of one component is kept essentially fixed, since the luminosity profile of bright ellipticals is known to obey the $R^{1/4}$ law. Dynamically, why should the two components respond to each other so that the luminous component has always a "nice" $R^{1/4}$ profile appears as part of a *new conspiracy*. Self-consistent, two-component, spherically symmetric models have been constructed by assuming that the relevant distribution functions are both of the f_∞-form (see Bertin, Saglia and Stiavelli 1988, 1989):

$$f_\infty = A(-E)^{3/2} exp[-aE - cJ^2/2], \tag{1}$$

where E and J^2 are respectively the energy and the angular momentum square per unit mass, and the distribution functions for the stars, f_L, and for the dark matter, f_D are identified by the six constants a_L, a_D, c_L, c_D, A_L, A_D. Such an assumption is not unreasonable since the dark halo was probably formed via dissipationless collapse, much like the luminous stellar system, and the f_∞ function appears to incorporate the essential features of such a process. Self-consistent solutions are obtained by solving the Poisson equation for the total potential Φ. Each solution is characterized by four dimensionless parameters: a_D/a_L, A_D/A_L, $\Psi = -a_L\Phi(0)$, and c_D/c_L. A model is then specified when its scale length and total mass are given. The 4-dimensional space of the parameters (a_D/a_L, c_D/c_L, A_D/A_L, Ψ) has been explored numerically. We have calculated 10,000 equilibrium models using an automatic procedure (see Saglia 1989).

Within this class of models a good indicator of the presence of dark matter is given by $\sigma_P(R)$. In particular we can determine the minimum amount of dark matter required to "explain" a given drop of $\sigma_P(R)$. This *minimum halo* solution is in reasonable agreement with the estimate of the mass of dark matter present inside the sphere of radius R_e which can be derived from the equations of "hydrostatic equilibrium":

$$(\frac{M_D}{M_L})_{R_e} \sim 3.5\frac{\sigma_P^2(R_e)}{\sigma_P^2(0)} - 1. \tag{2}$$

Dynamics and Interactions of Galaxies Editor: R. Wielen

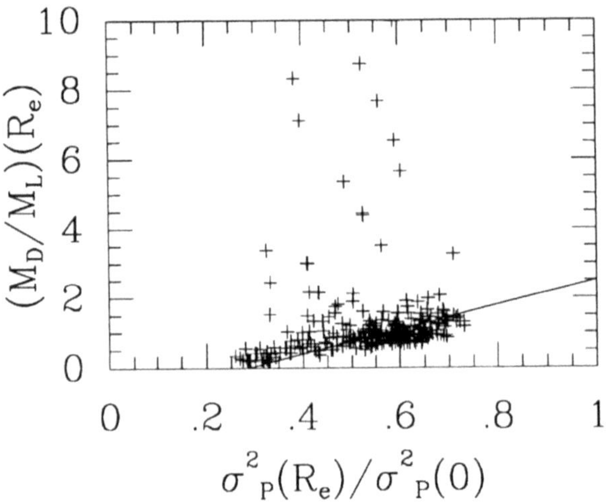

Figure 1: Crosses show the models described by Eq. (1) with a projected luminosity that follows the $R^{1/4}$ law. The line shows the prediction of Eq. (2).

This relation is based on the assumption that the dark component has a less concentrated and more extended density distribution than that of the luminous component and does not depend on the specific choice of the distribution function, provided the stellar pressure is nearly isotropic in the central parts of the model. In Fig. 1 we illustrate the properties of the two component models with a projected luminosity that follows with good accuracy the $R^{1/4}$ law, in the plane $\sigma_P^2(R_e)/\sigma_P^2(0) - (M_D/M_L)_{R_e}$. Models falling on the relation have a diffuse dark component ($R_D > R_L$, where R_D and R_L are the half-mass radii of the dark and luminous components), with a central density $\rho_D(0)$ much less than that of the luminous component. Models with M_D/M_L much larger than predicted by Eq. (2) have instead $\rho_D(0) \sim \rho_L(0)$ and $R_D > R_L$. Therefore Eq. (2) gives an estimate of the *minimum* amount of dark matter needed to account for the observed value of $\sigma_P^2(R_e)/\sigma_P^2(0)$. Models that follow Eq. (2) can be considered as the analogues for elliptical galaxies of the *maximum disk* models of spiral galaxies (van Albada and Sancisi 1986).

References

Bertin, G., Saglia, R.P., Stiavelli, M. 1988 *Ap. J.* **330**, 78

Bertin, G., Saglia, R.P., Stiavelli, M. 1989 in *Astronomy, Cosmology and Fundamental Physics* (Proceedings of the Third ESO-CERN Symposium, Bologna, May 1988) Kluwer Academic Publishers, Eds. M. Caffo et al., p. 303

Bertin, G., Bertola, F., Buson, L.M., Danziger, I.J., Dejonghe, H., Sadler, E.M., Saglia, R.P., Vietri, M., de Zeeuw, T., Zeilinger, W.W., 1989 *The Messenger* **56**, 19

Raimond, E., Faber, S.M., Gallagher III, S.J., Knapp, G.R., 1981 *Ap. J.* **246**, 708

Saglia, R.P. 1989, Ph. D. Thesis in preparation

Schweizer, F., van Gorkom, J.H., Seitzer, P., 1989 *Ap. J.* **263**, 770

van Albada, T.S., 1982 *MNRAS* **201**, 939

van Albada, T.S., Sancisi, R., 1986 *Phil. Trans. R. Soc.* London A **320**, 447

The Stability of Highly Flattened Isothermal Oblate Elliptical Galaxies

H.F. Levison[1], *M.J Duncan*[2], *and B.F. Smith*[3]

[1]US Naval Observatory, Flagstaff Station, Flagstaff, AZ 86002, USA
[2]Physics Department, Queens University
[3]Theoretical Studies Branch, NASA-Ames Research Center, MS 245-6, Moffett Field, CA 94035, USA

The stability of models of isothermal oblate E5 elliptical galaxies is studied using N-body techniques. The initial conditions of the runs are derived from models constructed by Levison & Richstone (1985) using linear programming techniques. These models have the same density distribution but differ in their internal dynamics. In particular, they can support their shapes by having either a large velocity dispersion in the ϖ direction, σ_ϖ, or by having a large velocity second moment in the ϕ direction, $\overline{v_\phi^2}$. Those of the latter case can either have all their stars in prograde orbits ($\overline{v_\phi}$ is large and σ_ϕ is small) or have some of their stars in retrograde orbits (which increases σ_ϕ and decreases $\overline{v_\phi}$). Thus, we must study a two dimensional parameter space; the shape of the velocity ellipsoid ($\Upsilon \equiv \overline{v_\phi^2}/\sigma_\varpi^2$) and the fraction of stars in retrograde orbits (β).

The activity observed in these runs is almost exclusively in the form of bar ($m = 2$) azimuthal modes. Even in runs where other modes are present, the $m = 2$ mode usually dominates. Therefore, we will concentrate most of our discussion on this mode. We find that all models for which $\beta \lesssim 0.25$ are unstable to a bar, independent of the value of Υ.

Models with large values of Υ and small values of β appear to go through the same instability as disk systems. We call this instability the 'streaming instability'. Figure 1 shows the distribution of particles of a model ($\beta = 0$, $\Upsilon = 13$) as a function of time viewed along the rotation axis. The system first forms a bar in the center. This bar causes spiral arms to form in the outer regions of the model. At intermediate times, the particles in these arms absorb some of the angular momentum of the bar. Thus, the bar slows down and the outer regions of the model puff up. At later times, the arms fade and we are left with a tumbling bar in the center of an approximately axisymmetric rotating system.

Models with small values of Υ and β are unstable to the radial orbit instability. Our results are the first indications that this instability occurs in flattened spheroidal systems. Figure 2 shows the distribution of particles of a model ($\beta = 0$, $\Upsilon = 0.07$) as a function of time. Note that the bar forms at all radii in the run, as opposed to the just the central regions, as in Figure 1. This instability is more violent than the streaming instability and produces the largest growth rates. Note that this system does have a small amount of angular momentum and that the bar is slowly tumbling.

In general, the strength of the bar mode instability decreases as the fraction of stars in retrograde orbits is increased. Systems for which $\Upsilon \gtrsim 2$ become stable to the $m = 2$ mode as

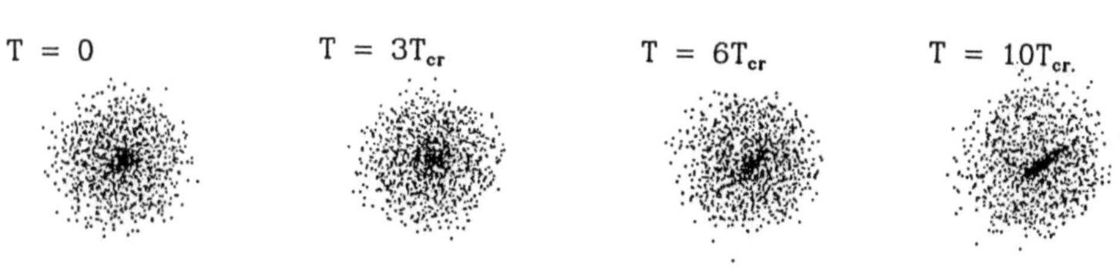

Figure 1

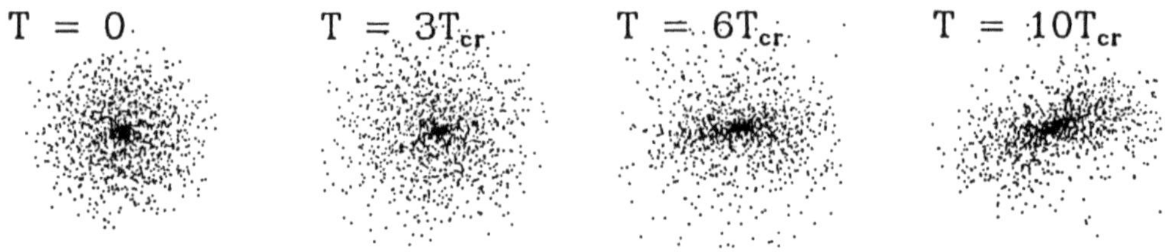

Figure 2

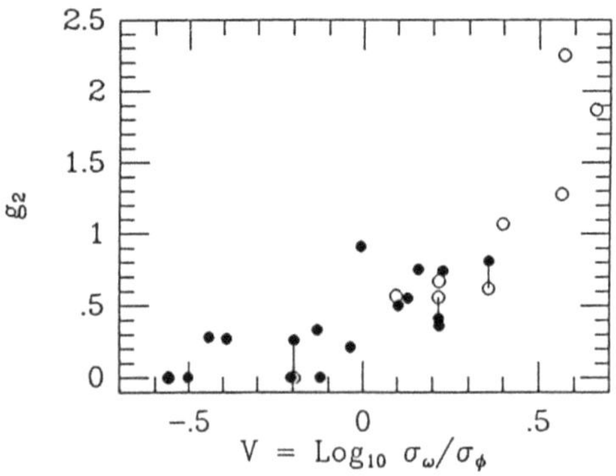

Figure 3: The $m = 2$ growth rate, g_2 as a function of $V \equiv \log_{10} \sigma_{\varpi}/\sigma_{\phi}$. The open symbols are runs done in a direct summation code by Duncan (1987). The closed symbols are runs done in an FFT code by Miller (1978). In addition, two symbols connected by a line represent the same load run in different codes. Although the two codes do not produce identical values g_2, we were surprised how well they agree.

$\beta \to 0.5$. Indeed these are the only runs that are stable. Systems with $\Upsilon \lesssim 2$ are unstable to the bar mode for all values of β. These systems tend to form counter-rotating bars!

In the previous few paragraphs we have shown that we are dealing with both the streaming and radial orbit instabilities. As a final part of our analysis, we attempt to find a global stability parameter. In this vein, we define $V \equiv \log \sigma_{\varpi}/\sigma_{\phi}$. Note that σ_{ϕ} is the random motions in the ϕ direction and is independent of rotation. Figure 3 shows the dependance of g_2 on V for all our runs. The symbols fall on a fairly well defined curve that shows g_2 increasing with V. This suggests that the radial orbit instability and the streaming instability are caused by the same process and that there is one parameter that determines the growth rates.

One of the most surprising result of this work is that the models that have the smallest value of Υ appear to be unstable to a $m = 1$ mode. There is no correlation between g_1 and β or between g_1 and V. The most significant question that must be answered is whether this is real. We see this in two very different N-body codes. Both codes produce similar growth rates and final distributions of the particles. Statler (private communication) has found $m = 1$ modes in a load with a similar distribution of orbits in a Steckel potential. Many authors have seen similar looking distributions in some barred spiral galaxies. Therefore, we must conclude that it indeed is real.

References:

Duncan, M. 1987, in *The Use of Supercomputers in Stellar Dynamics,* ed. P. Hut & S. McMillan (Springer-Verlag), p.162.

Levison, H. & Richstone D. 1985, *Ap.J.*, **295**, 349.

Miller, R. 1978, *Ap.J.*, **223**, 112.

Violent Relaxation: Difficulties with Maximum Entropy States

H. Wiechen and H.J. Ziegler

Ruhr-Universität Bochum, Theoretische Physik IV, Postfach 102148, D-4630 Bochum 1, Fed. Rep. of Germany

We prove that for all entropy-like concave functionals stationary points do not exist for arbitrary fixed total mass and energy, considering isolated self-gravitating systems in $\Re^3$. Only for a special choice for mass and energy, solutions of the variational problem are possible.

Introduction

We are concerned with the problem, whether it is possible to determine macroscopic final equilibrium states of collisionless self-gravitating systems by calculating stationary points of a suitable, entropy-like concave functional S under the constraints of mass and energy conservation. This problem has been posed by Tremaine et al. (1986), who consider concave functionals of the form

$$S = -\int C(f) d^3r d^3v, \tag{1}$$

where $C(f)$ is a convex function with $C(0) = 0$. A convex function can be characterized by the following definition:
If $C(f)$ is differentiable, then $C(f)$ is convex in $\Re^{\geq 0}$, if for all $f, f' \in \Re^{\geq 0}$

$$C(f') - C(f) \geq (f' - f)C'(f) \quad with \;\; C'(f) = \frac{dC}{df}. \tag{2}$$

Tremaine et al. state, that for any concave functional S of type (1), there exists a distribution function $f(h)$ with $df/dh < 0$, which is a stationary point of S and can be calculated by variation of S under the constraints of fixed total mass M and energy E. ($h = \frac{v^2}{2} + \psi$ denotes the energy per unit mass.) They ask whether final states of violent relaxation may be calculated by such a procedure with a suitable choice of S. This idea has stimulated some subsequent work (e.g. Dejonghe, 1987; Richestone and Tremaine, 1988). Also the theory of Lynden Bell (1967) using methods of statistical mechanics can be formulated as a problem similar to the type mentioned above.

Stationary Points of General Entropy-like Functionals

We prove that the following applies to all concave functionals $S = -\int C(f) d^3r d^3v$:
There exist no stationary points for arbitrary fixed total mass and energy, only for special choices of mass and energy stationary points may exist.
We make use of definition (2) for a convex function. In (2) we set f and f' respectively to zero and find

$$C'(f) \geq C'(0). \tag{3}$$

For a functional $S = -\int C(f) d^3r d^3v$ the first order variation under the constraints of mass and energy conservation yields

$$C'(f^*) = -\alpha - \beta h. \tag{4}$$

Dynamics and Interactions of Galaxies Editor: R. Wielen

α and β are Lagrange parameters, f^* denotes a solution of (4).Equation (3) together with (4) yields

$$-\alpha - \beta h \geq C'(0) \qquad \forall h. \tag{5}$$

This inequality must be fulfilled for all values of h, in particular for arbitrary large values. Thus

$$C'(0) = -\infty \tag{6}$$

must hold. Considering (4), we see that there are two possible cases to meet condition (6):
(i) $f^*(h) \to 0$ for $h \to \infty$ with $f^*(h) > 0$ for all $h < \infty$
However, it is easy to show, that it is necessary for a distribution function $f(h)$ that

$$f(h) \equiv 0 \qquad \forall h > 0 \tag{7}$$

if $f(h)$ should describe a system with finite total mass. Thus, in case (i), $f^*(h)$ describes a system with infinite total mass.
(ii) $f^*(h) \equiv 0$ for $h > h'$
If $h' \leq 0$, f^* meets the necessary condition for finite total mass, however, now condition (6) requires $\alpha = \infty$ or $\beta = \infty$. But this means that at least one Lagrange parameter is not free to fit f^* on arbitrary fixed mass or energy. In that case, solutions may exist only for a special choice of mass or energy, respectively. These problems also arise in Lynden-Bell's theory of violent relaxation. (A more detailed discussion will be published.)

Conclusions

The concept, to calculate final equilibrium states of violent relaxation as stationary points of a functional $S = -\int C(f) d^3r d^3v$ for arbitrary fixed mass and energy, fails.

An imaginable way out of this difficulty may lie in consideration of further ad hoc given constraints. Because of indications given by numerical studies of a stronger dependence of the final states on the initial states (e.g. Shlosman et al., 1979; Hoffman et al., 1979; van Albada, 1982; May and van Albada, 1984), such further constraints should be based on a more detailed knowledge about the dynamics of an individual system.

It seems more promising to develop a concept of violent relaxation, where the increase of the information entropy depends on the amount of nonequilibrium energy, which has to be converted into equilibrium energy by phase-mixing (Wiechen et al. 1988; Ziegler and Wiechen, 1989). This concept takes into consideration the mixing character of violent relaxation (Tremaine et al., 1986) and leads to globally stable final equilibrium states.

References

Dejonghe, H., 1987. *Astrophys. J.*, **320**, 477.
Hoffman, Y., Shlosman, I., & Shaviv, G., 1979. *Mon. Not. R. astr. Soc.*, **189**, 737.
Lynden-Bell, D., 1967. *Mon. Not. R. astr. Soc.*, **136**, 101.
May, A., & van Albada, T.S., 1984. *Mon. Not. R. astr. Soc.*, **209**, 15.
Richestone, D.O. & Tremaine, S. 1988. *Astrophys. J.*, **327**, 82.
Shlosman, I., Hoffman, Y., & Shaviv, G., 1979. *Mon. Not. R. astr. Soc.*, **189**, 723.
Tremaine, S., Hénon, M. & Lynden-Bell, D., 1986. *Mon. Not. R. astr. Soc.*, **219**, 285.
van Albada, T.S., 1982. *Mon. Not. R. astr. Soc.*, **201**, 939.
Wiechen, H., Ziegler, H.J. & Schindler, K., 1988. *Mon. Not. R. astr. Soc.*, **232**, 623.
Ziegler, H.J. & Wiechen, H., 1989. *Mon. Not. R. astr. Soc.*, **238**, 1261.

Exact Time Dependent Solutions – Violent Nonrelaxation

S. Sridhar and R. Nityananda*

Raman Research Institute, Bangalore – 560080, India
*Also Joint Astronomy Programme, Indian Institute of Science, Bangalore – 560012 India

1. INTRODUCTION. We present some exact oscillating solutions to the Collisionless Boltzmann Equation (CBE). The solutions all share the property that at any instant of time, the real space density is uniform over some (in general, ellipsoidal) region of space and zero outside. The shape and size of the region change with time. Then, the gravitational force within the region is linear in the spatial coordinates, though time dependent. The equations of motion of a particle are therefore linear. Under these conditions time dependent quadratic forms in the coordinates and momenta can be found which are constants of the motion. The phase space distribution function (f) is then chosen to be a function of these and solves the CBE by Jeans' theorem (see ref. 1). The functional form of f is so chosen that integrating over velocities gives uniform density ellipsoids.

2. BREATHING SPHERES AND WRITHING SPHEROIDS. The f's corresponding to these models are given in references 2 and 3. The equations governing the oscillations are given below.

2a. SPHERES[2]. The radius (R) obeys

$$\ddot{R} + \frac{a^2}{R^2} - \frac{1}{R^3} = 0 \tag{1}$$

where 'a' is a constant. So all spatially bound solutions are *time periodic.*

2b. SPHEROIDS[3]. The equations governing the oscillations are

$$\ddot{x} + \frac{\beta}{x^2}g - \frac{1}{x^3} = 0$$

$$\ddot{y} + \frac{2\beta}{x^2}(1-ug) - \frac{\alpha}{y^3} = 0 \tag{2}$$

where α and β are constants. y is the axis of symmetry and x is the other axis. u = axis ratio = y/x and ug(u) is the restoring force along the symmetry axis (see ref. 3). Surface of section studies reveal *periodic, quasiperiodic* and *chaotic* solutions.

Dynamics and Interactions of Galaxies Editor: R. Wielen

We conjecture that some of these models (especially the spheres) are stable. This implies that nearby solutions which do not have precisely uniform density are also non-relaxing.

3. A SOLVABLE MODEL OF GALAXY INTERACTIONS[4]. A model for interacting galaxies is proposed with the following features (i) The galaxies (before the encounter) are elliptic discs such as those proposed by Freeman[5]. (ii) The motion of their centres of mass is in the plane of the discs. (iii) The lowest order tidal approximation is made in modelling the interactions. In the course of the interactions, the discs acquire time dependence. We shall call these time dependent elliptic discs Generalized Freeman Discs (GFD). The properties of isolated GFDs are interesting in their own right, but the emphasis here is on their interactions. The surface density of a GFD is $\alpha\ (1 - \frac{x^2}{a^2} - \frac{y^2}{b^2})^{\frac{1}{2}}$. Its self gravity is linear in x and y. The tidal force on the disc due to a perturber (or another GFD) moving in the plane is also linear (in the lowest order tidal approximation) in x and y. Hence, there is an *invariant* (Q), quadratic in (x, y, V_x, V_y) - with time dependent coefficients - for motion in the combined potential of the GFD and the perturber. The distribution function for a GFD is of the form $f\ \alpha\ (1-Q)^{-\frac{1}{2}}$. On integrating f over the velocities, we get the surface density given above. Requiring self consistency, we get equations of motion for the ten parameters of Q. These parameters completely describe the response of the GFD without *any further approximations* (like impulse or adiabatic). In common with the models in section 2, there is no post encounter relaxation. However, energy and angular momentum transfer, heating and disruption of a model galaxy are issues relating to behaviour on at most a few dynamical time scales. We are optimistic that this model will prove to be a useful guide to the behaviour of more realistic systems.

REFERENCES

1. J. Binney, S. Tremaine : Galactic Dynamics (Princeton Univ. Press, Princeton, New Jersey) pp. 220.
2. S. Sridhar : Mon. Not. R. Astr. Soc. 238 (3), 1159 (1989).
3. S. Sridhar, R. Nityananda : to appear in Journ. Astrophys. Astron. (1989).
4. S. Sridhar, R. Nityananda : unpublished.
5. K.C. Freeman : Mon. Not. R. Astr. Soc. 134, 15 (1966).

Initial Conditions and Dynamical Evolution of Self-Gravitating Systems

G.C. Bierrenbach Lima Neto[1] *and H.V. Capelato*[2]

[1] Observatoire de Meudon/DAEC, F-92195 Meudon, France and
Instituto Astronômico e Geofísico/USP, São Paulo, S.P., Brazil
[2]Instituto de Pesquisas Espaciais/ASA, São José dos Campos, S.P., Brazil

I) Introduction

The time evolution of an expanding self-gravitating sphere (SGS) is derived, where the effect of pressure is included through a semi-phenomenological model. The results are compared with N-body simulations. The application to clusters of galaxies and the importance of statistical fluctuations on the initial conditions are discussed.

II) The Pressureless Sphere

We assume that the SGS remains homogeneous, spherically symmetric and pressureless at all times. The dynamical evolution of such a system is then given by an energy conservation equation (the problem is reduced to only one dimension) :

$$\frac{1}{2}\left(\frac{dr}{dt}\right)^2 - \frac{GM}{r} = E\,, \tag{1}$$

where E is the total mechanical energy of the sphere by unity of mass ($E < 0$) and M is the mass inside r. The above equation assumes that there are only radial motions. Also, this system will expand and contract isotropically and so any point within it will follow a Hubble-like law, $(dr/dt) \propto r$.

Let us now define some useful quantities. The collapsing time T_c is obtained by integrating (1) :

$$T_c = \sqrt{\frac{3\pi}{G}\frac{\zeta^3}{\rho_o}}\,, \tag{2}$$

where ζ, the expansion factor, is defined as the ratio between the maximum radius r_m and the initial one, r_0 . The initial density is ρ_0 . Finally, we define a quantity related to the virial theorem, $V = 2E_c/|E_p|$, where E_c and E_p are, respectively, the kinetic and potential energy of the system; for a bound SGS, V is limited between 0 and 2. The time evolution of V is independent of the initial conditions.

III) The Pressureness Sphere

We will now drop the unrealistic assumption that a SGS remains pressureless during its evolution. In an N-body system, the pressure is proportional to the square of the velocity dispersion of the system. This dispersion (or peculiar velocities) appears and develops due to random fluctuations in the density which induce non-radial accelerations. The total kinetic energy per unit of mass can be written as

$$E_c = \frac{1}{2}\left(\frac{dr}{dt}\right)^2 + P(t)\,, \tag{3}$$

The actual form of $P(t)$ would depend on the detailed evolution of the density and the density fluctuations. However, some mathematical properties of $P(t)$ can be infered through phenomenological considerations. First of all, we must have $P(t) \geq 0$ for $t > 0$, since the square of the velocity dispersion (temperature) is always a positive number. Secondly, $(dP(t)/dt) \geq 0$ for $t > 0$, because the perturbative effects should be of cumulative nature; a spontaneous decrease of the velocity dispersion would mean a decrease of entropy. Lastly, we have

$$\left(\frac{d^2}{dt^2} P(t)\right) \begin{cases} < 0 \ ; T_0 < t < T_{MAX} \\ > 0 \ ; T_{MAX} < t < T_c \end{cases} , \tag{4}$$

where T_{max} is the instant of maximum expansion of the sphere. When $P(t)$ vanishes identically , $T_{max} = T_c/2$. Equation (4) is a consequence of the $P(t)$ dependence on the density.

We do now a change of variable on $P(t)$, rewriting it as a function of a scale factor of the sphere, for instance, the radius r. It can be seen that if P is written as a function of r then P will have the form of a pseudo-potential. One of the simplest functions that can be written satisfying the above conditions is

$$P(r) = \begin{cases} A_1 r^{1/2} ; T_0 < t < T_{max} \rightarrow r_0 < r < r_{max} \\ A_2 r^{-2} \ ; T_{max} < t < T_c \rightarrow r_{max} > r > r_c \end{cases} , \tag{5}$$

and A_1 is given by the initial conditions. The evolution of V now depends on the initial values of ζ and A_1.

IV) The Real Sphere, the Initial Conditions and Conclusion

In order to compare our model of a pressureness sphere with a real SGS, we carried out a series of numerical simulations for N=800 bodies using Aarseth's N-body code.

Simply due to statistical fluctuations, important for small numbers, for each realization of the initial conditions a different value of ζ was obtained. The important point here is that the distribution of V_0 (and the quantities that depend on V_0) *are not* Gaussian . Our model cannot predict what happens after the collapse. We predict, however, that there will *never* be an ultimate collapse ($r \rightarrow 0$) when pression is taken into account. Also, it can be seen that as ζ gets bigger the collapse itself gets 'softer' and there is a critical point ζ_c, where there will be no more collapse at all. The value of ζ_c is evalueted from the condition $V(r_{max}) \geq 1$ and $v_r(r_{max})=0$. Such a system could be identified with clusters of galaxies in a cosmological scenario where galaxies are the biggest bound structures formed until $Z \approx 5 - 10$. For SGSs, ζ and γ could be calculated from cosmological models.

We acknowledge the support of the FAPESP and the CNPq. We gratefully acknowledge the hospitality at the Meudon Observatory (DAEC) where part of this work was done and S. Aarseth for the N-body simulation code.

Part XI

Groups and Clusters of Galaxies

Paradoxes of Tight Groups?

S.D.M. White

Steward Observatory, University of Arizona, Tucson, AZ85721, USA

1. Introduction

The first compact group of galaxies to be discovered, Stephan's Quintet, was identified more than a century ago (Stephan 1877). The second was probably Seyfert's Sextet, found in the 1940's (Seyfert 1948). (Nowadays this system is usually considered to consist of five galaxies with a massive tidal tail responsible for the sixth image.) These groups were remarkable for the close apparent proximity of the galaxies and for the fact that several of the images appear highly distorted. When they were observed spectroscopically it was a considerable surprise to find that each contained a galaxy with a redshift very different from that of the remaining members. An additional similar case, the compact galaxy chain VV 172, was found a few years later (Burbidge and Sargent 1971). As can be seen from the discussion of Burbidge and Sargent, it was appreciated immediately that the critical question is whether such discordant redshifts can be discarded as resulting from the superposition of unrelated systems. The very high surface density of the groups would seem to argue against this easy way out. However, the alternative implies radically new physics either in the form of explosive ejection of whole galaxies, or in the form of noncosmological redshifts.

Even if the discordant redshifts are indeed a result of projection effects, the remaining concordant galaxies form remarkable systems. Their outer isophotes overlap; substantial strong interaction not only seems likely *a priori*, it is also observed directly as dramatic distortions of several of the galaxies. Although such compact configurations are relatively rare, significant numbers of them have been catalogued by Rose (1977) and by Hickson (1982). If they are as dense as they appear, these systems must have short lifetimes. How were they formed (presumably quite recently) and what will they evolve into? Paradoxes arise if they appear to have no plausible precursors or descendants. As I discuss below, debate on these points has hinged on whether the groups are truly compact, and on how their evolution times should be estimated. The systematic properties of the galaxy population in compact groups may help sort out some of these questions, and and so are also discussed below.

Paradoxes which have been ascribed to compact groups are of three types. The most dramatic is associated with the discrepant redshifts. The second concerns the possible absence of precursors or descendants. The third results from unexpected population properties: a tendency for galaxies within a group to have similar morphology (Hickson, Kindl and Huchra 1988a); a tendency for them to be IR-loud (Hickson *et al.* 1989); and a tendency for elliptical galaxies to be radio-loud only if they are the brightest object in the group (Menon and Hickson 1985). The discussion which follows concludes that, in the current reviewer's reading of the literature, the paradoxical nature of all these effects is insufficiently demonstrated to *require* any departure from

Dynamics and Interactions of Galaxies Editor: R. Wielen

conventional ideas. This is not, of course, as strong a statement as suggesting that the observations *support* conventional ideas; in several cases the "conventional" predictions are too uncertain to be sure whether a paradox exists or not.

2. Catalogs

The first systematic survey of compact groups was carried out by Rose (1977). He searched 69 POSS plates (~6% of the sky) for compact configurations of three or more galaxies satisfying the conditions:

$$m_f \leq 17.5; \quad n(m_f)A_g \leq 0.0035, \tag{2.1}$$

where m_f is the blue magnitude of the faintest galaxies, $n(m_f)$ is the mean surface density of field galaxies brighter than m_f, and A_g is the solid angle subtended by the group. The limits in these inequalities were chosen so that the objects found would go at least as as faint and have at least as high a density contrast as Stephan's Quintet, Seyfert's Sextet and VV 172. Rose found about 170 triplets, 30 quartets and 2 quintets, which he estimated to contain about 1.4% of all galaxies brighter than 17.5. His groups seemed to have lower density halos which typically contained one or two nearby companions.

A catalog over a much larger region and with rather different selection criteria was compiled by Hickson (1982). He examined the whole of the POSS, effectively about 60% of the sky, and found all systems of four or more galaxies satisfying:

$$m_f - m_b \leq 3; \quad S_g \leq 26.0 \text{ mag/arsec}^2; \quad R_{nn} > 3R_g, \tag{2.2}$$

where m_b is the magnitude of the brightest galaxy, S_g is the mean surface brightness of the group averaged over area πR_g^2, the group radius, R_g, is defined as that of the smallest circle containing all members, and the nearest neighbor distance, R_{nn}, is measured from the center of that circle. The groups were thus required to have galaxies of roughly similar magnitude, to have a mean surface brightness similar to that at the outer edges of individual galaxies, and to be relatively isolated. After use of CCD photometry to revise the original catalog (Hickson, Kindl and Auman 1989) these criteria lead to a list of 60 quartets, 25 quintets, 8 sextets, 6 septets, and one octet. Notice that this fall-off in abundance is slower than that found by Rose, presumably because Rose's criteria require the density contrast of groups to increase as the number of galaxies increases. Hickson (1989) estimates that groups like those he has found contain roughly 1% of the light of the total galaxy population.

3. Discordant redshifts

Both the Rose and the Hickson catalogs have been used to estimate contamination probabilities, and so to assess whether systems like Stephan's Quintet and Seyfert's "Sextet" could be chance superpositions. Indeed the parameters for Rose's catalog were chosen expressly for this purpose. From (2.1) it is clear that a quartet in this catalog will lie on top of an unrelated field galaxy whose apparent brightness *exceeds* m_f with probability 0.0035. Thus one would expect $30 \times 0.0035 = 0.1$ discordant quintets in the catalog, or about 2 such systems in a similar catalog covering the whole sky. Since Rose's catalog contains 2 quintets in 6% of the sky, the probability that a randomly chosen quintet is discordant should be roughly 0.05. Thus it is quite surprising that the first two compact quintets studied should both be discordant, and that another such sytem (VV 172) should have been found in the handful of compact groups for which full velocity data are available. This conclusion is weakened by the fact that quartets can also be contaminated by field galaxies with magnitudes fainter than m_f but brighter than 17.5. This raises the expected number of discordant quintets. In addition, and perhaps more importantly, the relation between the classic discordant groups and those selected by (2.1) is far from clear (see, for example, Sulentic 1983).

A more rigorous argument has been given by Hickson, Kindl and Huchra (1988b; HKH) based on velocity and environment data for the Hickson groups themselves. Of 30 quartets with complete velocity information, these authors find four with a single discordant redshift and none with two. Of 10 quintets with full velocity data, four have a single discordant redshift and one has two. HKH obtain surface density and magnitude distributions for field galaxies in the vicinity of each of their 26 accordant quartets. They use these in Monte Carlo simulations to estimate the probability that a random repositioning of field galaxies would give rise to a discordant quintet satisfying (2.2). The mean value of this probability for the 26 systems is 0.16. Now out of the 60 quartets in entire catalog, 52 are expected to be accordant. Thus the expected number of discordant quintets is $52 \times 0.16 = 8.3$, or one third of all quintets. This is in very good agreement with the observation that four of the ten quintets with full velocity data contain a single discrepant galaxy.

Thus it seems at present that there is no compelling argument that the abundance of groups with discrepant redshifts is substantially higher than expected from random superpositions. Note, however, that a number of authors have presented evidence which may indicate direct physical interactions between discrepant redshift galaxies and other members of their groups (Arp 1973; Arp and Lorre 1976; Sulentic and Arp 1983; Sulentic and Lorre 1983). This should clearly make one uncomfortable with too easy an acceptance of the idea that all discordant redshifts result from superpositions.

4. Systematics of group populations

As a result of extensive programs mainly by P. Hickson and his collaborators, the properties of the Hickson compact groups are now known in considerable detail. Hickson, Kindl and Auman (1989) have obtained two color CCD photometry of all groups. This permits them to revise the group catalog slightly to ensure continued agreement with (2.2) and to obtain better Hubble types for all the group members. An ongoing program of optical spectroscopy (Hickson, Kindl and Huchra 1988a) has so far produced about 310 redshifts in 98 groups (67% completeness). Nearly 90 groups have been observed with the VLA resulting in about 50 detections of group members (Menon and Hickson 1985,1989), and a similar number of galaxies are found to coincide with sources in the IRAS Point Source Catalog (Hickson *et al.* 1989). HI observations of about half the catalog have been made by Williams and Rood (1987). Optical rotation curves for about 40 spirals in Hickson groups have been obtained by Rubin, Hunter and Ford (1990). Finally the immediate environment of the groups has been studied by Sulentic (1987), Williams and Rood (1988) and Hickson, Kindl and Huchra (1988b).

A detailed portrait of the groups emerges from all this work. The median redshift of the sample is ~ 9000 km/s, the most distant group being at 21,600 km/s. About half the galaxies are spirals or irregulars, and the rest are E or S0. This population fraction is the same for the first-ranked galaxies as it is for the population as a whole. The median galaxy luminosity is 2 $10^{10}h^{-2}$ $L_\odot$, suggesting a significant brightening relative to the field. Galaxies within a group tend to have similar luminosities (perhaps as a result of the group selection criteria) but the galaxy luminosity function for the sample as a whole is similar in shape to the field function (Hickson *et al.* 1984). The galaxies tend to have redder 60μ to optical colors than field objects, more than a third of them display clear optical distortions, and more than half of the measured rotation curves show some kinematic irregularities. All these properties seem consistent with enhanced star formation activity induced by interactions (see Larson and Tinsley 1978). The median projected separation within groups is $39h^{-1}$ kpc, their median velocity dispersion 207 km/s, and their median inferred M/L about $40h$. I discuss group dynamics further below, but it is worth noting here that the velocities are similar to the internal velocities within these relatively bright galaxies, and the typical separations are several times larger than the optical images.

The Hubble types of the galaxies show some interesting correlations with other properties of the groups. Hickson, Kindl and Huchra (1988a) find that groups containing a high fraction of E/S0 galaxies tend to be denser, more luminous, and to have higher velocity dispersions than groups with a high fraction of spirals. The correlation with velocity dispersion is particularly strong. These relations are reminiscent of the well known density-morphology relation discussed by Dressler (1980). They suggest that the apparent properties of the groups do, in fact, reflect their corresponding physical properties; apparent densities and velocity dispersions cannot be determined entirely by selection effects. Such relations can also explain the morphological concordance noted by Hickson, Kindl and Huchra (1988a): early-type galaxies prefer to form groups with other early-type galaxies, while spirals prefer to form groups with other spirals. A concordance consistent with the histograms of these authors is generated by a model in which the probability of of a group member being a spiral is correlated with some global property, X, of the group at a somewhat stronger level than that shown for the particular case where X is velocity dispersion.

5. The dynamical state of compact groups

The exact dynamical state of compact groups has long been a subject of controversy. The redshift concordance of most of the galaxies in most of the groups measured shows that these are certainly systems of associated objects rather than chance alignments. Of the Hickson quartets with complete redshift information, 12% turn out to be 3+1, *i.e.* systems with a single discordant redshift. Fully 50% of the quintets turn out to be 4+1 or 3+2. As discussed above, these numbers are probably consistent with chance projection, but in that case projection effects are not entirely negligible, particularly for the apparently richer systems.

A more subtle possibility is that the group members are associated because they are all part of the same diffuse group or cluster, but that the apparent compactness and isolation of the observed configuration is a result of chance alignments. Rose (1979) concluded from N-body simulations of "loose" groups that compact subgroupings appeared often enough to explain his observed frequency of compact systems. However, Hickson (1982; also Hickson *et al.* 1984) felt that the systematic properties of his groups showed them to be truly dense, a conclusion shared by Williams and Rood (1987), Sulentic (1987) and Hickson and Rood (1988). In contrast, Mamon (1986, 1987) claimed that most of Hickson's groups could be chance projections within larger aggregates. In a recent discussion of this issue, Walke and Mamon (1989) show that if the NBG group catalog of Tully (1987) is taken to represent the global properties of the loose group distribution, then projection effects can lead to a space density of spurious compact groups comparable to that found by Hickson. These "compact groups" should be preferentially located in the richer and higher density loose groups, and should often contain a "real" binary. It is not clear that this hypothesis is consistent with the rather modest density enhancements which Sulentic (1987) found in the neighborhood of the real Hickson groups. It also seems at odds with the low HI content of the groups, the high observed frequency of interactions and kinematic distortions, and with the observed correlations between their morphological composition and their apparent density, luminosity, and velocity dispersion. To this reviewer it seems that most of the Hickson groups do correspond to physically dense systems, but that many of them may be contaminated by superposed members of an associated loose group.

If compact groups really are physically dense, one must confront the question of their dynamical status. Because these groups are only a few times bigger than individual galaxies and have similar velocity dispersions, their inferred dynamical times and M/L values are only a few times those of galaxies. Thus either they have survived for many orbital times, or they only recently took up their present configuration. If they are bound and therefore will remain compact, one may wonder how long they can continue to survive. Rose (1979) has argued from simulations of 7-body "loose" groups that transient compact quartets occur frequently enough to explain the compact quartets in his catalog. Although Rose states that the transient quartets in his experiments are usually unbound, his tables show that this is not the case; in fact they are close to virial equilibrium, and must be transient because of the instability of the classical 4-body problem. Such transient bound configurations are quite rare in the somewhat more realistic simulations of Mamon (1986, 1987). This is due to the fact that pairs of his galaxies tend to merge before four of them can get together in a physically compact grouping. This effect is particularly important in simulations where the mass of the group is put in individual halos rather than the intergalactic medium, since the galaxies are then assigned large merger cross-sections. Most apparently compact subgroupings in Mamon's experiments turned out to be projection effects, and the true compact groupings rapidly merged into a single object. The main questions are then to determine accurate timescales for these processes, and to decide whether they are consistent with the observed statistics of compact groups, their progenitors and their progeny.

Direct interactions in compact groups are observed through the large number of optically distorted systems ($> 30\%$ according to Hickson 1989), of systems with kinematic peculiarities ($> 50\%$ according to Rubin, Hunter and Ford 1990) and through the presence in some groups of substantial intergalactic gas (Williams, McMahon and Van Gorkom 1990) and intergalactic stars (Sulentic 1987). (Note that Rose (1979) used the *absence* of intergalactic light in two groups to argue against significant interactions.) Of course, while these observations demonstrate that in many cases two or three galaxies in the group are interacting, they are not inconsistent with the possibility that one or two galaxies may often be superposed members of a surrounding loose group. For these reasons it is difficult to be sure exactly what multiplicity distribution to take for true compact systems. Although only 12% of Hickson's quartets have a discordant redshift, many more may involve accidental projection of a loosely associated galaxy. Half the observed quintets have discordant members, and the number of local projected galaxies may also be large. It is therefore possible that the number of truly compact systems containing four (and *a fortiori* five) galaxies is quite small. The record system of this type is V Zwicky 311 which has at least nine elliptical-like systems apparently embedded in a common envelope (Schneider and Gunn 1982).

The literature abounds with overly short estimates of the lifetimes of compact groups. The median separation of galaxies in the Hickson catalog is $40h^{-1}$ kpc in projection, corresponding to about $50h^{-1}$ kpc in three dimensions. The median group velocity dispersion is 200 km/s. Given that this is probably inflated by projection effects the typical orbital velocities of the galaxies are thus 300 km/s or less. At this velocity a circular orbit of radius $50h^{-1}$ kpc takes a tenth the Hubble time in an open universe and a seventh the Hubble time in a flat universe. Since the complete coalescence of two galaxies on a moderate eccentricity orbit cannot take place in much less than their orbital period, we see that a simple estimate of the minimum lifetime of compact groups is of order 10% of the age of the universe, rather than of order 1% as often assumed. In fact it is rather easy to lengthen this time, if it is still felt to be a problem, by assigning a major part of the mass of the group to a common background rather than attaching it to the individual galaxies (Barnes 1985, Mamon 1987). The extent to which this fix can be adopted is limited by the fact that the observed sizes of the galaxies are not very small compared to their separations; their total mass cannot, therefore, be a very small fraction of the group mass. In addition there is a problem because the mass density in compact groups is much higher than in loose groups. If the compact configuration is relatively young, the only reasonable way to achieve such a high background density of dark matter would seem to be to strip it off the galaxies as the compact group forms. However, the stripping and merging processes are so closely related, that fully self-consistent simulations offer the only realistic way to estimate group lifetimes in this situation.

6. N-body simulations of compact groups

The first program of fully self-consistent simulations of groups was that of Carnevali, Cavaliere and Santangelo (1981). These authors followed the evolution of virialised groups containing 10 or 20 galaxies, each made of 20 mass points. They found a kind of merging instability; once mergers started the increased mass and cross-section of the merger remnants produced a substantial enhancement of the merger rate. Their groups typically evolved into a single large remnant with a few remaining satellites, and resembled compact groups for roughly their first 7 dynamical times. (Following Barnes (1985) I define the group size, velocity dispersion and dynamical time through,

$$R = -GM^2/2E, \quad V^2 = -2E/M, \quad T = R/V, \tag{6.1}$$

where M and E are the group mass and energy *excluding* the internal binding energy of the galaxies.) Ishizawa *et al.* (1983) ran a single model of a collapsing 10 galaxy group, each galaxy represented by 100 particles. Very few of these galaxies survived the initial collapse. Almost all were incorporated into a single dominant remnant as they fell through it. This process was complete after the first two dynamical times, showing much more rapid evolution than the models of Carnevali *et al.* . The origin of this difference, together with many other aspects of the problem, was sorted out by the first truly systematic N-body study of small groups, the fine paper by Barnes (1985).

Barnes ran enembles of models of groups with 5 or 10 galaxies, both with and without significant amounts of generally distributed intergalactic matter, from virialised, from "maximum expansion", and from expanding initial conditions. In addition he studied a few examples in which both groups and galaxies formed "naturally" by hierarchical clustering in an expanding universe. The large number of models and the large variety of initial conditions allowed him to study both systematic trends and the scatter about those trends. In addition he presented his results in a manner designed to facilitate comparison with real groups. These experiments showed that merging occurs more slowly in groups with more galaxies, with more intergalactic matter, and with virialised, as opposed to collapsing, initial conditions. The central concentration of the "visible" parts of galaxies had little apparent influence on the merging rate. In groups of five galaxies it typically took only two to four initial dynamical times for merging to reduce the apparent population to two. However, during this evolution the *apparent* dynamical time derived from the galaxy positions and velocities was often much shorter than the true dynamical time. Thus the age of a group, measured in units of its apparent dynamical time, could be large even though the group was in fact evolving rapidly. Groups which had survived for more than 10 apparent dynamical times (and groups which would survive for at least 10 more) were easy to find. For comparison the median value of $(H_0 T_{app})^{-1}$ for the Hickson groups is about 40, corresponding to an age of 25 apparent dynamical times for $\Omega = 1$. Barnes's ensembles did contain a number of examples of groups with such extreme parameters (see his Figure 11).

More recently, Barnes (1989) has published a beautiful simulation of the evolution of a virialised group of six disk galaxies. His use of a total of more than 65,000 particles allowed him to make realistic models for the structure of the initial galaxies. As these galaxies interact and merge a variety of tidal streamers form and dissipate. In one case a few shells are formed as a small disk galaxy falls into an ellipsoidal merger remnant. In agreement with the results of many earlier experiments, such remnants are found to have radial structures similar to those of elliptical galaxies. However, because of substantial energy and angular momentum transfer to the unseen dark matter, they have substantially higher surface brightnesses and velocity dispersions than inferred by applying simple scaling arguments to the initial galaxies. As appreciated by the audience of the oral presentation of this talk, this is certainly a case where it is important to see the movie as well as to read the book.

7. Progeny and progenitors

It is clear that Hickson's selection criteria will induce a strong bias towards groups with short apparent dynamical times. Barnes's work shows that as a group evolves its apparent dynamical time is often as small as 10% to 20% of its remaining lifetime (Barnes 1985, Figures 7 and 10). Thus it seems likely that the typical lifetime of a Hickson group will be 5 to 10 times its apparent dynamical time. Since the median apparent dynamical time of the groups is 2.5% to 4% of the age of the universe, their typical remaining lifetime should be greater than 10%, and probably nearer to 20%, of t_0. Hickson (1989) estimates that his group population is responsible for 1% of the total observed luminosity density. Since the stellar population of at least the spirals in these groups is expected to fade substantially once merging is complete (see, for example, Veeraraghavan and White 1985), it would seem that 10% is a safe upper limit for the luminosity contribution of remnants of earlier compact groups, and that a few percent is a more likely estimate. There is thus no contradiction with the observed abundance of the elliptical galaxies which would presumably have to be identified with these remnants; the data presented by Schechter and Dressler (1987) suggest that at least 15% of the luminosity density comes from such galaxies. It remains to be proven, of course, that the systematic properties of bright elliptical galaxies are consistent with the formation of a significant fraction of them (perhaps about 20%) from compact groups. This is an extension of an old argument (*e.g.* Efstathiou and Jones 1980; Ostriker 1980; Tremaine 1981; White 1982; Veeraraghavan and White 1985; Barnes 1988; Schweizer 1983, this volume).

As most of the Hickson groups merge over the next 10% to 20% of a Hubble time, they will presumably be replaced by new compact configurations. These must form as existing loose groups collapse and amalgamate. Perhaps 50% of galaxies are part of loose groupings with collapse times comparable to the age of the universe (Nolthenius and White 1987, Table 3). If 1% are to be in compact groups, then on average a few percent of the evolution of a group must be spent with apparent dynamical times as short as those of the Hickson groups. This is at least roughly consistent with the diagrams presented by Barnes (1985); many of the groups in his ensemble H still have 4 or 5 remaining galaxies at epochs when their apparent dynamical times satisfy $T_{app}/t_0 < 0.05$ (compare his Figures 5c and 11b). Similarly, a different projection of Figure 1d of Barnes (1989) would clearly give a compact configuration of two disk and two elliptical galaxies with an apparent dynamical time considerably less than one tenth its age. Although the initial conditions of these simulations are idealised, they do not seem an unreasonable representation of the observed loose groupings from which future compact systems must form. I therefore conclude that there is no current reason to believe that the number of Hickson groups and their observed density are inconsistent with formation through a conventional hierarchical clustering process.

It is difficult to give a more rigorous analysis than the above without a specific model for the initial conditions; a particular example might be the Cold Dark Matter model (Blumenthal *et al.* 1984; Davis *et al.* 1985). Such a model makes specific predictions for the fraction of "Hickson groups" which result from the superposition of nearby galaxies on smaller interacting groups, and so removes one of the major uncertainties in interpreting the statistics of the observed systems. It also makes possible, a detailed comparison with the observed multiplicity, morphology, density and velocity dispersion distributions. This seems to me to be the most promising way to determine whether compact groups pose any serious challenge to our current ideas about the formation of structure in the Universe.

Acknowledgements

I thank Paul Hickson and Josh Barnes for very helpful discussions and for providing data and a movie before publication. This work was supported by NSF grant AST-8822297.

References

Arp, H. 1973, *Ap. J.*, **183**, 411.
Arp, H., and Lorre, J. J. 1976, *Ap. J.*, **210**, 58.
Barnes, J. E. 1985, *M.N.R.A.S.*, **215**, 517.
Barnes, J. E. 1988, *Ap. J.*, **331**, 699.
Barnes, J. E. 1989, *Nature*, **338**, 132.
Blumenthal, G. R., Faber, S. M., Primack, J. R., and Rees, M. J. 1984, *Nature*, **311**, 517.
Burbidge, E. M., and Sargent, W. L. W. 1971, in *Nuclei of Galaxies*, ed. D. O'Connell (North-Holland), p. 351.
Carnevali, P., Cavalere, A., and Santangelo, P. 1981, *Ap. J.*, **249**, 449.
Davis, M., Efstathiou, G., Frenk, C. S., and White, S. D. M. 1985, *Ap. J.*, **292**, 371.
Dressler, A. 1980, *Ap. J.*, **236**, 351.
Efstathiou, G., and Jones, B. J. T. 1980, *Comm. Ap. Sp. Sci.*, **8**, 169.
Hickson, P. 1982, *Ap. J.*, **255**, 382.
Hickson, P. 1989, private communication.
Huchra, P., Kindl, E., and Auman, J. 1989, preprint.
Hickson, P., Kindl, E., and Huchra, J. P. 1988a, *Ap. J.*, **331**, 64.
Hickson, P., Kindl, E., and Huchra, J. P. 1988b, *Ap. J. (Letters)*, **329**, L65.
Hickson, P., Menon, T. K., Palumbo, G. G. C., and Persic, M. Ap. J., *in press*, .
Hickson, P., Ninkov, Z., Huchra, J. P., and Mamon, G. A. 1984, in *Clusters and Groups of Galaxies*, ed. F. Mardirossian, G. Giuricin, and M. Mezzetti (Dordrecht: Reidel), p. 367.
Hickson, P., and Rood, H. J. 1988, *Ap. J. (Letters)*, **331**, L69.
Ishizawa, T., Matsumoto, R., Tajima, T., Kageyama, H., and Sakai, H. 1983, *Pub. Astr. Soc. Japan*, **35**, 61.
Larson, R. B., and Tinsley, B. M. 1978, *Ap. J.*, **219**, 46.
Mamon, G. A. 1986, *Ap. J.*, **306**, 426.
Mamon, G. A. 1987, *Ap. J.*, **321**, 622.
Menon, T. K., and Hickson, P. 1979, in preparation.
Menon, T. K., and Hickson, P. 1985, *Ap. J.*, **296**, 60.
Nolthenius, R., and White, S. D. M. 1987, *M.N.R.A.S.*, **235**, 505.
Ostriker, J. P. 1980, *Comments Ap.*, **8**, 177.
Rose, J. A. 1977, *Ap. J.*, **211**, 311.
Rose, J. A. 1979, *Ap. J.*, **231**, 10.
Rubin, V. C., Hunter, D.,a nd Ford, W. K. 1990, in preparation.
Schechter, P. L., and Dressler, A. 1987, *A. J.*, **94**, 563.
Schneider, D. P., and Gunn, J. E. 1982, *Ap. J.*, **263**, 14.
Schweizer, F. 1983, in *Internal Kinematics and Dynamics of Galaxies*, ed. E. Athanassoula (Dordrecht: Reidel), p. 319.
Seyfert, C. K. 1948, *A. J.*, **53**, 203.
Stephan, M. E. 1877, *M.N.R.A.S.*, **37**, 334.
Sulentic, J. W. 1983, *Ap. J.*, **270**, 417.
Sulenic, J. W. 1987, *Ap. J.*, **322**, 605.
Sulentic, J. W., and Arp. H. C. 1983, *A. J.*, **88**, 267.
Sulentic, J. W., and Lorre, J. J. 1983, *Astr. Ap.*, **120**, 36.
Tremaine, S. D. 1981, in *Struct. Evol. Norm. Gal.*, ed. S. M. Fall, and D. Lynden-Bell, p. 67.
Tully, R. B. 1987, *Ap. J.*, **321**, 280.
Veeraraghavan, S., and White, S. D. M. 1985, *Ap. J.*, **296**, 336.
Walke, D. G., and Mamon, G. A. 1989, *Astr. Ap.*, in press.
White, S. D. M. 1982, in *Morph. Dynam. Gal.*, ed. L. Martinet, and M. Mayor (Geneva Observatory).
Williams, B., and Rood, H. J. 1987, *Ap. J. Suppl.*, **63**, 265.

Rotation Curves of Spiral Galaxies in Groups and Clusters

B.C. Whitmore

Space Telescope Science Institute, 3700 San Martin Drive,
Baltimore, MD21218, USA

1. INTRODUCTION

The measurement of emission-line rotation curves is one of our primary tools for studying the structure and dynamics of galaxies. At the same time, it is becoming increasingly evident that interactions between galaxies, both now and in the distant past, may have played an important role in the evolution of galaxies. Such interactions would be most prevalent in regions of high galactic density such as groups and clusters. It is therefore rather surprising that it has only been during the last few years that a sample of extended, emission-line rotation curves has become available for galaxies in these dense environments. This article reviews some of the findings from these recent observations; more specifically, that rotation curves for spiral galaxies near the centers of clusters are falling, and a large fraction of rotation curves in compact groups are abnormal.

2. ROTATION CURVES FOR SPIRAL GALAXIES IN CLUSTERS

Based on limited optical material, Chincarini and de Souza [1] concluded that there was no qualitative difference between the rotation curves of cluster and field spirals. Guhathakurta et al. [2] supported this view based on 21 cm observations. However, Burstein et al. [3] reported that there was a difference in the distribution of mass types (i.e., the integral mass curve derived from the rotation curve) of cluster and field galaxies. Forbes and Whitmore [4] confirmed this result using a more objective method of determining mass types.

More recently, a set of extended H_α rotation curves for 21 spiral galaxies in four clusters was published by Rubin, Whitmore, and Ford [5]. This paper also examined the correlations between four global properties of the galaxies (luminosity, radius, maximum rotational velocity, and mass). While a variety of possible systematic biases make it difficult to make firm conclusions, they find tentative evidence that the Tully-Fisher relation between luminosity and rotational velocity is different in cluster galaxies than for field galaxies. While a much larger sample will be required to ascertain the reality of this effect, such a result would have important implications for studies which use the Tully-Fisher relation to determine distances to galaxies.

The measurement of the shape of the rotation curve is much less problematic than the determination of the global properties, since questions such as corrections for extinction, inclination, or large-scale motions in the universe are not relevant. Whitmore, Forbes,

Dynamics and Interactions of Galaxies Editor: R. Wielen

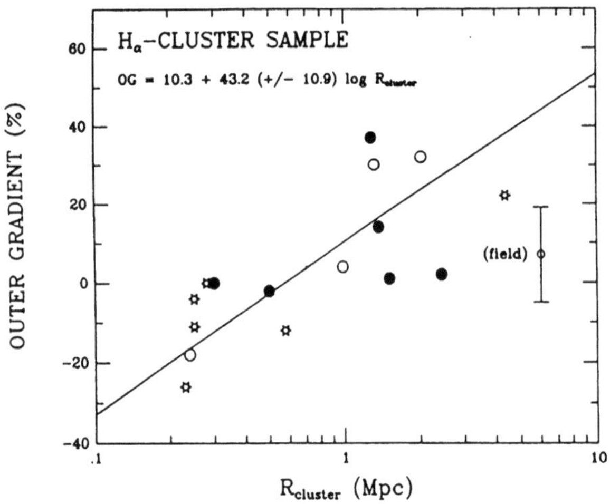

Figure 1 - Outer gradient of the rotation curve versus $R_{cluster}$ from Whitmore, Forbes, and Rubin [6]. Open circles are Sa galaxies; filled circles are Sb galaxies; and open stars are Sc galaxies. The average value of the outer gradient for the field sample is included with error bars showing the 1 σ scatter of the distribution.

and Rubin [6] examined the shape of rotation curves from Rubin, Whitmore, and Ford [5] using various methods of parameterizing the gradients. The most dramatic correlation is found using the outer gradient of the rotation curves. This is the percentage increase between 0.4 R_{25} and 0.8 R_{25}, normalized to the maximum rotational velocity. Figure 1 shows the correlation between the outer gradient and the radius from the center of the cluster ($R_{cluster}$). *The rotation curves for galaxies near the centers of clusters tend to have falling rotation curves in their outer regions, while galaxies farther from the cluster center, or in the field, tend to have flat or rising rotation curves.*

What physical mechanism is responsible for the correlation in Figure 1? Has the dark halo been stripped, either by tidal interactions with other galaxies or with the overall gravitational field of the cluster? Perhaps the inner cluster environment never allowed the galaxy to form a halo. Another possibility is that the inner cluster environment has modified the luminosity distribution, so that the decline in the rotation curves can be explained entirely in terms of the loss of luminous material. To address these questions we have calculated integral mass-to-light ($M/L = V^2R/L_{red}$) gradients for the cluster galaxies for which we have reliable photometry.

Figure 2 shows the M/L gradients for the cluster galaxies arranged in order of distance from the center of the clusters. We find that the inner cluster galaxies have flatter M/L gradients than the galaxies in the outer regions of the cluster. *If a strong gradient in M/L is indicative of a large fraction of dark matter in the outer regions of a galaxy, then the present sample provides evidence that the galaxies near the centers of clusters have a lower fraction of their mass in the form of dark matter than the galaxies in the outer regions of a cluster.*

Guhathakurta et al. [2] have obtained 21 cm observations at the VLA for 21 galaxies in the Virgo cluster. They conclude that the rotation curves for cluster galaxies are qualitatively similar to field galaxies. However, the rotation curves for the galaxies near the center of the Virgo clusters generally do not extend beyond a few tenths of R_{25} since they are HI deficient, making it difficult to make any meaningful comparison in this critical

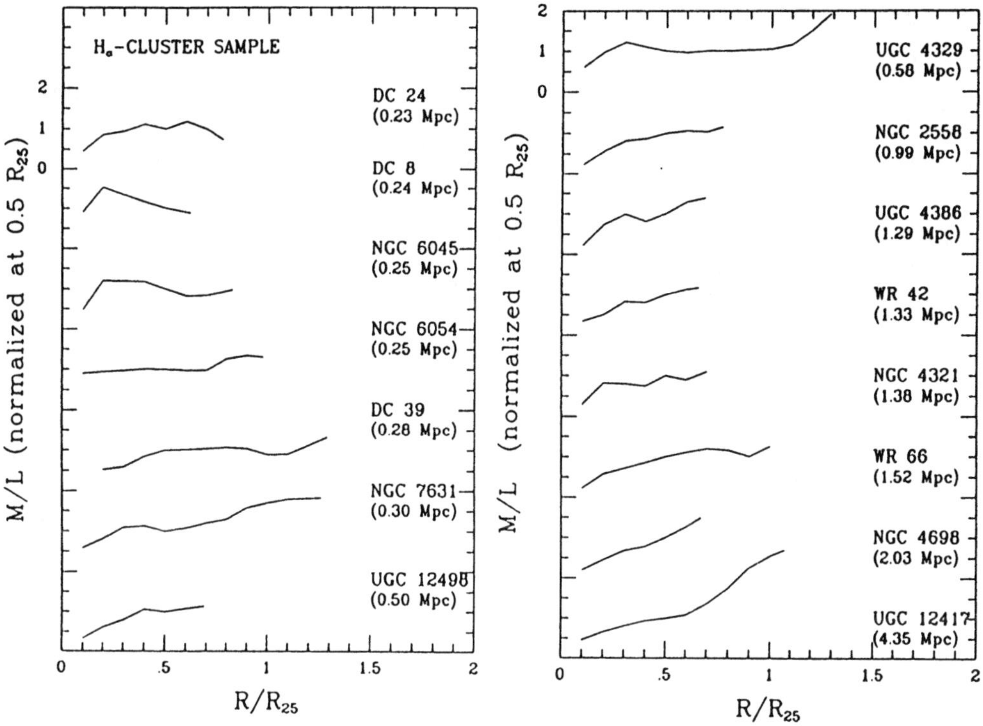

Figure 2 - Mass-to-light ratio gradients arranged in order of $R_{cluster}$ from Whitmore, Forbes, and Rubin [6]. Same symbol definition as Figure 1.

region. In addition, the poor spatial resolution of the radio observations (beamwidth = 45″ for most of the observations) makes the comparison uncertain. After eliminating the galaxies where resolution in questionable, Whitmore, Forbes, and Rubin find that the HI data follows essentially the same relationship between the outer gradient of the rotation curve and $R_{cluster}$ as the H_α data, although the lack of many galaxies near the center of the cluster makes the relationship much less compelling.

3. ROTATION CURVES FOR SPIRAL GALAXIES IN COMPACT GROUPS

The effect of the cluster environment on emission-line rotation curves is subtle, with the outer galaxies showing essentially no effect, and the inner galaxies showing outer gradients which are typically 10 - 30 % lower than the outer galaxies. The two sides of the rotation curves generally match relatively well, indicating the gas is in circular motion.

In compact groups, where the density is generally as high or higher than in the cores of clusters, and the relative velocities are lower, the effect of the environment on rotation curves is often much stronger. Rubin, Ford, and Hunter [7] have obtained spectra of 45 galaxies in 16 groups. Of the 33 spirals in the sample, 9 are too abnormal to classify, 13 are peculiar or irregular, and only 11 are normal. The rotation curves which are defined as peculiar generally fall into two categories: asymmetrical, with the two sides of the

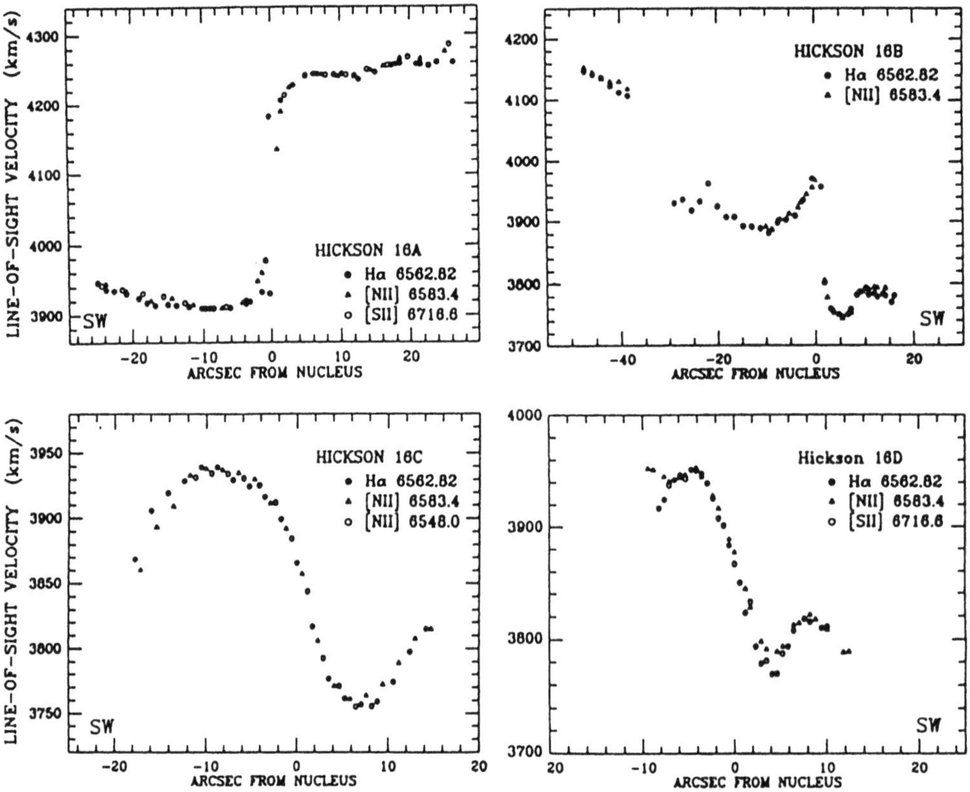

Figure 3 - Emission-line rotation curves for galaxies in Hickson 16 from Rubin, Ford, and Hunter [7]. Hickson 16A is an asymmetric curve while Hickson 16C is sinusoidal.

rotation curve having different outer gradients; and sinusoidal, with velocities actually going to zero or reversing at large radii. The famous merger candidate NGC 7252 also has a sinusoidal rotation curve (Schweizer [8]). *This provides support for the idea that these abnormal rotation curves are the result of recent interactions.* Figure 3 shows four of the rotation curves in Hickson 16. Hickson 16A is a good example of an asymmetric rotation curve, with one side rising and other falling, while Hickson 16C is a sinusoidal rotation curve. Both 16B and 16D are in the abnormal category.

Rubin, Ford, and Hunter also obtained spectra of 12 elliptical and S0 galaxies in the Hickson groups. They find a high incidence of extended emission in these galaxies, and in Hickson 23c the gas is counterrotating with respect to the stellar compoent. The presence of this gas in generally gas free systems is more evidence for the recent accretion of material from another galaxy during an interaction or merger event.

REFERENCES

1. Chincarini, G. and de Souza, R. 1985, Astr. Ap., **153**, 218.
2. Guhathakurta, P., van Gorkom, J. H., Kotanyi, C. G., and Balkowski, C. 1988, A. J., **96**, 851.

3. Burstein, D., Rubin, V. C., Ford, W. K., and Whitmore, B. C. 1986, Ap. J. (Letters), **305**, L11.
4. Forbes, D., and Whitmore, B. C. 1988, Ap. J., **339**, 657.
5. Rubin, V. C., Whitmore, B. C., and Ford, W. K. Jr. 1988, Ap. J., **333**, 522.
6. Whitmore, B. C., Forbes, D. A., and Rubin, V. C. 1988, Ap. J., **333**, 542.
7. Rubin, V. C., Ford, W. K., Jr., and Hunter, D. 1989, in Abstracts from 1989 STScI Workshop on Clusters of Galaxies, ed. by M. Fitchett and W. Oegerle.
8. Schweizer, F. 1982, Ap. J., **252**, 455.

The Origin of Central Cluster Galaxies

S. Tremaine

Canadian Institute for Theoretical Astrophysics, University of Toronto, 60 St. George Street, Toronto, Ontario M5S 1A1, Canada

Abstract The galaxies located at the centers of clusters of galaxies are the most luminous stellar systems in the universe, and explaining the origin of these remarkable objects is a major challenge for any comprehensive theory of galaxy formation. One possibility is that central galaxies are made by a series of mergers of cluster galaxies that have spiraled to the bottom of the cluster potential well through dynamical friction. However, the resulting rate of accumulation of luminosity at the cluster center appears to be too slow—by more than a factor of two—to produce central galaxies with the luminosities observed, at least in virialized clusters similar to those we see today. A related possibility is that central galaxies are formed before the cluster virializes, during the early stages of hierarchical clustering when the relative velocities of the galaxies are low enough that merging can occur in pairwise galaxy encounters. However, so far there are no convincing numerical models of hierarchical clustering in which the most massive stellar systems formed resemble observed central galaxies. The evidence for recent merging in central galaxies is still difficult to interpret: multiple nuclei in central galaxies do not provide direct evidence for merging, as most are simply projected cluster members, but there is strong circumstantial evidence that dumbbell galaxies are the ex-central galaxies from two recently merged clusters that will themselves merge in about 0.2 Hubble times.

1. Introduction

Wo viel licht is, ist starker Schatten. (Goethe, *Götz von Berlichingen*)

The bottoms of potential wells are sometimes interesting places. For example, the potential minima at the centers of galaxies are the sites of active galactic nuclei and may host supermassive black holes. The centers of star clusters are less spectacular, but sometimes exhibit cusps in the distribution of starlight that are believed to arise from core collapse. On the other hand, the centers of open clusters or of normal stars are not believed to contain any unusual features.

The potential wells associated with clusters of galaxies are $\gtrsim 10$ times deeper than the wells associated with galaxies, and it is natural to ask whether interesting phenomena occur at the bottoms of these potential wells. Unfortunately, it is difficult to locate the potential minimum in a given cluster, for at least two reasons: (i) Clusters are dynamically young and in many cases are still accreting new material; as a result their density distribution is often clumpy and irregular (Geller and Beers 1982, Dressler and Shectman 1988), and the potential surface may have several drainage basins, each with its own local minimum. (ii) The surface density is measured by counting galaxies, and bright galaxies are so rare that the density distribution can only be determined with limited accuracy. Because of the limited statistics, observers have traditionally assumed that the surface density distribution was azimuthally symmetric, so that the cluster center could be determined using strip counts and then the density distribution could be determined using ring counts; however, many clusters are so clumpy that this procedure does not yield an accurate picture of the cluster geometry.

Dynamics and Interactions of Galaxies Editor: R. Wielen

Our understanding of the spatial structure of clusters has improved dramatically in the last decade. The *Einstein* satellite has mapped the distribution of X-ray emitting gas in some 50 clusters, and the density distribution of this gas can be used to trace the cluster potential (Jones and Forman 1984). Meanwhile, optical observers have compiled extensive databases of galaxy positions, photometry, and radial velocities for many clusters. The locations of potential minima can be determined by finding the sites of maximum X-ray brightness or galaxy surface density (which coincide, in cases where comparison can be made; see, e.g. Gioia et al. 1982).

Given that we can locate the potential minima in clusters of galaxies, what do we find there? Jones and Forman (1984) show that in most cases, the potential minimum as measured by the maximum X-ray surface brightness coincides with the location of the brightest galaxy in the cluster core. Similarly, Beers and Geller (1983) find that the potential minimum as marked by the local maximum in galaxy surface density is often occupied by a bright galaxy. Moreover, the galaxies found at the potential minima are unusual in that they have extended luminous halos or envelopes (morphological type D or cD; see Tonry 1987 for a review of these galaxies). Over 80% of the D and cD galaxies in the sample examined by Beers and Geller are located at surface density maxima; and over 80% are the brightest galaxies in their cores. However, morphological type is a better predictor than brightness, since faint D and cD galaxies are still preferentially located at density maxima, while bright galaxies of other types are not (Beers and Geller 1983). Kinematic data confirm that D and cD galaxies tend to be located at potential minima, since their rms velocity with respect to the cluster mean is much smaller than that of typical galaxies (Quintana and Lawrie 1982, Smith et al. 1985).

According to the original definition, D galaxies are characterized by "an elliptical-like nucleus surrounded by an extensive envelope...The very large D galaxies observed in clusters are given the prefix 'c' " (Matthews, Morgan and Schmidt 1964). Unfortunately, there has never been universal agreement on the proper use of the terms D and cD. Some observers, starting with Morgan and Lesh (1965) modify the definition of a cD to include secondary properties (e.g., the cD is the brightest cluster member and is centrally located in the cluster), but strictly, these properties are neither necessary nor sufficient to identify a cD (Dressler 1984). An additional source of confusion is that most galaxies originally classified as D's turn out to be S0's (see discussion following Tonry 1987).

Despite the unsettled notation, the astronomical facts are clear, and perhaps they can best be stated by banishing the term "cD" wherever possible in this review and using the term "central galaxy" to denote a galaxy that is located at the bottom of a potential basin (as measured by maxima in the surface density of galaxies or X-ray surface brightness). Then: (i) most galaxies with extended envelopes (i.e. surface brightness profiles with relatively flat logarithmic slopes) are central galaxies; (ii) central galaxies are brighter than typical galaxies, and are usually brighter than any other galaxy in the local basin.

These striking correlations demand an explanation, and many theories for the formation of central galaxies have been advanced (see Dressler 1984 and Sarazin 1986 for reviews):

(i) The cooling time for gas in some clusters may be short enough that $\sim 10^{12}\,M_\odot$ of gas may accumulate at the cluster center over a Hubble time. Efficient conversion of the gas into stars could create a bright central galaxy. However, the colors of most central galaxies are inconsistent with a substantial present star formation rate unless most of the stars are low-mass and therefore undetectable; and in this case the mass-to-light ratio should be high, whereas in fact central galaxies have the same mass-to-light ratios as normal ellipticals.

(ii) Some special aspect of the galaxy formation process may produce bright galaxies with extended envelopes only in the high-density regions characteristic of cluster centers. A special formation process might also stimulate the production of globular clusters, which appear to be several times more numerous per unit luminosity in central galaxies relative to normal ellipticals (Harris 1987). This enhanced globular cluster population is difficult to explain in other models for the formation of central galaxies.

(iii) The central galaxy may consist of debris stripped from cluster galaxies by the cluster tidal field and by encounters with other galaxies. Stripped material may comprise the envelopes of central galaxies but the body of the galaxy must have a different origin since its velocity dispersion ($\sim 300\,\mathrm{km\,s^{-1}}$) is much smaller than the cluster dispersion ($\sim 800\,\mathrm{km\,s^{-1}}$).

(iv) The central galaxy may be made by a series of mergers of cluster galaxies that have spiraled to the bottom of the cluster potential well through dynamical friction.

These models are not necessarily mutually exclusive, since the cores and envelopes of central galaxies may be formed in different ways; also, not all central galaxies are necessarily formed by the same process. Nevertheless, for the sake of brevity, in this review I will discuss only the last model in the list, in which central galaxies are made by mergers.

2. Simple merger models

The idea that central galaxies are made by mergers dates back to papers by Lecar (1975), Ostriker and Tremaine (1975), and S. White (1976a). Although sophisticated numerical models of cluster evolution have been constructed more recently (notably by Merritt 1984a, 1985, Richstone and Malumuth 1983, and Malumuth and Richstone 1984), analytic models still provide a good introduction to the theory. In deriving these, I shall refer to the book by Binney and Tremaine (1987, hereafter BT) for some of the more standard results.

I first consider the dynamical evolution of galaxies after the cluster core has reached a quasi-steady state following collapse and virialization; evolution during cluster collapse is discussed in the next section.

A massive galaxy orbiting in a cluster is subject to a drag force arising from its gravitational interactions with the other mass in the cluster. This drag force is called "dynamical friction" (Chandrasekhar 1943; see also BT). For a galaxy of mass m travelling at speed $\mathbf{v}$ through an infinite homogeneous medium of density ρ, the drag force per unit mass is

$$\mathbf{a}_d \equiv \frac{d\mathbf{v}}{dt} = -\frac{4\pi G^2 m\rho \ln \Lambda F(v)}{v^3}\mathbf{v}. \qquad (1)$$

Here $F(v)$ is the fractional mass of objects in the medium with speeds less than $v = |\mathbf{v}|$, and $\Lambda = b_{\max}/b_{\min}$, where $b_{\max}$ and $b_{\min}$ are the maximum and minimum impact parameters of encounters contributing to the drag. The nature of the objects in the medium is irrelevant so long as they are both collisionless and of mass much less than m (so that there is no stochastic acceleration of the galaxy due to individual encounters).

Although (1) was derived for a homogeneous medium, it can also be used to compute the decay of the orbit of a galaxy in a cluster, with ρ now denoting the local density at the galaxy's position (I assume that most of the cluster mass is in objects with mass much less than a galaxy's). In this context, (1) has been tested against many numerical simulations and usually gives remarkably accurate predictions for the rate of orbital decay (S. White 1978, Lin and Tremaine 1983, Bontekoe and van Albada 1987, Zaritsky and S. White 1988).

I will model the cluster density distribution as an isothermal sphere (see BT), since this provides a satisfactory fit to the number density distribution in the inner parts of rich clusters. The isothermal sphere is specified by two parameters, the one-dimensional rms velocity dispersion σ, which is independent of radius, and the core radius r_c, which measures the extent of the constant-density core region.

The median dispersion of clusters containing a D or cD galaxy is $650\,\mathrm{km\,s^{-1}}$, similar to the dispersion for rich clusters generally (Geller 1988), but since I am concerned with the central rather than the average dispersion over the cluster I shall use the somewhat higher value $\sigma = 800\,\mathrm{km\,s^{-1}}$. The core radius r_c is less certain. Bahcall (1975) estimates $r_c = 125 \pm 20 h^{-1}$ kpc (h is the Hubble constant H_0 in units of $100\,\mathrm{km\,s^{-1}\,Mpc^{-1}}$), while Dressler (1978) estimates $r_c = 240 \pm 60 h^{-1}$ kpc, almost a factor of two larger. Much of the uncertainty arises because it is difficult to locate the exact cluster center. Beers and Tonry (1986; see also Yahil 1974) stress that centering errors usually lead to a measured core radius that is too large. They argue that in most clusters there is no evidence for any non-zero core radius, and show that in clusters whose centers are well-defined, by the presence of a cD galaxy or a local maximum in the X-ray surface brightness, the surface density of galaxies increases as r^{-1} right into the cluster center. (Indirect evidence from the frequency of multiple nuclei in central galaxies [see Lauer 1989 and §3] and from models of clusters as gravitational lenses [Grossman and Narayan 1989] also suggests that the core radius is much smaller than the Bahcall or Dressler estimates.)

The power-law density profiles found by Beers and Tonry and by Yahil closely resemble the profile of the singular (i.e. $r_c = 0$) isothermal sphere. Since the singular isothermal has the additional benefit of analytic simplicity, I shall adopt it as a crude model of the mass distribution in a cluster. The density, potential, enclosed mass, and fraction of objects with speed $< v$ are given by (see BT):

$$\rho(r) = \frac{\sigma^2}{2\pi G r^2}, \quad \Phi(r) = 2\sigma^2 \ln r + \text{const}, \quad M(r) = \frac{2\sigma^2 r}{G}, \quad F(v) = \text{erf}(x) - x\,\text{erf}'(x), \tag{2}$$

where $x = v/\sqrt{2}\sigma$ and erf denotes the error function.

I assume for simplicity that the galaxies move on circular orbits. Then the rate of change of orbital radius is related to the specific angular momentum $J(r) = \sqrt{r d\Phi/dr}$ and the drag force (eq. 1) by $dr/dt = dJ/dt(dJ/dr)^{-1} = -r|\mathbf{a}_d|(dJ/dr)^{-1}$. Specializing to the singular isothermal sphere and using (1) and (2) yields

$$\frac{dr}{dt} = -\frac{\text{erf}(1) - \text{erf}'(1)}{\sqrt{2}} \cdot \frac{Gm}{\sigma r} \ln \Lambda = -0.302 \frac{Gm}{\sigma r} \ln \Lambda. \tag{3}$$

The decay rate is proportional to the galaxy mass and hence is affected by the extent of massive halos in the cluster galaxies. I shall examine two limiting cases and compute for each the rate at which galaxies spiral into the center of the cluster:

(i) No massive halo I assume that the galaxy mass-to-light ratio Υ is independent of radius and equal to its value in the core of a typical nearby elliptical, $\Upsilon = 12h\Upsilon_\odot$ in the V band (cf. Lauer 1985). Tidal stripping by the cluster potential is small since the galaxy's mass is concentrated at small radii; thus I assume m is time-independent. I shall also neglect changes in $\ln \Lambda$ as the galaxy spirals in. Then (3) yields the initial orbital radius of galaxies that spiral into the cluster center in a time t:

$$r^2(t) = 0.604 \frac{Gmt}{\sigma} \ln \Lambda = (63h^{-1}\ \text{kpc})^2 \left(\frac{800\,\text{km}\,\text{s}^{-1}}{\sigma}\right) \left(\frac{\Upsilon}{12h\Upsilon_\odot}\right) \left(\frac{L}{L_\star}\right) (H_0 t) \ln \Lambda, \tag{4}$$

where I have written the galaxy luminosity L in units of the characteristic luminosity $L_\star = 1.0 \times 10^{10} h^{-2} L_\odot$.

Taking a Schechter luminosity function, and assuming that the density distribution of the galaxies is the same as that of the cluster mass, I can write the number density of galaxies per unit luminosity and volume in the form

$$n(L, r) dL d\mathbf{r} = n_\star(r) d\mathbf{r} \exp\left(-\frac{L}{L_\star}\right) \left(\frac{L}{L_\star}\right)^\alpha \frac{dL}{L_\star}, \tag{5}$$

where $\alpha = -1.25$ (e.g., Colless 1989). I assume that the cluster mass-to-light ratio Υ_c (only the light in galaxies is counted) is independent of position, so that $n_\star(r) \propto \rho(r)$; thus

$$\frac{\rho(r)}{n_\star(r)} = \Upsilon_c L_\star (\alpha + 1)!. \tag{6}$$

Combining (2), (4), (5) and (6) gives the total luminosity accreted to the cluster center:

$$\begin{aligned}\frac{L_{\rm acc}}{L_\star} &= 2\frac{(\alpha+\frac{3}{2})!}{(\alpha+1)!}\Upsilon_c^{-1}\left(\frac{0.604\sigma^3\Upsilon t\ln\Lambda}{GL_\star}\right)^{1/2} \\ &= 3.1\left(\frac{360h\Upsilon_\odot}{\Upsilon_c}\right)\left(\frac{\Upsilon}{12h\Upsilon_\odot}\right)^{1/2}\left(\frac{\sigma}{800\,{\rm km\,s^{-1}}}\right)^{3/2}\left(\tfrac{3}{2}H_0t\right)^{1/2}(\ln\Lambda)^{1/2}.\end{aligned} \tag{7}$$

The comparison value $360h\Upsilon_\odot$ used for the cluster mass-to-light ratio is the value obtained by Kent and Gunn (see Kent and Sargent 1983) for the Coma cluster, which is typical of rich clusters. The factor $\frac{3}{2}H_0t$ is unity if the cluster formed shortly after the Big Bang in a critical ($\Omega = 1$) Friedmann universe.

The appropriate value of $\ln\Lambda = \ln(b_{\rm max}/b_{\rm min})$ may be estimated by setting $b_{\rm max}$ to the typical orbital radius, here $60h^{-1}$ kpc by (4), and $b_{\rm min}$ to the median radius of a typical elliptical, $r_h \simeq 5h^{-1}$ kpc (see Bontekoe and van Albada 1987). (These values are approximate but only enter (7) in the square root of a logarithm.) Thus $\ln\Lambda = 2.5$ and I find

$$\frac{L_{\rm acc}}{L_\star} = 4.9\left(\frac{360h\Upsilon_\odot}{\Upsilon_c}\right)\left(\frac{\Upsilon}{12h\Upsilon_\odot}\right)^{1/2}\left(\frac{\sigma}{800\,{\rm km\,s^{-1}}}\right)^{3/2}\left(\tfrac{3}{2}H_0t\right)^{1/2}. \tag{8}$$

(ii) Massive halo Tidal forces from the cluster place a strong constraint on the extent of the massive halo (Gunn 1977, Merritt 1984a). Following Merritt, I locate the tidal radius r_t at the collinear Lagrange points (the locations where the effective potential—cluster plus galaxy plus centrifugal potential from the orbital motion—has a saddle point). For a galaxy of mass m on a circular orbit of radius r in a cluster potential given by (2) I find

$$r_t^3 = \frac{Gmr^2}{4\sigma^2}. \tag{9}$$

I approximate the galaxy as a singular isothermal sphere (2) with dispersion σ_g, which I identify with the central line-of-sight dispersion in the galaxy. I then write the galaxy mass as a fraction $f < 1$ of the mass contained in the isothermal sphere within the tidal radius r_t. (This is a plausible approximation since the dispersion is likely to decrease outwards, so that the actual galaxy mass inside any radius will be less than the mass in the isothermal sphere.) Thus

$$m = \frac{2\sigma_g^2 f r_t}{G}. \tag{10}$$

The fraction f is uncertain but a plausible value is $f = 0.5$. Then (9) and (10) yield

$$m = 2^{1/2}f^{3/2}\frac{\sigma_g^3 r}{G\sigma}, \qquad r_t = 2^{-1/2}f^{1/2}\frac{\sigma_g}{\sigma}r. \tag{11}$$

Then (3) yields the initial radius of galaxies that spiral into the cluster center in a time t:

$$r(t) = 0.428f^{3/2}\frac{\sigma_g^3\ln\Lambda t}{\sigma^2} = 25h^{-1}\,{\rm kpc}\left(\frac{800\,{\rm km\,s^{-1}}}{\sigma}\right)^2\left(\frac{L}{L_\star}\right)^{0.75}\left(\frac{f}{0.5}\right)^{3/2}(H_0t)\ln\Lambda, \tag{12}$$

where I have used the empirical Faber-Jackson law,

$$\sigma_g = \sigma_\star\left(\frac{L}{L_\star}\right)^{0.25}, \qquad \text{with} \quad \sigma_\star = 220\,{\rm km\,s^{-1}}. \tag{13}$$

Using (2), (5), and (6), I find the accreted luminosity to be

$$\frac{L_{\rm acc}}{L_\star} = 0.855 f^{3/2} \frac{(\alpha+0.75)!}{(\alpha+1)!} \Upsilon_c^{-1} \left(\frac{\sigma_\star^3 t \ln\Lambda}{GL_\star}\right)$$
$$= 2.0 \left(\frac{f}{0.5}\right)^{3/2} \left(\frac{360h\Upsilon_\odot}{\Upsilon_c}\right) \left(\tfrac{3}{2}H_0 t\right) \left(\frac{\sigma_\star}{220\,\mathrm{km\,s^{-1}}}\right)^3 \ln\Lambda. \tag{14}$$

I shall take $\ln\Lambda = \ln(r/0.5r_t) = \ln(2^{1/2}\sigma/0.5f^{1/2}\sigma_g)$ since the median radius is probably about half the tidal radius; for the standard values of the parameters $\ln\Lambda = 2.7$ and I have

$$\frac{L_{\rm acc}}{L_\star} = 5.4 \left(\frac{f}{0.5}\right)^{3/2} \left(\frac{360h\Upsilon_\odot}{\Upsilon_c}\right) \left(\tfrac{3}{2}H_0 t\right) \left(\frac{\sigma_\star}{220\,\mathrm{km\,s^{-1}}}\right)^3. \tag{15}$$

There are a number of uncertainties in the estimates (8) and (15) of the central luminosity produced by dynamical friction.

(i) The assumption of circular orbits is unrealistic. Our estimates of $L_{\rm acc}$ should probably be increased by about 10% to account for the fact that the galaxies are on isotropic rather than circular orbits. On the other hand, the assumption of circular orbits minimizes the mass stripping due to tidal forces, which are strongest at pericenter.

(ii) I have assumed that the luminous mass in the galaxy is concentrated at sufficiently small radii that it is not stripped by tidal forces. To check this, I use (9) to evaluate the orbital radius r_s at which the tidal radius r_t equals the median radius r_h for a galaxy with mass $m = \Upsilon L_\star$, $\Upsilon = 12h\Upsilon_\odot$, and dispersion $\sigma_\star = 220\,\mathrm{km\,s^{-1}}$. I find $r_s = 7h^{-1}$ kpc, small enough that stripping of the luminous mass should be unimportant. However, some stripping may have occurred—Strom and Strom (1979) find galaxies near cluster centers to be smaller than field galaxies—which would reduce the accreted luminosity.

(iii) Mass loss due to galaxy-galaxy collisions is probably unimportant (for rate estimates see Richstone 1975, Merritt 1983, 1984a), but galaxies on elongated orbits may suffer mass loss at pericenter due to impulsive tidal shocks from the central part of the cluster mass distribution (Aguilar, Hut and Ostriker 1988).

(iv) The fraction f in (10) is uncertain and may be less than 0.5; for example Merritt (1984a) argues for $f = 0.25$ (his α is $2\sqrt{f}$).

(v) In the model described here, central galaxies are formed from a small number of accreted cluster galaxies, and hence there is a statistical uncertainty in the luminosity of the central galaxy expected in a given cluster. Assuming that the luminosities and initial positions of the cluster galaxies are statistically independent, it is straightforward to show that the rms fractional uncertainty in $L_{\rm acc}$ is $[(\alpha+\frac{5}{2})L_\star/L_{\rm acc}]^{1/2}$ if no massive halos are present, and $[(\alpha+2.75)L_\star/L_{\rm acc}]^{1/2}$ if massive halos are present. Using the nominal parameters in (8) and (15) respectively, the fractional uncertainties are 0.5 in both cases.

(vi) Probably the main uncertainty in $L_{\rm acc}$ arises from the uncertain cluster core radius. Galaxy counts (Beers and Tonry 1986) yield core radii that are consistent with zero, but indirect arguments based on the frequency of multiple nuclei in central galaxies (see Lauer 1989 and §3 below) suggest that $r_c \gtrsim 20h^{-1}$ kpc, much smaller than the estimates of Bahcall (1975) and Dressler (1978) ($r_c = 0.13$ and $0.24h^{-1}$ kpc respectively), but still large enough to reduce substantially the number of bright galaxies accreted (eqs. 4 and 12). Moreover, it is likely that the core radius of the dark matter is larger than that of the galaxies, since it is not subject to dissipation or dynamical evolution; hence its density will be lower than in the singular isothermal and the drag will be reduced.

The conclusions from these simple models are consistent with more careful calculations made by Merritt (1984a, 1985). Using parameters similar to those in our "massive halo" calculation, except that the cluster core radius is $r_c = 120$ kpc, Merritt (1984a, Table 2) found that the luminosity accumulated in the core due to dynamical friction was only about $0.1L_\star$.

This is substantially smaller than our estimate of $5L_\star$ (eq. 15), and confirms that the accreted luminosity is a strongly decreasing function of core radius.

The typical luminosity of a cD galaxy is about $13L_\star$ (this is the median luminosity of a sample of 27 cD's from Schombert 1988, with the luminosity of the extended envelope removed, since it may have a different origin from the main body of the galaxy). Therefore the estimates (8) and (15) of the luminosity accreted onto the cluster center are more than a factor of two smaller than the actual luminosity of central galaxies, and the most important of the uncertainties described above are likely to make more realistic estimates smaller still. A similarly small estimate, $L_{\rm acc} \simeq 2L_\star$, has been obtained by Lauer (1988) using the fraction of close neighbors to the central galaxy that exhibit evidence of tidal interactions.

Thus, it appears that dynamical evolution of galaxies in a quasi-static, virialized cluster core similar to those we see today is not strong enough to produce central galaxies with luminosities comparable to those observed. This conclusion is originally due to Merritt (1984a, 1985), whose models are much more sophisticated than the ones here; the principal contribution of the present calculations is to show that Merritt's conclusion remains valid whatever the core radius may be.

3. Mergers during cluster formation

The formation of galaxies and clusters is likely to occur through hierarchical clustering (S. White and Rees 1978, Peebles 1980). The early universe contains density fluctuations on a wide range of scales. As the universe expands, bound density fluctuations eventually "break away" from the Hubble flow and collapse. The strongest density perturbations are on small scales, so that small systems generally collapse first; larger bound systems have lower overdensities and collapse only after the smaller systems they contain. Thus, small systems (e.g. galaxies, or, more properly, dark matter halos of galaxies) develop before larger ones (groups and clusters of galaxies). An important feature of hierarchical clustering is that most or all of these small subunits are generally disrupted by tidal forces during the collapse of the next larger structure in the hierarchy; thus, for example, the dark matter in a group or cluster of galaxies will be smoothly distributed rather than concentrated in distinct galaxy halos.

In hierarchical clustering models, the luminous material in galaxies forms at the bottoms of the potential wells of individual galaxy halos. We must therefore ask why the individual galaxies in a cluster have survived, when the halos have been disrupted. One possibility is that dissipation concentrates the luminous material so that its density is high enough to survive disruption (S. White and Rees 1978). Alternatively, it may be possible for the dense cores of the individual halos to survive, with luminous material at the center of each dark core. (These two alternatives correspond crudely to the two simple models—no massive halo and tidally limited halo—that were described in the previous section.)

Merritt (1985) has pointed out that since cD galaxies are found in groups as well as rich clusters (R. White 1978 and references therein) it is natural to ask whether the cD's in rich clusters might have formed while the cluster members were still distributed in groups.

The simple merger models of the previous section do not suggest that groups are more likely to form cD's than clusters: groups have lower velocity dispersion than clusters, and the accreted luminosity $L_{\rm acc}$ varies as $\sigma^{3/2}$ in models with no massive halos (eq. 8) and is independent of σ in models with halos (eq. 15). However, when the cluster dispersion σ becomes as small as the internal galaxy dispersion σ_g, a new effect appears: mergers can occur during galaxy-galaxy collisions at any location in the cluster. (The reason follows from the impulse approximation. In a high-speed head-on collision between two galaxies, the energy converted from relative motion of the two galaxies to internal energy is of order $\Delta E \approx \sigma_g^4/\sigma^2$ per unit mass, while the relative kinetic energy before the encounter is of order $E \approx \sigma^2$ per unit mass. Mergers can occur once $|\Delta E| \gtrsim E$, which requires $\sigma \lesssim \sigma_g$; of course in this limit the impulse approximation is not really valid, but it still gives the correct order-of-magnitude result.) A large merger remnant can be built up more rapidly by direct mergers between galaxies throughout the cluster than by merging only galaxies that have spiraled in to the cluster center.

Numerical simulations provide dramatic confirmation of the importance of merging in small groups (e.g. Ishizawa et al. 1983, Barnes 1989). These simulations typically follow a group of 5–10 spherical or disk galaxies. They find that within a few group crossing times most or all of the galaxies have merged to form a single amorphous system. Moreover, the material at the centers of the individual galaxies is concentrated at the center of the remnant system, suggesting that the luminous material will collect in a single structure at the center of the merged halos.

Thus, mergers in the early stages of hierarchical clustering can produce single galaxies that resemble observed central galaxies. However, the initial conditions for the simulations described above, though plausible, are not derived self-consistently by following the growth of linear perturbations with a given fluctuation spectrum. The challenge to numerical models of hierarchical clustering and galaxy formation is to yield enough mergers at early stages to produce $10 - 15L_\star$ galaxies and yet not so much that an entire rich cluster merges into a single enormous starpile.

4. Multiple nuclei and dumbbell galaxies

A remarkably large fraction of central galaxies have multiple nuclei or superimposed secondary galaxies, a fact already noted by Morgan and Lesh (1965). Hoessel and Schneider (1985) found that about 50% of brightest cluster galaxies had multiple nuclei brighter than about $0.05L_\star$ within an aperture of $10h^{-1}$ kpc radius; in contrast, the frequency of multiple nuclei in second or third brightest cluster galaxies is about an order of magnitude smaller (Schneider, Gunn and Hoessel 1983). The luminosity function of the secondary galaxies is similar to the luminosity function of cluster galaxies in general (Hoessel and Schneider 1985, Lauer 1989).

This concentration of galaxies near the central galaxy has often been cited as evidence that the central galaxy is presently consuming cluster galaxies at a rapid rate. However, I have never found this argument convincing. Consider a model in which the galaxies move on circular orbits, the galaxy number density is $n(r,t)$, and the rate of orbital decay is $dr/dt < 0$. Then the continuity equation implies that orbital decay leads to an increase in the number density ($\partial n/\partial t > 0$) if and only if $r^2 n|dr/dt|$ is increasing outwards. In a singular isothermal, $n \propto r^{-2}$, and $|dr/dt| \propto r^{-1}$ (for constant mass galaxies); hence the number density is depleted by friction. In a central core, $n \propto const$ and $|dr/dt| \propto r$ (see Merritt 1985); thus the number density is enhanced. However, the central galaxy is a strong source of drag (its density is higher and velocity dispersion lower than the cluster's) so that near the center the drag increases rapidly. If the increase is rapid enough so that $|dr/dt|$ increases inwards faster than r^{-2}, then the density will be depleted. Thus we expect an enhancement in the cluster galaxy density inside the core due to friction, but it is much less certain that there will be an enhancement in density within the central galaxy itself.

Recent observational evidence is consistent with this argument, in that it now appears that most multiple nuclei are not associated with the central galaxy at all, but are simply cluster galaxies projected in front of the central galaxy:

(i) The velocity dispersion of the nuclei is similar to that of the cluster (Tonry 1985, Smith et al. 1985 and references therein), which is expected if their orbits are those of typical cluster members, and inconsistent with orbits bound to the central galaxy, which has a much lower dispersion (e.g. Tonry 1984).

(ii) The numbers of multiple nuclei are about four times the number expected from chance projections for an assumed core radius of $150h^{-1}$ kpc (Schneider et al. 1983). However, Beers and Tonry (1986) argue that the actual core radii of clusters are quite uncertain and may be much smaller; assuming a simple model for the surface density

$$\Sigma(r) = \frac{\Sigma_0}{[1 + (r/r_c)^2]^\gamma}, \tag{16}$$

then for a given asymptotic surface density profile the central surface density varies as $\Sigma_0 \propto r_c^{-2\gamma}$. Thus, a core radius as small as $r_c = 75h^{-1}$ kpc ($\gamma = 1$) or $40h^{-1}$ kpc ($\gamma = \frac{1}{2}$) would imply that all the multiple nuclei could be chance projections.

Lauer (1989) has compared the total galaxy luminosities enclosed within $10h^{-1}$ kpc and $50h^{-1}$ kpc of the central galaxy center (excluding the central galaxy itself). If the multiple nuclei are due to chance projections and the core radius $r_c \gg 50h^{-1}$ kpc then the ratio of these luminosities should be 25. The ratios found by Lauer were 23 ± 8 (richness class 2), 13 ± 3 (class 1), and 8 ± 2 (class 0). These values are consistent with the hypothesis that all the multiple nuclei are chance projections if $r_c \gg 50h^{-1}$ kpc (richness class 2), $\simeq 30h^{-1}$ kpc (class 1), and $\simeq 20h^{-1}$ kpc (class 0). (I assume a surface density given by [16] with $\gamma = 1$.)

(iii) Lauer (1988) finds that roughly half of the nuclei show morphological disturbances not seen in isolated galaxies, implying recent direct interactions. To see whether this fraction is consistent with the assumption that the secondaries are simply typical cluster members, I use the surface brightness distribution (16) with $\gamma = 1$. This implies a space density distribution $n(r) \propto [1 + (r/r_c)^2]^{3/2}$, which implies that the fraction of galaxies projected on a central aperture of radius $r_{\rm ap} \ll r_c$ that are actually within a distance $r_{\rm int}$ small enough to be interacting is $f = r_{\rm int}/\sqrt{r_c^2 + r_{\rm int}^2}$. Assuming that the tidal disturbances damp on a crossing time $r_{\rm ap}/(\sqrt{3}\sigma_g)$, and that the typical encounter speed is $\sqrt{3}\sigma$, then $r_{\rm int} \approx r_{\rm ap}(\sigma/\sigma_g)$. Using the nominal parameters $\sigma_g = 220\,{\rm km\,s^{-1}}$, $\sigma = 800\,{\rm km\,s^{-1}}$, $r_{\rm ap} = 10h^{-1}$ kpc, $r_c = 75h^{-1}$ kpc, I find $r_{\rm int} = 36h^{-1}$ kpc and $f = 0.4$, consistent with the observed value of about 0.5.

(iv) Merritt (1984b) has pointed out that small core radii may be a natural consequence of dynamical friction, which tends to shrink the core radius. This may also provide a natural explanation for the smaller core radii deduced in (ii) for richness class 0 or 1 compared to richness class 2: the rate of shrinkage is likely to be faster in clusters of lower richness because their dispersion is lower and $dr/dt \propto \sigma^{-1}$ (eq. 3).

Thus multiple nuclei in central galaxies provide no direct support for the hypothesis that these galaxies have been made by mergers. The nuclei can be explained as typical cluster galaxies that happen to be projected on top of the central galaxy.

A related but distinct class is the binary supergiant or "dumbbell" (db) galaxies. Matthews et al. (1964) described the dumbbells as "a group allied to the D galaxies, in which two separated, approximately equal, nuclei are observed in a common envelope." In most dumbbells both components are very luminous—often brighter than any other galaxies in the cluster—and appear to belong to the class of D or cD galaxies.

There are no systematic surveys for dumbbell galaxies, although Leir and van den Bergh's (1977) study of the Abell catalog of rich clusters noted all brightest cluster galaxies that were "double or multiple", and Rood and Leir (1979) state that most of the galaxies classified as "multiple" in this way are dumbbells. They also stress that the fraction of clusters containing a multiple first-ranked galaxy is remarkably high—23% for Bautz-Morgan type I and I–II clusters.

Valentijn and Casertano (1988) have compiled a list of 44 clusters containing dumbbell galaxies. For most of these the magnitude difference between the components was less than 1 magnitude. (Rood 1988 estimates a median difference of 0.5 magnitude in the Rood and Leir sample.) The median separation of the centers of the two components is $10h^{-1}$ kpc and only 3 of the dumbbells have separations $\gtrsim 25h^{-1}$ kpc. The high frequency of dumbbells with such small separations is a striking result, although the absence of dumbbells with large separations is largely an artifact—by definition, "dumbbells" cannot have large separations.

The rms velocity difference between the dumbbell components in the Valentijn-Casertano sample is $650\,\mathrm{km\,s^{-1}}$, implying a single-object dispersion of $650/\sqrt{2} = 460\,\mathrm{km\,s^{-1}}$ if the velocities of the two components are uncorrelated. This is substantially smaller than the typical cluster dispersion of $800\,\mathrm{km\,s^{-1}}$, and smaller than the dispersion of multiple nuclei, which is similar to the cluster dispersion. The mean velocity difference increases with separation.

There is a simple picture for the formation of dumbbell galaxies that is consistent with much of this data. Consider two clusters that collapse and virialize, then fall towards each other and merge into a single cluster. Such structures arise commonly in cosmological simulations (e.g. S. White 1976b, Cavaliere et al. 1986) and strongly resemble structures that can be identified in both X-ray observations ("double clusters", see Forman et al. 1981) and galaxy counts (Geller and Beers 1982; see Geller 1988 for a review).

Now suppose that each of the two merging clusters had a central galaxy. The central galaxies will remain surrounded by their respective cores until the two cores merge. Thus the central galaxies, like most of the other collisionless material in the two original cores, will end up orbiting in or near the core of the merged cluster. (In other words, violent relaxation during a merger of this kind is not complete: material with high binding energy in the merging systems will have high binding energy in the merger remnant.)

This argument suggests that clusters containing two D or cD galaxies are the product of a merger of two clusters in which these were once the central galaxies, and that in the merged cluster the two galaxies will be separated by at most a couple of core radii, that is, by $\lesssim 200h^{-1}$ kpc. It remains to be shown why in most observed clusters the two D's or cD's are separated by a much smaller distance, $\lesssim 20h^{-1}$ kpc.

Once the ex-central galaxies are orbiting in the core of the merger remnant, dynamical friction will drag them toward the center of the remnant, where they will eventually merge to form a new central galaxy. The radius r of a circular orbit within the core decays exponentially, $r \propto \exp(-t/t_d)$, where (Merritt 1985)

$$t_d = \frac{\sqrt{2}\pi r_c^2 \sigma}{3Gm \ln \Lambda} = 0.44 \left(\frac{2}{3H_0}\right) \left(\frac{r_c}{75h^{-1}\,\mathrm{kpc}}\right)^2 \left(\frac{\sigma}{800\,\mathrm{km\,s^{-1}}}\right) \left(\frac{L_\star}{L}\right) \left(\frac{12h\Upsilon_\odot}{\Upsilon}\right), \tag{17}$$

and I have assumed $\ln \Lambda = 2.5$. I shall concentrate on the less luminous ex-central galaxy, since the more luminous (and presumably more massive) of the pair will settle to the center first and await its partner. Assuming a typical luminosity ratio $L_>/L_< = 1.6$ (i.e. 0.5 magnitude) and $L_< = 5L_\star$ (this gives a merged central galaxy with a typical cD luminosity $L_> + L_< = 13L_\star$), the e-folding time $t_d \simeq 0.1t_u$, where $t_u = \frac{2}{3}H_0^{-1}$ is the age of the universe. Probably of order one or two e-foldings are required before the ex-central galaxies reach the separation $r_{\mathrm{crit}} \approx 15h^{-1}$ kpc where their mutual drag begins to dominate the drag from the cluster core. Once this happens the two galaxies will rapidly merge. Since we have overestimated the drag somewhat by using the central density (even at one core radius the density is only 0.35 of the central value), it is likely that the two galaxies merge only after a total time $t_{\mathrm{tot}} \simeq 0.3t_u$.

This model explains many of the features of dumbbell galaxies. (i) The dumbbell components are D or cD galaxies because they used to be central galaxies. (ii) The strong concentration towards small separations arises because the frictional decay rate $dr/dt \propto r$ inside the core but outside r_{crit}; in a steady state this implies that the number density $n(r) \propto r^{-3}$ for $r \gtrsim r_{\mathrm{crit}}$. (iii) The rms velocity is lower than the cluster value because the dumbbell components are orbiting within the cluster core; within the core $v \propto r$, which explains why the velocities increase with separation. (iv) To estimate whether the numbers of dumbbell galaxies are consistent with the model, assume that there is a steady flux F of ex-central galaxies across radius r; then $F = 4\pi r^2 n(r)|dr/dt|$. Then the total number of such galaxies at a given time is $N = Ft_{\mathrm{tot}}$ and the total number of mergers over the age of the universe is $N_m = Ft_u$, assuming that the flux has been constant. Taking $t_{\mathrm{tot}} = 0.3t_u$, I find $N_m \approx 3N$. This is probably an underestimate, since the flux was surely higher in the past. For comparison, Rood and Leir (1979) estimate that 23% of Bautz-Morgan type I and I-II clusters have brightest galaxies that are multiple, and most of these are dumbbells; if these clusters have undergone an average of one transient dumbbell stage then $N_m = 4N$, consistent with the crude theoretical prediction.

Using similar arguments, Rood (1988) has reached a quite different conclusion: that the number of dumbbells is at least five times that predicted by simple hierarchical clustering models. The reason for the discrepancy is that Rood uses a much shorter decay time, $t_{tot} \simeq 3 \times 10^8 h^{-1}$ yr $= 0.05 t_u$. This is the decay time inside r_{crit}, where the drag is dominated by the galaxies; however, the steady-state population of dumbbells is mostly determined by the slowest part of the decay process, the decay through the cluster core outside r_{crit}, for which the much longer decay time of $0.3 t_u$ used above is the appropriate one.

The Coma cluster may be an example of a cluster that will eventually contain a dumbbell. Coma has two galaxies with many cD characteristics (Schombert 1988), with a present separation of $180h^{-1}$ kpc, and the distribution of galaxies both on the sky and in velocity space suggest that it consists of two merging subclusters (Fitchett and Webster 1987 and references therein).

It has often been argued that the presence of multiple nuclei in central galaxies provides strong evidence that merging is an important and ongoing process, while the existence of dumbbell galaxies suggests that dynamical friction and merging are less effective than theory would lead us to believe. It is interesting and a little ironic that present observations suggest strongly that multiple nuclei can be explained as nothing more than a projection effect, while it is the dumbbell galaxies that provide the most direct evidence that merging of giant galaxies is a common process in the centers of clusters.

5. Summary

The principal conclusions are summarized in the abstract.

I thank Josh Barnes, Dick Bond, Marc Davis, Gus Evrard, Jim Gunn, Nick Kaiser, Tod Lauer, Doug Richstone, Herb Rood, Alar Toomre, and Rachel Webster for helpful discussions, and Alar Toomre for access to unpublished notes. I am grateful to the Berkeley Astronomy Department for their hospitality while I prepared this review, and to NSERC for financial support through an operating grant.

References

Aguilar, L., Hut, P., and Ostriker, J. P. 1988. *Astrophys. J.* **335**, 720.
Bahcall, N. 1975. *Astrophys. J.* **198**, 249.
Barnes, J. 1989. *Nature* **338**, 123.
Beers, T. C., and Geller, M. J. 1983. *Astrophys. J.* **274**, 491.
Beers, T. C., and Tonry, J. L. 1986. *Astrophys. J.* **300**, 557.
Binney, J., and Tremaine, S. 1987. *Galactic Dynamics* (Princeton: Princeton University Press).
Bontekoe, Tj. R., and van Albada, T. S. 1987. *Mon. Not. Roy. Astron. Soc.* **224**, 349.
Cavaliere, A., Santangelo, P., Tarquini, G., and Vittorio, N. 1986. *Astrophys. J.* **305**, 651.
Chandrasekhar, S. 1943. *Astrophys. J.* **97**, 255.
Colless, M. 1989. *Mon. Not. Roy. Astron. Soc.* **237**, 799.
Dressler, A. 1978. *Astrophys. J.* **226**, 55.
Dressler, A. 1984. *Ann. Rev. Astron. Astrophys.* **22**, 185.
Dressler, A., and Shectman, S. A. 1988. *Astron. J.* **95**, 985.
Fitchett, M., and Webster, R. 1987. *Astrophys. J.* **317**, 653.
Forman, W., Bechtold, J., Blair, W., Giacconi, R., Van Speybroeck, L., and Jones, C. 1981. *Astrophys. J. Lett.* **243**, L133.
Geller, M. J. 1988. In *Large-Scale Structure in the Universe*, 17th Advanced Course of the Swiss Society of Astronomy and Astrophysics, ed. L. Martinet (to be published).

Geller, M. J., and Beers, T. C. 1982. *P.A.S.P.* **94**, 421.
Gioia, I. M., Geller, M. J., Huchra, J. P., Maccacaro, T., Steiner, J. E., and Stocke, J. 1982. *Astrophys. J. Lett.* **255**, L17.
Grossman, S. A., and Narayan, R. 1989. *Astrophys. J.* **xx**, xx.
Gunn, J. E. 1977. *Astrophys. J.* **218**, 592.
Harris, W. E. 1987. *Astrophys. J. Lett.* **315**, L29.
Hoessel, J. G., and Schneider, D. P. 1985. *Astron. J.* **90**, 1648.
Ishizawa, T., Matsumoto, R., Tajima, T., Kageyama, H., and Sakai, H. 1983. *P.A.S.Japan* **35**, 61.
Jones, C., and Forman, W. 1984. *Astrophys. J.* **276**, 38.
Kent, S. M., and Sargent, W. L. W. 1983. *Astron. J.* **88**, 697.
Lauer, T. R. 1985. *Astrophys. J.* **292**, 104.
Lauer, T. R. 1988. *Astrophys. J.* **325**, 49.
Lauer, T. R. 1989. Preprint.
Leir, A. A., and van den Bergh, S. 1977. *Ap. J. Supp.* **34**, 381.
Lecar, M. 1975. In *Dynamics of Stellar Systems*, IAU Symposium No. 69, ed. A. Hayli (Dordrecht: Reidel), 161.
Lin, D. N. C., and Tremaine, S. 1983. *Astrophys. J.* **264**, 364.
Malumuth, E. M., and Richstone, D. O. 1984. *Astrophys. J.* **276**, 413.
Matthews, T. A., Morgan, W. W., and Schmidt, M. 1964. *Astrophys. J.* **140**, 35.
Merritt, D. 1983. *Astrophys. J.* **264**, 24.
Merritt, D. 1984a. *Astrophys. J.* **276**, 26.
Merritt, D. 1984b. *Astrophys. J. Lett.* **280**, L5.
Merritt, D. 1985. *Astrophys. J.* **289**, 18.
Morgan, W. W., and Lesh, J. R. 1964. *Astrophys. J.* **142**, 1364.
Ostriker, J. P., and Tremaine, S. 1975. *Astrophys. J. Lett.* **202**, L113.
Peebles, P. J. E. 1980. *The Large-Scale Structure of the Universe* (Princeton: Princeton University Press).
Quintana, H., and Lawrie, D. G. 1982. *Astron. J.* **87**, 1.
Richstone, D. O. 1975. *Astrophys. J.* **200**, 535.
Richstone, D. O., and Malumuth, E. M. 1983. *Astrophys. J.* **268**, 30.
Rood, H. J. 1988. *P.A.S.P.* **100**, 1242.
Rood, H. J., and Leir, A. A. 1979. *Astrophys. J. Lett.* **231**, L13.
Sarazin, C. L. 1986. *Rev. Mod. Phys.* **58**, 1.
Schneider, D. P., Gunn, J. E., and Hoessel, J. G. 1983. *Astrophys. J.* **268**, 476.
Schombert, J. M. 1988. *Astrophys. J.* **328**, 475.
Smith, R. M., Efstathiou, G., Ellis, R. S., Frenk, C. S., and Valentijn, E. A. 1985. *Mon. Not. Roy. Astron. Soc.* **216**, 71P.
Strom, S. E., and Strom, K. M. 1979. *Astron. J.* **84**, 1091.
Tonry, J. L. 1984. *Astrophys. J.* **279**, 13.
Tonry, J. L. 1985. *Astron. J.* **90**, 2431.
Tonry, J. L. 1987. In *Structure and Dynamics of Elliptical Galaxies*, IAU Symposium No. 127, ed. T. de Zeeuw (Dordrecht: Reidel), 89.
Valentijn, E. A., and Casertano, S. 1988. *Astron. Astrophys.* **206**, 27.
White, R. A. 1978. *Astrophys. J.* **226**, 591.
White, S. D. M. 1976a. *Mon. Not. Roy. Astron. Soc.* **174**, 19.
White, S. D. M. 1976b. *Mon. Not. Roy. Astron. Soc.* **177**, 717.
White, S. D. M. 1978. *Mon. Not. Roy. Astron. Soc.* **184**, 185.
White, S. D. M., and Rees, M. J. 1978. *Mon. Not. Roy. Astron. Soc.* **183**, 341.
Yahil, A. 1974. *Astrophys. J.* **191**, 623.
Zaritsky, D., and White, S. D. M. 1988. *Mon. Not. Roy. Astron. Soc.* **235**, 289.

Cannibalism in Clusters' Cusps

T.R. Lauer

Princeton University Observatory, Peyton Hall, Princeton, NJ08544, USA

Galactic cannibalism has long been assumed to play a role in the luminosity evolution of the brightest galaxies in clusters, but whether is it important enough to grow the luminous ($L \sim 12L^*$) first-ranked galaxies completely over the age of the universe, or is rather just a modest source of luminosity growth has been a point of contention. I have been attempting to answer this question observationally through imaging investigations of multiple-nucleus first ranked galaxies. My work (Lauer 1986, 1988, and 1989) leads me to conclude that cannibalism may well be important for luminosity growth of first-ranked galaxies, but is probably insufficient to grow them completely — my best guess is that typical luminosity growth by cannibalism is $\sim 4L^*$. The high prevalence of multiple nucleus first-ranked cluster galaxies is often sited as strong evidence for cannibalism. Fully 48% of first-ranked galaxies in Abell clusters have a companion galaxy brighter and $M_B = -17.5$ within $20kpc$ of their centers; and if first-ranked galaxies are built by steady cannibalism, then it can be shown that a large fraction of these "secondary galaxies" are currently being cannibalized as we watch. For cluster galaxies initially distributed as $r^{-\gamma}$ in space density, the present epoch cannibalism rate is

$$\left.\frac{dL_T}{dt}\right|_{t=T} = \left(\frac{3-\gamma}{2T}\right) L_T, \tag{1}$$

where L_T is the total luminosity cannibalized, and T is the time since cluster formation. For typical first-ranked galaxy luminosity, $L_1 \approx 12L^*$, $T = 10^{10}yr$, and $\gamma = 2$, $dL_T/dt \approx 0.6L^*/10^9yr$. Absorption of the brightest secondary galaxies will take place in about an orbital period or typically $\Delta t \approx 3 \times 10^8 yr$. This gives the average luminosity actively being cannibalized within the metric radius at any time as $L_C \approx dL_T/dt\Delta t \approx 0.2L^*$. This number is close to the typical total luminosity of secondary galaxies within the metric radius averaged over all Abell clusters regardless of multiplicity, but the radial velocities of the secondary galaxies are high (Tonry 1984, 1985, Smith *et al.* 1985); the majority of these galaxies do not appear to be bound to the central galaxy in a merging event. Further, from my own morphological analysis of multiple-nucleus systems (Lauer 1988) I find that only $\sim$ 25% of multiple systems show interaction effects consistent with on going cannibalism at this epoch; the implied cannibalism rate, $dL_T/dt \approx 0.2L^*/10^9yr$, appears too low to support formation of first-ranked galaxies completely from cannibalism.

Regardless of whether or not there is current evidence for ongoing cannibalism, it can be asked if the spatial distribution of cluster galaxies is consistent with significant cannibalism over the cluster lifetime. The total luminosity cannibalized by a centrally located galaxy is just the luminosity contained within the volume defined by a luminosity and time dependent infall radius, giving

$$L_T = 4\pi n^* L^* \int_0^\infty \left(\frac{L}{L^*}\right)^{\alpha+1} e^{-L/L^*} \frac{dL}{L^*} \int_0^{r_i(L,T)} \rho(r) r^2\, dr. \tag{2}$$

While direct estimation of L_T requires knowledge of the initial galaxy distribution of the cluster, as well as its age and the form of its potential in order to estimate the infall radius, it is possible to infer the total luminosity cannibalized by comparing the present galaxy density distribution

Dynamics and Interactions of Galaxies Editor: R. Wielen

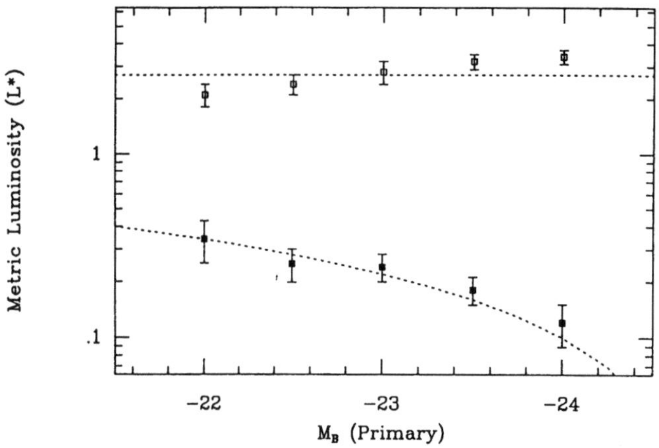

FIG. 1.—The lower solid symbols (■) represent the total luminosity of secondary galaxies within $20kpc$ of the center of the primary galaxies. The upper open symbols (□) represent the total luminosity of cluster galaxies within $100kpc$ of the primaries. The dashed lines plot the richness-corrected relationships between the metric luminosities L_{20} and L_{100} and primary galaxy luminosity.

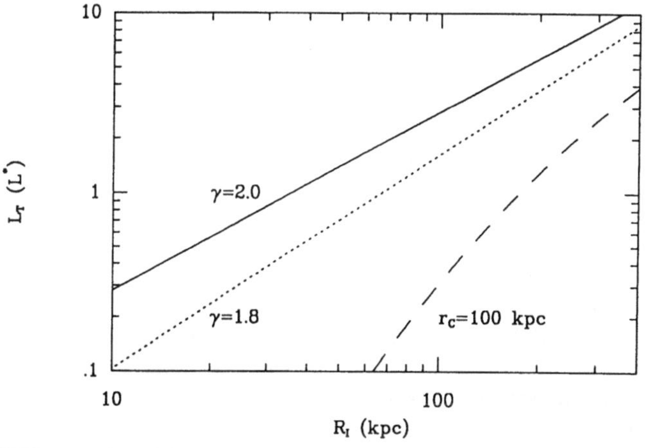

FIG. 2.—Total luminosity cannibalized as a function of dynamical friction infall radius for three models of the initial galaxy cluster distribution. The infall radius is normalized to a $1.0L^*$ galaxy; the primary galaxies are assumed to be at $r = 0$. Results are shown for two power-law density distributions with $\gamma = 2$ and $\gamma = 1.8$; and for the density distribution with a flat core with $r_C = 100kpc$.

$\rho'(r)$ to an initial assumed $\rho(r)$. In this case only the functional form of $r_i(L,T)$ is needed. At this epoch, $\rho'(r)$ gives the total luminosity of cluster galaxies seen in any metric aperture centered on the first-ranked galaxy, L_{r_m}, which is known from the line-of-sight integral

$$L_{r_m} = 4\pi n^* L^* \int_0^\infty \left(\frac{L}{L^*}\right)^{\alpha+1} e^{-L/L^*} \frac{dL}{L^*} \int_0^\infty \int_o^{r_m} \rho'(r) x \, dx \, dz. \tag{3}$$

The total luminosities of cluster galaxies within $20kpc$ and $100kpc$ of the first-ranked galaxies (averaged over all Abell clusters) are shown in Figure 1. Figure 2 shows the luminosity cannibalized as a function of infall radius for three models of the initial galaxy distribution for a

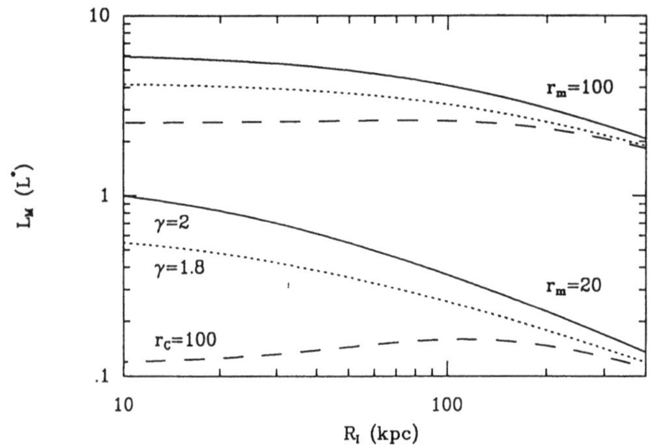

FIG. 3.—Metric luminosities of secondary galaxies as a function of dynamical friction infall radius for the three models of the initial galaxy cluster distribution given in Figure 2. The lower set of curves give L_{20} or the total luminosity of cluster galaxies within $20kpc$ of the primaries; the upper set of curves give L_{100}.

richness class 1 cluster. Figure 3 shows how the metric luminosities change as a function of the infall radius. The problem then is to pick a model, find the r_i in Figure 3 that matches the observations in Figure 1, and then deduce the cannibalized luminosity from Figure 2.

The $\gamma = 2$ model was motivated by Beers and Tonry (1986) who demonstrate that the central structure of clusters at $r > 100kpc$ appears to have a cusp of this form. Matching current cluster structure for $r < 100kpc$ with this model implies $100 \lesssim r_i \lesssim 400kpc$ and substantial cannibalism — 50% to 100% of the first-ranked galaxy luminosity would arise from accretion of cluster galaxies. This conclusion is extremely sensitive to the power-law slope, however. The second model shown assumes $\rho(r) \propto r^{-1.8}$ initially. This profile was suggested by the galaxy correlation function and is effectively indistinguishable from the $\gamma = 2$ over the central $500kpc$ of clusters. The estimate of the total luminosity cannibalized under this model is only $\sim 40\%$ of that estimated with the $\gamma = 2$ model; in this case, cannibalism would be an important effect on the luminosity and structural evolution of the first-ranked galaxies, but would be insufficient to form them initially. L_T estimated here is typically $\sim 3L^*$ or $\sim 25\%$ of L_1. The last model considered is a density distribution with $r_C = 100kpc$, which is about the largest core size for which any match to the present cluster structure is possible. The implied cannibalized is smaller yet than the $\gamma = 1.8$ model; the observations suggest $L_T \lesssim 2L^*$. The total luminosity cannibalized under the three models considered here thus spans the full range from coming close to forming first-ranked galaxies to being a mild source of luminosity growth for a preexisting galaxy. The points that emerge are that the distribution of galaxies around the primary galaxies at this epoch is cuspy or at least has a small core radius; and making the first-ranked galaxy completely by cannibalism requires $\gamma \gtrsim 2$.

The luminosity function of galaxies within $20kpc$ of the first-ranked galaxies appears more consistent with mild rather than strong cannibalism. Dynamical friction not only changes the central space density of the central population, but also effectively changes its luminosity function slope α by the factor $(1 - \gamma)/2$. For $\gamma \approx 2$, the central luminosity function should be depleted at the bright end. The observed luminosity function (Figure 4) shows no evidence for depletion at the bright end, and if anything is brighter and shallower than a normal cluster luminosity function. No effects will be apparent on the luminosity function for $\gamma = 1$, so the present results may be most consistent with $\gamma < 2$, and consequently mild rather than strong cannibalism.

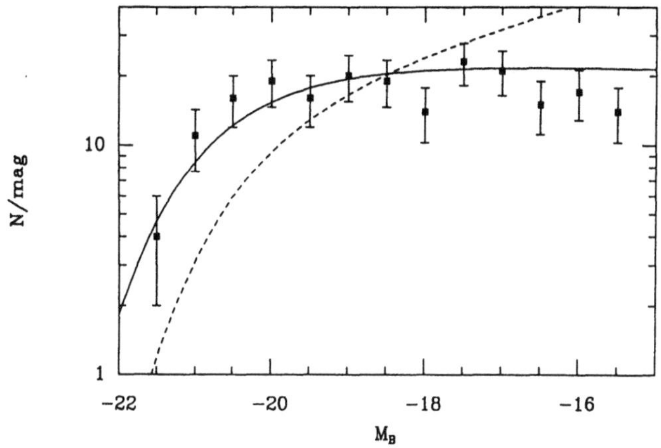

FIG. 4.—The luminosity function of secondary galaxies within 20*kpc* of the primary galaxy centers. The data point bins are one magnitude wide, but are shown at every half-magnitude step to provide proper sampling. The solid line is the best-fit Schechter function with $M_B^* = -21.0 \pm 0.5$ and $\alpha = -0.98 \pm 0.20$. The dashed line shows a "standard" Schechter function with $M_B^* = -20.6$ and $\alpha = -1.25$ normalized to the same number of total galaxies. The present data is incomplete for $M_B \gtrsim -17.0$, so the last four points were not used in the analysis.

Beers, T. C., and Tonry, J. L. 1986, *Ap. J.*, **300**, 557.

Lauer, T. R. 1986, *Ap. J.*, **311**, 34.

————. 1988, *Ap. J.*, **325**, 49.

————. 1989, *Ap. J.*, submitted.

Smith, R. M., Efstathiou, G., Ellis, R. S., Frenk, C. S., and Valentijn, E. A. 1985, *M. N. R. A. S.*, **216**, 71P.

Tonry, J. T. 1984, *Ap. J.*, **279**, 13.

————. 1985, *A. J.*, **90**, 2431.

Dynamics of the Local Group

B. Moore and C. Frenk
Department of Physics, University of Durham, UK

We have analysed in detail the properties of a binary system formed in an N-body simulation of a cold dark matter (CDM) Universe described by Frenk, White, Efstathiou and Davis, Nature, **317**, 595 1985, hereafter FWED. The internal structure and orbital parameters of this system resemble those of the Local Group. The simulation contains 32768 particles each of mass $6 \times 10^9 M_\odot$ of which 3000 belong to the "Local Group", the spatial resolution is 15kpc. We find that;

(i) Within the CDM model it is quite easy to find massive binary systems dynamically similar to the Local Group. Figure 1 shows the positions at various epochs of the particles which make up our model "Local Group" with their respective rotation curves plotted underneath. The left hand panel in figure 1 corresponds to a redshift $z = 0.2$. Even at this early time the rotation curves of the two halos are flat beyond ~ 50 kpc of the centre, the regions which we can confidently resolve in our calculation. Further in, the gravitational effects of the galaxy forming at the centre are likely to extend a flat rotation curve to the inner parts of the galaxy (Blumenthal *et. al.* Ap.J, **301**,27 1988, Ryden and Gunn Ap.J, **318**,15 1988).

The second panel in figure 1 corresponds to the epoch we identify with the present. The asymptotic values of the rotation curves of the two main clumps, 220 km s^{-1} and 260 km s^{-1}, are very similar to those of the Galaxy and M31 respectively. Furthermore, the relative velocity and separation of these two clumps lie within 3% of the values measured for the Galaxy and M31.

(ii) In our model, the dark matter halos around our galaxy and M31 are already interacting. The future evolution of the local group is illustrated in the last two panels of figure 1. The two halos will completely in $\sim 2 \times 10^9$yrs after the present, a timescale which is significantly shorter than the current crossing time of the system, $\sim 6 \times 10^9$ yrs. As the halos fall together, the rotation curves in the outer regions become distorted and, as the merging process proceeds, they rise steeply even in the inner regions, reaching an asymptotic value close to 400 km s^{-1}. The remnant is smooth and centrally concentrated; substructure is efficiently erased. Much of the orbital angular momentum is transferred to the outer halo and the inner regions of the remnant show little rotation.

(iii) The velocity field around our galaxy predicted by our model is consistent with that inferred from the kinematics of the outer satellites of the Milky Way. (Figure 2.)

(iv) Applying the timing argument to the Local Group gives a mass of $4.3 \times 10^{12} M_\odot$ for a universe that is 13 Gyrs old. The *actual* mass of our simulated "Local Group" (defined as the sum of the masses within spheres around each galaxy extending out to the centre of mass), is about a factor of 2 larger, $M_{tot} = 7.9 \times 10^{12} M_\odot$. This discrepancy is due to a significant transverse velocity of M31 in our model.

The dynamics of outlying satellites provide another method to estimate the mass of the Galaxy. Zaritsky *et. al.* (preprint, 1989), applied the timing argument to the orbit of Leo I around our galaxy and obtained a mass of $M_g(< 230\text{kpc}) = 1.3 \times 10^{12} M_\odot$. Applying a similar method to particles in a shell between 200 and 260 kpc of the centre of our simulated "Galaxy" gives a median mass within 20% of this value, but the actual mass is again, a factor of 2 larger.

Dynamics and Interactions of Galaxies Editor: R. Wielen

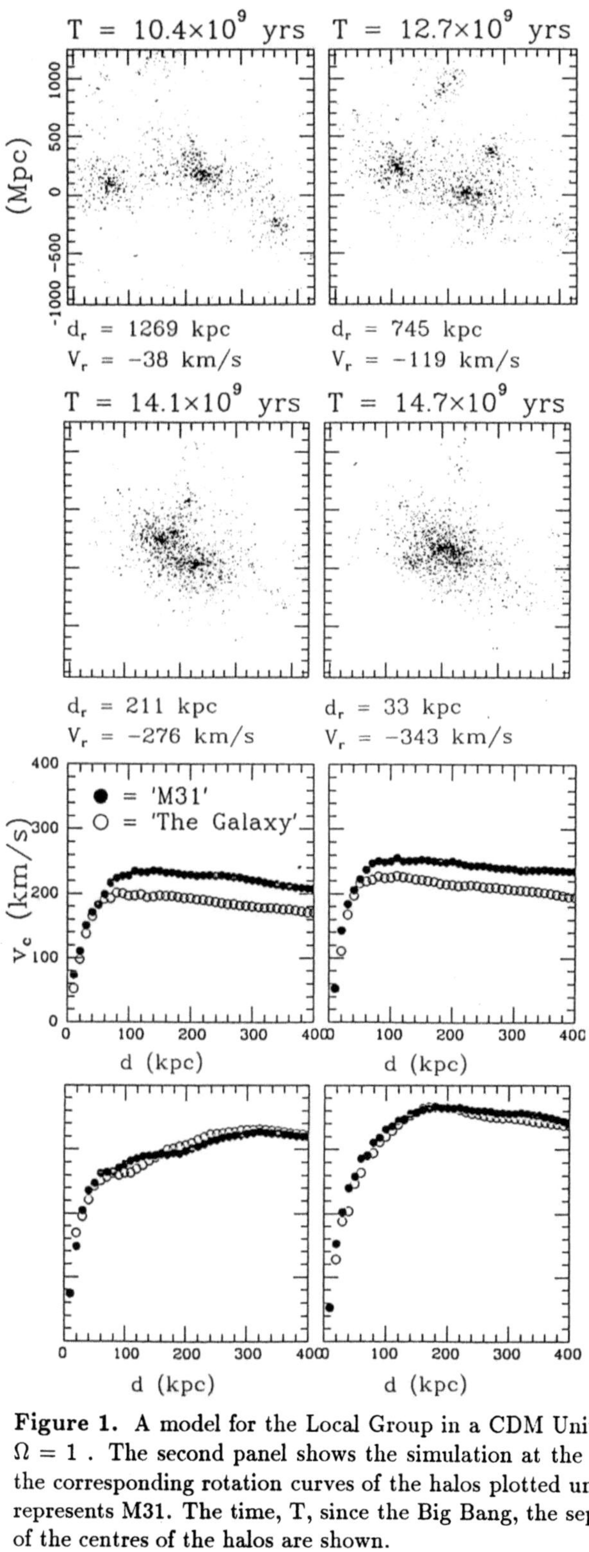

Figure 1. A model for the Local Group in a CDM Universe with $H_o = 50\ \mathrm{kms^{-1}Mpc^{-1}}$ and $\Omega = 1$. The second panel shows the simulation at the time we identify as the present, with the corresponding rotation curves of the halos plotted underneath. The larger halo in the plot represents M31. The time, T, since the Big Bang, the separation, d_r, and relative velocity, V_r, of the centres of the halos are shown.

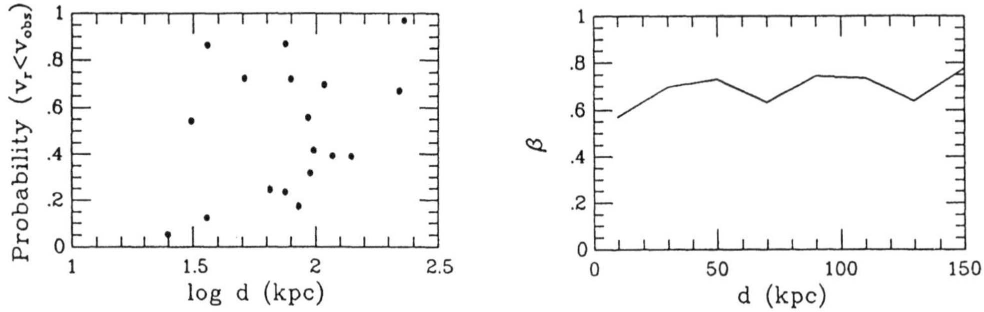

Figure 2.(left) Following Peebles *et. al.* (1989, preprint) we compare the velocity field around the "Galaxy" in the simulation with that around our own Galaxy by considering the measured radial velocities of 18 satellites within 250 kpc of the centre of the Milky Way. From the mean radial velocity of the simulation particles in a shell centred at the distance of each satellite, we calculate the probability of finding a velocity less than the observed value. A successful model would return a uniform distribution of probabilities with a mean of 0.5. For our model we find a mean of 0.50 with a standard deviation of 0.27.

Figure 2.(right) The velocity anisotropy parameter $\beta = \langle v_r^2\rangle/(\langle v_r^2\rangle + \langle v_t^2\rangle)$ for particles in successive radial shells centred around the "Galaxy" in our model. Here $\langle v_r^2\rangle$ and $\langle v_t^2\rangle$ denote the radial and tangential velocity dispersions respectively. The anisotropy parameter is fairly constant with radius and corresponds to orbits which are significantly radially biased.

Interactions Between Massive Galaxies and Their Satellites: Probing Galaxy Mass Distribution

J.E. Beckman[1], *J. Cepa*[1,2], *and A.M. Varela*[1]

[1]Instituto de Astrofísica de Canarias
[2]Observatorio del Roque de los Muchachos, La Palma

Using the measured common distance from the observer, angular separation of a dwarf satellite from its mother galaxy and their relative radial velocities, plus three criticable assumptions: circularity of the orbit, spherical symmetry of the gravitational potential, and that the radius of the dwarf galaxy is limited by tidal interaction with the mother, we show how to derive the orbital elements of the satellite. A trial of the method with the satellites of M31, M32 and NGC 205, gave results consistent with all three assumptions. Here we apply it to two systems, NGC 128 and NGC 1023, and their satellites, concluding that it is a useful exploratory tool, but that at its best it requires better data than the literature offers.

1. Introduction

In a previous article (Cepa and Beckman, [1]), we showed how to obtain the orbital elements of satellite galaxies round their massive parents, using the assumptions and observations outlined in the abstract, plus the rotation curve of the parent. In the cases of NGC 205 and M 32, we showed that they have retrograde orbits about M 31, at $\simeq$12 Kpc galactocentric radius, and are the probable cause of its star-forming ring. For numerical details the reader is referred to the publication cited.

2. New applications of the method

The problems associated with the technique are well known, when attempting to apply it to NGC 1023 and its satellites, illustrated in Figure 1. Satellite A has a differential radial velocity of 128 km s^{-1} and circular orbits are possible between 7.5 and 14.3 Kpc from the nucleus of NGC 1023.

Using the tidal method, taking the mass of A from HI line widths (Davis and Kinman, [2]) and its luminosity from Cappacioli et al. [3], we infer a minimum orbital radius of 23.8 Kpc. The problem is light contamination from NGC 1023, and CCD photometry is required. Satellite B has a differential radial velocity of -22 Km s^{-1}, so our technique cannot be applied. Satellite C, with a differential radial velocity of 288 Km s^{-1} is just consistent, within errors, with our three assumptions, and shows that the mass distribution of NGC 1023 is hardly flattened, given that C is gravitationally bound. For D, the velocity and distance give a circular orbit of radius 26.6 Kpc, and the tidal radius suggests $\simeq$21 Kpc. This is reasonably consistent, within the errors, and accepting 26.6 as the most reliable estimate, the circular velocity out of the plane is not measurably different from the in-plane rotation curve at the same radius. Taken together, C and D are at least consistent with a spherical massive halo for NGC 1023. Table 1 gives our schematic results for the orbital parameters (for definitions see Cepa and Beckman, [1]), and Table 2 gives the equivalent data for the satellites of NGC 128, *viz.* NGC 127 and NGC 130, for which lack of space prevents further description here.

Dynamics and Interactions of Galaxies Editor: R. Wielen

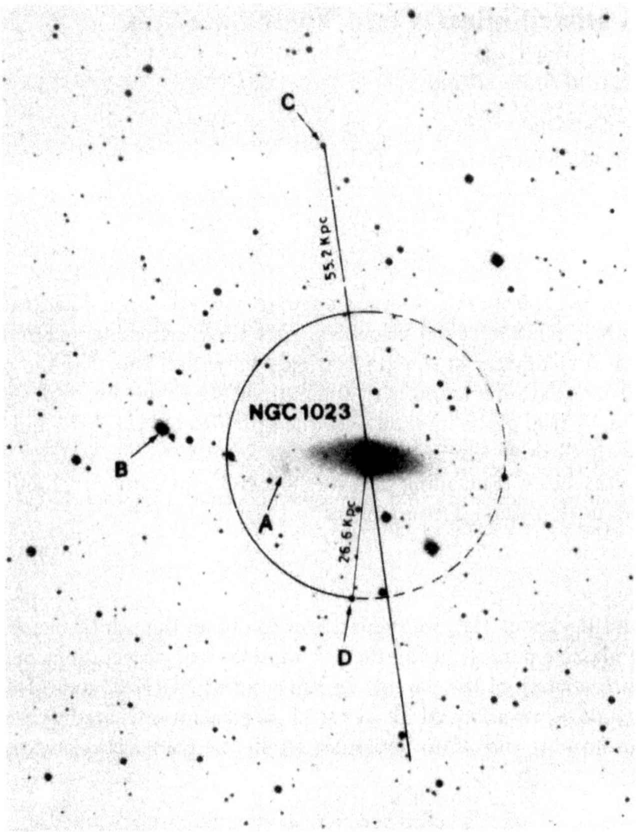

Fig. 1: Palomar plate of the NGC 1023 system with four satellites marked. Orbits for satellites C and D with reasonably reliable available data are indicated

Table 1: Set of permitted solutions for the orbital parameters of C and D companions of NGC 1023.
The correct solutions are in bold characters

Galaxy	R (Kpc)	V (Km/s)	α (^{0})	ϕ (^{0})	D (^{0})	γ (^{0})	β (^{0})
C	**55.16**	**288**	**90**	**2.6**	**257.3**	**-102.5**	**79.7**
D	26.6	245	19.1	**2.6**	**96.3**	**82.6**	**99.3**
				-177.4	-83.7	-78.5	92.7
				177.4	271.5	-80.1	88.1
				-2.6	**91.5**	**81.0**	**94.7**

Table 2: Set of permitted solutions for the orbital parameters of companions of NGC 128.
The correct solutions are in bold characters

Galaxy	R (Kpc)	V (Km/s)	α (^{0})	ϕ (^{0})	D (^{0})	γ (^{0})	β (^{0})
				141.3	227.7	-129.1	68.0
				-38.7	**47.7**	**47.6**	**102.9**
NGC 127	**26.0**	**228.0**	**84.5**				
				38.7	**56.5**	**56.2**	**102.2**
				-141.3	-123.5	-120.8	71.0
NGC 130	**22.6**	**228**	**90**	**0.0**	**-67.6**	**-68.2**	**75.6**

3. Conclusions

To some extent, the technique we have devised is not unique, in that a similar approach, using globular clusters, can be envisaged and would have striking statistical advantages. However, the present method, given the much wider range of orbital radii spanned by dwarf satellites, has the potential to explore the sphericity of galactic masses with higher sensitivity. It is clear, though, that literature data for the relative radial velocities, masses and luminosities of the satellites, and the rotation curves of the primaries are not yet, in general, precise enough to draw reliable conclusions.

References

1 J. Cepa and J. E. Beckman: *Astr. Ap.*, **200**, 21 (1988) *(and references there in)*
2 R. D. Davies and T. D. Kinman: *M.N.R.A.S.*, **207**, 173 (1984)
3 M. Capaccioli, H. Lorenz and V. L. Afanasjev: *Astr. Ap.*, **169**, 54 (1986)

The NGC 128 Group of Galaxies

B. Jarvis

European Southern Observatory, Casilla 19001, Santiago 19, Chile

I present preliminary results of a 2D photometric and kinematic study of the NGC 128 group of galaxies (NGC 127, NGC 128 and NGC 130). There is strong evidence that these three galaxies are currently suffering a strong dynamical interaction. Firstly, the tidal bridge between NGC 127 and NGC 128 is strong. The spread of systemic velocities between these galaxies is also less than 200 km s^{-1}. Secondly, the outer isophotes of NGC 130 are strongly box-shaped at the 23 mag $arcsec^{-2}$ level and fainter in R. Moreover, the major axis rotation curve of NGC 130 is also disturbed, most likely indicating a tidal origin. Examination of rotation and velocity dispersions along five cuts perpendicular and three cuts parallel to the disk major axis of NGC 128 show that the bulge of NGC 128 is in strong cylindrical rotation to at least 6kpc (assuming D=60Mpc) above and below the plane of the disk with no clear indication of a decrease at the largest z distances measured. The bulge is also one of the most pronounced peanut-shaped known. These data provide probably the most detailed kinematic and photometric picture of a peanut-shaped galaxy yet obtained and should now be sufficient to constrain several dynamical models recently proposed for their formation.

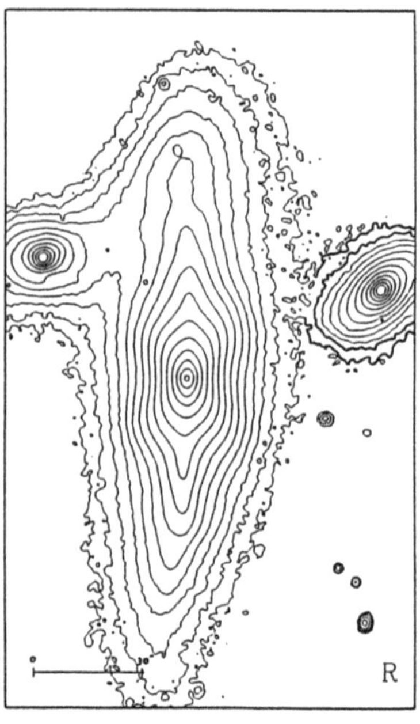

Surface brightness distribution of the NGC 128 group in R. NGC 130 is to the right and NGC 127 is to the left. The isophote spacing is 0.5 mag $arcsec^{-2}$ starting at 24.0 mag $arcsec^{-2}$. Note the boxy shape of the outer isophotes of NGC 130 (heavy black lines).

Dynamics and Interactions of Galaxies Editor: R. Wielen

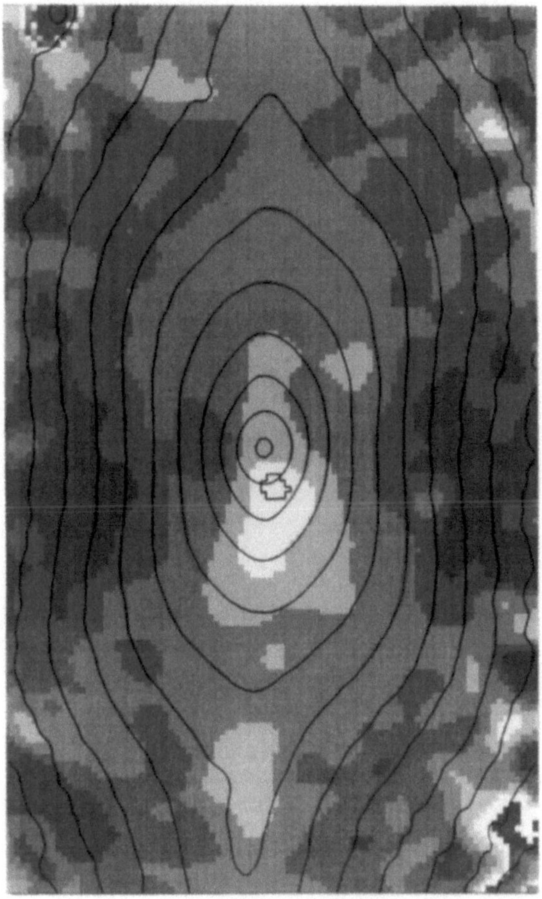

B-R colour distribution of the central 33" x 53" of NGC 128 with B isophotes superimposed (0.5 mag intervals). Note the non-coincidence of the colour distribution with the light distribution, probably indicating dust. N is up and E is to the right.

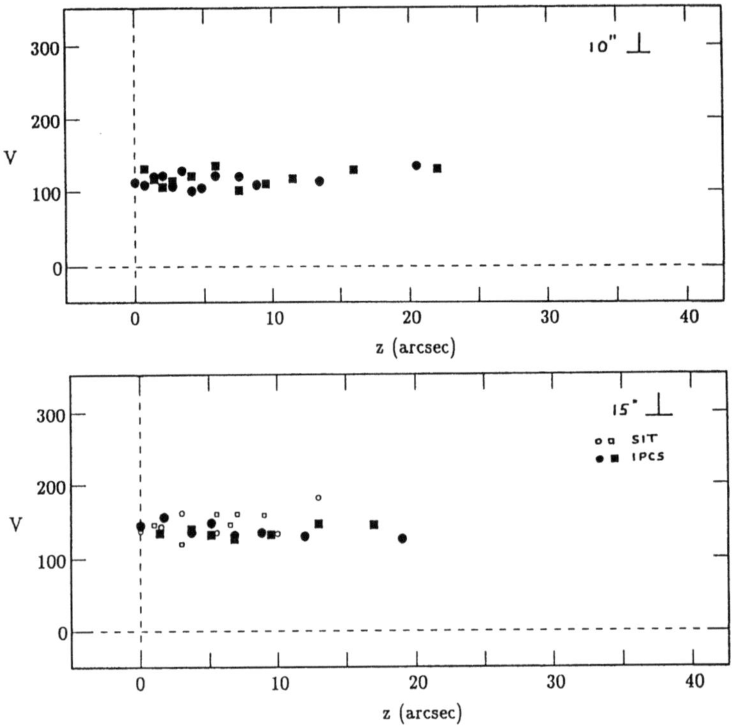

Observed rotation curves of NGC 128 perpendicular to the disk and offset from the minor axis by the amounts indicated in the two figures above. The velocities are measured relative to the systemic velocity of the galaxy. Note the almost constant velocity with increasing z, i.e. cylindrical rotation. The data have been folded about the photometric center.

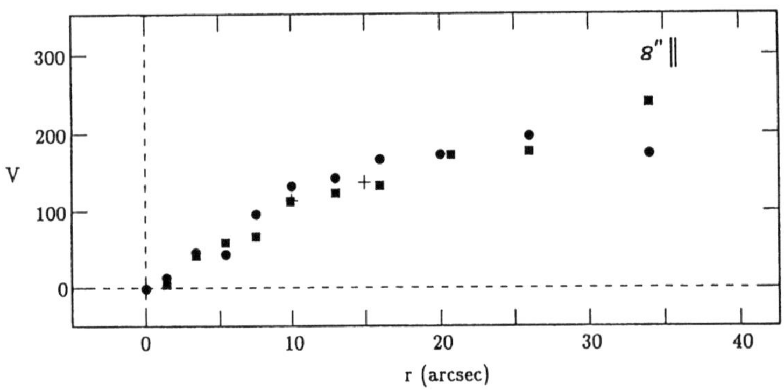

Observed rotational velocity parallel to the major axis of NGC 128 but offset 8". The crosses indicate the velocities from the intersecting slit positions from the top two panels. The data have been folded about the photometric center

Structural Properties of Early-Type Galaxies in Low Density Environments: Pairs

R. Rampazzo[1] *and L.M. Buson*[2]

[1]Osservatorio Astronomico di Brera, Milano, Italy
[2]Osservatorio Astronomico di Padova, Padova, Italy

Both formation conditions and subsequent environmental effects could have affected the present structure of 'normal' early-type galaxies. In the past few years it has been widely assumed that isophotal twists could be caused by tidal effects (*cf.*Kormendy, 1982). Recently, Fasano and Bonoli (1989) found that this phenomenon occurs with the same frequency and amplitude in their isolated galaxy sample as in samples analysed previously in the literature. They concluded that *"tidal effects cannot be invoked as a major cause for isophotal twisting in elliptical galaxies"*. In addition Rampazzo and Buson (1988) have found that the amount of measured twisting (θ_m) is independent of the density of galaxies ρ $[GalMpc^{-3}]$ (de Souza *et al.*1985) over four orders of magnitude from field ($log\rho < -2.5$) to clustered objects ($log\rho > -1.5$), concluding that the membership to the field, to a group or to a cluster does not seem to influence the structure. At the same time they did observe basically different geometrical properties in objects sharing the same environment. NGC 1549/1553, an isolated pair (following the Karachentsev's criteria) located in a small group (*cf.*Sadler and Sharp 1984), is an example of this phenomenon, (*cf.*Rampazzo 1987; *cf.*for internal kinematics Franx 1988 and Rampazzo 1988).

Since isolated pairs of galaxies differ from isolated single galaxies only in the effects induced by the companion, they represent the simplest model for testing the effectiveness of possible mutual tidal interactions. We have begun an observing program selecting a sample of pairs containing at least an early-type component from the compilation of Picchio & Tanzella–Nitti (1985).

Our program (still in progress) is primarily based upon data obtained with the 1.82m telescope of the Asiago Observatory including both B and R CCD photometry. The image analysis is being done with different packages at the Observatories of Padova and Brera in Milano.

Preliminary results, made without any claim of statistical significance at this phase in the program, are:

1) we confirm the general rule found by Galletta (1980), that flatter objects show smaller twisting effects.

2) Components in an E+E pairs seem to share the maximum ellipticity value in the sample we analysed, but they may show different twisting behaviour, in the sense that the two objects may have a comparable (*e.g.*NGC 1587/1588) or quite different (*e.g.*NGC 7033/7034; NGC 7557/7562) amounts of twisting. In Fig. 1 we report an example. This may be explained in term of a projection effect, but it is not unusual to find also pairs in wich one member is twisted while the other shows a negligible twisting. It should be remembered that, due to the small CCD field, in some cases we are dealing with only the inner part of the galaxies which, at least in principle, might be less affected by interaction.

3) (1) and (2), for what concern the twisting, seem true also for E/late–type pairs. The two cases presented here show that the early-type component may be twisted (*e.g.*NGC 2274 in pair with the spiral NGC 2275) or slightly twisted (*e.g.*NGC 3894 in pair with the spiral NGC 3895).

Dynamics and Interactions of Galaxies Editor: R. Wielen

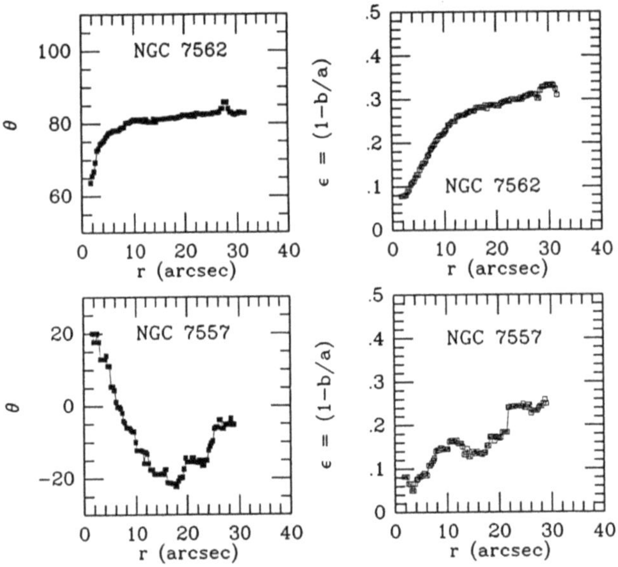

Fig. 1 Ellipticity and Twisting profiles of NGC 7557 and NGC 7562. Both are ellipticals with a $\Delta v = 126 Kms^{-1}$, a separation of 3.'1 and a B_T respectively of 15.0 and 13.0.

In this framework, twisting effects observed in 'normal' pairs cannot be easily ascribed to the mutual tidal interaction of the components. The present phenomenology may be sketched in terms of intrinsic structural features related to the formation process of the pair components. In some cases one should take into account possible merger events with small-size bodies (*cf.*the paper of Balcells and Quinn in these proceedings). In particular, the outstanding example of the nearby pair of shell galaxies NGC 1549/1553 in which only one component (NGC 1549) shows a large twisting, an asymmetric plume of material and very complex kinematics (including rotation along both minor and major apparent axes) is intriguing.

References

de Souza, R.E., Vettolani, G., Chincarini, G., 1985: *Astron. & Astrophys.*, **143**, 143.

Fasano, G., Bonoli, C. 1989: *Astron. & Astrophys. Supp. Ser.*, in press.

Franx, M. 1988: PhD Thesis Univ. Leiden, Netherland.

Galletta, G. 1980: *Astron. & Astrophys.*, **81**, 179.

Kormendy, J. 1982: in *Morphology and Dynamics of Galaxies* Twelfth Advanced Course of the Swiss Society of Astronomy and Astrophysics, eds. L. Martinet and M. Mayors, Geneva Observatory, 115.

Picchio, G., Tanzella-Nitti, G. 1985: *Astron. & Astrophys.*, **142**, 21.

Rampazzo, R. 1987: *I.V.S.L.A.*, Tomo CXLV, 81.

Rampazzo, R., Buson, L. 1988: in *Le Monde de Galaxies*, eds. L. Bottinelli and H.J. Corwin, Paris, in press.

Rampazzo, R. 1988: *Astron. & Astrophys.*, **204**, 81.

Sadler, E.M., Sharp, N.A. 1984: *Astron. & Astrophys.*, **133**, 216.

Optical Colors of Early-Type Galaxies in Compact Groups

S.E. Zepf[1] *and B.C. Whitmore*[2]

[1]Space Telescope Science Institute, 3700 San Martin Drive, Baltimore, MD21218, USA and the Johns Hopkins University, Baltimore, MD21218, USA
[2]Space Telescope Science Institute, 3700 San Martin Drive, Baltimore, MD21218, USA

1. INTRODUCTION Compact groups of galaxies with high spatial densities and low velocity dispersions are very unstable to merging. The mergers of spiral galaxies shoud be common in these environments, and if elliptical galaxies are the product of the merger of two spirals, then a population of compact group ellipticals with younger stellar populations than normally found in elliptical galaxies is expected. We have found one candidate elliptical galaxy which has the blue optical colors expected of an elliptical galaxy formed from the recent merger of two gas rich systems. We also have tentative evidence for similar signatures in several other compact group ellipticals. Models of the evolution of the number and color of early-type galaxies in compact groups do not fit the data very well unless it is assumed that the compact groups begin with a mix of galaxy types which has a greater fraction of old, red ellipticals than is observed in the loose groups from which compact groups are thought to originate.

2. OBSERVATIONS Our primary goal is to test the hypothesis that early-type galaxies are formed from the mergers of spirals by searching for bluish ellipticals in the regions where mergers are expected to be most common. At present, our observational data base consists of B - R photometry of all of the early-type galaxies in the Hickson groups (Hickson et al. [1]) as well as our own UBV observations of about 20 groups.

The narrow range in optical colors occupied by normal elliptical galaxies is clearly evident in the color-magnitude diagram of Sandage and Visvanathan [2] shown in Figure 1a. A comparison of this figure with the color-magnitude diagram of the elliptical galaxies in compact groups (Figure 1b., data from Hickson et al. [1]) indicates that the scatter is much larger for the galaxies in compact groups, especially towards the blue. However, much of this additional scatter is the result of the contribution from a few groups (noted as stars in Figure 1b.) in which all of the early types are much redder or bluer than expected. The one group for which all of the colors are red is at very low galactic latitude, so the unusual color is plausibly explained by insufficient correction for galactic reddening. For the groups with all blue galaxies, clouds affecting the red exposure remains a possibility.

Even when the galaxies in these strange groups are excluded, we find six ellipticals with blue colors suggestive of a past merger origin. These six appear to be reliable candidates (i.e. other early-types in the candidate's group have normal colors). For one of these ellipticals (Hickson 37e) we have UBV surface photometry which confirms both its classification as an early-type galaxy, and its unusually blue color. The strange concordance of unusual colors within a group dictate a cautious interpretation of the $B - R$ colors. However, **the B - R data provide at least tentative evidence for a population of abnormally blue ellipticals in compact groups.**

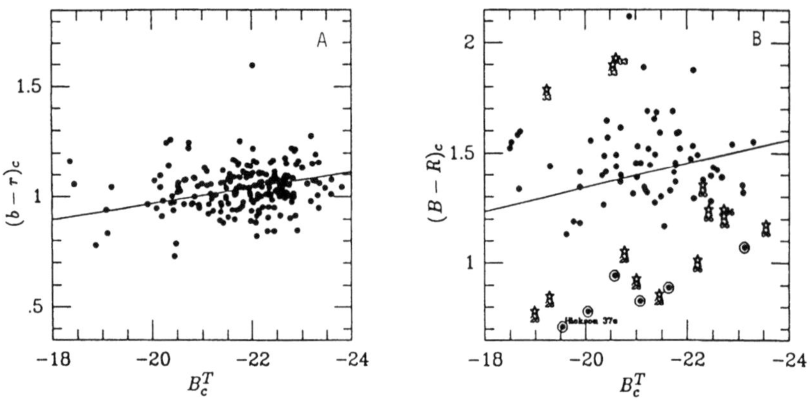

Figure 1. (a) The color-magnitude plot for all of the galaxies in Sandage and Visvanathan [2]. (b) The color-magnitude plot for all of the early-type galaxies in the Hickson compact groups [1]. The six candidate blue ellipticals are encircled.

3. MODELLING Is the distribution of the optical colors of compact group ellipticals as shown in Figure 1b. consistent with the prediction that the formation of young, blue ellipticals by the merging of spirals is a frequent process in compact groups? We have begun building simple models of the distribution of the optical colors of compact group ellipticals in an attempt to answer this question (Zepf and Whitmore [3]). The basis of the models is the combinination of Barnes' [4] dynamical simulations with the stellar population models of Larson and Tinsley [5], along with the assumption that the products of all of the mergers are ellipticals.

We find that it is difficult in the context of these models to explain both the observed colors and distribution of morphological types observed in compact groups. The lifetime of a compact group is not much longer (and possibly shorter) than the time it takes the product of a merger to evolve into a normal, red elliptical galaxy. Therefore, it is difficult to reproduce the greater fraction of early-type galaxies in compact groups compared to the field without violating the constraint that no more than 20% of them are abnormally blue. A slower evolution rate for the group and a relatively rapid recognition of a merger product as an elliptical brings the models into better agreement with the data. Of course, one can simply assume that the enhanced fraction of early-types was set by the initial conditions, and the few blue ellipticals are due to recent interactions with gas rich galaxies. We are undertaking photometric and spectroscopic observations which will distinguish these two hypotheses.

REFERENCES

1. Hickson, P., Kindl, E., and Aumann, J. 1989, *Astrophys. J. Suppl.*, in press.
2. Sandage, A., and Visvanathan, V. 1978, *Astrophys. J.*, **225**, 742.
3. Zepf, S.E. and Whitmore, B.C. 1989, *Astrophys. J.*, in preparation.
4. Barnes, J.E. 1989, *Nature*, **338**, 123.
5. Larson, R.B. and Tinsley, B.M. 1978, *Astrophys. J.*, **219**, 46.

Morphological Differences in the Nature of Radio Sources in High Density Groups

T.K. Menon

Department of Geophysics and Astronomy, University of British Columbia, Vancouver, B.C., V6T1W5 Canada

A VLA study of 393 galaxies in high density groups has shown that the absolute magnitude distributions of radio-loud and radio-quiet ellipticals in the groups are the same while the radio-loud spirals are significantly brighter than the radio-quiet spirals. Furthermore radio sources in ellipticals are found only among the brightest ellipticals in a group independent of their optical luminosity. These results suggest a fundamental difference in the ultimate consequences of interaction for ellipticals and spirals in a group and the subsequent triggering of nuclear activity.

1. Introduction

A large number of optical, infra-red, radio and x-ray studies of galaxies in different environments have demonstrated the important role of interactions in the interpretation of the diverse phenomena manifested at different wavelengths. These studies have also shown that the usually dominant elliptical galaxies in a group or cluster are much more likely to show evidence of activity than the less massive members. Since the cross-section for interaction depends directly on the local galaxy density an investigation of a sample of very high density groups can provide important information on the role of interaction in initiating activity in galaxies. I have investigated the radio radiation from a subsample of 85 Hickson (1982) Compact Groups containing 85 E, 119 SO and 189 spirals. The relevant optical data are given by Hickson, Kindl and Auman (1989). The radio observations were carried out with the VLA in various configurations at 1.4 GHZ and 5 GHZ (Menon and Hickson (1985) and Menon (1989)). Of the 87 galaxies detected so far, only 78 detected above a uniform sensitivity limit of 1 mJy are included in the present discussion.

2. Discussion

It has been known from earlier studies of Auriemma et al (1977) and Hummel (1980) that both in the case of E and S galaxies the radio luminosity is well correlated with optical luminosity. Hence the optically brighter members of a group may be expected to have a higher probability of being radio sources than the less bright members. An application of the Kolmogoroff-Smirnov test to the absolute magnitude distributions of the radio-loud and radio-quiet samples shows that in the case of E galaxies the optical luminosity distributions are statistically the same while in the case of spirals the radio-loud galaxies form an optically bright sample from the whole population. This result is particularly surprising since among the radio-loud galaxies the ellipticals show a higher degree of correlation with optical luminosity than the spirals which have a higher dispersion.

Another difference in the behaviour of the ellipticals and spirals in high density groups is seen in the ranking distribution of the radio-quick and radio-loud galaxies of the two morphological types. The present sample confirms the earlier preliminary conclusion of Menon and Hickson (1985) that, at least as far as the radio radiation is concerned, whenever an elliptical galaxy in a group is a radio source it is almost always only the brightest galaxy or the brightest elliptical in the group. This is in sharp contrast to the situation in the case of spirals where the first two ranks have almost the same probability of being radio sources and in addition it is fairly common to have two radio-loud spirals in a group. In the interaction scenario for the origin of radio sources in compact groups the fuelling of the central source is attributed to the accretion of gas on to a compact object at the center of the galaxy. In such a scenario the above discussed results imply that the physical processes, ending up with the accretion to the central object must be different for the ellipticals and spirals. It is significant that there are no extended radio sources among the 87 radio-loud galaxies. At the same time many of the spirals show optical and radio evidence of interaction. A comparison of the radio properties of the galaxies in the Hickson Groups and isolated galaxies of similar type (Menon (In Preparation)) shows that the group galaxies have a higher probability of being radio-loud than isolated galaxies. At the same time the enhancement is almost entirely in radio radiation from the central regions of galaxies. Hence it would appear that in the case of both ellipticals and spirals their membership in high density groups increases the probability of central accretion but in the case of ellipticals the cause of increased probability is related to the ascendency of the dominant elliptical while in the case of spirals interactions play an additional and possibly dominant role in their activity. A detailed discussion of the possible processes involved is in preparation (Menon and Hickson (1989)).

I should like to express my appreciation to P. Hickson for numerous discussions. VLA of NRAO is operated by AUI Inc. under contract with the National Science Foundation. This study is supported by grants from the Natural Science and Engineering Research Council of Canada.

Bibliography

Auriemma, C., Perola, C., Ekers, R., Fanti, R., Lari, C., Jaffe, W. and Ulrich, M.H., 1977, A&A, 57, 41
Hickson, P., 1982, Ap.J., 255, 382
Hickson, P., Kindl, E. and Auman, J., 1989, Ap.J. Suppl. (In Press)
Hummel, E., 1980, Ph.D. Thesis, University of Groningen
Menon, T.K., 1989 (In Preparation)
Menon, T.K., and Hickson, P., 1982, Ap.J., 296, 60
Menon, T.K., and Hickson, P., 1989, (In Preparation)

Are the Groups of Galaxies Flat?

U. Haud

Tartu Astrophysical Observatory, Institute of Astrophysics and Atmospheric Physics, Estonian Academy of Sciences, SU-202444 Tõravere, Estonia, USSR

In 1969 Holmberg studied 174 groups of galaxies centered on prominent spiral systems. He discovered that, in the case of spirals with an edgewise orientation, satellites have a peculiar distribution; most of them are found closer to the direction of the minor axis of the central system. We extended this study, using the catalogue of physical groups of galaxies compiled by Vennik (1984). To study the distribution of relatively smaller companions around giant galaxies we selected from the list all groups, where the brightest galaxy has the magnitude B_T less than -18^m and the second brightest galaxy is at least 0.9^m fainter. These groups were divided into two classes: 1) systems with a nearly face-on main galaxy, 2) all remaining groups. As a result, we had in the first class a total of 19 groups with 134 members and in the second class 32 groups with 185 members.

For the second class we constructed a histogram of relative numbers of the companions in position angle intervals of 15° from the plane of the main galaxy (Fig. 1). According to the chi-square test an observed distribution can be obtained, in the case of a spherically symmetric distribution of companions (dashed line in Fig. 1), only with a probability of 3.7 per cent. The inspection of the groups from our first sample demonstrates that most of them are considerably elongated. The situation is illustrated in Fig. 2, for which the position angles were measured from the direction of the minimal extent of the groups. In this case the observed distribution is consistent with a spherically symmetric distribution of companions (dashed line) only at a probability less than 0.1 per cent.

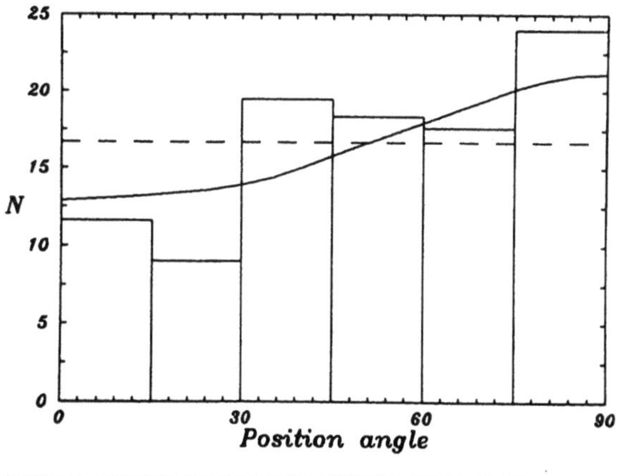

Figure 1. Histogram of the relative numbers of the companions of galaxies seen nearly edge on, in position angle intervals of 15° from the plane of the main galaxy. The dashed line represents the dependence for a spherically symmetric distribution, and the smooth solid line - the dependence in the case of a flat distribution

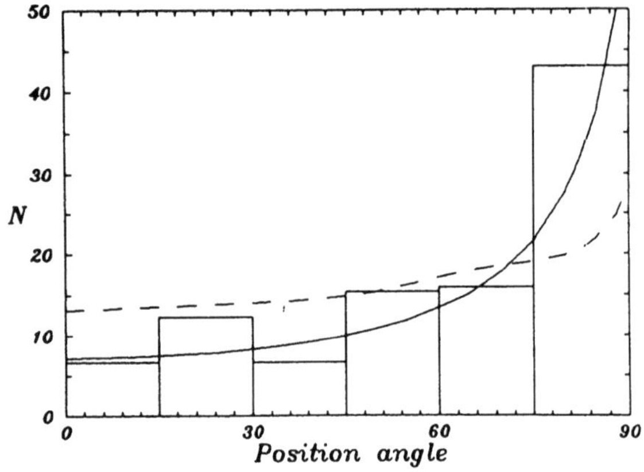

Figure 2. Histogram of the relative numbers of the companions of galaxies seen nearly face on, in position angle intervals of 15° from the direction of the minimal extent of the group. The dashed line represents the dependence for a spherically symmetric distribution, and the smooth solid line - the dependence in the case of a flat distribution

The described observational results seem to suggest that the companions of giant galaxies have a tendency to concentrate to the planes, nearly perpendicular to the main plane of these galaxies. We simulated observations of randomly orientated flat systems of companions, each of 3-15 members, where the distances of companions from the main galaxy were uniformly distributed between 20 kpc and 1 Mpc. The deviations of the disks of companions from perpendicularity to the main galaxy were generated according to the normal distribution with a zero mean and a dispersion of 26°. We generated randomly also the orientation of the group plane and the angle between the plane of the main galaxy and the line of sight. Absolute values of the sines of the latter angle were uniformly distributed between 0.75 and 1 for the models of the first sample, and between 0 and 0.75 for the models of the second sample. The obtained "observations" were processed as real ones and distributions of position angles are given with smooth solid lines in Figs. 1. and 2. They represent the actual observations satisfactorily.

Dr. E. Saar called our attention to a kinematical possibility to check the reality of the flat distribution of the companions. As in the first class of groups, we observe all disks of companions more or less edge on, and in the second class we have both the disks seen edge on and face on, then the mean observed velocity dispersion of the companions in the first class must be higher than that for the objects of the second class. From the bootstrap procedure we obtained the following estimates: **sample I: 155.4 ± 13.0 km/s, sample II: 124.8 ± 8.8 km/s.** With a probability of 97.2% the observed velocity dispersion in the groups of the first class is, indeed, higher than that in the groups of the second class.

REFERENCES

Holmberg, E.: 1969, *Ark. Astron.* **5,** 305
Vennik, J.: 1984, *Tartu Astrofüüs. Obs. Teated* **73,** 3

Formation of cD Galaxies and Dissipationless Collapse

N. Roos and G. Rhee

Sterrewacht Leiden, P.O. Box 9513, NL-2300 RA Leiden, The Netherlands

ABSTRACT. There is strong new evidence that first ranked galaxies are aligned with their parent cluster and with the direction of the nearest neighbour cluster (scale $15h^{-1}$Mpc) and that the average ellipticity of first ranked ellipticals is a strongly increasing function of radius. The alignment effect is limited to first ranked galaxies. In hierarchical clustering scenarios like the cold dark matter theory, galactic halos and clusters of galaxies are expected to have moderate asphericity. We present numerical results of a study of the dissipationless collapse of moderately aspherical systems. Our results indicate that the central part of the collapsed and virialized system does show the large scale elongation imposed by the initial conditions if the system is prolate. We also reproduce the increasing ellipticity as a function of radius that is observed for brightest cluster ellipticals.

Orientation and Ellipticity of First Ranked Galaxies in Clusters

During the last decades several alignment effects between galaxies and clusters have been reported and subsequently disputed (see Djorgovski, 1987 for a review). One alignment effect has survived and seems now well established: the alignment of first ranked galaxies with their parent cluster and even with the surrounding galaxy distribution on scales up to $15h^{-1}$Mpc (e.g. Rhee and Katgert, 1987, Lambas et al. 1988).

Porter (1988) has shown that the average ellipticity of brightest elliptical galaxies in 175 Abell clusters is a strongly increasing function of physical radius (see figure). This distinguishes brightest cluster ellipticals as a class from other elliptical galaxies. The outer isophotes are so flattened that it can be shown that the surface brightness contours of brightest cluster ellipticals at large radii cannot be drawn from a population of randomly oriented oblate spheroids.

Clusters are generally assumed to originate from overdense regions in the early universe which will generally not be spherically symmetric. Small deviations from sphericity tend to

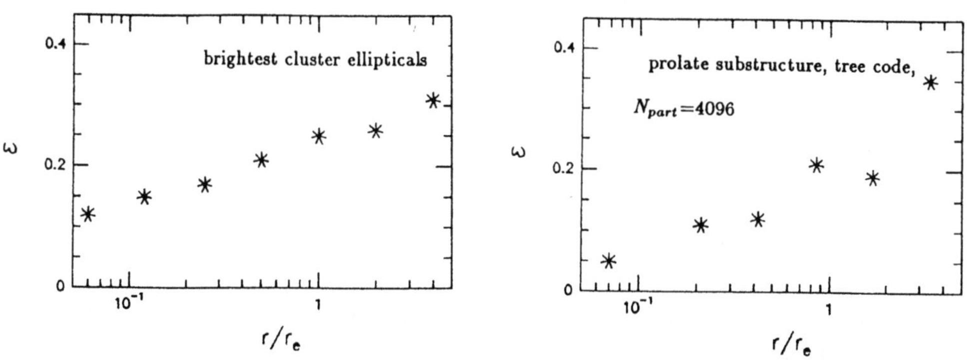

Figure 1 *Ellipticity as a function of de Vaucouleurs radius: left) Porter's sample of brightest cluster ellipticals; right) prolate ellipsoids 4096 particles with initial substructure*

Dynamics and Interactions of Galaxies Editor: R. Wielen

grow (Lynden-Bell, 1964, Icke,1984). In cold dark matter simulations the collapsed and virialized systems generally are triaxial having typical axial ratios of about 2 (Frenk et al. 1988).

In this paper we present results of a series of N-body experiments involving the collapse of dissipationless systems having moderate asphericity. A more complete account of our work will be published elsewhere (Rhee and Roos, 1989). The asphericity is similar to that of the virialized systems that form in the cold dark matter scenario. We will pay particular attention to the run of ellipticity over a large range in radius in the collapsed and virialized object. The question we wish to answer is: under what conditions will the initial large scale anisotropy be communicated to the innermost regions of the collapsed and virialized system.

Numerical Experiments

Initial conditions were generated by distributing the particles randomly inside prolate and oblate ellipsoids with axial ratios 2:1:1 and 2:2:1 respectively. (One prolate experiment was made with particles distributed in 16 spherical subclumps filling half the volume of, and randomly distributed within the ellipsoid). Particles have low initial random velocities such that the initial ratio of kinetic to potential energy is ~ 0.05. The evolution of these systems was followed over four collapse times.

Two different numerical integration schemes were used to check for consistency of the results. Firstly we used a hierarchical tree code (Barnes and Hut, 1986) and secondly a "Poisson solver" in which Poisson's equation is solved after expanding potential and density in Legendre functions (Bontekoe, 1988). We refer to the former code as the tree code and to the latter code as the van Albada code.

Results and Discussion

A position angle and associated uncertainty were calculated using a fourier analysis of the azimuthal particle distribution. Ellipticities were calculated by using the moment of inertia tensor. Results for one simulation are shown above.

We find that for prolate ellipsoid collapse, the ellipticity and orientation are communicated to the central part over a large range in radius. This is not the case for oblate ellipsoids. The result for prolate ellipsoids is not affected by distributing the particles in subclumps. The end result is not sensitive to the number of particles used (4096 and 20000) or the code used. Note that we reproduce the increasing ellipticity with radius that is observed in brightest cluster ellipticals.

We made a study of the dissipationless collapse of n-body systems having final asphericity comparable to virialized systems that form in cold dark matter scenarios. Our results provide a basis for a quantitative understanding of the shape and orientation of first ranked galaxies in clusters. The results indicate that the first ranked galaxies are formed during the collapse of the cluster as a whole and that they are generally sitting at the bottom of the cluster potential well.

References

Barnes, J. and Hut, P., 1986, *Nature,* **324**, 446

Bontekoe, T., 1988, Ph.D. Thesis, University of Groningen

Djorgovski, S.,1987, in the proceedings of the Santa Cruz Workshop *Nearly Normal Galaxies*, ed. S. Faber (Springer Verlag)

Frenk, C.S., White, S.D.M., Davis, M. and Efstathiou, G., 1988, *Ap.J.*, **327**, 507

Icke, V., 1984, *M.N.R.A.S.*, **206**, 1P.

Lambas, D.G., Groth, E.J. and Peebles, P.J.E. 1988, *Astron. J.*, **95**, 996
Lynden-Bell, D., 1964,*Ap.J.*, **139**, 1195
Porter, A.C., 1988, Ph.D. Thesis, California Institute of Technology
Rhee, G.F.R.N. and Katgert, P., 1987, *Astron. and Astrophys.*, **183**, 217
Rhee, G.F.R.N. and Roos, N., 1989, *M.N.R.A.S., submitted*

Encounters of Disk-Galaxies in Small Groups and the Stimulation of Activity in Galaxies

T. Schmutzler[1,*] *and P.L. Biermann*[1]

[1]Max-Planck-Institut für Radioastronomie, Auf dem Hügel 69, D-5300 Bonn 1, Fed. Rep. of Germany

*Present address: MPI für Kernphysik, Postfach 103980, and SFB 328, D-6900 Heidelberg, Fed. Rep. of Germany

The relationship between galaxy encounters and activity of galaxies, particularly in small groups, has been studied in model calculations. We have shown the important role of the interstellar gas which can dissipate the collisional energy. Gas exchange (loss or gain) is possible. The gain of gas is especially important, as it can initiate activity in galaxies.

Observational hints

The different types of activity in galaxies may be described in terms of enhanced star formation ranging up to bursts of star formation, infrared excess, X-ray halos and last not least active galactic nuclei. Observational evidence of a relationship between interactions and activity in galaxies has been reported by many authors, for example:

1. X-ray emission of compact groups of galaxies (e.g., Biermann *et al.* 1982, Biermann and Kronberg 1983, Bahcall *et al.* 1984). The X-ray halos of boxy ellipticals also can be interpreted as a relic of kinetic energy dissipation by the interstellar gas of merging (merged) galaxies (Bender *et al.* 1989).
2. Stimulated star formation inferred from a) enhanced radio emission in pairs of galaxies (e.g., Hummel 1981), b) infrared excess in interacting galaxies (e.g., Joseph *et al.* 1984) and c) the interacting systems M81 and M82 (starburst in M82, e.g., Rieke *et al.* 1980, Kronberg, Biermann and Schwab 1985).
3. A relationship between galaxy interactions and nuclear activity (e.g., Dahari 1985, Cutri and McAlary 1985, Sanders *et al.* 1988). Fricke *et al.* (1985) have shown a relationship between the class of activity of a group member and its distance to the most active galaxy for nine small groups. They found that all neighbouring galaxies within a projected distance of 250 kpc, and only these, are Seyferts, LINERs, or starburst galaxies.

Consequently the physics of interacting galaxies must consider more than pure stellar systems. The interstellar material plays an important role at least in colliding galaxies: stars will not touch each other but gas particles will do so.

Our model

Using the basic conservation laws of momentum, angular momentum and energy we calculate collisional interactions and the evolution of disk-galaxies in small groups, focussing our interest on the role of the interstellar gas. The model galaxies are represented by infinitely thin disks of finite size and mass with flat rotation curves. They move in a gravitationally bound group (*N-body calculation*). The rotation curve determines the surface mass density, the potential and the escape velocity as functions of radius within the disk. The gas content of each galaxy is time-dependent according to a simple evolution model (*gas consumption by star formation and gass loss of stars*). We assume that the stellar component is not significantly disturbed by interpenetrating collisions, which should be valid for reasonable high relative velocities. In the case of small relative velocities merging is possible. However, we concentrate on the role of the interstellar gas. From the exact collisional geometry and the time of collision we calculate the mass of gas affected by the inelastic impact. The resulting velocity of the gas, determined

Dynamics and Interactions of Galaxies Editor: R. Wielen

piecewise from the locally rotational and orbital velocity of both galaxies using the conservation law of momentum, is compared with the local escape velocity of each disk. Thus the gas may either be retained by one of the disks, or ejected from both.

Discussion of some results

1. The orbital kinetic energy can be dissipated and thermalized by the gas (*X-ray halo*). In the context of colliding galaxies Harwit *et al.* (1987) have proposed an explanation for the extremely luminous far-infrared sources (*infrared excess*). In contrast to the starburst model they suggest that dust grains are directly heated by UV-photons from shock heated gas produced by colliding molecular clouds in interpenetrating galaxies. However, the loss of kinetic energy implies that the galaxies move together in phase space. This leads to an increased rate of collisions.
2. The increased collision rate, the exchange of interstellar gas and the resulting shockwaves running through the interstellar medium may stimulate repeated bursts of star formation (*enhanced radio emission, infrared excess* and *extended X-ray emission*, vid. Schaaf *et al.* 1989). For example: one result of our model calculations is that the inner regions (1/30 of the disk radius) nearly always gain mass, typically 10^6 to $10^8\,M_\odot$ but sometimes up to $10^9\,M_\odot$, and even up to $10^{10}\,M_\odot$ within 1/3 of the disk radius. Thus assuming conservatively that $\sim 10\%$ of the gas is converted into stars in the sound travel time of $\sim 3\,10^7$ yr across the inner region of the disk, we obtain a star formation rate of $3\,10^{-2}$ to $30\,M_\odot\,yr^{-1}$. Using M82 as our prototype, with $4\,10^{10}\,L_\odot$ in the infrared and a gas mass of $\sim 3\,10^7\,M_\odot$ present within the starburst region, we can infer expected infrared luminosities stimulated by collisions. We estimate $L_{IR} = 1.3\,10^{12}(M_g/10^9\,M_\odot)L_\odot$. Using our result that the ratio of gas gain ΔM_g to gas content M_g in the inner part of a disk prior to collision (required to stop the incoming gas) ranges from 20% to 80%, we obtain $L_{IR} \simeq 5.4\,10^{12}((1 \pm .45)\Delta M_g/10^9\,M_\odot)L_\odot$. Thus we find infrared luminosities of up to $10^{12}\,L_\odot$. A more general relationship between gas mass and far-infrared luminosity was given by Krügel *et al.* (1988). M82 fits their result if the $2\,10^8\,M_\odot$ of gas surrounding the starburst region is also considered. Though this gas is not part of the starburst region so far, its dust reemits the UV-energy of the O and B stars in the mm- to IR-range.
3. The nuclei of galaxies can be fed, because the captured gas generally carries and transfers different local angular momenta in comparison to the stopping material.
$\star$ More details of our model and results can be found in Schmutzler and Biermann (1985).

References

Bahcall, N.A., Harris, D.E., Rood, H.R.: 1984, *Astrophys. J.* **284**, L29
Bender, R., Surma, P., Döbereiner, S., Möllenhoff, C., Madejsky, R.: 1989, *Astron. Astrophys.* **217**, 35
Biermann, P., Kronberg, P.P., Madore, B.F.: 1982, *Astophys. J.* **256**, L37
Biermann, P., Kronberg, P.P.: 1983, *Astrophys. J.* **268**, L69
Cutri, R.M., McAlary, C.W.: 1985, *Astrophys. J.* **296**, 90
Dahari, O.: 1985, *Astrophys. J. Suppl.* **57**, 643
Fricke, K.J., Kollatschny, W., Loose, H.H.: 1985, in *Structure and Evolution of Active Galactic Nuclei*; Eds.: Giuricin, G. *et al.*; *ASSL* **121**, 551 (Proceedings 1985; Reidel, Dordrecht)
Harwit, M., Houck, J.R., Soifer, B.T., Palumbo, G.G.C.: 1987, *Astrophys. J.* **315**, 28
Hummel, E.: 1981, *Astron. Astrophys.* **96**, 111
Joseph, R.D., Meikle, W.P.S., Robertson, N.A., Wright, G.S.: 1984, *Monthly Notices Roy. Astron. Soc.* **209**, 111
Kronberg, P.P., Biermann, P., Schwab, F.R.: 1985, *Astrophys. J.* **291**, 693
Krügel, E., Chini, R., Kreysa, E., Sherwood, W.A.: 1988, *Astron. Astrophys.* **193**, L16

Rieke, G.H., Lebofsky, M.J., Thompson, R.I., Low, F.J., Tokunaga, A.T.: 1980, *Astrophys. J.* **238**, 24

Sanders, D.B., Soifer, B.T., Elias, J.H., Madore, B.F., Mathews, K., Neugebauer, G., Scoville, N.Z.: 1988, *Astrophys. J.* **325**, 74

Schaaf, R., Pietsch, W., Biermann, P.L., Kronberg, P.P., Schmutzler, T.: 1989, *Astrophys. J.* **336**, 722

Schmutzler, T., Biermann, P.: 1985, *Astron. Astrophys.* **149**, 383

An Approximate Treatment of the Tidal Effects Between Galaxies in N-Body Simulations of Clusters of Galaxies

A. Goeres[1] *and R. Wielen*[2]

[1]Institut für Astronomie und Astrophysik, Technische Universität Berlin, Berlin, Germany
[2]Astronomisches Rechen-Institut, Mönchhofstr. 12–14, D-6900 Heidelberg, Fed. Rep. of Germany

1. Introduction

In N-body systems where each member consists of many gravitating particles, tidal effects frequently occur. They lead to a loss of kinetical energy of the N-body system in favour of the internal energies of its members. This process may have considerable consequences on the dynamical development of clusters of galaxies (Wielen 1978). Therefore these effects should be included in cosmological simulations of the formation and evolution of clusters of galaxies (Schwekendiek and Wielen 1989, this proceedings). In this paper we present a simple but efficient method of treating tidal effects in such simulations.

2. Equations of Motion

Energy conversion by tidal interactions is described by introducing

- a friction term into the equations of motion of the galaxies, and considering
- a second set of equations for the change of their mean radii.

The galaxies are treated as rigid plummer potentials (Plummer 1911) with variable plummer radii, each obeying softened Newtonian equations of motion. The total energy of the entire cluster is conserved.

The friction term represents a force antiparallel to the relative velocity between each pair of galaxies. The functional form is based on the impulsive approximation (Spitzer 1958) for distant encounters. It depends on mass m_i, radius ε_i, relative position vector $\vec{r}_{ij}$, relative velocity $\vec{v}_{ij}$ and the internal velocity dispersion $\sigma_i=\sqrt{3\pi G m/32\varepsilon}$. The equations of motion for a system of N galaxies are:

$$\ddot{\vec{r}}_i = -\sum_j^N \frac{G\, m_j}{\left(r_{ij}^2+\varepsilon_i^2+\varepsilon_j^2+2\zeta\varepsilon_i\varepsilon_j\right)^{3/2}}\left\{\vec{r}_{ij}+\gamma^2 G\left(\frac{m_j\,\varepsilon_i^2}{\left(r_{ij}^2+\eta(\vartheta\varepsilon_i^2+\varepsilon_j^2)\right)\left(v_{ij}^2+\lambda\sigma_i^2\right)^{3/2}}+\frac{m_i\,\varepsilon_j^2}{\left(r_{ij}^2+\eta(\varepsilon_i^2+\vartheta\varepsilon_j^2)\right)\left(v_{ij}^2+\lambda\sigma_j^2\right)^{3/2}}\right)\dot{\vec{r}}_{ij}\right\} \quad (1)$$

Energy conservation and the assumption of independent interactions of each pair of galaxies lead to a set of equations which describe the change of the plummer radii:

$$\dot{\varepsilon}_i = \gamma^2 G \frac{\displaystyle\sum_{i,j;i\neq j}^{N} \frac{v_{ij}^2\, m_j^2\, \varepsilon_i^2}{\left(r_{ij}^2+\varepsilon_i^2+\varepsilon_j^2+2\zeta\varepsilon_i\varepsilon_j\right)^{3/2}\left(r_{ij}^2+\eta(\vartheta\varepsilon_i^2+\varepsilon_j^2)\right)\left(v_{ij}^2+\lambda\sigma_i^2\right)^{3/2}}}{\displaystyle\frac{3\pi}{64}\frac{m_i}{\varepsilon_i^2}+\sum_{j;i\neq j}^{N}\frac{m_j\,(\varepsilon_i+\zeta\varepsilon_j)}{\left(r_{ij}^2+\varepsilon_i^2+\varepsilon_j^2+2\zeta\varepsilon_i\varepsilon_j\right)^{3/2}}} \quad (2)$$

The coefficients γ, η, λ, ϑ, ζ are dimensionless parameters which are fitted to numerical simulations of close encounters of two galaxies.

3. Discussion

The modified equations of motion allow to simulate tidal interactions of galaxies during the dynamical evolution of a cluster of galaxies in a simple way with a minimum of additional computing time.

- Mass loss is not considered. An inspection of tidal radii shows that mass loss depends mainly on the ratio of galaxy to cluster density. In simulation shown below with $N = 50$ cluster members the mean galaxy density/mean cluster density is ≥ 10 and the overall mass loss is expected $\leq 10\%$. In rich clusters with a higher density ratio the effect is entirely negligible.
- The model galaxies forming the central merger mainly constitute a background of stripped material. We checked the model simulations by calculating the dynamical friction experienced by a galaxy in a background with the distribution of a plummer model. No systematic differences were found.
- Simultaneous encounters of more than two galaxies require tensor addition of the deformations instead of vector addition of friction forces. Thus our model causes errors up to a factor two for correlated encounters. In rich clusters the errors are expected to compensate statistically except for merger remnants.
- We have treated merger remnants as a non-virialized superposition of plummer models. The error of the forces is smaller than a factor 2. For the most effective encounters with $r \simeq \varepsilon_i + \varepsilon_j$ the error is compensated by the error of vector addition of the friction forces. Merging causes no additional loss of computation time.

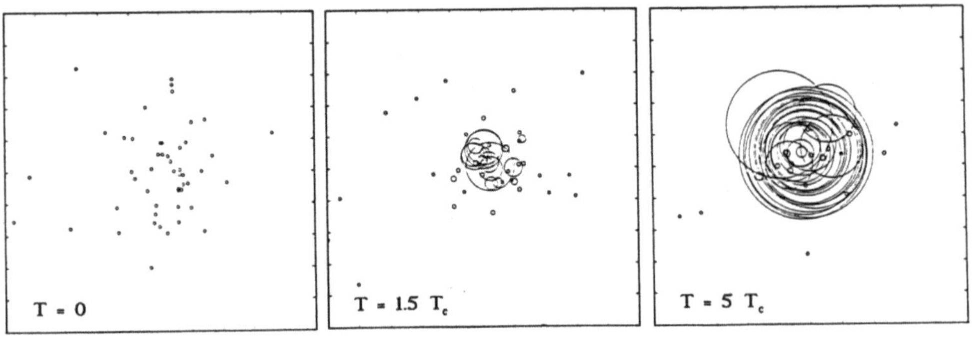

Fig.1: Dynamical evolution of a cluster of $N = 50$ galaxies:
- Parameters: $\gamma^2 = 10$, $\eta = \lambda = 3$, $\vartheta = 1$ according to Wielen(1978) and Biermann(1983).
- At $T=0$ all galaxies are equal and distributed according to a plummer model with a plummer radius r_c. The plummer radii of the galaxies are $\varepsilon = 0.05\ r_c$.

Each galaxy is represented by a circle with the radius ε_i. The figures show a two dimensional projection of the space distribution. Scale unit is r_c. T_c is the crossing time of the cluster.

REFERENCES

BIERMANN, P., 1983: private communication to R.Wielen.

GOERES, A.,1986:"Entwicklung und Untersuchung eines Modells zur Simulation inealstischer Wechselwirkungen zwischen Galaxien", Diplomarbeit, Freie Universität Berlin

PLUMMER, H.C., 1911: Monthly Not.R.Astron.Soc. **71**,460

SPITZER, L.J., 1958: Astrophys.J. **127**,17-27

WIELEN, R., 1978: in IAU Symposium79,"The Large scale structure of the Universe", Eds. M.S.Longair and J.Einasto; Dordrecht, Holland: Reidel, p.121

WIELEN, R., 1978: Mitt.Astron.Ges. Nr.45, p.16.

The Effect of Tidal Braking Between Galaxies on the Formation and Evolution of Clusters of Galaxies

P. Schwekendiek and R. Wielen

Astronomisches Rechen-Institut, Mönchhofstr. 12–14,
D-6900 Heidelberg, Fed. Rep. of Germany

The formation and development of clusters of galaxies in Euclidean expanding universes are investigated by N-body simulations. The dynamics of the galaxies is described by a continuously acting term of interaction representing the gravitational attraction, the tidal braking, and the increase of the galactic radii while conserving energy and angular momentum of the whole system. Models with purely elastic interactions and constant galactic radii serve as a reference.

The Models

We use a spherical test volume in an Euclidian universe ($\Omega = 1$), where 8000 galaxies are initially distributed in space at random. The galaxies are identical and defined by their effective radii ε and masses. The models start at $z + 1 = 10$ (i.e. $R(t_0) = 1$) at which time t_0 the effective radii of the galaxies are set to $\varepsilon = 0.01$. In the elastic case the effective radii remain constant.

The Terms of Interaction for Tidal Braking Between Galaxies

We use two terms described by GOERES & WIELEN (1989). The first term is a superposition of a softened Newtonian attractive force and a tidal braking term dependant upon the relative velocities and the distances between the interacting galaxies. The second term represents the change of the effective radii of the galaxies.

The Evolution of Groups

During the cosmological evolution the galaxies collect themselves in groups, while only a few field galaxies remain. In the elastic case we find that the galaxies stay well seperated within the virialized groups (Fig. 1). Under the effect of tidal braking, the effective galactic radii are increased by up to one order of magnitude and the galactic centers move closer together. The inner spatial region of a group often becomes filled with merged galaxies (Fig. 2). Note that the

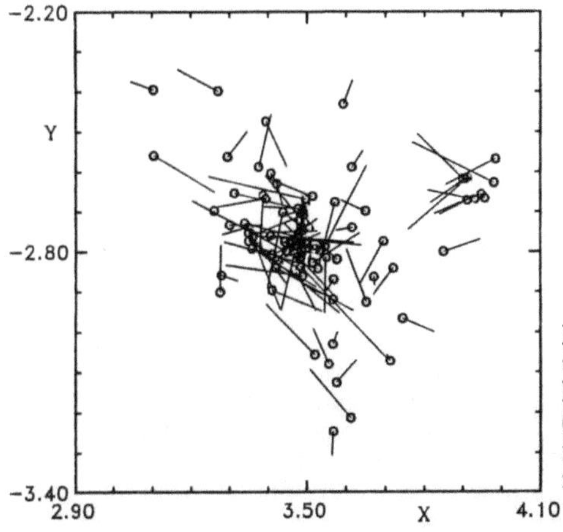

Fig. 1: A "typical" cluster in the pure elastic case at final expansion rate ($R = 10$). The lines represent the projected velocity vectors (the HUBBLE-flow is already subtracted). Notice the high velocity dispersion in the central region.

Dynamics and Interactions of Galaxies Editor: R. Wielen

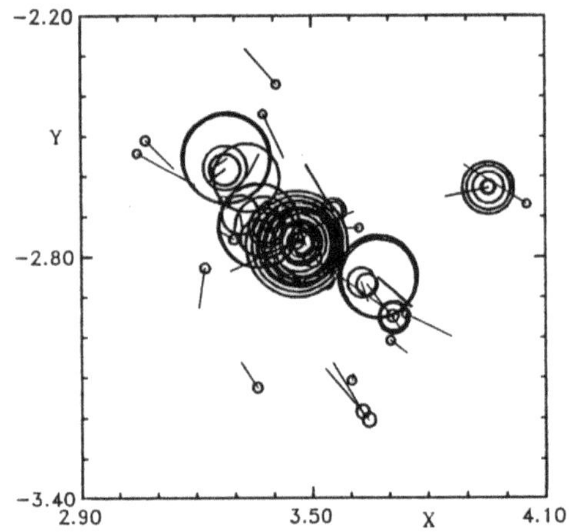

Fig. 2: A "typical" cluster in the tidal braking case at final expansion rate ($R = 10$). The lines represent the projected velocity vectors (the HUBBLE-flow is already subtracted). The circles represent the extended effective galactic radii. Notice the multiple mergers in the center.

effective radii of the model galaxies represent the extended dark galactic coronae rather than the visible inner parts of the galaxies. At the cluster centers merging naturally leads to a smaller number of more massive galaxies, and hence a decrease in the apparent number density. The velocity dispersions of the groups formed via tidal braking are significantly reduced. Hence the total mass of these groups can be up to twice the virial mass.

Global Structure

The global two-point-correlation function ξ of the spatial distribution of the galaxies in the simulations is a poor fit to the observed (GROTH & PEEBLES 1981) power law, and is only adequat for $\log \xi < 1$ (Fig. 3). This is because Newton's force is insufficient to pull the galaxies together during close encounters. However, with tidal braking the fit is significantly improved (Fig. 4) matching the power law well, all the way to the cluster core. Investigations of further models (not shown here) show that the final power law slope γ is strongly dependant on the initial distribution of the galaxies in space.

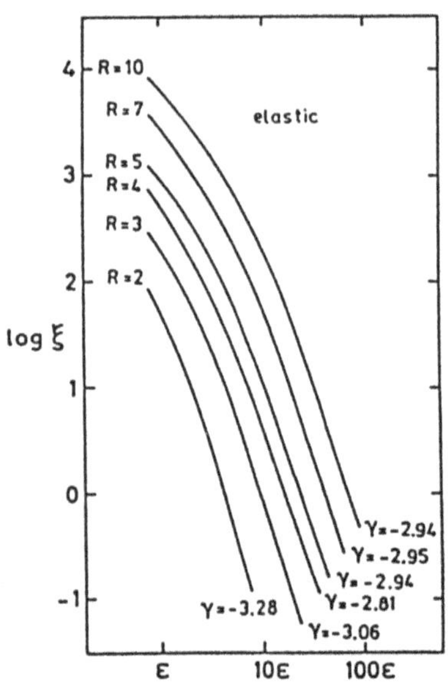

Fig. 3: The spatial two-point-correlation function ξ for different expansion rates $R = 10/(z+1)$ in the elastic case versus distance in units of the the effective galactic radii $\varepsilon = 0.01$. The power law γ can only be fitted for $\log \xi < 1$.

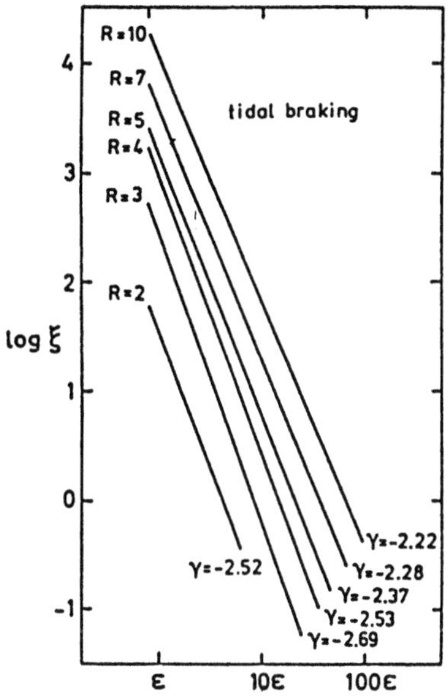

Fig. 4: The spatial two-point-correlation function ξ for different expansion rates $R = 10/(z+1)$ in the tidal braking case versus distance in units of the the start values of the effective galactic radii $\varepsilon = 0.01$. The power law γ can be fitted for all ξ and is closer to the observed value $\gamma \approx -1.8$.

References:

GOERES, A., WIELEN, R. 1989, An approximate treatment of the tidal effects between galaxies in N-body simulations of clusters of galaxies *(in these proceedings)*.

GROTH, E. J., PEEBLES, P. J. E. 1977, *Astrophys. J.* **217**, 385

Part XII

Nuclear Activity of Galaxies and Interactions

The QSO-Tidal Connection

A. Stockton

Institute of Astronomy, University of Hawaii at Manoa, 2680 Woodlawn Drive, Honolulu, HI96822, USA

This review examines the various kinds of observational evidence that have been cited to support the notion that interactions and mergers between galaxies are often responsible for triggering the nuclear activity seen in QSOs. The main concerns in evaluating the evidence that has been presented to date are (1.) that the criteria used to define the presence of an interaction or merger are generally subjective, and (2.) the absence of suitable control samples. However, the fact that several independent lines of evidence point to the importance of interactions in galaxies with active nuclei make the case fairly strong in spite of these worries.

I. Introduction

The suggestion that galaxy interactions or mergers might have something to do with the nuclear activity in QSOs and powerful radio galaxies has had at least one of its roots in the "Galactic Bridges and Tails" paper of Toomre and Toomre (1972), in a section near the end entitled "Stoking the Furnace?" Although the specific reference there was only to the tendancy of interacting galaxies to display vigorous nuclear star formation, the clear empirical lesson was that such encounters could have profound effects not only on the loosely-bound outer parts of galaxies, but also on processes occuring deep in their central regions.

The explicit connection to nuclear activity in the usual sense was made by Gunn (1979) in his well-known "Feeding the Monster" paper, in which he put forward one plausible means of channeling external gas into the nucleus of a galaxy. On the observational side, Adams (1977) had noticed an apparent excess of disturbed and interacting Seyfert galaxies. But in spite of these hints and suggestions, the climate of opinion a decade ago was essentially that, while one might occasionally find evidence for interactions in a few specific cases, these had little or nothing to do with the QSO phenomenon in general.

The current situation is quite different: that QSO activity is often, and perhaps always, triggered by interactions has achieved the status of conventional wisdom in many quarters. It seems an appropriate time to examine critically the observational bases of this belief. The questions we would really like to answer are of this sort:

- ▷ How do we recognize interactions at large redshifts?
- ▷ Can the observed features interpreted as evidence for interactions be accounted for in any other way?
- ▷ What would we see if we looked just as carefully at a sample of "normal" galaxies having similar luminosities and environments as QSO host galaxies?

We in fact are not yet able to give very good answers to questions like these, but we should keep them in mind as we now turn to the various kinds of evidence that have been put forward as support for interactions being common among QSOs.

Dynamics and Interactions of Galaxies Editor: R. Wielen

II. Types of Evidence for QSO Interactions

It is important to remember, in evaluating the evidence for interactions involving host galaxies of QSOs, that the original definition of QSOs meant that the host galaxies themselves were virtually undetectable by the best instrumentation available to astronomers two decades ago (this meant deep, large-image-scale photographs, not simply lack of visible extension on the Palomar Sky Survey prints), and that we are now staking our arguments on sometimes subtle features in these same galaxies. The formidible difficulties of these observations create pressures, firstly, to make the most of whatever observations one *can* obtain, with a consequent risk of overinterpreting the available data; and secondly, to concentrate on the more observationally tractable objects, which may give easier answers but not always to the original questions (in a parallel situation, some of the earlier studies aimed at determining whether QSOs were in galaxies effectively ended up demonstrating the not-so-astonishing fact that Seyfert nuclei were in galaxies). Appeals can and have been made to continuity arguments, but such appeals beg the question: the continuity in properties between low- and high-luminosity examples of active nuclei is one of the things being tested, and cannot be one of the assumptions. I shall try as nearly as possible, except when clearly indicated otherwise, to concentrate my discussion on objects that would have been called QSOs by the criteria generally accepted in the mid-1960s. This definition is admittedly arbitrary, but it is still a useful one in practice.

A. Close Companion Galaxies

Companion galaxies to QSOs are generally easier to detect than QSO host galaxies, but in the absence of additional morphological clues, there is no proof that a significant interaction has taken place in any individual instance. Strictly speaking, without redshift information, there is no way to exclude the possibility of a chance projection, although for low-redshift (*e.g.*, $z < 0.5$) QSOs, companion magnitudes not much below L_* (assuming the QSO redshift), and projected distances less than a few tens of kiloparsecs, the chance that the association is real has been shown to be quite high (Heckman *et al.* 1984). In any case, evidence for interactions based simply on the proximity of apparent companions has to be statistical in nature and involve sizable samples. Hutchings (1987) and Hutchings, Janson, and Neff (1989) report that roughly 70% of both radio-loud and radio-quiet QSOs have detected close companions. At least some of these also show other signs of interaction, which will be discussed in § II.B. These fractions are impressive and seem to leave little room for doubt, but no one has yet carried out an unbiased statistical survey of QSO companions matched to a control sample of normal galaxies having a similar luminosity and redshift distribution as the QSO host galaxies. Something of this sort is necessary to put claims of an enhanced probability of close companions around QSOs on a sound statistical footing.

B. Bridges, Tails, and Host Galaxy Properties

Over the past decade, several large programs have been carried out aimed at detecting and exploring the properties of QSO host galaxies (Gehren *et al.* 1984; Hutchings, Crampton, and Campbell 1984; Hutchings, Janson,and Neff 1989; Malkan, Margon, and Chanan 1984; Smith *et al.* 1986; see references in these papers for earlier work). In almost all cases, these surveys cover a wide range of luminosities, from luminous Seyferts to classical QSOs. There is a general consensus among the various groups on a number of points of importance. Firstly, all agree that current CCD surveys are capable of resolving the host galaxies of a large fraction (> 90%) of QSOs with redshifts < 0.5. Secondly, most agree that a large fraction of the resolved host galaxies show substantial distortion of one kind or another. Thirdly, steep-spectrum radio-loud QSOs appear to have larger and brighter host galaxies, on the average, than to radio-quiet QSOs. Finally, these surveys have produced some evidence to support (and little or none to contradict) the common belief that radio-quiet QSOs are an extension of the Seyfert sequence to higher luminosities, and therefore usually in spirals, while steep-spectrum radio-loud QSOs are correspondingly an extension of the broad-line radio galaxies and in (often perturbed) ellipticals.

Only Hutchings and his coworkers have produced much in the way of color information for significant samples of QSOs: Hutchings (1987) finds that the radio-loud host galaxies have quite blue colors, which he interprets as being due to extensive star formation. On the other hand, Hutchings, Janson, and Neff (1989) find the colors of radio-quiet host galaxies to be significantly redder, indicating relatively less star formation or perhaps more dust. Note that this implies that the host galaxies that are morphologically ellipticals typically have more vigorous star formation than those that are morphologically spirals.

The morphological distortions seen in QSO host galaxies range from large-scale, but unresolved, asymmetries to those that are clearly bridge-like or tail-like. Among classical, high-luminosity QSOs, good examples of probable tidal tails are the "one-armed spiral" features associated with 3CR 48 (Gehren *et al.* 1984; Stockton and MacKenty 1987) and Mkn 1014 (MacKenty and Stockton 1984; Stockton and MacKenty 1987; Sanders *et al.* 1988a). I suspect that the jet-like projection from OX 169 (Stockton 1978; Hutchings *et al.* 1981; Gehren *et al.* 1984) is a similar feature seen edge-on. Examples of QSOs with bridges connecting to companion galaxies are PKS 0735+178 (Hutchings, Johnson, and Pyke 1988) and 4C 37.43 (Stockton and MacKenty 1987).

C. Extended Emission

Extended ionized gas is found around many active galaxies, including Seyferts, radio galaxies, and QSOs (see Fosbury 1986 for a general review; van Breugel 1989 for a review of extended emission around radio galaxies; and Boroson, Persson, and Oke 1985, Stockton and MacKenty 1987, and references in each for discussions specifically of QSOs). In brief summary, extended emission-line regions (ELRs) are detected around roughly half of steep-radio-spectrum QSOs with at least moderately strong nuclear narrow-line regions; they occur with a much lower frequency around flat-radio-spectrum and radio-quiet QSOs. Their sizes range from a few kpc to a few tens of kpc. The spectra generally are of high ionization, although some of the weaker ELRs associated with flat-spectrum or radio-quiet QSOs show lower ionization. These ELRs are often highly structured, with narrow filaments sometimes a few tens of kpc long; however, unlike that of ELRs associated with many high-redshift radio galaxies, this structure rarely seems to be correlated with any extended continuum emission seen. The QSO ELRs also differ from the high-redshift radio-galaxy ELRs in that most show no clear association with the radio structure. The most important lesson from all of this is that these QSO ELRs are clearly non-equilibrium configurations, yet they are quite common: it therefore seems likely that they are in some way intimately involved with the same physical process that produces the QSO phenomenon.

Virtually all of the QSOs for which there is suspicious evidence for interaction, such as extremely close companions, large asymmetries in the continuum image, or tail-like or bridge-like structures, also seem to show ELRs. If it is legitimate to turn this correlation around and say that the ELRs usually are indicative of an interaction and are themselves a form of tidal debris, then it is possible to use them to make some inferences regarding the natures of the galaxies involved in the interaction. Just as substantial narrow or sheet-like structures of stars are indicative of pre-existing cold stellar disks (Toomre and Toomre 1972; Barnes 1988), the narrow gaseous structures we see in many of the ELRs point equally clearly to pre-existing cold gaseous disks in one or both galaxies.

However, the position that the ELRs are of tidal origin has generated some recent controversy, particularly from Crawford and Fabian (1989, and references therein). Their argument is briefly this: The ionized gaseous structures seen, if ejected from galaxies during an interaction would dissipate on timescales short compared with the interaction timescale unless they are confined in some way. The most reasonable means of confinement is a hot external medium. In the small groups to which QSOs in the relevant redshift range typically belong, the temperature of the gas confined in the group potential will be typically a few times 10^7 °K. Ionization equilibrium calculations based on measurments of the [O II] and [O III] line strengths and the estimated distance from the ionizing source (*i.e.*, the QSO nucleus) give typical densities of 10 cm^{-3}. The assumption of pressure equilibrium gives a cooling time for the assumed hot confining gas of less than a Hubble time for temperatures $< 5 \times 10^7$ °K, which is consistent with the ELRs being the results of cooling flows and obviates the need to depend on interactions as an origin of the emission-line-emitting gas.

I personally remain unconvinced by these arguments at present. We as yet have no *direct* evidence that substantial densities of hot gas exist in these small groups (there must, of course, be *some* external medium present to confine the radio lobes). The *Einstein* HRI did not have sufficient resolution and sensitivity to detect any extended hot gas in the presence of the strong nuclear x-ray emission from the QSO, but *Rosat* should be able to provide a definitive test for some objects. On the other hand, as mentioned above, we *do* have independent evidence that interactions are important for many of these objects. Furthermore, at least some QSOs with ELRs are clearly *not* at the centers of the groups to which they belong, in which case the presumed hot confining medium is not an intracluster medium, but the extended atmosphere of the QSO host galaxy itself. But such hot atmospheres are expected to have little rotation, and at least some QSO ELRs show radial velocities in the rest frame of the QSO of up to 400 km s^{-1} (Stockton, unpublished).

But what about the confinement problem? First of all, it may well be that confinement is not necessary. We are dealing with a non-equilibrium situation, and, while it is clear that the large structures we see have very small filling factors, and that the gas is in the form of relatively thin filaments, it is not clear that the individual filaments necessarily have to survive for an interaction time scale. What *is* necessary is that new filaments be formed to replace those that dissipate: essentially, that shocks continue to propagate through the relevant volume during the progress of the interaction, which, after all, does not seem unlikely. On the other hand, if some sort of confinement is required, gravitational confinement cannot yet be ruled out. I have emphasized the general lack of correlation between the distributions of the ELRs and the extended continuum material, but it is perhaps significant that for 4C 37.43, which has the highest surface brightness ELR and for which we have our deepest line-free continuum image, we do see continuum radiation underlying the brightest ELR cloud (Stockton and MacKenty 1987; see also Fig. 1. in §V.A). It is clear that this material must have a moderately-high mass-to-light ratio if it is to be called on to confine the gas, and deep continuum images are necessary for other objects to see if this is a general phenomenon, but at this point it remains a possibility.

D. How Frequent Were Interactions at $z \sim 0.3$?

The discussion up to now has centered on various kinds of evidence that many host galaxies of QSOs have recently undergone interactions or mergers. But, even if this evidence is taken at face value, it alone cannot demonstrate a causal link between interactions and nuclear activity. To take an extreme situation, suppose that virtually *all* major galaxies in environments like those in which QSOs are found were to have had major interactions within, say, 10^9 years prior to the epoch at which they were observed: then we could make no statement regarding a connection between interactions and QSO activity, precisely because the connection depends on the circumstantial evidence of a juxtaposition of two intrinsically rare events. An implicit assumption in most discussions has been that interaction rates at the epochs of the QSOs in the various surveys (typically corresponding to $z \sim 0.3$) were not too different from that of the present. Although it still seems unlikely that this assumption is so poor that it jeopardizes the argument, phenomena such as the Butcher-Oemler effect and the prevalence of galaxies with shells (see the papers by Lavery and Prieur, respectively, in this volume) give one pause. Also, one can't help wondering whether *most* galaxies might not look rather peculiar if scrutinized to the same degree as have been many QSO host galaxies. The problem, in both cases, is that we do not have a standard by which to judge the significance of the observations. Clearly, it would be desirable, though difficult, to define a control sample of galaxies having luminosities, redshifts, and environments that are similar to those of a sample of QSO host galaxies so that a direct comparison could be made.

III. Pitfalls

The following two examples serve as warnings that the interpretation of observations may not always be as straightforward as it seems, and that things we think we know may suddenly come unravelled. In both cases, the problems happen to have been caused by the presence of extended emission, but they serve as warnings of other, perhaps more subtle, traps of which we are not yet aware.

A. Compact Companions

Stockton (1982) pointed out that previously known features within about 3 arcseconds of three QSOs (two of which were classical, high-luminosity QSOs) were actually stellar or nearly stellar objects and suggested that they were tidally-limited companion galaxies. The redshifts of the objects near the high-luminosity QSOs were determined from emission lines in their spectra, while that near the remaining object showed stellar absorption lines. However, it is now known that the emission lines seen in the first two objects come from a general surrounding distribution of ionized gas, rather than specifically from the stellar objects (Stockton and MacKenty 1987), so the redshifts of these objects are no longer valid. Because a few similar objects with secure redshifts are known, I suspect that these particular objects probably *are* really associated with their respective QSOs, but until better spectroscopy is obtained we cannot be absolutely certain.

B. Emission-Line Structure in Broad-Band Images

The emission-line nebulosities around 3CR 48 and 4C 37.43 were originally discovered on broad-band images. This implies that it is not safe to assume that extended features seen on such images of QSOs necessarily reflect the distribution of stars in the host galaxy if a strong emission line falls within the bandpass. Unfortunately, the [O III] λ5007 line is usually the strongest in the optical spectra of low-redshift QSOs, and it falls within the r bandpass (which is often the one chosen for studying the morphology of QSO host galaxies) roughly for $0.2 < z < 0.4$. As I have noted in § II.C, the extended continuum emission and extended line emission often have unrelated morphologies, so a problem of this sort can lead to a significant misinterpretation of the data.

One example where I believe that this has happened is the r-band image of 4C 25.40 (1223+252) shown by Hutchings, Johnson, and Pyke (1988), where all of the non-radially-symmetric features seen in the immediate vicinity of the QSO (including the "companion" to the north) are due to emission-line gas (*cf.* Stockton and MacKenty 1987). The authors imply that similar structure is seen on a B image, but this can quite plausibly be attributed to [O II] λ3727 emission falling within that bandpass. While there might indeed be stars associated with this gas, continuum images obtained with bandpasses that include strong emission lines cannot be taken as evidence for their presence.

The potential for such confusion can only increase as attempts are made to reach fainter surface-brightness limits on QSO host galaxies, where even weaker lines will be a problem. It will also be crucially important to consider this problem in planning Hubble Space Telescope imaging of the inner regions of QSOs, where the surface brightness in many emission lines will likely be higher.

IV. Recent Related Developments

A. Ultraluminous IRAS Galaxies

One of the most impressive results from IRAS has been the identification of a class of infrared-bright galaxies undergoing global episodes of star formation. Many of these galaxies show unambiguous signs of interaction (Allen, Roche, and Norris 1985; Sanders *et al.* 1986; Armus, Heckman, and Miley 1987). From the standpoint of this review, the really interesting aspect of the IRAS galaxies are the claims that the most luminous of them (the so-called ultraluminous IRAS galaxies, with IR luminosities $> 10^{12}$ $L_\odot$) are not only virtually always interacting, but also usually contain active nuclei (Sanders *et al.* 1988*a,b*). Both of these claims are still controversial: *e.g.*, Lawrence *et al.* (1989), who tried to define reasonably objective criteria for identifying an interaction and who also observed a matched comparison sample, found a much lower rate of interaction for IRAS galaxies only slightly less luminous than the Sanders *et al.* samples; and Leech *et al.* (1989) find, from spectrophotometry of this same IRAS sample, that the fraction of galaxies having active nuclei is much lower than found by Sanders *et al.* and also that this fraction does not depend on IR luminosity. One possible explanation of these differences mentioned by Leech *et al.* is that, because of the greater distance of their sample, their slits covered a larger portion of the galaxies, possibly diluting any active nucleus component.

Because the ultraluminous IRAS galaxies potentially provide the clearest observational link between galaxy mergers and nuclear activity, it is extremely important to resolve these questions. But even assuming this link is clearly established, there is at least one other important question: What do the ultraluminous IRAS galaxies look like once they have aged sufficiently for the starburst to moderate and the dust to dissipate? If it could be shown that IR galaxies like these are the parent population of a large portion of QSOs in general, then we would have direct evidence of the importance of interactions and mergers for the QSO phenomenon. Soifer *et al.* (1986) made an estimate of the respective number densities of the ultraluminous IRAS galaxies and the PG QSOs and found the former a factor of ~ 3.5 times more numerous. Of course, in evaluating an evolutionary scenario, it is necessary to consider two additional factors: (1) What are the respective time scales for two stages? and (2) How does the nuclear luminosity change with time? At present, it isn't clear to me that we can answer these questions well enough to say whether the ultraluminous IR galaxies in the Saunders *et al.* sample ultimately will become what we would call low-luminosity QSOs (in which case they might contribute a substantial fraction of all such objects) or what we would be more likely to call Seyfert galaxies (in which case they would represent a relatively small contribution to the Seyfert galaxy population). In either case, however, and in spite of current uncertainties concerning the ubiquity of the phenomenon, the importance of the ultraluminous IRAS galaxies is that they vindicate the general idea that active nuclei can be triggered by interactions in a more direct and convincing way than has been possible by other means.

B. Structures of High-Redshift Radio Galaxies

The paper by Miley in these proceedings gives the current status of recent work on high-redshift radio galaxies, which I needn't repeat. In the present context, what I feel is important is that phenomenologically they are quite different from what we see in the low-redshift QSOs. In particular, the alignment of the line-emitting and continuum structures with the radio axis sets them apart, and it is difficult to imagine any scenario in which interactions would produce structure preferentially aligned along the radio axis. Another difference, already mentioned, is that for the luminous, high-redshift radio galaxies, there is a strong correlation between the extended line emission and the extended continuum emission, whereas this is not true for the low-redshift QSOs. In short, while interactions may conceivably be important in producing these radio galaxies, the extended continuum and line emission cannot generally be cited as support for this position.

V. Overview and Future Prospects

A. An Example

A number of the points mentioned in this review are illustrated in Figure 1, which shows three images of 4C37.43, a QSO with a redshift of 0.3708. At the top is a deep broad-band r image; to the lower left, a 30 Å bandpass interference filter image centered on [O III] λ5007 emission at the QSO redshift; and to the lower right is an image in the line-free continuum between 4675 Å and 4800 Å in the QSO rest frame. About 10 arcsec west of the QSO is a compact galaxy with a redshift of 0.3722 (Stockton 1978), which is connected to the QSO by a luminous continuum bridge (Stockton and MacKenty 1987). In the broad-band r image, much of the region around the QSO is clearly dominated by line emission from [O III] $\lambda\lambda$4959,5007. This extended emission region is the most luminous known among low-redshift QSOs. While there is no detailed correlation between the extended continuum emission and the extended line emission, the highest surface brightness line emission falls close to the region of highest continuum surface brightness in this case.

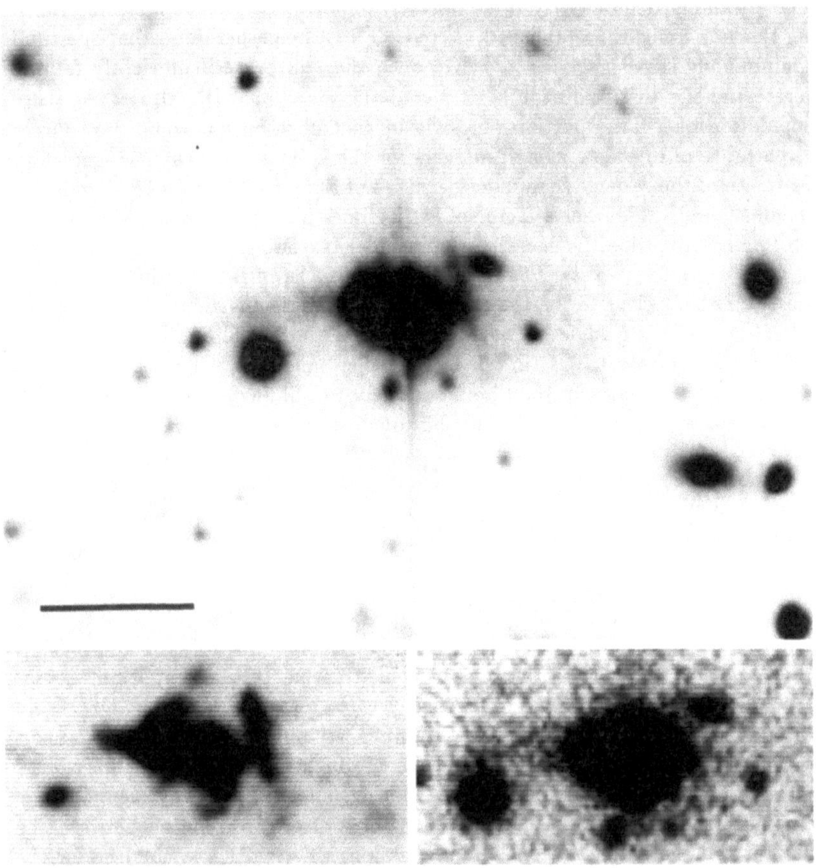

Figure 1.—Three images of 4C 37.43: broad-band r *(top)*, narrow-band [O III] λ5007 *(lower left)*, and line-free continuum just shortward of Hβ *(lower right)*. The bar indicates 10 arcsec.

B. HST Observations

The Hubble Space Telescope will clearly bring important new observations to bear on questions relating to QSO host galaxies in general and frequency of interactions in particular. Those of most value are likely to be (1.) high-angular-resolution imaging of the relatively high-surface-brightness inner regions of host galaxies and (2.) ultraviolet imaging to identify regions which have recently had high rates of star formation. On the other hand, pressures on HST time and the characteristics of the original complement of instruments make it unlikely that it will be used significantly in the next few years to push imaging of QSO host galaxies to lower surface brightnesses or to carry out more than very limited spectroscopic programs on host galaxies. Continued progress will depend on a carefully-planned coordination of HST and ground-based programs. Large ground-based telescopes at excellent sites will continue to be an essential part of this effort.

C. Where Do We Stand?

The question we want to answer is not whether some QSOs are in interacting or merging systems: the answer to that is clearly, "Yes." Instead, what we want to know is whether interactions and mergers are *important* mechanisms for triggering QSO activity. In my opinion, the evidence in favor of a "Yes" to this question as well is largely circumstantial, but still fairly strong. What makes me feel reasonably confident is that we have a number of different lines of evidence, none of which is completely compelling in itself, but which all point in the same direction:

▷ The number of QSOs that have very close companions is striking, particularly when one considers that many such companions will be difficult to detect if they are projected close to the QSO or in the process of merging. What keeps this observation from having great weight by itself is the lack of a suitable control sample.

▷ The major imaging surveys have produced a large number of objects with global asymmetries, bridges, or tails. Since a fair fraction of interactions and even mergers would be likely to have unfavorable interaction parameters or viewing angles and would not have obvious signatures at the redshifts involved, the observations are probably consistent with virtually all QSOs having undergone recent interactions. Again, one would feel slightly more secure in this statement if there were a control sample of galaxies of similar luminosities, environments, and redshifts that had been observed with just as much diligence. Nevertheless, I regard this evidence as quite strong.

▷ The spectra of the host galaxies of 3CR 48 and (less certainly) a few additional QSOs, together with the blue colors of many more QSO host galaxies, indicate fairly recent galaxy-wide star formation, consistent with recent interactions.

▷ In the ultraluminous IR galaxies, we have a sample of objects for which there may be a direct observational connection between strong interactions or mergers and an active nucleus, although both the fraction of IRAS galaxies that are interacting and the evidence for active nuclei are still controversial. I believe that the evidence presented by Sanders *et al.* 1988a for a high rate of interaction for the ultraluminous galaxies among the complete IRAS Bright Galaxy sample is quite strong but that there is room for dispute about the evidence for active nuclei, given the potential effect of the high and variable extinction in these galaxies on the diagnostic line ratios.

What needs to be done in the immediate future to resolve some of the remaining areas of uncertainty? Here is a highly personal and certainly incomplete list of priorities:

IRAS Galaxies: One would like to have a clear resolution of the present uncertainties regarding the luminous IRAS galaxies: What fraction are strongly interacting? What fraction have active nuclei? How do each of these fractions vary as a function of luminosity? How do the various time scales (*viz.*, of the high-luminosity IR source, of the active nucleus, of the interaction) affect these statistics, as well as discussions of possible evolutionary futures?

QSO Host Galaxies: HST imaging of the inner (*i.e.*, typically < 10 kpc radius) regions of QSO host galaxies should give us a much better impression of their morphologies and will allow us to evaluate evidence for interactions that is almost independent of, and complementary to, the ground-based data. Ideally, for both HST and ground-based surveys of host galaxies, one would like to have both well-defined criteria for what is or is not evidence for interaction and a comparison sample of galaxies with similar distributions of redshift, luminosity, and environment to those of QSO host galaxies.

Extended Emission-Line Regions: The principal question here is: What are the physical mechanisms responsible for the different types of extended emission seen around QSOs and other kinds of galaxies with active galactic nuclei, and how do these relate to the presence of the active nucleus? There is evidence that anisotropic emission due to shadowing close to the nucleus is largely responsible for determining the observed morphologies of emission regions associated with Seyfert galaxies and possibly also with powerful radio galaxies at high redshifts. Is the same true for QSOs? In any case, the question of the origin of the extended gas remains, and it is a tantalizingly difficult one to come to grips with observationally, although Rosat should provide better constraints than now exist on the presence of substantial densities of hot extended gas.

Acknowledgements

I thank Andy Fabian, John Hutchings, and Kieron Leech for sending me recent preprints, and Len Cowie, Bob Fosbury, and Peter Quinn for helpful discussions. I am also very grateful to Phil Crane not only for arranging for me to spend a week at the European Southern Observatory headquarters in Garching, where much of this paper was written, but also for allowing me to use both his office and his workstation in his absence. During my visit there, Joe Wampler several times kindly interrupted his schedule to help me find what I needed. I thank the U. S. National Science Foundation for support, most recently under grant AST87-15330.

References

Adams, T. A. 1977, *Astrophys. J. Suppl.*, **33** 19.

Allen, D. A., Roche, P. F., and Norris, R. P. 1985, *Monthly Notices Roy. Astron. Soc.*, **213**, 67P.

Armus, L., Heckman, T., and Miley, G. 1987, *Astron. J.*, **94**, 831.

Barnes, J. 1988, *Astrophys. J.*, **331**, 699.

Boroson, T. A., Persson, S. E., and Oke, J. B. 1985, *Astrophys. J.*, **293**, 120.

Crawford, C., and Fabian, A. 1989, *Monthly Notices Roy. Astron. Soc.*, **239**, 219.

Fosbury, R. 1986, in *Structure and Evolution of Active Galactic Nuclei*, ed. Giuricin *et al.* (Dordrecht: Reidel) p. 297.

Gehren, T., Fried, J., Wehinger, P. A., and Wyckoff, S. 1984, *Astrophys. J.*, **278**, 11.

Gunn, J. E. 1979, in *Active Galactic Nuclei*, ed. C. Hazard and S. Mitton (Cambridge: Cambridge Univ. Press) p. 213.

Heckman, T. M., Bothun, G. D., Balick, B., Smith, E. P. 1984, *Astron. J.*, **89**, 958.

Hutchings, J. B. 1987, *Astrophys. J.*, **320**, 122.

Hutchings, J. B., Crampton, D., Campbell, B., and Pritchet, C. 1981, *Astrophys. J.*, **247**, 743.

Hutchings, J. B., Crampton, D., and Campbell, B. 1984, *Astrophys. J.*, **280**, 41.

Hutchings, J. B., Johnson, I., and Pyke, R. 1988, *Astrophys. J. Suppl.*, **66**, 361.

Hutchings, J. B., Janson, T., and Neff, S., 1989, *Astrophys. J.*, **342**, 660.

Lawrence, A., Rowan-Robinson, M., Leech, K., Jones, D. H. P., and Wall, J. V. 1989, *Monthly Notices Roy. Astron. Soc., in press.*
Leech, K. J., Penston, M. V., Terlevich, R., Lawrence, A., Rowan-Robinson, M., and Crawford, J. 1989, *Monthly Notices Roy. Astron. Soc., in press.*
MacKenty, J. W., and Stockton, A. 1984, *Astrophys. J.*, **283**, 64.
Malkan, M. A., Margon, B., and Chanan, G. A. 1984, *Astrophys. J.*, **280**, 66.
Sanders, D. B., Scoville, N. V., Soifer, B. T., Young, J. S., Schloerb, F. P., Rice, W. L., and Daniels, G. E. 1986, *Astrophys. J. Letters*, **305**, L45.
Sanders, D. B., Soifer, B. T., Elias, J. H., Madore, B. F., Mathews, K., Neugebauer, G., Scoville, N. Z. 1988a, *Astrophys. J.*, **325**, 74.
Sanders, D. B., Soifer, B. T., Elias, J. H., Neugebauer, G., and Mathews, K. 1988b, *Astrophys. J. Letters*, **328**, L35.
Smith, E. P., Heckman, T. M., Bothun, G. D., Romanishin, W., and Balick, B. 1986, *Astrophys. J.*, **306**, 64.
Soifer, B. T., Sanders, D. B., Neugebauer, G., Danielson, G. E., Lonsdale, C. J., Madore, B. F., and Persson, S. E. 1986, *Astrophys. J. Letters*, **303**, L41.
Stockton, A. 1978, *Astrophys. J.*, **223**, 747.
Stockton, A. 1982, *Astrophys. J.*, **257**, 33.
Stockton, A., and MacKenty, J. W. 1987, *Astrophys. J.*, **316**, 584.
Toomre, A., and Toomre, J. 1972, *Astrophys. J.*, **178**, 623.
van Breugel, W. 1989, in *Hotspots in Extragalactic Radio Sources*, ed. H. J. Röser and K. Meisenheimer (Heidelberg: Springer-Verlag) p. 121.

Galaxy Interactions and the Radio Galaxy Phenomenon

E.P. Smith[1] *and T.M. Heckman*[2]

[1]Goddard Space Flight Center, Greenbelt, MD20771, USA
[2]Department of Physics and Astronomy, The Johns Hopkins University, Baltimore, MD21218, USA

1.0 Introduction

We have been studying powerful radio galaxies (which we define as having radio power at 178MHz, $P_{178} \geq 5\times10^{24}$ Watts/Hz for a cosmology of $H_o = 100$ km sec^{-1}Mpc^{-1} and $q_o = 0$) in order to determine their origins. There are two distinct classes of powerful radio galaxies (PRGs), the so called class A and B radio galaxies (Hine and Longair 1979; Heckman *et al.* 1986). The class A radio galaxies exhibit strong optical emission-lines ($L_{H\alpha} > 10^{41}$erg/s), "edge-brightened" Fanaroff-Riley (1974) type II radio morphologies, dominate the radio luminosity function at $L_{\rm Radio} > 10^{42}$ erg/s, and exhibit strong cosmological evolution. Two nearby examples are 3C 120 and NGC1275. The Class B radio galaxies typically have weak emission-lines or pure absorption line spectra, Fanaroff-Riley type I radio morphologies, dominate at $L_{\rm Radio} \sim 10^{40-42}$ erg/s, and exhibit weak cosmological evolution. Some nearby prototypes are M84 and M87. If we hypothesize that these, like other active galaxies, may be the products of interactions or mergers we must account for their striking differences given their similar triggering mechanism. We have acquired (in collaboration with S. Baum, W. van Breugel, G. Illingworth, and G. Miley) a large body of data on roughly equal numbers of class A and B radio galaxies (in the redshift range $0.01 < z < 0.3$) consisting of broad-band images (to study morphology, photometric structure, stellar content, dust, and the local environment), Narrow-band $H\alpha$ images (to study the ISM), spectroscopy (to examine the stellar dynamics and the ISM), and radio maps. These data were principally acquired with the CTIO 4m, KPNO 4m, KPNO 2.1m telescopes, and the VLA.

2.0 Results

Our investigations have yielded the following results for the class A and class B radio galaxies. Many of these findings are reported in Smith and Heckman 1989(a,b) and Baum *et al.* (1989). In the interest of brevity we summarize our findings and refer the reader to these articles for a more detailed description of the results and the methods used to determine them.

2.1 Class A PRGs (Strong Emission-line galaxies)

- 50% of the radio galaxies in our sample show large-scale (few kpc - 100 kpc) deviations from elliptical symmetry in their isophotes at surface brightness levels brighter than 25 V mag arcsec^{-2}. The deviations from elliptical symmetry are sharp or coherent structural distortions *e.g.*, tails, bridges, fans, shells; *not* mildly boxy isophotes, distorted envelopes, multiple nuclei, ellipticity gradients or isophotal twists.

Dynamics and Interactions of Galaxies Editor: R. Wielen

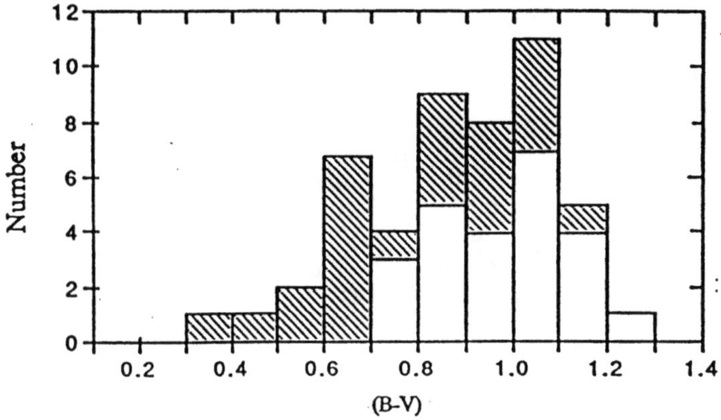

Figure 1: The mean $(B - V)$ color distributions for the SE (hashed) and WE/ABS (open) PRGs.

- Morphological distortions in the radio galaxies are usually stellar in composition and not merely the result of extended emission-line gas.
- 44% of the SE galaxies have *off nuclear* continuum colors which are blue (by at least 0.2 magnitudes) when compared to giant ellipticals at the same redshift ($<(B-V)_{SE}>= 0.79 \pm 0.04$ in the PRG rest-frame; see Figure 1).
- The (*B*-*V*) color maps of these galaxies are often quite complex, suggesting the presence of large amounts of dust and young stars.
- These galaxies have a strong ISM component seen by its $H\alpha$ and strong far infrared emission, and also through its obscuration of stellar light (dust).
- There is some evidence to suggest that the local environments ($r < 100$ kpc) of SE PRGs are not as dense as those of WE/ABS PRGs. Smith and Heckman (1990) found evidence indicating that the local galaxy density near SE PRGs is down by a factor of ~ 2 relative to WE/ABS galaxies.
- Although there are a number of these galaxies with unusual surface brightness profiles, on average SE PRGs have radial light distributions which are $r^{1/4}$-law in shape.
- SE PRGs have absolute magnitudes which, after removal of nonstellar light, are fainter than class B PRGs by $\sim 0.35V$ magnitudes. With $<M_V>= -21.42 \pm 0.10$.
- Peculiar stellar dynamics (particularly for those galaxies with peculiar optical morphologies; see figs[2,3]). The peculiar morphology galaxies have lower velocity dispersions ($\Delta \log \sigma \sim 0.15$), and appear to be more rapidly rotating - about 2/3 of the peculiar morphology galaxies lie near the oblate rotator line in the $v/\sigma - \epsilon$ plane.

2.2 Class B PRGs (**W**eak **E**mission-line/**ABS**orption-line galaxies)

- Sharp morphological peculiarities like those seen in SE PRGs are rare (< 10% of WE/ABS galaxies).

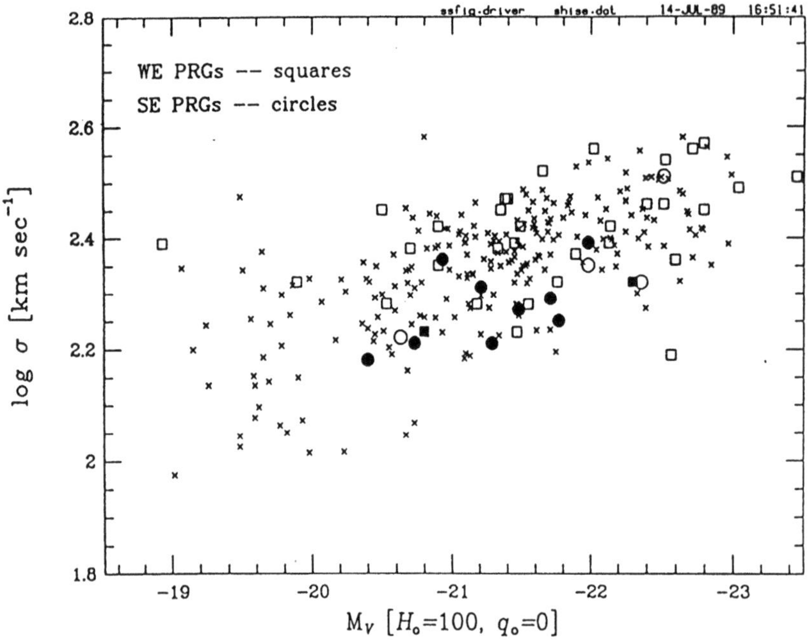

Figure 2: A plot of the absolute V magnitude *vs.* $\log\sigma$ for our PRG sample along with data for normal ellipticals from Faber *et al.* 1989. The x's are the normal ellipticals, squares are WE/ABS radio galaxies, circles are SE galaxies. Filled points mark the peculiar morphology galaxies.

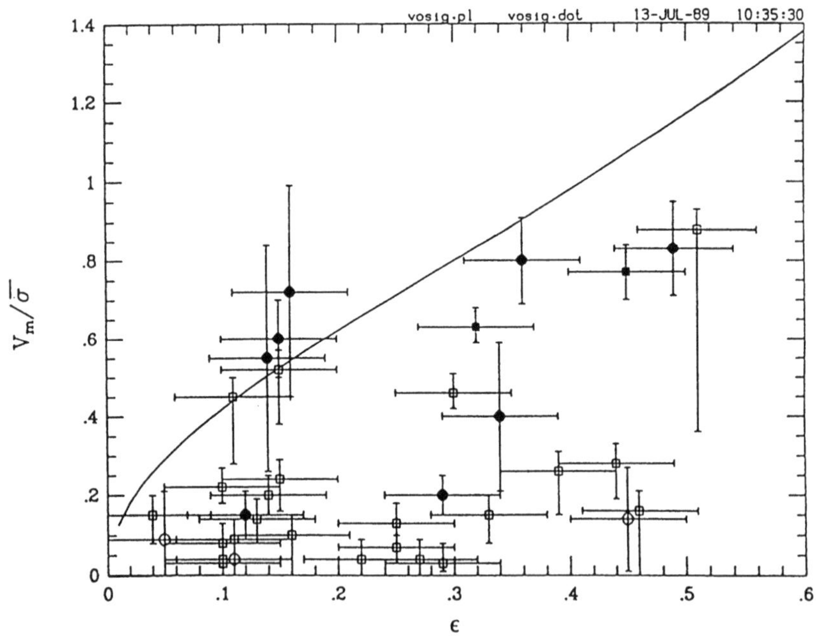

Figure 3: The maximum rotation velocity divided by the velocity dispersion within $r_e/2$ is plotted *vs.* the ellipticity of the outer isophotes. WE galaxies are again squares, SE galaxies are circles, and peculiar morphology galaxies are filled points. The line represents the expected relation for a rotationally supported oblate spheroid.

- WE/ABS galaxies are more likely to be found in often asymmetric common envelope systems, and have greater incidence of multiple nuclei. About 42% of the WE/ABS PRGs are in common envelope systems, and many (32%) have double nuclei (defined as nuclei within a projected distance of 9.6 kpc). In fact, ~20 – 25% are dumbbell, "db" galaxies with relative velocities of only a few hundred km/sec (Casertano and Valentijn, these proceedings).
- The color properties for WE/ABS galaxies ($< (B-V)_{WE/ABS} >= 0.99 \pm 0.02$) including color gradients appear indistinguishable from those of normal giant ellipticals and brightest cluster galaxies (BCGs).
- Surface brightness profiles are shallow in slope, similar to BCGs (Schombert 1986).
- Their absolute magnitudes are similar to those of BCGs ($< M_V >= -21.77 \pm 0.12$).
- These galaxies have normal elliptical galaxy stellar dynamics.

3.0 Summary and Conclusions

3.1 The Relationship Between the Two Classes of Radio Galaxies

The SE galaxies are not all just mildly damaged ellipticals, indeed their morphological differences with WE/ABS PRGs along with their other differences such as type of radio morphology, presence of optical emission-lines, continuum colors, local environments, and stellar dynamics confirm that these galaxies are *fundamentally* distinct populations.

3.2 Relationship of WE/ABS Radio Galaxies to Brightest Cluster Galaxies

In the course of our investigations we found that the WE/ABS powerful radio galaxies have many similarities to BCGs. (In fact some of the PRGs are themselves BCGs, e.g. 3C 264 [Abell 1367], 3C 317 [Abell 2052].)

1. Their mean absolute magnitudes are equivalent (though the PRGs have a larger dispersion in their distribution).
2. The global morphologies of the two classes of galaxies are quite similar.
   (a) Both have a high incidence of multiple nuclei and/or galaxies in common envelopes (~30 – 40%).
   (b) Surface brightness profiles for WE/ABS PRGs and BCGs appear distended relative to normal elliptical profiles.
3. The colors and color gradients of these two types of galaxies are quite similar.
4. Both are found in regions of very high galaxy density.

3.3 Galaxy Mergers/Collisions and Their Role in Formation of PRGs

In spite of their differences, one feature the SE and WE/ABS systems appear to have in common is the indication that one or more mergers (or strong interactions) have occurred in the system's past. In the case of the SE galaxies the strongest evidence is often the unmistakable presence of large, coherent *stellar* features reminiscent of the tidal interaction models of Toomre and Toomre (1972) or Quinn (1984). Heckman *et al.* (1986)

have suggested that the SE galaxies are the products of mergers of spiral-spiral or spiral-elliptical systems. The ISM of the disk systems supply the necessary fuel in the form of gas and dust for the large emission-line intensities and strong thermal infrared emission. The dynamically cold disk in addition allows the formation of extended coherent stellar features such as tails and fans. Our recent results from a stellar dynamical study of PRGs dovetail nicely with this picture. Rotation plays a more significant role in the structure of the morphologically peculiar galaxies. This would be expected if the merger remnant of two disk systems or a disk and elliptical retained a significant amount of the angular momentum of the disk(s). The blue colors then result either from the blue initial colors of the merger participants or from star formation stimulated by the merger. Moreover, the optically peculiar systems are preferentially found in regions of lower average density than PRGs with normal optical morphology. That is, the systems posited to be mergers involving disk galaxies are not found in dense clusters where large velocity dispersions inhibit merging and where few disk and/or gas-rich galaxies are found.

Although the implication of mergers in the WE/ABS class of galaxies is not as strong, we believe that the combination and frequency of double (or multiple) nuclei, common envelope systems (with distorted isophotes), and unusually shallow surface brightness profiles (see Aguiler and White 1986) tells us that mergers have played an important role here too. WE/ABS galaxies in this interaction scenario would result from mergers of elliptical type systems whose lack of a significant cool ISM would suppress the formation of emission-line features and whose "hot" dynamical state would decrease the chance of sharp tidally distended features being produced. In fact, it may be the case that many of the WE/ABS galaxies are undergoing a strong interaction, yet because of their (sometimes) large relative velocities will not actually merge. Future studies of the relative velocities of interacting systems combined with the increasingly superb galaxy interaction models will help to further our understanding of the importance merging plays in the formation of PRGs and other galaxies.

REFERENCES

Aguiler, L. A., and White, S. D. M. 1986, *Ap. J.*, **307**, 97.

Baum, S. A., Heckman, T. M., Bridle, A., van Breugel, W., and Miley, G. 1989, *Ap. J. Suppl.*, **68**, 643.

Faber, S. M., Wegner, G., Burstein, D., Davies, R. J., Dressler, A., Lynden-Bell, D., and Terlevich, R. J. 1989, *Ap. J. Suppl.*, **69**, 763.

Fanaroff, B. L, and Riley, J. M 1974, *M.N.R.A.S.*, **167**, 31P.

Heckman, T. M. Smith, E. P., Baum, S. A., van Breugel, W. J. M., Miley, G. K., Illingworth, G. D., Bothun, G. D., and Balick, B. 1986, *Ap. J.*, **311**, 526.

Hine, R. G., and Longair, M. S. 1979, *M.N.R.A.S.*, **188**, 111.

Quinn, P. J. 1984, *Ap. J.*, **279**, 596.

Schombert, J. M. 1986, *Ap. J. Suppl.*, **60**, 603.

Smith, E. P., and Heckman, T. M. 1989*a*, *Ap. J. Suppl.*, **69**, 365.

Smith, E. P., and Heckman, T. M. 1989*b*, *Ap. J.*, **341**, 658.

Smith, E. P., and Heckman, T. M. 1990, *Ap. J.*, **348**, in press.

Toomre, A., and Toomre, J. 1972, *Ap. J.*, **172**, 623.

Nuclear Activity: How Irritating Are Neighbours?

J.M. van der Hulst[1], *E. Hummel*[2], *W.C. Keel*[3], *and R.C. Kennicutt*[4]

[1]Kapteyn Astronomical Institute, Postbus 800,
NL-9700 AV Groningen, The Netherlands
[2]Nuffield Radio Astronomy Laboratories, Jodrell Bank, UK
[3]Dept. of Physics and Astronomy, University of Alabama, Tuscaloosa, AL35487, USA
[4]Steward Observatory, University of Arizona, Tucson, AZ85721, USA

1. Introduction

Many observational studies in the last decade have strongly suggested that galaxy - galaxy interactions do affect the properties of spiral galaxies, in particular the properties of their nuclei. The first indications came from radio surveys of galaxies (Stocke 1978, Hummel 1980, Condon et al. 1982). Later on numerous optical and IR studies of interacting galaxies and galaxy pairs (Heckman 1980, Heckman et al. 1980, Kennicutt and Keel 1984, Dahari 1984, Joseph et al. 1984, Lonsdale et al. 1984, Keel et al. 1985, Joseph and Wright 1985, Cutry and McAlary 1985, Bushouse 1987, Bushouse et al. 1988, Kennicutt et al. 1987) have provided insight in the physical processes causing the observed excess nuclear radio emission. These studies indicate that much of the inferred nuclear activity can be ascribed to either enhanced star formation in the central few hundred parsecs or enhanced non-thermal activity.

The samples chosen in most of the studies mentioned above are not very homogeneous and often incomplete. Though the conclusion that the presence of a companion galaxy does affect a galaxy's nuclear activity is almost always correct, there often is the question of selection effects. The study by Condon et al. (1982) for example focusses on a sample of galaxies with bright radio continuum nuclei, which in hindsight often appear to have companions. A comparison with a control sample of isolated galaxies, moreover, is often omitted.

To avoid this problem of selection effects we began an observing program comprising a complete sample of galaxies in pairs and groups selected from the unpublished catalogue of van Albada. This catalogue is drawn from the Uppsala General Catalogue (Nilson 1973) and provides for each galaxy a probability of being physically associated with the nearest neighbour of comparable brightness as seen projected onto the sky. This probability accounts for the the local surface density of galaxies of comparable brightness, and is a statistical estimate of whether two galaxies are a physical pair. The sample contains all spiral galaxies with $8^h < \alpha < 16^h$ and $\delta > 0°$ brighter than $B_T = 13.0$ and with an association probability of 0.9 or more. From this list we rejected pairs with radial velocity differences in excess of 600 km/s. The number of pairs rejected was consistent with the probability estimates. A detailed description of the sample can be found in Keel et al. (1985). As control sample we used the sample of bright galaxies studied by Keel (1983), omitting galaxies with physical companions as defined above. This control sample is complete for galaxies of type SO/a to Scd, $-15° < \delta < 40°$ and $B_T <$ 12.0.

For these samples we obtained spectrophotometry over the range 3700 - 6900 Å of the central 5" (corresponding to a mean linear size of 360 parsec) region, H_α and red continuum images, and high resolution (1.3") 20-cm radio continuum images. For a description of the observations see Keel et al. (1985), Kennicutt et al. (1987) and Hummel et al. (1987). Here we report our main results regarding the nuclei or central regions (typically a few hundred parsec). We want to emphasise that our study is based on observations of statistically complete samples and therefore unlikely to be hampered by major selection effects.

Dynamics and Interactions of Galaxies Editor: R. Wielen

2. Results.

We used the spectrophotometry to derive H_α luminosities and equivalent widths, and to classify the "nuclear activity". Following Baldwin et al. (1981) we used the O[III]/H_β and N[II]/H_α line ratios to distinguish HII region (star forming) nuclei, Seyfert nuclei and the so called Liners. It appears that in field galaxies the fraction in each of the three spectral classes is a strong function of Hubble type: the Liner nuclei (and also the Seyferts) occur preferentially in galaxies of early Hubble type (SO/a - Sab), while most of the HII nuclei are found in the later type galaxies (e.g. see figure 13 in Keel et al. 1985). This was already indicated by the studies of Heckman (1980), Heckman et al. (1980) and Stauffer (1982). Comparison of the interacting and control samples shows that the number of HII region nuclei increases at the expense of the Liners, especially in the galaxies of early Hubble type. The fraction of Seyfert nuclei is significantly higher in the interacting sample (8,5% compared to 5.6% in the control sample). The Seyfert activity appears to favour the very close pairs (Kennicutt and Keel 1984, Keel et al. 1985). We note in passing that we did not find other strong dependances on projected pair separation.

The relative decrease of Liner nuclei in the interacting sample could be a selection effect if strong star bursts tend to mask the presence of a Liner. The H_α luminosities of Liners are in general systematically lower than the H_α luminosities of HII region nuclei (Hummel et al. 1989) so that this effect may play an important role. A recent study by Kennicutt et al. (1989) indicates that a large fraction of HII region and star burst nuclei shows a secondary ionization component which could be due to an active Liner type nucleus or the presence of large scale shocks. We conclude that many nuclei may be composite in nature and in the following discussion one should keep in mind that we only consider the dominant process.

The H_α luminosity and H_α equivalent widths of the nuclei of the galaxies in the interacting sample are significantly higher than those of the nuclei in the control sample (Keel et al. 1985). This indicates a (circum)nuclear star formation rate which is about a factor three higher than in relatively isolated galaxies. In contrast to this we found that the star formation rates in the disks of the galaxies in the complete sample are not much higher than in the disks of isolated galaxies (Kennicutt et al. 1987). Bushouse (1987) reached the same conclusion on the basis of his observations of interacting galaxies. Whatever the mechanism is, the presence of a neighbour has more impact on the star formation in the central region than on the global star formation in a galaxy. This confirms the result of Hummel (1981) that the radio continuum emission from the disks of galaxies in pairs is not significantly higher than from the disks of isolated galaxies, while the central radio sources of galaxies in pairs do show a clear enhancement. The latter is also more clearly sown by the radio data we have on the nuclei of the galaxies in our samples (Hummel et al. 1989).

In the analysis thusfar we have made no clear distinction between the HII region nuclei and the Liners, and have only estimated the activity from the H_α luminosities and H_α equivalent widths. We used the radio continuum maps to measure the radio luminosity of the nuclear sources. If we separate the two classes, omitting the few Seyfert galaxies in the samples, we find the surprising result that the increase in radio luminosity in the interacting sample is almost entirely due to the HII region nuclei. This is clearly illustrated in figure 1 which shows the fractional luminosity functions (for a definition of FLF's see Hummel 1981) of the HII region nuclei and the Liners in the complete and the control samples. It is very clear that the luminosity of the Liners hardly increases for galaxies with companions, while the HII region nuclei show a very significant increase in luminosity. This strongly suggests that the Liner activity is hardly affected by the presence of companions and that "active" Liner nuclei are present in some fixed fraction of all galaxies, more notably the earlier types (Keel et al. 1985).

In addition the size of the radio emitting region is much smaller in the Liner nuclei (see figure 2). The median radio size of the HII region nuclei is 450 parsec, while the median size of

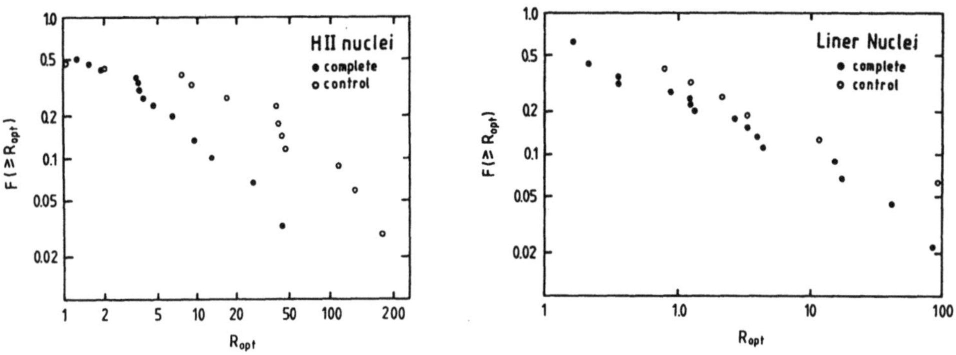

Figure 1. *The fractional luminosity fuctions for the HII region nuclei (left panel) and Liner nuclei (right panel) in the complete (filled circles) and control (open circles) samples.*

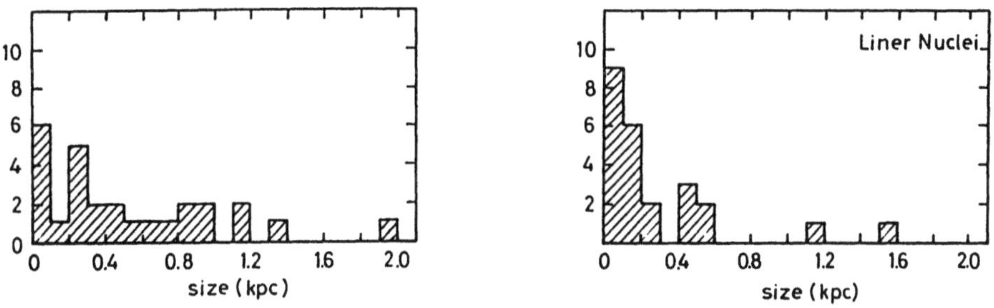

Figure 2. *Distribution of radio sizes for the HII region nuclei (left panel) and the Liner nuclei (right panel).*

the Liner nuclei is less than 100 parsec. In fact more than half the Liner nuclei are unresolved at our 1.3" resolution. To further verify the idea that Liner nuclei are very compact, as suggested by earlier work of Heckman (1980), we searched the literature for spiral with a compact nucleus detected in VLBI experiments. Spectra were available in the literature for most of the 12 spiral galaxies we found; the remainder we observed ourselves. Two of these 12 nuclei show Seyfert spectra and 9 have Liner characteristics. For one object no spectrum is yet available.

3. Concluding Remarks.

Detailed studies of the nuclear radio and H_α emission of galaxies in pairs and groups show that there is about a factor three or more enhancement in radio and H_α luminosity as compared with isolated galaxies. The spectrophotometry indicates that this enhancement is in most cases a result of enhanced massive star formation in the central few hundred parsecs of these galaxies.

About one quarter of the nuclei is dominated by Liner activity. It appears that the Liner nuclei in galaxies in pairs are not significantly more luminous that Liner nuclei in isolated galaxies, suggesting that the Liner phenomenon is independent of interactions between galaxies. Apparently a fixed fraction of all spiral galaxies, predominantly the galaxies of earlier Hubble type, have Liner nuclei whose activity does not critically depend on the presence of a neighbour. The lower fraction of Liner nuclei among the galaxies in pairs and groups is probably a result of enhanced star formation in the central region which tends to mask the less luminous Liners.

A comparison of spectrophotometry and VLBI detections for 19 galaxies suggests strongly that the Liner phenomenon is always restricted to a very small region in the nucleus. This supports the idea of the presence of a compact central engine such as a black hole, or perhaps a 'warmer' as proposed by Terlevich (1989). In the latter case, however, one has to understand why massive star formation is triggered in a very small volume, while the more widespread star formation like in the HII region nuclei does not occur.

References

Arp, H. C. 1966 *Atlas of Peculiar Galaxies* (Caltech) and *Astrophys. J., Suppl.* **14**, 1.

Baldwin, J., Phillips, M. and Terlevich, R., 1981, *Publ. Astron. Soc. Pac.* **93**, 5.

Bushouse, H. A., 1987, *Astrophys. J.,* **320**, 49.

Bushouse, H. A., Lamb, S. A. and Werner, M. W., 1988, *Astrophys. J.,* **335**, 74.

Condon, J. J., Condon, M. A., Gisler, G. and Puschell, J. J., 1982, *Astrophys. J.,* **252**, 102.

Cutri, R. M. and McAlari, C. W., 1985, *Astrophys. J.,* **296**, 90.

Dahari, O., 1984, *Astr. J.,* **89**, 966.

Heckman T. M., 1980, *Astr. Astrophys.,* **87**, 142.

Heckman, T. M., Balick, B, and Crane P. C., 1980 *Astr. Astrophys., Suppl.* **40**, 295.

Hummel, E., 1980, *Astr. Astrophys.,* **89**, L1.

Hummel, E., 1981, *Astr. Astrophys.,* **93**, 93 and **96**, 111.

Hummel, E., van der Hulst, J. M., Keel W. C. and Kennicutt, R. C., 1987, *Astr. Astrophys., Suppl* **70**, 517.

Hummel, E., van der Hulst, J. M., Keel W. C. and Kennicutt, R. C., 1989, *Astr. Astrophys.,* (in preparation)

Joseph, R. D., Meikle, W. P. S., Robertson, N. A. and Wright, G. S., 1984, *Mon. Not. R. astr. Soc.,* **209**, 111.

Joseph, R. D., and Wright, G. S., 1985, *Mon. Not. R. astr. Soc.,* **214**, 87.

Keel, W. C., 1983, *Astrophys. J., Suppl.* **52**, 229.

Keel, W. C., Kennicutt, R. C., van der Hulst, J. M. and Hummel, E., 1985, *Astr. J.,* **90**, 708.

Kennicutt , R. C. and Keel, W. C., 1984, *Astrophys. J.,* , **279**, L5.

Kennicutt, R. C., Keel, W. C., van der Hulst, J. M., Hummel. E. and Roettiger, K. A., 1987, *Astr. J.,* **93**, 1011.

Kennicutt, R. C., Keel, W. C. and Blaha, C. A. 1989, *Astr. J.,* **97**, 1022.

Lonsdale, C. J., Persson, S. E. and Matthews, K., 1984, *Astrophys. J.,* **287**, 95.

Nilson, P., 1973, *Uppsala General Catalogue of Galaxies* Uppsala Astron. Obs. Ann. **6**.

Stauffer, J. P., 1982, *Astrophys. J., Suppl.* **50**, 517.

Stocke, J. T., 1978, *Astr. J.,* **83**, 348.

Terlevich, R., 1989, IAU Coll. **120** *Structure and Dynamcis of the Interstellar Medium*

Terlevich, R., 1989, *Evolutionary Phenomena in Galaxies*, ed. Beckman, J. E. and Pagel, B. E., Cambridge University Press.

Ultraluminous Infrared Galaxies and the Origin of Quasars

D.B. Sanders, B.T. Soifer, and G. Neugebauer

Division of Physics, Mathematics and Astronomy,
California Institute of Technology, 320–47,
Pasadena, CA91125, USA

ABSTRACT. The Infrared Astronomical Satellite (IRAS) has discovered a class of ultraluminous infrared galaxies ($L_{ir} > 10^{12}\ L_{\odot}$) that are bolometrically as luminous, and slightly more numerous than optically selected quasars in the local universe ($z \lesssim 0.2$). Ground-based observations of a complete sample of 10 of these infrared galaxies suggest that they represent the initial dust-enshrouded stages of quasars. A second group of ultraluminous objects selected on the basis of 'warm' infrared colors are found to be preferentially Seyfert galaxies. We suggest that these ultraluminous 'warm' objects represent the transition phase between the infrared selected and optically selected quasars.

I. INTRODUCTION

One of the most important results of the IRAS survey has been the discovery of a significant number of ultraluminous galaxies ($L_{bol} > 10^{12}$ $L_{\odot}$, H_{o} = 75 km $s^{-1}Mpc^{-1}$, q_{o} = 0) that emit the bulk of their energy at infrared wavelengths (Soifer *et al.* 1986; Sanders *et al.* 1988a,b). These objects meet the generally accepted minimum bolometric luminosity criteria for quasars (Schmidt and Green 1983; Véron-Cetty and Véron 1985; Soifer *et al.* 1987).

Ground-based optical and infrared observations suggest an evolutionary connection between ultraluminous infrared galaxies and optical quasars. Below we summarise the evidence that suggests that ultraluminous infrared galaxies represent the initial, dust-enshrouded stages of quasars triggered by the strong interaction/merger of two gas-rich spiral galaxies.

II. PROPERTIES OF ULTRALUMINOUS INFRARED GALAXIES

Arp 220 (IC 4553) was the first galaxy from the IRAS survey to receive special attention due to it's extreme infrared luminosity ($L_{8-1000\mu m}$ = 1.6 × 10^{12} $L_{\odot}$; Soifer *et al.* 1984). However, Rieke and Low (1972) had discovered some years before that Mrk 231, a luminous Seyfert 1 galaxy with an absolute blue magnitude equal to that of optical quasars (M_B < -23.1; Véron-Cetty and Véron 1985), had "ultrahigh" near infrared luminosity (~4 × 10^{11} $L_{\odot}$ at 7.9-13.3μm), suggesting a total infrared luminosity equivalent to the total optical-UV luminosity of optical quasars. Arp 220 was unique in that virtually all (~ 98%) of it's luminosity was emitted in the far-infrared.

Searches of the IRAS data base have now revealed several score infrared selected objects with far-infrared luminosities in excess of 10^{12} $L_{\odot}$ (Lawrence *et al.* 1986; Soifer *et al.* 1986; Low *et al.* 1988; Sanders *et al.* 1988b). While the majority of these objects are known only by their IRAS name and have energy distributions more similar to Arp 220 than Mrk 231, the list spans a wide range of infrared colors, infrared/blue flux ratios, and optical spectral classes including starbursts, Seyfert II's, LINERS, and radio loud and radio quiet quasars. The most thoroughly studied of these objects to date are the ten ultraluminous galaxies from the IRAS Bright Galaxy (BG) Survey (Soifer *et al.* 1986, 1987), a flux limited sample (F_{ν} (60μm) > 5.2 Jy) of 320 galaxies observable from the northern hemisphere (δ > -30°, $|b|$ > 30°). As the brightest infrared objects in the sky the BG galaxies are the most amenable for detailed study at other wavelengths. The properties of the ten ultraluminous BG objects are summarized below.

Dynamics and Interactions of Galaxies Editor: R. Wielen

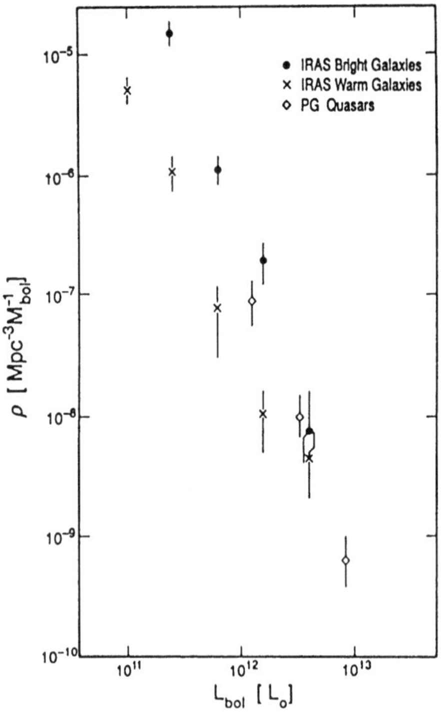

Figure 1. The luminosity functions of luminous infrared galaxies and optically selected quasars normalized to the same Hubble constant and plotted in units of bolometric luminosity. The (•) represents the IRAS Bright Galaxies (Soifer *et al.* 1987), the (×) represents IRAS 'warm' galaxies from the survey of Sanders *et al.* (1988b), and the (◇) represents optically selected quasars taken from the Palomar-Green Bright Quasar Survey (Schmidt and Green 1983).

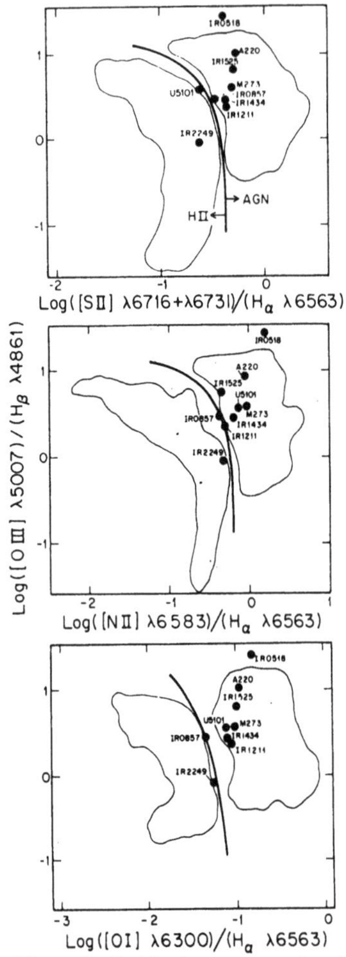

Figure 2. Reddening-corrected optical emission-line intensity ratios for ultraluminous infrared galaxies. In each panel the thick solid curve divides narrow-line AGN's from HII region-like galaxies, while the outlined regions represent the locus of narrow-line AGN and HII region-like galaxy spectra as summarized by Veilleux and Osterbrock (1988).

SPACE DENSITIES - The bolometric luminosity function for infrared selected galaxies is shown in Figure 1. At $L_{bol} = 10^{11}$ $L_{\odot}$ the space density of infared galaxies is comparable to that of optically selected Seyferts and starburst galaxies. At $L_{bol} > 10^{12}$ $L_{\odot}$ the space density of infrared objects exceeds by a factor of two that of the only other class of objects with comparable luminosity - the optically selected quasars.

OPTICAL SPECTRA - Two of the ten ultraluminous BG objects have very broad (>2000 km s^{-1}, FWZI) nuclear emission lines (IRAS 05189-2524, Mrk 231), three have emission linewidths in the range 1000 - 2000 km s^{-1} (UGC 5101, Mrk 273, Arp 220), and three (IRAS 12112+0305, IRAS 14348-1447, IRAS 15250+3609) have emission line ratios typical of active galactic nuclei (AGN; see Figure 2).

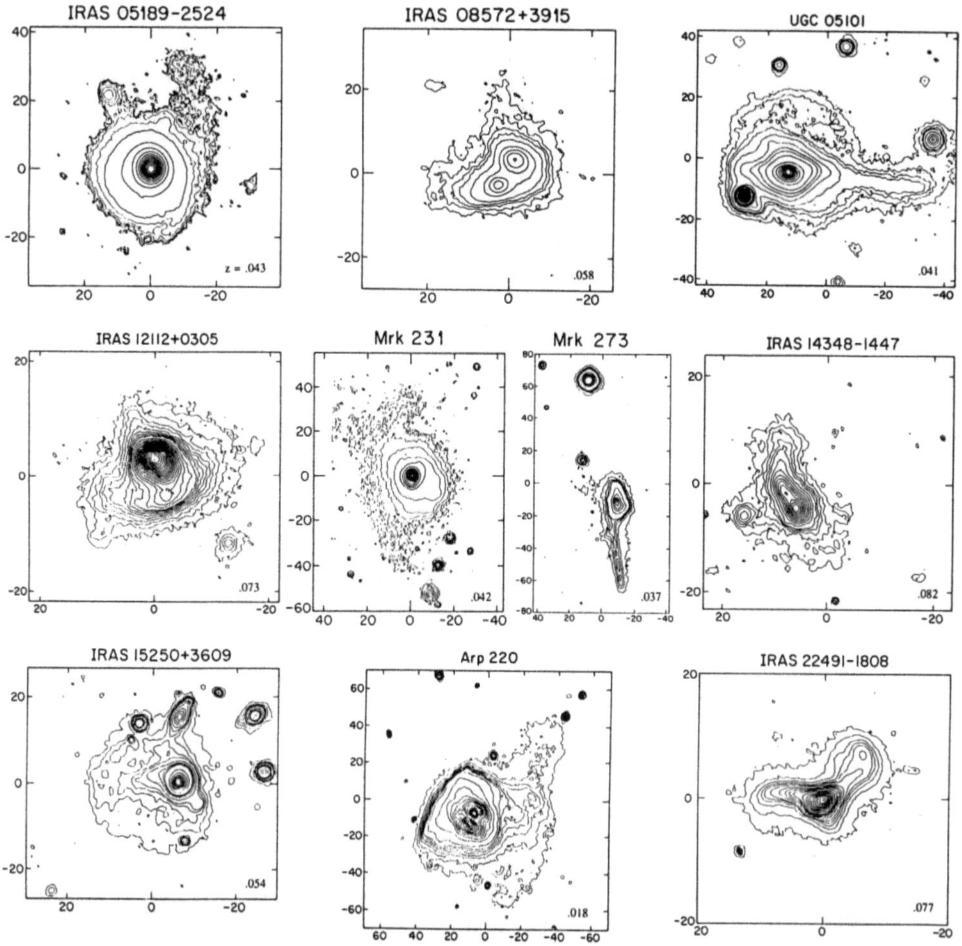

Figure 3. Contour maps from CCD images at 6500 Å (Gunn *r*) of the ten ultraluminous infrared galaxies from the IRAS Bright Galaxy Survey. The lowest four contours are linear with higher contours on a log scale. The axes are labeled in arcseconds.

MORPHOLOGY - The contour maps shown in Figure 3 suggest that all of the ultraluminous BG objects are strongly interacting galaxies. The majority are best described as ongoing mergers as evidenced by the seven objects that exhibit major tidal tails plus main bodies that appear to be overlapping disks. The remaining three objects appear to be strongly interacting systems as evidenced by prominant rings or tails and nearby companions.

GAS CONTENT - All of the ultraluminous BG objects are extremely rich in molecular gas. CO($1 \rightarrow 0$) observations yield H_2 masses in the range 1.5 - 6 $\times$ 10^{10} $L_{\odot}$, or 5 - 20 times larger than the value for the Milky Way (Sanders *et al.* 1989). Interferometer measurements of two objects (Mrk 231, Arp 220) show that both have abnormal concentrations of molecular gas at their centers.

The above properties of ultraluminous infrared galaxies have been interpreted as evidence that these objects are in fact dust-enshrouded quasars. Sanders *et al.* (1988a) suggest that the trigger for producing these quasar-like luminosities is the strong interaction/merger of two gas-rich spiral galaxies that results in the funneling of molecular clouds into the merger nucleus where they provide fuel for an intense starburst and an active nucleus. In the ultraluminous infrared phase it seems plausible that the active nucleus dominates the starburst which may already have begun to fade. Once the combined forces of radiation pressure and supernovae explosions begin to sweep dust clear of the nuclear region these objects should take on the appearance of optical quasars.

III. THE TRANSITION FROM GALAXY TO QUASAR

The ultraluminous BG galaxies with the warmest $25\mu m / 60\mu m$ color temperatures, e.g., Mrk 231, display the closest appearance to "classical" quasars, i.e., broad emission lines, dominant optical nucleus, etc. This suggests that the bridging of the "gap" between the ultraluminous BGS objects and the optically selected quasars might be found in the extragalactic objects selected to be "warmer" than

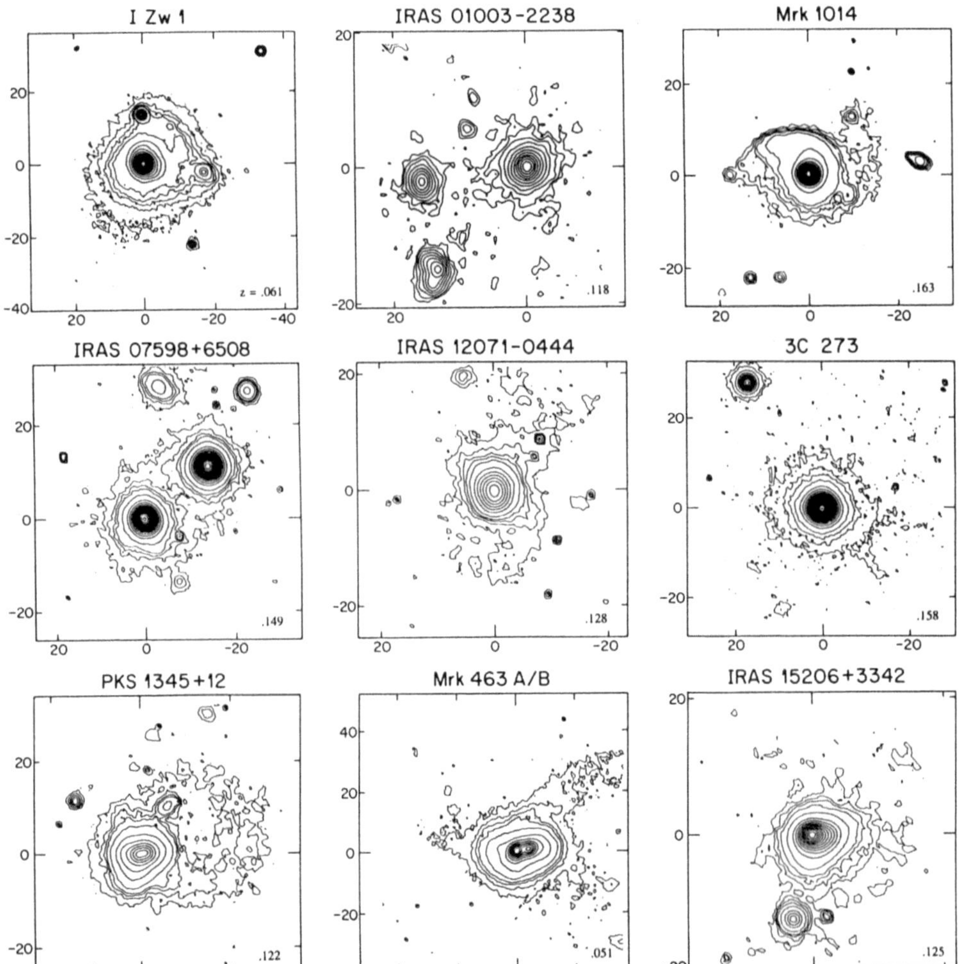

Figure 4. Contour maps from CCD images at 6500 Å (Gunn *r*) of nine of the 12 ultraluminous infrared galaxies from the IRAS 'warm' galaxy survey of Sanders *et al.* (1988b). The three remaining galaxies, IRAS 05189- 2524, IRAS 08572+3915, and Mrk 231, were displayed in Figure 3. The image of IRAS 01003-2238 was obtained at the CFHT by Hutchings and Neff (1988).

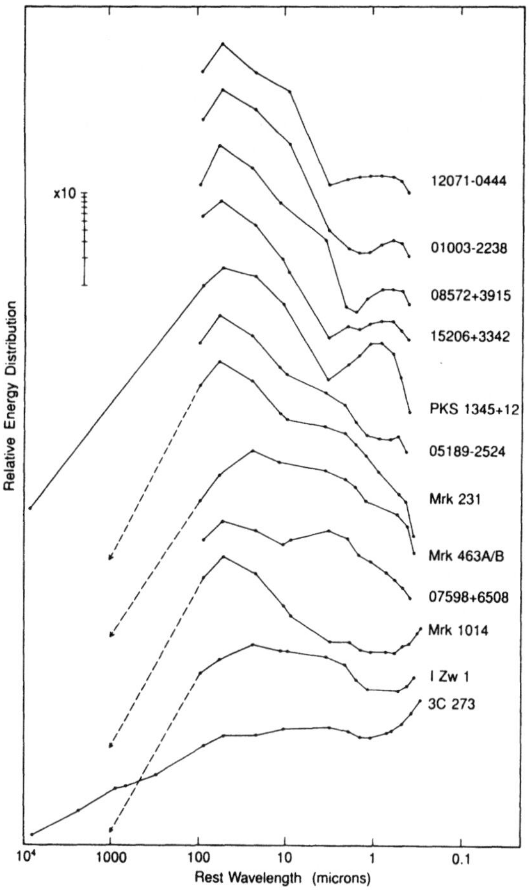

Figure 5. Spectral energy distributions for ultraluminous galaxies from the IRAS 'warm' galaxy survey of Sanders *et al.* (1988b). The data are arranged in order of decreasing infrared luminosity fraction, from the infrared extreme objects like IRAS 12071-0444 to the relatively unobscurred optical quasars like I Zw 1.

those selected by a strictly $60\mu m$ flux density criterion such as used for the BGS. This approach has successfully been used by, e.g. deGrijp, Miley, and Lub (1987); Vader and Simon (1987); and Low *et al.* (1988) to identify active galaxies in the IRAS database. A search for warm ultraluminous infrared objects was carried out by Sanders *et al.* (1988b) using criteria identical to the earlier BGS except for a lower limiting flux, f_ν $(60\mu m) > 1.5$ Jy, and an imposed color criterion, $0.2 < f_\nu\ (25\mu m)\ /\ f_\nu(60\mu m) < 2.0$. In addition to three objects already found in the BGS - IRAS 05189-2524, IRAS 08572+3915, Mrk 231 - nine additional ultraluminous objects were discovered; their images are shown in Figure 4. Five of these nine objects show clear evidence of interaction (double nuclei, tidal tails), one shows evidence of a distorted disk, while two are principally stellar. The remaining object, 3C 273, exhibits nebulosity and has at least one prominant jet. *All* twelve of the warm ultraluminous objects have Seyfert spectra, split equally between Seyfert 1's and Seyfert 2's.

Spectral energy distributions for all twelve warm ultraluminous objects are displayed in Figure 5. The fraction of total luminosity contributed by emission at $\lambda > 8\mu m$ ranges from ~90% for objects like IRAS 12071-0444 (similar to the majority of the ultraluminous BGS objects in Figure 3) to approximately 50% for IRAS 07598+6508. While the BGS contained no objects with strong UV excess, the warm ultraluminous sample contains a few such objects (e.g. Mrk 1014, I Zw 1, 3C 273).

In general the infrared energy distributions of the ultraluminous warm objects are flatter than those for the BGS galaxies.

The data for the warm ultraluminous objects appear to support the idea that ultraluminous infrared galaxies evolve into the classical quasars. The warm ultraluminous objects span many of the well-known classes of AGN, so the distinction between ultraluminous infrared, radio, of UV-loud AGN appears to be diminishing. It is plausible that such distinctions are predominantly related to the interstellar medium surrounding the luminosity source. Most of the ultraluminous warm objects contain luminous optical nuclei, so that the central powerhouse has apparently begun to break through the sussounding cloud of dust to reveal itself visibly. Qualatatively, the progression is from the objects with the greatest relative infrared luminosity being the most recognizable as galaxies to those with the least relative infrared luminosity having the brightest optical nucleus. The warm galaxy sample displays the transition from the cooler ultraluminous BGS galaxies to optical quasars by providing a significant sample of intermediate objects.

This research was supported by grants from the National Science Foundation and the IRAS Extended Mission Program.

REFERENCES

deGrijp, M. H. K., Miley, G. K., and Lub, J. 1988, *Astron. Ap.*, **182**, 362.

Hutchings, J. B., and Neff, S. G. 1988, *A. J.*, **96**, 1575.

Lawrence, A., Walker, D., Rowan-Robinson, M., Leech, K. J., and Penston, M. V. 1986, *M.N.R.A.S.*, **219**, 687.

Low, F. J., Huchra, J., Kleinmann, S. G., and Cutri, R. M. 1988, *Ap. J. (Letters)*, **327**, L41.

Mirabel, I. F., and Sanders, D. B. 1988, *Ap. J.*, **335**, 104.

Rieke, G. H., and Low, F. J. 1972, *Ap. J. (Letters)*, **176**, L95.

Sanders, D. B., Scoville, N. Z., Soifer, B. T., and Mirabel, I. F. 1989, *Ap. J.*, submitted.

Sanders, D. B., Soifer, B. T., Elias, J. H., Madore, B. F., Matthews, K., Neugebauer, G., and Scoville, N. Z. 1988a, *Ap. J.*, **325**, 74.

Sanders, D. B., Soifer, B. T., Elias, J. H., Neugebauer, G., and Matthews, K. 1988b, *Ap. J. (Letters)*, **328**, L35.

Schmidt, M., and Green, R. F. 1983, *Ap. J.*, **269**, 352.

Soifer, B. T., *et al.* 1984, *Ap. J. (Letters)*, **283**, L1.

Soifer, B. T., *et al.* 1986, *Ap. J. (Letters)*, **303**, L41.

Soifer, B. T., Sanders, D. B., Madore, B. F., Neugebauer, G., Persson, C. J., Persson, S. E., and Rice, W. L. 1987, *Ap. J.*, **320**, 238.

Vader, J. P., and Simon, M. 1987, *Nature*, **327**, 304.

Veilleux, S., and Osterbrock, D. E. 1988, *Ap. J. Suppl.*, **63**, 295.

Véron-Cetty. M.-P., and Véron, P. 1985, *ESO Sci. Rept., No. 4.*

Active Galactic Nuclei Without Black Holes

R. Terlevich

Royal Greenwich Observatory, Herstmonceux, BN27 1RP, UK

1. Introduction

Terlevich and Melnick (1985 ; hereafter TM85) investigated the properties of giant bursts of star formation with metallicities typical of those found in the nuclei of giant galaxies. We found that unlike the low metallicity case, metal rich massive stars have their evolution fundamentally affected by mass-loss in the form of stellar winds. Without exception all evolutionary computations find the same differences between conservative, $\dot{M}=0$, and mass-losing models found initially by Tanaka (1966); namely a change in the H and He burning time scale and an evolution towards high temperatures after hydrogen exhaustion. During the blueward evolution the star reaches effective temperatures well in excess of those typical of the ZAMS. Wolf-Rayet stars with massive progenitors ($M>60\,M_\odot$) are believed to be in the blueward evolutionary stage (Conti 1976, Chiosi 1981, Maeder 1983). The Wolf-Rayet stage is reached after a highly unstable phase, corresponding to the minimum effective temperature ever reached by the star (~15,000K) and roughly before the onset of helium burning. Luminous blue variables, like the Hubble-Sandage variables, η Carina and P-Cygni type stars are believed to be in this unstable phase. The most recent evolutionary star models for solar composition and incorporating mass loss and overshooting, indicate that by the end of the helium burning phase the effective temperature reaches up to 200,000K and their bolometric luminosity could be up to a factor of 2 larger than at the ZAMS. We called these extremely hot and luminous Wolf-Rayet stars WARMERS.

TM85 computed the changes that a population of WARMERS will introduce in to the emitted spectrum of a young metal rich cluster. Using Maeder's isochrones, we studied the evolution spectral of a burst of star formation using our computer code IMF. This program computes, from the theoretical isochrones and stellar atmospheres, the integrated spectrum of a star cluster as a function of age; then calculates the emission line spectrum of the associated HII region using the photoionization code CLOUDY developed by Gary Ferland. We found that the emitted spectrum of a metal rich HII region suffers a qualitative change after about 3Myr of evolution, when the most massive stars reach the WARMER phase. In a very short time the ionizing spectrum of the cluster is fundamentally modified by the appearance of the WARMER component. Consequently the emission line spectrum is transformed from that of a typical low excitation HII region into a high excitation Seyfert type 2. Following the evolution still further, shows that after 5Myr as the ionizing flux decreases and therefore the ionization parameter also decreases, the Seyfert type 2 nucleus becomes a Liner.

In brief, TM85 have shown that the 'traditional' method of using the Baldwin, Phillips and Terlevich (1981) diagnostic diagrams to classify emission line regions should be used with care. Power law type ionizing continuum can be the result of a few WARMERs modifying the emitted UV spectrum of a young cluster. The best evolutionary star models predict this behaviour for solar abundances.

This result raised the possibility of a direct relation between star formation and some forms of nuclear activity. This possibility has been discussed in the literature by several authors (Pronik 1973, Adams and Weedman 1975, Harwit and Pacini 1975, Osterbrock 1978, Weedman 1983, Perry and Dyson 1985). Some researchers have preferred the view that after the star formation event, the resulting cluster of compact remnants will eventually coalesce and form a massive blackhole (Rees 1978, 1984), while others suggest that the gas lost during the evolution of the massive stars is able to fall into the gravitational potential of the bulge, cool and form a blackhole (Norman and Scoville 1988). All these scenarios have a common point, they have to start the evolution with extremely compact clusters with large velocity dispersions to allow the formation of a blackhole.

Another possibility originally suggested by Shklovskii (1960), and further developed by Field (1964) and McCrea (1976) and more recently by Terlevich *et al.* (1987), Terlevich and Melnick (1987) and Terlevich (1989) is that no blackhole forms and that *all* the observed properties of at least some AGN are the *direct* consequence of a nuclear starburst.

In this paper I will argue that the observed variability of the Broad Line Region (BLR) in Seyfert type 1s and QSOs, the line profiles, line ratios and total luminosity of the BLR are consistent with the starburst scenario.

2. The BLR in the starburst scenario

Terlevich, Melnick and Moles (1987), extended the work of TM85 to include the supernova phase in the evolution of the nuclear starburst. In that paper we postulated that the observed properties of Seyfert type 1 nuclei may also be related to nuclear starburst activity. We distinguished two different supernova activity phases,

1 - a SN type Ib at the end of the life-time of the most massive stars ($M>25\,M_{\odot}$). These SNe have Wolf-Rayet/WARMER progenitors and they are predicted to be optically dim and radio loud. This stage is associated with the Seyfert 2 phase.

2 - a SN type II phase at the end of the life-time of intermediate mass stars ($8 < M < 25\,M_{\odot}$). These SNe have red supergiant progenitors. The SN ejecta after leaving the atmosphere of the star will presumably interact with dense ($n{\sim}10^{7}$) circumstellar/interstellar medium. This interaction will produce a hot and luminous remnant with a life time of about 2 years. The predicted velocities, dimensions, temperature, density and luminosity of these remnants are similar to those of the canonical BLR. The emitted spectrum of a fast shock ($V=1000\,km\,s^{-1}$) in a high density ($n=10^{7}$) environment has been studied by Daltabuit *et al.* (1978). They found that the emitted spectrum of the post shock cool dense shell, photoionized by the radiation of the shock, resembled that of a QSO. Thus in all aspects, we expect this rapidly evolving remnants to closely resemble the BLR.

Terlevich and Melnick (1988) provided evidence that this may be the case. They showed that the reported flare in the Seyfert type 1 NGC 5548, may have been the first detection of a SN in the nucleus of a galaxy, rather than an accretion event (Peterson and Ferland 1986). The spectrum of the flare looked very similar to that of SN 1983K the only supernova known with a probable Wolf-Rayet progenitor (Niemela *et al.* 1985). Also the luminosity and duration of the flare were similar to those of SN 1983K.

Perhaps the most compelling evidence comes from some recent observations of supernovae. Filippenko (1989) has reported the discovery of a supernova in an HII region in one of the spiral arms of the SBc galaxy NGC 4615. After maximum the spectrum was dominated by broad permitted emission lines of hydrogen, FeII and CaII; it has a striking resemblance to the spectrum of a Seyfert type 1 or a QSO. A second supernova reported in the same paper, SN 1988I, showed a similar spectrum. Even the luminosities of the SNe were comparable to those of Seyfert 1 nuclei.

Long term monitoring of radio-selected quasars has shown that many are variable on time scales of years and with variable amplitudes (Tritton and Selmes 1971, Barbieri and Romano 1981, Pica and Smith 1983; PS83), and some may not vary at all. A small and perhaps unrepresentative subset, the Optically Violently Variable (OVV) are the most variable ones. Work on optically selected QSOs, has indicated that most of them show only small amplitude (${\sim}0.2$mag) variability (Bonoli *et al.* 1979, PS83). Perhaps the largest dataset is that obtained at the Rosemary Hill Observatory (PS83). It includes 13 years of continuous monitoring of 130 AGN and OVV representing an average of 45 epochs per source. The average rms error of the photometry is 0.13 mag. PS83 found a trend with luminosity in the variability amplitude. This is illustrated in Figure 1 were I plotted the observed variability amplitude versus the absolute blue luminosity estimated for $H_o=50$ km s^{-1} Mpc^{-1}and $q_o=0$. As can be seen from the figure luminous QSOs are somewhat less variable than low luminosity QSOs and Seyfert nuclei. From a least-squares fit to the data, PS83 estimated the change in peak-to-peak amplitude with luminosity as 0.12 mag$\pm$0.02mag per magnitude for all sources and 0.18$\pm$0.03mag per magnitude for the radio quasars only. This trend is expected if the variability is produced by random flares, but Pica and Smith did not favour this explanation because the slope of the observed relation is much more shallow than the theoretically expected. The variability amplitude should decrease roughly as the inverse of the square root of the luminosity. In their simple analysis Pica and Smith did not include the observational errors; this would tend to flatten the relation.

To check this effect, and in collaboration with Alejandro Terlevich, I estimated the variability of a Starburst during the supernova stage using Montecarlo techniques. We

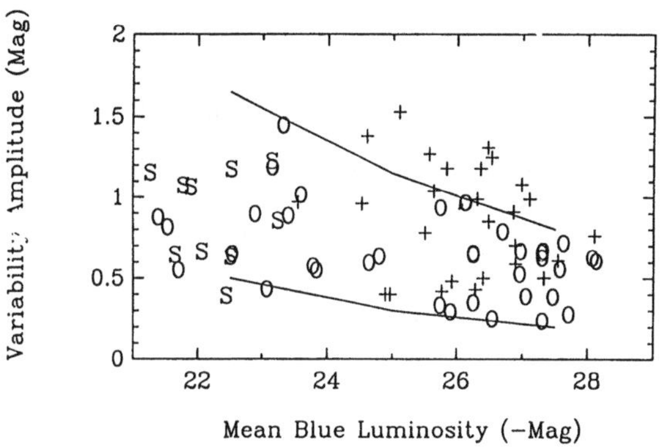

Figure 1 - Mean blue luminosity ($H_o = 50$, $q_o = 0$) versus peak-to-peak variability amplitude for all non-OVV sources from the survey by PS83. S = Seyfert galaxies; O = optically selected QSOs; + = radio loud Quasars. The lines represent the expected range of values for a starburst.

constructed the light curve by superposition of random events with random amplitude between $M_b = -18$ and $M_b = -20$. Gaussian noise with an rms scatter equal to the observational errors appropiate for the data set of PS83 (σ=0.13mg), was then added to this light curve. The lines in Figure 1 represent the result of the simulations. The region between the lines represent the range of values that we expect the observer to measure 95% of the time. We can see that most of the QSOs lie inside the expectations. Only some radio selected quasars have variability slightly over the predicted values.

We are lead to conclude that the range of variability observed in Seyfert galaxies, optically selected QSOs and in many radio selected quasars is well inside that expected in the starburst scenario. Also the variability time scale is *not* constraining the size of the BLR; in our scenario the BLR of a luminous QSO, can be and probably is distributed over few kpc and still will produce the required variability amplitude and observed time scales.

3. Supernova rates in young galaxies and starbursts

A typical L* elliptical galaxy has a mass of about 10^{12} $M_\odot$ and an effective radius (half mass radius) of 5 kpc and its present absolute blue luminosity is M_b=-22. The supernova rate (SNR) depends on the star-formation rate (SFR), and the details of the initial mass function (IMF). In the numerical examples that follow I assumed a Salpeter IMF. For this type of IMF one supernova explodes every 30-50 $M_\odot$ of new stars, depending on the adopted lower and upper mass limits. Following Larson's dissipative models for the formation of elliptical galaxies, during the formation of the inner kiloparsec at the end of the collapse, the SFR for a 10^{12} $M_\odot$ elliptical reaches a peak of about 3000 $M_\odot$/year corresponding to about 100 SN/year. If the blue luminosity of the SN and SN remnant is $M_b \sim 19.5$, and the remnants last for about 1 year, it is simple to estimate that the integrated blue luminosity of such a galaxy during the SN phase is about -25. A very massive elliptical may have a total mass up to 10^{13} $M_\odot$ and during the formation of the inner part will reach SN rates of about 1000 SN/year and M_b of -27 or -28.

A different approach is to estimate the supernova rate for a burst of star formation that represents 5% of the total mass of the galaxy. Norman and Scoville (1988) have recently estimated the flux of stars off the main sequence for a Salpeter IMF cluster of stars following stellar evolution models by Renzini and Buzzoni (1986) and Iben and Renzini (1986). They find that the flux of stars off the main sequence is between 20 and 10 stars/year for a cluster of 10^{10} $M_\odot$ and for stellar masses larger than 7 $M_\odot$, i.e. inside the lower limit for type II SNe. The SN rate is almost equal to the flux of stars off the main sequence, therefore, a burst of strength 5% in a 10^{12} $M_\odot$ galaxy will also produce about 100 SN/yr. This type of burst may be triggered by dynamical interactions, and presumably they are more common at high redshifts when galaxies had a larger gas content.

4. Concluding remarks

The ideas presented here demonstrate that the starburst model based on extremely simple assumptions is potentially able to reproduce the main observed properties of the BLR in Seyfert galaxies and QSOs.

The fact that the luminosities and variability of QSOs can be explained with the expected SN rates during the formation of a spheroidal galaxy leads naturally to the suggestion that perhaps most of the optically selected QSOs, representing the majority of the high redshift luminous objects, are young galaxies in the process of formation.

I like to thank Elena Terlevich for her comments that greatly improved this paper.

References

Adams, T.F. and Weedman, D.W., 1975. *Astrophys. J.* , **199,** 19.

Baldwin, J.A., Phillips, M.M. and Terlevich, R., 1981. *Publ. astr. Soc. Pacif.* , **93,** 5.

Barbieri, C. and Romano, G., 1981. *Astr. Astrophys. Suppl.* , **44,** 159.

Bonoli, F., Braccesi, A., Federici, L., Zitelli, V. and Formiggini, L., 1979. *Astr. Astrophys. Suppl.* , **35,** 391.

Chiosi, C., 1981. *in: The most massive stars, p.261, eds D'Odorico, S.Baade and Kjar,K. (Proceedings ESO workshop)*

Conti, P.S., 1976. *Mem.Soc.R.Sci.Liege*, **6eme Serie 9,** 193.

Daltabuit, E., MacAlpine, G. and Cox, D., 1978. *Astrophys. J.* , **219,** 372.

Field, G.B., 1964. *Astrophys. J.* , **140,** 1434.

Filippenko, A.V., 1989. *Astron. J.* , **97,** 726.

Harwit, M. and Pacini, F., 1975. *Astrophys. J. Lett.* , **200,** 1127.

Heckmann, T.M., Balick, B. and Crane, P.C., 1980. *Astr. Astrophys. Suppl.* , **40,** 295.

Iben, I. and Renzini, A., 1986. *Ann. Rev. Astr. Astrophys.* , **21,** 271.

Larson, R.B., 1974. *Mon. Not. R. astr. Soc.* , **166,** 585.

Maeder, A., 1983. *Astr. Astrophys.* , **120,** 113.

McCrea, W.H., 1976. *in: The galaxy and the local group, eds Dickens, R.J. and Perry, J.E. (RGO Bull 182).*

Moffat, J.W., Seggewiss, W. and Shara, M.M., 1985. *Astrophys. J.* , **295,** 109.

Norman, C. and Scoville, N., 1988. *Astrophys. J.* , **332,** 124

Osterbrock, D.E., 1978. *Phys. Scripta*, **17,** 285.

Perry, J.J. and Dyson, J.E., 1985. *Mon. Not. R. astr. Soc.* , **213,** 665.

Peterson, B.M. and Ferland, G., 1986. *Nature*, **324,** 345.

Pica, A.J. and Smith, A.G., 1983. *Astrophys. J.* , **272,** 11; (PS83).

Pronik, I.I., 1974. *Soviet astr.*, **16,** 628.

Rees, M.J., 1978. *Ann. N.Y. Acad. Sci.*, **320,** 613.

Rees, M.J., 1984. *Ann. Rev. Astr. Astrophys.* , **22,** 471.

Renzini, A. and Buzzoni, A., 1986. *in: Spectral evolution in Galaxies, eds. C.Chiosi and A. Renzini, p. 195, Reidel.*

Shklovskii, I.S., 1960. *Soviet astr.*, **4,** 885.

Tanaka, S., 1966. *Publ. astr. Soc. Japan* , **18,** 47.

Terlevich, R., 1988. *in The post recombination Universe, eds. N. Kaiser and A.N. Lasenby, Kluwer.*

Terlevich, R., 1989. *in Evolutionary phenomena in Galaxies, eds. J.E. Beckman and B.E.J. Pagel, Cambridge Univ. Press.*

Terlevich, R., Melnick, J. and Moles, M., 1987. *in Observational evidence for activity in galaxies, eds. E.Ye. Khachikyan, K.J. Fricke and J. Melnick, Reidel, Dordrecht; (TMM87).*

Terlevich, R and Melnick J., 1987. *in Starburst and galaxy evolution eds. T.X. Thuan, T. Montmerle and J. Tran Than Van, Frontiere.*

Terlevich, R. and Melnick, J., 1985. *Mon. Not. R. astr. Soc.* , **213,** 841; (TM85).

Terlevich, R. and Melnick, J., 1988. *Nature*, **333,** 239.

Terlevich, R. and Terlevich, A., 1990. *in preparation.*

Tritton, K.P. and Selmes, R.A., 1971. *Mon. Not. R. astr. Soc.* , **153,** 453.

Weedman, D.W., 1983. *Astrophys. J.* , **266,** 479.

Bar Formation in Interacting Galaxies and the Fueling of Nuclei

M. Noguchi

Tokyo Astronomical Observatory, Mitaka, Tokyo 181, Japan

1. Introduction

Recent observations suggest that interacting galaxies show a statistically higher incidence of nuclear starbursts than isolated galaxies (e.g., Keel et al. 1985). This means that the nuclear region of an interacting galaxy is somehow fuelled with the gas. Here the following mechanism for the gas fuelling is proposed based on the results of numerical simulations. Namely, the tidal force of the companion galaxy induces a bar structure in the stellar disc component of the perturbed galaxy (the parent galaxy), and this bar in turn induces infall of interstellar gas to the central disc region. Full description is given in Noguchi (1987,1988).

2. Numerical simulations

In the numerical simulation, a model for a disc galaxy has been constructed which consists of a halo and a disc. The halo was treated as a rigid spherical component which remains fixed during the interaction. The disc component contains both stars and gas. The stellar disc was constructed by 20K collisionless particles with the same mass. The gaseous disc was treated as an ensemble of 1000 cloud particles. Each cloud particle has a finite radius but no mass. It can be regarded as a giant molecular cloud. We assumed that the collision between two cloud particles is inelastic and produces an "OB star" particle which becomes a supernova at the end of its lifetime and gives kinetic energy to nearby cloud particles.

In the absence of a companion galaxy, the stellar disc is dynamically stabilized by the halo component and maintains nearly axisymmetric shape. In the gaseous component, the cloud-cloud collision rate stays nearly constant with the time. In interaction, a point mass perturber is introduced to disturb the parent galaxy.

Dynamics and Interactions of Galaxies Editor: R. Wielen

3. Results

1) Stellar component

Fig.1 shows the evolution of the standard model, i.e., Model A+A1 in Noguchi (1988). In this model, 20% of the total mass of the parent galaxy is contained in the disc component. The companion has the same mass as the parent galaxy and moves on a co-planar prograde parabolic orbit with the perigalactic distance twice the disc radius. Other parameters are given in Noguchi(1988).

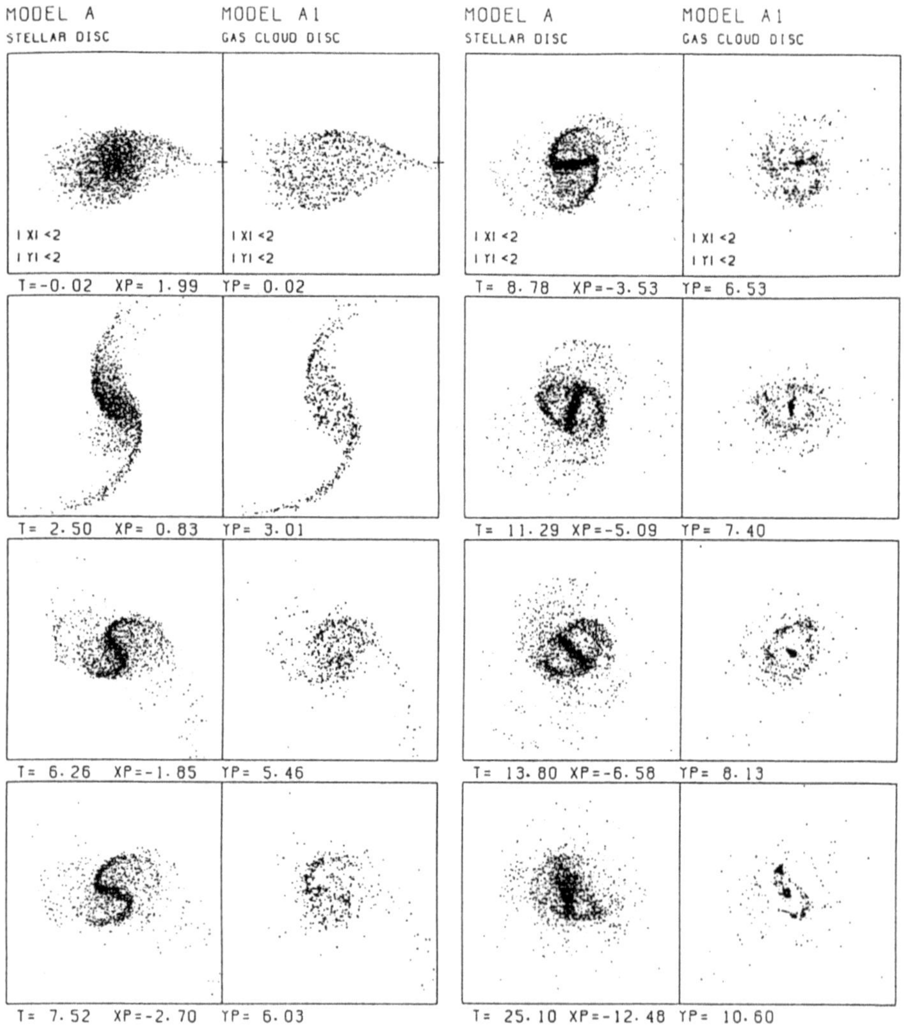

Fig.1. Evolution of the standard model. The perigalactic passage corresponds to T=0. One disc rotation period at the outer edge is 6.28. XP and YP indicate the coordinates of the companion galaxy in units of the initial disc radius

Left panels show the stellar disc component. Shortly after the perigalactic passage, a tail and a bridge form like in test-particle simulations (e.g., Toomre and Toomre 1972). About one rotation period after the perigalactic passage, a bar structure begins to evolve in the inner disc region. This point is quite different from the behavior of a test particle disc, in which only the outer part of the disc develops non-axisymmetric structure. It is evident that the self-gravity of the stellar disc plays an essential role in this bar formation; initial amplitude of tidal perturbation is smaller in the inner region than in the outer region. But the self-gravity is more dominant in the inner region. Therefore the growth of the perturbation is more efficiently accelerated in the inner region.

The bar lasts for one or two rotation periods. Although the detailed properties of the bar depend on the model parameters, the bar formation is a common behavior of a tidally-disturbed self-gravitating disc. One important observational implication of this result is that we should see barred galaxies more frequently in interacting systems than in isolated ones. It is interesting that we can see a slight hint of such effect in spiral galaxies in the Arp Atlas (Arp, 1966).

Another notable point is that the bar induced by tidal interaction can rotate very slowly. Detailed orbital analysis shows that the bar in Fig.1 lies inside the inner Lindblad resonance (ILR). It is suggested that the pattern angular velocity of the bar decreases as the disc mass fraction decreases (Noguchi, 1987). When the disc mass fraction is too large, the disc is dynamically unstable and develops a bar spontaneously (Ostriker and Peebles, 1973). For such massive disc, we can expect a rapidly rotating bar. Indeed several numerical studies show that such spontaneous bar usually ends near the co-rotation point (e.g., Miller and Smith 1979). On the other hand, when the disc mass is sufficiently small, the disc is stable. Nevertheless, such a disc can develop a bar under the tidal force of a companion galaxy as shown here. In this case, the pattern velocity is generally smaller than that of a spontaneous bar and can be very small depending on the disc mass fraction. Although such a slowly rotating bar may seem very strange, Telesco and Decher (1988) recently suggested that the central bar of NGC1068 is located within two ILRs. It is possible that the bar in NGC1068 has been formed in tidal interaction with a nearby galaxy NGC1055 (or 1072).

2)Gaseous component

The right panels of Fig.1 show the evolution of the gaseous component. Shortly after the perigalactic passage, the gas exhibits two-armed spiral structure similar to the one in the stellar component. However, after the stellar bar has been formed, a significant fraction of the gas begins to fall to the disc center and accumulates there. This is because the shock induced by the bar removes the kinetic energy and the angular momentum from the gas. The infall continues for about 10^9 years. Assuming that the total mass of the gas is 4 10^9 $M_{\odot}$, about 10^9 $M_{\odot}$ is gathered within 2 Kpc from the nucleus in this period. It is considered that active star formation (starburst) will result near the nucleus.

4. Discussion

We have seen that tidal interaction can induce active star formation in the galactic nucleus through bar formation and subsequent gas infall. It is observationally controversial whether Seyfert nuclei are likewise more frequent in interacting galaxies than in isolated galaxies (Dahari 1984,1985; Fuentes-Williams and Stocke 1988). In the case of Seyfert activity, the gas should fall to a far smaller region (<1pc). It should be clarified how such infall can be realized (cf, Lin et al. 1988; Shlosman et al. 1989).

References

Arp, H. 1966, Atlas of Peculiar Galaxies, Caltech, Pasadena.
Bushouse, H.A. 1987, Ap.J., **320**, 49.
Dahari, O. 1984, A.J., **89**, 966.
Dahari, O. 1985, A.J., **90**, 1772.
Fuentes-Williams, T., and Stocke, J.T. 1988, A.J., **96**, 1235.
Keel, W.C. et al. 1985, A.J., **90**, 708.
Lin, D.N.C. et al. 1988, Ap.J., **328**, 103.
Miller, R.H., and Smith, B.F. 1979, Ap.J., **227**, 785.
Noguchi, M. 1987, M.N.R.A.S., **228**, 635.
Noguchi, M. 1988, Astr.Ap., **203**, 259.
Ostriker, J.P., and Peebles, P.J.E. 1973, Ap.J., **186**, 467.
Shlosman, I. et al. 1989, Nature, **338**, 45.
Telesco, C.M., and Decher, R. 1988, Ap.J., **334**, 573.
Toomre, A., and Toomre, J. 1972, Ap.J., **178**, 623.

Interaction Versus Radio Source Generation: A CCD Survey of Galaxies with Radio Jets

L. Colina[1] *and I. Pérez-Fournon*[2]

[1]Space Telescope Science Institute, 3700 San Martin Drive, Baltimore, MD21218, USA
[2]Instituto de Astrofísica de Canarias, Tenerife, Spain

INTRODUCTION

We have started an observational program to cover a representative sample of radio galaxies containing radio jets in order to study the optical properties of the parent galaxy, their relationship to the radio properties as well as to detect optical emission associated with the radio structures.

The sample object of this study (47 galaxies) was selected from the Bridle and Perley (1984) list of galaxies containing well defined radio jets considering $0.01 \leq z \leq 0.15$ and $\delta \geq -15^0$. This sample, which is optical unbiased, can not be considered in any sense as complete although the presence of the same radio jet emission morphology can be see as evidence of same physical mechanism at work. Concerning radio luminosity at 1.4GHz, the sample is a low-intermediate luminosity sample, median $\mathrm{LogP}_{1.4}^{total}$= 24.83 Watt Hz^{-1}, with all but six galaxies with $P_{1.4}^{tot} \leq 10^{25} h_0^{-2}$ Watt Hz^{-1} ($h_0 = H_0/100$). If the radio jet morphology is considered, 80% of the sample shows a two side jet which is characteristic of Fanaroff-Riley type I radio sources.

We present some general results suggesting that ongoing strong interaction/merging between elliptical galaxies produces a well defined type of radio jet galaxies. A complete account of the observations, the galaxy sample, their optical morphological features and relation with the radio properties can be found elsewhere (Colina and Pérez-Fournon, 1989a,b; Colina et al., 1989). Throughout this paper H_0 = 75 km $s^{-1} Mpc^{-1}$ will be used.

OBSERVATIONS

All the observations were done at the Calar Observatory of the Max-Planck Society using an RCA CCD camera with a field of view of 112 arcsec by 180 arcsec. This instrument was attached to the 2.2m telescope and the filter used was the r of the Gunn-Thuan system. Typical exposure times were twenty to thirty minutes depending on weather conditions and redshift. The data reduction consisting of bias substraction, flat-field correction and background subtraction was done in the standard way.

RESULTS

We conclude that at least 50% of the galaxies in this sample present peculiarities like isophote twists, dust lanes, filaments or companion galaxies at very short distances.

A percentage as high as 13% of the sample (6 galaxies) are dumbbell galaxies while the 32% of the galaxies (15) appear with a small companion at distances shorter than 20 kpc with a mean center to center separation of 12.8 kpc and with a relative size 0.37 ± 0.18 (median value of 0.36) that of the radio galaxy. These two types of galaxies form the multicomponent sample.

The total number of multicomponent galaxies (21) corresponds to a fraction of 45% of the galaxies in the sample (only a small number of the multicomponent galaxies, 2-3 galaxies, could be due to projection effects). Moreover, almost all these galaxies excluding some dumbbell galaxies or galaxies sharing common envelopes appear with signs of distorted morphology indicating a real physical connection. Therefore, at least 35% of the observed galaxies with radio jets consist of galaxies interacting with its nearby companions.

The largest part of the sample, 55% or 26 galaxies, consists of objects without companions at distances shorter than 20 kpc.

Considering the importance of interstellar/intergalactic material in the radio jet phenomenon and the expected fuel supply to the nuclear engine over large periods of time due to tidal interactions between galaxies, one could ask how the presence of a companion galaxy at short distances, i.e. ongoing strong interaction, orbital motions or merging, could enhanced or produce a radio emission different from that of the more isolated galaxies where gravitational effects with more separated companions are less efective or where we see final stages of an interaction/merging process. Therefore, the observed sample have been divided into two major subgroups considering the existence or not of companion galaxies at distances shorter than 20 kpc. This division produces two samples with a similar number of objects: 21 and 26 for the multicomponent (dumbbell galaxies included) and isolated galaxies respectively.

The multicomponent galaxies form a very homogeneous sample with the following properties:
67% have $22.0 \leq LogP_5 \leq 23.0$ Watt$\cdot Hz^{-1}$
67% have $24.0 \leq LogP_{1.4} \leq 25.0$ Watt$\cdot Hz^{-1}$
67% have $10.0 \leq d(jet) \leq 40.0$ kpc
71% have $0.00 \leq P_5/P_{1.4} \leq 0.03$
while the isolated sample is more heterogeneous.

A quantitative study of the sample involving complete isophotal analysis of the galaxies is in progress. From the results of this analysis and its comparison with activity indicators like radio luminosity, spectral index, amount of ionized gas... we expect to tie the activity level with the interaction history of a particular galaxy.

References

Bridle, A.H., Perley, R.A., 1984, Ann. Rev. astron. Astrophys. **22**, 319.
Colina, L., Pérez-Fournon, I., 1989a, Ap. J. Suppl. in press.
Colina, L., Pérez-Fournon, I., 1989b, Ap. J. submitted.
Colina, L., Pérez-Fournon, I., Gonzalez-Serrano, J.I., 1989, in preparation.

Dusty Emission Nebulosity in NGC 1316 (Fornax A)

W.B. Sparks [1] *and F.D. Macchetto* [1,*]

[1]Space Telescope Science Institute**, 3700 San Martin Drive, Baltimore, MD21218, USA

*Affiliated with the Astrophysics Division, Space Science Department, European Space Agency (ESA)

**Operated by the Association of Universities for Research in Astronomy, Inc., under contract with the National Aeronautics and Space Administration

NGC1316 (Fornax A) is host to a powerful radio source, exhibiting characteristic jet structure on small scales and diffuse lobes on larger scales. The galaxy itself displays many of the features which we now believe are due to the merger of a companion galaxy (Schweizer 1980) — it has Malin-Carter shells, a semi-regular minor axis dust-lane with patches and filaments of dust elsewhere, associated emission-line gas and, of course, nuclear activity.

Here we present images of the galaxy taken in the light of $[NII] + H\alpha$ together with a new map of the dust distribution. Long-slit spectroscopy shows the emission is dominated by $[NII]$. There appears to be interaction between the optical line-emitting gas and the radio jet.

Description

We identify two regions where the jet and emission-line gas appear to be in contact. The first is at the position of the bend in the radio jet to the North-West. This coincides with a prominent dust patch and a region of enhanced optical emission. The second area is at the South-East end of the radio jet. We see a feature in the optical emission line gas reminiscent of a bow-shock and of much the same shape as the radio contours. A similar relationship was found on a much smaller scale in the Seyfert galaxy NGC1068 (Wilson & Ulvestad 1987). There are other areas of possible interaction which require confirmation — both very close to the nucleus and beyond the end of the NE radio jet. (The sharp-edged ellipse feature in the emission-line image is believed to be an artifact of the analysis procedure.)

The image of dust optical depth shows essentially identical features in the innermost regions to that published by Sparks (1987). The outer regions are much more clearly defined in the new image. We see prominent absorption features along the minor axis of the galaxy, an irregular arc following the isophotes beyond that, and a series of faint filaments or possibly a disk of dust along the major axis. Since the galaxy is known to be rotating fast, we may be seeing gas primarily in polar orbits in the process of settling to the principal plane of a rotationally flattened oblate spheroid (or mildly triaxial ellipsoid) as expected from models of infall.

Dynamics and Interactions of Galaxies Editor: R. Wielen

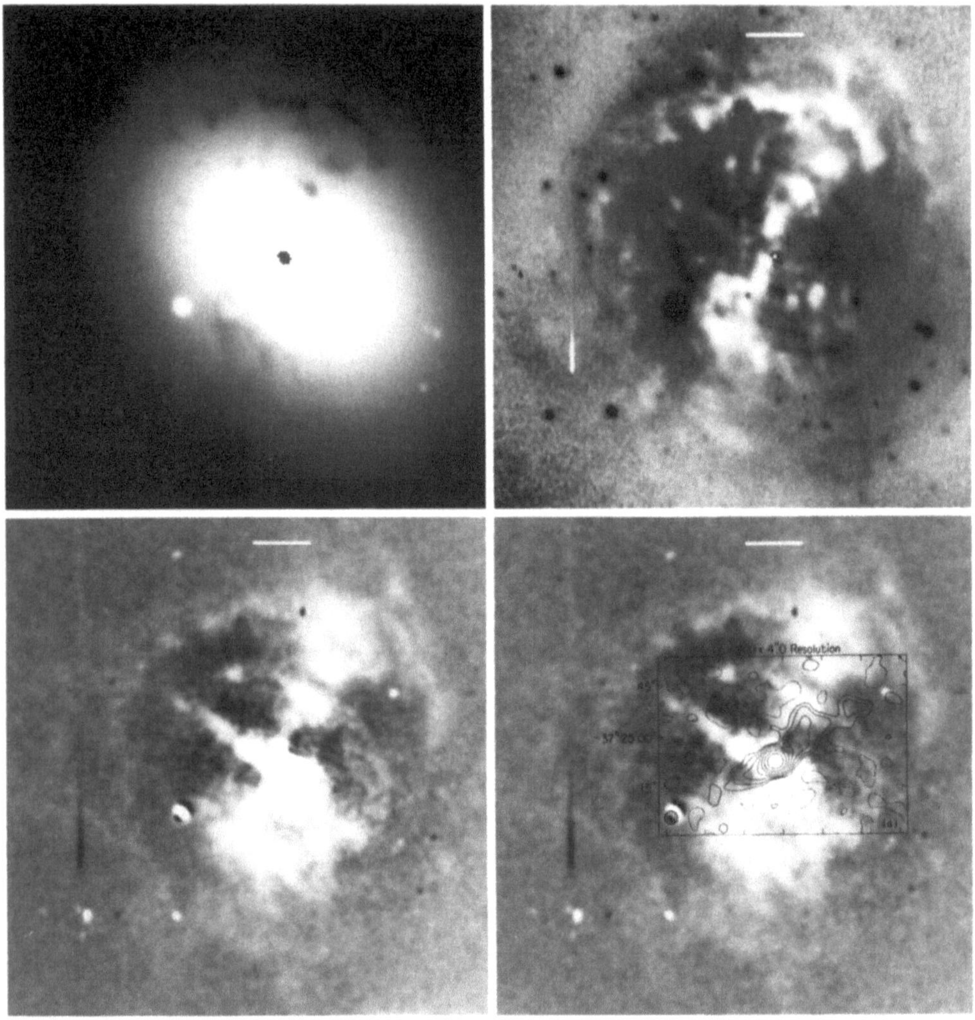

Fig. 1: NGC1316 (Fornax A) showing **top left**, direct B-band CCD frame taken with EFOSC on the ESO 3.6-m. telescope. **Top right,** image processed to show dust, with areas of high extinction showing bright. **Bottom left**, an image in $H\alpha + [NII]$ after continuum subtraction, showing extensive filamentary structure. **Bottom right,** overlay of the radio map of Geldzahler & Fomalont 1984 onto the optical emission-line image showing bright optical emission at the position of the bend in the radio jet. The scale bar is 18 arcsec in length.

References:

Geldzahler, B.J. & Fomalont, E.B., 1984, *Astr. J.* , **89**,1650.

Schweizer, F., 1980, *Ap. J.* , **237**,303.

Sparks, W.B., 1987, *in I.A.U.# 127 Structure and Dynamics of Elliptical Galaxies ed. T. de Zeeuw*.

Wilson, A.S. & Ulvestad, J.S., 1987, *Ap. J.* , **319**,105.

Strength of Interaction and Radio Emission in Seyfert Galaxies

G. Giuricin, F. Mardirossian, M. Mezzetti, and G. Bertotti

Department of Astronomy, via G.B. Tiepolo 11, I-34131 Trieste, Italy

We have found a significant, positive correlation between the total and central radio powers (per light unit) of a wide sample of Seyferts and the interaction class IAC assigned by Dahari (1985). Hence, the most interacting Seyferts appear to have enhanced total and central radio emission, in line with the general tendency displayed by normal spirals.

I.INTRODUCTION

An excess of total radio continuum emission in paired and interacting galaxies when compared to samples of relatively isolated objects has been often reported in the literature. In particular, the radio power (per unit light) of a sample of spiral galaxies selected from the "Atlas and Catalogue of Interacting Galaxies" of Vorontsov-Vel'yaminov (1959, 1977) was proved to show a positive correlation (Giuricin et al., 1988) with the interaction strength as defined by the interaction class IAC assigned by Dahari (1985) (e.g., IAC=1 is assigned to isolated, symmetric galaxies and IAC=6 to severely disordered galaxies or overlapping objects).

On the other hand, a correlation between the radio emission from Seyfert galaxies and the presence of a companion galaxy has been often unsuccessfully searched in the literature. In this study we have looked for a correlation between the logarithm of the radio-to-optical luminosity ratio (log R) and the interaction class IAC assigned by Dahari (1985) and Dahari and De Robertis (1988) for 116 Seyferts observed at radio wavelengths. We have generally taken the radio flux densities (or upper limits) from the compilation of Dahari and De Robertis (1988).

II.RESULTS

Applying the non-parametric methods of Cox regression and of the generalized Kendall's rank correlation to this sample with a fair number of upper limits on the radio fluxes (30), we have found a significant, positive log R - IAC class correlation. (The corresponding significance levels of independence of the two variables are 10^{-4} and $<10^{-4}$ for the two methods, respectively).

We have obtained similar results for the correlation between the central radio power (per unit galaxy light) and the interaction class IAC.

Dynamics and Interactions of Galaxies Editor: R. Wielen

Our results can be taken as indications that the most interacting Seyferts tend to display enhanced radio emission. Our finding turns out to be in line with the results of Giuricin et al. (1987, 1988), who noted similar positive log R-IAC class correlations for extensive samples of spiral galaxies.

REFERENCES

Dahari,O., 1985, Astrophys. J. Suppl. **57**,643.

Dahari,O. and De Robertis, M.M., 1988, Astrophys. J. Suppl. **67**,249.

Giuricin,G., Mardirossian,F., Mezzetti,M., 1987, Astron. Astrophys. **176**,175.

Giuricin,G., Mardirossian,F., Mezzetti,M., 1988, Astron. Astrophys. **195**,9.

Vorontsov-Vel'yaminov,B., 1959, Atlas and Catalogue of Interacting Galaxies, Part 1 (Moscow, Moscow State University).

Vorontsov-Vel'yaminov,B., 1977, Astron. Astrophys. Suppl. **28**,1.

Chaotic Orbits in Barred Structures with Central Mass Concentration

C. Norman[1] *and H. Hasan*[2]

[1]Space Telescope Science Institute, 3700 San Martin Drive, Baltimore, MD21218, USA and Johns Hopkins University, Baltimore, MD21218, USA
[2]Space Telescope Science Institute, 3700 San Martin Drive, Baltimore, MD21218, USA

The influence of a black hole, or central mass concentration, on stellar orbits in the plane of a barred galaxy is examined by analysing surfaces of section and computing phase space volumes. It is estimated that a black hole with mass equal to $\sim$ 17% of the total mass is required to destroy the B family of orbits, and hence the bar. Further, a very small black hole will destroy a thin bar while a thicker bar is far more robust to black holes. Finally, the orbits of stars which do not have enough energy to go far from the black hole get stochastic more readily than those with higher energy.

1 Formulation

The galactic potential is approximated by Plummer spheres representing the core and black hole, and an inhomogeneous prolate bar used in standard galaxy models (e.g. Anthanassoula et al,1983). The bar semi-major axis a was fixed at 9kpc and the total mass $M_T = 4.67X10^{10}M_O$ (Teuben and Sanders, 1985) resulting in a pattern speed $\Omega_p \sim$ 15km/sec/kpc so that $R_{cr}/a = 1$, where R_{cr} is the corotation radius. Orbits in the galactic plane corotating with the bar were considered. ***Phase space volumes were computed numerically*** using the formulation of Binney et al (1985).

2 Discussion of results

Variation of black hole mass. An examination of the surface of section (Fig. 1a) shows that in the absence of a black hole there is one family of retrograde orbits (not discussed further) and one family of direct orbits aligned with the bar (the B family). The orbits are all loops. The introduction of a central black hole with the black hole mass ratio $M_{bh}/M_T = .009$ causes the appearance at small radii of a minor family of loop orbits antialigned to the bar (Fig. 1b). We shall refer to them as "looplets" in analogy with the minor family of box orbits christened as "boxlets" by Miralda et al(1989). The looplets gradually join the major orbit family B as M_{bh}/M_T is increased to 0.013. As the black hole mass is increased further the outermost orbits start getting stochastic till at *mass ratio 0.17 all B family orbits have become stochastic and the bar dissolves completely (Fig. 1c).*

An estimate of stochasticity was obtained by computing the percentage P of the phase space volume occupied by the B family of orbits (Fig. 2). The phase space occupied by B orbits goes from $\sim$ 48% to zero as M_{bh}/M_T increases from 0 to 0.17. There is a dip in the curve for $M_{bh}/M_T = 0.01$ caused by the appearance of the looplets.

Physical insight may be gained by looking at the frequency curves for each case (e.g. Fig.3). Curves are plotted for the angular velocities Ω_x and Ω_y along x and y respectively and the corresponding curves for $\Omega_x - \kappa_x/2$ and $\Omega_y - \kappa_y/2$, where κ represents the "epicyclic frequency". In the absence of a black hole the pattern speed Ω_p is much higher than $\Omega - \kappa/2$ so that there are no Lindblad resonances and all direct orbits are of the B type. With the introduction of a black hole with increasing mass the peak in $\Omega - \kappa/2$ rises sharply causing the appearance of the inner Lindblad resonances (ILR's). The outer ILR moves further out with increase of black hole mass so that the region in phase space occupied by orbits of the B family decreases (Contopoulos,1980). For a black hole mass of 15% the outer ILR along the y-axis reaches the edge of the bar along this axis. The volume in phase space occupied by B orbits has gone down to $\sim$ 2% for this case (Fig. 2). When the outer ILR moves outside the bar all B orbits are destroyed.

Variation of bar axial ratio (b/a). The effect of a black hole of mass $0.03M_T$ was examined on three bars with $b/a = 0.25, 0.35, 0.45$ and all other parameters held constant. In each case $R_{cr}/a = 1$

Dynamics and Interactions of Galaxies Editor: R. Wielen

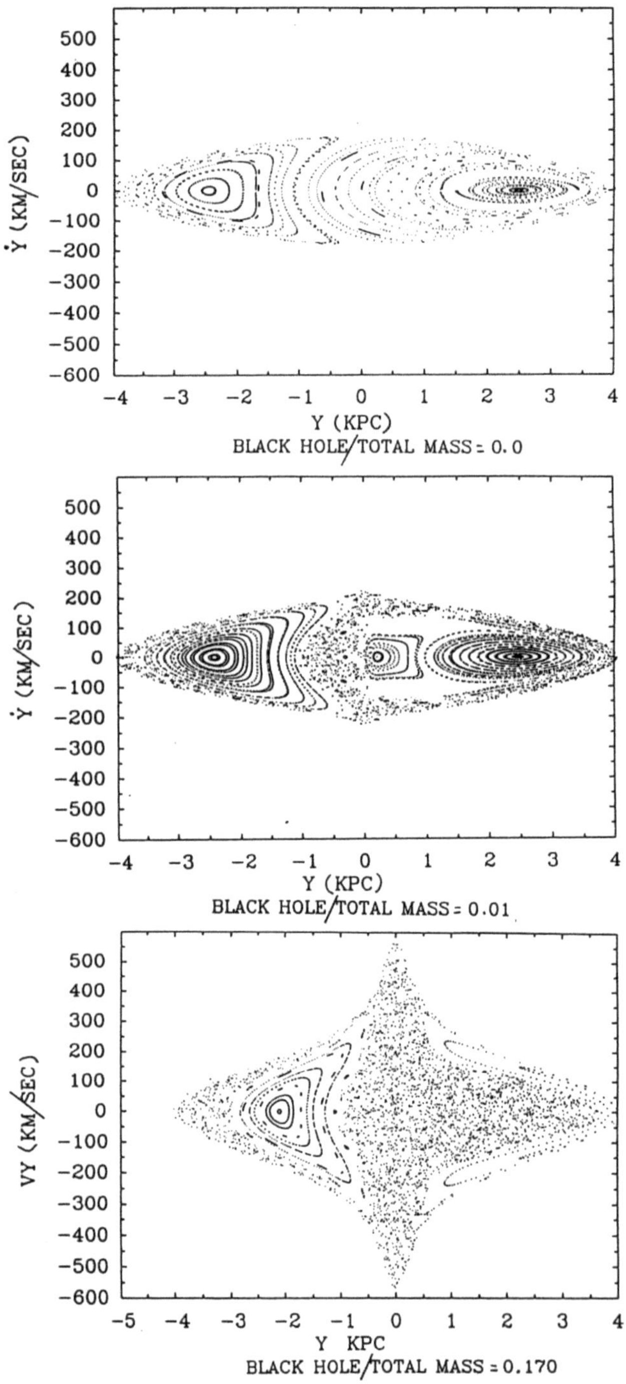

Fig. 1 Phase space plots for a rapidly rotating bar with (a) no black hole (b) $M_{bh}/M_T = 0.01$ (c) $M_{bh}/M_T = 0.17$.

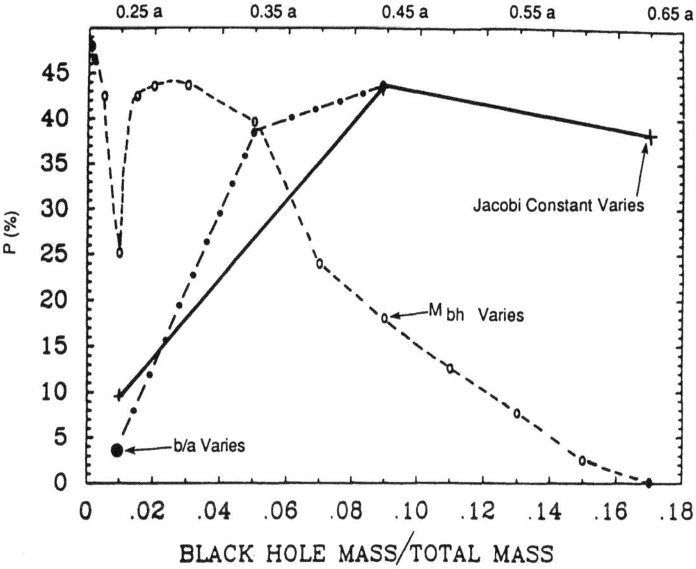

Fig. 2 **Percentage of phase space occupied by *B* orbits.**

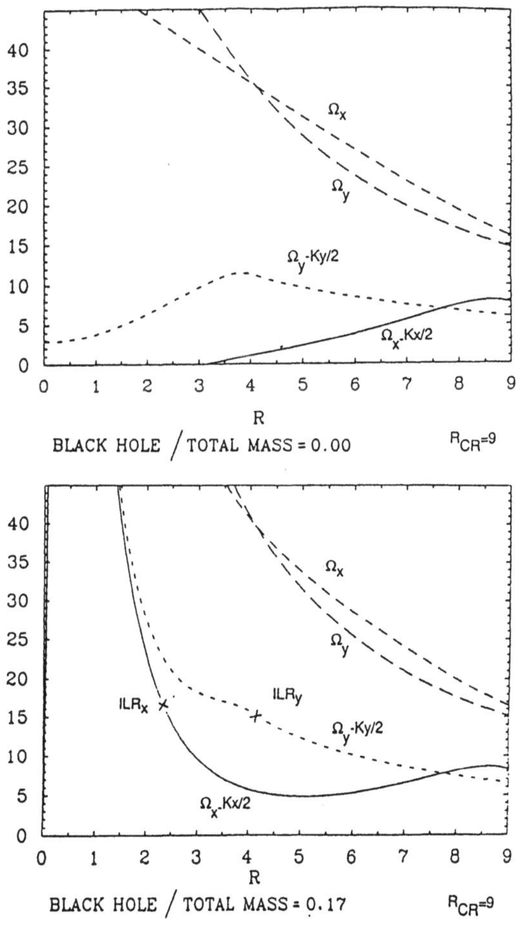

Fig. 3 **Frequency plots for (a) no black hole (b)** $M_{bh}/M_T = 0.17$.

and the Jacobi constant was fixed so that the star could go to maximum distance of $0.45a$. The phase space volume ratio, P, is plotted in Fig. 2. While P varies slowly from $\sim 44\%$ to $\sim 41\%$ as b/a decreases from 0.45 to 0.35, the effect is much more dramatic for $b/a = 0.25$ when P goes down to $\sim 3\%$. In each case a computation was done for a bar without a black hole and it was found that $P \sim 48\%$ in all cases. It seems quite clear that a *very small black hole will destroy a thin bar* while a thicker bar is far more robust to black holes.

Variation of Jacobi constant. A black hole of mass $0.03M_T$ causes greater ergodicity when the Jacobi constant corresponds to a maximum distance of $0.25a$ reached by the star than for distances $0.45a$ and $0.65a$. Thus the *orbits of stars* which do not have enough energy to go far from the *black hole get stochastic more readily* than those of stars with higher energy. The phase space volume ratio, P, is plotted in Fig. 2.

References

1. Anthanassoula, E., Bienayme, O., Martinet, L. and Pfenniger, D., 1983 *Astr. Astrophys.*,**127**,349.
2. Binney, J., Gerhard, O.E. and Hut, P., 1985 *Mon. Not. R. astr. Soc.*,**215**,59
3. Contopoulos, G. and Papayannopoulos, Th., 1980 *Astr. Astrophys.*, **92**,33.
4. Miralda-Escude, J. and Schwarzschild, M., 1989 *Atrophs. J.*, **339**,752.
5. Teuben, P.J. and Sanders, R.H.,1985 *Mon. Not. R. astr. Soc.*,**212**,257.

Fuelling AGN by Non-Axisymmetric Instabilities

J. Frank

Max-Planck-Institut für Astrophysik, Karl-Schwarzschild-Str. 2,
D-8046 Garching, Fed. Rep. of Germany

1. Introduction

Although galaxies contain sufficient interstellar material to power AGN by accretion on to a super-massive black hole ($M_h \gtrsim 10^8$ $M_\odot$), it is not possible to bring in this gas in from galactic scales down to the center via a turbulent viscous disk in less than a Hubble time. In most scenarios discussed in the literature which involve gas already present in the galaxy or gas captured from a merger, the main problem is not the total mass supply but how to get rid of the angular momentum in a sufficiently short time scale. A mechanism involving large scale non-axisymmetric dynamical instabilities is outlined below. A more detailed account is published elsewhere and references are given in the text.

2. Bars in Bars

In a recent paper (Shlosman, Frank & Begelman 1989) we suggested a two stage mechanism in which a large scale stellar bar sweeps the interstellar material inwards and under certain conditions becomes dynamically unstable again generating further inflow. The first stage can be due to the barred nature of the host galaxy or can be induced or triggered tidally by the passage of a companion galaxy. The second stage can occur when the mass of the rotationally supported disk formed at the center of the host as a consequence of the first stage of inflow exceeds a certain critical fraction of the mass of the host. As a first indicator of stability we apply the well known semiempirical criterion due to Ostriker & Peebles (1973) to this disk considering the gaseous disk to be the only "active" mass in the system (see also Bardeen 1975; Shlosman, Frank & Begelman 1989; and Frank & Shlosman 1989). We find that the critical mass fraction is $\sim 0.1 - 0.2$ of the host mass.
As it is further discussed in Begelman, Frank & Shlosman (1989) it is necessary to bring the material from galactic scales down to radii $\lesssim 10$ pc for turbulent viscous processes of the type operating in accretion disks in close binaries to become effective. This cannot in general be achieved in one step of instability in the gaseous disk. However, if the angular momentum transport and consequent inflow caused by the instability is *efficient* in the sense that most of the mass flows in, the new and smaller gaseous subdisk resulting from the inflow may go unstable again. Providing every stage occurs with high efficiency in the above sense, a runaway cycle of non axisymmetric instabilities at smaller and smaller scales could take place (see however Phinney, this volume). Such a runaway would cease when the potential of the central black hole becomes dominant.
The mechanism discussed here is not limited to spiral hosts. The first stage can assume quite different forms depending on the morphological type of the host, the presence/absence of tidal perturbers, and the gas richness of the galaxy. Some of these questions are addressed in Frank, Shlosman & Begelman (1989). For example: if the mass of the disk formed at the first stage is less than critical a nuclear starburst may follow. The presence/absence of resonances also affects the scale and shape of the region undergoing starburst activity. A summary of the different scenarios and their most likely outcomes is shown in Fig. 1.

Dynamics and Interactions of Galaxies Editor: R. Wielen

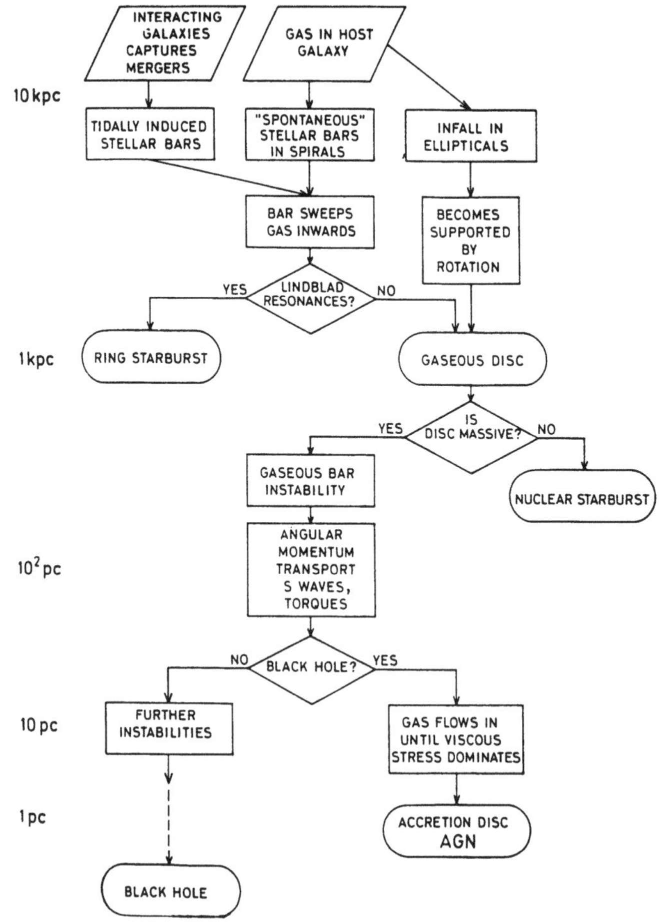

Figure 1. Different pathways by which host galaxies can acquire a massive black hole and/or become an AGN or a starburst. The approximate radial scales at which these processes take place are indicated on the left (adapted from Frank *et al.* 1989).

3. Discussion

We summarized here some recent work on the "bars in bars" mechanism for fuelling active galactic nuclei and we refer the reader to the cited papers for details. The coupling between different spatial scales of instability and the implications of differences in morphology and kinematics of the host on the first stage of inflow are currently under investigation.

REFERENCES

Bardeen, J. M. 1975, in *Dynamics of Stellar Systems*, ed. A. Hayli (Reidel: Dordrecht)

Begelman, M. C., Frank, J. & Shlosman, I. 1989. *Theory of Accretion Disks*, eds. F. Meyer, W. Duschl, J. Frank, and E. Meyer-Hofmeister, Dordrecht: Kluwer, in press.

Frank, J. & Shlosman, I. 1989. *Astrophys. J.* , in press.

Frank, J., Shlosman, I. & Begelman, M. C. 1989. *Theory of Accretion Disks*, eds. F. Meyer, W. Duschl, J. Frank, and E. Meyer-Hofmeister, Dordrecht: Kluwer, in press.

Ostriker, J. P. & Peebles, P. J. E. 1973. *Astrophys. J.* **186**, 467.

Shlosman, I., Frank, J. & Begelman, M. C. 1989. *Nature* **338**, 45.

Slow Dissipation in Bars and the Fuelling of Nuclei

D. Pfenniger[1] *and C. Norman*[2]

[1]Geneva Observatory, CH-1290 Sauverny, Switzerland
[2]Space Telescope Science Institute, 3700 San Martin Drive, Baltimore, MD21218, USA and Johns Hopkins University, Baltimore, MD21218, USA

The problem of multiphase gas flow in the central regions of barred galaxies, and the influence of a central mass on the rate of accretion are considered. In order to study the most essential factors, the complicated physics of interstellar clouds is simplified, such that gas clouds are represented by point mass particles moving in a fixed barred galaxy potential with various friction laws acting on them depending on the velocity in the rotating frame of the bar. The dissipation time-scale is fixed to be always much longer than the orbital period. This kind of system belongs to the class of dissipative systems in the limit of vanishingly small dissipation, and is close to the well studied Hamiltonian systems. The basic mechanisms are due to resonances with the main potential. These control how and why gas accretes in bars and can be studied without the distraction of additional effects, such as self-gravitation and collective effects.

1. DISSIPATION IN A BARRED POTENTIAL WITH A BLACK-HOLE

The barred galaxy model is standard (Pfenniger 1984). A small additional Plummer model represents the black-hole, its radius amounts to 10 pc, the bar semi-major axis to 6 kpc. Orbits start close to the Lagrangian points L_1 at very small initial velocities for various dissipation laws. The initial position in z is chosen slightly above the galactic plane (20 pc) in order to observe the effects of vertical instabilities. The dissipation laws considered here include dynamical friction, friction laws proportional to v^n, $n \geq 1$, where $\mathbf{v}$ is the velocity in the rotating frame, and more complicated laws. As long as dissipation is everywhere weak, i.e. much slower than the local orbital period, the orbits move throughout the basin of attraction associated with the potential, and *the results are verified to be qualitatively independent of the dissipation law.*

By choosing a sufficiently small friction coefficient, the dissipation time-scale without a black-hole is fixed to be typically of the order of 10 Hubble times, or approximately 1000 bar rotation periods. The total accretion time is not especially relevant, but the ratios of the times tell us the relative increase of accretion when the black-hole mass is increased. For a given friction coefficient α, the accretion rate increases systematically with the black-hole mass M_{bh}. Fig. 1a shows the evolution of the radii of typical orbits with identical initial conditions and an αv^2 friction law, the only difference is the black-hole mass. The orbits spend most of the time in the regions where stable periodic orbits can be found in the non-dissipative case. The plunging toward the center is dramatically increased by the presence of the black-hole. The steep contraction of the orbital radius occurs inside resonances, mainly the ILR, where chaos is the most pronounced. *A very general property of chaos is to accelerate the effects of dissipation.*

The orbital height above the plane is almost always *amplified by vertical resonances*, up to 1–2 kpc. This 3D effect is powerful enough to create a sizable part of the bulge. The excursion in z is only temporary here because the final attractor is the center.

Since a central density peak produces an ILR, the accretion onto the center is self-amplified. Once the clouds have crossed the ILR, they can accumulate onto the circular orbits around the central mass and form a new, small disk. As soon as this small disk is massive enough, it becomes gravitationally independent of the large bar, and may develop its own bar instability.

Dynamics and Interactions of Galaxies Editor: R. Wielen

2. DISSIPATION IN A DOUBLY BARRED POTENTIAL

Bars inside bars have been frequently observed (e.g. Jarvis *et al.* 1988). If a small bar develops, having its corotation located about at the ILR of the large bar, it provides a mechanism for regulating accretion of gas inside the small bar, because bars repel dissipating particles outside corotation. The previous central Plummer sphere is now replaced by a second rotating bar, the size and mass of which is smaller than the large bars by a factor 10. The small bar has its corotation close to the ILR of the large bar. The mass of the small bar amounts to 1% of the total galaxy mass. The small bar has its own, non-commensurable rotation frequency.

As before, we study the attractors in this time-dependent potential. In Fig. 1b, the orbital radius R is plotted as a function of the time T, for different friction coefficients α, and an αv^2 friction law. For small α's, R fluctuates in an erratic manner for very long times. *This is typical for a strange attractor.* For intermediate α's, the attractor may be an oval retrograde orbit (limit cycle) in the plane $z=0$. If dissipation is strong enough (or the bar strength weak enough), the repulsion due to the small bar is overcome by the attraction from the center. In those cases of

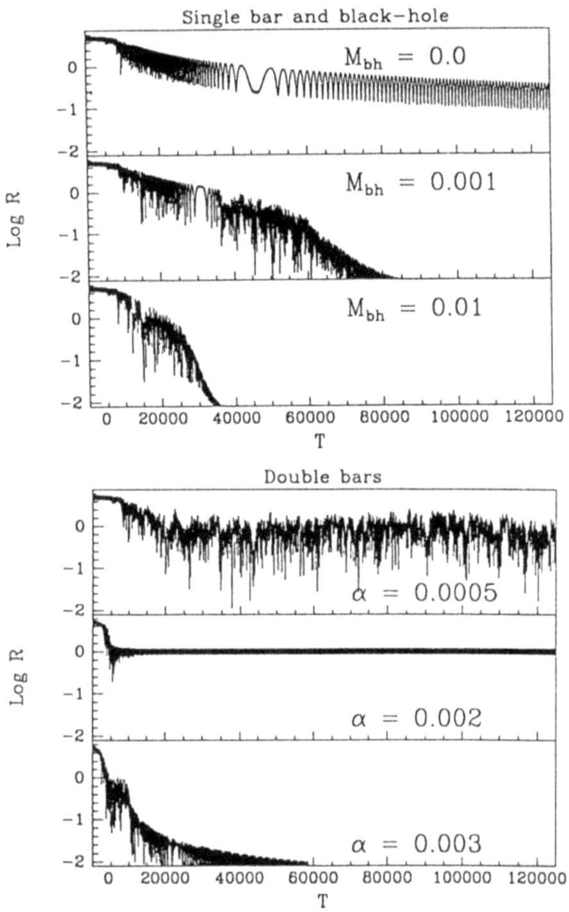

Fig. 1: *The orbit radius R as a function of time T,* **a)** *in the barred potential with a central black-hole of fractional mass $M_{\mathbf{bh}}$, and* **b)** *in the doubly barred galaxy potential with different friction coefficients α. Some aliasing in the curves is produced by the orbit sampling at constant time intervals $\Delta T = 25$*

strange attractors and limit cycles, the energy and angular momentum lost by dissipation have to be supplied by the small bar, which, in fully self-consistent systems, should limit the lifetime of the small bar.

As in the single bar potential, the orbits often achieve sufficiently large heights above the galactic plane (1–2 kpc) that they become part of the bulge. The strange attractors are essentially three-dimensional and have a fat spheroidal shape.

3. CONCLUSIONS

- ▷ In the limit of weak dissipation, the results are qualitatively independent of the dissipation law. The effects of dissipation are amplified by a chaotic region, around resonances.
- ▷ A black-hole at the center of a bar creates resonances, especially an ILR, which accelerates the accretion of gas clouds inside this ILR.
- ▷ If a secondary small bar rotates inside a much larger one, the possible attractors include the center, oval rings, and strange (i.e. chaotic) 3D attractors outside the small bar.
- ▷ Vertical resonances in bars are strong enough to feed bulges by disk heating.

References

Jarvis, B.J., Dubath, P., Martinet, L., Bacon, R.: 1988, *Astron. Astrophys. Suppl. Ser.* **74,** 513
Pfenniger, D.: 1984, *Astron. Astrophys.* **134,** 384

Bar-Driven Central Star-Bursts in SB0 and SBa Galaxies

L.L. Dressel[1], *C.M. Telesco*[2], *and R. Decher*[2]

[1]A.J. Dyer Observatory, Vanderbilt University, Nashville, TN 37235, USA
[2]NASA/Marshall Space Flight Center, Huntsville, AL 35812, USA

1. Introduction

Optical photographs of galaxies give little or no indication of on-going star formation at the early end of the Hubble sequence. Few or no star forming regions can be resolved in the disks of SO and early-type spiral galaxies. However, several lines of evidence indicate that a high rate of star formation is occurring at the centers of many SO and early-type spiral galaxies, particularly those with bars.

This hypothesis is supported by a study of the galaxies in the magnitude-limited Dressel and Condon (1978) Arecibo 2380 MHz survey. Statistical studies have been made of the radio flux densities and far infrared flux densities for each Hubble type, and more detailed observations have been made of individual galaxies. The evidence for vigorous star formation at the centers of early-type galaxies (especially SBO and SBa galaxies) is outlined below.

2. Data and Interpretation

In star-forming regions, dust absorbs much of the ultraviolet emission of the hot massive young stars and re-emits the energy as infrared radiation. This radiation is warmer than that produced by dust in the general interstellar medium. Powerful warm infrared sources can thus indicate sites of intense star formation. Globally warm powerful sources have been found to occur most frequently in SO and early-type spiral galaxies (Dressel 1988) in a study of the IRAS 60 and 100 micron flux densities (Lonsdale et al. 1985) of the Arecibo survey galaxies.

As shown by Helou, Soifer, and Rowan-Robinson (1985), radio continuum flux density and far infrared flux are extremely well correlated for both ordinary spiral disks and for star-burst galaxies. This correlation presumably results from a relationship between the numbers of luminous massive young stars (which heat the dust) and of recent supernovae (which contribute to the radio emission). The Arecibo SO + Sa sample (including barred galaxies) shows a similar correlation, consistent with the suggestion that the far infrared flux is produced by starlight-heated dust.

If warm bright infrared sources are indeed indicative of high rates of star formation, then high star formation rates occur much more frequently in barred galaxies than in galaxies without bars. Sources warmer than $F(100\mu)/F(60\mu) = 2.0$ and brighter than $F(60\mu) = 1.5$ Jy occur in <u>4%</u> of the 259 SO + Sa galaxies in the Arecibo sample and in <u>22%</u> of the 92 SBO + SBa galaxies.

Dynamics and Interactions of Galaxies Editor: R. Wielen

Ultraviolet and optical spectral observations confirm the presence of hot massive young stars in some of the infrared-bright early-type galaxies. In the Arecibo sample of non-Seyfert galaxies with warm bright infrared sources, 3 of the 11 SO + Sa galaxies are Markarian galaxies and 8 of the 15 SBO + SBa galaxies are Markarian galaxies. Optical spectrophotometric observations of the centers of early-type galaxies in the Arecibo sample (O'Connell and Dressel 1978) included several of the galaxies with warm bright infrared sources. These galaxies have conspicuous high order Balmer absorption lines characteristic of A stars and emission lines characteristic of HII regions.

VLA observations of several early-type galaxies with warm bright infrared sources have shown that the radio continuum emission is centrally concentrated (Dressel, in preparation). The radio sources are diffuse and occupy the central few kiloparsecs of the galaxies. They have a variety of morphologies, from ring-shaped to centrally peaked.

The infrared sources in infrared-bright early-type galaxies are also centrally concentrated. We have imaged six galaxies at 10.8 microns at the NASA Infrared Telescope Facility with a 5 x 4 array of 4.3" x 4.3" FWHM pixels. The sources are centered on the optical galaxies and are unresolved or marginally resolved. They are thus smaller in FWHM than the approximately 2 kpc FWHM pixels (for H_o = 50 km s^{-1} Mpc^{-1}).

3. Summary

In summary, radio continuum flux densities, far infrared flux densities, far infrared colors, and ultraviolet and optical spectra indicate that high rates of star formation are occurring in SO and early-type spiral galaxies, especially barred galaxies. Radio continuum and infrared images show that the star forming region is typically concentrated within the central few kpc of the galaxy. The high incidence of central star formation in early-type barred galaxies suggests that bar-driven inflow of gas is fueling the star-forming regions in these galaxies.

This research was supported by grant AST-8818606 from the National Science Foundation and by funds from NASA administered by the American Astronomical Society.

Reference

1. L. L. Dressel: Ap.J. (Letters) 329, L69 (1988)
2. L. L. Dressel, J. J. Condon: Ap.J. Suppl. 36, 53 (1978)
3. G. Helou, B. T. Soifer, M. Rowan-Robinson: Ap.J. (Letters) 298, L7 (1985)
4. C. J. Lonsdale, G. Helou, J. C. Good, W. Rice: Cataloged Galaxies and Quasars Observed in the IRAS Survey (Jet Propulsion Laboratory) (1985)
5. R. W. O'Connell, L. L. Dressel: Nature 276, 374 (1978)

Part XIII

Problems of Merger Scenarios

Galaxy Mergers: Views of an Appointed Skeptic

S. van den Bergh

Dominion Astrophysical Observatory, National Research Council of Canada, 5071 W. Saanich Road, R.R. 5, Victoria, B.C. V8X 4M6, Canada

If there were two earths in space, then they would both fall to the centre of the Universe and coalesce, because earth moves naturally to the centre.
Aristotle

1. INTRODUCTION

The conjecture about mergers that is cited above has been discussed in more detail by Chandrasekhar (1942) in his theory of dynamical friction. As Toomre (1977) points out: "It seems almost inconceivable that there wasn't a great deal of merging of sizable bits and pieces (including quite a few lesser galaxies) early in the career of every major galaxy, no matter what it looks like now." Observations by Zepf and Koo (1989) suggest that such interactions between galaxies may have been much more frequent in the past than they are now. If this view is correct then the galaxies we see today might bear little resemblance to the original population of primordial galaxies. It is noted in passing that most tidally distorted galaxies have nearby optical companions. It follows (van den Bergh 1969) that only a small fraction of all galaxies can consist entirely of dark matter.

2. ARGUMENTS IN FAVOUR OF MERGERS IN ELLIPTICALS

In recent years a considerable amount of evidence has accumulated favouring the idea that many ellipticals have accreted other galaxies during their lifetime. Spectroscopic and photometric observations of NGC5813 by Efstathiou, Ellis and Carter (1982) have been interpreted by Kormendy (1984) in terms of a model in which a low luminosity galaxy with a compact core merged with, and sank to the bottom of, the potential well of a more luminous elliptical with a larger core. Perhaps even more striking evidence in favour of the merger picture comes from observations of IC1459 by Franx and Illingworth (1988), which show this object to have a counter-rotating core. Observations of large samples of ellipticals by Bender (1988), and by Jedrzejewski and Schechter (1988) suggest that ~1/3 of all galaxies show some kinematic evidence for mergers.

Additional sources of information on mergers are provided by dust, and by the shells that have been discovered around some E galaxies. According to Quinn (1984) the shells observed in E galaxies are due to the phase wrapping of dynamically cold discs in the fixed potential wells of ellipticals. Surveys by Malin and Carter (1983), and by Schweizer and Ford (1985), show evidence for ripples and shells associated with 17% and 44%, respectively, of all field ellipticals. (The higher percentage of ripples in the latter investigation is due to the fact that these authors used a CCD detector and digital background subtraction techniques). Malin and Carter note that about half of all E galaxies with shells are either isolated, or located in small groups. In such groups (in which the velocity dispersion is relatively small) tidal interactions should be stronger than they are between fast-moving galaxies in rich clusters. A counter-example is, however, provided (Schweizer 1980) by NGC1316 (Fornax A), which is located in the rich Fornax I cluster. In summary it appears that both shells in the outer regions of ellipticals, and photometry and spectroscopy of their cores, shows that mergers

Dynamics and Interactions of Galaxies Editor: R. Wielen

play an important role in the evolutionary history of many elliptical galaxies. It does not, however, follow that all ellipticals were formed by mergers and that dissipative collapse played no role in their formation (Carlberg 1986).

Dense dwarf ellipticals, such as M32, provide a particularly severe test for merger scenarios that do not involve dissipation. This is so (Kormendy 1989) because such objects would have to form by merging faint low-density disc systems. Furthermore mergers involving many ancestral galaxies will have difficulty accounting for the colour gradients seen in most elliptical galaxies. Finally it appears likely that some (but not all!) "amorphous" galaxies (Sandage and Brucato 1979, Sandage and Tammann 1981, Noguchi 1988) are the remnants of such mergers.

Vader et al. (1988) have suggested that luminous ellipticals, which are not supported by rotation (Davies et al. 1983), might have formed by merger of less-luminous rotationally supported disc-like systems. This suggestion may, however, be difficult to reconcile with the observation (e.g. Visvanathan and Sandage 1977) that luminous ellipticals are redder than intrinsically fainter early-type galaxies.

Contrary to expectations from the merger scenario (Hickson, Kindl and Huchra 1988) compact groups with a high velocity dispersion (in which mergers should be rare) are predominantly composed of elliptical galaxies. On the other hand compact groups with a low velocity dispersion (in which mergers should be frequent) are mainly composed of spiral galaxies.

3. CONSTRAINTS ON MERGERS FROM GLOBULAR CLUSTER OBSERVATIONS

Globular clusters are tightly bound objects that should be able to survive mergers, collisions and all manner of other misfortunes that might befall a galaxy during the course of its evolution (Harris 1981, van den Bergh 1984). During tidal encounters and mergers some stars will escape. Since globular clusters are, on average, less tightly bound to their parent galaxies than typical stars (Harris 1988) they are more likely to escape. Mergers will therefore, in general, result in a slight lowering of the specific globular cluster frequency S, which is defined (Harris and van den Bergh 1981) as the number of globulars per $M_V = -15$ of parent galaxy luminosity.

Since only the brightest clusters in distant galaxies are usually visible the computation of S requires the assumption that all globular cluster systems have the same luminosity function. From an observational point of view the statement that S is high in Virgo cluster ellipticals is equivalent to saying that the globular clusters in Virgo have an above-average luminosity. However, a comparison of various distance determinations to the Virgo cluster (van den Bergh 1989) gives no evidence for a systematic under estimate of the Virgo distance modulus from distance estimates based on globular clusters. This indicates that Virgo globulars are probably not overluminous compared to their Local Group counterparts.

Studies of S in a few dozen galaxies lead to the following conclusions:

1. S is an order of magnitude lower in disc systems than it is in ellipticals. For the small number of disc systems that have so far been studied $S \approx 0.2$ (van den Bergh 1984). After the fading of the luminosity of intermediate-age populations is taken into account and, correcting for internal dust absorption, this figure would increase to $S \sim 0.4$ for an old dustless disc system. This value is an order of magnitude lower than that for typical ellipticals. This shows [but see Lake and Dressler (1986) for a contrary view] that most elliptical galaxies cannot have formed by merging disc-like galaxies resembling present-day spirals.

2. It has been suggested by Schweizer (1986) that the notion that E galaxies were formed by the merger of spirals might still be saved by making the ad hoc assumption that the globulars, that are presently observed in ellipticals, were formed as a direct consequence of the violent merger process itself. A major problem with this idea is, however, that the metallicity of the globulars in ellipticals should then be similar to that of the metal-enriched gas in spirals. Globular clusters in E galaxies would therefore be expected to be as rich, or metal-richer, than the bulk of the stellar population in elliptical galaxies.

This appears to conflict with the observation (Forte, Strom and Strom 1981, Strom et al. 1981, Cohen 1988) that the globular clusters associated with three giant ellipticals in Virgo are bluer (and hence presumably metal-poorer) than the background light of the galaxies on which they are superimposed. The alternative hypothesis that Virgo globulars are relatively blue because they are younger than the dominant population in giant ellipticals appears implausible because most merging in the core of the Virgo cluster must have taken place long ago before the cluster virialized i.e. when the relative velocities of galaxies were much lower than they are at the present time.

3. A few early-type galaxies, such as NGC5253, are observed (van den Bergh 1980, Caldwell and Phillips 1989) to contain large numbers of young star clusters. Such young clusters are easily distinguished from old ones by the fact that (1) their colours are blue and (2) they are distributed over the face of their parent galaxy in a very clumpy fashion. Both of these characteristics would, of course, be more difficult to observe in a galaxy in which a burst of star and cluster formation took place ~10 times as long ago as it did in NGC5253. The observations of Caldwell and Phillips (1989) show that the luminosity function of widely-distributed blue clusters in this galaxy starts at $M_V \approx -8$. After aging by a factor of 10 such clusters will have faded by almost 2 mag (Tinsley 1972). This would make the brightest widelydistributed clusters in NGC5253 slightly brighter than $M_V = -6$; which is well below the peak of the globular cluster luminosity function that occurs at $M_V = -7.5$. Old open clusters are therefore too faint to be confused with real globulars. Only the five blue nuclear knots in NGC5253, which have $-11.3 < M_V < -10.4$, might after they have aged, have been mistaken for globular clusters. The nuclear knots in NGC5253 are probably similar to the superassociations (van den Bergh 1971) discovered in the core of the active galaxy M82. The fact that such objects appear to be closely associated with nuclear starbursts makes it improbable that they would be confused with globular clusters, which are at least as widely dispersed as is the light of their parent galaxies. Only in the case of star-forming brightest cluster galaxies, such as NGC1275 and NGC6166, might confusion between old and young clusters be a problem.

4. Considering only the bulge light of 5 spirals Harris (1988) finds $\langle S \rangle \sim 4$ for the spheroidal componants of spiral galaxies. This value is comparable to that found for early-type galaxies, which are essentially all bulge. Aguilar, Hut and Ostriker (1988) have recently shown that bulge shocks destroy clusters much more efficiently than do disc shocks. For galaxies with M_V(elliptical) $\simeq$ M_V(spiral bulge) the globular cluster destruction rates should therefore be approximately the same.

5. The specific globular cluster frequencies in galaxies classified E and SO by Sandage and Tammann (1981) are indistinguishable. $\langle S \rangle = 4.5 \pm 0.8$ for 14 E's compared to $\langle S \rangle = 4.7 \pm 0.6$ for 9 SO's (van den Bergh 1984). The fact that ellipticals and SO's have similar S values is, most probably, due to the fact that most luminous SO galaxies are strongly bulge-dominated. The prototypical SO with a well-developed disc is NGC3115. For this object Hanes and Harris (1986a) find the relatively low value $S \approx 2$.

6. It has recently become clear that spheroidal galaxies are fundamentally different from ellipticals (Wirth and Gallagher 1984, Kormendy 1985, Sandage, Binggeli and Tammann 1985). Specific globular cluster frequencies for spheroidal galaxies are observed to exhibit a wide range from $S \sim 20$ for the Fornax dwarf to $S \sim 2$ in NGC205. It follows that ellipticals might have formed by merger of spheroidal galaxies. Kormendy (1989) has, however, given strong arguments against the notion that compact dwarf ellipticals, like M32, could have formed from mergers of less dense dwarf spheroidals. He therefore argues that dissipation must have played an important role in forming dwarf ellipticals.

7. For early-type galaxies (but excluding spheroidals) S is higher in rich clusters ($\langle S \rangle = 5.8 \pm 0.5$) than it is in the field and in small groups ($\langle S \rangle = 3.8 \pm 0.7$). This difference might be a consequence of the higher probability for mergers among field galaxies. Perhaps the velocity dispersion in presently rich clusters was only low enough to permit significant numbers of mergers before they virialized, whereas field galaxies continued to suffer mergers during most of their lifetimes. For early-type galaxies in the Virgo cluster $\langle S \rangle = 6.5 \pm 0.6$. A

lower value $\langle S \rangle = 3.9 \pm 1.0$ for 4 galaxies in the Fornax cluster is mainly due to the low value $S = 1.2 \pm 0.3$ in NGC1404.

8. Among ellipticals S does not appear to depend on apparent flattening $\varepsilon = 10(a - b)/a$ (van den Bergh 1984).

9. It would be interesting to see if there is a systematic difference between the S values of dusty ellipticals (which presumably suffered recent mergers) and transparent ellipticals, The presently available data samples are not yet large enough to study this problem.

10. S is significantly higher for the central galaxies in rich clusters than it is in other cluster ellipticals. Examples of this effect are M87 at the centre of the Virgo cluster, the D galaxy NGC3311 in the Hydra I cluster (Harris, Smith and Myra 1983), NGC1399 - which is the central galaxy of the Fornax I cluster (Dawe and Dickens 1976, Hanes and Harris 1986b), and NGC4874 in the Coma cluster (Harris 1987). From data compiled by Harris (1988) it is seen that $\langle S \rangle = 16.5 \pm 1.1$ for the 4 central cluster galaxies for which this information is presently available. This value is 2.8 ± 0.3 times that for other early-type galaxies in rich clusters.

Were the globular cluster systems of these central cluster galaxies always different from others, or did they become special because of their location? Computations by Muzzio (1986) appear to show that central galaxies cannot have acquired the bulk of their globular cluster swarm by capture. (M87 has an excess of ~10000 globular clusters). Note that cannibalism and merger with other cluster ellipticals and spirals (which have lower S values then the central galaxy) would have lowered the specific cluster frequency of the central object i.e. central galaxies must initially have had even higher S values than they do today. This suggests that central galaxies in rich clusters have always been special (van den Bergh 1983b). One might speculate that this is because they were the kernels around which giant clusters formed. Alternatively the early globular cluster forming phase of their evolution may have lasted longer (van den Bergh 1977) in central galaxies of rich clusters than it did for other cluster galaxies.

11. After excluding spheroidals and the central galaxies of rich clusters there is no evidence for a dependence of S on parent galaxy luminosity (Harris 1988). This result would be difficult to understand if globular clusters were only formed in "highly crunched gas" (Schweizer 1986). This is so because mergers of massive galaxies are expected to involve higher relative velocities than the gentle coalescence of dwarf systems.

12. Miller (1988) has suggested that galaxies resembling ellipticals may form as a result of violent shaking of spirals. Such shaking might result from collisions, mergers or tidal stresses. "Ellipticals" formed in this way would, however, have low S values, which is contrary to observation.

13. Morbey and McClure (1985) find an unusually low specific globular cluster frequency $S = 0.8 \pm 0.6$ for the fast-rotating elliptical NGC3557. The low specific globular cluster frequency, and high rotational velocity, in this object suggest that it might be a rare example of an elliptical that did form from merging spirals.

The discussion given above shows that more observations of the specific globular cluster frequency in galaxies might provide valuable insights into their evolutionary history.

4. DID THE GALAXY FORM BY MERGERS?

Observations of the flattening of globular clusters in the Galaxy, and of clusters in the Large Magellanic Cloud (van den Bergh 1983a, van den Bergh and Morbey 1984) show that clusters of all ages in the LMC are, on average, significantly more flattened than are those in the Galaxy. The high degree of flattening of the Small Cloud globular NGC121, and the noticeable flattening of the brightest globular cluster in NGC6822, suggests that the same conclusion probably also holds for other Magellanic irregular galaxies. The fact that the Galaxy contains few clusters as flattened as those in the Clouds indicates that our Milky Way system has not (yet) accreted any dwarfs similar to the Magellanic Clouds. The possibility that the Galaxy swallowed one or more very faint dwarf irregulars (such as IC1613) which appears to contain no star clusters at all (van den Bergh 1979b), cannot however, be excluded. The same conclusion follows for mergers during the last $\lesssim$10 Gyr from the observation that carbon stars, which are ubiquitous in relatively metal-poor galaxies such as the Magellanic Clouds, are rare in the Galactic halo. These arguments suggest that accretion of satellites similar to the Magellanic Clouds can only have have taken place before these objects started to form significant numbers of stars and clusters.

Van den Bergh (1975) has shown that the mean metallicity of globular cluster systems increases with increasing luminosity of their parent galaxy. This implies that the Galaxy cannot have formed from the merger of many lower luminosity galaxies, which would all have had very metal-poor cluster systems. The observation (Zinn 1985, Armandroff 1989) that metal-rich ([Fe/H] $\lesssim$ -0.8) clusters in the Milky Way system are confined to a thick disc also indicates that the Galaxy did not merge with a giant companion. This is so because some metal-rich clusters might have been expected to have survived if the Galaxy had, at some time in its history, merged with a massive metal-rich companion. Such clusters would not, in general, be expected to have orbits that lie in, or close to, the Galactic plane. The conclusion that the Galaxy has not merged with a massive companion conflicts with that of Carney, Latham and Laird (1989) who invoke merger with a relatively metal-rich satellite to account for a population of Galactic high-velocity stars with [Fe/H] $\sim$ -0.5.

Recently Larson (1986) has suggested that spiral galaxies form from the early merger of dense dwarf galaxies of a type that no longer exists at the present time. Possibly this type of scenario can explain how globular clusters and halo stars formed in what is now a region of exceedingly low average density. A model in which galaxies form at a rather late time, from what is essentially a very clumpy protogalaxy, has recently received some support from the observations (McCarthy et al. 1987) of the very distant possible protogalaxy 3C326.1.

SUMMARY AND CONCLUSIONS

The observation that globular clusters are an order of magnitude more frequent in ellipticals than they are in spirals shows that E galaxies have not formed by merger of disc-like systems. The possibility that ellipticals formed from merging spheroidals cannot, however, be excluded from observations of specific globular cluster frequencies. The fact that cluster ellipticals have a higher specific cluster frequency than field ellipticals might be understood if field E's have typically increased their luminosities by factors of ~50 percent as a result of accretion. Globular cluster-poor fast-rotating ellipticals, such as NGC3557, might be rare examples of elliptical galaxies that did form from merging discs.

The differences between Galactic globular clusters and clusters in the Magellanic Clouds suggest that the Galaxy did not form by merger of low-density ancestral galaxies. Formation of the Galaxy from a very clumpy proto-Galaxy does, however, appear to fit most of the presently known facts about Galactic evolution.

I thank John Kormendy for a number of useful suggestions and Yola Hurwitz for typing many versions of the manuscript of this paper.

REFERENCES

Aguilar, L., Hut, P., and Ostriker J.P. 1988, Astrophys.J., **335**, 720.
Armandroff, T.E. 1989, Astron.J., **97**, 375.
Bender, R. 1988, Astron. Astrophys., **202**, L5.
Caldwell, N., and Phillips, M.M. 1989, Astrophys.J., **338**, 789.
Carlberg, R.G. 1986, Astrophys.J., **310**, 593.

Carney, B.W., Latham, D.W., and Laird, J.B. 1989, Astron.J., **97**, 423.
Chandrasekhar, S. 1942, Principles of Stellar Dynamics, (New York: Dover Publications), p.295.
Cohen, J.G. 1988, Astron.J., **95**, 682.
Davies, R.L., Efstathiou, G., Fall, S.M., Illingworth, G., and Schechter, P.L. 1983, Astrophys.J., **266**, 41.
Dawe, J.A., and Dickens, R.J. 1976, Nature, **263**, 395.
Efstathiou, G., Ellis, R.S., Carter, D. 1982, Monthly Not. Roy. Astron.Soc., **201**, 975.
Forte, J.C., Strom, S.E., and Strom, K.M. 1981, Astrophys. J. Letters, **245**, L9.
Franx, M., and Illingworth, G.D. 1988, Astrophys. J. Letters, **327**, L55.
Hanes, D.A., and Harris, W.E. 1986a, Astrophys.J., **304**, 599.
Hanes, D.A., and Harris, W.E. 1986b, Astrophys.J., **309**, 564.
Harris, W.E. 1981, Astrophys.J., **251**, 497.
Harris, W.E. 1987, Astrophys. J. Letters, **315**, L29.
Harris, W.E. 1988, in Globular Cluster Systems = IAU Symp. No. 126, Eds. J.E. Grindlay and A.G.D. Philip, (Dordrecht: Kluwer), p.237.
Harris, W.E., Smith, M.G., and Myra, E.S. 1983, Astrophys.J., **272**, 456.
Harris, W.E., and van den Bergh, S. 1981, Astron.J., **86**, 1627.
Hickson, P., Kindl, E., and Huchra, J.P. 1988, Astrophys.J., **331**, 64.
Jedrzejewski, R., and Schechter, P.L. 1988, Astrophys. J. Letters, **330**, L87.
Kormendy, J. 1984, Astrophys.J., **287**, 577.
Kormendy, J. 1985, Astrophys.J., **295**, 73.
Kormendy, J. 1989, in preparation.
Lake, G., and Dressler, A. 1986, Astrophys.J., **310**, 605.
Larson, R.B. 1986, in Nearly Normal Galaxies, Ed. S.M. Faber, (Berlin: Springer Verlag), p.26.
Malin, D.F., and Carter, D. 1983, Astrophys.J., **274**, 534.
McCarthy, P.J., Spinrad, H., Djorgovski, S., Strauss, M.A., van Breugel, W., and Liebert, J. 1987, Astrophys. J. Letters, **319**, L39.
Miller, R.H. 1988, Comments on Astrophys., 13, 1.
Morbey, C.L., and McClure, R.D. 1985, Publ. Astron. Soc. Pacific, **97**, 110.
Muzzio, J.C. 1986, Astrophys.J., **301**, 23.
Noguchi, M. 1988, Astron. Astrophys., **201**, 37.
Quinn, P.J. 1984, Astrophys.J., **279**, 596.
Sandage, A., Binggeli, B. and Tammann, G.A. in The Virgo Cluster, eds. O.-G. Richter and B. Binggeli (Garching: ESO) p. 239.
Sandage, A., and Brucato, R. 1979, Astron.J., **84**, 472.
Sandage, A., and Tammann, G.A. 1981, A Revised Shapley-Ames Catalog of Bright Galaxies, (Washington: Carnegie Institution).
Schweizer, F. 1980, Astrophys.J., **237**, 303.
Schweizer, F. 1986, in Nearly Normal Galaxies, Ed. S.M. Faber, (Berlin: Springer Verlag), p.18.
Schweizer, F., and Ford, W.K. 1985, in New Aspects of Galaxy Photometry, Ed. J.-L. Nieto, (Berlin: Springer Verlag), p.145.
Strom, S.E., Forte, J.C., Harris, W.E., Strom K.E., Wells, D.C., and Smith, M.G. 1981, Astrophys.J., **245**, 416.
Tinsley, B.M. 1972, Astrophys. J. (Letters), **173**, L93.
Toomre, A. 1977, in The Evolution of Galaxies and Stellar Populations, Eds. B.M. Tinsley and R.B. Larson, (New Haven: Yale University Observatory), p.401.
Vader, J.P., Vigroux, L., Lachièze-Rey, M., and Souviron, J. 1988, Astron. Astrophys., **203**, 217.
van den Bergh, S. 1969, Nature, **224**, 891.
van den Bergh, S. 1971, Astron. Astrophys., **12**, 474.
van den Bergh, S. 1975, Ann. Rev. Astron. Astrophys., 13, 217.
van den Bergh, S. 1977, Vistas in Astron., **21**, 71.
van den Bergh, S. 1979a, Astrophys.J., **230**, L161.
van den Bergh, S. 1979b, Astrophys.J., **230**, 95.
van den Bergh, S. 1980, Publ. Astron. Soc. Pacific, **92**, 122.
van den Bergh, S. 1983a, Publ. Astron. Soc. Pacific, **95**, 839.

van den Bergh, S. 1983b, Comments Astrophys., **10**, 27.
van den Bergh, S. 1984, Publ. Astron. Soc. Pacific, **96**, 329.
van den Bergh, S. 1989, Astron. Astrophys. Rev., 1, (in press).
van den Bergh, S., and Morbey, C.L. 1984, Astrophys.J., **283**, 598.
Visvanathan, N., and Sandage, A. 1977, Astrophys.J., **216**, 214.
Wirth, A., and Gallagher, J.S. 1984, Astrophys.J., **282**, 85.
Zepf, S.E., and Koo, D.C. 1989, Astrophys.J., **337**, 34.
Zinn, R. 1985, Astrophys.J., **293**, 424.

Did Elliptical Galaxies Form by Mergers or by Dissipative Collapse?

J. Kormendy

Dominion Astrophysical Observatory, Herzberg Institute of Astrophysics,
5071 W. Saanich Road, Victoria, R.R. 5, B.C. V8X4M6, Canada

1. Introduction

Toomre (1977, 1985) has suggested that all bulges and elliptical galaxies formed by mergers of galaxy disks. It is now clear that mergers are fundamental for galaxy evolution. This meeting is likely to be remembered as an important watershed, at which the merger hypothesis completed the transition from heresy to paradigm. However, we may have overshot the truth. Several arguments suggest that dissipative collapse also was important (e. g., van den Bergh 1989). This paper (see also Kormendy 1989) makes two main points. First, the high central densities and small core radii of low-luminosity bulges and ellipticals imply that they formed by dissipative collapse, whether triggered by mergers or not. Second, I argue that the merger and dissipative collapse theories are evolving toward a common picture in which both processes are important.

2. Low-Luminosity Ellipticals Formed by Dissipative Collapse

The high densities of ellipticals have always been a problem for the merger picture (Ostriker 1980; Gunn 1987). Carlberg (1986) has shown that central phase-space densities f_c are much larger in small ellipticals than in galaxy disks; since f_c cannot increase without dissipation, such ellipticals did not form by dissipationless mergers. Dissipative mergers are not excluded.

Observed correlations between core parameters suggest that small ellipticals formed by dissipative collapse. Fig. 1 shows that low-luminosity ellipticals have small core radii and high central densities. They are very unlike disks. Also, a galaxy like M 32 is so faint that the only

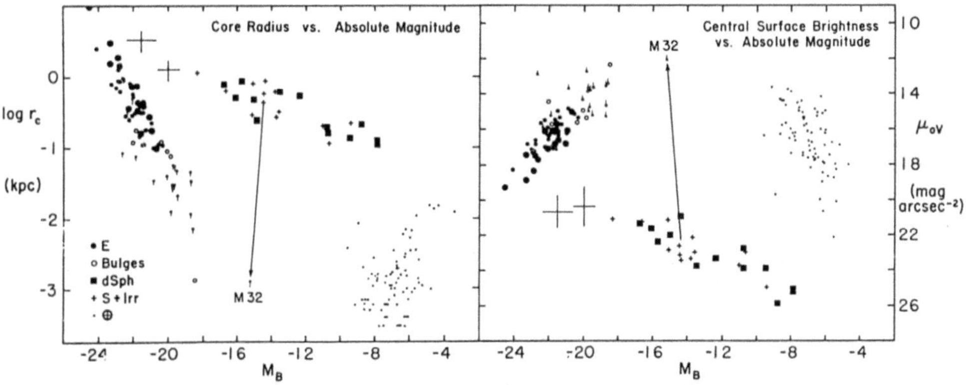

Fig. 1. – Correlations of core parameters with luminosity for various kinds of stellar systems (Kormendy 1989; $H_0 = 50$ km s^{-1} Mpc^{-1}). Larger points imply better resolution of the core; NGC 3379 has the smallest r_c of the best-resolved galaxies. The arrows indicate how core parameters changed if M 32 formed by a merger of two identical dwarf spiral or irregular galaxies.

Dynamics and Interactions of Galaxies Editor: R. Wielen

available progenitors are dwarf irregulars. To make M 32, we have to believe that the core radius decreased and the central density increased by factors of $\sim 10^3$ and $\sim 10^4$, respectively. This implies that M 32 formed by a dissipative collapse, whether triggered by a merger or not.

Galaxies as faint as M 32 are rare. Also, we cannot exclude the possibility that M 32 increased its central density by capturing gas from M 31. Therefore it is important to note that isolated and more common ellipticals like NGC 3377 are almost as extreme as M 32. Finally, while the progenitor disk population may in rich clusters not have been the same as the one we see today (Efstathiou 1989), this is less likely in the field, where the small galaxies in my sample are found. It is implausible in any case that *every* disk galaxy that could make a small elliptical was used up. Therefore, if mergers formed low-luminosity bulges and ellipticals, they did so via dissipative collapse. Possible examples include starbursting IRAS galaxies (Schweizer 1989).

The above argument is based on core properties. Since cores contain only a small fraction of the mass of an elliptical, we need to know how much of the mass had to collapse. Fig. 2

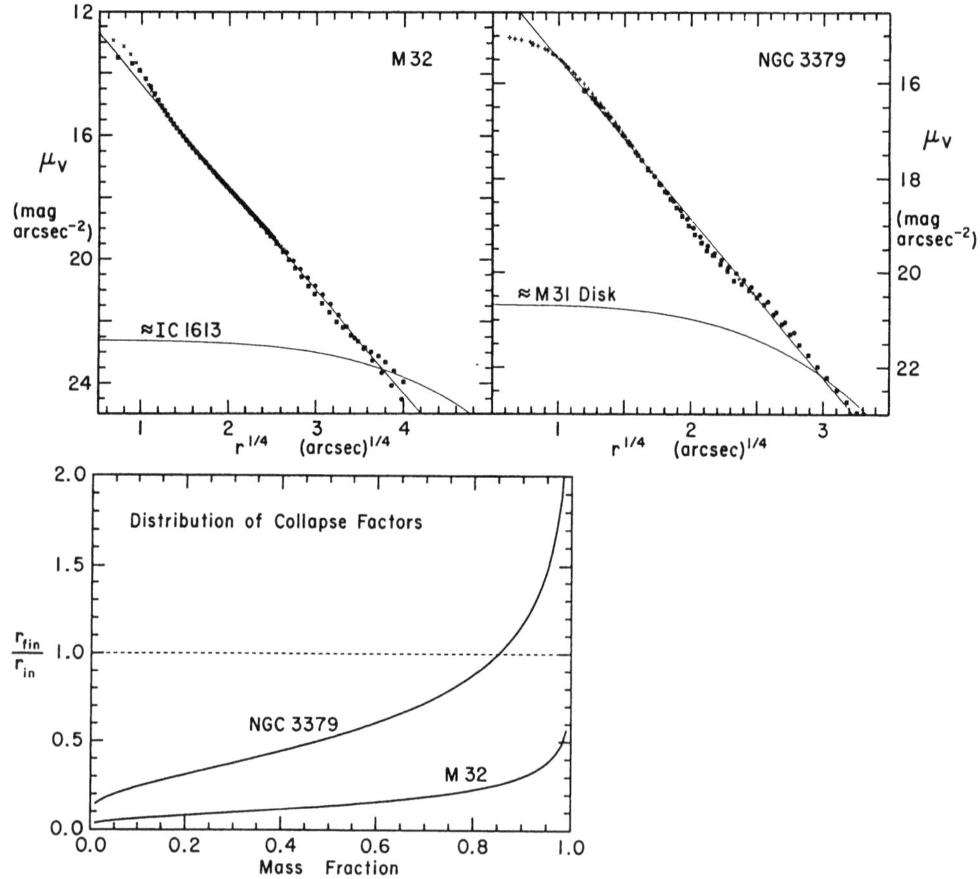

Fig. 2.– Brightness profiles and distribution of collapse factors for the hypothetical merger formation of M 32 and NGC 3379. Progenitor exponentials are also shown, and are labeled with the names of galaxies that have similar profiles. NGC 3379 represents the transition between faint ellipticals that require dissipative collapse and bright ones that can form from disks with little additional collapse. Perhaps not coincidentally, NGC 3379 is also near the transition between faint ellipticals that rotate rapidly and bright ones that are supported by velocity anisotropy (Davies *et al.* 1983). Still brighter ellipticals are relatively easy to make in mergers.

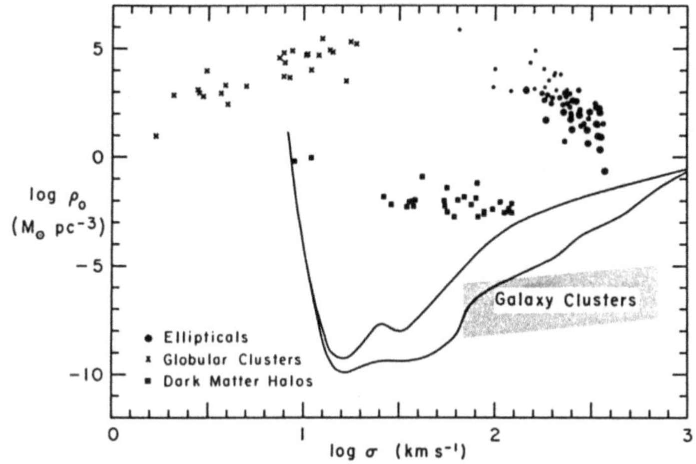

Fig. 3. – "Cooling diagram" for cores of various kinds of stellar systems and for dark matter halos (Kormendy 1989). Here ρ_0 is the central mass density and σ is the central velocity dispersion. M 32 is the highest-density elliptical. Halo parameters of Sc – Im galaxies are from decompositions of rotation curves into visible and dark components (Athanassoula, Bosma, and Papaioannou 1987). Protogalactic gas cooling curves are shown for solar metal abundance (lower) and for no metals (Blumenthal *et al.* 1984). Below the curves, the cooling time is longer than the dynamical collapse time; baryonic objects that have not dissipated lie here. Baryonic structures that formed with dissipation are above the curves, where the cooling time is shorter than the collapse time.

estimates the distribution of collapse factors $r_{\rm fin}/r_{\rm in}$ for various mass shells in the formation of two small ellipticals. I assume that the elliptical was made in a merger of two identical exponential disks whose parameters are taken from the ridge line of the S+Irr parameter correlations (Fig. 1) read at half of the luminosity of the elliptical. The elliptical is modeled by an $r^{1/4}$ law fitted to its mean profile. Now suppose that the elliptical was constructed in 1 % mass shells starting at the center and using up the disks from the center outward. I can then derive the ratio $r_{\rm fin}/r_{\rm in}$ of final to initial radii of each shell. This procedure minimizes the occurrence of extreme collapse factors. It is also conservative because it ignores gas and star formation; gas extends to larger radii than the stars. Fig. 2 shows that galaxies like NGC 3379 or smaller must have collapsed if they formed from disks like those observed today.

The amount of dissipation that took place during galaxy formation can be estimated using the cooling diagram (e. g., Rees and Ostriker 1977; Silk 1977); an observer's version is the correlation between density and velocity dispersion for a sample of galaxies (Fig. 3). The amount of dissipation is measured by how far cores are above and to the right of where they would be without dissipation. Suppose that bulges and ellipticals formed by mergers. Then, if they had formed with no dissipation, they would now appear below the distribution of their dark halos in Fig. 3 by an order of magnitude in ρ_0 (there is $\sim$10 % as much visible matter as dark matter). Dissipation during disk formation then moved these points upward; disks are $\sim$ 10 times as dense as their halos (Athanassoula, Bosma, and Papaioannou 1987, Fig. 11). (In Draco and UMi, the smallest halos plotted, the central visible matter densities are now 10 to 100 times smaller than those of their halos, but were presumably larger before most of the baryons were expelled.) Cores of ellipticals are far above the distribution of disks because of further dissipation. I conclude that if ellipticals formed by mergers of disks, most of the required dissipation took place after disk formation and during the merger. This shows again that even in the merger picture, small ellipticals formed by dissipative collapse. Of course, none of these data *require* that mergers had anything to do with the formation of elliptical galaxies.

3. Convergent Evolution of the Dissipative Collapse and Merger Theories

New theories are often based on oversimplified assumptions; this is a natural consequence of meager evidence and of a focusing of attention on essentials. Not surprisingly, research groups that emphasize different aspects of a problem are soon in conflict. Afterwards, continued emphasis on the original themes can prolong a debate even after new results lead to refinements in rival theories that substantially reduce the differences between them. I believe that this has happened to the dissipative collapse and merger theories of galaxy formation.

The merger picture has always included elements of the dissipative collapse picture. Toomre (1977) pointed out that most merging would take place early in the history of the universe. The need for dissipation and the fact that ellipticals are old imply that the main period of merging was essentially as long ago as in the dissipative collapse picture. Now § 2 shows that small ellipticals formed by dissipative collapse whether or not mergers were involved.

Meanwhile, the dissipative collapse picture has evolved toward the merger picture. In early calculations, the initial conditions were taken to be uniform density inside a sphere. But uniform-density protogalaxies were never more than a computational convenience. Clumpy initial conditions are more realistic, and produce final density profiles that agree better with observations (e. g., van Albada 1982). A clumpy protogalaxy is not physically very different from a collection of separate objects that is about to merge.

Also, it is clear that mergers are making some ellipticals now (e. g., Schweizer 1989).

The dissipative collapse and merger theories are therefore converging toward hybrid schemes that take both into account. The spectrum of initial conditions probably included both clumpy protogalaxies and ones where the clumps were distinct but still destined to merge. In both cases there was significant dissipation, more so in smaller galaxies. Hierarchical clustering, merging, and the accretion of small companions continue today. The difference between the merger and dissipative collapse pictures is shrinking; it is becoming more a matter of choice of emphasis than of taking sides in a controversy.

I thank R. Carlberg, G. Efstathiou, F. Schweizer, and A. Toomre for helpful and enjoyable conversations. This paper is based on observations obtained as a Visiting Astronomer, Canada-France-Hawaii Telescope, operated by the National Research Council of Canada, the Centre National de la Recherche Scientifique of France, and the University of Hawaii.

Athanassoula, E., Bosma, A., and Papaioannou, S. 1987, *Astr. Ap.*, **179**, 23.
Blumenthal, G. R., Faber, S. M., Primack, J. R., and Rees, M. J. 1984, *Nature*, **311**, 517.
Carlberg, R. G. 1986, *Ap. J.*, **310**, 593.
Davies, R. L., Efstathiou, G., Fall, S. M., Illingworth, G., and Schechter, P. L. 1983, *Ap. J.*, **266**, 41.
Efstathiou, G. 1989, this volume.
Gunn, J. E. 1987, in *Nearly Normal Galaxies: From the Planck Time to the Present*, ed. S. M. Faber (New York: Springer-Verlag), p. 455.
Kormendy, J. 1989, *Ap. J. (Letters)*, **342**, L63.
Ostriker, J. P. 1980, *Comments Ap.*, **8**, 177.
Rees, M. J., and Ostriker, J. P. 1977, *M. N. R. A. S.*, **179**, 541.
Schweizer, F. 1989, this volume.
Silk, J. 1977, *Ap. J.*, **211**, 638.
Toomre, A. 1977, in *The Evolution of Galaxies and Stellar Populations*, ed. B. M. Tinsley and R. B. Larson (New Haven: Yale University Observatory), p. 401.
Toomre, A. 1985, in *IAU Symposium 113, Dynamics of Star Clusters*, ed. J. Goodman and P. Hut (Dordrecht: Reidel), p. 500.
van Albada, T. S. 1982, *M. N. R. A. S.*, **201**, 939.
van den Bergh, S. 1989, this volume.

Observational Problems for Collision/Merger Scenarios

H. Arp

Max-Planck-Institut für Astrophysik, Karl-Schwarzschild-Str. 1,
D-8046 Garching, Fed. Rep. of Germany

Abstract. Infrared observations have shown interacting galaxies to be sites of enhanced star formation. The most extreme starbursting galaxies, however, tend to be smaller companions or isolated galaxies. It is argued that it is the recent formation of stars and newer galaxies from the internal activity of older galaxies, not the collision of previously separate galaxies, which is responsible for the current starbursts. It is proposed that material outflowing from active nuclei cools and condenses under the constraint of jets. The strong correlation between the intensity of non thermal radio emission and the infrared emission from star formation supports this model. Multi-wavelength observations of specific galaxies such as NGC 2777 are used to illustrate a recent ejection origin for starburst companions.

The general hypothesis of collision/mergers is considered. It is argued that collisional interaction generally inhibits rather than promotes star formation. It is pointed out that real peculiar velocities of galaxies are not sufficient to produce appreciable collisions and moreover that the necessary reservoir of potential interactors in groups is observationally absent.

1. The Spirit of the Atlas The publication of the *Atlas of Peculiar Galaxies* (Arp 1966) led to many hypotheses involving colliding galaxies. Actually the order of presentation in the Atlas had, after much study, been arranged so as to emphasize the empirical evidence for internal activity in galaxies. For example categories of jet galaxies were presented as well as small knots being apparently ejected, in a continuous gradation to larger companions attached to spiral arms or filaments.

Given contemporary belief that all galaxies were old, however, it was difficult for many astronomers to avoid the interpretation of disturbed adjacent galaxies as the recent collision of two previously well separated galaxies. In addition the advent of electronic computers furnished the opportunity to make n body models of galaxies interacting under the time honored law of Newtonian gravity. The essentially unlimited choice of initial conditions such as relative linear and rotational velocity, masses and densities and the distribution of these quantities between two colliding systems gave enormous range of possible results. Moreover a curved stream of stars could always be rotated to a special orientation where it would, for example, appear straight, or jet-like. Because many configurations in the Atlas of Peculiar Galaxies could be mimicked quite impressively by manipulation of n point representations, the belief grew that encounter scenarios and tidal star tails could, and therefore did, explain all observed configuration of peculiar galaxies. (Of course, this is clearly not true of such objects as Atlas 87 where helical spiral arms are observed, nos. 111 and 268 where loops are observed and no. 184 when multiple emerging streamers are observed. Naturally, there are also many examples of single objects where the disturbances must be internally generated.)

As for the recently popular hypothesis of mergers, it is difficult to disprove that a disturbed galaxy is a well mixed combination of two previously separate galaxies. But I would cite L. Woltjer's (1986) remarks at the Byurakan conference: "to every merger resulting for a short time in a galaxy with a double nucleus there must exist a whole variety of galaxies in other stages of merging and in premerger states. Very little work has been done to show that a statistically consistent picture can be made in which mergers are the dominant factor;...."

We will follow an inductive course here. First we examine a specific galaxy with energetic star production and identify conditions which appear to be responsible. Then we consider a larger class of star forming galaxies and test the two hypotheses of internal vs collisionally

Dynamics and Interactions of Galaxies Editor: R. Wielen

stimulated star formation against each other. Finally we consider the general proposition of collision/mergers and test that hypothesis against the basic, observed properties of galaxies.

2. NGC 2777, a galaxy undergoing rapid star formation. The luminous features in NGC 2777 are irregular and clearly not in rotational equilibrium (Arp and Sulentic 1988). Nor has there been sufficient time to interactively smooth the lumpy clouds into a relaxed configuration. The spectrum of NGC 2777 is quite startling. As Fig. 1 shows the blue wavelengths are dominated by only the Balmer series of hydrogen in absorption. The spectrum resembles that of a late B or early A star - a star that has an age of only 10^8 to 10^9 years. Is the galaxy this young?

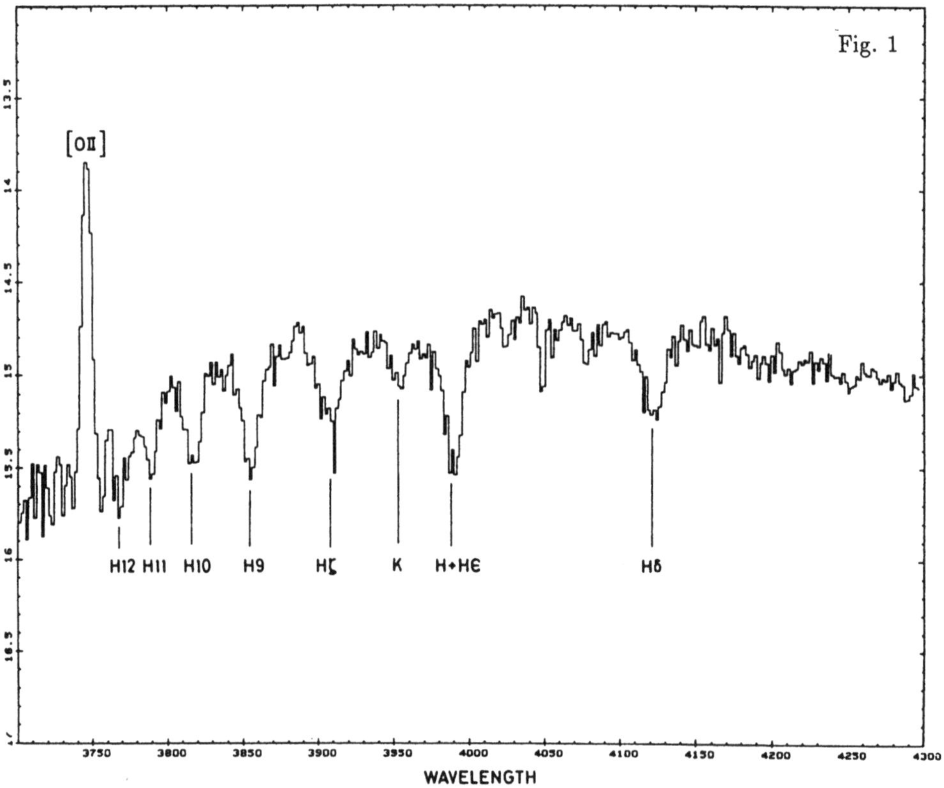

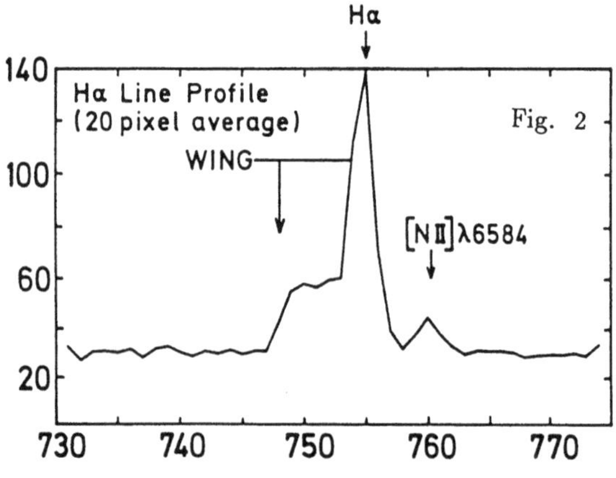

Fig. 2 shows a blue wing on the $H\alpha$ emission line which is most naturally interpretable as approaching outflow from the nucleus on the side nearest the observer. (The receding outflow on the side away from the observer is assumed to be obscured by the copious dust in the system). If this is outflow from the center at velocities reaching ~ 1000 km s^{-1} is must traverse dimensions of the galaxy in $\lesssim 10^7$ yrs. It is important to note that more than a decade ago (M.-H. Ulrich 1972; 1978) concluded from spectroscopy in NGC 1614 and NGC 253 that outflow of hy-

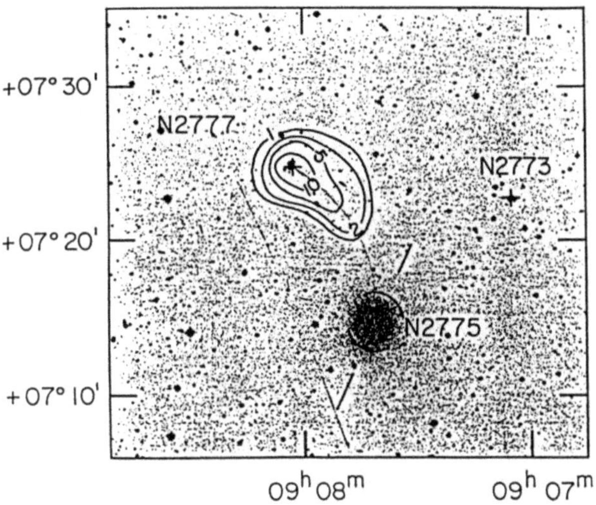

Fig. 3

drogen gas resulted from star formation processes in the interior of these galaxies. Table 1 shows NGC 2777 has the infrared characteristics of a strongly star forming galaxy ($R \gtrsim 0.5$).

2.1 Is star formation in NGC 2777 caused by interaction with NGC 2775? About 12 arc min away from NGC 2777 there is the larger Sa galaxy NGC 2775 (Fig. 3.) At + 139 km s^{-1} relative redshift, NGC 2777 is obviously a companion to NGC 2775. Fig. 3 shows a HI map of NGC 2775/2777 by Haynes (1981). We propose that a jet of ejected material was magnetically or ambiently constrained while it cooled into stars and dust. A clump of stars, forming the nucleus of NGC 2777 had enough velocity to escape and is of the order of 10^8 to 10^9 years old. The young companion, NGC 2777 is now ejecting from its own nucleus and is causing the current flow of excited hydrogen which is condensing into the very young stars and warm dust characteristic of a starbursting galaxy.

In support of this scenario, and in contradiction to a collision hypothesis we refer to Fig. 4 which shows a 1.5 GHz radio continuum tail pointing from NGC 2775 to the companion NGC 2777

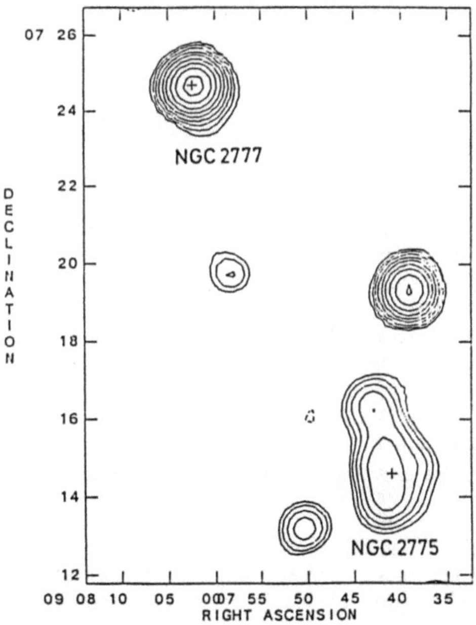

Fig. 4

Table 1 Well Known Galaxies With Properties Similar To NGC 2777

Galaxy	Type	OPTICAL Spectrum	IR FLUX RATIO 60/100μ	COMPANION STATUS $c\Delta z$ from main galaxy	
NGC 2777	Am	v. early abs. + em	0.58	+139 ± 9kms^{-1}	NGC 2775
M82	Am	early abs. + em	1.02	+ 286 ± 5	M81
NGC 3077	Am	early abs. + em	.59	+ 57 ± 6	M81
NGC 404	$S0_{pec}$	early abs. + em	.49	+ 228 ± 10	M31
NGC 1569	S_mIV(Am)	v. early abs. + em	.91	+ 157 ±8	M31
NGC 5195	SBO (Am)	A - F	1.7	+ 11 ± 8	M51
NGC 5253	Am	early abs. + em	1.06	(-104 ± 9)	NGC 5128
NGC 1510	Am	early abs. + em	.63	+69 ± 9	NGC 1512
NGC 520	Am	A - F	.66		uncertain
NGC 1808	SB_rpec	————	.72		uncertain

(Condon 1987). It would be important to study the age characteristics of this synchrotron, jet-like feature. For strongly star forming galaxies in general, radio observations support a model of star formation within magnetically constrained jets because the strongest correlation known in this subject is the one between non thermal radio emission and infrared emission associated with star formation (Wynn-Williams 1987). (In M82 and the galaxy supernovae are not sufficient to account for this relation). Fig. 4 suggests that the older ejection of NGC 2777 from NGC 2775 is marked by a now weaker radio jet whereas the currently strong star formation in NGC 2777 is marked by strong non thermal radio emission.

3. Are strong star bursts in companion galaxies caused by interaction?

Table 1 lists the brightest apparent magnitude, amorphous (i.e. high surface brightness) galaxies which have the high 60/100 μm infrared ratio of R = $\gtrsim$ 0.5 characteristic of strong star formation. They are predominantly companions to larger, more normal galaxies. M82 for example: If an interaction causes the starbursting in this companion, then the abundant star forming material in M81 should show star bursting characteristics. It does not. Neither is there any asymmetry to the observed star production in the arms of M81 or in the HI arm leading to NGC 3077. Most generally, to shock or tidally distort pre stellar material would put more kinetic energy into the material and therefore inhibit rather than enhance star formation. Even if we postulate a special orbital situation where a companion passes just within the distance range where disturbance in only the small galaxy is appreciable, it would be expected that the tidal forces would simply tend to disrupt the companion, lower its average density and retard its potential for star formation. Smashing galaxies together is the poorest way to form stars.

Another strong empirical argument against interaction caused star formation is that some completely isolated galaxies are star bursters. For example, the nearby galaxy NGC 253 is clearly in a strong star formation stage. Yet it is clearly an isolated galaxy with no signs of asymmetrical perturbation. The channeled flow from the nucleus (Ulrich 1978) is the only characteristic which could be responsible for its present high rate of star production. If star formation is caused by outflow from the nucleus in the case of the isolated NGC 253 then it certainly can be the cause in the starburst companions. The preference for starbursting in companions would be a result of their being younger and more active than most galaxies.

4. The General Case Against Collisions and Mergers The fact that extreme star bursters are common among companion galaxies is another argument against collisional encounters for galaxies in general. It is simply that if, in spite of the arguments that galaxy encounters inhibit

star formation, we postulate that there is presently a perturbing encounter of a companion with a larger parent galaxy, then we must look for the large number of companions which are presently not interacting. Since encounters last for a relatively short time, there must be a reservoir of companions lying outside the main galaxy, like an Oort cloud of comets waiting to feed an interacting companion to the main galaxy. But the optical observations show clearly that there are typically only two or three companions around a galaxy and not the reservoir of companions required. HI surveys are particularly telling in this regard because they have shown no appreciable population of even non luminous clouds. (Roberts 1987).

For galaxies in the field, the Hubble flow is too quiet to permit interaction with interlopers. But cosmological objects are characteristically double. If one galaxy in a coeval binary undergoes internal activity which stimulates star formation that galaxy would become the brighter of the pair. This would naturally explain the observation that in close and interacting pairs of galaxies it is the brighter which has the infrared excess characteristic of star formation. One example of the characteristic pairing of extragalactic objects is the class of blue dwarf galaxies often called "isolated, extragalactic HII regions". Ever since their discovery it has been noticed that they are characteristically double. They are conspicuous for their young stars and energetic events and, as systems, probably cannot be very old (Arp 1980; Bergvall and Olafsson 1984). It would surely be unreasonable to consider, for example, these blue, compact dwarf, double galaxies, isolated as they are in space, to be products of recent collisions.

Another independant indication that collisions can not be appreciable comes from the fact that real peculiar velocities must be much smaller than currently believed. In galaxy groups and clusters redshifts of less massive galaxies are systematically redshifted and quantized. (Arp 1987; 1988). Therefore the relative real motions, even in dense groupings, is reduced and collisions are rarer. The situation was summarized by a calculation made over 30 years ago: "...... collisions are, as has also been emphasized by Ambartsumian, exceedingly rare, of the order of 10 in the whole sky for galaxies down to 18th magnitude ...." (Minkowski 1958)

Arp, H. 1987 "Quasars, Redshifts and Controversies" publ. Interstellar Media.
Arp, H. 1988, Astron. Astrophys. *202*, 70.
Arp, H. and Sulentic J.W. 1988, "The Properties of NGC 2777 - Are Companion Galaxies Young?" MPA Preprint 388 and Ap. J. submitted.
Bergvall, N. and Olofsson, K. 1984, ESO Messenger No. 38, p 2.
Condon, J.J. 1987, Ap. J. 65, 485. Haynes, M.P. 1981, Astron. J. *86*, 1139.
Minkowski, R. 1958, Pub. Astr. Soc. Pacific *70*, 143.
Roberts, M.S. 1987, Symposium New Ideas in Astron., Venice 1987, Camb. Univ. Press p 65.
Ulrich, M.-H. 1972, Ap. J. 178, 113. Ulrich, M.-H. 1978, Ap. J. 219, 424.
Woltjer, L. 1986, I.A.U. Symposium 121, ed. E. Khachikian.
Wynn-Williams, C.G. 1987, in "Star Formation in Galaxies" NASA Conf. publ. 2466, p. 125.

Galactic Mergers: Major Issues Remaining

P.L. Schechter

Massachusetts Institute of Technology, Cambridge, MA 02139, USA

1. "Interactions" or the "M" Word?

The Scientific and Local Organizing Committees deserve our warmest thanks for what has been both a very productive and an exceedingly pleasant conference. But at the risk of seeming ungrateful, I think that some criticism is in order as well. The listener at this conference and the reader of these proceedings cannot help but have been struck by the frequency with which the word "merger" has been used, mostly, but not exclusively, in contexts supporting hypothesized mergers. It was surely evident to the scientific organizing committee that this was to be the case. It strikes me as unfortunate, therefore, that the word "merger" fails to appear in the conference title.

Even the chairman of the Scientific Organizing Committee, who has been in the merger vanguard, has been heard to emphasize that not *all* interactions are mergers. Minor collisions also produce a variety of interesting phenomena. But unless such interactions occur on hyperbolic orbits, it is likely that there will be later, more spectacular interactions, in which the distances of closest approach will be smaller. Such systems may have won brief reprieves, but they will ultimately merge.[1] The distinction between interactions and mergers therefore strikes me as unnecessarily fussy. The interesting question is not *whether* a merger will occur but *when* a merger will occur.

2. Astronomical Style: Case Studies or Censuses?

We have heard a convincing plea from Francois Schweizer for in-depth studies of specific cases of merging and merged systems. His own work on NGC 7252, Arnold Rots' and Lars Hernquist's work on M51, and Linda Sparkes' and Ulrich Schwarz' work on NGC 3718 testify to the power of such studies.

But there is a complementary approach, which emphasizes breadth rather than depth, and addresses the question of how important mergers have been for the *typical* galaxy rather than for some specific individual galaxy. My favorite contribution to this conference has been that of Pat Seitzer and Schweizer himself, whose complete survey of nearby ellipticals gives us the best statistical handle on the importance of recent mergers.[2] Honorable mentions in the census category go to Thijs van der Hulst's survey of radio continuum and optical emission in pairs and groups and to Albert Bosma and Lia Athanassoula's study of warps. The award for bravery goes to Ray Carlberg for his attempt to predict the luminosity function for quasars using a merger model.

3. *When* did mergers form ellipticals?

Seitzer and Schweizer have reported that 56% of their nearby ellipticals showed ripples, which they and others have taken as evidence for recent mergers. But if so many ellipticals had par-

[1] And of course if $\Omega = 1$, all galaxies are fated to merge.

[2] This work was carried out with a mere 0.9 m telescope at Kitt Peak. Our speakers have been somewhat stingy (perhaps only because of time constraints) in not crediting the telescopes and intruments used in obtaining the beautiful data they've shown us.

Dynamics and Interactions of Galaxies Editor: R. Wielen

ticipated in recent "major" mergers, one might expect to see more evidence for young stars. Seitzer and Schweizer do see stronger $H\beta$ absorption in the most disturbed of their systems, but $H\beta$ provides little discrimination in the less extreme cases. It may be that such ripples persist for more galactic years than one thought. Or perhaps the mergers weren't "major", and involved only low luminosity dwarfs, which are relatively abundant.

Some of the participants have taken the boxiness of isophotes as signatures of past mergers. But since diamond shaped isophotes are thought to indicate the presence of disks, one might ask whether there are processes other than mergers which might rid a galaxy of its disk, or prevent it from forming. Ralf Bender's correlations of radio and X-ray properties with boxiness are very impressive. The vehement chorus of objections which followed his talk convinces me that he has discovered something important. It is amusing to note that the same objectors who yesterday were content with a_2 are today saying that a_4 is not enough and that higher order moments must be considered.

This brings us to a related question: What *is* an elliptical galaxy? Strict constructionists might hold that objects with significantly non-zero a_4 parameters should not be called ellipticals. Bender has argued that those with $a_4 > 0$, the "disky" ellipticals, are just the high bulge-to-disk ratio end of a continuum which includes S0's and spirals. Many of the galaxies which were heretofore thought to be ellipticals, and even taken to be examples of the class, have now been banished from it. While under the merger hypothesis the number of ellipticals increases with time, the number of bona fide ellipticals has instead *decreased* dramatically over the last decade.

Kinematic substructure in ellipticals has also been taken as a merger marker. Aligned and concentric substructures (last year's great discovery) may be long lived fossils of ancient mergers, or even "primordial". But Stephan Wagner's non-concentric velocity dispersion contours for NGC 7626 ought to phase mix away in just a few galactic years.

There has been surprisingly little mention of the well known morphology-density relation. It is not obvious that mergers should occur more frequently in clusters than in the field. While the densities in clusters are higher, the cross sections are lower. Tod Lauer and Scott Tremaine have given us observational evidence and theoretical arguments that cD galaxies, long thought to be cannibalizing neighbors, may have ceased to do so. Could there have been more generations of mergers in clusters? Were the merging galaxies themselves more luminous than in low density regions? George Efstathiou took pains to point out that rich clusters grow out of a biased subset of early groups, and that these early groups may be responsible for the morphology we see today. Does the paucity of disks in clusters mean that most disks have been destroyed or that they never had the opportunity to form?

Sidney van den Bergh has argued that the specific frequency of globular clusters in ellipticals is a factor of ten higher than one would have expected if they had formed through mergers of today's spirals. Moreover, the specific frequency of globulars is higher in cluster ellipticals than in the field, and higher by a factor of 3 in cD galaxies than in non-cD's. Van den Bergh has challenged us to explain why, if mergers produce ellipticals, the mergers at the centers of clusters should involve galaxies with more globulars than those mergers that occur on the peripheries.

An amusing parallel can be drawn between globular cluster specific frequencies and the metallicities of galaxies. Marijn Franx has told us that the metallicity at a given point within an elliptical is closely correlated with the depth of the potential well at that point in the galaxy. Likewise the specific frequency of globular clusters appears to be correlated with the depth of the potential well at that point in the cluster of galaxies. But unlike the case of metal-rich stars, the globular-rich galaxies may have been the earliest to have formed, not the latest. On this argument the ellipticals and bulges near the centers of clusters would have undergone mergers first, and those in the field last. It remains to be explained why early forming systems would produce more globulars than late forming systems.

4. Do bulges *also* form via mergers?

Galaxies have two kinds of components – cool ones and hot ones, where cool and hot mean velocity dispersions which are respectively smaller than or roughly equal to the virial velocity. Since shape implies kinematics and *vice versa*, the hot components are to first order round and the cool components are to first order two-dimensional, usually flat. Ken Freeman has told us that the Milky Way has at least two interpenetrating cool components and three interpenetrating hot components.

So much has been written about the second order differences between ellipticals and the bulges of systems which also have disks that one tends to forget their first order similarity: both ellipticals and bulges are hot. This heat is produced by what Alar Toomre has called "mechanical violence". It may be primordial, generated in the initial collapse of the galaxy, or it may have been generated by the subsequent reheating of cool disks by mergers or instabilities within those disks. The Milky Way's metal rich bulge may well be an example of a component which formed by the latter process, but I doubt that it could have produced the metal weak halo.

An important paper by Barnes and White (1984) not mentioned thus far argues that bulges look *too* much like ellipticals. Adiabatic compression accompanying the accumulation of disks should have made them denser than their sister ellipticals.

5. Do mergers *destroy* disks or *make* them?

There is little doubt that when galaxies of comparable mass collide, their stellar disks are demolished. But in the cases where the colliding galaxy masses are not equal, the heating of the disk of the larger galaxy will vary inversely with the mass of the perturber. Moreover the perturber may be tidally disrupted before it has much opportunity to heat the disk. There has been little agreement among the theorists here on the fragility of disks, but I suspect that the theory of "disk bashing" will soon improve. Renzo Sancisi has, in the meantime, presented observational evidence that infall can be "episodic" without destroying disks.

The fate of gas in collisions has received only limited attention. Joshua Barnes finds that 60% of the gas in his experiments settles in the centers of his merger remnants, and presumably goes down the drain. But if sufficient energy is produced in the star formation which appears to accompany gas rich mergers, some or all of the gas may be blown out of the system. Somewhere in between the two extremes, gas may settle into a new disk.

One example of such a case is the galaxy MCG -5-7-1, a polar ring galaxy whose peculiar nature was first noticed by Arp and Madore (1986). Van Gorkom, Kristian and I (1987) obtained H I and deep CCD images of this galaxy and found that in addition to the pretty, narrow polar ring, there were ripples and extended H I, which appeared to coincide with each other. Some of the gas in this system evidently chose not to settle into the plane of the preexisting disk, but into a nearly perpendicular plane. But for every polar ring, there may be many more "equatorial" rings.[3] Werner Zeillinger and Giuseppe Galletta have told us about counter-rotating gas rings in S0's and ellipticals which may be precisely such cases. If the gas from the "secondary" event were *corotating*, it would be difficult to distinguish between a galaxy with such an "episodic" formation history and one which formed in a "coherent" or "monolithic" collapse.

James Binney has explained that there are two variants of the idea that disks are made by the raining down of gas clouds: the English drizzle model and the Brazillian downpour model. The cooling of a hot corona might produce the former, while the accumulation of a Magellanic companion might produce the later.

I argued earlier for censuses in addition to case studies. As proof that I practice what I preach, I would point to the results of a complete (which does not mean unbiased) survey

[3] I am indebted to Dr. V. Rubin for suggesting this locution.

of field spirals carried out by Alan Dressler and myself (1987) several years ago. We found a weak trend of increasing bulge-to-disk ratio with increasing luminosity, but with very large scatter (an rms scatter in $log_{10}[B/D]$ of 0.7) at all absolute magnitudes. In rich clusters, we see a yet weaker trend, but with more bulge and less disk. Overall, we found that the mass in the galaxies is approximately equally divided between bulges and disks. If bulges are indeed formed by collisions, than half of the stars we see in galaxies formed before and half have formed since the most recent collision.

Most galaxy formation models have disks forming later than bulges (although Freeman would appear to dissent on this point, and would have the halo of our Milky Way accumulate after the disk). But if bulges formed from the merging of galaxies which themselves had disks, then such disks would almost certainly have had to have been smaller, with shorter rotation periods, than today's disks. It may therefore be the case that there were not two distinct eras – the bulge formation era and the disk formation era – but two overlapping eras, with much of the early work of disk formation undone by subsequent mergers.

6. What *breaks* the merger hierarchy?

George Efstathiou, in our opening talk, reminded us that if one starts with a self-gravitating, expanding system with an initial power law density perturbation spectrum, and if one ignores the effects of dissipation, then as the system evolves, every epoch will look just like every other epoch, except that the scale changes with time. The cartoon version of such a scenario would be a hierarchy of binary mergers of equal mass companions, with merger epidemics occuring every time (1+z) changes by a factor of $\sqrt{2}$.

But substructure has a difficult time surviving in such a merger hierarchy. This was evident in White's experiments of 13 years ago (White 1976), as was emphasized by Rees at the 1977 Yale Symposium (Rees 1977). And yet we see clusters of galaxies, collapsed objects with persistent substructure. Why do these galaxies survive? The problem has been with us long enough to have been given a catchy code name – Ostriker called it "overmerging" at the I.A.U. Symposium on "Dark Matter in the Universe" – but none of the several mechanisms suggested to avoid the problem is entirely convincing.

Dissipation would make galaxies smaller, decreasing their cross sections and increasing their chances for survival. Fall and Efstathiou (1980) have argued that the material which presently forms the disks of spirals must have originally had rather low dimensionless angular momentum, and must have collapsed by a factor of 10 or more to support itself rotationally. That would be the entire story if we didn't think that galaxies extend beyond the disks that we see.

Many currently fashionable models have baryonic matter condensing and dissipating within dark matter halos which condense without dissipating. Such halos would seem to be doomed to merge when the clusters in which they are embedded collapse. White (1978) and van Albada (1982) have shown that while mergers do stir up galaxies, the mass at the centers of the merging systems ends up at the center of the merger remnant. It therefore seems hard to keep the galaxies embedded within halos from merging shortly after their halos merge. Barnes' three component experiments showed exactly this effect.

It is conceivable that halos are stripped in the formation of clusters, but that their cores survive. Core survival may be helped by the dissipative condensation of baryonic material within the core. Two effects work in concert, the added binding energy associated with such added mass and the adiabatic contraction of the core as the mass is added. The effect should be verifiable in N-body experiments. It should also manifest itself observationally in differences in the rotation curves of field and cluster spirals. The results presented by Brad Whitmore are especially intriguing in this regard.

Cluster galaxies might also be saved by an exceedingly shallow density perturbation spectrum. If there is enough power on large scales, then small and large scales will collapse almost simultaneously. If the velocities of the galaxies with respect to each other are sufficiently large, they fail to merge. Carlberg and Couchman (1989) have used both a shallow density perturbation spectrum and dissipative matter embedded within dark, non-dissipative matter to produce galaxies which survive cluster collapse.

7. Does the merging of clusters promote the merging of galaxies?

An idea which emerged in the course of this meeting which struck me as genuinely new was that of crosstalk between merging at the different levels of a clustering hierarchy.

Scott Tremaine has argued that cD galaxies cannot grow to their present size via cannibalism in a relaxed, slowly evolving cluster. But he and Richstone (1977) have argued that the brightest galaxies in clusters do not form a continuum with the vast majority of cluster members. cD galaxies do tend to locate themselves at the centers of clusters, enough so that the Matthews *et al.* (1964) designation "cD" has sometimes been mistaken for an abbreviation of the words "central dominant". Perhaps, Tremaine argues, it is in the merger of clusters that their central galaxies find the opportunity to merge.

Simon White has argued that the catastrophically short collapse times for dense groups are not quite so catastrophic. He nonetheless finds that such groups destroy themselves through mergers on a timescale which is a factor of 5 smaller than the Hubble time. Perhaps the supply of dense groups is replenished by the merging of smaller, less dense groups.

8. Theory or Observations?

Looking over the questions just asked, it seems to me that while a number of observations are suggested, the opportunities for theorists are richer than those for observers. The observers have given the theorists interesting but unexplained dimensionless numbers, including the fraction of ellipticals with ripples, the fraction of spirals with warps, and the relative proportions of spirals and ellipticals. The power of the theorist's second most important tool, the computer, is e-folding every few years, while optical observers are approaching the limit of 100% quantum efficiency and will only gain another factor of 4 in collecting area with the next generation of ground based optical telescopes. The greater gain for the observers will be in the increased resolution afforded by orbiting telescopes. Judging from the images Russell Lavery has shown us taken in Mauna Kea's good seeing, there is much to be learned from clusters at even the "modest" redshift of 0.5. It will be interesting to see how rapidly those dimensionless ratios evolve with lookback time. In case of bulge-to-disk ratio even the *sign* of the expected evolution is a matter of debate.

The skeptics among us might argue that many of these same questions were asked at the 1977 Yale Symposium, and have either resisted attack or have until now proved too daunting. The optimists, among whom I count myself, can point to the great strides since then, both in theory and observation, which have been in abundant evidence at this conference.

References

Arp, H. C. and Madore, B. M. 1986. *Catalogue of Southern Peculiar Galaxies and Associations* (Cambridge: Cambridge University Press).

Barnes, J. and White, S. D. M. 1984. *Mon. Not. Roy. Astron. Soc.* **211**, 253.

Carlberg, R. and Couchman, H. 1989. *Astrophys. J.* **340**, 47.

Fall, S. M., and Efstathiou, G. 1980. *Mon. Not. Roy. Astron. Soc.* **193**, 189.

Matthews, T. .A., Morgan, W. ,W., and Schmidt, M. 1964. *Astrophys. J.* **140**, 35.

Rees, M. J. 1977. In *The Evolution of Galaxies and Stellar Populations* ed. B. M. Tinsley and R. B. Larson (New Haven: Yale University Observatory).

Schechter, P. L., and Dressler, A. 1987. *Astron. J.* **94**, 563.

Tremaine, S. and Richstone, D. 1977. *Astrophys. J.* **212**, 311.
van Albada, T. S. 1982. *Mon. Not. Roy. Astron. Soc.* **201**, 939.
van Gorkom, J. V., Schechter, P. L., and Kristian, J. 1987. *Astrophys. J.* **314**, 457.
White, S. D. M. 1976. *Mon. Not. Roy. Astron. Soc.* **177**, 717.
White, S. D. M. 1978. *Mon. Not. Roy. Astron. Soc.* **184**, 185.

List of Participants

Afanasiev, V.L. (USSR)
Alladin, Saleh Mohammed (India)
Amram, Philippe (France)
Appenzeller, Immo (Fed. Rep. of Germany)
Arp, Halton C. (Fed. Rep. of Germany)
Athanassoula, E. (France)

Balcells, Marc (USA)
Barnes, Joshua (USA)
Barteldrees, Andreas (Fed. Rep. of Germany)
Beck, Rainer (Fed. Rep. of Germany)
Bender, R. (Fed. Rep. of Germany)
Bernlöhr, Konrad (Fed. Rep. of Germany)
Bertola, Francesco (Italy)
Bien, Reinhold (Fed. Rep. of Germany)
Biermann, Peter L. (Fed. Rep. of Germany)
Bierrenbach Lima Neto, Gastao C. (France)
Binney, James (UK)
Böhringer, Hans (Fed. Rep. of Germany)
Borne, Kirk D. (USA)
Bosma, Albert (France)
Briggs, Frank (Netherlands)
Brinks, Elias (UK)
Burbidge, Geoffrey (USA)
Bushouse, Howard A. (USA)
Buson, Lucio (Italy)
Byrd, Gene G. (USA)

Carlberg, Ray (Canada)
Carney, Bruce W. (USA)
Casertano, Stefano (Netherlands)
Centurioni, Sante (Italy)
Clowes, Roger (UK)
Colina, Luis (USA)
Combes, Francoise (France)
Cuddeford, P. (Italy)

Dahlem, Michael (Fed. Rep. of Germany)
Daniel, H.-U. (Fed. Rep. of Germany)
Davies, Roger L. (UK)
Davoust, Emmanuel (France)
de Jong, Teije (Netherlands)
Dejonghe, Herwig (Belgium)
Dettbarn, Christian (Fed. Rep. of Germany)
de Zeeuw, Tim (USA)
Dietrich, Matthias (Fed. Rep. of Germany)
Dressel, Linda L. (USA)

Efstathiou, George (UK)
Einasto, Jaan (USSR)
Elsässer, Hans (Fed. Rep. of Germany)
Engström, Stefan (Sweden)

Feitzinger, Johannes (Fed. Rep. of Germany)
Frank, Juhan (Fed. Rep. of Germany)
Franx, Marijn (USA)
Freeman, K.C. (Australia)
Frenk, C.S. (UK)
Fried, Josef (Fed. Rep. of Germany)
Fuchs, Burkhard (Fed. Rep. of Germany)

Galletta, Giuseppe (Italy)
Gerber, Richard (USA)
Gerhard, Ortwin (Fed. Rep. of Germany)
Giuricin, Giuliano (Italy)
Goeres, Achim (Fed. Rep. of Germany)

Haud, Urmas (USSR)
Heckman, Timothy M. (USA)
Hernquist, Lars (USA)
Hibbard, John E. (USA)
Hippelein, Hans (Fed. Rep. of Germany)
Hopp, Ulrich (Fed. Rep. of Germany)

Illingworth, Garth (USA)

Jahreiß, Hartmut (Fed. Rep. of Germany)
James, R.A. (UK)
Jarvis, Brian (Chile)
Joseph, R.D. (UK)

Kammel, Carl-Heinz (Fed. Rep. of Germany)
King, Ivan R. (USA)
Klaas, Ulrich (Fed. Rep. of Germany)
Kormendy, John (Canada)
Kraan-Korteweg, Renee C. (Switzerland)

Larson, Richard B. (USA)
Lauer, Tod R. (USA)
Laurikainen, Eija (Finland)
Lavery, Russell J. (Australia)
Leech, Kieron (UK)
Lesch, H. (Fed. Rep. of Germany)
Levison, Harold F. (USA)
Longo, Giuseppe (Italy)
Loose, H.-H. (Fed. Rep. of Germany)

Lorenz, Hilmar (German Democratic Republic)
Lutz, Dieter (Fed. Rep. of Germany)

Macchetto, F.D. (USA)
Madejsky, Rainer (Fed. Rep. of Germany)
Martinet, Louis (Switzerland)
Marziani, Paolo (Italy)
McCarthy, Patrick J. (USA)
McGlynn, Thomas A. (USA)
Menon, T.K. (Canada)
Messina, Antonio (Italy)
Miley, George (Netherlands)
Möllenhoff, Claus (Fed. Rep. of Germany)
Moore, Benjamin (UK)
Morin, Stephane (France)
Münch, Guido (Fed. Rep. of Germany)

Nieto, Jean-Luc (France)
Noguchi, Masafumi (Japan)
Norman, Colin (USA)

Palouš, Jan (Czechoslovakia)
Peletier, Reynier F. (Netherlands)
Pfenniger, Daniel (Switzerland)
Pfleiderer, J. (Austria)
Phinney, Sterl (USA)
Prieur, Jean-Louis (Fed. Rep. of Germany)
Prugniel, Philippe (France)

Quinn, Peter J. (Australia)

Radons, Gunnar (Fed. Rep. of Germany)
Rampazzo, Roberto (Italy)
Reimers, Dieter (Fed. Rep. of Germany)
Renz, Wolfgang (Fed. Rep. of Germany)
Richter, Otto-G. (USA)
Rix, Hans-Walter (USA)
Rodgers, Alex W. (Australia)
Roos, Nico (Netherlands)
Rosa, Michael (Fed. Rep. of Germany)
Rots, Arnold H. (USA)

Saar, Enn (USSR)
Sackett, Penny D. (Netherlands)
Saglia, Roberto Philip (Italy)
Salmon, John (USA)
Salo, Heikki (Finland)
Sancisi, R. (Netherlands)
Sanders, David B. (USA)
Schechter, Paul (USA)
Schindler, Sabine (Fed. Rep. of Germany)
Schmidt-Kaler, Th. (Fed. Rep. of Germany)
Schmutzler, Thomas (Fed. Rep. of Germany)
Schulz, H. (Fed. Rep. of Germany)
Schweizer, Francois (France)
Schwekendiek, Peter (Fed. Rep. of Germany)
Scorza, Cecilia (Venezuela)
Seitzer, Patrick (USA)
Shaw, Martin A. (UK)
Shuter, W.L.H. (Canada)
Smith, Eric P. (USA)
Sopp, Hazel M. (UK)
Sparke, Linda (USA)
Sparks, W.B. (USA)
Spies, W. (Fed. Rep. of Germany)
Sridhar, S. (India)
Stockton, Alan (USA)
Sundelius, Bjoern (Sweden)
Sundin, Maria (Sweden)
Surma, P. (Fed. Rep. of Germany)
Sweatman, Winston (UK)

Tammann, G.A. (Switzerland)
Terlevich, Roberto Juan (UK)
Teuben, Peter (USA)
Tonry, John (USA)
Toomre, Alar (USA)
Toomre, Juri (USA)
Tremaine, Scott (Canada)

van Albada, Tjeerd S. (Netherlands)
van den Bergh, Sidney (Canada)
van den Broek, A.C. (Netherlands)
van der Hulst, J.M. (Netherlands)
van der Marel, R.P. (Netherlands)
van Driel, Wim (Netherlands)
van Gorkom, Jaqueline (USA)
van Woerden, Hugo (Netherlands)
Varela Perez, Antonia (Spain)
Völk, Heinrich (Fed. Rep. of Germany)

Wagner, Stefan (Fed. Rep. of Germany)
White, S.D.M. (USA)
Whitmore, Brad (USA)
Wiechen, Heinz (Fed. Rep. of Germany)
Wielen, Roland (Fed. Rep. of Germany)
Wilkinson, Althea (UK)
Wolstencroft, Ramon D. (UK)

Zeilinger, Werner W. (Italy)
Zepf, Stephen E. (USA)
Zhang, Yun (China)

Index of Contributors

MIX
Papier aus verantwortungsvollen Quellen
Paper from responsible sources
FSC® C105338

If you have any concerns about our products,
you can contact us on
ProductSafety@springernature.com

In case Publisher is established outside the EU,
the EU authorized representative is:
Springer Nature Customer Service Center GmbH
Europaplatz 3, 69115 Heidelberg, Germany

Printed by Libri Plureos GmbH
in Hamburg, Germany